TRANSLATIONS OF MATHEMATICAL MONOGRAPHS

VOLUME 55

Systems of Quasilinear Equations and Their Applications to Gas Dynamics

by B. L. ROŽDESTVENSKIĬ
N. N. JANENKO

American Mathematical Society · Providence · Rhode Island

СИСТЕМЫ КВАЗИЛИНЕЙНЫХ УРАВНЕНИЙ И ИХ ПРИЛОЖЕНИЯ К ГАЗОВОЙ ДИНАМИКЕ

ИЗДАНИЕ ВТОРОЕ

Б. Л. РОЖДЕСТВЕНСКИЙ
И Н. Н. ЯНЕНКО

«НАУКА»
ГЛАВНАЯ РЕДАКЦИЯ
ФИЗИКО-МАТЕМАТИЧЕСКОЙ ЛИТЕРАТУРЫ
МОСКВА 1978

Translated from the Russian by J. R. Schulenberger

1980 *Mathematics Subject Classification.*
Primary 35L60, 35L45, 76N15, 65M10; Secondary 35D99.

ABSTRACT. This book is essentially a new edition, revised and augmented by results of the last decade, of the work of the same title published in 1968 by "Nauka". It is devoted to mathematical questions of gas dynamics.

In Chapter 1 the theory of systems of quasilinear equations, the basic mathematical apparatus of gas dynamics, is presented. Chapter 2 contains a consideration of the main problems of one-dimensional gas dynamics, while Chapter 3 is an account of difference methods. The last, fourth chapter is devoted to the theory of discontinuous solutions of systems of quasilinear equations.

Bibliography: 325 titles.

Library of Congress Cataloging in Publication Data

Rozhdestvenskiĭ, B. L. (Boris Leonidovich)
Systems of quasilinear equations and their applications to gas dynamics.
(Translations of mathematical monographs, ISSN 0065-9282; v. 55)
Translation of: Sistemy kvazilineĭnykh uravneniĭ i ikh prilozheniia͡ k gazovoĭ dinamike.
Bibliography: p.
1. Differential equations, Partial. 2. Gas dynamics. I. I͡Anenko, N. N. (Nikolaĭ Nikolaevich) II. Title. III. Series.
QA374.R8213 1983 533′.2′01515353 82-24488
ISBN 0-8218-4509-8

TABLE OF CONTENTS

PREFACE TO THE SECOND EDITION

Ten years have passed since the appearance of the first edition of our book. In this time many new and interesting results on the questions considered in the book have been obtained, especially on difference methods of solving problems of mathematical physics and gas dynamics.

Some corrections of the text have been inserted in the second edition, and also a considerable number of additions to reflect the progress that has taken place. But as in the first edition, the general plan and style of the exposition affects the choice of material. This applies particularly to the new material in Chapter 3, which is devoted to difference methods. The great number of papers on difference methods, together with the necessarily small size of Chapter 3, have compelled us to omit a number of theoretical questions already covered in sufficient detail in accessible monographs and textbooks.

In preparing the second edition we have again made use of the help of our friends and colleagues, as well as our students. We express our profound gratitude to all of them.

FROM THE PREFACE TO THE FIRST EDITION

In writing the book the authors interacted with various collectives of Soviet mathematicians. Among them we mention those headed by M. V. Keldyš, A. N. Tihonov and A. A. Samarskiĭ, and I. M. Gel′fand. The interaction with friends and professional colleagues inevitably affected our opinions and points of view; a number of results became known to us through conversations with them.

During the course of a number of years each of us gave special courses for students on the theme of this book. As a result of work on the book, several new results were obtained which are published here for the first time.

The present book arose from many years of work during which time we constantly enjoyed the help of many of our friends and professional colleagues and also of many of our students.

We are grateful to A. N. Tihonov, whose advice we constantly followed.

The help of L. V. Ovsjannikov was especially valuable to us; he not only looked through the manuscript of the entire book and made a number of valuable remarks, but also placed at our disposal materials which we used in writing §13 of Chapter 1.

A. A. Samarskiĭ read the manuscript for Chapter 3 and made a number of valuable suggestions.

N. N. Kuznecov helped us a great deal; he read the entire manuscript, made a number of valuable suggestions, and as editor of this book greatly contributed to its improvement.

We express our gratitude to all of them.

INTRODUCTION

To describe the behavior of a continuous medium (a gas, liquid, or solid) theoretical physics uses various models which in most cases lead to nonlinear partial differential and integro-differential equations.

The mechanics of continuous media is the main area of practical application of systems of nonlinear partial differential equations, but it is not the only one. In describing the majority of real physical processes we arrive at nonlinear equations, and only essential additional assumptions regarding the smallness of the amplitudes of waves of the field or the amplitudes of oscillations of the medium, the amplitudes of the deviation for the equilibrium state, etc. lead to linear equations which have been studied more extensively. In Chapter 4 of the book a number of examples are presented of problems of physics, chemistry, and mathematics which are connected with nonlinear equations.

The study of general properties of nonlinear equations and methods of solving them is a rapidly developing area of modern mathematics.

For all the interesting facts and the variety of original and clever methods for investigating and solving nonlinear equations, this area of mathematics does not yet have as solid a theoretical foundation as the theory of linear equations. This is related primarily to the fact that the principle of superposition of solutions is not applicable to nonlinear differential equations, so that the manifold of solutions is not linear.

Systems of quasilinear equations are the simplest hyperbolic systems of nonlinear partial differential equations. Systems of equations with two independent variables have been most studied; these systems describe, in particular, nonstationary one-dimensional and supersonic two-dimensional stationary flows of compressible gases and liquids. However, even for these systems there is presently no sufficiently complete theory; there are no general existence and uniqueness theorems for a solution of the initial value problem.

This is due to the fact that for hyperbolic systems of nonlinear equations the solution of the Cauchy problem involves considerable complication both in the very formulation of the problem and the methods of solving it. Moreover, almost all the basic difficulties that arise here are present already for the case of two independent variables, and it may be expected that solutions of multidimensional equations of gas dynamics locally have essentially the same features as solutions of one-dimensional equations.

The study of hyperbolic systems of nonlinear equations in two independent variables thus constitutes an altogether necessary and so far incomplete stage in the investigation of more general nonlinear equations.

On the basis of these considerations the authors decided to restrict attention mainly to the theory of hyperbolic systems in two independent variables and the study of one-dimensional, nonstationary flows of compressible liquids and gases. Therefore, as a rule, we speak of one of the variables as the time and denote it by the letter t.

Here we shall briefly describe the present state of the question of the solvability of the Cauchy problem for hyperbolic systems of quasilinear equations and the difficulties arising in the attempt to construct a global solution of this problem. The basic method of solving hyperbolic systems of quasilinear equations is the method of characteristics, which is expounded in detail in Chapter 1. The existence, uniqueness, and the continuous dependence on the initial data of a classical solution of the Cauchy problem are proved by means of this method. The results obtained are highly satisfactory in the sense that the classical solution is constructed in the entire domain of the variables t and x where it exists. We observe that the domain of existence of a classical solution is, in general, limited, since solutions of nonlinear equations, in contrast to solutions of linear equations, possess the property of unbounded growth of the magnitude of derivatives which is called the "gradient catastrophe".

The meaning of this property is that even for arbitrarily smooth initial data the first derivatives of a solution become unbounded, in general, in finite time. For some $t_0 > 0$ they become unbounded, and for $t > t_0$ a classical solution of the Cauchy problem does not exist.

From the point of view of gas dynamics this corresponds to the formation of a shock wave (a condensation jump) from the compression wave. Thus, if we wish to determine a solution of the Cauchy problem for any $t \geq 0$, i.e. globally (and this is precisely the problem, for example, in gas dynamics), then we must first find a definition of the solution, since, as already mentioned, a solution of the system of equations in the usual sense—a classical solution—does not exist for $t > t_0$.

In the majority of physical problems and, in particular, in gas dynamics the definition of a generalized solution is dictated by the very formulation of the problem. Thus, for example, in gas dynamics the basic physical laws from which we derive all consequences are the laws of conservation of mass, momentum, and energy. These conservation laws have the character of integral relations, and they are applicable not only to smooth (differentiable) flows. On the other hand, the differential equations of gas dynamics are obtained from these conservation laws under the assumption that the flow is smooth.

We thus define a generalized solution of the equations of gas dynamics as a flow (possibly even with discontinuous parameters) satisfying the basic conservation laws of mass, momentum, and energy. To this we add the thermodynamic requirement that the entropy increase in each thermodynamically closed system. There is a broadly held opinion, which has so far not been contradicted by a single example, that a solution thus defined exists, is unique, and satisfies all reasonable requirements.

The requirement of thermodynamics regarding the increase of entropy is very essential: it indicates the possible direction of the process of rapid variation of the state of the gas. This requirement does not enter in considering classical solutions of the equations of gas dynamics for a gas without viscosity or thermal conductivity, since in smooth flows the entropy of the system is conserved by virtue of the same basic conservation laws.

In gas dynamics another approach to generalized (discontinuous) flows of an ideal gas without viscosity or thermal conductivity is well known. Since a gas without dissipation is an idealization of a gas possessing dissipative processes, it is natural to consider its discontinuous flow as a "limit flow" of a viscous, conducting gas as the coefficients of viscosity and thermal conductivity tend to zero. Here it is assumed that the viscous flows are always described by classical solutions of the differential equations, while the limit as the dissipative coefficients tend to zero exists and is unique in a reasonable sense. This assumption has, in fact, so far not been contradicted by a single example, although precise proofs have presently been obtained only for the very special case of a stationary shock wave.

Here it should be observed that in many cases real gases possess rather small dissipation, so that they can be "approximated" by nondissipative gases. However, the presence of dissipative processes, even small ones, leads to an increase of the entropy of the system. Thus, the requirement of the increase of entropy in the discontinuous flow of an ideal gas is related to the representation of this flow as a "limit" flow of a viscous, thermally conducting gas.

We note that from a mathematical point of view the requirement of the increase of entropy is the condition guaranteeing the uniqueness of a generalized solution and its stability with respect to perturbations.

Although this formulation of the problem of the flow of compressible gases has been known for more than a century (Riemann studied the simplest discontinuous flows), the progress in studying general properties of generalized solutions of the equations of gas dynamics has been relatively minor. Thus, as we have already mentioned, there are still no satisfactory existence and uniqueness theorems.

On the other hand, practical requirements occasioned by the pressing necessity of the practical study of discontinuous flows and also new computing capabilities related to the application of rapid computing technology have led to the situation that, in spite of our inadequate information regarding general properties of discontinuous flows, various numerical algorithms have been created and used which make it possible to satisfactorily compute flows with shock waves. It should be mentioned that in creating these numerical algorithms the majority of the conjectures that we mentioned above were accepted as true.

In view of the fact that a direct and rigorous justification of various assumptions regarding generalized solutions in gas dynamics is a difficult problem, it is natural to hope to verify our views for model equations and systems of equations which to some extent imitate the equations of gas dynamics.

A consequence of this desire was the appearance in the last decade of the so-called *theory of generalized solutions of systems of quasilinear equations* or, more briefly, *the theory of systems of quasilinear equations* (here systems of hyperbolic type are usually meant). This theory poses the problem of introducing in analogy with gas dynamics the concept of a generalized solution for an "arbitrary" system of quasilinear partial differential equations of hyperbolic type, proving its existence, uniqueness, and continuous dependence on the initial data of the problem, and studying the properties of such solutions. At least formally, this theory is more general than one-dimensional gas dynamics and includes the latter as a special case.

It has attracted the attention of many mathematicians, and the number of results obtained through the efforts of Soviet and foreign scientists make it possible to expect further development of the theory.

On the basis of this view of the development of the theory of generalized (discontinuous) solutions of systems of quasilinear equations, the authors limited their attention to the case of only two independent variables and included in the book the following basic questions:

1. Methods of constructing classical solutions of systems of quasilinear equations; proofs of existence and uniqueness theorems and continuous dependence of classical solutions; analytic methods of constructing solutions of systems of nonlinear equations; and conditions for the formation of discontinuities in solutions of arbitrary systems of quasilinear equations. These questions are discussed in Chapter 1. Results obtained for classical solutions of systems of quasilinear equations during recent years are presented there.

2. Classical and generalized solutions of the equations of gas dynamics for

one-dimensional, nonstationary flows. This question is discussed in Chapter 2. The authors considered it expedient to consider in detail certain questions of gas dynamics which are discussed in many textbooks. The foundations of thermodynamics, a derivation of the equations of gas dynamics for various symmetries of a one-dimensional flow, the Hugoniot conditions, generalized properties of flows, the theory of the shock transition, and self-similar and analytic solutions of gas dynamics are presented. The inclusion in the book of these traditional questions of gas dynamics makes it possible to expound from a unified point of view certain mathematical problems which arise in gas dynamics; moreover, the majority of numerical methods in gas dynamics are actually based on this material. The basic problem of the theory of discontinuous solutions of the equations of gas dynanics—the problem of the decay of an arbitrary discontinuity—and also the interaction of shock waves with one another, with travelling waves, and with a contact boundary are considered in detail.

3. Chapter 3 is devoted to difference methods for solving the equations of gas dynamics. In our time these methods have become a basic means of studying problems of gas dynamics, and therefore progress in the study of discontinuous flows is to a considerable extent connected with difference methods.

In this chapter we have had to present the basic concepts of the theory of difference methods. Unfortunately, the majority of assertions of this theory pertain only to the case of linear equations.

The present situation regarding the justification of difference methods applied to the numerical solution of problems of gas dynamics is briefly as follows. Classical solutions (smooth flows) can be computed with almost arbitrary accuracy. The basic method—the numerical method of characteristics—for classical solutions is sufficiently justified. On the other hand, numerical methods applied to compute discontinuous flows have not been rigorously justified, and in the majority of cases certain conjectures regarding the behavior of solutions, the approximation of certain equations by others, etc. are used. Simple equations for which the behavior of the discontinuous solution is well known are most frequently used to verify the various assumptions. It is no accident that in this chapter in most cases each scheme is subject to verification on a simple quasilinear equation with a solution which can be written out explicitly.

This situation regarding the justification of difference methods indicates that progress in this area is to a considerable extent connected with progress in the study of general properties of generalized solutions of systems of quasilinear equations and, in particular, the equations of gas dynamics. On the other hand, difference methods provide experimental material and strongly stimulate the development of the theory of generalized solutions.

4. Chapter 4 is devoted to the theory of generalized solutions of systems of

quasilinear equations of hyperbolic type, and contains the basic results obtained in this area in recent years. A major success here is the construction of a theory of a generalized solution of a single quasilinear equation, which may be considered near completion. For this equation existence, uniqueness, and continuous dependence of the generalized solution of the initial data are proved, and the equivalence of the definitions of a generalized solution from the point of view of the conservation law on the one hand and as a limit of "viscous solutions" on the other is demonstrated.

On the other hand, as in gas dynamics, the study of generalized solutions of systems of equations encounters major difficulties, and very meager results have so far been obtained here. The basic problem, which is currently being subjected to thorough investigation, is the problem of the decay of an arbitrary discontinuity. By means of this simple problem it is possible to study the structure of the generalized solution, and on the basis of this structure it is even possible to construct generalized solutions for the case of a system of two equations.

It should be noted that in recent years more general problems for systems of quasilinear equations have also been studied intensively.

In Chapter 4 the basic results obtained for a single quasilinear equation are presented, the problem of the decay of a discontinuity for an arbitrary hyperbolic system of quasilinear equations is considered, and some results pertaining to more general cases are presented. To conclude this chapter a number of problems are described in various areas of science which are related to the theory of systems of quasilinear equations and, in particular, to discontinuous solutions of such equations.

From what has been said above it should be clear that the mathematical theory of discontinuous solutions of systems of quasilinear equations and, in particular, of the equations of gas dynamics, although it contains many remarkable results and achievements, is far from complete. We hope that our book will afford the reader an idea of the modern methods of solving and studying systems of quasilinear equations and will at the same time spur him to further investigations in this interesting and rapidly developing area of applied mathematics.

The book is subdivided into chapters, sections, and subsections. Formulas are numbered independently in each subsection, and hence references contain the number of the section and subsection in addition to the number of the formula, so that formula (2.7.18) signifies formula (18) in Subsection 7 of Section 2 (i.e., §2.7) of the given chapter. The number of the formula alone is given only in the case where the reference does not go beyond the confines of the given subsecton.

CHAPTER 1

FOUNDATIONS OF THE THEORY OF SYSTEMS OF QUASILINEAR EQUATIONS OF HYPERBOLIC TYPE IN TWO INDEPENDENT VARIABLES

§1. Basic definitions

In this book we restrict our attention to differential equations for functions depending only on two independent variables.

A system of relations

$$\mathscr{F}_i\left(x, t, u_1, \ldots, u_n, \frac{\partial u_1}{\partial x}, \ldots, \frac{\partial u_n}{\partial x}, \frac{\partial u_1}{\partial t}, \ldots, \frac{\partial u_n}{\partial t}\right) = 0$$
$$(i = 1, \ldots, m), \qquad (1)$$

connecting the values of the unknown functions $u_1(x, t), \ldots, u_n(x, t)$ and their first derivatives $\partial u_1/\partial x, \ldots, \partial u_n/\partial x$ and $\partial u_1/\partial t, \ldots, \partial u_n/\partial t$ is called a *system of differential equations of first order* for the functions $u_1, \ldots, u_n$. The system (1) is called *determinate* in the case $m = n$. We limit consideration to only this case.

Introducing the vectors

$$u = \{u_1, \ldots, u_n\},$$
$$\frac{\partial u}{\partial x} = \left\{\frac{\partial u_1}{\partial x}, \ldots, \frac{\partial u_n}{\partial x}\right\}, \qquad \frac{\partial u}{\partial t} = \left\{\frac{\partial u_1}{\partial t}, \ldots, \frac{\partial u_n}{\partial t}\right\},$$

we can write the system (1) more compactly as

$$\mathscr{F}_i\left(x, t, u, \frac{\partial u}{\partial x}, \frac{\partial u}{\partial t}\right) = 0 \qquad (i = 1, 2, \ldots, n). \qquad (2)$$

Functions $u_i = u_i(x, t)$ possessing continuous first derivatives and satisfying the equations of the system (2) constitute a *solution* of this system of equations.

If the system of nonlinear differential equations (2) can be represented in a form solved for the derivatives of the functions $u_1, \ldots, u_n$ with respect to some variable (for example, t),

$$\frac{\partial u_i}{\partial t} = \varphi_i\left(x, t, u, \frac{\partial u}{\partial x}\right) \qquad (i = 1, 2, \ldots, n), \tag{3}$$

then this form of the system (2) is called the *normal form*. The system (3) is called a system of *Cauchy-Kovalevskaja type*. We remark that in reducing the system (2) to normal form transformation of the variables x, t is admissible.

The system (2) is called a *system of quasilinear equations* if the functions $\mathscr{F}_i$ are linear in $\partial u/\partial x$ and $\partial u/\partial t$; if the functions $\mathscr{F}_i$ are linear in all the variables u, $\partial u/\partial x$ and $\partial u/\partial t$, then the system (2) is called *linear*.

A system of quasilinear equations of first order can be written in the form

$$\sum_{j=1}^{n} a_{ij}\frac{\partial u_j}{\partial t} + b_{ij}\frac{\partial u_j}{\partial x} = c_i \qquad (i = 1, 2, \ldots, n), \tag{4}$$

where the coefficients a_{ij}, b_{ij}, and c_i depend on x, t, and u. If the coefficients a_{ij} and b_{ij} do not depend on u, then the system (4) is called *semilinear* (in this case if the c_i depend linearly on u, then it is linear). It is possible to simplify somewhat the notation for the system (4) if we introduce the vectors u, $\partial u/\partial x$ and $\partial u/\partial t$ considered above, the vector $c = \{c_1, \ldots, c_n\}$, and the matrices

$$A = \left\| \begin{matrix} a_{11} & \ldots & a_{1n} \\ a_{n1} & \ldots & a_{nn} \end{matrix} \right\| = ((a_{ij})), \qquad B = ((b_{ij})).$$

In using matrix notation we assume that the symbols Au and uA represent vectors with components which are computed according to the rules

$$(Au)_k = \sum_{j=1}^{n} a_{kj}u_j, \qquad (uA)_k = \sum_{j=i}^{n} a_{jk}u_j = (A'u)_k \qquad (k = 1, 2, \ldots, n), \tag{5}$$

where A' is the transposed matrix.

If the matrix A is symmetric, then $A' = A$, and $Au = uA$. Below we shall consider real matrices and vectors. The inner product of vectors u and v is defined by

$$uv = (u, v) = \sum_{i=1}^{n} u_i v_i.$$

From (5) it thus follows that

$$v(Au) = (vA)u = vAu.$$

The *norm of a vector* u is the quantity

$$\|u\| = \sqrt{(u, u)} = \sqrt{\sum_{i=1}^{n} u_i^2}.$$

The *norm of a matrix A* is the smallest number $\|A\|$ which for any vectors u satisfies the inequality

$$\|Au\| \leq \|A\|\,\|u\|.$$

It is not hard to see that $\|A\| = \sqrt{\lambda}$, where λ is the largest eigenvalue of the matrix AA' (or, equivalently, $A'A$). Since $\lambda \leq \operatorname{Tr} AA'$, it follows that $\|A\| \leq (\sum a_{ij}^2)^{1/2}$.

We recall some further definitions from linear algebra. A vector $l = \{l_1, \ldots, l_n\}$ and a number ξ are called respectively a *left eigenvector* and an *eigenvalue of the matrix A if*

$$lA = \xi l, \qquad \|l\| \neq 0. \tag{6}$$

Similarly, a vector r is called a *right eigenvector of the matrix A* if

$$Ar = \xi r, \qquad \|r\| \neq 0. \tag{7}$$

According to (5) and (6), an eigenvalue ξ of a matrix A is a root of the characteristic equation

$$\operatorname{Det}((a_{ij} - \xi\delta_{ij})) = 0, \tag{8}$$

where δ_{ij} is the Kronecker symbol ($\delta_{ij} = 0$ for $i \neq j$ and $\delta_{ij} = 1$ for $i = j$). To each eigenvalue ξ of a matrix A there corresponds a linear space of left eigenvectors l and right eigenvectors r. The dimension of these spaces is equal to $n - \beta$, there β is the rank of the matrix

$$A - \xi E = ((a_{ij} - \xi\delta_{ij})). \tag{9}$$

It is known that the rank of the matrix (9) is not less than $n - \alpha$, where α is the multiplicity of the root ξ of (8).

We assume that the eigenvalues ξ of A are real. We enumerate them in increasing order, i.e., we assume that

$$\xi_1 \leq \xi_2 \leq \cdots \leq \xi_n. \tag{10}$$

The equality sign in (10) is admissible in view of the possibility of multiple roots of (8), while each multiple root ξ is repeated in (10) as many times as its multiplicity.

If for any eigenvalue ξ of the matrix A of multiplicity α the rank of the matrix

(9) is equal to $n - \alpha$, then both the left eigenvectors l and the right eigenvectors r corresponding to all eigenvalues form a basis of the space E_n.

Thus, in this case we may assume that there exist eigenvectors $l^1, \ldots, l^n$ which form a basis of the space E_n, i.e., they satisfy the condition

$$\operatorname{Det} \Lambda = \operatorname{Det}((l_i^k)) = \operatorname{Det} \begin{Vmatrix} l_1^1 \ldots l_n^1 \\ \cdot \;\cdot \;\cdot \;\cdot \\ l_1^n \ldots l_n^n \end{Vmatrix} \neq 0. \tag{11}$$

Here the index of the left eigenvector l^k corresponds to the index of the eigenvalue ξ_k; the latter are ordered by (10).

If $\xi_k \neq \xi_j$, then l^k and r^j are orthogonal. Indeed, suppose that

$$l^k A = \xi_k l^k, \qquad A r^j = \xi_j r^j. \tag{12}$$

Taking the inner product of the first of the equalities (12) with r^j and of the second with l^k and subtracting, we obtain

$$(\xi_k - \xi_j) l^k r^j = (l^k A) r^j - l^k (A r^j) = l^k (A r^j) - l^k (A r^j) = 0. \tag{13}$$

Since $\xi_k \neq \xi_j$, this implies that l^k and r^j are orthogonal. In the case where all eigenvalues of A are simple, there can be no equality sign in (10), and the left and right eigenvectors form a biorthogonal system, i.e.,

$$l^k r^j = \sum_{\alpha=1}^{n} l_\alpha^k r_\alpha^j = 0 \quad \text{for } k \neq j. \tag{14}$$

If the matrix A is symmetric, then it may be assumed that $r^k = l^k$. We require that the left eigenvectors l^k of A satisfy the normalization condition

$$\|l^k\| = 1 \qquad (k = 1, \ldots, n). \tag{15}$$

Then, if for any eigenvalue $\xi = \xi_k$ the rank of the matrix (9) is equal to $n - \alpha$, the eigenvectors l^k form a normalized basis in E_n and, of course, satisfy the condition (11); moreover, if A is symmetric, then the basis $\{l^k\}$ can be chosen to be orthonormal.

A matrix A is called *nonsingular* if $\xi = 0$ is not an eigenvalue, and *singular* otherwise.

Having stated these brief facts from linear algebra, we write the system (4) in the form

$$A \frac{\partial u}{\partial t} + B \frac{\partial u}{\partial x} = c. \tag{16}$$

In the case where the matrix A is nonsingular the system (16) reduces to the normal form (3), and after transformation it can be written in the form

$$\frac{\partial u}{\partial t} + A_1 \frac{\partial u}{\partial x} = b, \tag{17}$$

where $A_1 = A_1(x, t, u)$ and $b = b(x, t, u)$ are a new matrix and a new vector respectively. We henceforth restrict our study to systems (16) which reduce to the normal form (17).

Above we have made an assumption regarding the matrix $A(x, t, u)$ of the system of equations (16). However, A depends on u, i.e., on the solution, which, as a rule, is unknown to us. We therefore state in what sense we make assumptions regarding the coefficients of the systems (16) and (17); this is done in one of the following two ways.

1) We assume that a solution $u = u(x, t)$ of the systems (16) and (17) is given as a function of the variables x and t. The satisfaction of the various conditions imposed on the matrices A, B, and A_1 and the vectors c and b can then be verified directly.

2) We assume that these conditions are satisfied identically (for any values of $u = \{u_1, \ldots, u_n\}$) in some simply connected region of the space (x, t, u) in which the system of quasilinear equations and its solutions are considered.

In this chapter we shall mainly impose conditions in the second sense.

§2. Characteristic directions of a system of quasilinear equations

1. *The directional derivative.* Let $f(x, t)$ be a differentiable function of its variables. At some point (x_0, t_0) we consider the expression

$$A\frac{\partial f}{\partial t} + B\frac{\partial f}{\partial x}, \tag{1}$$

assuming that A and B are not simultaneously equal to zero.

For any A and B which are continuous in some neighborhood of the point (x_0, t_0) it is possible to find a smooth curve Γ passing through this point and such that for a suitable parametrization of it the expression (1) is proportional to the derivative of $f(x, t)$ on Γ with respect to the parameter τ.

Indeed, suppose that the curve Γ is given by the equations

$$\Gamma: x = x(\tau), \qquad t = t(\tau), \qquad x(\tau_0) = x_0, \qquad t(\tau_0) = t_0. \tag{2}$$

On Γ the function $f(x, t)$ is then a function of the single variable τ: $f(x(\tau), t(\tau)) = F(\tau)$. We require that the expression (1) be proportional to $F'(\tau)$ for any function f.

This will be satisfied if

$$\frac{dx}{d\tau} = \alpha B, \qquad \frac{dt}{d\tau} = \alpha A, \tag{3}$$

where α is any continuous function of τ. It is clear that the essential condition which uniquely determines the direction of Γ at the point (x_0, t_0) is the equation

$$\frac{dx}{B} = \frac{dt}{A}, \tag{4}$$

while formulas (3) determine the corresponding parameterization.

The *derivative of the function f in the direction of* Γ is the derivative $F'(\tau)$ for the natural parameterization of Γ, where $\alpha = 1/\sqrt{A^2 + B^2}$. In this case the parameter τ is the arc length of Γ. For $\alpha = 1$ we call (1) the derivative of f with respect to the parameter τ in the direction of Γ. This simple concept finds important applications in the theory of partial differential equations.

We consider the simplest differential equation

$$A(x, t)\frac{\partial u}{\partial t} + B(x, t)\frac{\partial u}{\partial x} = 0, \tag{5}$$

assuming that the functions A and B are continuously differentiable. The equations

$$\frac{dt}{d\tau} = A(x, t), \qquad \frac{dx}{d\tau} = B(x, t) \tag{6}$$

or (4) determine a one-parameter family of curves Γ. The parameter τ is uniquely determined along each of these curves if along some (arbitrarily chosen) curve γ_0 intersecting the curves Γ we set $\tau = \tau_0$ (Figure 1.1).

To each curve of Γ we assign the value of some parameter ω (for example, the arc length of a curve γ_0 reckoned from any point on it to the point of its intersection with a given curve of Γ). Then to each point (x, t) there corresponds one and only one pair of numbers τ, ω.

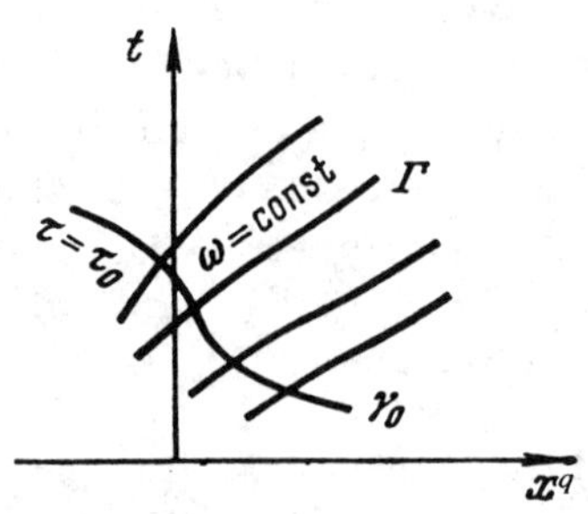

FIGURE 1.1

The function $u(x, t)$ may thus be considered a function of the variables τ and ω; the equations of the lines Γ have the form $\omega = \text{const}$, while (5), according to the foregoing, can be written in the form

$$\frac{\partial u(\tau, \omega)}{\partial \tau} = 0. \tag{7}$$

This implies that $u = F(\omega)$ is the general solution of (5), and the function $u(x, t)$ is constant along the curves Γ.

The directions of the curves Γ defined by the vector $\{B, A\}$ are called the *characteristic directions* of equation (5), and the curves Γ are called *characteristics.*

We observe that the form (7) of equation (5) does not assume the existence of the derivatives $\partial u/\partial x$ and $\partial u/\partial t$: (7) is satisfied by any function $F(\omega)$, in particular, even by a discontinuous function. The function $u = F(\omega)$ in this case can be treated as a solution of (5) in a generalized sense.

2. *Hyperbolic systems of quasilinear equations.* We consider the system of quasilinear equations

$$\frac{\partial u}{\partial t} + A\frac{\partial u}{\partial x} = b. \tag{1}$$

Multiplying it on the left by the vector l, we obtain the scalar equation

$$l\frac{\partial u}{\partial t} + lA\frac{\partial u}{\partial x} = lb. \tag{2}$$

If l is a left eigenvector of the matrix A, then (2) can be written in the form

$$l\left(\frac{\partial u}{\partial t} + \xi\frac{\partial u}{\partial x}\right) = lb, \tag{3}$$

where ξ is the corresponding eigenvalue of A.

In (3) all the components u_i of the vector u are differentiated in the same direction. Indeed, writing (3) in terms of components, we obtain

$$\sum_{i=1}^{n} l_i\left(\frac{\partial u_i}{\partial t} + \xi\frac{\partial u_i}{\partial x}\right) = \sum_{i=1}^{n} l_i b_i.$$

Denoting by

$$\left(\frac{du_i}{dt}\right) = \frac{\partial u_i}{\partial t} + \xi\frac{\partial u_i}{\partial x}$$

the derivative of $u_i(x, t)$ with respect to t in the direction $dx/dt = \xi$, we see that (3) contains a linear combination of the derivatives (du_i/dt). The equation $dx/dt = \xi$ defines the common direction of differentiation in (3) for all functions

$u_i(x, t)$, which is called a *characteristic direction* of the system of equations (1).

We call a system of quasilinear equations (1) *hyperbolic* in some simply connected domain D of the space of the variables x, t and u if at each point of this domain the following conditions are satisfied:

1) All the eigenvalues $\xi = \xi_k(x, t, u)$ of the matrix $A = A(x, t, u)$ are real.

2) There exists a basis $\{l^1(x, t, u), \dots, l^n(x, t, u)\}$ in E_n composed of left eigenvectors of A subject to the normalization condition, i.e., there exist normalized eigenvectors $l^1, \dots, l^n$ satisfying the condition (1.11).

We note that if the system (1) is semilinear, then the eigenvalues ξ_k and the left eigenvectors l^k do not depend on u. Therefore, the property of hyperbolicity for semilinear systems is defined in a domain of the variables (x, t) for arbitrary u (in a cylindrical region).

In connection with the definition of hyperbolicity we note that the condition of a particular smoothness of the eigenvectors l^k and the eigenvalues ξ_k is frequently added to conditions 1) and 2). Thus, for example, in the book of I.G. Petrovskiĭ (1961) the system (1) is called hyperbolic if conditions 1) and 2) are satisfied, and, moreover, $\xi_k(x, t, u)$ and $l^k(x, t, u)$ possess the same smoothness as the elements of the matrix $A(x, t, u)$.

Of course, we must subsequently resort to assumptions regarding the smoothness of ξ_k and l^k. We shall impose them as necessary. In connection with this we note that, whatever the smoothness of l^k and ξ_k, it does not always follow from an assumption regarding the same smoothness of A.

We shall demonstrate this with the following example of a system of two quasilinear equations:

$$\frac{\partial u_1}{\partial t} + u_1 \frac{\partial u_1}{\partial x} = 0, \qquad \frac{\partial u_2}{\partial t} + a(u_1, u_2)\frac{\partial u_1}{\partial x} + u_2 \frac{\partial u_2}{\partial x} = 0.$$

The matrix A in this case has the form

$$A = A(u) = \begin{Vmatrix} u_1 & 0 \\ a(u_1, u_2) & u_2 \end{Vmatrix},$$

and its eigenvalues ξ_1 and ξ_2 are defined from the equation

$$(\xi - u_1)(\xi - u_2) = 0,$$

whence $\xi_1 = u_1$ and $\xi_2 = u_2$. The eigenvectors l^1 and l^2 are determined from the equations

$$0 \cdot l_1^1 + a(u_1, u_2) l_2^1 = 0, \qquad 0 \cdot l_1^1 + (u_2 - u_1) l_2^1 = 0, \tag{4}$$

$$(u_1 - u_2) l_1^2 + a(u_1, u_2) l_2^2 = 0, \qquad 0 \cdot l_1^2 + 0 \cdot l_2^2 = 0. \tag{5}$$

Using the normalization condition, from this we determine l^1 and l^2 in the region $u_1 \neq u_2$:

$$\left.\begin{aligned} l^1 &= \{1, 0\}, \\ l^2 &= \left\{ \frac{\dfrac{a(u_1, u_2)}{u_2 - u_1}}{\sqrt{1 + \left(\dfrac{a}{u_2 - u_1}\right)^2}}, \frac{1}{\sqrt{1 + \left(\dfrac{a}{u_2 - u_1}\right)^2}} \right\}. \end{aligned}\right\} \tag{6}$$

In the plane of the variables (u_1, u_2) we consider the line $u_1 = u_2$. If $a(u_1, u_1) \neq 0$, then from (5) we find that $l^2 = \{1, 0\}$, and the condition 2) of hyperbolicity is violated on the line $u_1 = u_2$. We shall consider the case $a(u_1, u_1) = 0$ in more detail. In this case (4) and (5) become identities on the line $u_1 = u_2$ and for l^1 and l^2 it is possible to choose any two noncollinear unit vectors; the system is thus hyperbolic.

If we require that the vectors l^1 and l^2 possess some particular smoothness, this is not guaranteed by the same smoothness of the function $a(u_1, u_2)$. Suppose, for example, we require that the vectors l^1 and l^2 be continuous on the line $u_1 = u_2$.

It is obvious that the only vector l^1 continuous on the line $u_1 = u_2$ is the vector

$$l^1(u_1, u_2) = \{1, 0\};$$

as concerns l^2, the continuity of the function $a(u_1, u_2)(u_2 - u_1)^{-1}$ is necessary for its continuity. For example, if

$$a(u_1, u_2) = |u_1 - u_2|^\alpha\, b(u_1, u_2) \qquad (b(u_1, u_1) \neq 0),$$

then for $\alpha > 1$ the desired continuous vector l^2 is given by (6) (for $u_1 \neq u_2$) and by

$$l^2(u_1, u_1) = \{0, 1\}.$$

For $\alpha < 1$ no continuous vector l^2 exists.

It is not hard to verify that in this example to ensure the continuity of the mth derivatives of l^2 it is necessary to require that $a(u_1, u_2)$ have m continuous derivatives and on the line $u_1 = u_2$ it also have $(m + 1)$st derivatives.

The situation is much simpler in the very important special case of hyperbolic systems which we call strictly hyperbolic systems.

A system of equations (1) is called *strictly hyperbolic* in a simply connected domain D of the variables (x, t, u) if at each point of this domain the eigenvalues $\xi_1, \ldots, \xi_n$ of the matrix A are real and distinct.

In this case the eigenvalues can be ordered, and we assume that everywhere in D

$$\xi_1(x, t, u) < \xi_2(x, t, u) < \cdots < \xi_n(x, t, u).$$

It was shown in §1 that then the eigenvectors $l^k(x, t, u)$ are linearly independent. It is easy to verify that in this case $l^k(x, t, u)$ and $\xi_k(x, t, u)$ possess the same degree of smoothness as the elements of $A(x, t, u)$.

Thus, by multiplication by left eigenvectors l^k the system (1) in the domain D can be brought to the form

$$l^k(x, t, u)\left[\frac{\partial u}{\partial t} + \xi_k(x, t, u)\frac{\partial u}{\partial x}\right] = f_k(x, t, u) \qquad (k = 1, \ldots, n), \tag{7}$$

where $f_k = l^k b = \sum_1^n l_\alpha^k b_\alpha$. From condition 2) of the definition of hyperbolicity of the system (1) it follows that the system (7) is equivalent to the original system (1).

We call (7) the *characteristic form* of the system (1).

The expanded form of this system in terms of components is

$$\sum_{\alpha=1}^{n} l_\alpha^k(x, t, u)\left[\frac{\partial u_\alpha}{\partial t} + \xi_k(x, t, u)\frac{\partial u_\alpha}{\partial x}\right] = f_k(x, t, u) \qquad (k = 1, \ldots, n).$$

We shall sometimes write it in the form

$$l^k(x, t, u)\left(\frac{du}{dt}\right)_k = f_k(x, t, u) \qquad (k = 1, \ldots, n),$$

where $(df/dt)_k$ denotes the quantity

$$\left(\frac{df}{dt}\right)_k = \frac{\partial f}{\partial t} + \xi_k\frac{\partial f}{\partial x}.$$

3. ***A hyperbolic system of nonlinear equations.*** We consider a system of nonlinear equations written in normal form:

$$\frac{\partial v_i}{\partial t} + \varphi_i\left(x, t, v, \frac{\partial v}{\partial x}\right) = 0 \qquad (i = 1, \ldots, n). \tag{1}$$

Setting

$$\partial v_i/\partial x = \omega_i, \tag{2}$$

we write (1) in the form

$$\frac{\partial v_i}{\partial t} + \varphi_i(x, t, v, \omega) = 0 \qquad (i = 1, 2, \ldots, n). \tag{3}$$

Suppose that $\varphi_k(x, t, v, \omega) \in C_2$. We denote by $A(x, t, v, \omega)$ the square matrix of order n

$$A = \left(\left(\frac{\partial \varphi_k(x, t, v, \omega)}{\partial \omega_i}\right)\right) = \left(\left(\frac{\partial \varphi}{\partial \omega}\right)\right). \tag{4}$$

We call the system of nonlinear equations (1) *hyperbolic in some domain* of variation of the variables x, t, v and ω if at each point of this domain the eigenvalues $\xi = \xi_k(x, t, v, \omega)$ are real and from the left eigenvectors $l^k(x, t, v, \omega)$ of the matrix A it is possible to construct a basis of E_n.

The hyperbolic system of nonlinear equations (1) reduces to a system of quasilinear equations of hyperbolic type. Differentiating each of the equations (1) with respect to x and recalling the notation (2), we obtain

$$\begin{aligned}\frac{\partial}{\partial x}\frac{\partial v_i}{\partial t} &+ \frac{\partial}{\partial x}\varphi_i(x, t, v, \omega)\\ &= \frac{\partial \omega_i}{\partial t} + \frac{\partial \varphi_i}{\partial \omega_\alpha}\frac{\partial \omega_\alpha}{\partial x} + \frac{\partial \varphi_i}{\partial v_\alpha}\omega_\alpha + (\varphi_i)'_x = 0.\end{aligned} \tag{5}$$

In (5) summation is implied on the Greek index α from 1 to n.[1] To simplify notation we shall subsequently often use this convention.

Combining (3) and (5), we obtain a system of $2n$ equations

$$\left.\begin{aligned}\frac{\partial v}{\partial t} &= -\varphi(x, t, v, \omega),\\ \frac{\partial \omega}{\partial t} + A(x, t, v, \omega)\frac{\partial \omega}{\partial x} &= \bar{f}(x, t, v, \omega),\end{aligned}\right\} \tag{6}$$

where

$$\left.\begin{aligned}&\bar{f}_i = -[\varphi_i(x, t, v, \omega)]'_x - \frac{\partial \varphi_i}{\partial v_\alpha}\omega_\alpha, \qquad v = \{v_1, \dots, v_n\},\\ &\omega = \{\omega_1, \dots, \omega_n\} \qquad \varphi = \{\varphi_1, \dots, \varphi_n\}, \qquad \bar{f} = \{\bar{f}_1, \dots, \bar{f}_n\},\end{aligned}\right\} \tag{7}$$

which we may consider a system of quasilinear equations for the $2n$ unknowns

$$\begin{aligned}u = \{u_1, \dots, u_{2n}\} &= \{v_1, \dots, v_n; \omega_1, \dots, \omega_n\}\\ &= \left\{v_1, \dots, v_n; \frac{\partial v_1}{\partial x}, \dots, \frac{\partial v_n}{\partial x}\right\}.\end{aligned}$$

[1]To avoid confusion we emphasize that summation is taken only over Greek indices. For example,

$$l^k_\alpha \xi_k \frac{\partial u_\alpha}{\partial x} = \sum_{\alpha=1}^{n} l^k_\alpha \xi_k \frac{\partial u_\alpha}{\partial x},$$

while there is no summation over the Latin index k.

We shall show that the system of $2n$ quasilinear equations (6) is hyperbolic. Multiplying the second group of equations (6) by the left eigenvector $l^k(x, t, v, \omega)$ of the matrix $A(x, t, v, \omega)$, we obtain

$$l^k\left[\frac{\partial \omega}{\partial t} + \xi_k \frac{\partial \omega}{\partial x}\right] = f_k \qquad (k = 1, \ldots, n), \tag{8}$$

$$f_k = l^k \bar{f}. \tag{9}$$

Thus, the system (6) can be brought to the form

$$l^k\left[\frac{\partial \omega}{\partial t} + \xi_k \frac{\partial \omega}{\partial x}\right] = f_k, \quad \frac{\partial v_k}{\partial t} = -\varphi_k(x, t, v, \omega) \qquad (k = 1, \ldots, n), \tag{10}$$

which implies its hyperbolicity.

If $\varphi_k = \varphi_k(x, t, \omega)$, i.e., $\partial\varphi_k/\partial v_i \equiv 0$, then the first group of equations (10) can be considered independently as a hyperbolic system of n quasilinear equations for the n unknowns $\omega_1, \ldots, \omega_n$.

Of course, it is not possible to speak of the equivalence of the systems (10) and (1).

First of all, not every solution v_k, ω_k of (10) gives a solution $v_k(x, t)$ of (1). Indeed, a solution v_k, ω_k of (10) need not, generally speaking, satisfy (2).

We shall show in §9.3 that conditions (2) are satisfied if they are satisfied on the line $t = 0$. Thus, solutions of (10) lead to a solution $v(x, t)$ of (1) only when the conditions (2) are satisfied. On the other hand, the nonequivalence of (10) and (1) is manifest also in the fact that a solution of (10) satisfying (2) requires that $v(x, t) \in C_2$. On the other hand, the definition of a solution $v(x, t)$ of (1) requires only its continuous differentiability. Therefore, equivalence of (10) and (1) can occur only for solutions $v(x, t)$ of (1) of class C_2.

§3. The Riemann invariants

1. *Invariants of semilinear systems of equations.* In each equation of the characteristic form (2.2.7) the functions $u_i(x, t)$ are differentiated in a single direction. In some cases further simplification of the characteristic form is possible: by a change of variables one can arrange that in each of the equations (2.2.7) only one function of the variables x, t and u is differentiated.

We first consider the case of a semilinear system. Equations (2.2.7) can then be written in the form

$$\left(\frac{dr_k}{dt}\right)_k = \frac{\partial r_k}{\partial t} + \xi_k \frac{\partial r_k}{\partial x} = g_k \qquad (k = 1, \ldots, n), \tag{1}$$

where

$$r_k = l^k u = l^k_\alpha u_\alpha, \qquad g_k = f_k + u_\alpha\left(\frac{dl^k_\alpha}{dt}\right)_k = f_k + u_\alpha\left[\frac{\partial l^k_\alpha}{\partial t} + \xi_k \frac{\partial l^k_\alpha}{\partial x}\right]. \quad (2)$$

Since, for a hyperbolic system, Det Λ = Det$((l^k_i)) \neq 0$, it follows that

$$\frac{\partial[r_1, \ldots, r_n]}{\partial[u_1, \ldots, u_n]} = \operatorname{Det} \Lambda \neq 0 \quad (3)$$

and the quantities $r_k(x, t, u)$ can be taken as new unknown functions. We express $u_1, \ldots, u_n$ in terms of $r_1, \ldots, r_n$ from (2):

$$u_k = \lambda^k_\alpha r_\alpha = \lambda^k r, \quad (4)$$

where by r we denote the vector $\{r_1, \ldots, r_n\}$, while λ^k_α are the coefficients of the matrix Λ^{-1} inverse to the matrix Λ:

$$\Lambda = ((l^k_i(x, t))), \qquad \Lambda^{-1} = ((\lambda^k_i(x, t))).$$

Substituting (4) into the right sides of (2), we arrive at the system

$$\frac{\partial r_k}{\partial t} + \xi_k \frac{\partial r_k}{\partial x} = g_k(x, t, r) \qquad (k = 1, \ldots, n), \quad (5)$$

which we call a *system written in invariants*.

We illustrate the concept of invariants for the example of the wave equation

$$\frac{\partial^2 u}{\partial t^2} = a^2 \frac{\partial^2 u}{\partial x^2} \qquad (a = \text{const}).$$

It reduces to the hyperbolic system

$$\frac{\partial u}{\partial t} + a^2 \frac{\partial v}{\partial x} = 0, \qquad \frac{\partial v}{\partial t} + \frac{\partial u}{\partial x} = 0$$

with characteristic form

$$\left(\frac{\partial u}{\partial t} - a \frac{\partial u}{\partial x}\right) - a\left(\frac{\partial v}{\partial t} - a \frac{\partial v}{\partial x}\right) = 0,$$

$$\left(\frac{\partial u}{\partial t} + a \frac{\partial u}{\partial x}\right) + a\left(\frac{\partial v}{\partial t} + a \frac{\partial v}{\partial x}\right) = 0.$$

Hence, the invariants defined by (2) are $r_1 = u - av$ and $r_2 = u + av$. (We remark that we have here used unnormalized vectors l^1 and l^2.)

The system written in invariants,

$$\frac{\partial r_1}{\partial t} - a \frac{\partial r_1}{\partial x} = 0, \qquad \frac{\partial r_2}{\partial t} + a \frac{\partial r_2}{\partial x} = 0,$$

shows that the invariant r_1 is constant along lines $x + at = \text{const}$, while r_2 is constant along lines $x - at = \text{const}$; therefore, $r_1 = f(x + at)$ and $r_2 = g(x - at)$, where f and g are arbitrary functions.

Returning to the function u, we obtain the well-known general solution of the wave equation

$$u = \frac{f(x + at) + g(x - at)}{2}.$$

In the case of a system of quasilinear equations the vectors l^k depend on x, t and u. We consider the differential forms

$$\omega_k(x, t, u, du) = l^k(x, t, u)du = l^k_\alpha(x, t, u)du_\alpha. \tag{6}$$

Suppose that each of these forms considered for fixed values of the variables x and t has an integrating factor $\mu_k = \mu_k(x, t, u)$, so that for any $k = 1, \ldots, n$ we have

$$\mu_k(x, t, u)\omega_k(x, t, u, du) = \mu_k l^k_\alpha\, du_\alpha = \frac{\partial r_k(x, t, u)}{\partial u_\alpha} du_\alpha \tag{7}$$

(we recall that summation goes only over Greek indices; in (7) the index k is fixed). After multiplication by μ_k, equations (2.2.7) take the form

$$\left(\frac{dr_k(x, t, u)}{dt}\right)_k = \frac{\partial r_k}{\partial t} + \xi_k \frac{\partial r_k}{\partial x} = \bar{g}_k(x, t, u) \qquad (k = 1, \ldots, n) \tag{8}$$

and

$$\bar{g}_k = \mu_k f_k + r'_{kt} + \xi_k r'_{kx}. \tag{9}$$

In (9) the quantities r'_{kx} and r'_{kt} are the partial deritavives of $r_k(x, t, u)$ with respect to x and t respectively for fixed values of the variables $u = \{u_1, \ldots, u_n\}$.

Using the independence of the functions $r_k(x, t, u)$ once again, we can express the variables u in terms of them, after which we obtain from (8) the system of quasilinear equations

$$\frac{\partial r_k}{\partial t} + \xi_k \frac{\partial r_k}{\partial x} = g_k(x, t, r) \qquad (k = 1, \ldots, n). \tag{10}$$

The quantities r_k are called *invariants* (the Riemann invariants), while (10) is called a *system in invariants*. B. Riemann first introduced the concept of invariants in his classic work (1860). If the systems (2.2.1) and (2.2.7) are homogeneous and do not depend explicitly on x and t ($A = A(u), f \equiv 0$), then the equations (10) are also homogeneous:

$$\frac{\partial r_k}{\partial t} + \xi_k(r)\frac{\partial r_k}{\partial x} = 0 \qquad (k = 1, \ldots, n), \tag{11}$$

i.e., the functions $r_k(x, t)$ are constant along integral curves of the equations

$$\frac{dx}{dt} = \xi_k(r(x, t)), \tag{12}$$

which are called the *characteristics* of the system (11).

2. ***Systems of two and three quasilinear equations.*** It is known that not every differential form $\omega_k(u, du)$ has an integrating factor. An exception is the case $n = 2$, when this factor always exists. In this case the Riemann invariants can be defined as follows. Suppose that the equations

$$\omega_k(x^0, t^0, u, du) = 0 \qquad (k = 1,\ 2)$$

have the integrals

$$\Phi_k(x^0, t^0, u) = \text{const} \qquad (k = 1, 2).$$

As the Riemann invariants it is then obviously possible to take the functions

$$r_k = \Phi_k(x, t, u).$$

We now consider the case $n = 3$. It is known (see V. V. Stepanov (1958)) that in this case any differential form

$$\omega_k(x, t, u, du) = l^k(x, t, u)\, du \tag{1}$$

(x and t are fixed) can be represented in one of the forms

$$\text{a) } dU, \qquad \text{b) } VdU, \qquad \text{c) } dU + VdW,$$

where U, V, and W are functions of x, t and u. The cases a), b), and c) follow one another in order of generality.

If the forms ω_k belong for $k = 1, 2, 3$ to the types a) and b), then this implies the existence of an integrating factor μ_k for each form ω_k, i.e., the possibility of reducing the system of quasilinear equations to invariants.[2]

In the general case the forms ω_k belong to the type c), and the system can be brought to the form

$$\left(\frac{dU_k}{dt}\right)_k + V_k\left(\frac{dW_k}{dt}\right)_k = \bar{g}_k \qquad (k = 1, 2, 3), \tag{2}$$

[2]It is known (see V. V. Stepanov (1958)) that a form $l^k_\alpha du_\alpha$ belongs to type a) if curl $l^k = 0$ and to type b) if l^k curl $l^k = 0$, and these conditions are necessary and sufficient (the operation curl is taken over the variables u_1, u_2 and u_3).

where U_k, W_k and V_k are functions of the variables u_1, u_2, u_3, x and t. Suppose that $V_3 = 0$, i.e., the form $\omega_3(x, t, u, du)$ has an integrating factor, and the system of quasilinear equations has one Riemann invariant $r_3 = U_3$. Then (2.2.7) can be brought to the form

$$l^k\left(\frac{\partial u}{\partial t} + \xi_k \frac{\partial u}{\partial x}\right) = f_k \qquad (k = 1,\ 2), \tag{3}$$

$$\frac{\partial U_3}{\partial t} + \xi_3 \frac{\partial U_3}{\partial x} = \bar{g}_3, \tag{4}$$

where $U_3 = U_3(x, t, u)$.

If the forms $\omega_k(x, t, u, du)$ $(k = 1, 2)$ are considered on the surface $U_3 = U_3(x, t, u) = \text{const}$, $x = x_0$, $t = t_0$, then they have an integrating factor. This implies that

$$\mu_k(x, t, u)\omega_k(x, t, u, du)$$
$$= dU_k(x, t, u) - \frac{\partial U_k}{\partial U_3} dU_3 - \frac{\partial U_k}{\partial t} dt - \frac{\partial U_k}{\partial x} dx \qquad (k = 1, 2),$$

where it is assumed that

$$U_k(x, t, u) = U_k(x, t, u_1, u_2, U_3(u_1, u_2, u_3, x, t)).$$

In this case the system of quasilinear equations can be written in the following characteristic form:

$$\left.\begin{aligned} \left(\frac{dU_k}{dt}\right)_k + \left(\eta_k - \frac{\partial U_k}{\partial U_3}\right)\left(\frac{dU_3}{dt}\right)_k &= \bar{g}_k \qquad (k = 1, 2), \\ \left(\frac{dU_3}{dt}\right)_3 &= \bar{g}_3, \end{aligned}\right\} \tag{5}$$

where η_k and $\bar{g}_k$ are functions of the variables $U_k(x, t, u)$, x and t.

§4. Transformations of systems of quasilinear equations

1. *Transformation of systems via a solution.* By means of a transformation of independent and dependent variables

$$x' = x'(x, t), \qquad t' = t'(x, t), \qquad v = v(x, t, u), \tag{1}$$

having an inverse, i.e., such that

$$\frac{\partial[x', t']}{\partial[x, t]} \neq 0, \qquad \frac{\partial[v_1, v_2, \ldots, v_n]}{\partial[u_1, u_2, \ldots, u_n]} \neq 0,$$

a system of quasilinear equations of hyperbolic type goes over into a new system of quasilinear equations of hyperbolic type. The characteristic directions of a hyperbolic system are invariants of the transformation (1). This means that if the direction $dx/dt = \xi_k$ is characteristic for the original system, then after the transformation (1) the direction

$$\frac{dx'}{dt'} = \xi_k' = \frac{\partial x'/\partial t + (\partial x'/\partial x)\xi_k}{\partial t'/\partial t + (\partial t'/\partial x)\xi_k}$$

is also characteristic.

We shall consider a transformation of the independent variables used in gas dynamics which we call a *transformation of independent variables via a solution.* Suppose that the new variables x', t' are related to the old variables x, t by the formulas

$$\left.\begin{aligned} dx' &= \varphi_1(x, t, u)dx - \psi_1(x, t, u)\, dt, \\ dt' &= \varphi_2(x, t, u)dx - \psi_2(x, t, u)\, dt. \end{aligned}\right\} \tag{2}$$

In order that the lines $x' = \text{const}$ and $t' = \text{const}$ form a regular mesh for any solutions $u = u(x, t)$ of the original system, i.e., in order that to each point x, t there correspond one and only one point x', t', it is sufficient that in the simply connected domain of the variables x, t in question the following conditions be satisfied:

$$\frac{\partial \varphi_i(x, t, u)}{\partial t} + \frac{\partial \psi_i(x, t, u)}{\partial x} = 0 \qquad (i = 1, 2), \tag{3}$$

$$\Delta = \varphi_1\psi_2 - \varphi_2\psi_1 \neq 0. \tag{4}$$

In (3) $u = u(x, t)$ is an arbitrary solution of the original system; in the differentiation the dependence of u on x and t must be considered.

Since the equations (3) must hold for any solution $u = u(x, t)$ of the original system, they themselves must be consequences of it. We suppose that this is so and (4) is satisfied. From (2) we obtain the differentiation formulas

$$\frac{\partial u}{\partial x} = \frac{\partial u}{\partial x'}\varphi_1 + \frac{\partial u}{\partial t'}\varphi_2, \qquad \frac{\partial u}{\partial t} = -\frac{\partial u}{\partial x'}\psi_1 - \frac{\partial u}{\partial t'}\psi_2. \tag{5}$$

From (5) the derivatives $\partial u/\partial t$ and $\partial u/\partial x$ can be linearly expressed in terms of $\partial u/\partial t'$ and $\partial u/\partial x'$, and after substitution into the original system a system of quasilinear equations is obviously again obtained.

The transformation (2) is more general than ordinary transformations of the independent variables (1). In order than it be applicable, however (in the case where φ_i and ψ_i depend on u), it is necessary that the system of quasilinear equations havespecial equatio ns as consequences. As we shall see in §5, not every system of quasilinear equations has as a consequence even one equation of the type (3).

We shall present an example of a transformation of independent variables via a solution. The system of equations of gas dynamics (see Chapter 2, §3)

$$\left.\begin{aligned} \frac{\partial \rho}{\partial t} + \frac{\partial \rho u}{\partial x} = 0, \\ \frac{\partial u}{\partial t} + u\frac{\partial u}{\partial x} + \frac{1}{\rho}\frac{\partial p}{\partial x} = 0, \qquad p = p(\rho, S), \\ \frac{\partial S}{\partial t} + u\frac{\partial S}{\partial x} = 0 \end{aligned}\right\} \tag{6}$$

contains three equations with three unknowns ρ, u, S. We consider the transformation

$$dx' = \rho dx - \rho u dt, \qquad dt' = dt. \tag{7}$$

In this case $\varphi_1 = \rho$, $\psi_1 = \rho u$, $\varphi_2 = 0$ and $\psi_2 = -1$; conditions (3) are satisfied by virtue of the first of the equations (6), while condition (4) leads to the condition $\rho > 0$. Thus, if $\rho > 0$, then

$$\frac{\partial}{\partial x} = \rho\frac{\partial}{\partial x'}, \qquad \frac{\partial}{\partial t} = \frac{\partial}{\partial t'} - \rho u\frac{\partial}{\partial x'} = \frac{\partial}{\partial t'} - u\frac{\partial}{\partial x},$$

and (6) goes over under the transformation (7) into the new system

$$\frac{\partial}{\partial t'}\left(\frac{1}{\rho}\right) - \frac{\partial u}{\partial x'} = 0, \qquad \frac{\partial u}{\partial t'} + \frac{\partial p}{\partial x'} = 0, \qquad \frac{\partial S}{\partial t'} = 0. \tag{8}$$

In gas dynamics the variables x and t are called *Euler variables*, while the variables $q = x'$ and $t' = t$ are called the *Lagrange variables*.

2. ***The hodograph transformation.*** For a homogeneous system of quasilinear equations with coefficients that do not depend explicitly on x and t, and for $n = 2$,

$$\partial u/\partial t + A(u)\partial u/\partial x = 0, \qquad u = \{u_1, u_2\}, \tag{1}$$

we interchange the roles of the dependent and independent variables, i.e., we assume that $x = x(u_1, u_2)$ and $t = t(u_1, u_2)$. Simple computations lead to the result

$$\frac{\partial u_1}{\partial x} = \Delta\frac{\partial t}{\partial u_2}, \qquad \frac{\partial u_2}{\partial x} = -\Delta\frac{\partial t}{\partial u_1}, \qquad \frac{\partial u_1}{\partial t} = -\Delta\frac{\partial x}{\partial u_2}, \qquad \frac{\partial u_2}{\partial t} = \Delta\frac{\partial x}{\partial u_1}, \tag{2}$$

where

$$\Delta = \begin{vmatrix} \dfrac{\partial u_1}{\partial x} & \dfrac{\partial u_1}{\partial t} \\ \dfrac{\partial u_2}{\partial x} & \dfrac{\partial u_2}{\partial t} \end{vmatrix} = \frac{\partial[u_1, u_2]}{\partial[x, t]}. \tag{3}$$

If $\Delta \neq 0$, then, by substituting (2) into (1), we arrive at the linear system of two equations

$$\left.\begin{aligned} \frac{\partial x}{\partial u_2} - a_{11}(u)\frac{\partial t}{\partial u_2} + a_{12}(u)\frac{\partial t}{\partial u_1} = 0, \\ \frac{\partial x}{\partial u_1} - a_{21}(u)\frac{\partial t}{\partial u_2} + a_{22}(u)\frac{\partial t}{\partial u_1} = 0, \end{aligned}\right\} \tag{4}$$

where $a_{ij}(u)$ are the elements of the matrix $A(u)$.

Such a transformation of the variables is called a *hodograph transformation*; it is used in gas dynamics.

3. *The augmented system.* We write a system of quasilinear equations of hyperbolic type in the characteristic form

$$l^k(x, t, u)\left[\frac{\partial u}{\partial t} + \xi_k(x, t, u)\frac{\partial u}{\partial x}\right]$$
$$= l^k_\alpha\left[\frac{\partial u_\alpha}{\partial t} + \xi_k\frac{\partial u_\alpha}{\partial x}\right] = f_k(x, t, u) \qquad (k = 1, \ldots, n). \tag{1}$$

In many investigations together with (1) it is useful to consider the system in which the derivatives of the solution $u(x, t)$ are also unknowns. This system is obtained by differentiating (1) and is its differential corollary.

We shall call the system (1) and its differential corollaries the *augmented system.*

We set

$$\frac{\partial u}{\partial x} = p, \quad \frac{\partial u}{\partial t} = q \qquad \left(\frac{\partial u_i}{\partial x} = p_i, \quad \frac{\partial u_i}{\partial t} = q_i\right); \tag{2}$$

the system (1) can then be written in the form

$$l^k(q + \xi_k p) = l^k_\alpha(q_\alpha + \xi_k p_\alpha) = f_k \qquad (k = 1, \ldots, n). \tag{3}$$

Differentiating each equation of (3) with respect to the variables t and x, we obtain

$$l^k\left(\frac{\partial q}{\partial t} + \xi_k\frac{\partial p}{\partial t}\right) = \mathscr{G}_k, \qquad l^k\left(\frac{\partial q}{\partial x} + \xi_k\frac{\partial p}{\partial x}\right) = \mathscr{F}_k, \tag{4}$$

where

$$\left.\begin{aligned}\bar{\mathscr{G}}_k &= \frac{\partial f_k}{\partial t} + \frac{\partial f_k}{\partial u_\alpha} q_\alpha \\ &\quad - (q_\alpha + \xi_k p_\alpha)\left(\frac{\partial l^k_\alpha}{\partial t} + \frac{\partial l^k_\alpha}{\partial u_\beta} q_\beta\right) - l^k_\alpha\left(\frac{\partial \xi_k}{\partial t} + \frac{\partial \xi_k}{\partial u_\beta} q_\beta\right) p_\alpha, \\ \bar{\mathscr{F}}_k &= \frac{\partial f_k}{\partial x} + \frac{\partial f_k}{\partial u_\alpha} p_\alpha \\ &\quad - (q_\alpha + \xi_k p_\alpha)\left(\frac{\partial l^k_\alpha}{\partial x} + \frac{\partial l^k_\alpha}{\partial u_\beta} p_\beta\right) - l^k_\alpha\left(\frac{\partial \xi_k}{\partial x} + \frac{\partial \xi_k}{\partial u_\beta} p_\beta\right) p_\alpha.\end{aligned}\right\} \tag{5}$$

From (2) we obtain as integrability conditions the equations

$$\frac{\partial q}{\partial x} = \frac{\partial p}{\partial t}, \tag{6}$$

and hence (4) can be rewritten in the form

$$l^k\left(\frac{\partial q}{\partial t} + \xi_k \frac{\partial q}{\partial x}\right) = \bar{\mathscr{G}}_k, \qquad l^k\left(\frac{\partial p}{\partial t} + \xi_k \frac{\partial p}{\partial x}\right) = \bar{\mathscr{F}}_k, \tag{7}$$

where $\bar{\mathscr{G}}_k$ and $\bar{\mathscr{F}}_k$, according to (5), are functions of x, t, u, p and q.

We call the equations (2) and (7) the *augmented system*. The augmented system can be written in another form. The equations

$$\frac{\partial u}{\partial t} = q, \qquad l^k\left(\frac{\partial q}{\partial t} + \xi_k \frac{\partial q}{\partial x}\right) = \bar{\mathscr{G}}_k \tag{8}$$

are an augmented system of $2n$ equations if the variables p contained in $\bar{\mathscr{G}}_k$ are eliminated by means of (3). It is here assumed that $\xi_k \neq 0$.[3]

The augmented system was introduced in this form by Courant and Lax (1949).

We shall consider a remarkable property of the augmented system. It has been shown that a hyperbolic, semilinear system is reducible to invariants. For systems of quasilinear equations this is not the case, in general. However, the augmented system of any system of quasilinear equations of hyperbolic type does possess this property, i.e., it is reducible to invariants.

Indeed, denoting

$$\mathscr{P}_k = l^k p = l^k_\alpha p_\alpha, \qquad \mathscr{Q}_k = l^k q \tag{9}$$

and passing in (7) to the variables $\mathscr{P}_k$ and $\mathscr{Q}_k$, we obtain

[3]The condition $\xi_k \neq 0$ is not essential. By a change of variables it is possible to arrange that $\xi_k \neq 0$ for all $k = 1, \ldots, n$.

$$\left.\begin{aligned}\frac{\partial \mathscr{P}_k}{\partial t} + \xi_k \frac{\partial \mathscr{P}_k}{\partial x} &= \mathscr{F}_k(x, t, u, \mathscr{P}, \mathscr{Q}),\\ \frac{\partial \mathscr{Q}_k}{\partial t} + \xi_k \frac{\partial \mathscr{Q}_k}{\partial x} &= \mathscr{G}_k(x, t, u, \mathscr{P}, \mathscr{Q}),\end{aligned}\right\} \tag{10}$$

where

$$\left.\begin{aligned}\mathscr{F}_k &= \bar{\mathscr{F}}_k + p_\alpha\left[\frac{\partial l_\alpha^k}{\partial t} + \xi_k \frac{\partial l_\alpha^k}{\partial x} + \frac{\partial l_\alpha^k}{\partial u_\beta}(q_\beta + \xi_k p_\beta)\right],\\ \mathscr{G}_k &= \bar{\mathscr{G}}_k + q_\alpha\left[\frac{\partial l_\alpha^k}{\partial t} + \xi_k \frac{\partial l_\alpha^k}{\partial x} + \frac{\partial l_\alpha^k}{\partial u_\beta}(q_\beta + \xi_k p_\beta)\right].\end{aligned}\right\} \tag{11}$$

Since Det $((l_\alpha^k)) \neq 0$, the quantities p and q can be uniquely expressed in terms of $\mathscr{P}$ and $\mathscr{Q}$, and can be eliminated from $\mathscr{F}_k$ and $\mathscr{G}_k$.

By adjoining to (10) the equations

$$\frac{\partial u_k}{\partial x} = p_k = \lambda_\alpha^k \mathscr{P}_\alpha, \qquad \frac{\partial u_k}{\partial t} = q_k = \lambda_\alpha^k \mathscr{Q}_\alpha, \tag{12}$$

we obtain a system of $4n$ quasilinear equations written in invariants.

It is possible to reduce the number of equations to $2n$ if, for example, to the first groups of equations in (10) we adjoin the first group of equations in (12), i.e., we consider the system of $2n$ equations in invariants

$$\frac{\partial \mathscr{P}_k}{\partial t} + \xi_k \frac{\partial \mathscr{P}_k}{\partial x} = \mathscr{F}_k, \qquad \frac{\partial u_k}{\partial x} = p_k = \lambda_\alpha^k \mathscr{P}_\alpha, \tag{13}$$

and assume that in the functions $\mathscr{F}_k$ the quantities q have been eliminated by means of (3) and the quantities p by means of (12).

However, the second group of equations in (13) is not convenient for investigation. We transform it. From (3) we have

$$q_k = \lambda_\alpha^k f_\alpha - \lambda_\alpha^k l_\beta^\alpha \xi_\alpha p_\beta = \lambda_\alpha^k f_\alpha - \lambda_\alpha^k \xi_\alpha \mathscr{P}_\alpha. \tag{14}$$

Therefore, in place of (13) it is possible to consider the system

$$\frac{\partial \mathscr{P}_k}{\partial t} + \xi_k \frac{\partial \mathscr{P}_k}{\partial x} = \mathscr{F}_k, \qquad \frac{\partial u_k}{\partial t} = q_k = \lambda_\alpha^k f_\alpha - \lambda_\alpha^k \xi_\alpha \mathscr{P}_\alpha, \tag{15}$$

which is also written in invariants, and the functions $\mathscr{F}_k$ are functions of x, t, u and $\mathscr{P}$.

We write the augmented system (15) in the final form

$$\frac{\partial \mathscr{P}_k}{\partial t} + \xi_k \frac{\partial \mathscr{P}_k}{\partial x} = \mathscr{F}_k(x, t, u, \mathscr{P}), \tag{16}$$

$$\frac{\partial u_k}{\partial t} = F_k(x, t, u, \mathscr{P}). \tag{17}$$

From (15), (11), (5), and (3) it follows that

$$\mathscr{F}_k = \mathscr{F}^k(x, t, u) + \mathscr{F}^k_\alpha(x, t, u)\mathscr{P}_\alpha + \mathscr{F}^k_{\alpha\beta}(x, t, u)\mathscr{P}_\alpha\mathscr{P}_\beta, \tag{18}$$

$$F_k = F^k(x, t, u) + F^k_\alpha(x, t, u)\mathscr{P}_\alpha, \tag{19}$$

where $\mathscr{F}^k$, $\mathscr{F}^k_\alpha$, $\mathscr{F}^k_{\alpha\beta}$, F^k and F^k_α are functions depending only on x, t and u. The formulas for these quantities are rather involved, and we shall not write them out here. We note, however, that F^k and F^k_α can be expressed in terms of the coefficients of the original system, while $\mathscr{F}^k$, $\mathscr{F}^k_\alpha$ and $\mathscr{F}^k_{\alpha\beta}$ can be expressed in terms of the coefficients and their first derivatives with respect to the variables x, t and u.

We shall subsequently use the augmented system (16), (17) to estimate the growth of a solution of a system of quasilinear equations and its derivatives (see §8).

According to §2, a system of n nonlinear equations of hyperbolic type can be reduced to a system of $2n$ quasilinear equations. The augmented system for any system of quasilinear equations of hyperbolic type can be reduced in turn to equations in invariants. Therefore, a system of n nonlinear equations of hyperbolic type can be reduced to a system of no more than $4n$ quasilinear equations in invariants by forming an augmented system.

§5. Conservative systems of quasilinear equations

1. *Definitions.* If an equation

$$\frac{\partial\varphi(x, t, u)}{\partial t} + \frac{\partial\psi(x, t, u)}{\partial x} = F(x, t, u) \tag{1}$$

is a consequence of a system of quasilinear equations

$$\frac{\partial u}{\partial t} + A\frac{\partial u}{\partial x} = b \tag{2}$$

for any solutions of (2), then we call it a *conservation law of the system* (2).

Suppose that the system (2) has m conservation laws (1) corresponding to functions $\varphi_1, \ldots, \varphi_m$ and $\psi_1, \ldots, \psi_m$. We call these conservation laws *independent in a domain D* if the functions $1, \varphi_1(x_0, t_0, u), \ldots, \varphi_m(x_0\ t_0, u)$ are linearly independent for all x_0, t_0, u of the domain D in question.

If $\varphi = \varphi(x, t)$, then, according to this definition, (1) is not an independent conservation law.

If the system (2) has n independent conservation laws satisfying the condition

$$\frac{\partial[\varphi_1, \ldots, \varphi_n]}{\partial[u_1, \ldots, u_n]} \neq 0,$$

then we call it *conservative*; otherwise it is called *nonconservative*.

Thus, a conservative system (2) can be brought to the form

$$\frac{\partial\varphi(x, t, u)}{\partial t} + \frac{\partial\psi(x, t, u)}{\partial x} = F(x, t, u), \tag{3}$$

where by φ, ψ and F we now mean vectors with n components. We remark that a system of the type (3) is often called a "divergence" system; sometimes this term refers only to the case $F = 0$.

We shall obtain equations which serve to determine all conservation laws of the system (2), i.e., the functions φ and ψ. To this end we multiply (2) by the vector $\alpha = \alpha(x, t, u) = \{\alpha_1, \ldots, \alpha_n\}$, and we require that the result have the form (1).

We arrive at the equations

$$\alpha_i = \frac{\partial\varphi}{\partial u_i}, \qquad \sum_{j=1}^n \alpha_j a_{ji} = \frac{\partial\psi}{\partial u_i} \quad (i = 1, \ldots, n),$$

$$F = \sum_{i=1}^n \alpha_i b_i + \varphi_t' + \psi_x'. \tag{4}$$

Eliminating the quantities α_i from these equations, we obtain a system containing only the two unknown functions $\varphi(x, t, u)$ and $\psi(x, t, u)$:

$$\frac{\partial\psi}{\partial u_i} = \sum_{j=1}^n a_{ji}(x, t, u)\frac{\partial\varphi}{\partial u_j} \qquad (i = 1, 2, \ldots, n). \tag{5}$$

The variables x and t are contained in the coefficients of this system as parameters.

The set of linearly independent solutions of (5) determines the set of independent conservation laws of (2).

If the system (2) is linear or semilinear, then it is conservative. Indeed, in this case $A = A(x, t)$, and (5) has the n independent solutions

$$\varphi_k(x, t, u) = u_k, \quad \psi_k(x, t, u) = a_{k\alpha}(x, t)u_\alpha \qquad (k = 1, \ldots, n).$$

For $n \leq 2$ the system (5) is either underdetermined or determinate and has an infinite number of solutions. For $n \geq 3$ the system (5) is overdetermined and, generally speaking, does not have a single solution φ, ψ which depends in an essential way on u. A proof of this assertion can be obtained with the example of the system

$$\frac{\partial u_1}{\partial t} + u_2\frac{\partial u_1}{\partial x} = 0, \qquad \frac{\partial u_2}{\partial t} + u_3\frac{\partial u_2}{\partial x} = 0, \qquad \frac{\partial u_3}{\partial t} + u_1\frac{\partial u_3}{\partial x} = 0,$$

for which it is easily established that the system (5) has no nontrivial solutions (see Roždestvenskiĭ (1959a)).

2. ***The conservation laws of gas dynamics.*** As an example (see Roždestvenskiĭ (1957)) we consider the system of equations of gas dynamics in Lagrangian coordinates (Chapter 2, §2.8) which we write in the form

$$\frac{\partial V}{\partial t} - \frac{\partial u}{\partial x} = 0, \qquad \frac{\partial u}{\partial t} + \frac{\partial p(V, S)}{\partial x} = 0, \qquad \frac{\partial S}{\partial t} = 0. \tag{1}$$

We pose the problem of finding all conservation laws of this system (it is obvious that (1) is already written in the form of conservation laws and is therefore conservative).

We assume that the pressure $p = p(V, S)$ is a twice continuously differentiable function of the specific volume V and the entropy S.

For (1) we write out the corresponding system (5.1.5) for $\varphi = \varphi(V, u, S)$ and $\psi = \psi(V, u, S)$:

$$\frac{\partial \psi}{\partial V} = p'_V(V, S)\frac{\partial \varphi}{\partial u}, \tag{2}$$

$$\frac{\partial \psi}{\partial u} = -\frac{\partial \varphi}{\partial V}, \tag{3}$$

$$\frac{\partial \psi}{\partial S} = p'_S(V, S)\frac{\partial \varphi}{\partial u}. \tag{4}$$

We shall seek twice continuously differentiable functions φ and ψ satisfying the system (2)–(4). Combining (2) and (4), we obtain

$$p'_S(V, S)\frac{\partial \psi}{\partial V} - p'_V(V, S)\frac{\partial \psi}{\partial S} = 0,$$

which implies the functional dependence for fixed u of the quantities ψ and p, i.e.,

$$\psi = \Psi(p, u). \tag{5}$$

Using (5), we reduce the system (2)–(4) to the two equations

$$\frac{\partial \Psi}{\partial p}(p, u) = \frac{\partial \varphi(V, u, S)}{\partial u}, \tag{6}$$

$$\frac{\partial \Psi}{\partial u}(p, u) = -\frac{\partial \varphi(V, u, S)}{\partial V}. \tag{7}$$

Differentiating (6) with respect to V and using (7), we obtain

$$p'_V(V, S) \cdot \frac{\partial^2 \Psi(p, u)}{\partial p^2} = - \frac{\partial^2 \Psi(p, u)}{\partial u^2}. \tag{8}$$

We shall consider the case where the quantity p'_V is functionally independent (as a function of the variables V and S) of the pressure $p(V, S)$, i.e.,

$$p'_V p''_{VS} - p'_S \cdot p''_{VV} \neq 0. \tag{9}$$

If this is not the case,[4] then the left side of (8) contains a function of three variables, while the right side contains a function of only two variables. Therefore, (8) is possible only in the case

$$\frac{\partial^2 \Psi(p, u)}{\partial p^2} = \frac{\partial^2 \Psi(p, u)}{\partial u^2} = 0. \tag{10}$$

From this we have the general solution of the system (2)–(4) for $\psi(V, u, S)$:

$$\psi(V, u, S) = \Psi(p, u) = C_1 u + C_2 p + C_3 up, \tag{11}$$

where C_1, C_2, and C_3 are constants.

Substituting (11) into (6) and (7), we obtain

$$\frac{\partial \varphi}{\partial V} = - C_3 p - C_1; \qquad \frac{\partial \varphi}{\partial u} = C_3 u + C_2. \tag{12}$$

From the second law of thermodynamics (Chapter 2, (1.1.16)) we have $p = - \partial \varepsilon(V, S)/\partial V$, and hence

$$\frac{\partial \varphi}{\partial V} = C_3 \frac{\partial \varepsilon(V, S)}{\partial V} - C_1; \qquad \frac{\partial \varphi}{\partial u} = C_3 u + C_2. \tag{13}$$

Integrating (13), we find the general solution of (2)–(4) relative to $\varphi = \varphi(V, u, S)$:

$$\varphi(V, u, S) = - C_1 V + C_2 u + C_3 \left[\varepsilon(V, S) + \frac{u^2}{2} \right] + f(S), \tag{14}$$

where $f(S)$ is an arbitrary function of the entropy.

The quantity $\varepsilon(V, S)$ is the internal energy of the gas, and according to the second law of thermodynamics it is defined by the equality

$$d\varepsilon = T\, dS - p\, dV. \tag{15}$$

[4]The condition (9) excludes the case $p = F(V + f(S))$ where F and f are arbitrary smooth functions.

Thus, if (9) is satisfied, then we have found the general representation of all conservation laws of the system considered:

$$\varphi = -C_1V + C_2u + C_3\left[\varepsilon + \frac{u^2}{2}\right] + f(S), \tag{16}$$

$$\psi = C_1u + C_2p + C_3up + 0\cdot f(S). \tag{17}$$

It is easy to see that (16) and (17) contain both the conservation laws of volume, momentum, and entropy expressed by the equations of (1) as well as the independent conservation law

$$\frac{\partial}{\partial t}\left(\varepsilon + \frac{u^2}{2}\right) + \frac{\partial pu}{\partial x} = 0,$$

which corresponds to the case $C_3 = 1$, $C_1 = C_2 = f(S) = 0$ and is called the *law of conservation of energy*. We observe that when (9) is violated the system (1) may have additional conservation laws.

From our proof it follows that if (9) is satisfied the equations of gas dynamics (1) have no other conservation laws except the known laws of conservation of mass, momentum, energy, and entropy.

As still another example we consider the system

$$\frac{\partial}{\partial t}\frac{\partial \mathscr{L}}{\partial u_i} + \frac{\partial}{\partial x}\frac{\partial \mathscr{L}^1}{\partial u_i} = 0 \qquad (i = 1, \ldots, n), \tag{18}$$

where $\mathscr{L} = \mathscr{L}(u_1, \ldots, u_n)$ and $\mathscr{L}^1 = \mathscr{L}^1(u_1, \ldots, u_n)$ are scalar functions (see S. K. Godunov (1961a)). The system (18) is hyperbolic if the matrix

$$\mathscr{L}_{uu} = \left(\left(\frac{\partial^2 \mathscr{L}}{\partial u_i \partial u_j}\right)\right)$$

is definite.

The system (18) is obviously conservative. We can easily verify that it has still another conservation law which is independent of the conservation laws (18) if $\mathscr{L}_{uu}$ is a nonconstant matrix:

$$\varphi = u_\alpha \mathscr{L}_{u_\alpha} - \mathscr{L}, \qquad \psi = u_\alpha \mathscr{L}^1_{u_\alpha} - \mathscr{L}^1.$$

It is interesting to note that the equations of gas dynamics and also certain other systems of equations of mathematical physics reduce to the form (18).

3. *The potential of a solution of a conservative system of quasilinear equations.* We consider the conservative system of n quasilinear equations

$$\frac{\partial \varphi}{\partial t} + \frac{\partial \psi}{\partial x} = F.$$

According to the definition of conservativity,

$$\frac{\partial[\varphi_1, \ldots, \varphi_n]}{\partial[u_1, \ldots, u_n]} \neq 0.$$

Therefore, as new dependent variables it is possible to take the quantities $v_i = \varphi_i(x, t, u)$ and to consider only conservative systems of the special form

$$\frac{\partial u}{\partial t} + \frac{\partial \varphi(x, t, u)}{\partial x} = F(x, t, u). \tag{1}$$

Suppose that a solution $u(x, t) \in C_1$ of (1) is known. We find a vector $\mathscr{F}(x, t)$ such that

$$\frac{\partial \mathscr{F}(x, t)}{\partial x} = F(x, t, u(x, t)).$$

It is obvious that $\mathscr{F}(x, t)$ is not uniquely determined; to be specific we set

$$\mathscr{F}(x, t) = \int_{x_0(t)}^{x} F(\xi, t, u(\xi, t))\, d\xi, \tag{2}$$

where $x = x_0(t)$ is a smooth curve projecting uniquely onto the axis $x = 0$. The system (1) can be rewritten in the form

$$\frac{\partial u}{\partial t} + \frac{\partial}{\partial x}[\varphi(x, t, u) - \mathscr{F}(x, t)] = 0.$$

Integrating this equation over a domain $\mathscr{G}_C$ bounded by a contour C, we conclude that the contour integral

$$\oint_C u\, dx - [\varphi(x, t, u) - \mathscr{F}(x, t)]\, dt$$

vanishes for any piecewise smooth, closed contour C. Therefore, the curvilinear integral

$$\Phi(x, t) = \int_{(x_0, t_0)}^{(x, t)} u\, dx - [\varphi - \mathscr{F}]\, dt \tag{3}$$

does not depend on the path of integration and defines a vector $\Phi(x, t) \in C_2$ if $\varphi \in C_1$ and $F \in C_0$.

From (3) we obtain

$$\frac{\partial \Phi}{\partial x} = u, \qquad \frac{\partial \Phi}{\partial t} = -\varphi(x, t, u) + \mathscr{F}(x, t).$$

Eliminating u and using (2), we find that

$$\frac{\partial\Phi}{\partial t}+\varphi\left(x,t,\frac{\partial\Phi}{\partial x}\right)=\int_{x_0(t)}^{x}F\left(\xi,t,\frac{\partial\Phi(t,\xi)}{\partial\xi}\right)d\xi. \tag{4}$$

The system of nonlinear integrodifferential equations (4) can now be considered in its own right independently of (1). If a solution $\Phi(x, t)\in C_2$ of (4) is known, then $u=\partial\Phi/\partial x\in C_1$ is a solution of (1). By reducing the system (1) to (4), we can thus consider less smooth solutions $u(x, t)$ of (1) as derivatives of solutions $\Phi(x, t)$ of (4) having greater smoothness.

For this reason the above technique finds application in considering generalized (for example, discontinuous) solutions of systems of quasilinear equations.

We call the vector $\Phi(x, t)$ a *potential for the solution* $u(x, t)$ of the system (1) (see Roždestvenskiĭ (1958a)).

We note some special cases. If $F\equiv 0$, then (4) becomes a nonlinear system of Cauchy-Kovalevskaja type. The reduction of (1) to (4) in this case should be compared with the reverse technique of reducing a nonlinear system to a system of quasilinear equations (§2). If the system (1) is linear, so is (4). Therefore, this process of raising the smoothness of a solution can be further applied to systems of linear equations.

§6. Formulation of the Cauchy problem for a system of quasilinear equations of hyperbolic type

1. *Formulation of the problem.* For the system of quasilinear equations of hyperbolic type

$$\frac{\partial u}{\partial t}+A\frac{\partial u}{\partial x}=b, \tag{1}$$

which we shall also write in characteristic form

$$l^k\left[\frac{\partial u}{\partial t}+\xi_k\frac{\partial u}{\partial x}\right]=f_k \qquad (k=1,\ldots,n),$$

we consider the following problem.

In some neighborhood of an arc $a\le\tau\le b$ of a curve $\mathscr{L}$ $(x=x(\tau),\ t=t(\tau))$, find a solution $u(x, t)$ of the system (1) which on $\mathscr{L}$ assumes prescribed values

$$u(x(\tau),t(\tau))=u^0(\tau),\qquad a\le\tau\le b. \tag{2}$$

The conditions (2) are called the *initial conditions*, the vector-valued function u^0 is called the *initial function*, and the curve $\mathscr{L}$ is called the *initial curve*.

The problem (1), (2) is called an initial value problem or a *Cauchy problem*.

The Cauchy problem for equation (1) can be interpreted geometrically as the problem of constructing in the $(n+2)$-dimensional space of the variables (x, t, u) a two-dimensional surface $u = u(x, t)$ passing through the given curve

$$x = x(\tau), \qquad t = t(\tau), \qquad u = u^0(\tau),$$

which we shall also call the initial curve.

To precisely formulate the Cauchy problem it is necessary to indicate:

a) the smoothness of the matrix $A(x, t, u)$, the vector $b(x, t, u)$ (or l^k, ξ_k and f_k), the initial curve, and the function $u^0(\tau)$ (we call these quantities the initial data of the Cauchy problem); and

b) the domain G of the variables x and t in which a solution of the Cauchy problem is sought.

These questions will be considered in subsequent sections in constructing solutions of the Cauchy problem.

We remark that a solution $u(x, t)$ of (1) is continuously differentiable ($u \in C_1$) by definition. If $u(x, t)$ possesses less smoothness but satisfies (1), then $u(x, t)$ is called a *generalized solution* of (1).

In this chapter we construct a solution $u(x, t) \in C_1$ for a system of quasilinear equations of hyperbolic type; for linear and semilinear systems a generalized solution $u(x, t) \in C$ will be constructed. Following K. O. Friedrichs, we call the latter a *solution* of the Cauchy problem *in the broad sense*.

2. *Solvability of the Cauchy problem. Characterisitics.* Suppose that $x(\tau)$, $t(\tau)$, $u^0(\tau) \in C_1$, $l^k(x, t, u)$, $\xi_k, f_k \in C$, and that a vector-valued function $u(x, t) \in C_1$ assumes the value $u^0(\tau)$ on the curve $\mathscr{L}$ while its derivatives p and q satisfy (6.1.1) on $\mathscr{L}$.

We pose the problem of the possibility of determining (on the line $\mathscr{L}$) the derivatives p, q on the basis of these data, i.e., the question of the existence of a function $u(x, t) \in C_1$ satisfying these conditions.

On the line $\mathscr{L}$ we have the equalities

$$u(x(\tau), t(\tau)) = u^0(\tau), \qquad l^k[q + \xi_k p] = f_k, \tag{1}$$

where l^k, ξ_k, f_k are obviously known functions of the variable τ on $\mathscr{L}$.

Differentiating $u(x(\tau), t(\tau)) = u^0(\tau)$ with respect to τ, we obtain

$$t'(\tau)q + x'(\tau)p = \frac{du^0(\tau)}{d\tau} = \varphi(\tau), \tag{2}$$

and hence $\varphi(\tau) \in C$.

Equations (1) and (2) form a system of 2n equations for determining (on $\mathscr{L}$) the derivatives p and q. Since the matrix $((l^k_\alpha))$ is nonsingular, by eliminating the vector q from (1) and (2), we obtain

$$[t'(\tau)\xi_k - x'(\tau)]l^k p = [t'(\tau)\xi_k - x'(\tau)]l^k_\alpha p_\alpha = \bar{f}_k, \tag{3}$$

where $\bar{f}_k$ is a continuous function of τ. The determinant $D(\tau)$ of the system (3) can easily be computed:

$$D(\tau) = [\mathrm{Det}((l^k_\alpha))] \prod_{k=1}^{n} [t'(\tau)\xi_k - x'(\tau)] = \mathrm{Det}\, \Lambda \prod_{k=1}^{n} [t'(\tau)\xi_k - x'(\tau)].$$

It is nonzero if for all $k = 1, \ldots, n$

$$\left.\frac{dx}{dt}\right|_{\mathscr{L}} = \frac{x'(\tau)}{t'(\tau)} \neq \xi_k = \xi_k(x(\tau), t(\tau), u^0(\tau)). \tag{4}$$

We assume that $|x'(\tau)| + |t'(\tau)| \neq 0$. If $t'(\tau) = 0$, then clearly $D(\tau) \neq 0$, since the ξ_k are bounded.

Thus, if conditions (4) are satisfied, then the system (3) has a unique solution $p = p(\tau)$, and hence the derivatives p and q of $u(x, t)$ are uniquely determined on $\mathscr{L}$ from conditions (1).

Suppose now that $u(x, t) \in C_2$, $x(\tau)$, $t(\tau)$, $u^0(\tau) \in C_2$, and l^k, ξ_k, $f_k \in C_1$. If $u = u^0$ on $\mathscr{L}$, the first derivatives of $u(x, t)$ satisfy (6.1.1) on $\mathscr{L}$, and the second derivatives of $u(x, t)$ satisfy the differential corollaries of (6.1.1) on $\mathscr{L}$ (i.e., the equations obtained by formal differentiation of (6.1.1) with respect to x and t), then, subject to conditions (4), the second derivatives of $u(x, t)$ are also uniquely determined on $\mathscr{L}$.

In the same way, if conditions (4) are satisfied it is possible to determine on $\mathscr{L}$ the derivatives of any order m of the function $u(x, t)$ provided that conditions (1) are satisfied and, moreover, all derivatives of u through order m satisfy all differential corollaries of (6.1.1) through order m on $\mathscr{L}$. Here, of course, the initial data must be sufficiently smooth.

We remark that, as is easily seen, in these arguments it suffices to consider not all differential corollaries but only those which are obtained by differentiation of (6.1.1) in any fixed direction not coinciding with the direction of $\mathscr{L}$ (a so-called *outgoing* direction). The direction of the normal, for example, is such a direction.

This procedure of determining derivatives can be continued as far as desired if the initial data are analytic; and this makes it possible to construct an analytic solution of problem (6.1.1), (6.1.2) for such data. This circumstance forms the analytic basis for the Cauchy-Kovalevskaja method.

If conditions (4) are satisfied on $\mathscr{L}$ for all $k = 1, \ldots, n$, then the Cauchy problem is called *normal*.

A curve $\mathscr{L}$ defined in the space of the $n + 2$ variables x, t, u by the equations

$$x = x(\tau), \qquad t = t(\tau), \qquad u = u^0(\tau), \tag{5}$$

is called a *characteristic of index* k_0 of the system (6.1.1) if on this curve

$$\left.\frac{dx}{dt}\right|_{\mathscr{L}} = \frac{x'(\tau)}{t'(\tau)} = \xi_{k_0}(x(\tau), t(\tau), u^0(\tau)). \tag{6}$$

In the case where several characteristic values ξ_k coincide on $\mathscr{L}$ the curve may be simultaneously a characteristic of several indices k.

Sometimes we shall also call the projection of the curve (5) onto the plane of the variables (x, t) a characteristic, bearing in mind, however, that (6) is satisfied on it.

Suppose that the curve $\mathscr{L}$ in the conditions of the Cauchy problem is a characteristic of index k_0. The left side of the equation of (3) corresponding to $k = k_0$ then vanishes. If the right side $\bar{f}_{k_0}(\tau)$ of this equation does not vanish identically, then the system (3) has no solutions. Therefore, there exists no function $u(x, t) \in C_1$ assuming on $\mathscr{L}$ the prescribed values $u^0(\tau)$ which satisfies (6.1.1) on $\mathscr{L}$. There thus exists no solution $u(x, t) \in C_1$ of the Cauchy problem (6.1.1), (6.1.2).

Thus, if the initial curve is characteristic, then the initial conditions (6.1.2) and the system (6.1.1) are, in general, inconsistent, and the Cauchy problem is not solvable.[5]

In order that the Cauchy problem be meaningful in this case we must require that $\bar{f}_{k_0}(\tau) \equiv 0$.

Thus, if the initial curve is characteristic of index k_0, then the initial data cannot be prescribed arbitrarily; they must satisfy the condition

$$\bar{f}_{k_0}(\tau) = t'(\tau) f_{k_0}(x(\tau), t(\tau), u^0(\tau)) - l^{k_0}(x(\tau). t(\tau), u^0(\tau)) \frac{du^0(\tau)}{dt} \equiv 0, \tag{7}$$

which is called the *solvability condition.*

Suppose that the initial curve $\mathscr{L}$ is a characteristic of index k_0, and the solvability condition (7) is satisfied. Then the system (3) is consistent but has an infinite number of solutions. Thus, the solution $u(x, t)$ of the Cauchy problem is not uniquely determined by the initial conditions (6.1.2), and there exist infinitely many solutions of (6.1.1) satisfying the same initial conditions. We thus arrive at the general definition of a characteristic of the system (6.1.1).

A *characteristic* is a curve $\mathscr{L}$ for which the Cauchy problem is either not solvable or is solvable but not uniquely.

To uniquely determine a solution $u(x, t)$ in the case where the curve $\mathscr{L}$ is

[5]From this consideration there follows another definition of a characteristic as a curve $\mathscr{L}$ on which some linear combination of the equations of the system being considered contains only inner derivatives, i.e., derivatives with respect to the parameter τ in the direction of $\mathscr{L}$.

characteristic and the solvability conditions are satisfied one may impose certain additional conditions. Examples of such problems are considered in §11.

The question of continuing a solution $u(x, t)$ across the curve $\mathscr{L}$ is related to the Cauchy problem. Suppose that a solution $u(x, t) \in C_1$ is known on one side of $\mathscr{L}$, and it is required to continue it to the other side. This problem reduces to a Cauchy problem with $\mathscr{L}$ as the initial curve.

If $\mathscr{L}$ is not characteristic, then this Cauchy problem is normal, and the question of extending the solution can be solved uniquely. Moreover, the condition of the continuity of the extension implies the continuity of all derivatives of $u(x, t)$ that exist on $\mathscr{L}$; in particular, it implies that $u(x, t) \in C_1$. If the curve $\mathscr{L}$ is characteristic, then the corresponding Cauchy problem is still solvable, since the solvability condition (7) is obviously satisfied (since $u(x, t)$ is a solution on one side of $\mathscr{L}$). However, it is not uniquely solvable.

We shall consider, for example, a continuous extension. As we have seen, the values of $u(x, t)$ on $\mathscr{L}$ do not uniquely determine its first derivatives; therefore, a continuous extension of the solution with a discontinuity of the first derivatives on the characteristic $\mathscr{L}$ is possible in an infinite number of ways. If continuity of the first derivatives on $\mathscr{L}$ is required, then derivatives of higher order may still suffer a discontinuity, so that even in this case the continuation is not uniquely determined.

Thus, a characteristic is a line through which a solution can be continued nonuniquely.

The problem of continuing a solution $u(x, t)$ through a characteristic $\mathscr{L}$ is uniquely solvable only in the case of analytic solutions (as is, obviously, any problem of analytic continuation of a function).

We shall now devote our main attention to the normal Cauchy problem. By a transformation of the variables $x' = x'(x, t)$, $t' = t'(x, t)$ taking the curve $\mathscr{L}$ into a segment of the axis $t' = 0$, the general problem reduces to a Cauchy problem of special type with initial conditions prescribed on the axis $t = 0$: find a solution $u(x, t)$ of (6.1.1) which satisfies the initial conditions

$$u(x, 0) = u^0(x), \qquad a \leq x \leq b. \tag{8}$$

From the boundedness of the quantities ξ_k it follows that this Cauchy problem is normal. We shall solve the Cauchy problem (6.1.1), (8) only in the half plane $t \geq 0$. The construction of a solution $u(x, t)$ for $t \leq 0$ is carried out in similar fashion.

3. ***The domain of dependence and the domain of determinacy. The concept of correctness of the Cauchy problem.*** Suppose a solution $u(x, t)$ of the system (6.1.1) is known that assumes the initial values (6.28). Through a point

$\mathscr{M} = (x^0, t^0)$ of the half plane $t > 0$ we draw the characteristics $x = x_k(x^0, t^0, \tau)$ defined by the equations

$$\frac{dx_k}{d\tau} = \xi_k(x_k, \tau, u(x_k, \tau)) \qquad (k = 1, 2, \ldots, n),$$

to their intersection with the axis $t = 0$. Suppose that they intersect this axis at certain points, and call the two extreme intersection points a' and b' $(a' < b')$ (Figure 1.2). The segment $a' \leq x \leq b'$ of the initial axis $t = 0$ is called the *domain of dependence of the solution* $u(x, t)$ at the point $\mathscr{M}$.

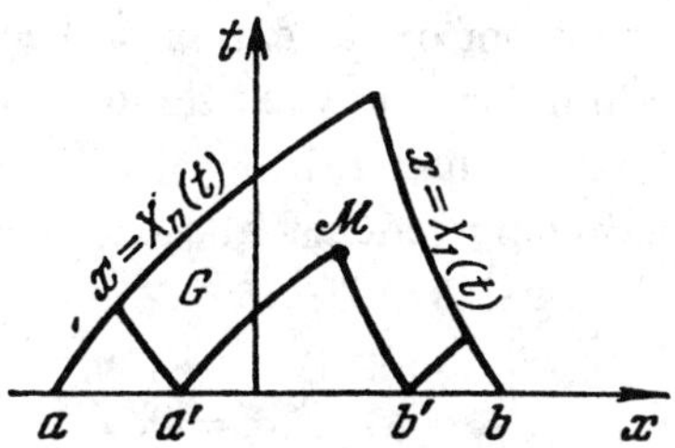

FIGURE 1.2

The *domain of determinacy* G of a solution of the Cauchy problem is the part of the half plane $t \geq 0$ consisting of all points (x, t) for which the domain of dependence $a' \leq x \leq b'$ belongs to the initial segment $[a, b]$, i.e., $[a', b'] \subset [a, b]$.

Finally, the *domain of influence of a segment* $a' \leq x \leq b'$ of the initial axis is the domain G' of the half plane $t \geq 0$ consisting of all points (x, t) with domain of dependence having nonempty intersection with $[a', b']$.

Inasmuch as the characteristics of the system can only be found simultaneously with a solution $u(x, t)$, the determination of the domain of determinacy is a difficult problem. The situation is considerably simpler in the case of a semilinear system, where $\xi_k = \xi_k(x, t)$. In this case the domain of determinacy G is given by the conditions

$$G: t \geq 0, \qquad X_n(t) \leq x \leq X_1(t),$$

where $X_1(t)$ and $X_n(t)$ denote the solutions of the differential equations

$$\frac{dX_n(t)}{dt} = \max_{k=1,\ldots,n} \{\xi_k(X_n(t), t)\}, \qquad \frac{dX_1(t)}{dt} = \min_{k=1,\ldots,n} \{\xi_k(X_1(t), t)\},$$

which assume for $t = 0$ the values $X_n(0) = a$ and $X_1(0) = b$.

In the case of quasilinear equations an a priori determination of G is difficult. However, if it is known that $\|u(x, t)\| \leq U$, then $\bar{G} \subseteq G$, where

$$\bar{G}: t \geq 0, \qquad \bar{X}_n(t) \leq x \leq \bar{X}_1(t),$$

$$\frac{d\bar{X}_n}{dt} = \max_{k=1,\ldots,n} \max_{\|u\|\leq U} \{\xi_k(\bar{X}_n(t), t, u)\}, \qquad \bar{X}_n(0) = a,$$

$$\frac{d\bar{X}_1}{dt} = \min_{k=1,\ldots,n} \min_{\|u\|\leq U} \{\xi_k(\bar{X}_1(t), t, u)\}, \qquad \bar{X}_1(0) = b.$$

A Cauchy problem is called *correct* or *correctly posed* if it has a unique solution $u(x, t)$ which depends continuously on the initial data. Of course, the question of the metric in which continuous dependence occurs depends on the classes of solutions and initial data considered and is resolved differently in each of these classes. In the proof of the existence theorems for a solution of the Cauchy problem we will indicate in what sense the continuous dependence of solutions on the initial data is intended.

We shall clarify the concepts introduced above with the example of a linear system with constant coefficients:

$$\frac{\partial u}{\partial t} + A\frac{\partial u}{\partial x} = 0. \tag{1}$$

The invariants

$$r_k = l^k u = l^k_\alpha u_\alpha \tag{2}$$

satisfy the equations

$$\frac{\partial r_k}{\partial t} + \xi_k \frac{\partial r_k}{\partial x} = 0 \qquad (k = 1, \ldots, n),$$

i.e., the system (1) decomposes into n independent equations. The characteristics of the system (1) are the lines

$$x = x_k = x_k^0 + \xi_k t.$$

Hence, the domain of determinacy of a solution of the Cauchy problem for the system (1) is the triangle

$$G: t \geq 0, \qquad a + \xi_n t \leq x \leq b + \xi_1 t,$$

while the domain of dependence at a point $\mathscr{M}(x, t)$ is the segment $[a', b']$ of the axis $t = 0$, where $a' = x - \xi_n t$ and $b' = x - \xi_1 t$ (Figure 1.3).

The functions $r_k(x, t)$ are easily found:

$$r_k(x, t) = f_k(x - \xi_k t),$$

where the f_k are arbitrary functions.

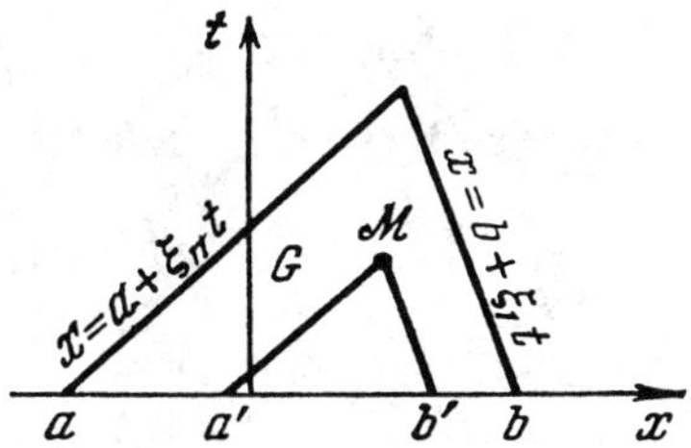

FIGURE 1.3

If we impose the initial conditions $u(x, 0) = u^0(x)$, then, according to (2),

$$r_k(x, 0) = l^k u^0(x) = r_k^0(x).$$

Hence

$$r_k(x, t) = r_k^0(x - \xi_k t) = l^k u^0(x - \xi_k t) = l_\alpha^k u_\alpha^0(x - \xi_k t).$$

Returning via (2) to the functions u, we obtain

$$u_k(x, t) = \lambda_\alpha^k l_\beta^\alpha u_\beta^0(x - \xi_\alpha t).$$

This immediately implies the continuous dependence of a solution of the Cauchy problem for the system (1) with constant coefficients on the initial data—the matrix A and the initial functions $u^0(x)$.

4. ***The method of characteristics and a survey of results.*** The application of the method of characteristics to the proof of the basic theorems on the solvability of the Cauchy problem will be expounded in detail in §§7 and 8. Here we only briefly describe the notion of the method of characteristics, so that the reader not interested in the details may get some idea of the method of characteristics and the results it yields without reading the proofs of the corresponding theorems.

Suppose that for the system

$$l^k(x, t, u)\left[\frac{\partial u}{\partial t} + \xi_k(x, t, u)\frac{\partial u}{\partial x}\right] = f_k(x, t, u) \tag{1}$$

we pose the initial conditions

$$u(x, 0) = u^0(x). \tag{2}$$

For simplicity we assume that $a = -\infty$ and $b = \infty$; we assume that the quantities l^k, ξ_k, f_k, and u^0 are sufficiently smooth functions of their arguments.

We suppose that in the strip $0 \leq t \leq t_0$ a smooth solution $u(x, t)$ of problem (1), (2) is known. The functions

$$\bar{l}^k(x, t) = l^k(x, t, u(x, t)), \qquad \bar{\xi}_k = \xi_k(x, t, u(x, t)),$$
$$\bar{f}_k = f_k(x, t, u(x, t))$$

may then be considered functions of x and t, while (1) may be considered as a system of linear equations

$$\bar{l}^k\left[\frac{\partial u}{\partial t} + \bar{\xi}_k \frac{\partial u}{\partial x}\right] = \bar{f}_k$$

and it can be written in invariants:

$$\frac{\partial r_k}{\partial t} + \bar{\xi}_k \frac{\partial r_k}{\partial x} = \bar{g}_k(x, t). \tag{3}$$

Here the functions $\bar{g}_k(x, t)$ can be expressed in terms of $u(x, t)$, $\bar{f}_k$, $\bar{l}^k$, $\bar{\xi}_k$, and the first derivatives of $\bar{l}^k$.

Each of the equations (3) can be integrated. Indeed, if we denote by $x = x_k(x^0, t^0, \tau)$ the solution of the problem

$$\frac{dx_k}{d\tau} = \bar{\xi}_k(x_k, \tau), \qquad x_k|_{\tau=t^0} = x^0,$$

then $\partial/\partial t + \bar{\xi}_k \partial/\partial x$ is the operator of differentiation with respect to t in the direction of the characteristic $x = x_k(x^0, t^0, \tau)$; therefore,

$$r_k(x, t) = r_k^0(x_k(x, t, 0)) + \int_0^t \bar{g}_k(x_k(x, t, \tau), \tau)\, d\tau, \tag{4}$$

where, in correspondence with (3.1.2), the r_k^0 are equal to

$$r_k^0(x) = \bar{l}^k(x, 0)u^0(x) = l^k(x, 0, u^0(x))u^0(x).$$

In view of the hyperbolicity of the system (1) the matrix $((\bar{l}_\alpha^k))$ is nonsingular, and hence $u(x, t)$ can be found from (4):

$$u_k(x, t) = \bar{\lambda}_\alpha^k(x, t) r_\alpha(x, t), \qquad \bar{\lambda}_\alpha^k = \lambda_\alpha^k(x, t, u(x, t))$$

(see (3.1.4)).

However, we do not know the solution $u(x, t)$, and hence $\bar{l}^k$, $\bar{\xi}_k$ and $\bar{f}_k$ are unknown. Therefore, with the exception of the simplest cases the construction of the solution $u(x, t)$ does not reduce to the indicated simple procedure but requires application of the method of successive approximations.

Suppose that in the strip $0 \le t \le t_0$ an approximate value $\overset{(s)}{u}(x, t)$ of the solution of the Cauchy problem (1), (2) is known. We can then determine the quantities $\overset{(s)}{l^k}$, $\overset{(s)}{\xi}$ and $\overset{(s)}{f_k}$, and find the next approximation $\overset{(s+1)}{u}(x, t)$ in the manner indicated above.

Thus, the approximation $\overset{(s+1)}{u}(x, t)$ can be considered as the result of applying to $\overset{(s)}{u}(x, t)$ a certain operator T:

$$\overset{(s+1)}{u}(x, t) = T\overset{(s)}{u}(x, t).$$

This operator is nonlinear and contains the operations of differentiation with respect to x and t and integration along the characteristics. In this approach the solution $u(x, t)$ of problem (1), (2) satisfies the equation $u(x, t) = Tu(x, t)$, which obviously describes the Cauchy problem (1), (2) symbolically. To prove the convergence of the successive approximations $\{\overset{(s)}{u}(x, t)\}$ it is necessary to first establish their uniform boundedness in some strip $0 \leq t \leq t_0$. Thereafter the proof of convergence reduces to establishing the complete continuity of the operator T (i.e., the fact that it maps any bounded set into a compact set). Finally, it is shown that the limit possesses the required smoothness and is a solution of our problem. This last step is usually connected with an investigation of the sequence of derivatives $\partial\overset{(s)}{u}/\partial x$. These questions are studied in detail in the next two sections.

The first results on the existence and uniqueness of a solution of the Cauchy problem were obtained by the Cauchy-Kovalevskaja method for systems of equations of Cauchy-Kovalevskaja type assuming analyticity of the initial data of the Cauchy problem. These restrictive conditions limit the value of the results, since it is desirable to consider the Cauchy problem for equations of hyherbolic type with minimal smoothness conditions on the initial data.

H. Lewy (1927) showed that the solution of a system of linear equations of hyperbolic type with two independent variables reduces essentially to solving a Cauchy problem for systems of ordinary differential equations. This work formed the basis for the classical method of characteristics.[6]

Here we briefly discuss results obtained recently on the question of the solvability of the Cauchy problem for systems of equations of hyperbolic type with two independent variables.

K.O. Friedrichs (1948) considered the question of the existence and uniqueness of a solution of problem (1), (2) for systems of linear, semilinear, and quasilinear equations. For a linear system Friedrichs assumed continuous differentiability of $l^k(x, t)$, Lipschitz continuity of $\xi_k(x, t)$, and continuity of

[6]For a single equation the method of characteristics had been developed earlier.

$f_k(x, t, u) = f^k(x, t) + f^k_\alpha(x, t)u_\alpha$. Under these conditions, he established the existence of a solution in the broad sense assuming only continuity of $u^0(x)$ (the concept of a solution in the broad sense will be considered in §7). For a system of quasilinear equations Friedrichs required that l^k_α, $\xi_k \in C_2$, $f_k \in C_1$ and $u^0(x) \in C_2$. Under these conditions the existence of a solution $u(x, t) \in C_2$ is proved.

Courant and Lax (1949) use the representation of the prolonged system in invariants. In spite of the elegance of the proofs, they imposed more restrictive assumptions on the smoothness of the initial data. Thus, for example, the existence of three derivatives of the initial functions $u^0(x)$ is required.

In the work of Douglis (1952) and Hartman and Wintner (1952) a solution of problem (1), (2) is constructed assuming continuous differentiability of the initial data. Douglis first construct a solution for smoother initial data; the original initial data are considered by means of a limiting procedure. Hartman and Wintner constructed a solution under the assumptions indicated directly by the method of characteristics. Lemma 2, presented in the next subsection, plays an essential role in their construction.

Thus, the minimal requirement on the initial data under which the existence and uniqueness of a solution $u(x, t) \in C_1$ have so far been proved is the condition of their continuous differentiability. To this it should be added that the simplest examples show that a solution $u(x, t) \in C_1$ does not exist if the initial data are not differentiable.

We remark that some existence theorems can be obtained as a specialization to the case of two independent variables of more general theorems. Thus, for example, Petrovskiĭ (1937) and Hristianovič (1937) obtained general results which, in particular, imply existence theorems for the case of interest to us. We note, however, that here the conditions on the initial data are naturally more stringent.

Finally, we make several remarks on the exposition of these questions given in the present book. The Cauchy problem for a linear system is studied on the basis of the work of Friedrichs; the existence theorem is proved under the same assumptions.

The method of characteristics as presented by Hartman and Wintner is taken as the basis for systems of quasilinear equations. Some essential details, however, distinguish our exposition. We mention some of them. The proof of the boundedness of the successive approximations and their derivatives is usually very cumbersome in the method of characteristics. The growth estimates by means of a "majorant system" that we use considerably simplify and generalize these estimates while avoiding the necessity of arithmetic computation. Another difference in our presentation is in the proof of the uniform convergence in the domain of determinacy G not only of the successive approximations but also of the sequence of their first derivatives.

5. ***Two lemmas.***

LEMMA 1. *Suppose that a vector-valued function* $u(t) = \{u_1, \ldots, u_n\}$ *which is continuous on the segment* $0 \leq t \leq t_0$ *satisfies the inequality*

$$\|u(t)\| \leq U_0 + \int_0^t [A(\tau) + B(\tau) \max_{0 \leq \xi \leq \tau} \|u(\xi)\|]\, d\tau, \tag{1}$$

and suppose that for $0 \leq t \leq t_0$

$$|A(t)| \leq A, \qquad |B(t)| \leq B, \qquad U_0 \geq 0.$$

Then for $0 \leq t \leq t_0$

$$\|u(t)\| \leq \max_{0 \leq \tau \leq t} \|u(\tau)\| \leq U_0 e^{Bt} + \frac{A}{B}(e^{Bt} - 1). \tag{2}$$

For $B = 0$ *this becomes the obvious inequality*

$$\|u(t)\| \leq \max_{0 \leq \tau \leq t} \|u(\tau)\| \leq U_0 + At.$$

PROOF. We set $U(t) = \max_{0 \leq \tau \leq t} \|u(\tau)\|$. Let $t \geq 0$ be any number in the segment $[0, t_0]$, and suppose that $U(t) = \|u(t')\|$ $(0 \leq t' \leq t)$. Writing (1) for the point t', we obtain

$$\begin{aligned}\|u(t')\| = U(t) &\leq U_0 + \int_0^{t'} [A + BU(\tau)]\, d\tau \\ &\leq U_0 + \int_0^t [A + BU(\tau)]\, d\tau = U_0 + At + B\int_0^t U(\tau)\, d\tau.\end{aligned}$$

By repeated application of this estimate we have

$$\begin{aligned}U(t) \leq U_0\left[1 + Bt + \cdots + \frac{(Bt)^s}{s!}\right] \\ + A\left[t + \frac{Bt^2}{2!} + \cdots + \frac{(Bt)^{s+1}}{B(s+1)!}\right] + U(t_0)\frac{(Bt)^{s+1}}{(s+1)!},\end{aligned}$$

whence (2) follows.

COROLLARY. *Suppose that a vector* $u(x, t) \in C_1$ *satisfies*

$$\left\| \frac{du}{dt} \right\| \le \|f(t)\| + \|f^k_\alpha u_\alpha(t)\|.$$

Then (2) *holds with*

$$U_0 = \|u(0)\|, \qquad A = \max_t \|f(t)\|, \qquad B = \max_t \|((f^k_\alpha(t)))\|.$$

LEMMA 2. *Let* $u(x, t, \tau)$, $v(x, t, \tau) \in C_1$. *Then the function*

$$I(x, t) = \int_0^t u(x, t, \tau) \frac{\partial v(x, t, \tau)}{\partial \tau} \, d\tau$$

is continuously differentiable ($I(x, t) \in C_1$).

PROOF. We compute, for example, the derivative $\partial I/\partial x$. We have

$$\frac{1}{\Delta x}[I(x + \Delta x, t) - I(x, t)]$$

$$= \frac{1}{\Delta x} \int_0^t \left\{ u(x + \Delta x, t, \tau) \frac{\partial v(x + \Delta x, t, \tau)}{\partial \tau} - u(x, t, \tau) \frac{\partial v(x, t, \tau)}{\partial \tau} \right\} d\tau$$

$$= \int_0^t \frac{u(x + \Delta x, t, \tau) - u(x, t, \tau)}{\Delta x} \cdot \frac{\partial v(x + \Delta x, t, \tau)}{\partial \tau} \, d\tau$$

$$+ \frac{1}{\Delta x} \int_0^t u(x, t, \tau) \frac{\partial}{\partial \tau} [v(x + \Delta x, t, \tau) - v(x, t, \tau)] \, d\tau.$$

Integrating by parts in the last integral and passing to the limit as $\Delta x \to 0$, we obtain

$$\frac{\partial I(x, t)}{\partial x} = u(x, t, t) v_x(x, t, t) - u(x, t, 0) v_x(x, t, 0)$$
$$+ \int_0^t [u_x(x, t, \tau) v_\tau(x, t, \tau) - u_\tau(x, t, \tau) v_x(x, t, \tau)] \, d\tau. \qquad (3)$$

This proves Lemma 2 and simultaneously provides a rule for computing the derivatives of $I(x, t)$.

§7. The Cauchy problem for linear and semilinear systems

1. ***Existence and uniqueness of the solution to the Cauchy problem in the broad sense.*** We consider the semilinear system

$$\frac{\partial u}{\partial t} + A(x, t) \frac{\partial u}{\partial x} = b(x, t, u), \qquad (1)$$

and we suppose that the system

$$\frac{\partial r_k}{\partial t} + \xi_k(x, t)\frac{\partial r_k}{\partial x} = g_k(x, t, r) \tag{2}$$

is the form of the system (1) in invariants.

Suppose that on some segment $[a, b]$ of the axis $t = 0$ for the system (1) there are given the initial conditions

$$u(x, 0) = u^0(x). \tag{3}$$

We remark that the segment $[a, b]$ may be unbounded.

Setting $r_k^0(x) = l^k(x, 0)u^0(x)$, we obtain the initial conditions

$$r(x, 0) = r^0(x) \tag{4}$$

for the system (2).

We assume that A, ξ_k, $l^k \in C_1$ in G (we recall that G denotes the domain of determinacy of our problem), f_k, $\partial f_k/\partial u_\alpha \in C_0$ in the region $\bar{G} \times \{u\colon -\infty < \|u\| < \infty\}$, and $u^0(x) \in C_0$ on $[a, b]$. Then g_k, $\partial g_k/\partial r_k \in C_0$ in the region $G \times \{r\colon -\infty < \|r\| < \infty\}$ and $r^0(x) \in C_0$ on $[a, b]$.

Functions $r(x, t)$ continuous in G are called *a solution of the Cauchy problem* (2), (4) *in the broad sense* if $r(x, 0) = r_0(x)$ and each of the functions $r_k(x, t)$ is continuously differentiable in t along the corresponding characteristic $x = x_k(\xi, \tau, t)$, and

$$\frac{d}{dt} r_k(x_k(\xi, \tau, t), t) = g_k(x_k, t, r(x_k, t)). \tag{5}$$

The vector $u(t, x)$ obtained from the vector $r(x, t)$ according to (3.1.4) is called *a solution in the broad sense of the Cauchy problem* (1), (3).

We shall prove the *uniqueness* of a solution in the broad sense. Suppose that in G there exist two solutions $u(x, t)$ and $\bar{u}(x, t)$ in the broad sense of problem (1), (3). To them there correspond two solutions in the broad sense $r(x, t)$ and $\bar{r}(x, t)$ of problem (2), (4).

We introduce the difference

$$v(x, t) = r(x, t) - \bar{r}(x, t) \qquad (v(x, 0) \equiv 0). \tag{6}$$

Subtracting from equation (5) written for $r(x, t)$ the same equation written for $\bar{r}(x, t)$, we obtain

$$\frac{dv_k(x_k(\xi, \tau, t), t)}{dt} = g_\alpha^k(x_k, t)v_\alpha(x_k, t), \tag{7}$$

where $g^k(x, t)$ denote the quantities

$$g_\alpha^k(x, t) = \int_0^1 \frac{\partial g_k}{\partial r_\alpha}(x, t, r(x, t) - \lambda v(x, t))\, d\lambda.$$

According to the definition of a solution in the broad sense, the functions r, $\bar{r}$, and v are continuous in G. Therefore, the functions g_α^k are continuous in G and are bounded in any strip $0 \leq t \leq t_0$.

Integrating (7) with respect to τ from t to 0 and using (6), we obtain

$$v_k(\xi, \tau) = \int_0^\tau g_\alpha^k(x_k, t) v_\alpha(x_k, t)\, dt \qquad (k = 1, \ldots, n).$$

Since the matrix $((g_\alpha^k))$ is bounded, after applying Lemma 1 of §6 we obtain from this that $\|v(x, t)\| = 0$ everywhere in G, and hence $r(x, t) \equiv \bar{r}(x, t)$ and $u(x, t) \equiv \bar{u}(x, t)$. The proof of the uniqueness is complete.

From this uniqueness theorem, of course, it also follows that a classical (i.e., continuously differentiable) solution of the Cauchy problem is unique.

We shall prove the *existence* of a solution in the broad sense for a linear system. We construct the solution of problem (2), (4) for a linear system by the method of successive approximations.

Let

$$g_k(x, t, r) = g^k(x, t) + g_\alpha^k(x, t) r_\alpha.$$

According to (5), the solution r_k satisfies

$$r_k(x, t) = r_k^0(x_k(x, t, 0)) + \int_0^t g^k(x_k(x, t, \tau), \tau)\, d\tau$$
$$+ \int_0^t g_\alpha^k(x_k(x, t, \tau), \tau) r_\alpha(x_k(x, t, \tau), \tau)\, d\tau. \qquad (8)$$

Applying the method of successive approximations, we set

$$\overset{(s+1)}{r_k}(x, t) = \overset{(0)}{r_k}(x, t) + \int_0^t g_\alpha^k(x_k(x, t, \tau), \tau) \overset{(s)}{r_\alpha}(x_k(x, t, \tau), \tau)\, d\tau$$
$$(s = 0. 1, \ldots), \qquad (9)$$

where

$$\overset{(0)}{r_k}(x, t) = r_k^0(x_k(x, t, 0)) + \int_0^t g^k(x_k(x, t, \tau), \tau)\, d\tau.$$

From this it is evident that $\overset{(s+1)}{r_k}(x, t)$ are defined and continuous in G and have a continuous derivative with respect to t in the corresponding characteristic direction. We shall prove the uniform convergence of the sequence $\{\overset{(s)}{r}(x, t)\}$.

From (9) it follows that

$$\left\|\overset{(s+1)}{r}(x, t) - \overset{(s)}{r}(x, t)\right\| \leq B \int_0^t \sup_{G_\tau} \left\|\overset{(s)}{r} - \overset{(s-1)}{r}\right\| d\tau, \tag{10}$$

where G_τ denotes the intersection of G with the strip $0 \leq t \leq \tau$, and B is a constant bounding the norm of the matrix $((g_\alpha^k))$ in G.

Setting

$$\overset{(s)}{V}(t) = \sup_{(\xi, \eta) \in G_t} \left\|\overset{(s+1)}{r_k}(\xi, \eta) - \overset{(s)}{r_k}(\xi, \eta)\right\|,$$

we write (10) in the form

$$\overset{(s)}{V}(t) \leq B \int_0^t \overset{(s-1)}{V}(\tau)\, d\tau.$$

Then

$$\overset{(s)}{V}(t) \leq \frac{(Bt)^s}{s!} \overset{(0)}{V}(t) \to 0 \quad \text{as } s \to \infty,$$

and hence the sequence $\{\overset{(s)}{r}(x, t)\}$ converges uniformly to a continuous function $r(x, t)$ if the domain G is finite in the variable t. If G is infinite in t, then the sequence $\{\overset{(s)}{r}(x, t)\}$ converges uniformly on any finite (with respect to t) subdomain G_{t_0}. Passing to the limit as $s \to \infty$ in (9), we find that $r(x, t)$ satisfies (8). Since $x_k(x, t, \tau) = x_k(x_0, t_0, \tau)$, if the point (x_0, t_0) lies on this characteristic, then (8) can be rewritten in the form

$$r_k(x_k(x_0, t_0, t), t) = r_k^0(x_k(x_0, t_0, 0)) + \int_0^t g^k(x_k(x_0, t_0, \tau), \tau)\, d\tau$$

$$+ \int_0^t g_\alpha^k(x_k(x_0, t_0, \tau), \tau) r_\alpha(x_k(x_0, t_0, \tau), \tau)\, d\tau. \tag{11}$$

The continuity of the integrands implies the continuous differentiability of the right side of (11) with respect to t. Hence, $r_k(x, t)$ is continuously differentiable along the characteristic $x = x_k$; moreover, (5) is satisfied. Thus, $r_k(x, t)$ is a solution of problem (2), (4) in the broad sense. The solution $u(x, t)$ of problem (1), (3) can be obtained from (3.1.4).

In the case of a semilinear system the successive approximations are defined as in (9):

$$\overset{(s+1)}{r_k}(x, t) = \overset{(0)}{r_k}(x, t) + \int_0^t g_k(x_k(x, t, \tau), \tau)\overset{(s)}{r}(x_k(x, t, \tau), \tau)\, d\tau,$$

and they converge in a subdomain G_{t_0} of G in which they are uniformly bounded.

Thus, the construction of a solution in the broad sense for a semilinear system differs from the linear case only in that the domain of convergence of the successive approximations is a domain G_{t_0} in which the solution of the semilinear system remains bounded. We defer the question of boundedness of the solution to §8.

2. ***Existence of a classical solution of the Cauchy problem for a linear system.*** Suppose that the system (7.1.1) is linear: $b(x, t, u) = b_1(x, t) + B(x, t)u$. Then in the system (7.1.2)

$$g_k(x, t, r) = g^k(x, t) + g^k_\alpha(x, t)r_\alpha,$$

and $g^k(x, t) = l^k b_1$, while

$$g^k_\alpha(x, t) = \gamma^k_\alpha(x, t) + \left[\frac{\partial l^k_\beta}{\partial t} + \xi_k \frac{\partial l^k_\beta}{\partial x}\right]\lambda^\beta_\alpha(x, t), \tag{1}$$

where $\gamma^k_\alpha = l^k_\beta B_{\beta\delta}\lambda^\delta_\alpha$ (see (3.1.2)).

We assume that l^k, ξ_k, $B_{\alpha\beta} \in C_1$ in G. Then obviously also λ^α_β, $\gamma^k_\alpha \in C_1$ in G. Suppose also that $u^0(x) \in C_1$ and $r^0(x) \in C_1$.

We shall show that under these assumptions the solution constructed above is continuously differentiable in G and hence delivers a solution of problem (7.1.2), (7.1.4) in the usual sense.

In the case of (1), (7.1.9) takes the form

$$\overset{(s+1)}{r_k}(x, t) = \overset{(0)}{r_k}(x, t) + \int_0^t \gamma^k_\alpha(x_k(x, t, \tau), \tau)\, \overset{(s)}{r_\alpha}(x_k(x, t, \tau), \tau)\, d\tau$$

$$+ \int_0^t \left[\frac{d}{d\tau} l^k_\beta(x_k(x, t, \tau), \tau)\right]\lambda^\beta_\alpha(x_k(x, t, \tau), \tau)\overset{(s)}{r_\alpha}(x_k(x, t, \tau), \tau)\, d\tau. \tag{2}$$

It is obvious that our assumptions imply the continuous differentiability in G of the first two terms on the right side of (2) if the approximation $\overset{(s)}{r}(x, t)$ is continuously differentiable in G. As concerns the last term of (2), it is also continuously differentiable in G if $\overset{(s)}{r}(x, t) \in C_1$. This follows from Lemma 2 of §6.

Differentiating (2) with respect to the variable x (the last term is differentiated according to (6.5.3)), we obtain

$$\frac{\partial \overset{(s+1)}{r}_k}{\partial x} = \frac{\partial \overset{(0)}{r}_k}{\partial x} + \int_0^t \frac{\partial x_k(x, t, \tau)}{\partial x} \left[\frac{\partial \gamma_\alpha^k}{\partial x} \overset{(s)}{r}_\alpha + \gamma_\alpha^k \frac{\partial \overset{(s)}{r}_\alpha}{\partial x} \right] d\tau$$

$$+ \left[\frac{\partial}{\partial x} l_\beta^k(x_k(x, t, \tau), \tau) \right]_{\tau=t} \lambda_\alpha^\beta(x, t) \overset{(s)}{r}_\alpha(x, t)$$

$$- \left[\frac{\partial}{\partial x} l_\beta^k(x_k(x, t, \tau), \tau) \right]_{\tau=0} \lambda_\alpha^\beta(x_k(x, t, 0), 0) \overset{(s)}{r}_\alpha(x_k(x, t, 0), 0)$$

$$+ \int_0^t \left\{ \frac{\partial}{\partial x} \left[\lambda_\alpha^\beta(x_k(x, t, \tau), \tau) \overset{(s)}{r}_\alpha(x_k(x, t, \tau), \tau) \right] \left[\frac{\partial}{\partial \tau} l_\beta^k(x_k(x, t, \tau), \tau) \right] \right.$$

$$\left. - \frac{\partial}{\partial \tau} \left[\lambda_\alpha^\beta(x_k(x, t, \tau), \tau) \overset{(s)}{r}_\alpha(x_k(x, t, \tau), \tau) \right] \left[\frac{\partial}{\partial x} l_\beta^k(x_k(x, t, \tau), \tau) \right] \right\} d\tau.$$

This implies that

$$\left\| \frac{\partial \overset{(s+1)}{r}}{\partial x} - \frac{\partial \overset{(s)}{r}}{\partial x} \right\| \le A^1(x, t) \left\| \overset{(s)}{r}(x, t) - \overset{(s+1)}{r}(x, t) \right\|$$

$$+ \int_0^t A^2(x, t, \tau) \max_{G_\tau} \left\| \overset{(s)}{r} - \overset{(s-1)}{r} \right\| d\tau + \int_0^t A^3(x, t, \tau) \max_{G_\tau} \left\| \frac{\partial \overset{(s)}{r}}{\partial x} - \frac{\partial \overset{(s-1)}{r}}{\partial x} \right\| d\tau,$$

while the functions A^1, A^2, and A^3 are continuous and bounded in G.

Since the sequence $\{\overset{(s)}{r}(x, t)\}$ converges in G and A^1, A^2, and A^3 are bounded in G, the proof of uniform convergence in G of the sequence $\{\partial \overset{(s)}{r}_k/\partial x\}$ is analogous to the foregoing. Hence, the solution in the broad sense constructed above is continuously differentiable with respect to x. The continuous differentiability of $r(x, t)$ with respect to t is proved similarly. Actually, this already follows from the continuity in G of the combination of derivatives

$$\frac{\partial r_k}{\partial t} + \xi_k \frac{\partial r_k}{\partial x} = g_k(x, t, r).$$

Thus, under the conditions formulated above the solution $r(x, t)$ in the broad sense is also a solution of the Cauchy problem (7.1.2), (7.1.4) in the usual sense. Passing from r to u by (3.1.4), we conclude that u is a solution of (7.1.1) in the usual sense.

For a semilinear system the following assertion is proved in an altogether analogous way: if in the domain G_{t_0} the solution $u(x, t)$ of problem (7.1.1),

(7.1.3) is bounded and ξ_k, $l^k \in C_1$, $f_k(x, t, u) \in C_1$, and $u^0(x) \in C_1$, then the solution in the broad sense constructed above is continuously differentiable, i.e., $u(x, t) \in C_1$.

By direct differentiation of the system (7.1.2) we see that under our assumptions the derivatives p and q satisfy the equations of the prolonged system in the broad sense.

3. *Some properties of solutions of linear and semilinear systems.* The solution of the Cauchy problem in the broad sense for linear and semilinear systems is uniformly continuous in G if the domain G is finite; if G is an unbounded domain, then uniform continuity holds in any finite part of G. Further, in the case of a semilinear system uniform continuity of the solution in the broad sense holds only in the domain G_{t_0} of boundedness of the solution. These properties are easily derived from the formulas (7.1.8) determining the solution $r(x, t)$.

In the case of a classical solution ($u(x, t) \in C_1$) the derivatives p and q satisfy the equations of the augmented system in the broad sense (see §4.3). Noting that in the case of a semilinear system the augmented system is linear in the derivatives p and q, we conclude that the derivatives of the solution of a semilinear system remain bounded in the domain G_{t_0} in which the solution itself is bounded.

Suppose now that $u(x, t)$ and $\bar{u}(x, t)$ are two solutions in the broad sense of the Cauchy problem with initial data l^k, ξ_k, f_k, $u^0(x)$ and $\bar{l}^k$, $\bar{\xi}_k$, $\bar{f}_k$, $\bar{u}^0(x)$ respectively. We shall assume that the initial data satisfy the conditions formulated in the proof of the existence theorem for a solution in the broad sense.

It is easy to see that for ξ_k, $\bar{\xi}_k \in C_1$ we have $\bar{x}_k(x, t, \tau) \to x_k(x, t, \tau)$ if $\bar{\xi}_k \to \xi_k$; therefore $\bar{G} \to G$ as $\bar{\xi}_k \to \xi_k$. From (7.1.8) it follows that if $\bar{\xi}_k \to \xi_k$, $\bar{l}^k \to l^k$, etc., then

$$\bar{r}_k(x, t) \to r_k(x, t), \qquad \bar{u}_k(x, t) \to u_k(x, t).$$

Thus, the solution of the Cauchy problem in the broad sense depends continuously (in the norm of C) on the initial data of this problem. The Cauchy problem is thus correctly posed in this formulation.

If the initial data are continuously differentiable, then, as we have seen, the solution possesses continuous derivatives and is a classical solution. Of course, what has been said above regarding continuous dependence extends also to these solutions.

If not only the initial data but also the derivatives of the initial data tend to one another uniformly, then not only $\bar{u} \to u$ but also

$$\frac{\partial \bar{u}}{\partial x} \to \frac{\partial u}{\partial x}, \qquad \frac{\partial \bar{u}}{\partial t} \to \frac{\partial u}{\partial t}.$$

This follows from the fact that the derivatives of the solution satisfy the equations of the augmented system in the broad sense, and these equations are linear in the derivatives. Analogous conclusions can be drawn regarding the derivatives of higher order if the conditions on the smoothness of the initial data are strengthened correspondingly.

The construction of the solution of the Cauchy problem reduces to the construction of a mapping taking the initial function into the solution of the Cauchy problem at time t. To be specific, we consider the linear system (7.1.1) with $b = Bu$. Then the solution $u(x, t)$ is given by the formula $u = Su^0$, where the operator S is linear. The domain of this operator is obviously the set of continuously differentiable functions. We shall consider S as a mapping from C to C. The operator S is then defined on a dense set (the set of continuously differentiable functions is dense in C), and it is bounded on this set. According to a well-known theorem of functional analysis, S admits a continuous extension with preservation of norm to the entire space C. Let S^* be this extension. Then $u = S^*u^0$, $u^0 \in C$, is the solution of the Cauchy problem in the broad sense whose existence we have proved independently.

This implies that in order to construct a generalized (in the sense of extensions of operators) solution it suffices to approximate the initial function $u^0(x)$ by an element of a dense set $u^0_\delta(x)$ (i.e., to "smooth" the initial function), construct a smooth solution $u_\delta = Su^0_\delta$, and pass to the limit as $\delta \to 0$ using the corresponding metric.

Analogous arguments go through in the semilinear case as well. They are only complicated by the fact that the operator S delivering the solution of the Cauchy problem, $u = S(u^0)$, is now nonlinear. It is defined, bounded, and continuous as an operator from C to C on the set of continuously differentiable functions satisfying the inequality

$$\|u^0\| < U_{t_0}, \qquad U_{t_0} = \text{const} \tag{1}$$

($0 \leq t < t_0$ is the strip in which the solution is considered). Just as in the preceding case, S can be extended to a continuous, bounded operator defined on the entire set of elements of C satisfying (1). The result of this extension is a solution in the broad sense, just as in the linear case.

We emphasize that generalized solutions of linear and semilinear equations are thus limits of classical solutions in some metric.

§8. The Cauchy problem for a system of quasilinear equations

1. ***A growth estimate of the solution and its derivatives. The majorant system.*** For a system of quasilinear equations we have $l^k = l^k(x, t, u)$ and $\xi_k = \xi_k(x, t, u)$. In this case the construction of a solution of the Cauchy problem is more complicated than for a linear system. We indicate the basic differences in this case:

1) For a system of quasilinear equations, due to the lack of invariants it is not possible to introduce the concept of a solution in the broad sense.

2) The domain of determinacy G of the solution of the Cauchy problem is determined simultaneously with the solution $u(x, t)$ and, generally speaking, cannot be indicated beforehand.

3) The solution $u(x, t)$ and its derivatives do not remain bounded.

We must thus first establish growth estimates for the solution and its derivatives and find a domain $\bar{G} \subseteq G$ of the variables x and t in which the solution and its derivatives clearly remain bounded.

We introduce the domain $G_0(U)$ of the space (x, t, u) defined by the conditions

$$G_0(U) = \{a_0 \le x \le b_0, 0 \le t \le T_0; \|u\| \le U\}, \qquad a_0 < a, \qquad b_0 > b.$$

Let $l^k(x, t, u)$, $\xi_k(x, t, u)$, $f_k(x, t, u) \in C_1(G_0(U))$ for any $U > 0$; let $u^0(x) \in C_1(a, b)$.

According to §4.3 the prolonged system for a hyperbolic system of quaslinear equations has the form

$$\frac{\partial \mathscr{P}_k}{\partial t} + \xi_k \frac{\delta \mathscr{P}_k}{\delta x} = \mathscr{F}_k; \qquad \frac{\partial u_k}{\partial t} = F_k, \tag{1}$$

where

$$\mathscr{F}_k = \mathscr{F}^k(x, t, u) + \mathscr{F}^k_\alpha(x, t, u)\mathscr{P}_\alpha + \mathscr{F}^k_{\alpha\beta}(x, t, u)\mathscr{P}_\alpha\mathscr{P}_\beta,$$

$$F_k = F^k(x, t, u) + F^k_\alpha(x, t, u)\mathscr{P}_\alpha, \qquad \mathscr{P}_k = l^k_\alpha(x, t, u)p_\alpha.$$

The quantities $\mathscr{F}^k$, $\mathscr{F}^k_\alpha$, $\mathscr{F}^k_{\alpha\beta}$, F^k and F^k_α can be expressed in terms of l^k_α, ξ_k, f_k and their derivatives. These quantities are therefore continuous in the domain $G_0(U)$ for any $U > 0$. We introduce the notation

$$\mathscr{F}_0(U) = \max_{G_0(U)} \|\mathscr{F}(x, t, u)\|, \qquad \mathscr{F} = \{\mathscr{F}^k\},$$

$$\mathscr{F}_1(U) = \max_{G_0(U)} \|\mathscr{F}^k_\alpha(x, t, u)\|, \qquad \mathscr{F}_2(U) = \max_{G_0(U)} \|\mathscr{F}^k_{\alpha\beta}(x, t, u)\|,$$

$$F_0(U) = \max_{G_0(U)} \|F(x, t, u)\|, \qquad F_1(U) = \max_{G_0(U)} \|F^k_\alpha(x, t, u)\|$$

and we consider the system of two ordinary differential equations

$$\frac{d\mathscr{P}}{dt} = \mathscr{F}_0(U) + \mathscr{F}_1(U)\mathscr{P} + \mathscr{F}_2(U)\mathscr{P}^2, \tag{2}$$

$$\frac{dU}{dt} = F_0(U) + F_1(U)\mathscr{P}, \tag{3}$$

which we also call a *majorant system*. We denote by U_0 and $\mathscr{P}_0$ the quantities

$$U_0 = \max_{a \le x \le b} \|u^0(x)\|, \qquad \mathscr{P}_0 = \max_{a \le x \le b} \|l^k_\alpha(x, 0, u^0(x))\, du^0_\alpha(x)/dx\|.$$

For the system (2), (3) we prescribe the initial conditions

$$\mathscr{P}(0) = \mathscr{P}_0, \qquad U(0) = U_0. \tag{4}$$

From a comparison of (1), (2), and (3) it follows that if

$$\|u(x, t)\| \le U(t), \qquad \|\mathscr{P}(x, t)\| \le \mathscr{P}(t), \tag{5}$$

then

$$\left\|\frac{\partial u(x, t)}{\partial t}\right\| \le \frac{dU(t)}{dt}, \qquad \left\|\left\{\frac{\partial \mathscr{P}_k(x, t)}{\partial t} + \xi_k \frac{\partial \mathscr{P}_k(x, t)}{\partial x}\right\}\right\| \le \frac{d\mathscr{P}(t)}{dt}.$$

Since it follows from (4) that (5) is satisfied for $t = 0$, for any $t \ge 0$

$$\|u(x, t)\| \le U(t), \qquad \|\mathscr{P}(x, t)\| \le \mathscr{P}(t).$$

Thus, the functions $U(t)$ and $\mathscr{P}(t)$ majorize the growth of the solution $u(x, t)$ and its first derivatives.

Suppose that for $0 \le t \le t_0$ the solution $U(t)$, $\mathscr{P}(t)$ of the majorant system satisfying the initial conditions (4) remains bounded. Then for $0 \le t \le t_0$ the solution $u(x, t)$ and its derivatives $p(x, t)$ clearly also remain bounded. We can therefore define the domain $\bar{G} \subseteq G$ in the following manner:

$$\bar{G} = \{0 \le t \le t_0, \bar{X}_1(t) \le x \le \bar{X}_2(t)\},$$

$$\frac{d\bar{X}_1}{dt} = \max_{k=1,\dots,n} \max_{\|u\| \le U(t)} \{\xi_k(\bar{X}_1(t), t, u)\}, \qquad \bar{X}_1(0) = a,$$

$$\frac{d\bar{X}_2}{dt} = \min_{k=1,\dots,n} \min_{\|u\| \le U(t)} \{\xi_k(\bar{X}_2(t), t, u)\}, \qquad \bar{X}_2(0) = b.$$

We shall construct a solution of the Cauchy problem for a system of quasilinear equations in the domain $\bar{G} \subseteq G$ (Figure 1.4).

We write out a majorant system for some examples.

1) *A linear system.* Let $l^k_\alpha = l^k_\alpha(x, t)$, $\xi_k = \xi_k(x, t)$ and $f_k = f^k(x, t) + f^k_\alpha(x, t)u_\alpha$. From the formulas of §4.3 it then follows that

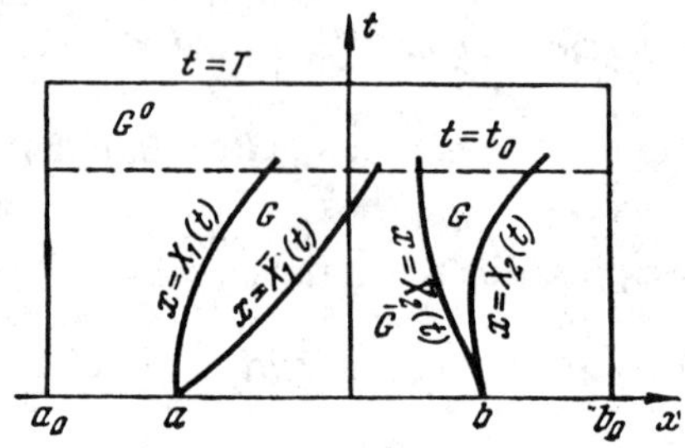

FIGURE 1.4

$$\mathscr{F}_0(U) = \mathscr{F}_0 + \bar{\mathscr{F}}_0 U, \qquad \mathscr{F}_1(U) = \mathscr{F}_1, \qquad \mathscr{F}_2(U) = 0,$$

$$F_0(U) = F_0 + \bar{F}_0 \cdot U, \qquad F_1(U) = F_1,$$

where $\mathscr{F}_0$, $\bar{\mathscr{F}}_0$, $\mathscr{F}_1$, F_0, $\bar{F}_0$ and F_1 are constants depending only on the domain G of the variables x and t. The majorant system for a system of linear equations is also linear. Hence, the solution of the majorant system and thus also the solution $u(x, t)$ and its derivatives $p(x, t)$ remain bounded in any finite part of the half plane $t \geq 0$. The quantity t_0 in this case is arbitrary, while the domain $\bar{G}$ coincides with the domain of determinacy G.

2) *A semilinear system.* Let $l_\alpha^k = l_\alpha^k(x, t)$, $\xi_k = \xi_k(x, t)$ and $f_k = f_k(x, t, u)$. Then $\mathscr{F}_2(U) = 0$. The majorant system (2), (3) takes the form

$$d\mathscr{P}/dt = \mathscr{F}_0(U) + \mathscr{F}_1(U)\mathscr{P},$$

$$dU/dt = F_0(U) + F_1(U)\mathscr{P}.$$

From this it follows that $\mathscr{P}(t)$ is bounded if $U(t)$ is bounded. This situation expresses a general property of solutions of a semilinear system: *the derivatives of the solution remain bounded as long as the solution itself is bounded.*

2. *The uniqueness and existence theorems for a solution.* Under the assumptions made in subsection 1 we consider the Cauchy problem

$$l^k(x, t, u)\left[\frac{\partial u}{\partial t} + \xi_k(x, t, u)\frac{\partial u}{\partial x}\right] = f_k(x, t, u), \tag{1}$$

$$u(x, 0) = u^0(x). \tag{2}$$

We first prove a *uniqueness theorem.* Suppose that in the domain G there exist two solutions $u(x, t)$ and $\bar{u}(x, t)$ of the Cauchy problem (1), (2). Then the difference $v(x, t) = u - \bar{u}$ in G satisfies the system of linear equations

$$\bar{l}^k_\alpha\left[\frac{\partial v_\alpha}{\partial t} + \bar{\xi}_k \frac{\partial v_\alpha}{\partial x}\right] = \bar{f}^k_\alpha v_\alpha \qquad (k = 1, \ldots, n)$$

and the zero initial conditions $v(x, 0) \equiv 0$. Here we have used the notation

$$\left.\begin{aligned} &\bar{l}^k_\alpha = l^k_\alpha(x, t, \bar{u}(x, t)), \qquad \bar{\xi}_k = \xi_k(x, t, \bar{u}(x, t)),\\ &\bar{f}^k_\alpha = \int_0^1 \left\{\frac{\partial f_k}{\partial u_\alpha}(x, t, \bar{u} + \lambda v(x, t)) - \frac{\partial u_\beta(x, t)}{\partial t}\frac{\partial l^k_\beta}{\partial u_\alpha}(x, t, \bar{u} + \lambda v(x, t))\right.\\ &\qquad\qquad \left. - \frac{\partial u_\beta(x, t)}{\partial x}\frac{\partial}{\partial u_\alpha}(l^k_\beta \cdot \xi_k)(x, t, \bar{u} + \lambda v)\right\} d\lambda. \end{aligned}\right\} \quad (3)$$

The quantities $\bar{l}^k_\alpha$ and $\bar{\xi}_k$ are obviously continuously differentiable in G, and $\bar{f}^k_\alpha$ are continuous.

By the uniqueness theorem for a solution of the Cauchy problem for a system of linear equations established in §7, we fine that $v(x, t) \equiv 0$ in G, i.e., $u(x, t) \equiv \bar{u}(x, t)$. The proof of the uniqueness theorem is complete.

To prove the *existence theorem* we apply the method of successive approximations. Suppose that an approximation $\overset{(s)}{u}(x, t) \in C_1$ has been constructed. We define $\overset{(s+1)}{u}(x, t)$ as the solution of the Cauchy problem for the linear system

$$l^k(x, t, \overset{(s)}{u}(x, t))\left[\frac{\partial \overset{(s+1)}{u}}{\partial t} + \xi_k(x, t, \overset{(s)}{u}(x, t))\frac{\partial \overset{(s+1)}{u}}{\partial x}\right] = f_k(x, t, \overset{(s)}{u}(x, t)), \qquad (4)$$

which assumes the initial values (2):

$$\overset{(s+1)}{u}(x, 0) = u^0(x). \qquad (5)$$

From the existence theorem for a solution of a linear system of equations established in §7, it follows that in the domain of determinacy $\overset{(s+1)}{G}$ of the Cauchy problem (4), (5) there exists a solution $\overset{(s+1)}{u}(x, t) \in C_1$, so that all successive approximations are defined and continuously differentiable in the domains $\overset{(s)}{G}$.

The first step is to prove the existence of some domain $\tilde{G}$, belonging to all the domains $\overset{(s)}{G}$, such that the successive approximations and their first derivatives are uniformly bounded in this domain.

We set

$$\overset{(s)}{\mathscr{P}}_k = l^k_\alpha(x, t, \overset{(s-1)}{u})\frac{\partial \overset{(s)}{u}_\alpha}{\partial x}$$

and write out the prolonged system for the linear system (4). It has the form

$$\left.\begin{aligned} \frac{\partial \overset{(s+1)}{\mathscr{P}_k}}{\partial t} + \xi_k(x, t, \overset{(s)}{u}) \frac{\partial \overset{(s+1)}{\mathscr{P}_k}}{\partial x} &= \mathscr{F}^k(x, t, \overset{(s)}{u}) + \Phi^k_\alpha(x, t, \overset{(s)}{u}, \overset{(s-1)}{u}) \overset{(s)}{\mathscr{P}_\alpha} \\ &\quad + \Psi^k_\alpha(x, t, \overset{(s)}{u}, \overset{(s-1)}{u}) \overset{(s+1)}{\mathscr{P}_\alpha} + \mathscr{F}^k_{\alpha\beta}(x, t, \overset{(s)}{u}, \overset{(s-1)}{u}) \overset{(s)}{\mathscr{P}_\alpha} \overset{(s+1)}{\mathscr{P}_\beta}, \\ \frac{\partial \overset{(s+1)}{u_k}}{\partial t} &= F^k(x, t, \overset{(s)}{u}) + F^k_\alpha(x, t, \overset{(s)}{u}) \overset{(s+1)}{\mathscr{P}_\alpha}. \end{aligned}\right\} \tag{6}$$

We shall not write out explicit expressions for the functions contained in this system, since they are obtained in a manner altogether analogous to (4.3.16)–(4.3.19). We note only that they are related to the functions contained in (8.1.1) by the obvious formulas

$$\left.\begin{aligned} \mathscr{F}^k_\alpha(x, t, u) &= \Phi^k_\alpha(x, t, u, u) + \Psi^k_\alpha(x, t, u, u), \\ \mathscr{F}^k_{\alpha\beta}(x, t, u) &= \mathscr{F}^k_{\alpha\beta}(x, t, u, u). \end{aligned}\right\} \tag{7}$$

Together with (6) we consider the system of ordinary differential equations

$$\left.\begin{aligned} d\tilde{P}/dt &= \mathscr{F}_0(\tilde{U}) + \Phi_1(\tilde{U})\tilde{P} + \Phi_2(\tilde{U})\tilde{P}^2, \\ d\tilde{U}/dt &= F_0(\tilde{U}) + F_1(\tilde{U})\tilde{P}, \end{aligned}\right\} \tag{8}$$

where the functions $\mathscr{F}_0$, F_0 and F_1 are defined in subsection 1, and

$$\Phi_1(\tilde{U}) = \Phi'(\tilde{U}) + \Psi(\tilde{U}), \qquad \Phi'(\tilde{U}) = \max_{\substack{\|u\| \le \tilde{U} \\ \|v\| \le \tilde{U}}} \|\Phi^k_\alpha(x, t, u, v)\|,$$

$$\Psi(\tilde{U}) = \max_{\substack{\|u\| \le \tilde{U} \\ \|v\| \le \tilde{U}}} \|\Psi^k_\alpha(x, t, u, v)\|, \qquad \Phi_2(\tilde{U}) = \max_{\substack{\|u\| \le \tilde{U} \\ \|v\| \le \tilde{U}}} \|\mathscr{F}^k_{\alpha\beta}(x, t, u, v)\|.$$

For the system (8) we prescribe the initial conditions $\tilde{P}(0) = P_0$ and $\tilde{U}(0) = U_0$ (cf. (8.1.4)).

Since it follows from (7) that

$$\mathscr{F}_1(U) \le \Phi_1(U), \qquad \mathscr{F}_2(U) \le \Phi_2(U),$$

the solution of (8) majorizes the solution of the majorant system (8.1.2), (8.1.3):

$$U(t) \le \tilde{U}(t), \qquad \mathscr{P}(t) \le \tilde{P}(t).$$

Hence, if the domain $\tilde{G}$ is constructed on the basis of $\tilde{U}(t)$ in the same way that $\bar{G}$ was constructed on the basis of $U(t)$ in subsection 1, then $\tilde{G} \subseteq \bar{G}$.

We now suppose that all successive approximations $\overset{(k)}{u}$, $k = 1, \ldots, s$, satisfy the inequalities

$$\|\overset{(k)}{u}\| \le \tilde{U}(t), \qquad \|\overset{(k)}{\mathscr{P}}\| \le \tilde{P}(t). \tag{9}$$

Setting

$$U_{s+1}(t) = \sup_x \|\overset{(s+1)}{u}\|, \qquad \mathscr{P}_{s+1}(t) = \sup_x \|\overset{(s+1)}{\mathscr{P}}\|,$$

from (6) we get

$$\frac{d\mathscr{P}_{s+1}}{dt} \leq \mathscr{F}_0(\tilde{U}) + \Phi'(\tilde{U})\tilde{P} + \Psi(\tilde{U})\mathscr{P}_{s+1} + \Phi_2(U)\tilde{P}\mathscr{P}_{s+1},$$

$$\frac{dU_{s+1}}{dt} \leq F_0(\tilde{U}) + F_1(\tilde{U})\tilde{P},$$

so that obviously

$$U_{s+1}(t) \leq \tilde{U}(t), \qquad \mathscr{P}_{s+1}(t) \leq \tilde{P}(t).$$

Since the initial approximation can be chosen so that (9) is satisfied, we have thus proved that all successive approximations satisfy (9). This implies the existence of a domain $\tilde{G}$ belonging both to $\bar{G}$ and to all the domains $\overset{(s)}{G}$ in which the inequalities (9) are satisfied.

The second step of our proof is the proof of the uniform convergence in the domain $\tilde{G}$ of the sequence $\{\overset{(s)}{u}\}$. Let

$$\overset{(s)}{r}_k = l^k(x, t, \overset{(s-1)}{u})(\overset{(s)}{u} - \overset{(s-1)}{u}).$$

Then from (4) we obtain

$$\frac{\partial \overset{(s+1)}{r}_k}{\partial t} + \xi_k(x, t, \overset{(s)}{u})\frac{\partial \overset{(s+1)}{r}_k}{\partial x} = \overset{(s)}{g}{}^k_\alpha \overset{(s+1)}{r}_\alpha + \overset{(s)}{f}{}^k_\alpha \overset{(s)}{r}_\alpha, \tag{10}$$

where the functions $\overset{(s)}{g}{}^k_\alpha$ and $\overset{(s)}{f}{}^k_\alpha$ are computed from formulas analogous to (3). These formulas contain $\overset{(s)}{u}$, $\overset{(s-1)}{u}$ and their derivatives; moreover, by (9), everywhere in G we have

$$\|\overset{(s)}{f}{}^k_\alpha\| \leq B, \qquad \|\overset{(s)}{g}{}^k_\alpha\| \leq B.$$

Since the domain of determinacy of the system (10) contains $\tilde{G}$, by integrating along characteristics we find that for each point of $\tilde{G}$

$$|\overset{(s+1)}{r}_k| \leq \int_0^t |\overset{(s)}{f}{}^k_\alpha \overset{(s)}{r}_\alpha + \overset{(s)}{g}{}^k_\alpha \overset{(s+1)}{r}_\alpha| \, d\tau,$$

so that

$$R_{s+1}(t) \leq B\int_0^t [R_s(\tau) + R_{s+1}(\tau)]\, d\tau,$$

where

$$R_s(t) = \max_{(x,\tau)\in\tilde{G},\, \tau\leq t} \|\overset{(s)}{r}\|.$$

Applying Lemma 1 of §6, we obtain

$$R_{s+1}(t) \leq C\int_0^t R_s(\tau)\, d\tau,$$

or

$$R_{s+1}(t) \leq \text{const}\,\frac{(Ct)^s}{s!},$$

which proves the uniform convergence in $\tilde{G}$ of the sequence $\{\overset{(s)}{u}\}$.

Finally, we proceed to the last step of the proof. We shall show that the derivatives of the successive approximations converge uniformly in $\tilde{G}$. This is obviously equivalent to the proof of the uniform convergence of the sequence $\{\overset{(s)}{\mathscr{P}}\}$.

We first prove the equicontinuity of this sequence in x. In other words, we show that there exists a function $M(\delta)$, $M(\delta) \to 0$ as $\delta \to 0$, such that everywhere in $\tilde{G}$ and for all s

$$\|\overset{(s)}{\mathscr{P}}(x'', t) - \overset{(s)}{\mathscr{P}}(x', t)\| \leq M(\delta) \quad \text{as } |x'' - x'| \leq \delta.$$

It was shown above that the sequence $\{\overset{(s)}{\mathscr{P}}\}$ was uniformly bounded. This implies that functions $\overset{(s)}{u}(x, t)$ and hence also the functions $\xi_k(x, t, \overset{(s)}{u}(x, t))$, $\mathscr{F}^k(x, t, \overset{(s)}{u})$, $\Phi^k_\alpha(x, t, \overset{(s)}{u}, \overset{(s-1)}{u})$, $\Psi^k_\alpha(x, t, \overset{(s)}{u}, \overset{(s-1)}{u})$, and $\mathscr{F}^k_{\alpha\beta}(x, t, \overset{(s)}{u}, \overset{(s-1)}{u})$ (see (6)) are equicontinuous in $\tilde{G}$. Morever, the equicontinuity of the functions $\xi_k(x, t, \overset{(s)}{u})$ implies the analogous property for the functions $x_k^{(s)}(x, t, \tau)$ defining the characteristics of the system (6).

Therefore, setting

$$M_s(t, \delta) = \max_{k\leq s} \sup_{\substack{|x'-x''|\leq\delta \\ \tau\leq t}} \|\overset{(k)}{\mathscr{P}}(x'', \tau) - \overset{(k)}{\mathscr{P}}(x', \tau)\|$$

and integrating (6) along the characteristics, we obtain

$$M_{s+1}(t, \delta) \leq (t+1)N(\delta) + C\int_0^t M_{s+1}(\tau, \delta)\, d\tau,$$

where $N(\delta) \to 0$ as $\delta \to 0$. Applying Lemma 1 of §6, we conclude that the required function $M(\delta)$ exists, so that the sequence $\{\overset{(s)}{\mathscr{P}}\}$ is equicontinuous.

Since by Arzelà's lemma from any uniformly bounded and equicontinuous sequence it is possible to extract a (uniformly) convergent subsequence, some sequence $\{\overset{(s_k)}{\mathscr{P}_\alpha}\}$ and hence also $\{\overset{(s_k)}{p_\alpha}\}$ converges uniformly in G to a continuous function p_α. By a well-known theorem of analysis, this implies that the function $u_\alpha = \lim_{s\to\infty} \overset{(s)}{u_\alpha}$ is continuously differentiable in $\tilde{G}$, and $\partial u_\alpha/\partial x = p_\alpha$.

From this it follows that the sequence $\{\overset{(s)}{p_\alpha}\}$ has only one limit point, and hence the sequence $\{\overset{(s)}{p_\alpha}\}$ is not only compact but also convergent. The proof of the convergence of the sequence $\overset{(s)}{p}$ (and hence also of $\overset{(s)}{q}$) is complete.

Passing to the limit in (4), we conclude that u is a solution of the problem (1), (2). The existence theorem is proved.

3. ***Some properties of solutions of the Cauchy problem for systems of quasilinear equations.*** Let $u(x, t)$ and $\bar{u}(x, t)$ be two solutions of the Cauchy problem for a system of quasilinear equations corresponding to the respective initial data $l_\alpha^k, \xi_k, f_k, u^0$ and $\bar{l}_\alpha^k, \bar{\xi}_k, \bar{f}_k, \bar{u}^0$. We shall assume that the initial data of these two Cauchy problems satisfy the conditions of subsection 2, i.e., they are continuously differentiable.

In the intersection G of the domains of determinacy of the solutions $u(x, t)$ and $\bar{u}(x, t)$ the difference $v(x, t) = u(x, t) - \bar{u}(x, t)$ satisfies the system of linear equations

$$\bar{l}^k\left[\frac{\partial v}{\partial t} + \bar{\xi}_k \frac{\partial v}{\partial x}\right] = \gamma_\alpha^k v_\alpha + \Delta f_k,$$

where γ_α^k and Δf_k are bounded and continuous in G, and the Δf_k tend to zero as $\bar{l}_\alpha^k, \bar{\xi}_k, \bar{f}_k \to l_\alpha^k, \xi_k, f_k$.

As we saw in §7, solutions of systems of linear equations depend continuously on the initial data of the Cauchy problem. This implies that, in G, $v(x, t) \rightrightarrows 0$ as $\bar{l}_\alpha^k \rightrightarrows l_\alpha^k$, $\bar{\xi}_k \rightrightarrows \xi_k$ and $\bar{f}_k \rightrightarrows f_k$, and $v(x, 0) = u^0(x) - \bar{u}^0(x) \rightrightarrows 0$.

Here, however, a clarification must be made. It is essential that the initial data of each Cauchy problem have bounded derivatives.

In contrast to the case of a semilinear system, the strip $0 \leq t \leq t_0$ in which the solution $u(x, t)$ (and its derivatives) remains bounded depends on the derivatives of the initial functions, and $t_0 \to 0$ if $\|du^0/dx\| \to \infty$. Therefore, continuous dependence of solutions of the Cauchy problem for a system of quasilinear equations occurs only in the class of initial data with uniformly bounded derivatives.

If, as previously, we symbolically write the procedure for solving the Cauchy problem as the equality $u(x, t) = Su^0(x)$, then the nonlinear operator S determines the solution $u(x, t)$ only in the domain $\bar{G}$ of the half plane $t \geq 0$. The width of the strip $0 \leq t \leq t_0$ in which $\bar{G}$ is contained depends on the derivatives du^0/dx and tends to zero as $\|du^0/dx\| \to \infty$.

Therefore, in contrast to the case of a linear system, the operator S does not admit an extension to the class of continuous initial functions $u^0(x)$. For this reason a generalized solution of a system of quasilinear equations cannot be defined formally by means of an extension of the space of classical solutions. The concept of a generalized solution of a system of quasilinear equations must thus be introduced independently of the concept of a solution of this system. Generalized solutions will be studied in detail in Chapter 4.

We note a minor extension of the class of initial data for which the existence of a solution of the Cauchy problem follows from the material presented above.

We call a function $\xi_k(x, t, u)$ *Lipschitz continuous* in a domain $G_0(U)$ in the set of variables x, t, u if there exists a constant $C > 0$ such that

$$|\xi_k(\bar{x}, \bar{t}, \bar{u}) - \xi_k(x, t, u)| \leq C\{|\bar{x} - x| + |\bar{t} - t| + \|\bar{u} - u\|\}$$

for any $(x, t, u), (x, t, u) \in G_0(U)$.

If we consider the class of Lipschitz continuous initial data characterized by a Lipschitz constant K, then it can be considered as the closure of the class of initial data with first derivatives bounded uniformly by the same constant K. Therefore, a solution of the Cauchy problem with Lipschitz continuous initial data may be considered as the limit of classical solutions $u(x, t) \in C_1$, since the latter form a family with uniformly bounded first derivatives.

Of course, this limit is not a solution of the Cauchy problem in the usual sense, since it does not possess continuous first derivatives. However, it is a Lipschitz continuous function of x and t, and possesses derivatives almost everywhere in $\bar{G}$. These derivatives satisfy the system of quasilinear equations almost everywhere in $\bar{G}$.

The class of Lipschitz continuous solutions $u(x, t)$ of the Cauchy problem is an example of a formal extension of the operator S defined on the class C_1 to the class of Lipschitz continuous initial data.

However, even this minor extension of the class of classical solutions contains interesting solutions which are broadly studied. An example of such solutions are solutions with so-called weak discontinuities where the solution itself is continuous but its derivatives suffer discontinuities while remaining bounded (see §10.1).

§9. The Cauchy problem for a single equation

1. ***A single quasilinear equation.*** The results of §8 hold unconditionally for the case of a single quasilinear equation. However, they are too general in application to this case, in which there are important simplifying details.

We shall therefore consider in detail the Cauchy problem for a single quasilinear equation

$$\left(\frac{du}{dt}\right) = \frac{\partial u}{\partial t} + \xi(x, t, u)\frac{\partial u}{\partial x} = f(x, t, u) \tag{1}$$

with the initial condition

$$u(x, 0) = u_0(x), \qquad a \leq x \leq b. \tag{2}$$

The integration of (1) reduces to solving a system of two ordinary differential equations

$$\frac{dX}{dt} = \xi(X, t, U), \qquad \frac{dU}{dt} = f(X, t, U), \tag{3}$$

which form the *characteristic system* of equation (1). Each solution $x = X(t)$, $u = U(t)$ defines a characteristic in the space of the variables x, t, u.

We assume that the functions ξ and f are continuously differentiable. Then through any point (x_0, t_0, u_0) there passes one and only one characteristic.

The Cauchy problem (1), (2) can be interpreted geometrically as the problem of constructing an integral surface of equation (1) passing through a prescribed initial curve: $t = 0$, $u = u_0(x)$. Since we wish to obtain a single-valued, differentiable function $u(x, t)$ of the variables x and t, this surface should obviously project injectively onto the plane $u = 0$ of the variables x and t.

Since the solution u is uniquely determined along each characteristic passing through the point (x_0, t_0, u_0), this problem consists in constructing a surface consisting of characteristics passing through the initial curve and projecting injectively onto the plane $u = 0$.

We denote by $X = X(t, x_0, u_0)$, $U = U(t, x_0, u_0)$ the solution of the characteristic system (3) satisfying the initial conditions

$$X(0, x_0, u_0) = x_0, \qquad U(0, x_0, u_0) = u_0. \tag{4}$$

Then the solution $u(x, t)$ of the Cauchy problem (1), (2) is given by

$$u(X(t, x_0, u_0(x_0)), t) = U(t, x_0, u_0(x_0)). \tag{5}$$

Formula (5) implicitly defines a function $u(x, t)$ which in the case $u_0(x) \in C_1$ is continuously differentiable at all points x, t for which the equation

$$x = X(t, x_0, u_0(x_0)) \tag{6}$$

is uniquely solvable for the parameter x_0 and the right-hand side of (5) remains bounded.

Suppose that at these points

$$x_0 = X^{-1}(t, x) = x_0(x, t) \tag{7}$$

is the result of solving (6) for x_0. From (5) we then obtain an explicit formula for the solution $u(x, t)$ of problem (1), (2):

$$u(x, t) = U(t, x_0(x, t), u_0(x_0(x, t))). \tag{8}$$

We explain graphically the construction of the solution of the Cauchy problem (1), (2). Through any point $x_0 \in [a, b]$ we draw the projection of the characteristic onto the plane $u = 0$ (the (x, t) plane), setting $u_0 = u_0(x_0)$ (Figure 1.5). This projection (6) we also call a *characteristic*.

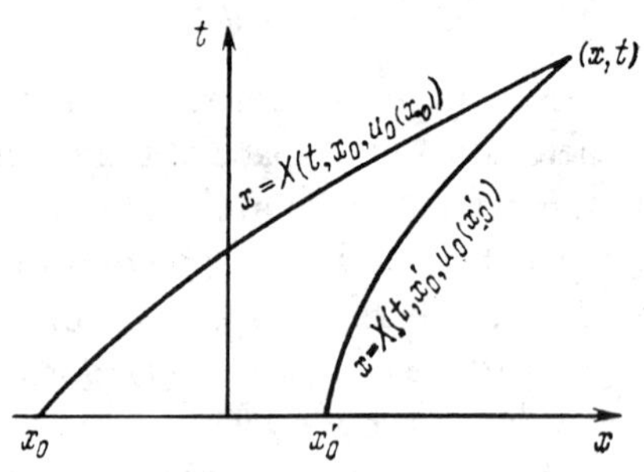

FIGURE 1.5

On the characteristic (6) there is given a continuously differentiable function $U(t, x_0, u_0(x_0))$ of the variable t which gives the solution $u(x, t)$ on the line $x = X(t, x_0, u_0(x_0))$.

It may happen that at certain points (x, t) for $t > 0$ two or more lines $x = X(t, x_0, u_0(x_0))$ corresponding to distinct values of the parameter x_0 intersect (Figure 1.5). At these points equation (6) relative to x_0 has more than one solution, and the formulas (5) and (8) define some multivalued function of x and t. In this case there does not exist a continuous solution $u(x, t)$ of problem (1), (2).

We shall show that for $0 \leq t \leq t_0$ and sufficiently small $t_0 > 0$ through any point $(x, t) \in G$ there passes a unique characteristic (6), i.e., (6) has a unique solution relative to x_0.

To justify the possibility of unique solvability of (6) for x_0 it suffices to show that

$$\dot{X} = \frac{\partial X(t, x_0, u_0(x_0))}{\partial x_0} > 0, \tag{9}$$

since $X(0, x_0, u_0(x_0)) \equiv x_0$ and

$$\dot{X}(0, x_0, u_0(x_0)) = 1. \tag{10}$$

Setting

$$\dot{U} = \frac{\partial U(t, x_0, u_0(x_0))}{\partial x_0} \qquad (\dot{U}(0, x_0, u_0(x_0)) \equiv u_0'(x_0)) \tag{11}$$

and differentiating (3), we obtain

$$\frac{d\dot{X}}{dt} = \xi_u'\dot{U} + \xi_x'\dot{X}, \qquad \frac{d\dot{U}}{dt} = f_u'\dot{U} + f_x'\dot{X}, \tag{12}$$

where the arguments of all quantities are omitted for brevity. Thus, $\dot{X}$ and $\dot{U}$ satisfy the system of linear ordinary differential equations (12) and the initial conditions (10) and (11). From this it is clear that (9) holds for sufficiently small t_0.

Thus, there exists $t_0 > 0$ such that for $0 \leq t \leq t_0$ the formulas (7) and (8) define a function $u(x, t) \in C_1$ satisfying (1) and (2).

The Cauchy problem (1), (2) requires, as we have already noted, the existence of a unique function $u(x, t) \in C_1$ of x and t satisfying (1) and (2). On the other hand, the more general problem of determining an integral surface S passing through the initial curve does not require that this surface project uniquely onto the plane of the variables x and t and may have (and, as a rule, has) a solution in a larger domain of the variables x and t than the Cauchy problem (1), (2).

For example, we shall seek an equation for the surface S in the form

$$\varphi(x, t, u) = 0. \tag{13}$$

Any characteristic (4) of equation (1) must lie on S; therefore,

$$\varphi(X, t, U) = 0. \tag{14}$$

Differentiating (14) with respect to t and using (3), we obtain

$$\frac{\partial \varphi}{\partial t} + \xi(x, t, u)\frac{\partial \varphi}{\partial x} + f(x, t, u)\frac{\partial \varphi}{\partial u} = 0, \tag{15}$$

which is a linear differential equation of first order for the function φ depending on the three variables (x, t, u).

The surface S is uniquely determined by (13) and (14) for any x_0 and t for which X and U are finite, and it is a smooth surface ($\varphi \in C_1$) if $u_0(x) \in C_1$ and $f \in C_1$.

A function $u(x, t) \in C_1$ is determined from (13) that gives a solution of problem (1), (2) only in the range of values of x and t for which (6) is uniquely solvable for x_0.

Thus, the difference in the formulations of the Cauchy problem (1), (2) and the problem of determining the surface S is that in the first case an integral surface $u = u(x, t)$ is sought which projects injectively onto the plane $u = 0$; in the second case this surface may be arbitrary.

For sufficiently small values of t_0 in the case $u_0(x) \in C_1$ both these Cauchy problems are equivalent: globally (i.e. for any $t > 0$) the geometric formulation of the Cauchy problem is more general and admits a solution even when problem (1), (2) is not solvable.

If it is supposed that $u(x, t)$ describes some physical quantity in the space of the variables x and t, then it is natural that this quantity should be a single-valued function of x and t. Therefore, problems of physics leading to the Cauchy problem require the determination of a single-valued function $u = u(x, t)$. As we have seen, the Cauchy problem (1), (2) in this formulation is solvable in the class of continuous solutions $u(x, t) \in C$ only in a sufficiently small strip $0 \le t \le t_0$.

Figure 1.6 shows a typical form of the surface S. In this figure it is clear that $|\partial u(x, t)/\partial x| \to \infty$ at those points (x, t) near which S does not project injectively onto the plane of the variables (x, t).

We shall illustrate what has been said with the simplest quasilinear equation

$$\frac{\partial u}{\partial t} + u \frac{\partial u}{\partial x} = 0, \tag{16}$$

for which we pose the initial condition

$$u(x, 0) = u_0(x) = \begin{cases} u^- = \alpha a + \beta & \text{for } x \le a, \\ \alpha x + \beta & \text{for } a \le x \le b, \\ u^+ = \alpha b + \beta & \text{for } x \ge b. \end{cases} \tag{17}$$

The initial function $u_0(x)$ is continuous for $-\infty < x < \infty$; the derivative $u_0'(x)$ has discontinuities at $x = a$ and $x = b$. We shall construct a solution of problem (16), (17) satisfying (16) in the broad sense at points where $\partial u/\partial t$ and $\partial u/\partial x$ do not exist.

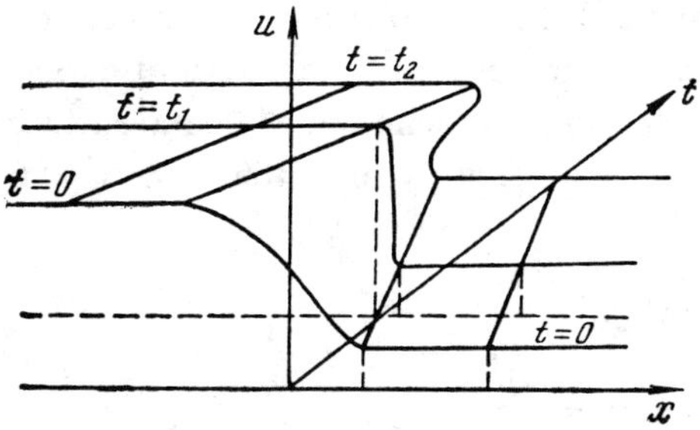

FIGURE 1.6

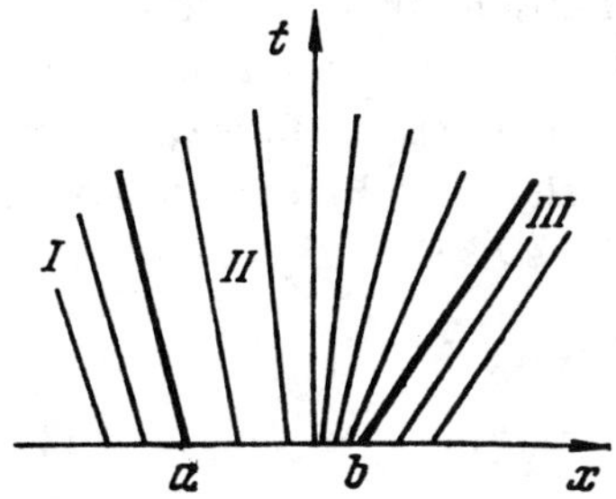

FIGURE 1.7

The characteristic system (3) for equation (16) has the solution

$$X(t, x_0, u_0) = x_0 + u_0 t, \qquad U(t, x_0, u_0) = u_0, \tag{18}$$

which remains bounded for any values of t, x_0 and u_0. Suppose that $\alpha \geq 0$. The projections of the characteristics (18) onto the plane $u = 0$ have the form shown in Figure 1.7. In this case through each point (x, t) of the half plane $t \geq 0$ there passes a unique characteristic $x = X(t, x_0, u_0(x_0))$, i.e., (6) has a unique solution relative to x_0. The function $u(x, t)$ is constant along the characteristics (6); therefore, in zone I, i.e., for $x \leq a + u^- t$,

$$u(x, t) = u^- = a\alpha + \beta,$$

and in zone III, for $x \geq b + u^+ t$,

$$u(x, t) = u^+ = \alpha b + \beta.$$

In zone II, for $a + u^- t \leq x \leq b + u^+ t$, equation (6) is solvable for x_0:

$$x_0 = \frac{x - \beta t}{1 + \alpha t}.$$

We determine the solution $u(x, t)$ in zone II on the basis of (8):

$$u(x, t) = u_0(x_0(x, t)) = \alpha \cdot \frac{x - \beta t}{1 + \alpha t} + \beta = \frac{\alpha x + \beta}{1 + \alpha t}.$$

Thus, the solution of the Cauchy problem (16), (17) for $\alpha \geq 0$ is given by

$$u(x, t) = \begin{cases} u^- = \alpha a + \beta & \text{for } x \leq a + u^- t, \\ \dfrac{\alpha x + \beta}{1 + \alpha t} & \text{for } a + u^- t \leq x \leq b + u^+ t, \\ u^+ = \alpha b + \beta & \text{for } x \geq b + u^+ t. \end{cases} \tag{19}$$

The solution (19) is continuously differentiable for $t \geq 0$ everywhere except on the lines $x = a + u^- t$, $x = b + u^+ t$ where the first derivatives are discontinuous.

In the space of the variables x, t, u the solution (19) defines the integral surface S shown in Figure 1.8. This surface projects injectively onto the plane $u = 0$ for $t \geq 0$.

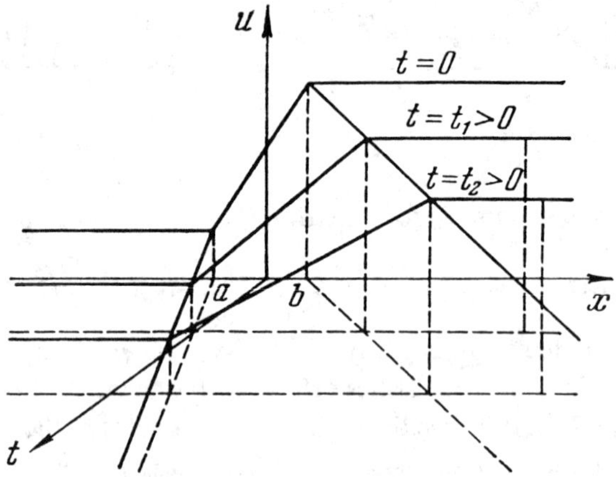

FIGURE 1.8

In the case $\alpha < 0$ we have $u^- > u^+$, and the picture of the characteristics in the projection onto the (x, t) plane has the form shown in Figure 1.9. All the characteristics (6) for $a \leq x_0 \leq b$ converge at the point $x = x_0 = -\beta/\alpha$, $t = t_0 = -1/\alpha > 0$. In zone I we have $u(x, t) = u^-$; in zone II, $u(x, t) = u^+$. In zone III we have $u(x, t) = (\alpha x + \beta)/(1 + \alpha t)$, but since $\alpha < 0$ this formula does not define the solution $u(x, t)$ for $t = -1/\alpha$. Finally, in zone IV the function

$$U(t, x_0(x, t), u_0(x_0(x, t)))$$

is triple-valued and assumes the values

$$u_{\mathrm{I}}(x, t) = u^-, \qquad u_{\mathrm{II}}(x, t) = \frac{\alpha x + \beta}{1 + \alpha t}, \qquad u_{\mathrm{III}}(x, t) = u^+.$$

Thus, in the case $\alpha < 0$ a continuous solution $u(x, t)$ of the Cauchy problem (16), (17) exists only for $t < -1/\alpha$, while the integral surface S is defined for all $t \geq 0$ (Figure 1.10); however, for $t \geq -1/\alpha$ it does not project injectively onto the plane $u = 0$.

2. ***A single nonlinear equation.*** The Cauchy problem for the nonlinear equation

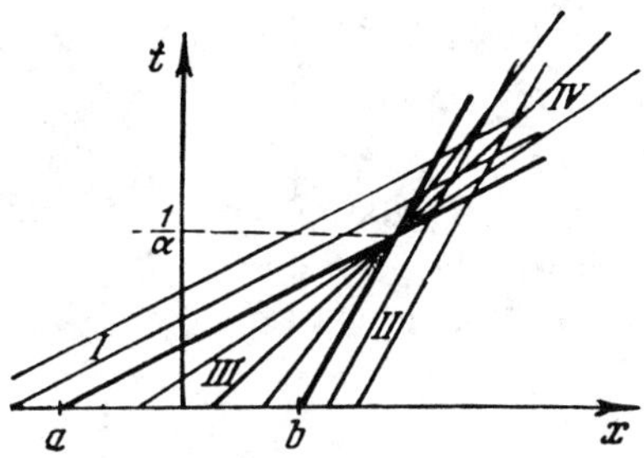

FIGURE 1.9

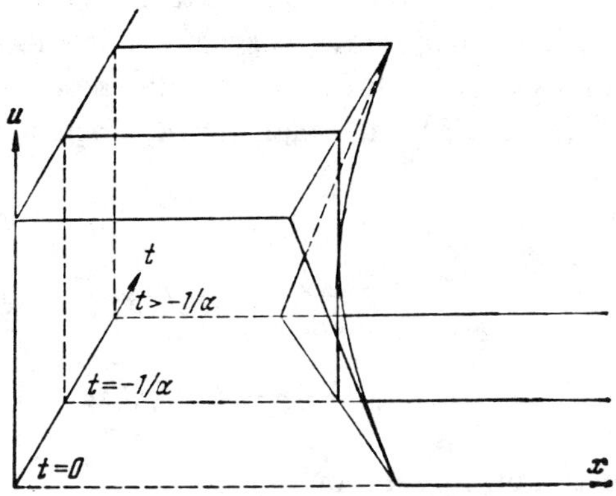

FIGURE 1.10

$$\frac{\partial v}{\partial t} + \varphi(x, t, v, \omega) = 0, \qquad \omega = \frac{\partial v}{\partial x}, \qquad v(x, 0) = v_0(x) \tag{1}$$

in the case $\varphi \in C_2$, $v_0(x) \in C_2$ reduces, after we differentiate (1) with respect to x, to the Cauchy problem for a system of two quasilinear equations

$$\left.\begin{aligned} \frac{\partial \omega}{\partial t} + \varphi'_\omega(x, t, v, \omega)\frac{\partial \omega}{\partial x} &= -\varphi'_v\omega - \varphi'_x, \\ \frac{\partial v}{\partial t} &= -\varphi(x, t, v, \omega) \end{aligned}\right\} \tag{2}$$

with the initial conditions $v(x, 0) = v_0(x)$ and $\omega(x, 0) = v_0'(x)$.

The last condition ($\omega(x, 0) = v_0'(x)$) guarantees that for any t for which there exists a solution of (2) the following condition is satisfied:

$$\omega(x, t) = \frac{\partial v(x, t)}{\partial x}.$$

A proof of this will be given in the next subsection for a more general case.

Using this, we rewrite (2) in the form

$$\frac{\partial v}{\partial t} + \varphi'_\omega \frac{\partial v}{\partial x} = -\varphi + \varphi'_\omega \cdot \omega, \tag{3}$$

$$\frac{\partial \omega}{\partial t} + \varphi'_\omega \frac{\partial \omega}{\partial x} = -\varphi'_v \cdot \omega - \varphi'_x. \tag{4}$$

The system (3), (4) is characterized by the fact that in each of the equations the functions v and ω are differentiated in the same direction, i.e., the characteristic direction $dx/dt = \varphi'_\omega$ for this system has multiplicity two.

Systems of this type are called *systems with identical principal part*.

This circumstance makes it possible to reduce the problem to the integration of a system of three ordinary differential equations.

Indeed, if for the characteristic system of (1)

$$\left.\begin{aligned} \frac{dX}{dt} &= \varphi'_\omega(X, t, V, \Omega), \\ \frac{dV}{dt} &= \Omega\,\varphi'_\omega(X, t, V, \Omega) - \varphi(X, t, V, \Omega), \\ \frac{d\Omega}{dt} &= -\,\Omega\varphi'_v(X, t, V, \Omega) - \varphi'_\tau(X, t, V, \Omega), \end{aligned}\right\} \tag{5}$$

we known a solution

$$X = X(t, x_0, v_0, \omega_0), \qquad V = V(t, x_0, v_0, \omega_0), \qquad \Omega = \Omega(t, x_0, v_0, \omega_0),$$

satisfying the initial conditions

$$X(0, x_0, v_0, \omega_0) = x_0, \qquad V(0, x_0, v_0, \omega_0) = v_0, \qquad \Omega(0, x_0, v_0, \omega_0) = \omega_0, \tag{6}$$

then the solution $v(x, t)$ of the Cauchy problem (1) is given by the formula

$$v\left(X\left(t, x_0, v_0(x_0), \frac{dv_0(x_0)}{dx_0}\right), t\right) = V\left(t, x_0, v_0(x_0), \frac{dv_0(x_0)}{dx_0}\right), \tag{7}$$

which defines $v(x, t)$ parametrically. If $x_0 = x_0(x, t)$ is the result of the single-valued solution of the equation

$$x = X\left(t, x_0, v_0(x_0), \frac{dv_0(x_0)}{dx_0}\right) \tag{8}$$

with respect to the parameter x_0, then from (7) we obtain an explicit formula for the solution $v(x, t)$ of the Cauchy problem (1):

$$v(x, t) = V(t, x_0(x, t), v_0(x_0(x, t)), v_0'(x_0(x, t))). \tag{9}$$

This determines a single-valued function $v(x, t) \in C_2$ only at those points (x, t) for which (8) is uniquely solvable for x_0.

We have obtained a solution of the Cauchy problem (1) only for the case $v_0(x) \in C_2$. Simple examples show if $v_0'(x)$ is only continuous, then, in general, there does not exist a solution $v(x, t) \in C_1$ of problem (1). The formulation of the Cauchy problem (1) in the case of initial functions $v_0(x) \in C_1$ must be made more precise.

3. ***A hyperbolic system of nonlinear equations.*** In the case of the Cauchy problem for a system of nonlinear equations of hyperbolic type

$$\frac{\partial v}{\partial t} + \varphi(x, t, v, \omega) = 0, \tag{1}$$

$$v(x, 0) = v^0(x), \tag{2}$$

$$\omega = \frac{\partial v}{\partial x} = \left\{\frac{\partial v_1}{\partial x}, \ldots, \frac{\partial v_n}{\partial x}\right\} \tag{3}$$

we assume that $\varphi \in C_2$ and $v^0 \in C_2$. According to §2.3, the functions v and ω are then a solution of the system

$$\frac{\partial v}{\partial t} = -\varphi(x, t, v, \omega), \qquad \frac{\partial \omega}{\partial t} + A(x, t, v, \omega)\frac{\partial \omega}{\partial x} = \bar{f} \tag{4}$$

(see §3.2), where the matrix $A = \partial\varphi/\partial\omega = ((\partial\varphi_i/\partial\omega_j))$. If for the system (4) we pose the initial conditions

$$v(x, 0) = v^0(x), \qquad \omega(x, 0) = \omega^0(x) = \frac{dv^0(x)}{dx}, \tag{5}$$

then the question of whether the function $v(x, t)$ determined as a result of solving the Cauchy problem (4), (5) is a solution of the original problem (1), (2) is the question of whether (3) is satisfied for any $(x, t) \in G$.

Differentiating the first group of equations in (4) with respect to x, subtracting the second group of equations in (4), and noting that $\bar{f} = -\varphi'_x - (\partial\varphi/\partial v)\omega$ (see (2.3.7)), we find that

$$\frac{\partial}{\partial t}\left(\frac{\partial v_i}{\partial x} - \omega_i\right) = -\frac{\partial \varphi_i}{\partial v_\alpha}\left(\frac{\partial v_\alpha}{\partial x} - \omega_\alpha\right) \qquad (i = 1, \ldots, n) \tag{6}$$

and, according to (5),

$$\frac{\partial v_i}{\partial x}(x, 0) = \omega_i(x, 0). \tag{7}$$

On the basis of the corollary of Lemma 1 (§6.5) it follows from (6) and (7) that

$$\frac{\partial v_i}{\partial x}(x, t) \equiv \omega_i(x, t), \tag{8}$$

i.e., the equality (3) is satisfied identically.

Since we have proved the existence of a solution of a system of quasilinear equations only in the class C_1, for systems of nonlinear equations of hyperbolic type the solutions constructed belong to the class C_2.

§10. The behavior of the derivatives of a solution of a system of quasilinear equations

1. ***A weak discontinuity. The transport equation.*** In §§7 and 8 a solution of the Cauchy problem for a system of quasilinear equations was constructed having continuous first derivatives. In considering Lipschitz continuous initial data, we arrived at a certain generalization of classical solutions—Lipschitz continuous solutions of the Cauchy problem possessing first derivatives almost everywhere in the domain where defined.

We shall consider a more special class of generalized solutions $u(x, t)$ of a system of quasilinear equations: the class of continuous functions $u(x, t)$ possessing piecewise continuous first derivatives. We assume that a vector-valued function $u(x, t)$ is continuous and possesses continuous first derivatives everywhere except for certain piecewise differentiable lines on which the first derivatives p and q have a discontinuity of first kind; we suppose that away from the lines of discontinuity of the first derivatives $u(x, t)$ satisfies the system of quasilinear equations

$$l_\alpha^k\left[\frac{\partial u_\alpha}{\partial t} + \xi_k \frac{\partial u_\alpha}{\partial x}\right] = f_k, \tag{1}$$

where $l_\alpha^k(x, t, u)$, $\xi_k(x, t, u)$, $f_k(x, t, u) \in C_1$. Let $x = x(t)$ be the equation of a line of discontinuity of the first derivatives of $u(x, t)$; we set

$$p_k^\pm = p_k^\pm(x(t), t) = p_k(x(t) \pm 0, t) = \frac{\partial u_k}{\partial x}(x(t) \pm 0, t),$$

$$q_k^\pm = q_k^\pm(x(t), t) = q_k(x(t) \pm 0, t) = \frac{\partial u_k}{\partial t}(x(t) \pm 0, t).$$

If $p^+ \neq p^-$ but the solution $u(x, t)$ is continuous on the line $x = x(t)$, then such a singularity of the solution is called a *weak discontinuity*, and the line $x = x(t)$ is called a *line of weak discontinuity*.

From the condition that $u(x, t)$ be continuous on the line of weak discontinuity $x = x(t)$ it follows that

$$x'(t)p^- + q^- = x'(t)p^+ + q^+,$$

i.e.,

$$[p]x'(t) = -[q], \tag{2}$$

where $[p] = p^+ - p^-$ and $[q] = q^+ - q^-$. By hypothesis, $u(x, t)$ satisfies (1) to the left and right of the line $x = x(t)$; therefore, at the points of this line,

$$l^k(q^+ + \xi_k p^+) = f_k, \qquad l^k(q^- + \xi_k p^-) = f_k \tag{3}$$

(the quantities l^k, ξ_k and f_k are continuous on the line $x = x(t)$).

Subtracting the second group of equations (3) from the first, we obtain

$$l^k_\alpha(q^+_\alpha - q^-_\alpha) + \xi_k(\mathscr{P}^+_k - \mathscr{P}^-_k) = 0,$$

where $\mathscr{P}^\pm_k = l^k_\alpha\, p^\pm_\alpha$ $(k = 1, \ldots, n)$. Eliminating $(q^+_\alpha - q^-_\alpha)$ by means of (2), we obtain

$$[\xi_k - x'(t)][\mathscr{P}_k] = [\xi_k - x'(t)](\mathscr{P}^+_k - \mathscr{P}^-_k) = 0 \qquad (k = 1, \ldots, n). \quad (4)$$

We set

$$\mathscr{P}^+_k - \mathscr{P}^-_k = [\mathscr{P}_k] = \eta_k.$$

If $x'(t) \neq \xi_k$ for all $k = 1, \ldots, n$, then $\mathscr{P}^+_k = \mathscr{P}^-_k$. Since the eigenvectors l^k are linearly independent, in this case $p^-_i = p^+_i$ for all $i = 1, \ldots, n$, i.e., there is no discontinuity of the derivatives. Therefore,

$$x'(t) = \xi_s(x(t), t, u(x(t), t)). \quad (5)$$

This equation implies that a line of weak discontinuity $x = x(t)$ is a characteristic of the system (1). This conclusion is, of course, consistent with the definition of a characteristic as a line across which a solution of the system (1) extends in a nonunique fashion (see §6.2).

Suppose that (5) is satisfied on the considered segment of the line of weak discontinuity $x = x(t)$, and suppose that $x = x(t)$ is an m-fold characteristic of the system (1), i.e., (5) holds for $s = j, j+1, \ldots, j+m-1$.

From (4) and (5) it then follows that $\eta_k = \mathscr{P}^+_k - \mathscr{P}^-_k = 0$ for $k < j$ and $k \geq j+m$. We shall derive equations which are satisfied by the η_k characterizing the magnitude of the weak discontinuity. Since, to the left and right of the line of weak discontinuity $x = x(t)$, $u(x, t)$ is a classical solution of (1), on the line $x = x(t)$ the quantities $\mathscr{P}^\pm_k$ satisfy the equations of the augmented system (§4.3) written for the characteristic $x = x(t)$ in the broad sense:

$$\left(\frac{d}{dt}\mathscr{P}^+_k\right)_k = \frac{\partial \mathscr{P}^+_k}{\partial t} + \xi_k \frac{\partial \mathscr{P}^+_k}{\partial x} = \mathscr{F}^k + \mathscr{F}^k_\alpha \mathscr{P}^+_\alpha + \mathscr{F}^k_{\alpha\beta}\mathscr{P}^+_\alpha \mathscr{P}^+_\beta, \quad (6)$$

$$\left(\frac{d}{dt}\mathscr{P}^-_k\right)_k = \frac{\partial \mathscr{P}^-_k}{\partial t} + \xi_k \frac{\partial \mathscr{P}^-_k}{\partial x} = \mathscr{F}^k + \mathscr{F}^k_\alpha \mathscr{P}^-_\alpha + \mathscr{F}^k_{\alpha\beta}\mathscr{P}^-_\alpha \mathscr{P}^-_\beta$$

$$(k = j, j+1, \ldots, j+m-1). \quad (7)$$

The coefficients of (6) and (7) are continuous on the line $x = x(t)$, and we therefore do not provide them with the indices $\pm$.

Subtracting (7) from (6), we obtain

$$\left(\frac{d\eta_k}{dt}\right)_k = \frac{\partial\eta_k}{dt} + \xi_k\frac{\partial\eta_k}{\partial x} = \mathscr{F}^k_\alpha\eta_\alpha + \mathscr{F}^k_{\alpha\beta}[\mathscr{P}^+_\alpha\mathscr{P}^+_\beta - \mathscr{P}^-_\alpha\mathscr{P}^-_\beta]$$
$$(k = j, j+1, \ldots, j+m-1). \tag{8}$$

Since

$$\mathscr{P}^+_\alpha\mathscr{P}^+_\beta - \mathscr{P}^-_\alpha\mathscr{P}^-_\beta = \mathscr{P}^-_\alpha\mathscr{P}^+_\beta + \eta_\alpha\mathscr{P}^+_\beta - \mathscr{P}^-_\alpha\mathscr{P}^-_\beta = \mathscr{P}^-_\alpha\eta_\beta + \mathscr{P}^+_\beta\eta_\alpha$$

and $\eta_\alpha = 0$ for $\alpha \neq j, j+1, \ldots, j+m-1$, the equations (8) can be written in the form

$$\left(\frac{d\eta_k}{dt}\right)_k = \sum_{\alpha=j}^{j+m-1} \psi^k_\alpha\eta_\alpha \qquad (k = j, j+1, \ldots, j+m-1), \tag{9}$$

where

$$\psi^k_\alpha = \mathscr{F}^k_\alpha + \sum_{\beta=1}^{n}(\mathscr{F}^k_{\alpha\beta}\mathscr{P}^+_\beta + \mathscr{F}^k_{\beta\alpha}\mathscr{P}^-_\beta).$$

The linearity of the system of ordinary differential equations (9) implies an important conclusion: *a weak discontinuity of a solution of a system of quasilinear equations of hyperbolic type can neither appear nor disappear while propagating along a characteristic, provided that the solution and its first derivatives remain bounded.*

In the case of a strictly hyperbolic system (§2.2) the characteristic $x = x(t)$ is simple. Therefore, the system (9) becomes a single differential equation. The equations (9) are called the *transport equations* for the weak discontinuity.

Since

$$\mathscr{P}^+_\alpha\mathscr{P}^+_\beta - \mathscr{P}^-_\alpha\mathscr{P}^-_\beta = \eta_\alpha\eta_\beta + \mathscr{P}^-_\alpha\eta_\beta + \mathscr{P}^-_\beta\eta_\alpha,$$

the system (9) can be written in the form

$$\left(\frac{d\eta_k}{dt}\right)_k = \mathscr{F}^k_\alpha\eta_\alpha + \mathscr{F}^k_{\alpha\beta}[\eta_\alpha\eta_\beta + \mathscr{P}^-_\alpha\eta_\beta + \mathscr{P}^-_\beta\eta_\alpha]$$
$$(k = j, j+1, \ldots, j+m-1). \tag{10}$$

The system (10) is nonlinear. From this equation we can conclude that the magnitude of a weak discontinuity η can become infinite in finite time. Indeed, for strictly hyperbolic systems, for example, the system (10) becomes a single equation of Ricatti or Bernoulli type. However, the quantities η_k only become infinite simultaneously with $\mathscr{P}^-_k$ and $\mathscr{P}^+_k$. Therefore, this effect is not a specific property of a weak discontinuity of a solution but is rather a consequence of the general property of unbounded growth of the derivatives of a solution of a system of quasilinear equations of hyperbolic type.

For a system of two quasilinear equations which are strictly hyperbolic the transport equation in the form (10) was obtained by Nitsche (1953).

We have established that for a strictly hyperbolic system a weak discontinuity is nonzero at all points of the characteristic. Hence, a weak discontinuity of a solution of the Cauchy problem occurs only when the initial functions possess a discontinuity of the first derivatives.

An arbitrary discontinuity of the derivatives of the initial functions decomposes into weak discontinuities which, generally speaking, propagate along all characteristics issuing from the point of discontinuity of the derivatives of the initial functions while satisfying the conditions (9) on each characteristic. This effect is sometimes called the *decay of an arbitrary weak discontinuity*.

2. *Unboundedness of derivatives. The gradient catastrophe.* According to §8.1, the growth of a solution $u(x, t)$ and its first derivatives with increasing t can be estimated by means of a solution of the majorant system (8.1.2), (8.1.3). This system is a nonlinear system of two ordinary differential equations, and from this system it follows immediately that for a sufficiently large value of $t > 0$ the quantities $\mathscr{P}(t)$ and $U(t)$, in general, become infinite simultaneously. Thus, the estimate of the growth of a solution and its derivatives by means of a solution of the majorant system leads to the conclusion that for an arbitrary system of quasilinear equations of hyperbolic type the solution $u(x, t)$ and its derivatives $p(x, t)$, in general, tend to infinity with increasing t.

This conclusion pertains to an arbitrary system of quasilinear equations. However, special classes of systems of quasilinear equations are also of interest, for example, systems with solutions which remain bounded for any values of t.

This property is possessed, for example, by systems of linear equations and also systems reducing to invariants, i.e., representable in the form

$$\frac{\partial r_k}{\partial t} + \xi_k \frac{\partial r_k}{\partial x} = f_k(x, t, r) \qquad (k = 1, \ldots, n), \tag{1}$$

if here f_k does not grow too rapidly with increasing r, for example, if for any x, t and r

$$\left| \frac{\partial f_k}{\partial r_j} \right| \leq C \qquad (k, j = 1, \ldots, n).$$

It is not hard to see that solutions $r(x, t)$ of systems of this type remain bounded for any values of t, but their derivatives nevertheless increase without bound in absolute value if $\xi_k(x, t, r)$ depend in an essential way on $r = \{r_1, \ldots, r_n\}$.

The effect of the formation of unbounded derivatives while the solution of a system of quasilinear equations remains bounded is called the *gradient catastrophe.*

We shall illustrate this with a simple example. We consider a homogeneous system of two quasilinear equations with coefficients which do not depend explicitly on x and t. It reduces to invariants and can be written in the form

$$\frac{\partial r_k}{\partial t} + \xi_k(r)\frac{\partial r_k}{\partial x} = 0 \qquad (k = 1, 2). \tag{2}$$

We assume that $\partial\xi_k/\partial r_k > 0$, and we consider for the system (2) the Cauchy problem with initial conditions posed on the entire axis $t = 0$:

$$r_1(x, 0) = r_1^0(x), \qquad r_2(x, 0) = r_2^0(x) = r_2^0 = \text{const}. \tag{3}$$

Suppose that $r_1^0(x) \in C_1$ and $|r_1^0(x)| < M$. The Cauchy problem (2), (3) reduces to the Cauchy problem for a single quasilinear equation:

$$\frac{\partial r_1}{\partial t} + \xi_1(r_1, r_2^0)\frac{\partial r_1}{\partial x} = 0, \qquad r_1(x, 0) = r_1^0(x).$$

According to §9, the solution $r_1(x, t)$ of this problem is given implicitly by the formula $r_1(x, t) = r_1^0(x - \xi_1(r_1(x, t), r_2^0)\cdot t)$.

We compute the derivative $\partial r_1/\partial x$:

$$\frac{\partial r_1}{\partial x} = \frac{(dr_1^0(x_0))/dx_0}{1 + (\partial\xi_1(r_1(x, t), r_2^0)/\partial r_1)(dr_1^0(x_0))/dx_0\cdot t},$$

where $x_0 = x - \xi_1(r_1, r_2^0)\cdot t$. This implies that for $dr_1^0(x_0)/dx_0 < 0$ the derivative $\partial r_1/\partial x$ decreases monotonically with increasing t on the characteristic $x = x_0 + \xi_1(r_1, r_2^0)\cdot t$, and for

$$t = t_m = \frac{-1}{d\xi_1(r_1^0(x_0), r_2^0)/dx_0} > 0$$

it becomes unbounded.

Thus, if $\partial\xi_k/\partial r_k \neq 0$, then, as a rule, the derivatives of a solution $r(x, t)$ of the system (2) grow without bound (in modulus) with increasing t.

For the system (2) these same conclusions can be obtained in a more formal way. Setting $p_k = \partial r_k/\partial x$, by differentiation of (2) we obtain

$$\begin{aligned}\frac{\partial p_1}{\partial t} + \xi_1(r_1, r_2)\frac{\partial p_1}{\partial x} &= -\frac{\partial\xi_1}{\partial r_1}(r_1, r_2)\, p_1^2 - \frac{\partial\xi_1}{\partial r_2}(r_1, r_2)p_1 p_2,\\ \frac{\partial p_2}{\partial t} + \xi_2(r_1, r_2)\frac{\partial p_2}{\partial x} &= -\frac{\partial\xi_2}{\partial r_2}(r_1, r_2)p_2^2 - \frac{\partial\xi_2}{\partial r_1}(r_1, r_2)p_1 p_2.\end{aligned} \tag{4}$$

Introducing the functions

$$\varphi_1(r_1, r_2) = \exp\left[-\int_0^{r_2} \frac{\partial \xi_1}{\partial r_2}(r_1, r_2) \frac{1}{\xi_2(r_1, r_2) - \xi_1(r_1, r_2)}\, dr_2\right],$$

$$\varphi_2(r_1, r_2) = \exp\left[\int_0^{r_1} \frac{\partial \xi_2}{\partial r_1}(r_1, r_2) \frac{1}{\xi_2(r_1, r_2) - \xi_1(r_1, r_2)}\, dr_1\right], \tag{5}$$

from (4), (5), and (2) we obtain

$$\frac{\partial}{\partial t}(\varphi_1 p_1) + \xi_1 \frac{\partial}{\partial x}(\varphi_1 p_1) = \frac{d}{dt_1}(\varphi_1 p_1) = -\kappa_1(r_1, r_2)(\varphi_1 p_1)^2,$$

$$\frac{\partial}{\partial t}(\varphi_2 p_2) + \xi_2 \frac{\partial}{\partial x}(\varphi_2 p_2) = \frac{d}{dt_2}(\varphi_2 p_2) = -\kappa_2(r_1, r_2)(\varphi_2 p_2)^2, \tag{6}$$

where $\kappa_i = (\partial \xi_i / \partial r_i)/\varphi_i$.

The prolonged system (6) of the original system (2) makes it possible to easily determine the occurrence of the gradient catastrophe and also conditions under which there exists a global classical solution of (2). For example, suppose that $\partial \xi_i / \partial r_i > 0$ for all r_1 and r_2; from (6) it then follows that if

$$\min_x \min_i \{p_i(x, 0)\} < 0,$$

then the derivatives of a solution of (2) become unbounded for a finite value of $t > 0$. If, under the same conditions on (2),

$$\min_x \min_i \{p_i(x, 0)\} \geq 0,$$

then the derivatives of the solution remain bounded for any $t > 0$. Hence, in this case there exists a classical solution of the Cauchy problem for any $t > 0$.

3. *Strongly and weakly nonlinear systems of quasilinear equations.* We call the system of quasilinear equations (10.2.1) *weakly nonlinear* in a region of the space of the variables x, t, r if in this region

$$\frac{\partial \xi_k(x, t, r)}{\partial r_k} \equiv 0 \qquad (k = 1, 2, \ldots, n); \tag{1}$$

otherwise we call the system (10.2.1) *strongly nonlinear*.

According to the definition, a weakly nonlinear system of two quasilinear equations can be written in the form

$$\left.\begin{aligned} \frac{\partial r_1}{\partial t} + \xi_1(x, t, r_2) \frac{\partial r_1}{\partial x} &= f_1(x, t, r_1, r_2), \\ \frac{\partial r_2}{\partial t} + \xi_2(x, t, r_1) \frac{\partial r_2}{\partial x} &= f_2(x, t, r_1, r_2). \end{aligned}\right\} \tag{2}$$

We note that if $\partial\xi_1/\partial r_2 \neq 0$ and $\partial\xi_2/\partial r_1 \neq 0$, then (2) can be brought to the form

$$\frac{\partial r_1}{\partial t} + r_2 \frac{\partial r_1}{\partial x} = f_1, \qquad \frac{\partial r_2}{\partial t} + r_1 \frac{\partial r_2}{\partial x} = f_2.$$

We shall now show that the derivatives of a solution of the weakly nonlinear system (2) remain bounded for any values of t provided that the solution $r(x, t)$ itself remains bounded.

THEOREM. *Suppose that the solution $r(x, t)$ of the system* (2) *is bounded*[7] *for* $0 \leq t \leq T$,

$$|r_k(x, t)| \leq R, \tag{3}$$

that the system (2) *is strictly hyperbolic,*[8] *i.e.,*

$$\xi_2(x, t, r_1(x, t)) - \xi_1(x, t, r_2(x, t)) \geq \varepsilon > 0 \quad \text{for } 0 \leq t \leq T, \tag{4}$$

and the functions ξ_k, $f_k \in C_1$. Then the derivatives $\partial r_k/\partial x$ and $\partial r_k/\partial t$ are bounded for $0 \leq t \leq T$ if they are bounded for $t = 0$.

PROOF. We consider $r(x, t)$ to be a solution of the Cauchy problem for the system (2) with the initial conditions $r_k(x, 0) = r_k^0(x)$, $-\infty < x < \infty$. According to the hypotheses of the theorem,

$$|r_k^0(x)| \leq R, \qquad \left|\frac{dr_k^0}{dx}(x)\right| \leq P_0.$$

Let $x = x_1(t, x_0)$ be the equation of the characteristic of the system (2) of the first family which passes through the point $x = x_0$ of the axis $t = 0$. We rewrite the first equation of (2) in the form

$$\left(\frac{dr_1}{dt}\right)_1 = \frac{\partial r_1}{\partial t} + \xi_1(x, t, r_2(x, t)) \frac{\partial r_1}{\partial x} = f_1(x, t, r_1, r_2).$$

If $r_2(x, t)$ is assumed known, then the determination of $r_1(x, t)$ reduces to the solution of a Cauchy problem for a system of two ordinary differential equations

$$\frac{dx_1}{dt} = \xi_1(x_1, t, r_2(x_1, t)), \tag{5}$$

$$\frac{d\bar{r}_1}{dt} = f_1(x_1, t, \bar{r}_1, r_2(x_1, t)) \qquad (\bar{r}_1 = \bar{r}_1(t, x_0)) \tag{6}$$

[7] As we have noted, condition (3) is automatically satisfied if $|\partial f_k/\partial r_i| \leq C$.

[8] For us it is sufficient that (4) be satisfied for $r(x, t)$. Of course, the system (2) may be strictly hyperbolic identically, i.e., for any r_1 and r_2.

with the initial data

$$x_1(0, x_0) = x_0, \qquad \bar{r}_1(0, x_0) = r_1^0(x_0). \tag{7}$$

If $\bar{r}_1(t, x_0)$ is the solution of problem (5)–(7), then the formula

$$r_1(x_1(t, x_0), t) = \bar{r}_1(t, x_0) \tag{8}$$

defines the solution $r_1(x, t)$.

We set

$$\dot{x}_1(t, x_0) = \frac{\partial}{\partial x_0} x_1(t, x_0)$$

and differentiate (5) with respect to the parameter x_0. We obtain

$$\frac{d\dot{x}_1}{dt} = \left\{\frac{\partial}{\partial x} [\xi_1(x, t, r_2(x, t))]\right\}_{x=x_1(t, x_0)} \dot{x}_1(t, x_0). \tag{9}$$

Since

$$\begin{aligned}\left(\frac{d\xi_1(x, t, r_2)}{dt}\right)_2 &= \frac{\partial \xi_1}{\partial t} + \xi_2 \frac{\partial \xi_1}{\partial x} \\ &= \frac{\partial \xi_1}{\partial r_2}(x, t, r_2)\left[\frac{\partial r_2}{\partial t} + \xi_2 \frac{\partial r_2}{\partial x}\right] + \xi'_{1t} + \xi_2 \xi'_{1x},\end{aligned}$$

it follows that

$$\frac{\partial \xi_1}{\partial t} + \xi_2 \frac{\partial \xi_1}{\partial x} = \bar{f}_2(x, t, r), \tag{10}$$

where

$$\bar{f}_2 = \frac{\partial \xi_1}{\partial r_2} f_2 + \xi'_{1t} + \xi_2 \xi'_{1x}.$$

Subtracting from (10) the equality

$$\frac{\partial \xi_1}{\partial t} + \xi_1 \frac{\partial \xi_1}{\partial x} = \left(\frac{d\xi_1}{dt}\right)_1,$$

we find that

$$\frac{\partial \xi_1}{\partial x} = \frac{\bar{f}_2 - \left(\frac{d\xi_1}{dt}\right)_1}{\xi_2 - \xi_1}. \tag{11}$$

In a similar way,

$$\frac{\partial \xi_2}{\partial t} + \xi_1 \frac{\partial \xi_2}{\partial x} = \left(\frac{d\xi_2}{dt}\right)_1 = \bar{f}_1.$$

We transform (11) by means of the identities

$$\frac{\partial \xi_1}{\partial x} = \frac{\bar{f}_2 - \bar{f}_1}{\xi_2 - \xi_1} + \frac{\bar{f}_1 - \left(\frac{d\xi_1}{dt}\right)_1}{\xi_2 - \xi_1} = \frac{\bar{f}_2 - \bar{f}_1}{\xi_2 - \xi_1} + \frac{\left(\frac{d\xi_2}{dt}\right)_1 - \left(\frac{d\xi_1}{dt}\right)_1}{\xi_2 - \xi_1}$$
$$= \frac{\bar{f}_2 - \bar{f}_1}{\xi_2 - \xi_1} + \left(\frac{d}{dt} \ln[\xi_2(x, t, r_1) - \xi_1(x, t, r_2)]\right)_1.$$

Substituting this into (9), we bring it to the form

$$\frac{d}{dt} \ln\left[\frac{\dot{x}_1(t, x_0)}{\xi_2(x, t, r_1(x, t)) - \xi_1(x, t, r_2(x, t))}\right]_{x=x_1(t, x_0)} = \left[\frac{\bar{f}_2 - \bar{f}_1}{\xi_2 - \xi_1}\right]_{x=x_1(t, x_0)}. \tag{12}$$

From the initial condition (7) we have $\dot{x}_1(0, x_0) = 1$; therefore, integrating (12) from 0 to t, we obtain

$$\frac{\dot{x}_1(t, x_0)[\xi_2(x_0, 0, r_1^0(x_0)) - \xi_1(x_0, 0, r_2^0(x_0))]}{[\xi_2(x, t, r_1(x, t)) - \xi_1(x, t, r_2(x, t))]_{x=x_1(t, x_0)}} = \exp \int_0^t \frac{\bar{f}_2 - \bar{f}_1}{\xi_2 - \xi_1} d\tau. \tag{13}$$

From the hypotheses (3) regarding the boundedness of the solution and the continuity of f_k, $\bar{f}_k$ and ξ_k it follows that there exists a number $M > 0$ such that

$$|\bar{f}_k| \leq \frac{M}{2}, \qquad |\xi_2 - \xi_1| \leq M, \qquad |\bar{f}_2 - \bar{f}_1| \leq M.$$

Moreover, by hypothesis (4) of the theorem we have $\xi_2 - \xi_1 \geq \varepsilon > 0$. Therefore, from (13) we obtain the estimate

$$\frac{\varepsilon}{M} e^{-Mt/\varepsilon} \leq \dot{x}_1(t, x_0) \leq \frac{M}{\varepsilon} e^{Mt/\varepsilon}, \tag{14}$$

which shows that the field of characteristics of the first family $x = x_1(t, x_0)$ is differentiable with respect to x_0 for all t in $[0, T]$.

This implies that the characteristics $x = x_1(t, x_0)$ do not intersect one another for $0 \leq t \leq T$. It is now easy to finish the proof of our theorem. Setting

$$\dot{\bar{r}}_1 = \dot{\bar{r}}_1(t, x_0) = \frac{\partial \bar{r}_1(t, x_0)}{\partial x_0}$$

and differentiating (6) with respect to x_0, we obtain

$$\frac{d\dot{\bar{r}}_1}{dt} = \frac{\partial f_1}{\partial r_1} \dot{\bar{r}}_1 + \dot{x}_1\left[\frac{\partial f_1}{\partial r_2} \cdot \frac{\partial r_2}{\partial x}\bigg|_{x=x_1} + f'_{1x}\right]. \tag{15}$$

We set

$$P(t) = \sup_{\substack{0\le\tau\le t\\ -\infty<x<\infty}} \max_{k=1,2} \left| \frac{\partial r_k(x, \tau)}{\partial x} \right|$$

and assume that the constant M is so large that for all $0 \le t \le T$ and $-\infty < x < \infty$

$$\left| \frac{\partial f_i}{\partial r_j} \right| \le M, \qquad |f'_{ix}| \le M \quad (i, j = 1, 2). \tag{16}$$

Integrating (15), we obtain

$$\dot{r}'_1(t, x_0) = \dot{r}'_1(0, x_0)\exp\left\{\int_0^t \frac{\partial f_1}{\partial r_1}\, d\tau\right\}$$

$$+ \int_0^t \left[\frac{\partial f_1}{\partial r_2} \cdot \frac{\partial r_2}{\partial x} + f'_{1x}\right] \dot{x}_1(\tau, x_0) \exp\left\{\int_\tau^t \frac{\partial f_1}{\partial r_1}\, d\zeta\right\} d\tau.$$

Substituting (14) and (16) and using the initial condition (7), we obtain

$$|\dot{r}'_1(t, x_0)| \le P_0 e^{Mt} + \frac{M^2}{\varepsilon}\, e^{(2M\varepsilon+M)t/\varepsilon} \int_0^t [P(\tau) + 1]\, d\tau. \tag{17}$$

From (8) we have

$$\frac{\partial r_1(x, t)}{\partial x} = \frac{\dot{r}'_1(t, x_0)}{\dot{x}_1(t, x_0)},$$

so that from (17) and (14) it follows that

$$\left| \frac{\partial r_1(x, t)}{\partial x} \right| \le \frac{M}{\varepsilon}\, e^{Mt+Mt/\varepsilon} P_0 + \frac{M^3}{\varepsilon^2}\, e^{2Mt(\varepsilon+1)/\varepsilon} \int_0^t [P(\tau) + 1]\, d\tau.$$

It is easy to see that $\partial r_2/\partial x$ can be estimated in a similar way; therefore,

$$\left| \frac{\partial r_k}{\partial x} \right| \le P_0 \frac{M}{\varepsilon}\, e^{Mt+Mt/\varepsilon} + \frac{M^3}{\varepsilon^2}\, e^{2Mt(\varepsilon+1)/\varepsilon} \int_0^t [P(\tau) + 1]\, d\tau \qquad (k = 1, 2).$$

From these inequalities we obtain

$$P(t) \le AP_0 + B \int_0^t [P(\tau) + 1]\, d\tau,$$

where

$$P(0) = P_0, \qquad A = \frac{M}{\varepsilon} e^{Mt(1+1/\varepsilon)}, \qquad B = MA^2,$$

which is valid for $0 \leq t \leq T$.

Applying Lemma 1 of §6, we have

$$P(t) \leq [AP_0 + Bt]e^{Bt},$$

whence it follows that the derivatives $\partial r_k(x, t)/\partial x$ are bounded on the entire segment $0 \leq t \leq T$. The proof of the theorem is complete.

Since T is arbitrary, the derivatives $\partial r_k/\partial x$ of a solution of the weakly nonlinear system (2) remain bounded in any strip of the variable t in which conditions (3) and (4) of the theorem are satisfied.

From the proof of the theorem we obtain the following result.

COROLLARY. *The Cauchy problem for a weakly nonlinear system of two quasilinear equations which is strictly hyperbolic is solvable in the domain of determinacy G if the solution $r(x, t)$ remains bounded in this domain.*[9]

We shall discuss this corollary in more detail. For an arbitrary system of quasilinear equations the derivatives become unbounded even when the solution itself is bounded. If we consider the Cauchy problem with initial conditions prescribed, for example, on the entire initial axis $t = 0$, then for a strongly nonlinear system the derivatives become infinite for a finite value of $t_0 > 0$, and for $t > t_0$ there exists no (classical) solution of this Cauchy problem.

For a weakly nonlinear system which is weakly hyperbolic with a solution which remains bounded (for example, when $|\partial f_k/\partial r_i| \leq C$ $(i, k = 1, 2)$), the derivatives remain bounded for all $t > 0$. Therefore, a solution of the Cauchy problem can be constructed in any finite strip $0 \leq t \leq T$ by the method set forth in §8. Thus, for a weakly nonlinear system there exists a global solution of the Cauchy problem, i.e., for all finite values of t. This circumstance makes weakly nonlinear systems comparable to linear systems.

On the other hand, this shows that any singularity of the initial data which is smoothed for $t = 0$ does not reappear for $t > 0$. Hence, generalized solutions of a weakly nonlinear system which is strictly hyperbolic may be considered the limits of smooth solutions in the entire half plane $t \geq 0$ in a manner analogous to the way in which this is done for linear equations.

As an example, we consider the weakly nonlinear system of two equation[10]

---

[9]This property of weakly nonlinear systems has been proved for any system (10.2.1) with $f_k \equiv 0$ (see Roždestvenskiĭ and Sidorenko (1967)).

[10]The general integral of the system (18) was obtained by Janenko (1955).

$$\frac{\partial r_k}{\partial t} + \xi_k(r_{3-k})\frac{\partial r_k}{\partial x} = 0 \qquad (k = 1, 2). \tag{18}$$

Here $f_1 = f_2 = \bar{f}_1 = \bar{f}_2 = 0$; therefore, (13) becomes the equality

$$\dot{x}_k(t, x_0^k) = \frac{\xi_2(r_1(x, t)) - \xi_1(r_2(x, t))}{\xi_2(r_1^0(x_0^k)) - \xi_1(r_2^0(x_0^k))} \qquad (k = 1, 2),$$

and for the derivative $\partial r_k/\partial x$ we have

$$\frac{\partial r_k(x, t)}{\partial x} = \frac{dr_k^0(x_0^k)}{dx_0^k} \cdot \frac{\xi_2(r_1^0(x_0^k)) - \xi_1(r_2^0(x_0^k))}{\xi_2(r_1(x, t)) - \xi_1(r_2(x, t))}. \tag{19}$$

This implies a more precise estimate for the derivatives of a solution of the system (18): $|\partial r_k(x, t)/\partial x| \leq P_0 M/\varepsilon$.

We note an interesting corollary of (19): if $\xi_2(r_1(x, t)) = \xi_1(r_2(x, t))$, then $\partial r_k(x, t)/\partial x = \infty$.

REMARK. Then definition of weakly nonlinear systems of quasilinear equations was given only for systems reducting to invariants. In general, this is possible only for $n \leq 2$ (see §3). The theorem on the boundedness of the derivatives has hence been proved only for $n = 2$.

The question of distinguishing a class of systems which do not lead to unbounded derivatives in the case $n \geq 3$ remains open. It is possible that the derivatives of a solution of a strictly hyperbolic system remain bounded if the following conditions are satisfied:

$$r_\alpha^k(x, t, u)\frac{\partial \xi_k(x, t, u)}{\partial u_\alpha} \equiv 0 \qquad (\alpha, k = 1, \ldots, n) \tag{20}$$

(see P. Lax (1957)). It is easy to see that the conditions (20) and (1) coincide in the case where invariants exist.

If the conditions (20) are satisfied, then it is easy to verify that the coefficients $\mathscr{F}_{\alpha\alpha}^k$ in the augmented system (4.3.16)–(4.3.19) vanish identically. In combination with the condition of strict hyperbolicity it is possible that this will make it possible to prove the boundedness of the derivatives of a solution of such systems as a consequence of the boundedness of the solution itself.

Another interesting question which has so far not been satisfactorily solved is that of distinguishing a class of systems with solutions which remain bounded even when the derivatives become infinite. Obviously the interesting case here is $n \geq 3$ when the system does not have invariants.

§11. Remarks on the mixed problem

1. *Formulation of the mixed problem for a linear system.* We shall consider a typical mixed problem: find a solution $u(x, t)$ of a system of linear equations of hyperbolic type which for $t = 0$ assumes prescribed values

$$u(x, 0) = u^0(x), \qquad a \le x \le b, \tag{1}$$

and satisfies certain boundary conditions

$$c^i_\alpha(x, t)u_\alpha(x, t)|_{\Gamma_1} = c^i u|_{\Gamma_1} = c_i(x, t) \qquad (1 \le i \le n_1), \tag{2}$$

$$d^i_\alpha(x, t)u_\alpha(x, t)|_{\Gamma_2} = d^i u|_{\Gamma_2} = d_i(x, t) \qquad (1 \le i \le n_2), \tag{3}$$

which are given on lines Γ_1 and Γ_2 issuing respectively from the end points $x = a$ and $x = b$ of a segment $[a, b]$ of the axis $t = 0$ (Figure 1.11).

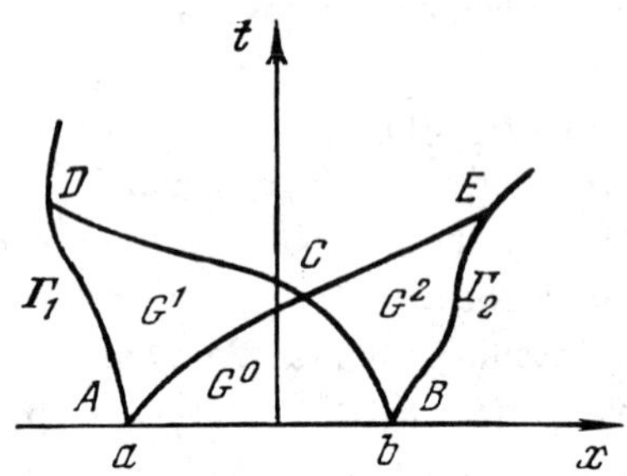

FIGURE 1.11

We shall assume that the curves Γ_1 and Γ_2 have continuously varying tangents and that $l^k_\alpha(x, t)$, $\xi_k(x, t)$, $f^k(x, t)$, $f^k_\alpha(x, t) \in C_1$ in the domain bounded by the curves Γ_1, Γ_2 and the axis $t = 0$; c^i_α, $c_i \in C_1$ on Γ_1, and d^i_α, $d_i \in C_1$ on Γ_2.

Suppose that the initial conditions (1) and the boundary conditions (2) and (3) satisfy the following consistency conditions:

$$\left.\begin{aligned} c^i_\alpha(a, 0)u^0_\alpha(a) &= c_i(a, 0) \qquad (1 \le i \le n_1), \\ d^i_\alpha(b, 0)u^0_\alpha(b) &= d_i(b, 0) \qquad (1 \le i \le n_2). \end{aligned}\right\} \tag{4}$$

If the conditions (4) are not satisfied a solution $u(x, t)$ of the mixed problem is discontinuous and must be interpreted as a generalized solution.

We expand the vectors c^i and d^i in terms of the vectors $l^k(x, t)$:

$$c^i(x, t) = \mu^i_\alpha(x, t)l^\alpha(x, t), \qquad d^i(x, t) = \nu^i_\alpha(x, t)l^\alpha(x, t).$$

The boundary conditions (2) and (3) can then be written in the form

$$\mu_\alpha^i l_\beta^\alpha u_\beta|_{\Gamma_1} = c_i(x, t) \qquad (i = 1, \dots, n_1),$$
$$\nu_\alpha^i l_\beta^\alpha u_\beta|_{\Gamma_2} = d_i(x, t) \qquad (i = 1, \dots, n_2),$$

or, in invariants,

$$\mu_\alpha^i r_\alpha|_{\Gamma_1} = c_i(x, t) \qquad (i = 1, \dots, n_1), \tag{5}$$

$$\nu_\alpha^i r_\alpha|_{\Gamma_2} = d_i(x, t) \qquad (i = 1, \dots, n_2). \tag{6}$$

Suppose that the conditions (5) and (6) are compatible and linearly independent, i.e., the rank of the matrix $((\mu_\alpha^i))$ is equal to n_1, and the rank of $((\nu_\alpha^i))$ is n_2.

Let G^0 be the domain of determinacy of the solution of the Cauchy problem with the initial condition (1).

It is obvious that the curves Γ_1 and Γ_2 must lie outside the domain G^0, since the solution of a linear system of equations is uniquely determined in G^0 by the initial condition (1) and, in general, it does not satisfy (5) and (6) there.

We consider the case in which the curve $x = X_1(t)$ intersects Γ_1 at a point D, while the curve $x = X_n(t)$ (see §6.3) intersects Γ_2 at a point E.

The solution $u(x, t)$ is uniquely determined in G^0 and can be constructed in this domain by the method of successive approximations (see §7). It therefore suffices to consider the problem of constructing $u(x, t)$ in the domain ACD; in BCE the solution is constructed in a similar way.

To determine conditions for the solvability of the mixed problem it is essential to know which families of characteristics $x = x_k(\xi, 0, t)$ issuing from points of the segment $[a, b]$ of the initial axis ($a \le \xi \le b$) intersect the arcs AD and BE of the curves Γ_1 and Γ_2.

Suppose that for $k = 1, \dots, p$ the characteristics $x = x_k(\xi, 0, t)$ for $\xi \in [a, b]$ intersect the segment AD of Γ_1, while for $k = n - s + 1, n - s + 2, \dots, n$ they intersect the segment BE of Γ_2.

Denoting by $u^0(x, t)$, $r^0(x, t)$ the solution of the Cauchy problem with the initial condition (1) in G^0, we arrive at the following problem in the domain ACD (G^1): find a solution $r(x, t)$ of the linear system

$$\frac{\partial r_k}{\partial t} + \xi_k \frac{\partial r_k}{\partial x} = g^k(x, t) + g_\alpha^k(x, t) r_\alpha, \tag{7}$$

which on the line AC satisfies the conditions

$$r_k(x, t)|_{AC} = r_k^0(x, t)|_{AC} \qquad (k = 1, \dots, n), \tag{8}$$

and on AD satisfies the conditions (5).

According to the definition of G^0, its boundary consists of segments of characteristics of the system (7).

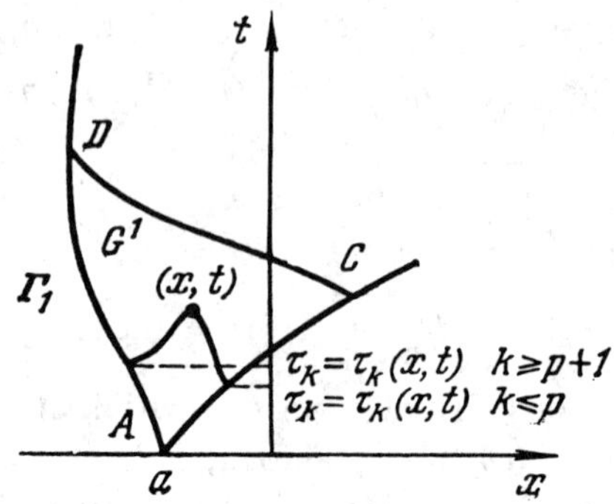

FIGURE 1.12

Thus, the determination of a solution in G^1 has reduced to finding a solution of (7) assuming prescribed values on the characteristic AC and satisfying the conditions (5) on AD (Figure 1.12). Since a solution of (7) exists in G^0, the values of $r^0(x, t)$ on the line AC satisfy the solvability conditions (§6.2). We thus encounter for the first time a problem with the initial values given on a characteristic. A problem with data on a characteristic is usually called a *Goursat problem.*

We consider an arbitrary point (x, t) of G^1 (Figure 1.12), and through it we draw all characteristics $x = x_k(x, t, \tau)$ of the system (7). According to our assumptions, the characteristics $x = x_k(x, t, \tau)$ intersect the curve AC for $1 \leq k \leq p$ at certain points $(x_k(x, t, \tau_k), \tau_k)$ with $\tau_k = \tau_k(x, t) \leq t$.

Similarly, the characteristics $x = x_k$ intersect the curve AD at points $(x_k(x, t, \tau_k), \tau_k)$ for $k \geq p + 1$ and $\tau_k = \tau_k(x, t) \leq t$.

We call the boundary conditions (5) *correct* if:

1) the number n_1 of boundary conditions (5) is equal to the number $q = n - p$ of characteristics $x = x_k(x, t, \tau)$ incident from an arbitrary point $(x, t) \in G^1$ on the line AD; and

2) the conditions (5) can be uniquely solved for $r_k(x, t)$ for $k \geq p + 1$.

Suppose that these conditions are satisfied, i.e., $n_1 = q = n - p$ and for $(x, t) \in AD$

$$\operatorname{Det} \left\| \begin{matrix} \mu^1_{p+1} & \cdots & \mu^1_{p+q=n} \\ \cdots & \cdots & \cdots \\ \mu^q_{p+1} & \cdots & \mu^q_n \end{matrix} \right\| \neq 0.$$

The conditions (5) can then be rewritten in the form

$$r_i(x, t)|_{AD} = \bar{c}_i(x, t) + \sum_{\alpha=1}^{p} \bar{\mu}^i_\alpha(x, t) r_\alpha(x, t)|_{AD}$$
$$(i = p + 1, p + 2, \ldots, p + q = n).$$

We shall henceforth drop the bar over c_i and μ^i_α. In G^1 the solution $r(x, t)$ satisfies

$$r_k(x, t) = r^0_k(x_k(x, t, \tau_k), \tau_k) + \int\limits_{\tau_k(x,t)}^{t} \{g^k(x_k(x, t, \tau), \tau) + g^k_\alpha(x_k(x, t, \tau), \tau) r_\alpha(x_k(x, t, \tau), \tau)\} \, d\tau \qquad (k \leq p),$$

$$r_k(x, t) = c_k(x_k(x, t, \tau_k), \tau_k) + \sum_{\alpha=1}^{p} \mu^k_\alpha(x_k(x, t, \tau_k), \tau_k) r_\alpha(x_k(x, t, \tau_k), \tau_k) + \int\limits_{\tau_k(x,t)}^{t} \{g^k(x_k(x, t, \tau), \tau) + g^k_\alpha(x_k(x, t, \tau), \tau) r_\alpha(x_k(x, t, \tau), \tau)\} \, d\tau \qquad (k \geq p + 1).$$

We shall seek a solution $r(x, t)$ by the method of successive approximations,

$$\overset{(s+1)}{r_k}(x, t) = r^0_k(x_k(x, t, \tau_k), \tau_k) + \int\limits_{\tau_k(x,t)}^{t} \{g^k(x, \tau) + g^k_\alpha(x, \tau) \overset{(s)}{r_\alpha}(x, \tau)\}_{x = x_k(x, t, \tau)} \, d\tau \qquad (k \leq p),$$

$$\overset{(s+1)}{r_k}(x, t) = \left[c_k(\zeta, \eta) + \sum_{\alpha=1}^{p} \mu^k_\alpha(\zeta, \eta) \overset{(s)}{r_\alpha}(\zeta, \eta) \right]_{\substack{\zeta = x_k(x, t, \tau_k) \\ \eta = \tau_k}} + \int\limits_{\tau_k(x,t)}^{t} \{g^k(x, \tau) + g^k_\alpha(x, \tau) \overset{(s)}{r_\alpha}(x, \tau)\}_{x = x_k(x, t, \tau)} \, d\tau \qquad (k \geq p + 1),$$

choosing a suitable initial approximation possessing continuous first derivatives in G^1.

If we set

$$\overset{(s)}{V}(t) = \max_{G^1_t} \max_{k=1,\dots,n} |\overset{(s)}{r_k}(\zeta, \tau) - \overset{(s-1)}{r_k}(\zeta, \tau)|,$$

then as in §7 we obtain the estimate

$$\overset{(s+1)}{V}(t) \leq B \int\limits_0^t \overset{(s)}{V}(\tau) \, d\tau + pn\mu \int\limits_0^t \overset{(s)}{V}(\tau) d\tau = (B + pn\mu) \int\limits_0^t \overset{(s)}{V}(\tau) \, d\tau.$$

Here we assume that in G^1

$$B \geq n|g^k_\alpha(x, t)|, \qquad \mu \geq |\mu^k_\alpha(x, t)| \,|_{\Gamma_1}.$$

The estimate obtained implies the uniform convergence in G^1 of the sequence $\{\overset{(s)}{r}(x, t)\}$ to a solution $r(x, t)$ of the mixed problem.

The solution $r(x, t)$ $(u(x, t))$ constructed in G^1 possesses there all the properties of a solution of the Cauchy problem enumerated in §7; it is continuously differentiable and depends in a continuous fashion on the initial data of the mixed problem including the curve Γ_1 if it satisfies properties 1) and 2).

We note, however, that the line AC is, in general, a line of discontinuity of the derivatives of the solution of the mixed problem. In order that the solution of the mixed problem have continuous first derivatives in the domain $G^0 + G^1$ it is necessary that the initial and boundary conditions satisfy consistency conditions for the derivatives. These conditions will be obtained below for a more general situation.

Above we assumed for simplicity that for fixed k all characteristics of the kth family $x = x_k(x, t, \tau)$ passing through any point $(x, t) \in G^1$ intersect for $\tau < t$ either only the line AC or only the line AD.

Of course, it may happen that this is not the case.

Suppose, for example, that through the point A there passes a characteristic $x = x_k(a, 0, \tau)$ dividing the domain G^1 into two parts (Figure 1.13). In this case a solution is constructed in a manner analogous to the foregoing with obvious changes.

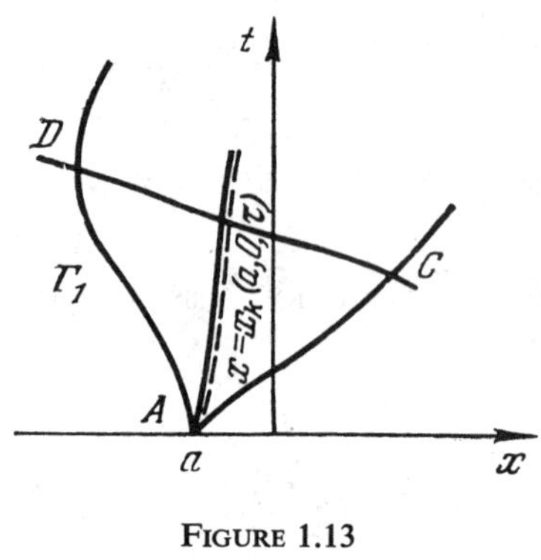

FIGURE 1.13

We note that this characteristic is also a line of discontinuity of the first derivatives.

In general, derivatives of a solution $r(x, t)$ are discontinuous on characteristics issuing from the point A unless consistency conditions for derivatives are satisfied at this point.

2. *Correctness of the boundary conditions for systems of quasilinear equations.* For a system of quasilinear equations of hyperbolic type we pose the initial conditions

$$u(x, 0) = u^0(x) \tag{1}$$

and the boundary conditions

$$c_i(x, t, u)\,|_{\Gamma_1} = 0 \qquad (i = 1, \ldots, n_1), \tag{2}$$

$$d_i(x, t, u)\,|_{\Gamma_2} = 0 \qquad (i = 1, \ldots, n_2). \tag{3}$$

We shall assume that the coefficients of the system satisfy the conditions imposed in §8 in proving the existence theorem for solutions of the Cauchy problem.

Let $u^0(x) \in C_1$, suppose that Γ_1 and Γ_2 are curves with continuous tangents lying in the half plane $t \geq 0$ and passing respectively through the points $(a, 0)$ and $(b, 0)$, and suppose that $c_i(x, t, u)$ and $d_i(x, t, u)$ are continuously differentiable functions of their arguments.

We say that at the point $(a, 0)$ the consistency conditions (the conditions for the continuity of the solution) are satisfied if

$$c_i(a, 0, u^0(a)) = 0 \qquad (i = 1, \ldots, n_1).$$

We shall establish consistency conditions for the derivatives at $(a, 0)$. Let $x = X(t)$ be the equation of the curve Γ_1. We assume the existence of a solution $u(x, t) \in C_1$ of the system of quasilinear equations which satisfies (1) and (2).

Differentiating the boundary conditions (2) with respect to the variable t on the line $x = X(t)$, we obtain

$$\frac{\partial c_i}{\partial u_\alpha}(q_\alpha + X'(t)p_\alpha) + \frac{\partial c_i}{\partial t} + \frac{\partial c_i}{\partial x} X'(t) = 0 \qquad (i = 1, \ldots, n_1).$$

The derivatives $p_\alpha = \partial u_\alpha / \partial x$ for $t = 0$ are determined from the initial conditions (1), and therefore from the system of equations

$$l_\alpha^k[q_\alpha + \xi_k p_\alpha] = f_k$$

it is possible to determine the derivatives $q_\alpha = \partial u_\alpha / \partial t$ on the initial axis $t = 0$:

$$q_k(x, 0) = -\lambda_\alpha^k \xi_\alpha l_\beta^\alpha p_\beta(x, 0) + \lambda_\alpha^k f_\alpha = q_k^0(x),$$

$$p_k(x, 0) = p_k^0(x) = du_k^0(x)/dx.$$

In these formulas the quantities λ_α^k, ξ_α, l_β^α and f_α are known functions of x; for example, $\lambda_\alpha^k = \lambda_\alpha^k(x, 0, u^0(x))$, etc.

We say that at the point $A(a, 0)$ of the curve Γ_1 the *consistency conditions for derivatives* are satisfied if

$$\frac{\partial c_i}{\partial u_\alpha}[q_\alpha^0(a) + X'(0)p_\alpha(a)] + c'_{it} + X'(0)c'_{ix} = 0 \qquad (i = 1, \ldots, n_1);$$

whereby the functions $\partial c_i/\partial u_\alpha$, $\partial c_i/\partial t$ and $\partial c_i/\partial c$ are evaluated at the point $(a, 0)$; for example, $\partial c_i/\partial u_\alpha = (\partial c_i/\partial u_\alpha)(a, 0, u^0(a))$.

Assuming, as before, the existence of a solution $u(x, t) \in C_1$ of the mixed problem, we shall establish conditions which the boundary conditions must satisfy. In general, the conclusion that Γ_1 and Γ_2 should lie outside the domain of determinacy G^0 of the solution of the Cauchy problem is not valid for essentially nonlinear equations ($\partial\xi_k/\partial u_j \neq 0$). There are correctly-posed mixed problems where Γ_1 and Γ_2 may lie in G^0. An example of such a problem is the piston problem in gas dynamics (see Chapter 2, §3). However, solutions of these problems are discontinuous. Restricting attention to classical solutions, we henceforth eliminate this case by assuming that Γ_1 and Γ_2 lie outside G^0. We set

$$\bar{l}_\alpha^k(x, t) = l_\alpha^k(x, t, u(x, t)),$$
$$\bar{\xi}_k = \xi_k(x, t, u(x, t)) \quad \text{etc.},$$

and we consider our problem as a mixed problem for the linear system

$$\bar{l}_\alpha^k\left[\frac{\partial u_\alpha}{\partial t} + \bar{\xi}_k\frac{\partial u_\alpha}{\partial x}\right] = \bar{f}_k$$

with the initial and boundary conditions (1)–(3).

The functions $\bar{r}_k = \bar{l}_\alpha^k u_\alpha$ satisfy the linear system of equations

$$\frac{\partial \bar{r}_k}{\partial t} + \bar{\xi}_k\frac{\partial \bar{r}_k}{\partial x} = \bar{g}^k(x, t) + \bar{g}_\alpha^k(x, t)\bar{r}_\alpha.$$

We consider the domain G^1 (Figure 1.14), and we suppose that the point $(X(t), t)$ lies on Γ_1.

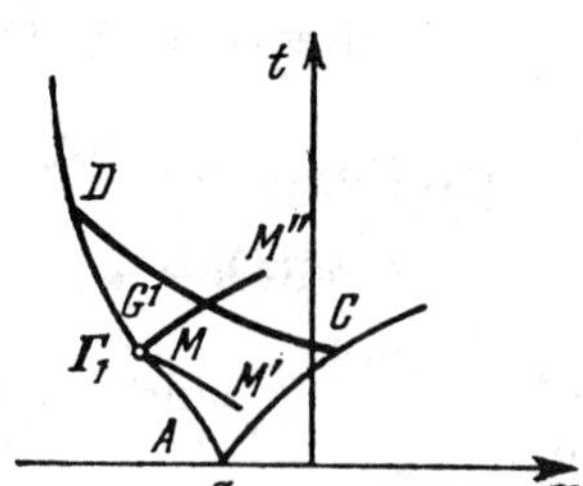

FIGURE 1.14

We call a characteristic $x = x_k(X(t), t, \tau)$ *incoming* at the point $(X(t), t)$ of Γ_1 if it lies in G^1 for $\tau \leq t$ and *outgoing* if it lies in G^1 for $\tau \geq t$. In Figure 1.14, MM' is an incoming characteristic, and MM'' is an outgoing characteristic.

Suppose that at each point of Γ_1 the characteristics $x = x_k$ for $k = 1, \ldots, p$ are incoming, while for $k = p + 1, \ldots, n$ they are outgoing.

In analogy to the case of a linear system we require that

1) the number of conditions (2) be equal to $q = n - p$, and

2) the equations (2) are uniquely solvable for $\bar{r}_{p+1}, \ldots, \bar{r}_n$ with known $\bar{r}_1, \ldots, \bar{r}_p$.

If we rewrite (2) in the form

$$c_i(x, t, \bar{\lambda}_\alpha^k \bar{r}_\alpha) = 0 \qquad (i = 1, 2, \ldots, n_1 = q),$$

then condition 2) is satisfied if

$$\operatorname{Det}\left(\left(\frac{\partial c_i}{\partial u_\alpha} \lambda_k^\alpha\right)\right) \neq 0 \qquad (i = 1, \ldots, q; k = p + 1, \ldots, n = p + q), (x, t) \in \Gamma_2.$$

We remark, however, that it is very difficult to verify the correctness of boundary conditions for a system of nonlinear equations, since conditions 1) and 2) depend on the solution $u(x, t)$, which is unknown to us. In solving a mixed problem it is nevertheless possible to proceed in the following manner.

On a sufficiently small segment of the curve Γ_1 next to the point A the solution $u(x, t)$ (if it exists) will be close to the value $u^0(a)$. This affords the possibility of verifying the correctness conditions in the formulation of the mixed problem for sufficiently small values of t. If they are satisfied, then we solve the problem for this small interval and consider the values of the solution $u(x, t)$ at the end of the interval as new initial values. In this way we are able to consider the solution of the mixed problem in the entire domain of the variables x and t where it exists.

We consider an example. For the system of two quasilinear equations

$$\frac{\partial r_1}{\partial t} + \xi_1(r_1, r_2)\frac{\partial r_1}{\partial x} = 0, \qquad \frac{\partial r_2}{\partial t} + \xi_2(r_1, r_2)\frac{\partial r_2}{\partial x} = 0$$

we pose the initial conditions

$$r_i(x, 0) = r_i^0 = \text{const}, \qquad |x| \leq a,$$

and the boundary conditions

$$c(t, r_1(-a, t), r_2(-a, t)) = 0, \qquad d(t, r_1(a, t), r_2(a, t)) = 0.$$

Suppose that $\xi_1 < 0$ and $\xi_2 > 0$. The consistency conditions are satisfied if

$$c(0, r_1^0, r_2^0) = 0, \qquad d(0, r_1^0, r_2^0) = 0. \tag{4}$$

If these conditions are satisfied, the solution $r(x, t)$ is continuous in some neighborhood of the axis $t = 0$. If in addition to (4) the consistency conditions for derivatives are satisfied,

$$\frac{\partial c}{\partial t}(0, r_1^0, r_2^0) = 0, \qquad \frac{\partial d}{\partial t}(0, r_1^0, r_2^0) = 0,$$

then $r(x, t)$ possesses continuous derivatives.

The solution $r(x, t)$ is constant in G^0 (Figure 1.15):

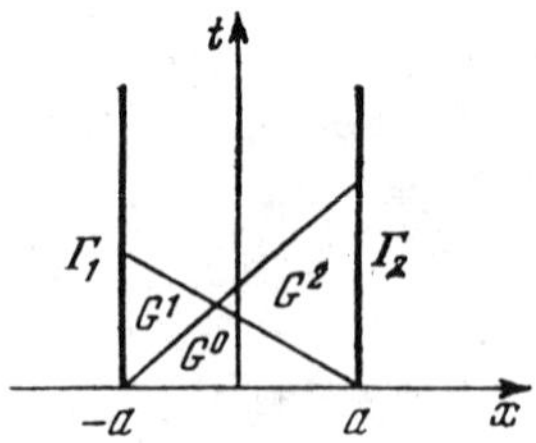

FIGURE 1.15

$$r_i(x, t) = r_i^0.$$

In G^1, $r_1(x, t) \equiv r_1^0$. The boundary conditions are correct if

$$\partial c(t, r_1^0, r_2)/\partial r_2 \neq 0, \qquad \partial d(t, r_1, r_2^0)/\partial r_1 \neq 0.$$

§12. Analytic methods of finding solutions of systems of differential equations in two independent variables

1. ***Investigation of the compatibility of overdetermined systems of equations.*** In most cases analytic methods for solving systems of quasilinear equations lead to the construction of more restricted classes of solutions than the general solution. As was shown in the preceding sections, a general solution of a system of n quasilinear equations in two variables depends on n arbitrary functions of a single variable. One can attempt to construct more restricted classes of solutions depending on fewer arbitary functions or even on parameters by requiring that a solution of the original system satisfy additional conditions, for example, additional equations.

As a result, we arrive at overdetermined systems of equations, i.e., systems in which the number of equations exceeds the number of unknown functions. Overdetermined systems may have no solutions at all; therefore, to establish the existence of solutions and the extent to which they are arbitrary an analysis of the compatibility of overdetermined systems is required. Here it is frequently possible to assume that certain coefficients of the original system which can be varied also remain to be defined. This approach means that we seek special cases of the original system for which this method of constructing solutions yields results.

The most universal method of analyzing the compatibility of systems of equations is Cartan's method of exterior forms (see É. Cartan (1945) and P. K. Raševskiĭ (1947)). However, here we present a simpler method, antedating Cartan's, which suffices for our purposes. To simplify the exposition we illustrate the method with several characteristic examples.

EXAMPLE 1. We consider the overdetermined system of equations

$$\partial u_i/\partial x_j = f_{ij}(x, u) \qquad (i = 1, \ldots, n; j = 1, \ldots, m), \tag{1}$$

where $u = \{u_1, \ldots, u_n\}$ and $x = \{x_1, \ldots, x_m\}$, for n functions $u_1, \ldots, u_n$ of m independent variables $x_1, \ldots, x_m$. Suppose that the functions f_{ij} are continuously differentiable in some domain G of the variables x and u. If for any $i = 1, \ldots, n$ and $j, k = 1, \ldots, m$ in G the conditions

$$\frac{\partial f_{ij}}{\partial x_k} + \sum_{\alpha=1}^{n} \frac{\partial f_{ij}}{\partial u_\alpha} f_{\alpha k} = \frac{\partial f_{ik}}{\partial x_j} + \sum_{\alpha=1}^{n} \frac{\partial f_{ik}}{\partial u_\alpha} f_{\alpha j} \tag{2}$$

ensuring the equality of the mixed derivatives $\partial^2 u/\partial x_j \partial x_k$ and $\partial^2 u/\partial x_k \partial x_j$ are identically satisfied, then the system (1) is called *completely integrable* in G.

In this case for any $(x^0, u^0) \in G$ there exists a neighborhood $|x - x^0| < r$ of x^0 in which (1) determines a unique solution $u(x) \in C_2$ such that $u(x^0) = u^0$ (see V. I. Smirnov (1958)). The determination of $u(x)$ hereby reduces to solving a system of ordinary differential equations

$$\frac{du_i}{dt} = \sum_{\alpha=1}^{m} f_{i\alpha}(x(t), u) \cdot \frac{dx_\alpha}{dt} \tag{3}$$

with the initial condition $u(0) = u^0$ and $x(0) = x^0$. Here $x = x(t) \in C_2$ is a line joining the point x^0 and the point x in question, and $u_i = u_i(x(t))$.

Thus, a solution $u(x)$ of a completely integrable system (1) depends on $n + m$ constants $u_1^0, \ldots, u_n^0$ and $x_1^0, \ldots, x_m^0$. It is therefore said that a solution of a completely integrable system has "constant arbitrariness".

EXAMPLE 2. We consider an overdetermined system of linear, homogeneous partial differential equations of first order

$$L_i u = \sum_{\alpha=1}^{m} a_{i\alpha} \frac{\partial u}{\partial x_\alpha} \qquad (i = 1, \ldots, p) \tag{4}$$

with one unknown function $u(x_1, \ldots, x_m)$, and we suppose that the coefficients a_i are sufficiently smooth functions of $x_1, \ldots, x_m$.

An algorithm is known for investigating the compatibility of systems of the type (4); it reduces to successively forming the so-called Poisson brackets (see Raševskiĭ (1947) and Smirnov (1958)).

For linear operators L_i and L_j we form the commutator

$$[L_i, L_j] = L_iL_j - L_jL_i = \sum_{\alpha=1}^{n} b_{ij\alpha} \frac{\partial}{\partial x_\alpha},$$

$$b_{ij\alpha} = \sum_{\beta=1}^{m} \left(a_{i\beta} \frac{\partial a_{j\alpha}}{\partial x_\beta} - a_{j\beta} \frac{\partial a_{i\alpha}}{\partial x_\beta} \right).$$

The operator $[L_i, L_j]$ is called the *Poisson bracket*. If $u(x_1, \ldots, x_m) \in C_2$ is a solution of (4), then it also satisfies the linear, homogeneous equations of first order

$$[L_i, L_j]u = 0 \qquad (i, j = 1, \ldots, p). \tag{5}$$

By adjoining the equation (5) for all $i, j = 1, \ldots, p$ to the system (4), we obtain a system of equations of first order of the same type as the original system—the so-called *extended system*. We call this cycle of operations a *prolongation*. After a finite number of prolongations we arrive at a linear system for which the adjoining of Poisson brackets yields no new equations, i.e., the commutators of the differential operators of the system are linear combinations of these operators.

Such systems are called *complete*. Thus, by definition, the system (4) is called complete if

$$[L_i, L_j] = \sum_{\alpha=1}^{p} C_{ij}^{\alpha}(x_1, \ldots, x_m) \cdot L_\alpha \qquad (i, j = 1, \ldots, p). \tag{6}$$

Suppose that the rank of the matrix $((a_{i\alpha}(x_1, \ldots, x_m)))$ in G is equal to p. For a complete system (4) there are two possible cases:

a) $p = m$: then the system (4) admits only the trivial solution $u = \text{const}$;

b) $p < m$: it is then possible to show (see Goursat (1929) or Smirnov (1958)) that the system (4) by means of a change of variables reduces to a single linear, homogeneous equation for a single unknown function of $m - p + 1$ arguments $y_1, \ldots, y_{m-p+1}$. Hence, the general solution of a complete system (4) in this case depends on one arbitrary function of $m - p$ arguments.

Thus, the investigation of the compatibility of the system (4) consists in extending it to a complete system and in computing the rank of the matrix of coefficients of the complete system.

EXAMPLE 3. We consider an overdetermined system of two nonlinear equations, one of which is the Monge-Ampère equation,

$$b\begin{vmatrix} \dfrac{\partial^2 u}{\partial x_1^2} & \dfrac{\partial^2 u}{\partial x_1\,\partial x_2} \\ \dfrac{\partial^2 u}{\partial x_1\,\partial x_2} & \dfrac{\partial^2 u}{\partial x_2^2} \end{vmatrix} + \sum_{\alpha,\beta=1}^{2} a_{\alpha\beta}\frac{\partial^2 u}{\partial x_\alpha \partial a_\beta} + a = 0, \tag{7}$$

while the other equation of first order is

$$\varphi\left(x_1, x_2, u, \frac{\partial u}{\partial x_1}, \frac{\partial u}{\partial x_2}\right) = 0. \tag{8}$$

Here b, $a_{\alpha\beta}$, and a are smooth functions of x_1, x_2, u, $\partial u/\partial x_1$ and $\partial u/\partial x_2$; $a_{\alpha\beta}=a_{\beta\alpha}$.

We shall consider conditions under which the overdetermined system (7), (8) admits a family of solutions depending on one arbitrary function of a single argument.

Setting

$$du = p dx_1 + q dx_2, \qquad dp = r dx_1 + s dx_2, \qquad dq = s dx_1 + t dx_2, \tag{9}$$

we write (7) and (8) in the form

$$b(rt - s^2) + a_{11}r + 2a_{12}s + a_{22}t + a = 0, \tag{10}$$

$$\varphi(x_1, x_2, u, p, q) = 0. \tag{11}$$

For fixed x_1, x_2, u, p and q equation (10) in the three-dimensional space of the quantities r, s and t determines a surface of second order (a quadric) having, in general, two families of rectilinear generators.

It is known (see Goursat (1929)) that these families are determined by the equations

$$\left.\begin{aligned} br - b\mu s + a_{22} - \mu\lambda_i &= 0,\\ bs - b\mu t + \lambda_{3-i} - \mu a_{11} &= 0, \end{aligned}\right\} \quad (i = 1, 2), \tag{12}$$

where λ_1 and λ_2 are roots of the characteristic equation

$$\lambda^2 + 2a_{12}\lambda + a_{11}a_{22} - ab = 0.$$

The quantity μ determines a specific generator in each of the family of lines (12); the pair μ, s gives a point on the generator. By varying μ and s independently we obtain the original quadric (10).

Differentiating (11) with respect to x_1 and x_2, we find that

$$\left.\begin{aligned}\frac{\partial\varphi}{\partial p}r+\frac{\partial\varphi}{\partial q}s+\frac{\partial\varphi}{\partial x_1}+\frac{\partial\varphi}{\partial u}p=0,\\ \frac{\partial\varphi}{\partial p}s+\frac{\partial\varphi}{\partial q}t+\frac{\partial\varphi}{\partial x_2}+\frac{\partial\varphi}{\partial u}q=0.\end{aligned}\right\}\tag{13}$$

In order that the system (7), (8) have a solution depending on an arbitrary function it is necessary that the algebraic system of equations (10), (13) for the variables r, s and t admit an infinite set of solutions. Indeed, otherwise r, s, and t would define specific functions of x_1, x_2, u, p and q which when substituted into (9) would yield a system of equations in total differentials. This system would give all derivatives of u, p and q with respect to x_1 and x_2 as functions of u, p, q, x_1 and x_2, and in the best of cases would admit only constant arbitrariness, as we have seen in Example 1.

The requirement of an infinite number of solutions of the system (10), (13) means that the line (13) is one of the generators (12). This leads to the equations

$$\left.\begin{aligned}\frac{\partial\varphi/\partial p}{b}=\frac{\partial\varphi/\partial q}{-b\mu}=\frac{\partial\varphi/\partial x_1+(\partial\varphi/\partial u)p}{a_{22}-\mu\lambda_i},\\ \frac{\partial\varphi/\partial p}{b}=\frac{\partial\varphi/\partial q}{-b\mu}=\frac{\partial\varphi/\partial x_2+(\partial\varphi/\partial u)q}{\lambda_{3-i}-\mu a_{11}},\end{aligned}\right\}\tag{14}$$

where $i=1, 2$ is the index of the family of generators (12) to which the line (13) belongs.

Eliminating μ from (14), we get a system of equations for $\varphi(x_1, x_2, u, p, q)$:

$$\left.\begin{aligned}b\left(\frac{\partial\varphi}{\partial x_1}+p\frac{\partial\varphi}{\partial u}\right)-a_{22}\frac{\partial\varphi}{\partial p}-\lambda_i\frac{\partial\varphi}{\partial q}=0,\\ b\left(\frac{\partial\varphi}{\partial x_2}+q\frac{\partial\varphi}{\partial u}\right)-\lambda_{3-i}\frac{\partial\varphi}{\partial p}-a_{11}\frac{\partial\varphi}{\partial q}=0.\end{aligned}\right\}\tag{15}$$

We shall show that if (15) is satisfied for some $i=1, 2$, then the original system (7), (8) has a solution depending on an arbitrary function. Indeed, suppose φ is a solution of (15) and $u(x_1, x_2)$ is a solution of (8); we shall show that $u(x_1, x_2)$ is a solution of (7). To this end we differentiate (8); we obtain (13). Equations (15) imply that it is possible to introduce the parameter μ so that (14) is satisfied. Equations (14) imply that the line (13) lies on the quadric (10), i.e., $u(x_1, x_2)$ satisfies (7). Since $u(x_1, x_2)$ is any solution of (8), it depends, in general, on one arbitrary function of a single argument. The assertion has thus been proved.

To find φ we must solve the overdetermined system (15) for $i=1$ or $i=2$. Here we can use the method of investigating compatibility considered in Example 2 for a linear, homogeneous system in one unknown function such as the system (15).

M.H. Martin (1953b), C.S. Ludford (1955), and Ju. S. Zav'jalov (1956) analyzed the compatibility of the system (7), (8) for the special case

$$b = 1, \qquad a_{\alpha\beta} = 0, \qquad a = f^2(x_1, x_2), \tag{16}$$

which finds application in gas dynamics.

We shall present their result. In order that the system (7), (8) under conditions (16) be compatible and its solution depend on an arbitrary function of a single variable it is necessary and sufficient that $f(x_1, x_2)$ be representable in one of the two forms

$$f = \sqrt{F'(\alpha_1 x_1 + \alpha_2 x_2)}, \tag{17}$$

$$f = \frac{1}{(x_2 + \alpha_2)^2} \sqrt{F'\left(\frac{x_1 + \alpha_1}{x_2 + \alpha_2}\right)}, \tag{18}$$

where F is an arbitrary function of a single argument and α_1 and α_2 are arbitrary constants.

For these $f(x_1, x_2)$ the $\varphi(x_1, x_2, u, p, q)$ are found:

$$\varphi = \alpha_2 p - \alpha_1 q \pm g(\alpha_1 x_1 + \alpha_2 x_2),$$

if f is given in the form (17), and

$$\varphi = u - p(x_1 + \alpha_1) - q(x_2 + \alpha_2) \pm g\left(\frac{x_1 + \alpha_1}{x_2 + \alpha_2}\right),$$

if f has the form (18). In both cases $g(\theta)$ is related to $F(\theta)$ by $g'(\theta) = \sqrt{F'(\theta)}$. Finally, if $f = 0$, then $\varphi = \varphi(p, q)$.

These results were used by Martin, Ludford and Zav'jalov to obtain generalized Riemann waves (see Chapter 2, §9. 3).

EXAMPLE 4. We shall also consider the case $b = 0$ in problem (7), (8) when (7) is replaced by a linear equation of second order. In the notation of (9) it has the form

$$a_{11} r + 2a_{12} s + a_{22} t + b_1 p + b_2 q + c = 0. \tag{19}$$

We shall also consider the case where the coefficients a_{ik}, b_i, and c depend only on x_1 and x_2.

The added equation (8) or (11) restricts the class of solutions of (19); we wish to find functions $\varphi(x_1, x_2, u, p, q)$ such that the system (19), (11) is compatible and admits functional arbitrariness in the solution. For this it is necessary that the system (13), (19), which is linear in the second derivatives r, s and t, contain no more than two linearly independent equations. This condition leads to two homogeneous equations for φ:

$$\eta\varphi_p - \varphi_q = 0, \tag{20}$$

$$a_{11}\eta\varphi_{x_1} + a_{22}\varphi_{x_1} + (a_{11}\eta p + a_{22}q)\varphi_u - (b_1 p + b_2 q + c)\varphi_q = 0, \tag{21}$$

where $\eta = \eta(x_1, x_2)$ is a root of the characteristic equation

$$a_{11}\eta^2 - 2a_{12}\eta + a_{22} = 0.$$

We consider (20) and (21) as an overdetermined system for φ. The compatibility of this system can be investigated by forming Poisson brackets (see Example 2). The Poisson bracket for the operators of (20) and (21) leads to an equation for φ:

$$(a_{22} - a_{11}\eta^2)\varphi_u + (a_{11}\eta\eta_{x_1} + a_{22}\eta_{x_2})\varphi_p + (b_1\eta - b_2)\varphi_q = 0. \tag{22}$$

The Poisson bracket for the operators of (20) and (22) is identically zero, while for the operators (21) and (22) it leads to still another equation,

$$\nu_1\varphi_u + \nu_2\varphi_p + \nu_2\varphi_q = 0, \tag{23}$$

in which the quantities ν_i are determined only in terms of the coefficients of (19) and their derivatives.

In order that the system (20), (21), (22) be complete it is necessary that (23) be an algebraic consequence of (20) and (22). This condition leads to an equation connecting the coefficients $a_{ij}(x_1, x_2)$, $b_i(x_1, x_2)$ and their derivatives. If it is satisfied, then for φ we have a complete system of three equations which has a solution depending on one arbitrary function of two arguments.

If $\varphi(x_1, x_2, u, p, q)$ is a solution of this complete system, then the equation $\varphi = 0$ is called an intermediate integral of equation (19), since any solution of $\varphi = 0$ is simultaneously a solution of (19). Therefore, for a certain class of equations (19) the search for an intermediate integral in the form (8) leads to the construction of special solutions of (19).

We shall compare the class of solutions obtained above with the class of so-called functionally invariant solutions (see Smirnov and Sobolev (1932)). A solution $u = u(x_1, x_2)$ of (19) is called *functionally invariant* if, for any function $F(u) \in C_2$, $z = F(u(x_1, x_2))$ is also a solution of (19). Considering the case $c = 0$, we find that if $u = u(x_1, x_2)$ is a functionally invariant solution of (19), then

$$a_{11}\left(\frac{\partial u}{\partial x_1}\right)^2 + 2a_{12}\frac{\partial u}{\partial x_1}\cdot\frac{\partial u}{\partial x_2} + a_{22}\left(\frac{\partial u}{\partial x_2}\right)^2 = 0. \tag{24}$$

The study of the compatibility of (19) and (24) leads to the conclusion that the class of equations (19) having functionally invariant solutions is more restricted than the class of equations possessing an intermediate integral.

2. *Solutions characterized by differential constraints.* To distinguish a class of special solutions of a particular differential equation or system of equations we introduced additional algebraic or differential equations and required that a solution of the original problem satisfy these additional conditions. Following N.N. Janenko (1964), we call the additional differential equations *differential constraints*, and we call the highest order of the derivatives they contain the *order* of the constraint.

Suppose there is given a system of two quasilinear equations written in invariants:

$$\frac{\partial r_i}{\partial t} + \xi_i(x, t, r_1, r_2)\frac{\partial r_i}{\partial x} = f_i(x, t, r_1, r_2) \qquad (i = 1, 2). \tag{1}$$

We seek a solution of (1) satisfying a *differential constraint of first order*

$$F(x, t, r_1, r_2, p_1, p_2) = 0, \tag{2}$$

where $p_i = \partial r_i/\partial x$. It is obvious that the equations (1) make it possible to eliminate the derivatives $q_i = \partial r_i/\partial t = f_i - \xi_i p_i$ from the equation of a constraint, and it therefore suffices to consider only the case (2).

We shall find conditions under which a family of solutions of (1) satisfying the differential constraint (2) has functional arbitrariness. We have the equalities

$$dr_i = p_i dx + q_i dt = p_i dx + (f_i - \xi_i p_i)\, dt \qquad (i = 1, 2) \tag{3}$$

and the conditions for the integrability of (3)

$$\frac{\partial p_i}{\partial t} + \xi_i \frac{\partial p_i}{\partial x} = g_i \qquad (i = 1, 2),$$

$$g_i = f'_{ix} - p_i \xi'_{ix} + \sum_{\alpha=1}^{2}\left(\frac{\partial f_i}{\partial r_\alpha} - p_i \frac{\partial \xi_i}{\partial r_\alpha}\right) p_\alpha. \tag{4}$$

The equations (4) are obviously the equations of the prolonged system of the original system (1). Differentiating the equation of the constraint (2), we obtain

$$F'_x + \sum_{\alpha=1}^{2}\left(F'_{r_\alpha} \cdot p_\alpha + F'_{p_\alpha}\frac{\partial p_\alpha}{\partial x}\right) = 0,$$

$$F'_t + \sum_{\alpha=1}^{2}\left(F'_{r_\alpha} \cdot q_\alpha + F'_{p_\alpha}\frac{\partial p_\alpha}{\partial t}\right) = 0. \tag{5}$$

In order that a solution have functional arbitrariness it is necessary that in the system of four equations (4), (5) there are no more than three linearly independent equations, since otherwise four derivatives are determined from it as functions of x, t, r_α and p_α, and, according to subsection 1, a solution has no more than constant arbitrariness. We thus require that the rank of the extended matrix of the system (4), (5), which is linear in $\partial p_\alpha/\partial x$ and $p_\alpha/\partial t$, be equal to 3. This leads to two conditions, the first of which has the form

$$F'_{p_1} \cdot F'_{p_2} \cdot (\xi_2 - \xi_1) = 0.$$

Suppose that $\xi_1 \neq \xi_2$; there are then two possible cases: $F'_{p_1} \equiv 0$ and $F'_{p_2} \equiv 0$. We shall consider only the case $F'_{p_2} \equiv 0$; the second condition can then be written in the form

$$A + Bp_2 = 0, \tag{6}$$

where

$$A = \xi_1 F'_x + F'_t + f_1 \cdot F'_{r_1} + f_2 \cdot F'_{r_2} + F'_{p_1}[f'_{1x} + (f'_{1r_1} - \xi'_{1x})p_1 - \xi'_{1r_1}p_1^2], \tag{7}$$

$$B = (\xi_1 - \xi_2)F'_{r_2} + (f'_{1r_2} - \xi'_{1r_2} \cdot p_1)F'_{p_1}. \tag{8}$$

If (6) is not satisfied identically, this equality makes it possible to determine p_2 or p_1; in this case we again arrive only at constant arbitrariness. Hence, for functional arbitrariness the following linear equations for F must be satisfied:

$$A = 0, \qquad B = 0. \tag{9}$$

The problem of the compatibility of these two linear equations and the degree of arbitrariness of their solutions can again be solved by forming Poisson brackets. Since in the present case ($F'_{p_2} = 0$) the function F depends only on five arguments, for the existence of a solution of (9) with functional arbitrariness it suffices that the rank of the complete system formed by adjoining the Poisson brackets to (9) be equal to 4.

The case of a differential constraint of any order when the constraint equation contains derivatives of higher orders of the solution of (1) can be considered in a similar fashion.

It is possible to consider a conservative system of quasilinear equations for which the constraint equation contains the values of potentials $\Phi_i(x, t)$ of a solution of this system (see §6.3 of this chapter). An example of a constraint of this type is considered in Chapter 2, §9.3.

To conclude this subsection we apply the concept of a differential constraint to the linear Darboux equation

$$\frac{\partial^2 u}{\partial x_1\, \partial x_2} = f(x_1, x_2)u, \tag{10}$$

which is encountered in problems of gas dynamics.

We have already considered the differential constraint of first order

$$\varphi(x_1, x_2, u, p, q) = 0, \qquad p = \partial u/\partial x_1, \qquad q = \partial u/\partial x_2, \tag{11}$$

in the preceding subsection (see Example 4).

In this case the differential constraint provides an intermediate integral of the Darboux equation. The compatibility conditions for (10) and (11) lead to the condition $\varphi_p \cdot \varphi_q = 0$. Suppose, for example, that $\varphi_q = 0$. Then

$$f(x_1, x_2) = 0; \qquad \varphi = \varphi(x_1, p) = \varphi(x_1, \partial u/\partial x_1),$$

i.e., the Laplace invariant vanishes, and the Darboux equation becomes the wave equation.

We shall consider the compatibility of a differential constraint of second order

$$\partial p/\partial x_1 = \varphi(x_1, x_2, p) \tag{12}$$

with the Darboux equation which we write in the form

$$\partial p/\partial x_2 = f(x_1, x_2)u. \tag{13}$$

The compatibility condition for (12) and (13) has the form

$$\varphi'_{x_2} + \varphi'_p \frac{\partial p}{\partial x_2} = \varphi'_{x_2} + \varphi'_p \cdot f \cdot u = f'_{x_1} u + fp,$$

and it is satisfied identically if

$$\varphi'_{x_2} = f \cdot p; \qquad \varphi'_p \cdot f = f'_{x_1}. \tag{14}$$

As a condition for the compatibility of (14), from this it follows that

$$f = \frac{\partial^2 \ln f}{\partial x_1 \partial x_2} \tag{15}$$

and

$$\varphi = p \frac{\partial \ln f}{\partial x_1} + F(x_1), \tag{16}$$

where $F(x_1)$ is an arbitrary function.

We see that in this case the Darboux equation can be reduced by a Laplace transformation (see Ovsjannikov (1960)) to the wave equation

$$\frac{\partial^2 u}{\partial x_1 \partial x_2} = 0.$$

We show how to construct the Riemann function for the Darboux equation (10) by means of the method of differential constraints. The Riemann function $R(x_1, x_2; \xi_1, \xi_2)$ is a solution of (10) (in the variables x_1 and x_2) and satisfies additional conditions on the two characteristics $x_1 = \xi_1$ and $x_2 = \xi_2$ of this equation: $R = 1$ for $x_1 = \xi_1$ and for $x_2 = \xi_2$.

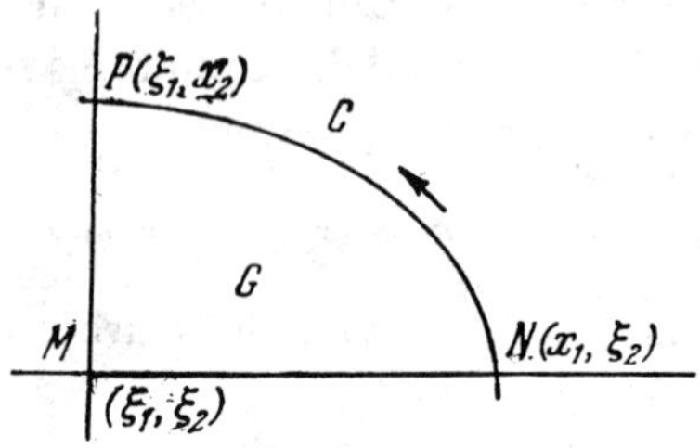

FIGURE 1.16

In other words, $R(x_1, x_2; \xi_1, \xi_2)$ is a solution of (10) which is equal to 1 on the characteristics PM and MN (Figure 1.16). Suppose that $u(x_1, x_2)$ is any solution of (10); application of Green's formula to the functions $u = u(x_1, x_2)$ and $v = R(x_1, x_2; \xi_1, \xi_2)$ then leads to the formula

$$u(M) = \frac{u(P) + u(N)}{2} - \frac{1}{2}\left[\int_N^P \left(u\frac{\partial R}{\partial x_1} - R\frac{\partial u}{\partial x_1}\right)dx_1 + \left(R\frac{\partial u}{\partial x_2} - u\frac{\partial R}{\partial x_2}\right)dx_2\right], \tag{17}$$

where the integral is taken along any smooth curve joining the points P and N on the axes $x_1 = \xi_1$ and $x_2 = \xi_2$.

Knowledge of the Riemann function R makes it possible to solve the Cauchy problem for (10) with prescribed values of u, u'_{x_1} and u'_{x_2} on the curve PN.

The case of a differential constraint of first order is obvious, since then $f \equiv 0$, $R \equiv 1$, and (17) is the familiar d'Alembert formula.

In the case of a differential constraint of second order we have (15) and (16); by $u(x_1, x_2)$ we now mean the Riemann function.

Integrating (12) using (16), we obtain

$$p = C(x_1, x_2)f,$$
$$C(x_1, x_2) = \int_{\xi_1}^{x_1} \frac{F(\tau)}{f(\tau, x_2)}\,d\tau + g(x_2), \tag{18}$$

where $g(x_2)$ is an arbitrary function.

We determine $F(x_1)$ from the conditions for the Riemann function

$$u(x_1, \xi_2) = 1; \qquad \frac{\partial u}{\partial x_1}(x_1, \xi_2) = p(x_1, \xi_2) = 0,$$

whence it follows that

$$C(x_1, \xi_2) = 0. \tag{19}$$

The equality (19) is possible only if $F(x_1) = 0$ and $g(\xi_2) = 0$, and therefore

$$p(x_1, x_2) = g(x_2) f'(x_1, x_2). \tag{20}$$

Integrating, we find that

$$u(x_1, x_2) = g(x_2) \int_{\xi_1}^{x_1} f(\tau, x_2)\, d\tau + K(x_2). \tag{21}$$

Since $u(\xi_1, x_2) = 1$, it follows that $K(x_2) \equiv 1$, and for $u(x_1, x_2)$ we obtain the representation

$$u(x_1, x_2) = g(x_2) \int_{\xi_1}^{x_1} f(\tau, x_2)\, d\tau + 1. \tag{22}$$

By satisfying the Darboux equation (13), we obtain an equation for $g(x_2)$:

$$\frac{dg}{dx_2} + g \frac{\partial \ln f}{\partial x_2} = g \int_{\xi_1}^{x_1} f(\tau, x_2)\, d\tau + 1. \tag{23}$$

Noting the condition (15), we bring (23) to the form

$$g'(x_2) + l(\xi_1, x_2) g(x_2) = 1, \tag{24}$$

where

$$l(x_1, x_2) = \frac{\partial \ln f}{\partial x_2}, \qquad f = \frac{\partial l}{\partial x_1}. \tag{25}$$

From this we obtain an expression for $g(x_2)$,

$$g(x_2) = g(x_2; \xi_1, \xi_2) = \left[\frac{1}{f(\xi_1, x_2)}\right] \int_{\xi_2}^{x_2} f(\xi_1, \tau)\, d\tau,$$

and for the Riemann function

$$u(x_1, x_2) = R(x_1, x_2; \xi_1, \xi_2) = \frac{\int_{\xi_2}^{x_2} f(\xi_1, \tau)\, d\tau \int_{\xi_1}^{x_1} f(\tau, x_2)\, d\tau}{f(\xi_1, x_2)} + 1. \tag{26}$$

If we observe that a function satisfying (15) has the form (see Ovsjannikov (1960))

$$f = 2\,\frac{\alpha_1'(x_1)\alpha_2'(x_2)}{[\alpha_1(x_1) + \alpha_2(x_2)]},$$

we finally obtain

$$R(x_1, x_2; \xi_1, \xi_2) = 2\,\frac{[\alpha_2(x_2) - \alpha_2(\xi_2)][\alpha_1(x_1) - \alpha_1(\xi_1)]}{[\alpha_1(\xi_1) + \alpha_2(\xi_2)][\alpha_1(x_1) + \alpha_2(x_2)]} + 1.$$

3. ***Solutions of systems of quasilinear equations with degenerate hodograph.*** One of the basic problems of the analytic theory of partial differential equations is to obtain special solutions and to construct by means of them a broader class of solutions. For linear equations the principle of superposition makes it possible to extend the class of solutions; in particular, this is the basis for the Fourier method which is widely used in mathematical physics. However, for systems of nonlinear equations there is no superposition principle, and the problem of constructing broader classes of solutions on the basis of particular solutions is thus considerably more complicated. There is the known method of envelopes, which makes it possible to construct solutions depending on an arbitrary function (a general integral) from solutions depending on two arbitrary parameters (a complete integral); however, generally speaking, this method is applicable only to the case of a single nonlinear equation.

The method of envelopes is especially effective for the case of equations not depending directly on the unknown functions but only on their derivatives. For example, the nonlinear equation

$$\frac{\partial \Phi}{\partial t} + \varphi\left(\frac{\partial \Phi}{\partial x}\right) = 0 \tag{1}$$

has the obvious solution

$$\Phi = ax - \varphi(a)\cdot t + b, \tag{2}$$

where a and b are arbitrary numbers. We assume that $b = \psi(a)$ is a function of the parameter a; then the family (2) can have an envelope defined by the equations

$$\begin{aligned} \Phi &= ax - \varphi(a)\cdot t + \psi(a), \\ 0 &= x - \varphi'(a)\cdot t + \psi'(a). \end{aligned} \tag{3}$$

It is known that (3) implicitly defines a solution of (1) depending on an arbitrary function ψ, i.e., a general integral.

To (3) there corresponds a general solution of the single quasilinear equation

$$\frac{\partial u}{\partial t} + \frac{\partial \varphi(u)}{\partial x} = 0, \tag{4}$$

for which $\Phi(x, t)$ is a potential (see §5. 3), and it is given by

$$u = \frac{\partial}{\partial x}[ax - \varphi(a)\cdot t + \psi(a)] = a = a(x, t), \tag{5}$$

where $a = a(x, t)$ is determined from the second equation of (3).

However, for $n \geq 2$ the method of envelopes leads only to the construction of more restricted classes of solutions. This circumstance has a simple geometric interpretation.

For $n = 1$ the space of (Φ, x, t) is threedimensional, and the integral surfaces (2) corresponding to different values of the parameter a intersect along a line (a characteristic). The one-parameter family of characteristics then forms an enveloping surface which is an integral manifold of equation (1). If $n = 2$, then $\Phi_i = \Phi_i(x, t, a)$ are two-dimensional surfaces in the four-dimensional space of the variables (x, t, Φ_1, Φ_2), and the intersection of these surfaces corresponding to two distinct values of the parameter a occurs, generally speaking, not along a line but at a point.

We consider the class of solutions to which the method of envelopes leads for a system of nonlinear equations

$$\begin{gathered} \frac{\partial \Phi_i}{\partial t} + \varphi_i\left(\frac{\partial \Phi}{\partial x}\right) = 0 \qquad (i = 1, 2, \ldots, n), \\ \frac{\partial \Phi}{\partial x} = \left\{\frac{\partial \Phi_1}{\partial x}, \ldots, \frac{\partial \Phi_n}{\partial x}\right\}. \end{gathered} \tag{6}$$

We have a family of solutions of the system (6):

$$\left.\begin{aligned} \Phi_i &= a_i x - \varphi_i(a)\cdot t + b_i \\ \varphi_i(a) &= \varphi_i(a_1, \ldots, a_n) \end{aligned}\right\} \qquad (i = 1, \ldots, n). \tag{7}$$

We shall consider the family of integral surfaces (7) assuming that the quantities a_i and b_i are functions of a single parameter τ. The envelope of this family is then determined by (7) and the n further equations

$$0 = \dot{a}_i x - t\sum_{j=1}^{n} \frac{\partial \varphi_i(a)}{\partial a_j}\dot{a}_j + \dot{b}_i \qquad (i = 1, \ldots, n), \tag{8}$$

which serve to determine τ.

In order that there exist a characteristic (a common line of intersection of the surfaces (7) and (8)) it is necessary and sufficient that in (8) there be only one independent equation, i.e., that the following conditions be satisfied:

$$\sum_{j=1}^{n} \frac{\partial \varphi_i(a)}{\partial a_j} \dot{a}_j = \mu(\tau)\dot{a}_i, \qquad \dot{b}_i = \nu(\tau)\dot{a}_i \quad (i = 1, \ldots, n). \tag{9}$$

Suppose that the system (6) is strictly hyperbolic, i.e., the matrix $A = ((\partial \varphi_i(a)/\partial a_j))$ has real and distinct eigenvalues $\xi_1(a) < \xi_2(a) < \cdots < \xi_n(a)$ to which there correspond the right eigenvectors $r^1(a), \ldots, r^n(a)$. Equations (9) then have n distinct solutions

$$\mu(\tau) = \xi_k(a); \qquad \dot{a}_i(\tau) = \lambda_k(\tau) r_i^k(a(\tau)), \qquad \dot{b}_i(\tau) = \nu_k(\tau)\dot{a}_i(\tau), \tag{10}$$

corresponding to each eigenvalue $\xi_k(a)$. Since the parameter τ is undetermined, we may, for example, assume that $\lambda_k(\tau)$ is a given function of τ which corresponds to a particular parameterization of a_i and b_i. In (10) there then remains one arbitrary function $\nu_k(\tau)$ on which the envelope of the family (7) depends in an essential way.

If the conditions (10) are satisfied, then the equations (8) reduce to the single equation

$$x - \xi_k(a(\tau))t + \nu_k(\tau) = 0, \tag{11}$$

which together with (10) implies that the quantities a_i and b_i are constant along the kth characteristic (11).

To the solution $\Phi_i(x, t)$ there corresponds a solution $u_i(x, t) = \partial \Phi_i/\partial x$ of the conservative system of quasilinear equations

$$\frac{\partial u_i}{\partial t} + \frac{\partial \varphi_i(u)}{\partial x} = 0 \qquad (i = 1, 2, \ldots, n),\ u = \{u_1, \ldots, u_n\}. \tag{12}$$

Morever, by (7) and (8) we have

$$u_i(x, t) = a_i(\tau(x, t)), \tag{13}$$

where $\tau = \tau(x, t)$ is determined from (11). A constant value τ defines a characteristic—the line (11)—along which by (13) the solution $u(x, t)$ is constant. The method of envelopes has thus led us to construct a solution of a conservative system of quasilinear equations depending on an arbitrary function of the single variable $\nu(\tau)$. The class of solutions of (12) we have constructed is called the class of *simple waves*.

A simple wave is a solution $u = u(x, t)$ of the system of equations (12) satisfying the condition

$$\operatorname{rank}\begin{Vmatrix} \dfrac{\partial u_1}{\partial t} & \dfrac{\partial u_2}{\partial t} & \cdots & \dfrac{\partial u_n}{\partial t} \\ \dfrac{\partial u_1}{\partial x} & \dfrac{\partial u_2}{\partial x} & \cdots & \dfrac{\partial u_n}{\partial x} \end{Vmatrix} = 1. \tag{14}$$

This condition implies that the functions $u_i(x, t)$ can be represented in the form

$$u_i(x, t) = u_i(\tau), \qquad \tau = \tau(x, t), \tag{15}$$

where $\tau(x, t)$ is a smooth function.

The solution (13) constructed above is thus a simple wave. Just as above, it follows from the representation (15) that the lines (11) are characteristics of a simple wave, while $\tau(x, t)$ is constant along these characteristics.

It is easy to see that a simple wave is completely determined by the matrix $A(u) = ((\partial\varphi_i/\partial u_j))$. A simple wave is therefore defined analogously for a non-conservative system

$$\frac{\partial u}{\partial t} + A(u)\frac{\partial u}{\partial x} = 0; \qquad u = \{u_1, \ldots, u_n\} \tag{16}$$

by the equalities

$$\dot{u}(\tau) = \lambda_k(\tau)\cdot r^k(u(\tau)), \qquad \frac{\partial \tau}{\partial t} + \xi_k(u(\tau))\cdot\frac{\partial \tau}{\partial x} = 0. \tag{17}$$

A special case of a simple wave is the *centered wave*, where the characteristics of the kth family are lines intersecting at a single point x_0, t_0. In this case it is possible to take for the function $\tau(x, t)$ in (15) the slope of these characteristics, i.e.,

$$\tau(x, t) = \frac{x - x_0}{t - t_0} = \xi_k(u(\tau)). \tag{18}$$

From (15) it follows that in a simple wave of kth type there is the $(n - 1)$-functional dependence:[11]

$$v_i^k(u_1, \ldots, u_n) = c_i \qquad (i = 1, 2, \ldots, n - 1). \tag{19}$$

We shall consider a one-parameter family of centered waves (18), (19) in which the constants c_i are fixed, while x_0 and t_0 are connected by a relation. We then have a one-parameter family of integral surfaces of equation (16), which has an envelope. This envelope is a simple wave, but it is not centered. Hence, simple waves are envelopes of centered waves.

[11]The collection $\{v_1^k(u), \ldots, v_{n-1}^k(u)\}$ is called an $(n - 1)$-dimensional vectorial Riemann invariant (see Chapter 4, §3.2).

Each solution $u = u(x, t)$ of (16) may be considered as the parametric definition in the space of the variables $(u_1, \ldots, u_n)$ of some manifold S in terms of the variables (x, t). If (14) is not satisfied in some domain D of the variables x and t then this is a two-dimensional manifold. If (14) is satisfied in D, then this manifold is one dimensional, and the formulas $u_i = u_i(x, t)$ define a mapping of D onto some line. Thus, simple waves are represented in the space of the hodograph $(u_1, \ldots, u_n)$ as lines $u_i = u_i(\tau)$, and simple waves are spoken of as *solutions with degenerate hodograph.* As we have seen above, for homogeneous systems of quasilinear equations (16) to each such line in the space of the hodograph there corresponds a family of solutions (simple waves) with single-function arbitrariness.

It is possible to obtain equations defining the surface S in the hodograph space (see Janenko (1955)) and also to show that, as a rule, to the surface S there corresponds a family of solutions with no more than constant arbitrariness. To the surface S there corresponds a family of solutions with functional arbitrariness only in exceptional cases.

For example, for a system of three quasilinear equations the surface S is determined by one quasilinear equation of third order (see also Chapter 2, §9).

In conclusion we present examples of inhomogeneous systems of quasilinear equations which have solutions of simple-wave type. Thus, the system of two quasilinear equations

$$\frac{\partial r_i}{\partial t} + \xi_i(r_1, r_2)\frac{\partial r_i}{\partial x} = f_i(r_1, r_2) \qquad (i = 1, 2)$$

has the solution $r_i = r_i(\tau) = r_i(x - ct)$ with a degenerate hodograph and constant arbitrariness ($c =$ const); the functions $r_i(\tau)$ are determined by solving the system of ordinary differential equations

$$\frac{dr_i}{d\tau} = \frac{f_i(r_1(\tau), r_2(\tau))}{\xi_i(r_1(\tau), r_2(\tau)) - c} \qquad (i = 1, 2).$$

Some systems of equations go over after a change of variables into systems having a solution with degenerate hodograph. For example, the system

$$\frac{\partial r_i}{\partial t} + \xi_i(r_1, r_2)\frac{\partial r_i}{\partial x} = \frac{g_i(r_1, r_2)}{x^\alpha} \qquad (i = 1, 2) \tag{20}$$

with functions ξ_i and g_i homogeneous in the variables r_1 and r_2,

$$\left.\begin{aligned} \xi_i(\theta r_1, \theta r_2) &= \theta^{\beta_1}\cdot\xi_i(r_1, r_2), \\ g_i(\theta r_1, \theta r_2) &= \theta^{\beta_2}\cdot g_i(r_1, r_2) \end{aligned}\right\} \qquad (i = 1, 2)$$

for $\nu\beta_1 = 1$, $\alpha = 1$ and $\gamma\beta_2 - \gamma - \alpha = 0$ goes over under the substitution

$$r_i = \left(\frac{x}{t}\right)^\nu R_i \quad (i = 1, 2), \qquad \tau = \ln t, \qquad z = \ln x,$$

into a system of the type (20) but with $\alpha = 0$, and it again has solutions with degenerate hodograph; this family also has only constant arbitrariness.

The system (2) for $\alpha \neq 1$ reduces by the change of variables

$$r_i = x^\nu \cdot R_i \qquad (i = 1, 2),$$
$$z = x^{-\nu\beta_1+1}, \qquad \nu = \frac{\alpha - 1}{\beta_2 - 1 - \beta_1}$$

again to a system of the type (20) but with the value $\alpha' = 1$, and hence it has solutions with degenerate hodograph.

§13. Group properties of differential equations

The problem of finding and increasing the number of solutions is closely related to group properties of differential equations. The fact that simple waves are the envelopes of centered waves, and the latter are *self-similar*, i.e., invariant under the homothetic transformation $x^1 \to kx^1$, $x^2 \to kx^2$ in the x^1, x^2 plane, has group character.

The possibility of reducing quasilinear equations to a form convenient for obtaining simple waves can be discovered after analysis of their group properties.

The long-known self-similar solutions of one-dimensional gas dynamics were essentially obtained by means of group analysis in the specific form of dimension theory.

Group analysis makes it possible to construct regular algorithms for finding particular solutions without invoking additional considerations but proceeding only from the given system of differential equations.

1. *The one-parameter Lie group.* In the n-dimensional space X_n of the variables $x^1, \ldots, x^n$ we consider the system of ordinary differential equations

$$\frac{dx^i}{dt} = \xi^i(x^1, \ldots, x^n), \tag{1}$$

for which it is possible to pose the Cauchy problem

$$x^i(t_0) = x_0^i. \tag{2}$$

For sufficiently smooth functions $\xi^i(x^1, \ldots, x^n)$ problem (1), (2) has a unique solution

$$x^i(t) = f^i(x_0^1, \ldots, x_0^n, t), \tag{3}$$

which is a sufficiently differentiable function of the initial values x_0^i and the parameter t in some region of variation of the variables x_0^i and t.

The solution (3) of problem (1), (2) possesses the property of invariance under translation in the parameter t: if in problem (1), (2) the initial values x_0^i are retained and t_0 is replaced by $t_0 + \tau$, then it has the solution

$$x^i(t) = f^i(x_0^1, \ldots, x_0^n, t - \tau). \tag{4}$$

The formula (3) can be written symbolically in the form

$$x(t) = S(t, t_0)x(t_0), \tag{5}$$

where $S(t, t_0)$ is the operator taking $x(t_0)$ into $x(t)$. Because of the invariance of the solution with respect to translation in t the operator $S(t, t_0)$ possesses the property

$$S(t, t_0) = S(t - t_0, 0) = S(t - t_0). \tag{6}$$

By successively solving the Cauchy problem with initial times $t = t_0, t_1, t_2$, we arrive at the composition property

$$S(t_1 + t_2) = S(t_2)S(t_1) = S(t_1)S(t_2). \tag{7}$$

To this property we add the property of continuous approach of the solution to the initial data

$$S(\tau) \to E \qquad (\tau \to 0) \tag{8}$$

and the property of invertibility

$$S(\tau)S(-\tau) = E, \tag{9}$$

where E is the identity operator.

A collection of operators (transformations) $S(t)$ possessing properties (7)–(9) forms, by definition, a *one-parameter continuous group* (*a Lie group*).

We call $\xi = \{\xi^1, \ldots, \xi^n\}$ the *direction vector* of the one-parameter Lie group and denote the group with direction vector ξ by the symbol $G_1(\xi)$.

2. *Group invariants.* A scalar-valued function $F(x) = F(x^1, \ldots, x^n)$ is called an *invariant* of the group $G_1(\xi)$ if

$$F(Sx) \equiv F(x) \tag{1}$$

for any transformation $S \in G_1(\xi)$.

We introduce the concept of the Lie derivative of a function $F(x)$. For x^i the Lie derivative is, by definition, the quantity

$$\delta x^i/\delta t = \xi^i \qquad (i = 1, \dots, n). \tag{2}$$

The *Lie derivative* of a function $F(x)$ is defined from the rule for the differentiation of a composite function

$$\frac{\delta F(x)}{\delta t} = \frac{\partial F}{\partial x^\alpha}\frac{\delta x^\alpha}{\delta t} = \xi^\alpha \frac{\partial F}{\partial x^\alpha} \qquad (\alpha = 1, \dots, n). \tag{3}$$

The Lie derivative $\delta F(x)/\delta t$ is nothing other than the derivative of the function $F(x)$ along a trajectory with respect to the parameter t.

The differential operator

$$L(\xi) = \xi^\alpha \frac{\partial}{\partial x^\alpha} \qquad (\alpha = 1, \dots, n) \tag{4}$$

is called the *infinitesimal operator* (henceforth simply the *operator*) of the group $G_1(\xi)$, and the operator $L(\xi)$ is said to generate the transformation group $G_1(\xi)$.

The Lie derivative of a function $F(x)$ is thus none other than the result of applying the operator $L(\xi)$ to $F(x)$. It is not hard to see that $F(x)$ is an invariant of $G_1(\xi)$ if and only if the Lie derivative of $F(x)$ is equal to zero:

$$\frac{\delta F}{\delta t} = L(\xi)F = 0. \tag{5}$$

The expression

$$\delta F = \left(\xi^\alpha \frac{\partial F}{\partial x^\alpha}\right)\delta t \tag{6}$$

is called the *Lie differential* of F. In particular, $\delta x^i = \xi^i \delta t$ is the Lie differential of the function x^i.

The *differential* dx^i is the expression

$$dx^i = \frac{\partial f^i}{\partial x_0^\alpha} dx_0^\alpha \qquad (\alpha = 1, \dots, n), \tag{7}$$

computed for fixed t. The differential of a function $F(x)$ is defined similarly:

$$dF = \frac{\partial F}{\partial x^\alpha} dx^\alpha = \frac{\partial F}{\partial x^\alpha}\cdot\frac{\partial f^\alpha}{\partial x_0^\beta} dx_0^\beta \qquad (\alpha, \beta = 1, \dots, n). \tag{8}$$

It is not hard to see that the operators δ and d commute, so that

$$\delta dx^i = d\delta x^i. \tag{9}$$

A manifold Φ defined by the equations

$$F^i(x^1, \ldots, x^n) = C^i \qquad (i = 1, \ldots, q), \tag{10}$$

is called *invariant* under $G_1(\xi)$ if the transformations $S \in G_1$ take points of the manifold again into points of the manifold.

For the invariance of Φ it is necessary and sufficient that the vector ξ be tangent to Φ at points of Φ, i.e., that on the manifold (1) the following conditions be satisfied:

$$\xi^\alpha \frac{\partial F^i}{\partial x^\alpha} = LF^i = 0 \qquad (i = 1, \ldots, q;\ \alpha = 1, \ldots, n). \tag{11}$$

If an arbitrary point of Φ is not invariant under G_1, but the manifold Φ itself is invariant, we say that the group G_1 *induces* a continuous transformation group on Φ. We define Φ by the parametric equations

$$x^i = f^i(v^1, \ldots, v^p) \qquad (p = n - q;\ i = 1, \ldots, n). \tag{12}$$

The group $G_1(\xi)$ then induces in the space of $v = \{v^1, \ldots, v^p\}$ a group $g_1(\eta)$, where the direction vector $\eta = \{\eta^1, \ldots, \eta^p\}$ is related to the direction vector $\xi = \{\xi^1, \ldots, \xi^n\}$ by the equations

$$\xi^i = \frac{\delta x^i}{\delta t} = \frac{\partial f^i}{\partial v^\beta} \cdot \frac{\delta v^\beta}{\delta t} = \frac{\partial f^i}{\partial v^\beta} \eta^\beta \qquad (i = 1, \ldots, n;\ \beta = 1, \ldots, p), \tag{13}$$

which follow from (12).

Let

$$\Omega(x, dx) = A_\alpha(x^1, \ldots, x^n)\, dx^\alpha \tag{14}$$

be a linear differential form. Then the Lie differential of Ω is, by definition, the expression

$$\delta\Omega = \delta(A_\alpha dx^\alpha) = \delta A_\alpha dx^\alpha + A_\alpha \delta\, dx^\alpha. \tag{15}$$

Using the commutativity of the operators d and δ, we have

$$\delta\Omega = \delta A_\alpha dx^\alpha + A_\alpha d\delta x^\alpha = \left[\frac{\partial A_\alpha}{\partial x^\beta}\xi^\beta dx^\alpha + A_\alpha d\xi^\alpha\right]\delta t \quad (\alpha, \beta = 1, \ldots, n). \tag{16}$$

The Lie derivative of Ω is, by definition, the expression

$$\Lambda(x, \xi, dx, d\xi) = \frac{\partial A_\alpha}{\partial x^\beta}\xi^\beta\, dx^\alpha + A_\alpha\, d\xi^\alpha, \tag{17}$$

which is itself a linear differential form.

A form Ω is called *invariant* under $G_1(\xi)$ if its Lie derivative, i.e., the form $\Lambda(x, \xi, dx, d\xi)$, is equal to zero.

Suppose that the group $G_1(\xi)$ leaves the forms

$$\Omega^i = A^i_\alpha(x)\,dx^\alpha \qquad (i = 1, \ldots, r;\ \alpha = 1, \ldots, n) \tag{18}$$

invariant and also the manifold Φ defined either by the equations (10) or the parametric equations (12).

The forms Ω^i on Φ go over into forms

$$\omega^i = a^i_\beta dv^\beta, \qquad a^i_\beta = A^i_\alpha \frac{\partial x^\alpha}{\partial v^\beta}$$
$$(i = 1, \ldots, r;\ \alpha = 1, \ldots, n;\ \beta = 1, \ldots, p), \tag{19}$$

which remain invariant under the group $g_1(\eta)$ induced by the group $G_1(\xi)$ on Φ.

By definition, the Lie derivatives of the forms ω^i are equal to zero, i.e.,

$$\Lambda^i(v, \eta;\ dv, d\eta) = \frac{\partial a^i_\beta}{\partial v^\gamma}\eta^\gamma dv^\beta + a^i_\beta d\eta^\beta = 0$$
$$(i = 1, \ldots, r;\ \beta, \gamma = 1, \ldots, p). \tag{20}$$

Let $G_1(\xi), \ldots, G_p(\xi)$ be one-parameter groups of transformations in the space X_n. We seek the common invariants of these groups. If $F(x) = F(x^1, \ldots, x^n)$ is an invariant of $G_1(\xi), \ldots, G_p(\xi)$, then the following relations must be satisfied:

$$\left.\begin{array}{l} L_1F = \xi^\alpha_1 \dfrac{\partial F}{\partial x^\alpha} = 0, \\ L_2F = \xi^\alpha_2 \dfrac{\partial F}{\partial x^\alpha} = 0, \\ \cdots\cdots\cdots\cdots\cdots\cdots \\ L_pF = \xi^\alpha_p \dfrac{\partial F}{\partial x^\alpha} = 0 \\ (\alpha = 1, \ldots, n). \end{array}\right\} \tag{21}$$

Suppose that the system (21) is complete, i.e., the operators $[L_i, L_j]$ are linear combinations of the L_i $(i = 1, \ldots, p)$:

$$[L_i, L_j] = c^\alpha_{ij} L_\alpha \qquad (i, j, \alpha = 1, \ldots, p).$$

If the coefficients c_{ij} are constant and satisfy the conditions

$$c^\alpha_{ij} = -c^\alpha_{ji}, \qquad c^\alpha_{ij}c^l_{\alpha k} + c^\alpha_{ki}c^l_{\alpha j} + c^\alpha_{jk}c^l_{\alpha i} = 0,$$

then the collection of operators $L_1, \ldots, L_p$ generates a p-parameter Lie group in such a way that any operator

$$L(\xi) = \lambda_1 L_1(\xi_1) + \cdots + \lambda_p L_p(\xi_p)$$

with constant coefficients $\lambda_1, \ldots, \lambda_p$ generates a one-parameter group (see Ovsjannikov (1962)).

3. ***The extended group.*** Until now all the variables x^α have enjoyed equal status. We now define a space $X_n = \{x^\alpha\}$ which corresponds to distinguishing in the point space $X_{m+2}(x^1, x^2, u^1, \ldots, u^m)$ the two-dimensional manifolds

$$u^i = u^i(x^1, x^2) \qquad (i = 1, \ldots, m). \tag{1}$$

A point $x = \{x^1, \ldots, x^n\} \in X_n$ is determined by the coordinates

$$x^s = x^s, \qquad x^{2+i} = u^i, \tag{2a}$$

$$x^{2+m+i} = \frac{\partial u^i}{\partial u^1} = p^i_1, \qquad x^{2+2m+i} = \frac{\partial u^i}{\partial x^2} = p^i_2$$

$$(s = 1, 2;\ i = 1, \ldots, m;\ n = 3m + 2). \tag{2b}$$

Among all transformations in the space X_n we distinguish transformations leaving the following forms invariant:

$$\Omega^i(x, dx) = dx^{2+i} - x^{2+\beta m+i}dx^\beta \qquad (i = 1, \ldots, m;\ \beta = 1, 2). \tag{3}$$

Such transformations we call *tangential transformations.* For these transformations the relations (2b) remain unchanged.

Among the tangential transformations we distinguish the subclass of *extended* transformations which are characterized by the fact that the subspace $X_{m+2} = \{x^1, \ldots, x^{m+2}\} \subset X_n$ remains invariant. We call the transformations in X_{m+2} *point* transformations.

We shall show that *the point transformations completely determine the extended transformations.*

Let $\xi = \{\xi^1, \ldots, \xi^{m+2}\}$ be the direction vector of a point transformation of $G_1(\xi)$ in X_{m+2}, and let $\bar{\xi} = \{\bar{\xi}^1, \ldots, \bar{\xi}^{m+2}, \bar{\xi}^{m+3}, \ldots, \bar{\xi}^{3m+2}\}$ be the direction vector of the corresponding extended transformation of $\bar{G}_1(\xi)$.

From the invariance of X_{m+2} it follows that

$$\bar{\xi}^i = \xi^i \qquad (i = 1, \ldots, m + 2). \tag{4}$$

From the condition of invariance of Ω^i under $\bar{G}_1(\xi)$ we find that

$$d\xi^{2+i} - x^{2+\beta m+i}\, d\xi^\beta - \bar{\xi}^{2+\beta m+i}\, dx^\beta = 0 \qquad (i = 1, \ldots, m;\ \beta = 1, 2). \tag{5}$$

This implies that

$$\bar{\xi}^{2+sm+i} = \frac{\partial \xi^{2+i}}{\partial x^s} + \frac{\partial \xi^{2+i}}{\partial u^\alpha} p^\alpha_s - p^i_\beta \left[\frac{\partial \xi^\beta}{\partial x^s} + \frac{\partial \xi^\beta}{\partial u^\alpha} p^\alpha_s\right]$$

$$(s, \beta = 1, 2;\ i, \alpha = 1, \ldots, m). \tag{6}$$

The assertion is proved.

Of course, it is possible to consider further extensions of the group in the space of components of derivatives of second and higher order.

4. *Groups of transformations admitted by a system of differential equations.* For simplicity we restrict attention to quasilinear equations. Using the notation (13.3.2), we write the system of quasilinear equations

$$\frac{\partial u^i}{\partial x^2} + a^i_\alpha(x^1, x^2; u^1, \ldots, u^m)\frac{\partial u^\alpha}{\partial x^1} = f^i(x^1, x^2; u^1, \ldots, u^m)$$

$$(i, \alpha = 1, \ldots, m) \qquad (1)$$

in the form of equations in differentials

$$\Omega^i(x, dx) = dx^{2+i} - x^{2+\beta m+i}dx^\beta = 0 \qquad (i = 1, \ldots, m;\ \beta = 1, 2) \qquad (2)$$

and finite relations

$$F^i(x^1, \ldots, x^n) = x^{2m+2+i} + a^i_\alpha x^{m+2+\alpha} - f^i = 0 \qquad (i, \alpha = 1, \ldots, m). \qquad (3)$$

An integral manifold of the system (2), (3) can then be defined as a surface in the space $X_n = X_{3m+2}$ of the variables $x^1, \ldots, x^{3m+2}$

$$x^i = x^i(x^1, x^2) \qquad (i = 1, \ldots, n = 3m + 2), \qquad (4)$$

on which the forms Ω^i vanish and which itself lies on the manifold Φ defined by (3).

The system (1) *admits* the group $G_1(\xi)$ in the space X_{m+2} of the variables x^i $(i = 1, \ldots, m + 2)$ if the extended group $\bar{G}_1(\xi)$ corresponding to it leaves the manifold Φ invariant, i.e., the group $G_1(\xi)$ takes an integral surface of equation (1) into an integral surface. The collection of groups $G_1(\xi)$ forms the set of admissible transformations of the system (1).

The condition of invariance of the manifold Φ has the form

$$\bar{\xi}^{2m+2+i} + a^i_\alpha\bar{\xi}^{m+2+\alpha} + \frac{\partial a^i_\alpha}{\partial x^\beta}\xi^\beta x^{m+2+\alpha} - \frac{\partial f^i}{\partial x^\beta}\xi^\beta = 0$$

$$(\beta = 1, \ldots, m + 2;\ i, \alpha = 1, \ldots, m). \qquad (5)$$

Substituting into (5) the expressions for $\bar{\xi}^{2+\beta m+i}$ from (13.3.6) and expressing $x^{2m+2+i} = p^i_2$ in terms of $x^{m+2+i} = p^i_1$ by means of (3), we arrive at the system

$$A^i + A^i_\alpha p^\alpha_1 + A^i_{\alpha_1\alpha_2}p^{\alpha_1}_1 p^{\alpha_2}_1 = 0 \qquad (i, \alpha, \alpha_1, \alpha_2 = 1, \ldots, m), \qquad (6)$$

where

$$\left.\begin{aligned} A^i &= b^{i\alpha}_{\beta}(x^1, \ldots, x^{m+2})\partial\xi^\beta/\partial x^\alpha + b^i_\beta(x^1, \ldots, x^{m+2})\xi^\beta, \\ A^i_j &= b^{i\alpha}_{j\beta}(x^1, \ldots, x^{m+2})\partial\xi^\beta/\partial x^\alpha + b^i_{j\beta}(x^1, \ldots, x^{m+2})\xi^\beta, \\ A^i_{j_1j_2} &= b^{i\alpha}_{j_1j_2\beta}(x^1, \ldots, x^{m+2})\partial\xi^\beta/\partial x^\alpha + b^i_{j_1j_2\beta}(x^1, \ldots, x^{m+2})\xi^\beta \\ &(i, j, j_1, j_2 = 1, \ldots, m;\ \alpha, \beta = 1, \ldots, m+2); \end{aligned}\right\} \tag{7}$$

the coefficients $b(x)$ depend on

$$a^i_\alpha(x),\quad f^i(x),\quad \partial a^i_\alpha(x)/\partial x^\beta,\quad \partial f^i(x)/\partial x^\beta \quad (i, \alpha = 1, \ldots, m;\ \beta = 1, \ldots, m+2).$$

On the $(2m+2)$-dimensional manifold Φ the quantities $x^1, \ldots, x^{m+2}$ and $p^1_1, \ldots, p^m_1$ are independent parameters. Since the relations (6) must be satisfied identically with respect to these parameters, it follows from (6) that

$$A^i = 0, \qquad A^i_\alpha = 0, \qquad A^i_{\alpha_1\alpha_2} + A^i_{\alpha_2\alpha_1} = 0 \qquad (i, \alpha, \alpha_1, \alpha_2 = 1, \ldots, m). \tag{8}$$

The equations (8) form a linear homogeneous system of equations for $\xi^1, \ldots, \xi^{m+2}$; this system is called the *defining system* and is, generally speaking, overdetermined.

In order that the group $G_1(\xi)$ be admissible for the given system (1) it is necessary and sufficient that (8) be satisfied.

We shall consider a number of examples.

On the basis of the indicated algorithm, Ovsjannikov (1962) investigated the transformation group for the system of one-dimensional equations of gas dynamics and a polytropic equation of state:

$$\left.\begin{aligned} \frac{\partial u}{\partial t} + u\frac{\partial u}{\partial x} + \frac{1}{\rho}\frac{\partial p}{\partial x} &= 0, \\ \frac{\partial \rho}{\partial t} + u\frac{\partial \rho}{\partial x} + \rho\frac{\partial u}{\partial x} &= 0, \\ \frac{\partial p}{\partial t} + u\frac{\partial p}{\partial x} + \gamma p\frac{\partial u}{\partial x} &= 0. \end{aligned}\right\} \tag{9}$$

For any value of γ the system (9) admits the operators

$$\left.\begin{aligned} &L_1 = \partial/\partial t, \qquad L_2 = \partial/\partial x, \qquad L_3 = t\partial/\partial x + \partial/\partial u, \\ &L_4 = t\partial/\partial t + x\partial/\partial x, \qquad L_5 = -t\partial/\partial t + u\partial/\partial u - 2\rho\partial/\partial\rho, \\ &L_6 = \rho\partial/\partial\rho + p\partial/\partial p. \end{aligned}\right\} \tag{10}$$

For $\gamma = 3$ the system (9) admits a further independent operator

$$L_7 = t^2\partial/\partial t + tx\partial/\partial x + (x - tu)\partial/\partial u - t\rho\partial/\partial\rho - 3tp\partial/\partial p. \tag{11}$$

The operators L_1 and L_2 are operators of translation along the axes x and t, the operator L_3 corresponds to the Galilean transformation, and the operator L_4 is the operator of similarity transformation (homothety) in the x, t plane. These operators occur for any equation of state. The operators L_5 and L_6 correspond to the polytropic equation of state, and L_7 corresponds to the specific value $\gamma = 3$.

The Euler-Poisson equation

$$\frac{\partial^2 u}{\partial x_1 \partial x_2} + \frac{a}{(x_1 - x_2)^2} u = 0 \qquad (a = \text{const}) \tag{12}$$

admits the operators

$$\left.\begin{aligned} L_1 = \partial/\partial x_1 + \partial/\partial x_2, \qquad L_2 = x_1\partial/\partial x_1 + x_2\partial/\partial x_2, \\ L_3 = x_1^2\partial/\partial x_1 + x_2^2\partial/\partial x_2, \qquad L_4 = u\partial/\partial u, \qquad L_5 = \varphi(x_1, x_2)\partial/\partial u, \end{aligned}\right\} \tag{13}$$

where $\varphi(x_1, x_2)$ is an arbitrary solution of (12). L_1 is the operator of coordinate translation along the axes x_1 and x_2; L_2 is the operator of homothety in the x_1, x_2 plane; and L_3 is the inversion operator. In its coefficients the operator L_5 contains two arbitrary functions of a single argument and together with the operators L_1, L_2, L_3 and L_4 generates an infinite-dimensional Lie group.

Following Ovsjannikov, we shall show how knowledge of the transformation group admitted by the Euler-Poisson equation (12) makes it possible to determine the Riemann function for it.

As is known (see §12.2), the Riemann function is a solution $u(\xi_1, \xi_2; x_1, x_2)$ of (12) that satisfies the conditions

$$u(\xi_1, \xi_2; \xi_1, x_2) = u(\xi_1, \xi_2; x_1, \xi_2) = 1. \tag{14}$$

The operators L_1, L_2 and L_3 generate a three-parameter group of consistent linear fractional transformations

$$\bar{x}_1 = \frac{ax_1 + b}{cx_1 + d}, \qquad \bar{x}_2 = \frac{ax_2 + b}{cx_2 + d}, \tag{15}$$

where a, b, c and d are arbitrary constants with $ad - bc \neq 0$. Since (12) admits the operators L_1, L_2 and L_3, the transformations (15) leave it unchanged. Using this fact, we choose a transformation (15) so that to the values $x_1 = \xi_1$ and $x_2 = \xi_2$ there correspond $\bar{x}_1 = 0$ and $\bar{x}_2 = \infty$. This is achieved by the transformation

$$\bar{x}_1 = \frac{x_1 - \xi_1}{x_1 - \xi_2}, \qquad \bar{x}_2 = \frac{x_2 - \xi_1}{x_2 - \xi_2}. \tag{16}$$

As a result, the solution $u(\xi_1, \xi_2; x_1, x_2)$ of problem (12), (14) goes over into a solution $v(\bar{x}^1, \bar{x}^2)$ of the problem

$$\left.\begin{aligned} &\frac{\partial^2 v}{\partial \bar{x}_1 \partial \bar{x}_2} + \frac{a}{(\bar{x}_1 - \bar{x}_2)^2} v = 0, \\ &v(0, \bar{x}_2) = 1, \qquad v(\bar{x}_1, \infty) = 1. \end{aligned}\right\} \tag{17}$$

The problem (17) admits the dilation operator L_2 and is thus self-similar. Its solution can be sought in the form

$$v(\bar{x}_1, \bar{x}_2) = v(\xi), \qquad \xi = \frac{\bar{x}_1}{\bar{x}_2}. \tag{18}$$

Substituting (18) into the first of the relations (17), we get

$$\xi v'' + v' - \frac{a}{(\xi - 1)^2} v = 0, \tag{19}$$

where the prime denotes differentiation with respect to ξ. The desired solution $v(\bar{x}_1, \bar{x}_2) = v(\xi)$ satisfies the second and third conditions of (17) if

$$v(0) = 1. \tag{20}$$

We represent a in the form

$$a = m(m - 1). \tag{21}$$

Equation (19) reduces to the hypergeometric equation by the substitution

$$v = (1 - \xi)^m w \tag{22}$$

and takes the form

$$\xi(\xi - 1)w'' + [-1 + (2m + 1)\xi]w' + m^2 w = 0. \tag{23}$$

The parameters α, β and γ of this equation are found from the relations

$$\gamma = 1, \qquad \alpha + \beta + 1 = 2m + 1, \qquad \alpha\beta = m^2 \tag{24}$$

and have the values

$$\alpha = \beta = m, \qquad \gamma = 1. \tag{25}$$

Recalling (20), we thus find that

$$w = F(m, m, 1, \xi), \tag{26}$$

where $F(\alpha, \beta, \gamma, \xi)$ is the hypergeometric function. Returning now to the functions $v(\xi)$, $v(\bar{x}_1, \bar{x}_2)$ and $u(\xi_1, \xi_2; x_1, x_2)$, we have

$$v(\xi) = (1 - \xi)^m F(m, m; 1, \xi), \tag{27}$$

$$v(\bar{x}_1, \bar{x}_2) = \left(1 - \frac{\bar{x}_1}{\bar{x}_2}\right)^m F\left(m, m; 1, \frac{\bar{x}_1}{\bar{x}_2}\right), \tag{28}$$

$$u(\xi_1, \xi_2; x_1, x_2) = \left[\frac{(\xi_1 - \xi_2)(x_1 - x_2)}{(\xi_1 - x_2)(x_1 - \xi_2)}\right]^m F\left(m, m; 1, \frac{(x_1 - \xi_1)(x_2 - \xi_2)}{(x_1 - \xi_2)(x_2 - \xi_1)}\right). \tag{29}$$

In the case of conservative systems it is possible to construct transformations not only in the space of the extended system but also in the space of potentials. By introducing the potential φ^i (see §5.3), the conservative system

$$\frac{\partial u^i}{\partial x^2} + \frac{\partial v^i(u^1, \ldots, u^n)}{\partial x^1} = \frac{\partial u^i}{\partial x^2} + \frac{\partial v^i}{\partial u^\alpha} \cdot \frac{\partial u^\alpha}{\partial x^1} = 0 \qquad (i, \alpha = 1, 2, \ldots, n) \tag{30}$$

can be transformed to the system of equations in differentials

$$\omega^i = d\varphi^i - u^i dx^1 + v^i(u^1, \ldots, u^n) dx^2 = 0. \tag{31}$$

In the space of x^1, x^2, u^i and φ^i ($i = 1, \ldots, n$) we consider the operator

$$L = \xi^\beta \frac{\partial}{\partial x^\beta} + \xi^{2+\alpha} \frac{\partial}{\partial u^\alpha} + \xi^{n+2+\alpha} \frac{\partial}{\partial \varphi^\alpha} \qquad (\alpha = 1, \ldots, n; \ \beta = 1, 2). \tag{32}$$

To the operator L there correspond admissible transformations if the Lie derivative of the forms ω^i is equal to zero:

$$\delta\omega^i / \delta\tau = L\omega^i = 0. \tag{33}$$

From (33) it follows that

$$d\xi^{n+2+i} - u^i \, d\xi^1 + v^i \, d\xi^2 - \xi^{2+i} \, dx^1 + a^i_\alpha \xi^{2+\alpha} \, dx^2 = 0. \tag{34}$$

We thus arrive at equations for the ξ^i ($i = 1, \ldots, 2n + 2$)

$$\left.\begin{aligned} \frac{d\xi^{n+2+i}}{dx^1} - u^i \frac{d\xi^1}{dx^1} + v^i \frac{d\xi^2}{dx^1} &= \xi^{2+i}, \\ \frac{d\xi^{n+2+i}}{dx^2} - u^i \frac{d\xi^1}{dx^2} + v^i \frac{d\xi^2}{dx^2} &= -a^i_\alpha \xi^{2+\alpha}, \end{aligned}\right\} \tag{35}$$

where

$$\left.\begin{aligned} \frac{d}{dx^1} &= \frac{\partial}{\partial x^1} + p^\alpha_1 \frac{\partial}{\partial u^\alpha} + u^\alpha \frac{\partial}{\partial \varphi^\alpha}, \\ \frac{d}{dx^2} &= \frac{\partial}{\partial x^2} + p^\alpha_2 \frac{\partial}{\partial u^\alpha} - v^\alpha \frac{\partial}{\partial \varphi^\alpha}, \\ p^\alpha_s &= \frac{\partial u^\alpha}{\partial x^s} \qquad (\alpha = 1, \ldots, n; \ s = 1, 2). \end{aligned}\right\} \tag{36}$$

By virtue of (30) the quantities p_s^i are connected by the relations

$$p_2^i = -a_\alpha^i p_1^\alpha, \qquad a_\alpha^i = \partial v^i/\partial u^\alpha \quad (i = 1, \ldots, n). \tag{37}$$

Substituting (36) into (35), we find that

$$\left.\begin{aligned}
&\frac{\partial\xi^{n+2+i}}{\partial x^1} + u^\alpha\frac{\partial\xi^{n+2+i}}{\partial\varphi^\alpha} - u^i\left(\frac{\partial\xi^1}{\partial x^1} + u^\alpha\frac{\partial\xi^1}{\partial\varphi^\alpha}\right)\\
&\quad + v^i\left(\frac{\partial\xi^2}{\partial x^1} + u^\alpha\frac{\partial\xi^2}{\partial\varphi^\alpha}\right) + p_1^\alpha\left(\frac{\partial\xi^{n+2+i}}{\partial u^\alpha} - u^i\frac{\partial\xi^1}{\partial u^\alpha} + v^i\frac{\partial\xi^2}{\partial u^\alpha}\right) = \xi^{2+i},\\
&\frac{\partial\xi^{n+2+i}}{\partial x^2} - v^\alpha\frac{\partial\xi^{n+2+i}}{\partial\varphi^\alpha} - u^i\left(\frac{\partial\xi^1}{\partial x^2} - v^\alpha\frac{\partial\xi^1}{\partial\varphi^\alpha}\right)\\
&\quad + v^i\left(\frac{\partial\xi^2}{\partial x^2} - v^\alpha\frac{\partial\xi^2}{\partial\varphi^\alpha}\right) + p_2^\alpha\left(\frac{\partial\xi^{n+2+i}}{\partial u^\alpha} - u^i\frac{\partial\xi^1}{\partial u^\alpha} + v^i\frac{\partial\xi^2}{\partial u^\alpha}\right)\\
&\qquad = -a_\alpha^i\xi^{2+\alpha}.
\end{aligned}\right\} \tag{38}$$

By requiring that (38) be satisfied identically in p_1^α, we arrive at the system of defining equations

$$\left.\begin{aligned}
&\frac{\partial\xi^{n+2+i}}{\partial x^1} + u^\alpha\frac{\partial\xi^{n+2+i}}{\partial\varphi^\alpha} - u^i\left(\frac{\partial\xi^1}{\partial x^1} + u^\alpha\frac{\partial\xi^1}{\partial\varphi^\alpha}\right)\\
&\qquad + v^i\left(\frac{\partial\xi^2}{\partial x^1} + u^\alpha\frac{\partial\xi^2}{\partial\varphi^\alpha}\right) = \xi^{2+i},\\
&\frac{\partial\xi^{n+2+i}}{\partial x^2} - v^\alpha\frac{\partial\xi^{n+2+i}}{\partial\varphi^\alpha} - u^i\left(\frac{\partial\xi^1}{\partial x^2} - v^\alpha\frac{\partial\xi^1}{\partial\varphi^\alpha}\right)\\
&\qquad + v^i\left(\frac{\partial\xi^2}{\partial x^2} - v^\alpha\frac{\partial\xi^2}{\partial\varphi^\alpha}\right) = -a_\alpha^i\xi^{2+\alpha},\\
&\frac{\partial\xi^{n+2+i}}{\partial u^\alpha} - u^i\frac{\partial\xi^1}{\partial u^\alpha} + v^i\frac{\partial\xi^2}{\partial u^\alpha} = 0.
\end{aligned}\right\} \tag{39}$$

5. ***Partially invariant and invariant solutions.*** On the manifold Φ given by (13.4.3) we seek submanifolds $\varphi \in \Phi$ invariant under some collection of admissible transformations. Without going into details, we shall sketch a manner of finding invariant submanifolds and the corresponding admissible transformations.

Let

$$x^i = \psi^i(x^1, \ldots, x^{2m+2}) \qquad (i = 1, \ldots, 3m+2) \tag{1}$$

be a parametric representation of Φ, let

$$x^j = \varphi^j(v^1, \ldots, v^p) \qquad (j = 1, \ldots, 2m+2) \tag{2}$$

be a parametric representation of the desired manifold φ, and suppose $\bar{G}_1(\xi)$ induces on φ the group $G_1(\eta)$, $\eta = \{\eta^1, \ldots, \eta^p\}$. The relations (1) and (2) take the forms of (13.4.2) into forms

$$\omega^i = c^i_\alpha(v)\, dv^\alpha \qquad (i = 1, \ldots, m;\ \alpha = 1, \ldots, p). \tag{3}$$

Here the coefficients c^i_α are known functions of φ^j ($j = 1, \ldots, 2m + 2$) and are linear functions of $\partial\varphi^j/\partial v^k$ ($k = 1, \ldots, p$). The conditions for the invariance of the forms ω^i under $G_1(\eta)$ have the form

$$\frac{\partial c^i_\alpha}{\partial v^\beta}\eta^\beta\, dv^\alpha + c^i_\alpha d\eta^\alpha = 0 \qquad (i = 1, \ldots, m;\ \alpha, \beta = 1, \ldots, p). \tag{4}$$

By virtue of (1) and (2), $\bar{\xi}$ and η are related by

$$\bar{\xi}^i = \frac{\partial x^i}{\partial x^\alpha} \cdot \frac{\partial \varphi^\alpha}{\partial v^\beta}\eta^\beta \qquad (i = 1, \ldots, 3m + 2;$$

$$\alpha = 1, \ldots, 2m + 2;\ \beta = 1, \ldots, p), \tag{5}$$

where

$$\partial x_\alpha/\partial x^\beta = \delta^\alpha_\beta \qquad (\alpha, \beta = 1, \ldots, 2m + 2). \tag{6}$$

In particular, for $i = 1, \ldots, m + 2$ we have

$$\bar{\xi}^i = \xi^i = \frac{\partial \varphi^i}{\partial v^\beta}\eta^\beta. \tag{7}$$

Substituting (7) into the defining system (13.4.8), we obtain a linear, homogeneous system of equations of first order for η:

$$B^{(k)}_{\beta\gamma}\frac{\partial \eta^\beta}{\partial v^\gamma} + B^{(k)}_\beta \eta^\beta = 0 \qquad (\beta = 1, \ldots, p;\ \gamma = 1, \ldots, p), \tag{8}$$

with coefficients B which depend on φ^i and are linear in $\partial\varphi^i/\partial v^\gamma$ ($i = 1, \ldots, 2m + 2;\ \gamma = 1, \ldots, p$).

From (4) we find that

$$\frac{\partial c^i_j}{\partial v^\beta}\eta^\beta + c^i_\beta\frac{\partial \eta^\beta}{\partial v^j} = 0 \qquad (i = 1, \ldots, m;\ j, \beta = 1, \ldots, p). \tag{9}$$

The compatibility conditions for the system (8), (9) lead to certain equations for the functions $\varphi^j(v^1, \ldots, v^p)$. Solutions (integral manifolds) lying in φ are called *partially invariant* (see Ovsjannikov (1962)).

In the special case where φ is an integral manifold we have an *invariant solution*. Invariant solutions are a generalization of the well-known self-similar solutions. We see that the theory of partially invariant and invariant solutions is closely related to the method of differential constraints, since the equations defining the manifolds φ are nothing other than differential constraints of first order.

L. V. Ovsjannikov has indicated a relation between simple and binary waves known in gas dynamics and partially invariant solutions.

CHAPTER 2

CLASSICAL AND GENERALIZED SOLUTIONS OF ONE-DIMENSIONAL GAS DYNAMICS

§1. General remarks on the mathematical description of the motion of compressible gases

1. ***A gas as a continuous medium.*** A gas is a collection of a large number of particles (molecules, atoms, ions) in a continuous state of motion. In order to characterize the state of a gas at a given time it is necessary to give the position and velocity of each particle of the gas. The problem of considering the interactions and motion of each particle of the gas is extremely difficult; a statistical approach is therefore adopted in describing the state of a gas.

In the statistical description of the state of a gas it is convenient to assume that its component particles continuously fill out the volume they occupy. Here, of course, only volumes are considered which are sufficiently large as compared with the distances between particles of the gas. The expressions "small volume" and "infinitely small volume" used below are thus to be understood as sufficiently large in the sense just indicated.

The motion of particles of a gas can be characterized as a collection of particles of a given sort located at a given plane in space and having a given velocity of motion. This collection is proportional to a distribution function which satisfies integrodifferential transport equations (the so-called *kinetic* equations). The Boltzmann equation is the simplest example of a kinetic equation for gases (see Chapman and Cowling (1952), and Patterson (1956)).

The description of the state of a gas by means of distribution functions and the solution of the kinetic equations are also very difficult problems. On the other hand, it is known that there exist flows that can be described with good accuracy by means of certain specific distribution functions. This description is achieved by means of the concept of the state of thermodynamic equilibrium as a state in which the distribution functions have an altogether determinate character.

We briefly recall some facts from thermodynamics (see Leontovič (1950)). Among the parameters characterizing the state of a gas, some are determined only by bodies external to the mass of gas considered and do not depend in any way on the gas itself. These parameters are called *external* parameters. For example, the volume occupied by the gas, the intensity of external electromagnetic and gravitational fields, etc. are external parameters. In contrast to external parameters, the *internal* parameters are determined by the state of the gas itself (for example, its energy, temperature, and pressure).

The state of a gas is called an *equilibrium* state if it does not change in time, and there is also no exchange of energy with external bodies. We emphasize that simply the invariability of the state (a stationary state) does not imply that the gas is in equilibrium. The equilibrium state is a state which the gas cannot spontaneously leave. If a gas is in an arbitrary state, then, left to itself (i.e., without exchange of energy with surrounding bodies and with fixed external parameters), it will pass to an equilibrium state within a certain time (the so-called *relaxation time*).

Exchange of energy between the gas and external bodies occurs primarily by means of heat transfer, and, secondarily, when work is performed on the gas (or work is performed by the gas on external bodies). Work is done by the gas only with variation of the external parameters a_i, and for sufficiently small changes in the latter the work is given by

$$\delta W = \sum_i A_i \, da_i, \tag{1}$$

where A_i are the so-called *generalized forces*.

If δQ is the amount of heat transferred to the gas, then the change in the internal energy E of the gas (the kinetic energy of the motion of the molecules plus the potential energy of interaction of the molecules) according to the law of conservation of energy can be written in the form

$$dE = \delta Q - \delta W = \delta Q - \sum_i A_i \, da_i. \tag{2}$$

Regarding the quantities A_i, we remark that for an arbitrary state of the gas, in addition to the a_i, they depend furthermore on the position and velocities of individual molecules of the gas, i.e., on the microscopic state of the gas.

The situation simplifies if we consider equilibrium states of the gas and infinitely small deviations from them. According to the fundamental theorem of thermodynamics all the internal parameters including the A_i are single-valued functions of the external parameters a_i and the energy (or temperature T) of the gas.

The generalized forces A_i themselves are usually taken as the equilibrium internal parameters. Thus, in a state of equilibrium

$$A_i = A_i(T, a_1, \ldots, a_n), \tag{3}$$

$$E = E(T, a_1, \ldots, a_n), \tag{4}$$

and (2) takes the form

$$dE = \delta Q - \sum_i A_i(T, a)\, da_i. \tag{5}$$

The relations (3) and (4) are determined by the microscopic structure of the gas, and are called the *equations of state*. The relations (3) are called the *thermic equations of state of the gas*, while (4) is called the *caloric equation of state.*

We shall consider the case in which the gas is physically and chemically homogeneous with respect to its microscopic composition and does not interact with any fields (i.e., there are no gravitational forces, electromagnetic fields, etc.). The only external parameter of the gas is then the volume V that it occupies, and the force A is the pressure p, so that the manifold of thermodynamic states is two-dimensional. Therefore,

$$d\varepsilon = \delta Q - p\, dV. \tag{6}$$

If we consider a unit mass of gas, then the quantity $V = 1/\rho$ is called the *specific volume*, ρ is the *density* of the gas, and the quantity ε is the *specific internal energy* of the gas.

According to the second law of thermodynamics, the quantity

$$dS = \frac{\delta Q}{T} = \frac{1}{T}(d\varepsilon + p\, dV) \tag{7}$$

is the total differential of the function $S = S(V, T)$, which is called the *entropy* of unit mass of the gas. Thus, the second law of thermodynamics can be written in the form

$$T\, dS = d\varepsilon + p\, dV, \tag{8}$$

where p, ε and S are given by the equations

$$p = p(V, T), \tag{9}$$

$$\varepsilon = \varepsilon(V, T), \tag{10}$$

$$S = S(V, T). \tag{11}$$

It follows from the second law of thermodynamics (8) that the equations of state (9) and (10) cannot be independent, since the integrability condition for (8) imposes the following restriction on them:

$$\frac{\partial}{\partial V}\left(\frac{1}{T}\frac{\partial \varepsilon}{\partial T}\right) = \frac{\partial}{\partial T}\left(\frac{1}{T}\frac{\partial \varepsilon}{\partial V} + \frac{p}{T}\right), \quad \text{or} \quad \frac{\partial \varepsilon}{\partial V} = T^2 \frac{\partial}{\partial T}\left(\frac{p}{T}\right). \tag{12}$$

If, for example, the equation of state (9) is given, then (12) determines the caloric equation (10) up to an additive function of the temperature.

Thus, in a state of thermodynamic equilibrium the gas is described by the following quantities: the density ρ, the mass per unit volume; the specific volume $V = 1/\rho$; the pressure p, the force acting on unit area; ε, the internal energy per unit mass of gas; T, the temperature of the gas; and S, the entropy per unit mass of gas.

According to (9)–(11), among all these thermodynamic variables there are only two that are independent. From (8) it follows that

$$\frac{\partial S}{\partial T} = \frac{1}{T}\frac{\partial \varepsilon}{\partial T} = \frac{c_V}{T}, \qquad \frac{\partial S}{\partial V} = \frac{1}{T}\left(\frac{\partial \varepsilon}{\partial V} + p\right). \tag{13}$$

Thus, for given equations of state (9) and (10) the entropy S is determined up to an additive constant which can be eliminated if the entropy is normalized by the Nernst relation: $S \to 0$ as $T \to 0$. The equations of state of the gas (9)–(11) can also be given with another choice of independent parameters, for example,

$$p = p(V, S), \qquad \varepsilon = \varepsilon(V, S), \qquad T = T(V, S) \tag{14}$$

or

$$\varepsilon = \varepsilon(p, V), \qquad S = S(p, V), \qquad T = T(p, V). \tag{15}$$

With this prescription of the equations of state the second law of thermodynamics (8) requires that

$$\left.\begin{aligned} \frac{\partial \varepsilon(V, S)}{\partial V} = -\,p(V, S), \qquad \frac{\partial \varepsilon(V, S)}{\partial S} = T(V, S), \\ \frac{\partial T(V, S)}{\partial V} + \frac{\partial p(V, S)}{\partial S} = 0 \end{aligned}\right\} \tag{16}$$

or, respectively,

$$\left.\begin{aligned} T(p, V)\frac{\partial S(p, V)}{\partial V} = p + \frac{\partial \varepsilon(p, V)}{\partial V}; \qquad T\frac{\partial S(p, V)}{\partial p} = \frac{\partial \varepsilon(p, V)}{\partial p}, \\ \frac{\partial T(p, V)}{\partial p}\frac{\partial S(p, V)}{\partial V} - \frac{\partial T(p, V)}{\partial V}\frac{\partial S(p, V)}{\partial p} = \frac{\partial[T, S]}{\partial[p, V]} = 1. \end{aligned}\right\} \tag{17}$$

2. ***Nonequilibrium states and processes in gases.*** The basic concepts of thermodynamics—temperature, pressure, and entropy—lose their meaning in a nonequilibrium state of a gas. A nonequilibrium state cannot be fully described in terms of thermodynamic (i.e., macroscopic) concepts at all but requires a microscopic analysis. However, for the purposes of classical gas dynamics the generally accepted approach of nonequilibrium thermodynamics suffices. We suppose that the mass of gas in question is subdivided into a large number of elementary parts of very small size each of which we assume to be in a state of thermodynamic equilibrium. This assumption is justified by the fact that the relaxation time of a system decreases with its decreasing size, so that for a small portion of gas it is close to zero.

Understanding by a "point" in a gas an "infinitely small" volume in the sense indicated in subsection 1, we can thus introduce for each point of the gas at each time the concepts of pressure, temperature, and energy. They are now interpreted as functions of the coordinates of the point and the time:

$$p = p(x_1, x_2, x_3, t); \qquad T = T(x_1, x_2, x_3, t); \qquad S = S(x_1, x_2, x_3, t).$$

As concerns the density $\rho(x_1, x_2, x_3, t)$ and the energy $\varepsilon = \varepsilon(x_1, x_2, x_3, t)$, these quantities are obviously meaningful even without our assumption. Thus, a nonequilibrium state of a gas is interpreted to mean the absence of equilibrium between individual particles of the gas, each of which in itself is in equilibrium.

From the assumption of the equilibrium of small portions of the gas it follows that the functions $p(V, T)$, $\varepsilon(V, T)$ and $S(V, T)$ satisfy the equations of state (1.1.9)–(1.1.11). Thus, as they change, the thermodynamic parameters satisfy the equations of state of a gas which are defined for an equilibrium gas. This process is called an *equilibrium* or *reversible* process.

The above considerations concerning the relaxation time, however, give no idea of the limits of applicability of thermodynamic concepts. Such an idea can be obtained on the basis of a more general model of a gas, such as the statistical model.

Microscopic consideration leads to the conclusion that the thermodynamic concepts of temperature and entropy are meaningful if the changes of parameters characterizing the state of the gas over lengths of the order of the mean free path of molecules of the gas in space and over times of the order of the time between collisions of the molecules are small as compared with these quantities themselves.

We consider, for example, the one-dimensional, nonequilibrium flow of a gas. Let $u(x, t)$ be the flow speed. One of the approximations reducing the Boltzmann equation to the equations of gas dynamics assumes that the temperature T and the pressure p satisfying the equations of state (1.1.9), (1.1.10) can be introduced in the gas, but the flux of momentum and energy is also determined by nonequilibrium components connected with molecular diffusion. For a materially fixed particle of the gas in place of the pressure p the quantity

$$\bar{p} = p - \mu \operatorname{div} \boldsymbol{u} = p - \mu\, \partial u/\partial x, \tag{1}$$

is introduced in the flux of momentum, and the flux of energy is determined by the quantity

$$\bar{p}u - \kappa \operatorname{grad} T = pu - \mu u\, \partial u/\partial x - \kappa\, \partial T/\partial x. \tag{2}$$

In this approximation the function $p(V, T)$ satisfies the equation of state (1.1.9); μ and κ are respectively the coefficients of viscosity and thermal conductivity which are proportional to the length of the mean free path of molecules of the gas.

This implies that if over lengths of the mean free path and for times of the order of the time between collisions the variation of the thermodynamic quantities is small, then the approximation of equilibrium thermodynamics is valid. Indeed, for a large number of problems of the motion of gases and liquids this condition may be considered to be satisfied. In this case the process has a quasiequilibrium character, and it is possible to introduce the temperature and entropy, which to a high degree of accuracy will satisfy all the thermodynamic relations. On the other hand, as we shall see in this chapter, zones of abrupt and rapid variation of the quantities characterizing the flow occur in flows of gases and liquids. In these regions it is not possible to neglect nonequilibrium components in the flows of momentum and energy. However, these zones have dimensions of the order of the mean free path of molecules of the gas. Therefore, if this length is small compared with the characteristic dimensions of the problem, we can represent the nonequilibrium zone as a surface of discontinuity separating zones of smooth variation of the flow parameters. In this approach we assume that these parameters satisfy the thermodynamic relations everywhere, while on surfaces of discontinuity conditions for the continuity of the flux of mass, momentum, and energy must be satisfied. We thus consider discontinuous flows in gases and liquids with viscosity and thermal conductivity that are sufficiently small. In the present chapter we consider mainly just this case.

Finally, in the case of sufficiently large coefficients of viscosity and thermal conductivity it is necessary to consider nonequilibrium components in the flux of momentum and energy. It may happen that even this consideration is insufficient, and it is necessary to invoke the intergrodifferential Boltzmann equation.

There are thus flows in which the temperature and entropy retain their thermodynamic meaning and satisfy all thermodynamic relations; zones of abrupt variation are considered as surfaces of discontinuity of the flow parameters. The present chapter is devoted to the study of this type of flows.

In an equilibrium process there is the relation

$$\frac{dS}{dt} = \frac{1}{T}\left(\frac{d\varepsilon}{dt} + p\,\frac{dV}{dt}\right) \tag{3}$$

or

$$\frac{dS}{dt} = \frac{1}{T}\,\frac{dQ}{dt}, \tag{4}$$

where dQ/dt is the rate of flow of heat into the portion of gas considered.

If the portion of gas considered is thermally insulated ($dQ = 0$), then the equilibrium process is called adiabatic. For an adiabatic process

$$dS/dt = 0. \tag{5}$$

For a nonequilibrium process the relation (4) does not hold, but according to the second law for a thermally insulated system

$$dS/dt \geq 0. \tag{6}$$

Suppose that a mass of gas participates in a nonequilibrium process while exchanging heat with external bodies. In this case the second law of thermodynamics requires that

$$\frac{dS}{dt} + \frac{dS_e}{dt} > 0, \tag{7}$$

where S_e is the entropy of the external bodies. The quantity dS_e/dt may be interpreted as the flux of entropy from external bodies to the mass of gas.

We demonstrate the flux of entropy dS_e/dt for the example of the exchange of heat of the gas with a thermostat at constant temperature T_0. In this case

$$\frac{dS_e}{dt} = -\frac{1}{T_0}\,\frac{dQ}{dt}, \tag{8}$$

where dQ/dt is the quantity of heat flowing from the external bodies into the portion of gas being considered. Therefore, in the case of a constant-temperature thermostat constituting the external bodies the second law of thermodynamics requires that

$$\frac{dS}{dt} > \frac{1}{T_0}\,\frac{dQ}{dt}. \tag{9}$$

We shall apply these considerations to the analysis of an isothermic gas in §4.

3. ***Various means of describing flows. Euler and Lagrange variables.*** The motion of a continuous medium can be described in two distinct ways. In the

first method, at each time t we define the parameters of the state of the gas as functions of the coordinates x_1, x_2, x_3 of a point in some fixed coordinate system. Thus, in this method $\boldsymbol{u} = \boldsymbol{u}(x_1, x_2, x_3, t)$ means the velocity of a particle located at the point (x_1, x_2, x_3) at time t. Similarly, all other quantities characterize the state of a particle of the gas located at the point (x_1, x_2, x_3) at time t. This method of describing the motion of a continuous medium is called the *Euler method*, and the coordinates x_1, x_2, x_3 are called the *Euler coordinates*.

Another method of describing the motion, which is called the *Lagrange method*, assumes the definition of the thermodynamic quantities and the velocity $\boldsymbol{u}$ of the gas for each particle as functions of the time t. Suppose that we distinguish one particle of the gas from others by certain parameters y_1, y_2, y_3. We then seek all quantities characterizing the flow as functions of the variables y_1, y_2, y_3 and t. In this manner of description, for example, the vector $\boldsymbol{u}(y_1, y_2, y_3, t)$ for fixed y_1, y_2, y_3 denotes the velocity of spatial displacement of a particular particle of the gas. The coordinates y_1, y_2, y_3 are called the *Lagrange coordinates*.

In most cases the Euler coordinates of the point at which a particle of the gas is found at some specific moment of time (for example, at $t = 0$) are taken as the coordinates y_1, y_2, y_3. If we adopt this choice of Lagrange coordinates y_1, y_2, y_3, then it is easy to compute the position of the particle at time $t \neq 0$. Since for fixed y_1, y_2, y_3 the velocity $\boldsymbol{u}(y_1, y_2, y_3, t)$ is the velocity of the particle, it follows that

$$x_i = y_i + \int_0^t u_i(y_1, y_2, y_3, \tau)\, d\tau \qquad (i = 1, 2, 3). \tag{1}$$

Here $x_i = x_i(y_1, y_2, y_3, t)$ are the coordinates at time t (the Euler coordinates) of a particle which at time $t = 0$ is located at the point with coordinates $x_i = y_i$; $u_i(y_1, y_2, y_3, t)$ are the components of the velocity vector $\boldsymbol{u}(y_1, y_2, y_3, t)$. The formulas (1) thus establish the relation between the Lagrange coordinates y_i of a particle and its Euler coordinates x_i.

Suppose that $f(x_1, x_2, x_3, t)$ is any function in Euler coordinates, while $\bar{f}(y_1, y_2, y_3, t)$ is representation of the same function in Largrange coordinates. Then, according to (1),

$$f\left(y + \int_0^t \boldsymbol{u}(y, \tau)\, d\tau, t\right) = \bar{f}(y, t), \tag{2}$$

where to simplify notation we have written

$$y = \{y_1, y_2, y_3\}, \qquad x = \{x_1, x_2, x_3\}, \qquad \boldsymbol{u} = \{u_1, u_2, u_3\}.$$

Differentiating (2) with respect to t (it is naturally here assumed that f is differentiable), we obtain

$$\frac{\partial \bar{f}(y, t)}{\partial t} = \frac{\partial f(x, t)}{\partial t} + \boldsymbol{u}(y, t) \operatorname{grad} f(x, t), \tag{3}$$

since, according to (1), $x = y + \int_0^t \boldsymbol{u}\, d\tau$. By the very meaning of the concept of the velocity of a particle of the gas we have $\boldsymbol{u}(y, t) = \boldsymbol{u}(x, t)$ (here we denote different relations by the same letter $\boldsymbol{u}$: $\boldsymbol{u}(y, t)$ is the velocity in Lagrange variables, and $\boldsymbol{u}(x, t)$ is the velocity in Euler variables); we can therefore rewrite (3) in the form

$$\frac{\partial \bar{f}(y, t)}{\partial t} = \frac{\partial f(x, t)}{\partial t} + \boldsymbol{u}\nabla f(x, t), \tag{4}$$

where $\boldsymbol{u}\nabla$ is the operator of differentiation in space in the direction of the vector $\boldsymbol{u}$:

$$\boldsymbol{u}\nabla = u_1 \frac{\partial}{\partial x_1} + u_2 \frac{\partial}{\partial x_2} + u_3 \frac{\partial}{\partial x_3}. \tag{5}$$

In gas dynamics all quantities are usually denoted by the same letters both in the Euler and Lagrange representations. To avoid confusion the quantity $\partial \bar{f}(y, t)/\partial t$ is therefore written $df(y, t)/dt$. In this notation (3) takes the form

$$\frac{df}{dt} = \frac{\partial f}{\partial t} + \boldsymbol{u}\nabla f = \frac{\partial f}{\partial t} + u_\alpha \frac{\partial f}{\partial x_\alpha}. \tag{6}$$

The quantity df/dt is equal to the total derivative with respect to time of the function f along the trajectory of the particle $x = y + \int_0^t \boldsymbol{u}\, d\tau$ and is called the *substantive derivative.*

According to (1), to the Lagrange coordinates y, t there correspond particular Euler coordinates, and the mapping $y, t \to x, t$ is single-valued. However, the mapping $x, t \to y, t$ is not always defined, in general. Indeed, it is easy to conceive of a case in which at time $t = 0$ a gas occupies the entire space y_1, y_2, y_3, but as a result of motion at $t > 0$ some part of the space x_1, x_2, x_3 is free of gas. This means that in this part of space for the given $t > 0$ to the coordinates x_1, x_2, x_3 there correspond no values y_1, y_2, y_3. As we shall see below, the condition that the mapping $x, t \to y, t$ be one-to-one is expressed in the form $\rho(x, t) \neq 0$.

4. *Equations of state of gases. A perfect gas. A van der Waals gas. A normal gas.* Having made these brief remarks concerning the methods of describing the motion of gases and liquids, we shall briefly consider some of the simplest equations of state of gases.

A perfect (ideal) gas is a gas for which Clapeyron's law holds:

$$pV = RT, \tag{1}$$

where R is the gas constant. From (1.1.12) it then follows that

$$\partial\varepsilon(V, T)/\partial V = 0, \quad \text{or} \quad \varepsilon = \varepsilon(T), \tag{2}$$

i.e., the internal energy of the gas is a function only of the temperature; moreover, the specific heat capacity of the gas $c_V = \partial\varepsilon/\partial T = c_V(T)$ is also a function of the temperature only.

A gas is called *polytropic* if c_V does not depend on the temperature. Then

$$\varepsilon = c_V T, \tag{3}$$

i.e., the internal energy is proportional to the temperature of the gas.

The kinetic theory of gases leads to the equations of state of a polytropic gas under the following assumptions.

1. The potential energy of the interaction of molecules is negligibly small as compared with their kinetic energy. As a consequence of this assumption, the energy of a given mass of gas is the sum of the kinetic energies of the molecules of the gas comprising it.

2. Only pairwise collisions between molecules are possible, and these occur according to the laws of elastic collision. The internal structure of the molecules remains unchanged, and their total momentum and kinetic energy are preserved.

The second assumption implies, in particular, that the proper volume of the molecules is small compared with the volume occupied by the gas.

Kinetic theory gives the following expressions for the coefficients c_V, c_p and R of the equations of state (1) and (2) of an ideal gas:

$$c_V = \frac{f}{2} k \frac{N}{M}, \tag{4}$$

$$R = c_p - c_V, \tag{5}$$

where f is the number of degrees of freedom of a molecule ($f = 3$ for a monatomic gas, $f = 5$ for a diatomic gas, etc.), k is the Boltzmann constant, N is Avogadro's number, and M is the molecular weight.

Taking account of the equations of state (1) and (3), from the basic relation (1.1.3) we have

$$\begin{aligned} S &= c_V \ln T + R \ln V + \text{const} \\ &= c_V \ln T + c_p \ln V - c_V \ln V + \text{const} \\ &= c_V \ln p + c_p \ln V + \text{const}. \end{aligned} \tag{6}$$

The simplest correction to the equation of state of an ideal gas, which involves considering the molecular volume and the forces of molecular coupling, is given by the van der Waals equation

$$p = \frac{RT}{V-b} - \frac{a}{V^2}. \tag{7}$$

Here a is a quantity proportional to the coupling force of molecules of the gas, and b is a quantity proportional to the proper volume of the molecules.

In a similar way we obtain expressions for ε and S:

$$\varepsilon = \int c_V(T)\, dT - \frac{a}{V}, \tag{8}$$

$$S = \int \frac{c_V(T)}{T}\, dT + R \ln(V - b) + \text{const}. \tag{9}$$

If some element of gas is subject to slow expansion or compression so that no heat exchange occurs with the surrounding medium, then this element undergoes an adiabatic transition from one thermodynamic state to another. The slow process is here invertible, and the entropy of the element remains unchanged. Such a transition is therefore called *isentropic*. All thermodynamic states through which the given element of gas passes lie on a curve

$$S = \text{const}, \tag{10}$$

which is called the *Poisson adiabat*.

For an ideal polytropic gas it follows easily from (6) that the equation of a Poisson adiabat has the form

$$p = \frac{A^2}{\gamma} V^{-\gamma}, \tag{11}$$

where

$$\gamma = \frac{c_p}{c_V} = 1 + \frac{R}{c_V} > 1, \qquad A^2 = A^2(S) = a^2 e^{S/c_V} = \text{const}.$$

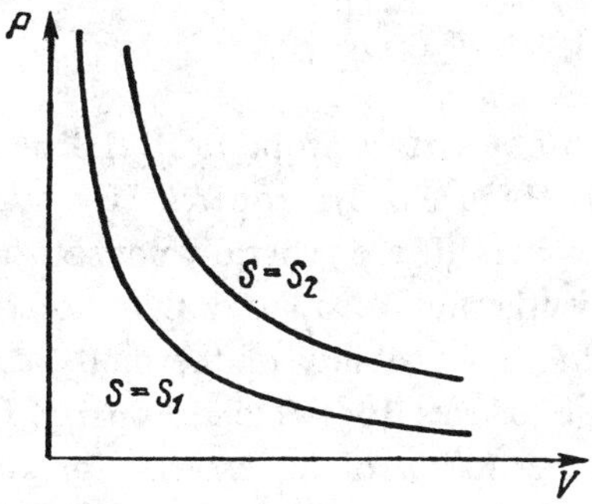

FIGURE 2.1.

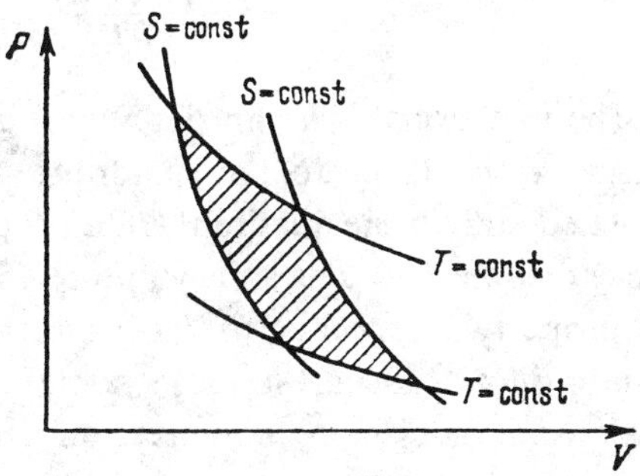

FIGURE 2.2

It is easy to see that for a polytropic gas along a Poisson adiabat (Figure 2.1)

$$\frac{dp}{dV} = \frac{\partial p(V, S)}{\partial V} < 0 \quad \text{(property I)}, \tag{12}$$

$$\frac{d^2p}{dV^2} = \frac{\partial^2 p(V, S)}{\partial V^2} > 0 \quad \text{(property II)}. \tag{13}$$

Thus, a Poisson adiabat $p = p(V, S_0)$ is a curve which is monotonically decreasing with respect to V and is convex upward. It is easy to see that $V = 0$ and $p = 0$ are asymptotes of a Poisson adiabat, i.e., $p \to 0$ as $V \to \infty$ and $p \to \infty$ as $V \to 0$ (property III).

If some element of gas is subject to compression or expansion so that the temperature of the element remains unchanged, then the transition of the gas is called an *isothermic* process.

In an ideal gas undergoing an isothermic process V and p are connected by the relation

$$p = c^2 \frac{1}{V} = c^2 \rho, \qquad c^2 = (c_p - c_V)\, T = RT = \text{const.} \tag{14}$$

Therefore, in some cases an ideal isothermic gas can formally be considered as a polytropic gas with exponent $\gamma = 1$.

The curve (14) is called an *isotherm*. It satisfies all three properties of a Poisson adiabat. An isotherm T and a Poisson adiabat A intersect in the V, p plane as shown in Figure 2.2. Thus, isotherms and Poisson adiabats of an ideal gas form a regular mesh in the V, p plane.

It is clear that the properties of an ideal gas ensure that the following properties hold:

$$\left.\begin{aligned} \frac{\partial p(V, S)}{\partial S} &> 0 \quad \text{(property IV)}, \\ c_V = \frac{\partial \varepsilon(V, T)}{\partial T} &> 0 \quad \text{(property V)}. \end{aligned}\right\} \tag{15}$$

The isotherms of a van der Waals gas do not satisfy properties I–III in the entire V, p plane. In the quadrant $A'CB'$ (Figure 2.3) property II may be violated, and in the quadrant BCA properties I and II are generally not satisfied, since the line $V = b$ is an asymptote of the isotherms. We remark that violation of property I expressed by (12) indicates the impossibility of thermodynamic equilibrium. Indeed, suppose some volume of gas for which property I is violated is subject to compression under the action of an external pressure which we assume to be constant and exceeding the initial pressure of the gas. If the compression occurs slowly and if $\partial p(V, S)/\partial V > 0$, then the pressure

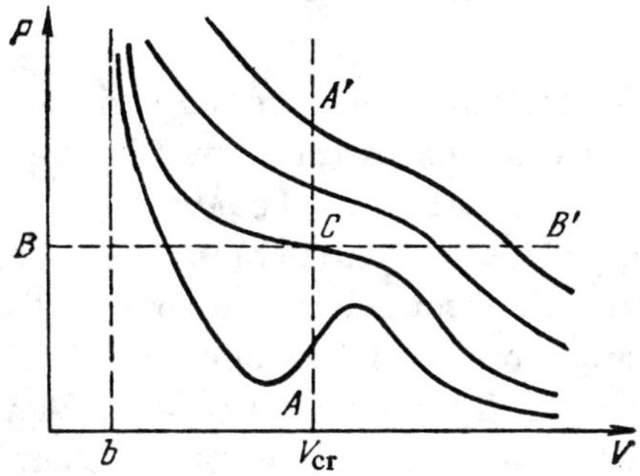

FIGURE 2.3.

in the gas drops, i.e., its resistence to the external pressure decreases. As a result, total collapse of the volume of gas occurs. On the other hand, a volume of gas possessing a pressure in excess of that of the surrounding medium must increase without bound.

If we imagine now a gas for which $\partial p(V, S)/\partial V > 0$ in thermodynamic equilibrium at constant pressure, then this equilibrium is absolutely instable, since the slightest deviation from equilibrium entails that some parts of the gas will contract while others expand without returning to the original state.

The situation is otherwise for the case of a gas satisfying condition I. Under adiabatic compression it will increase its pressure and impede compression. In a deviation from equilibrium oscillations occur in the gas which return it to the initial state.

Property I is thus a condition for the stability of thermodynamic equilibrium and is satisfied for all real substances in a stable state.

We shall henceforth restrict attention to gases with equations of state which satisfy a number of assumptions not following from the requirements of thermodynamics in addition to the thermodynamic relations of subsection 1 which are valid for any substances. Namely, we require that properties I–V be satisfied:

I. $$\frac{\partial p(V, S)}{\partial V} < 0. \tag{16}$$

II. $$\frac{\partial^2 p(V, S)}{\partial V^2} > 0. \tag{17}$$

III. $$p(V, S) \to \infty \quad \text{as } V \to 0. \tag{18}$$

IV. $$\frac{\partial p(V, S)}{\partial S} > 0. \tag{19}$$

V. $$c_V = \frac{\partial \varepsilon(V, T)}{\partial T} > 0. \tag{20}$$

We additionally require that the region of the variables V and T in which properties I–V are satisfied be convex (property VI). A gas with equations of state which satisfy properties I–VI is called a *normal gas*.[1]

The property I obviously means that the pressure p monotonically decreases with increasing V along a Poisson adiabat A; property II requires that this curve be convex upward; property III requires that $V = 0$ be an asymptote of any Poisson adiabat; finally, property IV means that a Poisson adiabat corresponding to greater entropy lies higher in the plane of the variables V, p (see Figure 2.1), i.e.,

$$\frac{\partial S(V, p)}{\partial p} > 0, \qquad \frac{\partial S(V, p)}{\partial V} > 0. \tag{21}$$

We shall consider the behavior of the surface $S = S(V, p)$ in the three-dimensional space of the variables V, p, S assuming that the equations of state satisfy properties I–V. In correspondence with properties I–IV each horizontal section or "contour" of the entropy is a monotone, convex curve with asymptote $V = 0$. Each section by a plane $V = \text{const}$ or $p = \text{const}$ is a monotone curve with respect to p or V respectively. Thus, any upper horizontal section projects inside a lower section (Figure 2.4).

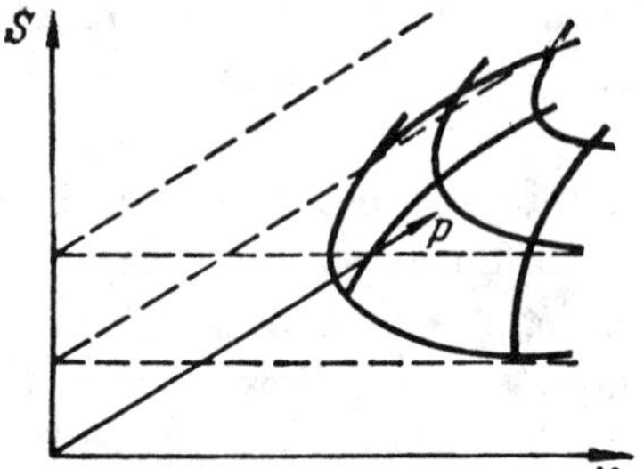

FIGURE 2.4.

From the basic relation (1.1.7)

$$T\, dS = d\varepsilon + p\, dV$$

it follows by property V that

[1] The relations I-V were formulated by Bethe and Weyl (see Weyl (1949)).

$$\frac{\partial S(V, T)}{\partial T} = \frac{1}{T}\frac{\partial \varepsilon(V, T)}{\partial T} = \frac{c_V}{T} > 0. \tag{22}$$

Hence

$$\frac{dp(V, T)}{dT} = \frac{\partial p(V, S)}{\partial S}\frac{\partial S(V, T)}{\partial T} > 0. \tag{23}$$

Using (21) and (22), we find that

$$\frac{\partial S(V, T)}{\partial V} > 0. \tag{24}$$

The last inequality means that the isotherms and adiabats A form a regular mesh of the same type as in the case of an ideal gas (see Figure 2.2).

Differentiating the identity $S = S(V, p) = S(V, p(V, S))$ with respect to V and S, we obtain

$$\frac{\partial S(V, p)}{\partial p}\frac{\partial p(V, S)}{\partial S} = 1, \qquad \frac{\partial S(V,p)}{\partial V} + \frac{\partial S(V, p)}{\partial p}\frac{\partial p(V, S)}{\partial V} = 0, \tag{25}$$

and, differentiating the second equation of (25) with respect to V, we get

$$S_{VV} + 2S_{Vp}p_V + S_{pp}p_V^2 = -S_p p_{VV} < 0. \tag{26}$$

Substituting (25) into (26), we obtain

$$S_{VV}S_p^2 - 2S_{Vp}S_VS_p + S_{pp}S_V^2 = -S_p^3 p_{VV} < 0. \tag{27}$$

We consider the nature of the variation of the entropy S along the lines

$$p = p_0 + at, \qquad V = V_0 + bt \tag{28}$$

in the plane of the variables V, p.

The following assertions hold:

a) *If* $a, b \geq 0$ $(a, b \leq 0)$, *then*

$$\frac{dS}{dt} \geq 0 \quad \left(\frac{dS}{dt} \leq 0\right). \tag{29}$$

This follows from (21).

b) *If* $(dS/dt)_{t=t_0} = 0$, *then*

$$\left(\frac{d^2S}{dt^2}\right)_{t=t_0} = \frac{\partial^2 S}{\partial p^2}a^2 + 2\frac{\partial^2 S}{\partial p \partial V}ab + \frac{\partial^2 S}{\partial V^2}b^2 < 0. \tag{30}$$

Indeed, from $(dS/dt)_{t=t_0} = (\partial S/\partial p)a + (\partial S/\partial V)b = 0$ we have $\partial S/\partial p = kb$ and $\partial S/\partial V = -ka$, where k is some factor of proportionality. Substituting $b = k^{-1}(\partial S/\partial p)$ and $a = -k^{-1}(\partial S/\partial V)$ into (30), we obtain (27).

c) *If* $dS/dt \leq 0$ *for* $t = 0$, *then* $dS/dt < 0$ *for* $t > 0$.

We suppose otherwise. Then there exist $t_1 \geq 0$ and $\varepsilon > 0$ such that $dS/dt < 0$ in the interval $0 \leq t < t_1$, $dS/dt = 0$ for $t = t_1$, and $dS/dt > 0$ for $t_1 < t < t_1 + \varepsilon$. Hence, $d^2S/dt^2 \geq 0$ for $t = t_1$, which contradicts property b). The proof of the assertion is complete.

We remark that properties a), b) and c) follow easily from the character of the "contours" of the function $S(V, p)$ (Figure 2.4).

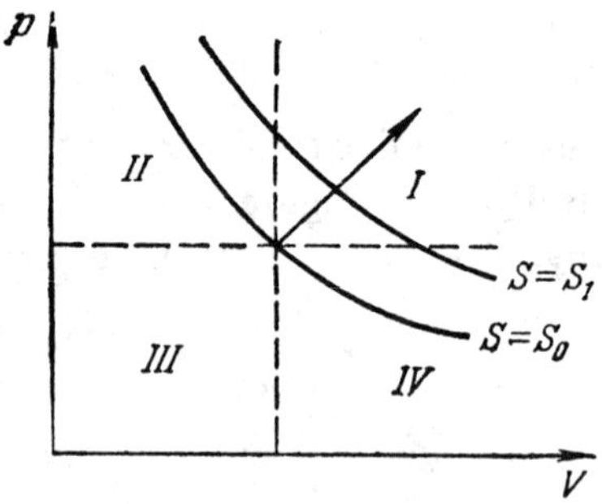

FIGURE 2.5.

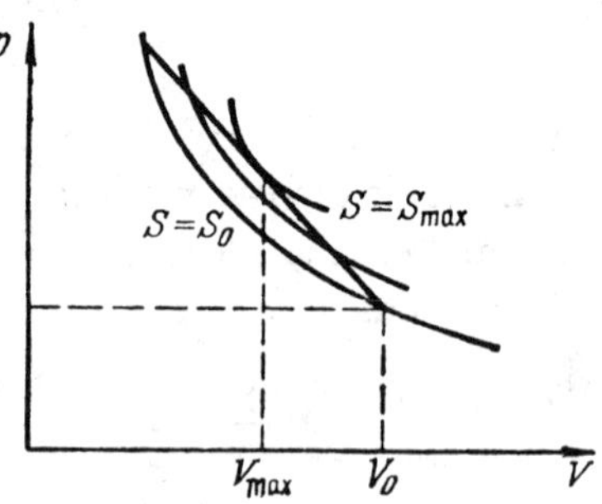

FIGURE 2.6.

Because of properties I and II of Poisson adiabats a ray passing through a point V_0, p_0 of the adiabat $S = S_0$ does not intersect it in more than this point if it lies in the zones I and III (Figure 2.5) and intersects it in still another point if it lies in the zones II and IV (we include here the case where the ray is tangent to the adiabat A, and we thus assume that a point of tangency is a double point).

In the first case S is a monotone function of the parameter t; in the second case S has a unique maximum. The point of the maximum is the point of tangency of the ray to some adiabat. We here note the following: the point $S = S_{max}$ on the ray divides it into two halves, so that an arbitrary adiabat intersects it in the upper half with its upper branch and in the lower half with its lower branch (Figure 2.6). In the upper half of the ray we have

$$\frac{\partial p(V, S)}{\partial V} < \frac{p - p_0}{V - V_0} \qquad (V < V_{max}); \tag{31}$$

in the lower half of the ray we have the reverse inequality

$$\frac{\partial p(V, S)}{\partial V} > \frac{p - p_0}{V - V_0} \qquad (V > V_{max}). \tag{32}$$

§2. Integral conservation laws. The hydrodynamic equations of one-dimensional flows

1. *General assumptions regarding the flow of compressible gases.* We picture the motion of a gas as the motion of a continuous medium in three-dimensional space x_1, x_2, x_3. In correspondence with §1 the motion is completely determined if the relations $\boldsymbol{u} = \boldsymbol{u}(x, t)$, $\rho = \rho(x, t)$, $p = p(x, t)$, $\varepsilon = \varepsilon(x, t)$, and $S = S(x, t)$ are known. We shall assume almost everywhere in this chapter that there is no internal friction or thermal conductivity in the gas, i.e., the particles of the gas are thermally insulated from one another.

Under these conditions no element of the fluid participates in heat exchange or loses its energy to friction. Therefore, for smooth changes of state the entropy of each element must remain constant. However, we especially wish to focus attention on the fact that this corollary is valid only for slow, smooth changes of the parameters of the gas particles. Therefore, if a particle of the gas abruptly changes its thermodynamic parameters its entropy does not remain constant in time.

In correspondence with §1 we assume that the flow is sufficiently smooth (a quasiequilibrium flow) everywhere except for surfaces of discontinuity. Therefore, a particle changes its entropy only by intersecting a surface of discontinuity; away from surfaces of discontinuity the concepts of temperature, pressure, and entropy are defined, and together with the equations of state they satisfy all thermodynamic relations.

Thus, we shall consider a fluid without internal friction and thermal conductivity; however, particles of the fluid may change their entropy by intersecting a surface of discontinuity. Actually a surface of discontinuity of the flow parameters is a narrow zone (a zone on nonequilibrium) in which the effect of viscosity and thermal conductivity is considerable however small they are. Indeed, as we have already mentioned in §1, a change of the entropy is determined by the quantities $\mu \operatorname{div} \boldsymbol{u}$ and $\kappa \operatorname{grad} T$, which in this narrow zone are finite.

In our consideration of these zones of large gradients as surfaces of discontinuity of the thermodynamic parameters we do not consider in detail the nonequilibrium flow in these zones, but we do consider this nonequilibrium globally, which leads to an increase of the entropy of a particle when it intersects a surface of discontinuity (zone of nonequilibrium).

We shall see (§5) that consideration of the viscosity and thermal conductivity in a fluid and subsequent passage to the limit as μ, $\kappa \to 0$ lead precisely to this picture of the flow. Thus, the fluid flows which we shall mainly consider here are limit flows of a viscous and thermally conducting fluid as the coefficients of viscosity and thermal conductivity tend to zero. On the other hand, such a description of the flows requires the application of only general laws which are valid both for equilibrium and nonequilibrium processes. These are the basic conservation laws of physics: the law of conservation of the mass of the fluid, the law of conservation of momentum, and the law of conservation of energy; we proceed to derive these laws in the next subsection.

2. ***The laws of conservation of mass, momentum, and energy in three-dimensional space.*** We shall study one-dimensional flows of compressible gases. However, in order to obtain the equations for various types of flow symmetry (plane, spherical, and cylindrical symmetry) in a unified way we first obtain the equations for an arbitrary three-dimensional, nonstationary flow, and from these equations we then derive the equations of interest to us for one-dimensional flows.

Thus, suppose that a gas moves in three-dimensional space with Cartesian coordinates (x_1, x_2, x_3) taken in some inertial system. We call these coordinates the *Euler coordinates*. We shall assume that there are no external forces acting on the gas and also that in the space occupied by the gas there are no sources of mass, momentum, and energy.

Let G be some portion of the space (x_1, x_2, x_3) bounded by a smooth, closed surface Σ_G. The amount of gas at time t in the volume G is equal to

$$\iiint_G \rho\, dx_1\, dx_2\, dx_3 = \int_G \rho(P, t)\, dG. \tag{1}$$

Here we denote the volume element by dG, and P is the point with coordinates (x_1, x_2, x_3). The quantity

$$\int_G [\rho(P, t_2) - \rho(P, t_1)]\, dG \tag{2}$$

is equal to the increment of the mass of gas in G from time $t = t_1$ to time $t = t_2$. Since there are no sources in G, the increment of mass of the gas (2) must be equal to the mass of gas entering G through the surface Σ_G during the time from $t = t_1$ to $t = t_2$. We denote by $\boldsymbol{d\Sigma}$ the vector having the direction of the outer normal to Σ and equal in magnitude to the area of a small piece of Σ. Through the element $d\Sigma$ during unit time a mass of gas equal to $-\rho \boldsymbol{u}\, \boldsymbol{d\Sigma}$ enters G. Therefore, equating the quantity (2) to the amount of gas entering G, we obtain

$$\int_G [\rho(P, t_2) - \rho(P, t_1)]\, dG + \int_{t_1}^{t_2} \left[\int_{\Sigma_G} \rho \boldsymbol{u}\, \boldsymbol{d\Sigma} \right] dt = 0. \tag{3}$$

Equation (3) expresses the law of conservation of mass.

The change of momentum in the volume G during the time from $t = t_1$ to $t = t_2$ is equal to the quantity

$$\int_G \left[\rho \boldsymbol{u}|_{t=t_2} - \rho \boldsymbol{u}|_{t=t_1} \right] dG = \int_G \rho \boldsymbol{u} \Big|_{t_1}^{t_2} dG \tag{4}$$

and is caused by gas flowing through the surface Σ_G carrying momentum

$$-\int_{t_1}^{t_2} \left[\int_{\Sigma_G} p\boldsymbol{u}(\boldsymbol{u}\, \boldsymbol{d\Sigma}) \right] dt, \tag{5}$$

and also by pressure forces p due to the remaining mass of gas acting on the gas in G along the normal to Σ_G. The total momentum of pressure forces acting on the gas in G is equal to the quantity

$$-\int_{t_1}^{t_2} \left[\int_{\Sigma_G} p\, \boldsymbol{d\Sigma} \right] dt. \tag{6}$$

Equating (4) to the sum of (5) and (6), we obtain an integral relation expressing the law of conservation of momentum:

$$\int_G \rho \boldsymbol{u} \Big|_{t_1}^{t_2} dG + \int_{t_1}^{t_2} \left[\int_{\Sigma_G} \rho\, \boldsymbol{d\Sigma} + \rho \boldsymbol{u}(\boldsymbol{u}\, \boldsymbol{d\Sigma}) \right] dt = 0. \tag{7}$$

In deriving (7) we have used the fact that there are no sources of momentum in G. The momentum in G is a vector; therefore, (7) contains three scalar equations for each of the components of the momentum.

The energy per unit volume is equal to $\rho(u^2/2 + \varepsilon)$. The quantity

$$\int_G \rho\left(\frac{u^2}{2} + \varepsilon\right) \Big|_{t_1}^{t_2} dG \tag{8}$$

is equal to the increment of the total energy in the volume G during the time from $t = t_1$ to $t = t_2$. This increment is caused by the transport of energy by the moving gas in the amount

$$-\int_{t_1}^{t_2} \left[\int_{\Sigma_G} \rho\left(\varepsilon + \frac{u^2}{2}\right) \boldsymbol{u}\, \boldsymbol{d\Sigma} \right] dt \tag{9}$$

and by the work of pressure forces p. A force $-p\, \boldsymbol{d\Sigma}$ acts from the side of the surrounding gas on the gas in G through the surface element $d\Sigma$; the gas particles on Σ_G move with velocity $\boldsymbol{u}$. Therefore, in each unit of time the pressure forces do work on the gas equal to $-\int_{\Sigma_G} p\boldsymbol{u}\, \boldsymbol{d\Sigma}$, while the total work of pressure forces during the time from $t = t_1$ to $t = t_2$ is equal to

$$-\int_{t_1}^{t_2} \left[\int_{\Sigma_G} p\boldsymbol{u}\, \boldsymbol{d\Sigma} \right] dt. \tag{10}$$

Equating the quantity (8) to the sum of (9) and (10), we obtain the integral relation expressing the law of conservation of energy:

$$\int_G \rho\left(\varepsilon + \frac{u^2}{2}\right)\Big|_{t_1}^{t_2} dG + \int_{t_1}^{t_2} \left[\int_{\Sigma_G} \rho\left(\varepsilon + \frac{u^2}{2} + \frac{p}{\rho}\right) \boldsymbol{u}\, \boldsymbol{d\Sigma} \right] dt = 0. \tag{11}$$

In deriving (11) we assumed that the gas has no thermal conductivity. The formulas (3), (7), and (11) express mathematically the laws of conservation of mass, momentum, and energy for gases free of friction and thermal conductivity, and they are the basic equations determining the motion of the gas.

To (3), (7), and (11) we add the equations of state of the gas. For example, if we prescribe the relation

$$p = p(V, \varepsilon), \qquad V = \frac{1}{\rho}, \tag{12}$$

then the problem of determining the flow reduces to finding five quantities: three components of the velocity $\boldsymbol{u}$ and two thermodynamic quantities: ρ and ε, since according to (12) p is a function of ρ and ε. The five scalar relations (3), (7), and (11) serve this purpose.

We remark that the equations of state can also be given in any of the other forms considered in §1, rather than in the form (12).

The vector $\rho\boldsymbol{u}$ is called the *vector of mass flow*; $\rho\boldsymbol{u}(\varepsilon + p/\rho + u^2/2)$ is the *energy flux vector*; the tensor $\Pi_{ik} = \delta_{ik}\, p + \rho u_i u_k$ is the *tensor of momentum flux*.

We now proceed to a more detailed study of one-dimensional flows, i.e., flows in which the quantities $\boldsymbol{u}$, ρ, p, and ε depend only on one spatial variable x and the time t. We shall consider three cases of one-dimensional flows:

1. A plane one-dimensional flow where the quantities $\boldsymbol{u}$, ρ, p and ε are constant on planes $x = x_1 = \text{const}$; $\boldsymbol{u} = \{u, 0, 0\}$. We call this case the case of plane symmetry.

2. A cylindrical one-dimensional flow (the case of cylindrical symmetry). In this case $\boldsymbol{u}$, ρ, p, and ε are constant on the surfaces of cyclinders $x = \sqrt{x_1^2 + x_2^2} = \text{const}$ (for fixed t). We set

$$\boldsymbol{u} = u(x, t)\left\{\frac{x_1}{x}, \frac{x_2}{x}, 0\right\}; \qquad \rho = \rho(x, t); \qquad p = p(x, t), \ldots. \tag{13}$$

3. A spherically symmetric flow (the case of spherical symmetry) is obtained if the following formulas hold:

$$\boldsymbol{u} = u(x, t)\left\{\frac{x_1}{x}, \frac{x_2}{x}, \frac{x_3}{x}\right\}, \qquad x = \sqrt{x_1^2 + x_2^2 + x_3^2}\,;$$

$$\rho = \rho(x, t); \qquad p = p(x, t), \ldots .$$

3. ***Integral conservation laws for one-dimensional flows in Euler coordinates.*** In the case of a plane one-dimensional flow all quantities depend only on x and t. The integral conservation laws (2.2.3), (2.2.7), and (2.2.11) can be rewritten in the form[2]

$$\int_{x_1}^{x_2} \rho \Big|_{t_1}^{t_2} dx + \int_{t_1}^{t_2} \rho u \Big|_{x_1}^{x_2} dt = 0 \tag{1}$$

(the law of conservation of mass),

$$\int_{x_1}^{x_2} \rho u \Big|_{t_1}^{t_2} dx + \int_{t_1}^{t_2} [p + \rho u^2] \Big|_{x_1}^{x_2} dt = 0 \tag{2}$$

(the law of conservation of momentum), and

$$\int_{x_1}^{x_2} \left[\rho\left(\varepsilon + \frac{u^2}{2}\right)\right] \Big|_{t_1}^{t_2} dx + \int_{t_1}^{t_2} \rho u\left(\varepsilon + \frac{p}{\rho} + \frac{u^2}{2}\right) \Big|_{x_1}^{x_2} dt = 0 \tag{3}$$

(the law of conservation of energy). In (1)–(3), x_1 and x_2 fix the distinguished volume G, and t_1 and t_2 are arbitrary times.

In the case of cylindrical symmetry we obtain the integral relations by writing the conservation laws (2.2.3), (2.2.7), and (2.2.11) for the volume G shown in Figure 2.7. Since all quantities are constant on the cylindrical surfaces $x =$ const, the law of conservation of mass can be written for G in the form

$$2\alpha l \int_{x_1}^{x_2} \rho \Big|_{t_1}^{t_2} x\, dx + 2\alpha l \int_{t_1}^{t_2} (\rho u x) \Big|_{x_1}^{x_2} dt = 0$$

or, after cancelling $2\alpha l$, in the final form

$$\int_{x_1}^{x_2} \rho \Big|_{t_1}^{t_2} x\, dx + \int_{t_1}^{t_2} (\rho u x) \Big|_{x_1}^{x_2} dt = 0. \tag{4}$$

The derivation of the integral conservation law for the momentum is somewhat

[2] By x_1 and x_2 we henceforth mean two values of a single coordinate x rather than the distinct Cartesian coordinates considered in the preceding subsection.

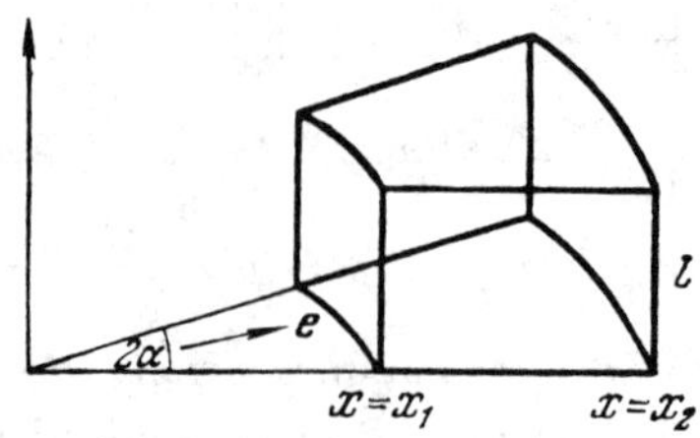

FIGURE 2.7

more involved. We write out the equality (2.2.7) in application to the volume G (Figure 2.7) only for the component of the momentum in the direction of the vector $\boldsymbol{e}$. We note immediately that for the two directions orthogonal to $\boldsymbol{e}$ the equality (2.2.7) leads to an identity. Because of (2.2.13) the integral $\int \rho u_e\, dG$ can be written in the form

$$\int_G \rho u_e\, dG = 2l \int_0^\alpha \cos\varphi\, d\varphi \int_{x_1}^{x_2} \rho u x\, dx$$
$$= 2l \sin\alpha \int_{x_1}^{x_2} \rho u x\, dx.$$

The integrand in the integral $\int_{\Sigma_G} \rho\, u_e(\boldsymbol{u}\, \boldsymbol{d\Sigma})$ is nonzero only on the parts $x = x_1$ and $x = x_2$ of the surface Σ_G; therefore,

$$\int_{\Sigma_G} \rho u_e(\boldsymbol{u}\, \boldsymbol{d\Sigma}) = 2l \sin\alpha\, [\rho u^2 x]\Big|_{x_1}^{x_2}.$$

Finally, the component of the integral $\int_{\Sigma_G} p\, \boldsymbol{d\Sigma}$ in the direction of the vector $\boldsymbol{e}$ consists of two terms: the integrals over the parts $x = x_1$ and $x = x_2$ of the surface Σ_G,

$$2l \int_0^\alpha \cos\varphi\, d\varphi \Big[px\, \Big|_{x_1}^{x_2} \Big] = 2l \sin\alpha \Big[px\, \Big|_{x_1}^{x_2} \Big],$$

and the integrals over the parts $\varphi = \pm\,\alpha$ of the surface Σ_G (Figure 2.7),

$$- 2l \sin \alpha \int_{x_1}^{x_2} p\, dx.$$

Substituting all these expressions into (2.2.7), we obtain after cancelling $2l \sin \alpha$ the final formula

$$\int_{x_1}^{x_2} [\rho u]\Big|_{t_1}^{t_2} x\, dx + \int_{t_1}^{t_2} [(p + \rho u^2)x]\Big|_{x_1}^{x_2} dt = \int_{t_1}^{t_2}\int_{x_1}^{x_2} p\, dx\, dt. \tag{5}$$

In application to the volume G the law of conservation of energy (2.2.11) for a flow with cylindrical symmetry can be written in the form

$$\int_{x_1}^{x_2} \left[\rho\left(\varepsilon + \frac{u^2}{2}\right)\right]\Big|_{t_1}^{t_2} x\, dx + \int_{t_1}^{t_2} \left[\rho u\left(\varepsilon + \frac{p}{\rho} + \frac{u^2}{2}\right) x\right]\Big|_{x_1}^{x_2} dt = 0. \tag{6}$$

Thus, (4), (5), and (6) give the representation of the laws of conservation of mass, momentum, and energy for a flow with cylindrical symmetry.

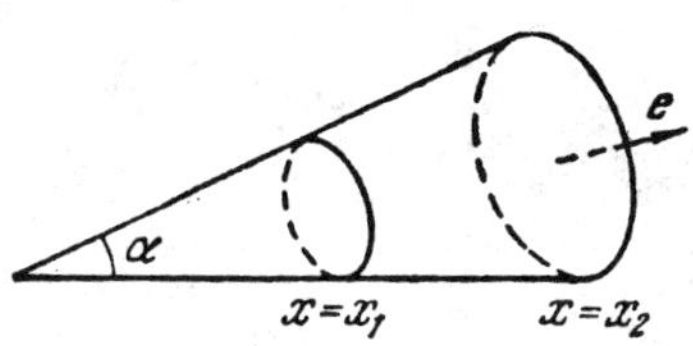

FIGURE 2.8

For a spherically symmetric flow the laws of conservation of mass, momentum, and energy can be written for the volume G cut out from a circular cone with aperture angle α by the spheres $x = x_1$ and $x = x_2$ (Figure 2.8). Computations analogous to the foregoing lead to the following equations:

$$\int_{x_1}^{x_2} \left(\rho\Big|_{t_1}^{t_2}\right) x^2\, dx + \int_{t_1}^{t_2} (\rho u x^2)\Big|_{x_1}^{x_2} dt = 0, \tag{7}$$

$$\int_{x_1}^{x_2} \left(\rho u\Big|_{t_1}^{t_2}\right) x^2\, dx + \int_{t_1}^{t_2} [(p + \rho u^2)x^2]\Big|_{x_1}^{x_2} dt = \int_{t_1}^{t_2}\int_{x_1}^{x_2} 2px\, dx\, dt, \tag{8}$$

$$\int_{x_1}^{x_2} \left[\rho\left(\varepsilon + \frac{u^2}{2}\right)\Big|_{t_1}^{t_2}\right] x^2\, dx + \int_{t_1}^{t_2} \left[\rho u\left(\varepsilon + \frac{p}{\rho} + \frac{u^2}{2}\right) x^2\right]\Big|_{x_1}^{x_2} dt = 0. \tag{9}$$

It is now easy to see that for all three cases of one-dimensional flows the laws of conservation of mass, momentum, and energy can be described by the common formulas

$$\int_{x_1}^{x_2} \rho \Big|_{t_1}^{t_2} x^\nu\, dx + \int_{t_1}^{t_2} (\rho u x^\nu)\Big|_{x_1}^{x_2} dt = 0, \tag{10}$$

$$\int_{x_1}^{x_2} \rho u \Big|_{t_1}^{t_2} x^\nu\, dx + \int_{t_1}^{t_2} [(p + \rho u^2)x^\nu]\Big|_{x_1}^{x_2} dt = \int_{t_1}^{t_2}\int_{x_1}^{x_2} \nu p x^{\nu-1}\, dx\, dt, \tag{11}$$

$$\int_{x_1}^{x_2} \left[\rho\left(\varepsilon + \frac{u^2}{2}\right)\right]\Big|_{t_1}^{t_2} x^\nu\, dx + \int_{t_1}^{t_2} \left[\rho u\left(\varepsilon + \frac{p}{\rho} + \frac{u^2}{2}\right)x^\nu\right]\Big|_{x_1}^{x_2} dt = 0. \tag{12}$$

In (10)–(12) we must set $\nu = 0$ in the case of plane symmetry, $\nu = 1$ in the case of cylindrical symmetry, and $\nu = 2$ in the case of spherical symmetry of the flow.

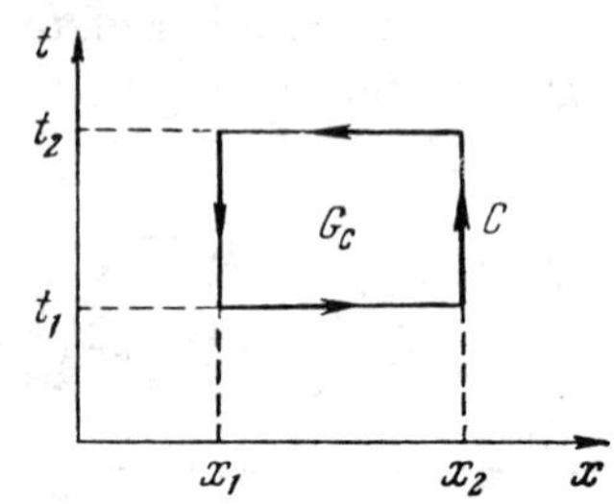

FIGURE 2.9

In the plane of the variables x and t we consider a contour C and the domain G_C bounded by it (Figure 2.9). It is obvious that (10)–(12) can be rewritten in the form

$$\oint_C \rho x^\nu\, dx - \rho u x^\nu\, dt = 0, \tag{13}$$

$$\oint_C \rho u x^\nu\, dx - (p + \rho u^2)x^\nu\, dt = -\iint_{G_C} \nu p x^{\nu-1}\, dx\, dt, \tag{14}$$

$$\oint_C \rho\left(\varepsilon + \frac{u^2}{2}\right)x^\nu\, dx - \rho u\left(\varepsilon + \frac{p}{\rho} + \frac{u^2}{2}\right)x^\nu\, dt = 0, \tag{15}$$

if the contour has the form indicated in Figure 2.9.

In their physical interpretation the quantities $\int_{x_1}^{x_2} \rho x^\nu\, dx$, $\int_{x_1}^{x_2} \rho u x^\nu\, dx$, and $\int_{x_1}^{x_2} \rho(\varepsilon + u^2/2)x^\nu\, dx$ are continuous functions of t, x_1 and x_2, while $\int_{t_1}^{t_2} \rho u x^\nu\, dt$, $\int_{t_1}^{t_2} (p + \rho u^2)x^\nu\, dt$ and $\int_{t_1}^{t_2} \rho u(\varepsilon + p/\rho + u^2/2)\, dt$ are continuous functions of x, t_1 and t_2. Therefore, assuming that u, ρ, ε and p are bounded and piecewise continuous functions[3] of x and t, we conclude that (13)–(15) are satisfied for any closed piecewise smooth contour C and domain G_C bounded by it.

4. ***Integral conservation laws in Lagrange coordinates.*** Suppose that r denotes the initial position of a particle of gas, for example, at time $t = 0$; let $x = x(r, t)$ be the position of this same particle at time t. As we saw in §1.3, the Lagrange coordinate r and the Euler coordinate x are related by

$$x = r + \int_0^t u(r, \tau)\, d\tau = x(r, t). \tag{1}$$

In (1) the velocity $u = u(r, t)$ is given as a function of the Lagrange variables. For a known velocity $u(r, t)$, equation (1) determines the trajectory of the particle. It is obvious that the mass of gas enclosed in the volume bounded by the sections $x = x(r_0, t)$ and $x = x(r, t)$ remains constant in time. We can therefore write[4]

$$\int_{x(r_0, t)}^{x(r, t)} \rho(x, t) x^\nu\, dx = \int_{r_0}^{r} \rho_0(r) r^\nu\, dr = M = \text{const}, \tag{2}$$

where $\rho_0(r)$ denotes the density at time $t = 0$.

Differentiating (2) with respect to r, we obtain

$$\frac{\partial x(r, t)}{\partial r} = \frac{r^\nu}{x^\nu} \frac{\rho_0(r)}{\rho(r, t)}, \tag{3}$$

since obviously $\rho(x(r, t), t) = \rho(r, t)$. Formula (3) shows that the mapping of Lagrange coordinates r onto Euler coordinates x is one-to-one under the condition $\rho(r, t) \neq 0$. In regions where $\rho = 0$ (regions of vacuum) to the points x, t there correspond no Lagrange coordinates r, t, i.e., through these points there pass no trajectories of the flow.

According to (1),

$$\frac{\partial x(r, t)}{\partial t} = u(r, t). \tag{4}$$

Therefore, from (3) and (4) we conclude that the passage from Euler coordinates x, t to Lagrange coordinates r, t is given by the relation

$$dx = \frac{r^\nu}{x^\nu} \frac{\rho_0(r)}{\rho(r, t)}\, dr + u(r, t)\, dt. \tag{5}$$

[3] In the case of spherical and cylindrical symmetry the boundedness of these quantities may be violated on the line $x = 0$. This, however, is of no consequence for what follows.

[4] The quantity M has the dimension of mass only in the case of spherical symmetry.

The substitution of (5) into the law of conservation of mass (2.3.13) reduces it to an identity. However, from (5) we obtain the integral relation

$$\oint_C x^\nu\, dx = \int_C \frac{\rho_0(r)}{\rho(r, t)} r^\nu\, dr + u(r, t)x^\nu\, dt = 0, \tag{6}$$

which is equivalent to the law of conservation of mass, so that it is a consequence of (2). In (6) C is an arbitrary piecewise smooth, closed contour in the plane of the variables r, t; $x = x(r, t)$ is determined by means of (1). We remark that (6) is called the integral conservation law of the volume occupied by the gas.

Passing from the variables x, t to Lagrange variables r, t in (2.3.14), we obtain

$$\begin{aligned}\oint_C u\rho_0(r)r^\nu\, dr - px^\nu(r, t)\, dt &= -\iint_{G_C} \nu p x^{\nu-1}\, dt \frac{\partial x(r, t)}{\partial r}\, dr \\ &= -\iint_{G_C} \nu p x^{\nu-1} \frac{r^\nu}{x^\nu} \frac{\rho_0(r)}{\rho(r, t)}\, dt\, dr \\ &= -\iint_{G_C} \nu p \frac{r^\nu}{x(r, t)} \frac{\rho_0(r)}{\rho(r, t)}\, dt\, dr \end{aligned}\tag{7}$$

which is the equation for conservation of momentum. Here C is a contour in the r, t plane, and G_C is the domain of these variables bounded by it. Finally, the law of conservation of energy (2.3.15) can be written in Lagrange variables in the form

$$\int_C \left(\varepsilon + \frac{u^2}{2}\right) \rho_0(r)r^\nu\, dr - upx^\nu(r, t)\, dt = 0. \tag{8}$$

Equations (6)–(8) constitute the laws of conservation of volume, momentum, and energy in Lagrange variables. We remark that they can be obtained directly by applying the conservation laws to a fixed mass of gas without passing from Euler coordinates to Lagrange coodinates.

The conservation laws (6)–(8) simplify somewhat if we introduce the notation

$$V(r, t) = V = \frac{1}{\rho(r, t)}; \qquad q = q(r) = \int_0^r \rho_0(r)r^\nu\, dr. \tag{9}$$

The quantity q coincides with M and is called the *Lagrange mass coordinate*. Going over to these variables, we obtain

$$\oint_C V\, dq + x^\nu(q, t)u\, dt = 0, \tag{10}$$

$$\oint_C u\,dq - px^\nu\,dt = -\iint_{G_C} \frac{\nu p V}{x}\,dq\,dt, \tag{11}$$

$$\oint_C \left(\varepsilon + \frac{u^2}{2}\right) dq - upx^\nu\,dt = 0. \tag{12}$$

In (10)–(12) C is an arbitrary piecewise smooth, closed contour in the plane of the variables q, t.

5. ***Differential equations for one-dimensional flows.*** We suppose that in some domain of the variables x and t the functions u, ρ, p and ε describing the gas flow are continuously differentiable. Then by Green's formula the contour integrals in (2.3.13)–(2.3.15) can be transformed into integrals over the domain G_C; the integrands will contain the first derivatives of u, ρ, p and ε. Since G_C is arbitrary, these integrands must vanish. Therefore, for smooth flows (u, ρ, p, $\varepsilon \in C_1$) the integral conservations laws (2.3.13)–(2.3.15) yield the differential equations

$$\frac{\partial}{\partial t}(x^\nu \rho) + \frac{\partial}{\partial x}(x^\nu \rho u) = 0, \tag{1}$$

$$\frac{\partial}{\partial t}(x^\nu \rho u) + \frac{\partial}{\partial x}[x^\nu(p + \rho u^2)] = \nu x^{\nu-1} p, \tag{2}$$

$$\frac{\partial}{\partial t}\left[x^\nu \rho\left(\varepsilon + \frac{u^2}{2}\right)\right] + \frac{\partial}{\partial x}\left[x^\nu \rho u\left(\varepsilon + \frac{p}{\rho} + \frac{u^2}{2}\right)\right] = 0. \tag{3}$$

The differential equations (1)–(3) are written in Euler coordinates x, t and are valid for smooth flows. Similarly, from the conservation laws (2.4.10)–(2.4.12) we obtain the differential equations in Lagrange coordinates q, t

$$\frac{\partial V}{\partial t} - \frac{\partial}{\partial q} x^\nu u = 0, \tag{4}$$

$$\frac{\partial u}{\partial t} + \frac{\partial}{\partial q} x^\nu p = \frac{\nu p V}{x}, \tag{5}$$

$$\frac{\partial}{\partial t}\left(\varepsilon + \frac{u^2}{2}\right) + \frac{\partial}{\partial q}(upx^\nu) = 0. \tag{6}$$

In (4)–(6) the Euler coordinate of the particle x must be considered as a function of the Lagrange coordinate q and the time t, i.e., $x = x(q, t)$; from (2.4.4) we obtain a differential equation for $x(q, t)$:

$$\frac{\partial x(q, t)}{\partial t} = u(q, t). \tag{7}$$

Equations (4)–(7) described a smooth, one-dimensional gas flow in Lagrange coordinates.

Equations (1)–(3) can be transformed to the form

$$\frac{\partial \rho}{\partial t} + \frac{\partial \rho u}{\partial x} + \frac{\nu \rho u}{x} = 0, \tag{8}$$

$$\frac{\partial u}{\partial t} + u\frac{\partial u}{\partial x} + \frac{1}{\rho}\frac{\partial p}{\partial x} = 0, \tag{9}$$

$$\frac{\partial S}{\partial t} + u\frac{\partial S}{\partial x} = 0, \tag{10}$$

where S is the entropy defined by $TdS = d\varepsilon + pdV$. Similarly, combining (4)–(7) in Lagrange coordinates, we obtain

$$\frac{\partial S(q, t)}{\partial t} = 0.$$

This equation shows that the entropy S of each particle of the gas remains constant in time in the entire region of smoothness of the flow. This implies that if the flow is smooth, then the entropy of each particle of the gas remains constant.

We observe that if the gas possesses finite viscosity and thermal conductivity, then equations (1)–(3) for a viscous, thermally conducting gas are replaced by

$$\frac{\partial}{\partial t}(x^\nu \rho) + \frac{\partial}{\partial x}(x^\nu \rho u) = 0, \tag{11}$$

$$\frac{\partial}{\partial t}(x^\nu \rho u) + \frac{\partial}{\partial x}\left[x^\nu\left(p + \rho u^2 - \mu\frac{\partial u}{\partial x}\right)\right] = \nu x^{\nu-1}\left(p - \mu\frac{\partial u}{\partial x}\right), \tag{12}$$

$$\frac{\partial}{\partial t}\left[x^\nu \rho\left(\varepsilon + \frac{u^2}{2}\right)\right] + \frac{\partial}{\partial x}\left[x^\nu \rho u\left(\varepsilon + \frac{p}{\rho} + \frac{u^2}{2}\right) - \mu x^\nu u\frac{\partial u}{\partial x}\right] = \frac{\partial}{\partial x}x^\nu \kappa\frac{\partial T}{\partial x}, \tag{13}$$

where $\mu > 0$ and $\kappa > 0$ are respectively the coefficients of viscosity and thermal conductivity, and T is the temperature of the gas. The corresponding equations in Lagrange coordinates have the form

$$\frac{\partial V}{\partial t} - \frac{\partial}{\partial q}x^\nu u = 0, \tag{14}$$

$$\frac{\partial u}{\partial t} + \frac{\partial}{\partial q}\left[x^\nu\left(p - \mu\rho x^\nu\frac{\partial u}{\partial q}\right)\right] = \frac{\nu V}{x}\left[p - \mu\rho x^\nu\frac{\partial u}{\partial q}\right], \tag{15}$$

$$\frac{\partial}{\partial t}\left(\varepsilon + \frac{u^2}{2}\right) + \frac{\partial}{\partial q}\left[ux^\nu\left(p - \mu\rho x^\nu\frac{\partial u}{\partial q}\right)\right] = \frac{\partial}{\partial q}\left[x^{2\nu}\rho\kappa\frac{\partial T}{\partial q}\right]. \tag{16}$$

Combining (11)–(13) in analogy to the foregoing, we find that

$$\frac{\partial S}{\partial t} + u\frac{\partial S}{\partial x} = \frac{\mu}{T\rho}\left(\frac{\partial u}{\partial x}\right)^2 + \frac{1}{T}\frac{1}{x^\nu \rho}\frac{\partial}{\partial x}\kappa x^\nu \frac{\partial T}{\partial x}. \tag{17}$$

or in Lagrange coordinates

$$\frac{\partial S(q, t)}{\partial t} = \frac{\mu\rho x^{2\nu}}{T}\left(\frac{\partial u}{\partial q}\right)^2 + \frac{1}{T}\frac{\partial}{\partial q}\kappa x^{2\nu}\rho\frac{\partial T}{\partial q}. \tag{18}$$

From (17) and (18) it follows that for any thermally insulated mass of gas possessing viscosity and thermal conductivity its total entropy does not decrease with increasing t. Indeed, writing (18) in the form

$$\frac{\partial S(q, t)}{\partial t} = \frac{\mu\rho x^\nu}{T}\left(\frac{\partial u}{\partial q}\right)^2 + \frac{\partial}{\partial q}\kappa x^{2\nu}\rho\frac{1}{T}\frac{\partial T}{\partial q} + \frac{\kappa x^{2\nu}\rho}{T^2}\left(\frac{\partial T}{\partial q}\right)^2 \tag{19}$$

and integrating it between the limits $q = q_1$ and $q = q_2$, we obtain

$$\frac{\partial}{\partial t}\int_{q_1}^{q_2} S(q, t)\, dq = \frac{\kappa x^{2\nu}\rho}{T}\frac{\partial T}{\partial q}\bigg|_{q_1}^{q_2} + \int_{q_1}^{q_2}\left[\frac{\mu\rho x^\nu}{T}\left(\frac{\partial u}{\partial q}\right)^2 + \frac{\kappa x^{2\nu}\rho}{T^2}\left(\frac{\partial T}{\partial q}\right)^2\right] dq. \tag{20}$$

Denoting by $S_s = \int_{q_1}^{q_2} S(q, t)\, dq$ the total energy of the given mass of gas, from (20) we find that

$$\frac{dS_s}{dt} \geq -\frac{W}{T}\bigg|_{q_1}^{q_2}, \tag{21}$$

where $W = -\kappa x^{2\nu}\rho(\partial T/\partial q)$ is the heat flux; W/T is the entropy flux. Thus, (21) shows that for any mass of a viscous, thermally conducting gas the production of entropy exceeds its influx through the boundary of this mass of gas. If the mass of gas contained between the sections $q = q_1$ and $q = q_2$ is thermally insulated, then $\kappa(\partial T/\partial q)x^{2\nu}\rho = 0$ for $q = q_1$ and $q = q_2$. Therefore, it follows from (20) that

$$\frac{dS_s}{dt} \geq 0. \tag{22}$$

Thus, if we consider the motion of a gas free of viscosity and thermal conductivity as the limit motion of a viscous and thermally conducting gas as $\mu, \kappa \to 0$, then from (17), (18), and (20) it follows that $S(q, t) = S(q, 0)$ only if $\sqrt{\mu}(\partial u/\partial q) \to 0$ and $\sqrt{\kappa}(\partial T/\partial q) \to 0$ as $\mu, \kappa \to 0$. As we shall see below, the motion of a gas free of viscosity and thermal conductivity is not smooth, since regions form in the gas where the gradients $\partial u/\partial q$ and $\partial T/\partial q$ are unbounded. For this reason if we consider the flow of a nonviscous gas which is not thermally conducting as the limit as $\mu, \kappa \to 0$, then conservation of entropy S for each particle of the gas occurs only so long as the trajectory of the particle lies in a region of smoothness of the flow.

If the trajectory of the particle passes through a zone where the gradients are unbounded or through a surface of discontinuity of the hydrodynamic variables, then the entropy of this particle increases. This principle of the increase of entropy when the particle passes through a surface of discontinuity (a zone of nonequilibrium) we shall subsequently use to select stable discontinuities in the study of the flow of a gas without viscosity or thermal conductivity.

We now observe that to uniquely determine the flow the equations of state of the gas must be added to equations (1)–(3) or (4)–(6) for a nonviscous and nonconducting gas and also to equations (11)–(13) or (14)–(16) for a gas with viscosity and thermal conductivity. We shall henceforth prescribe the equations of state in one of the following forms:

$$p = p(\rho, T), \qquad \varepsilon = \varepsilon(\rho, T); \tag{23}$$

$$\varepsilon = \varepsilon(\rho, p) \quad \text{or} \quad p = p(\rho, \varepsilon); \tag{24}$$

$$\varepsilon = \varepsilon(V, S), \quad p = p(V, S) \quad \text{or} \quad p = p(\rho, S). \tag{25}$$

The equations of state can be given in any of these forms, but in the case of a gas with viscosity and thermal conductivity it is especially convenient to apply them in the form (23), since this makes it possible to choose the temperature T as a basic thermodynamic variable.

In the case where various gases participate in the motion, we must assume that these relations are different in the regions occupied by different gases. Since these regions are unknown beforehand, we cannot, in general, prescribe these relations in the form of functions of the Euler coordinates x, t.

In this connection we note the advantage of the Lagrange coordinates in which the equations of state (23)–(25) can be given as functions of q and t, for example,

$$\varepsilon = \varepsilon(V, p, q). \tag{26}$$

In this case the relation (26) is taken to be discontinuous in the variable q at points $q = q_i = \text{const}$ which form the boundaries separating different gases.

6. *Study of the equations in Euler coordinates. The characteristic form. Characteristics.* We assume that the equation of state is given in the form (2.5.25):

$$p = p(\rho, S).$$

Then

$$\frac{\partial p}{\partial x} = \frac{\partial p}{\partial \rho}\frac{\partial \rho}{\partial x} + \frac{\partial p}{\partial S}\frac{\partial S}{\partial x}.$$

Substituting this into the system (2.5.8)–(2.5.10), we obtain the equations of one-dimensional flow in Euler coordinates in the form

$$\frac{\partial \rho}{\partial t} + \frac{\partial \rho u}{\partial x} = -\frac{\nu \rho u}{x}, \tag{1}$$

$$\frac{\partial u}{\partial t} + u\frac{\partial u}{\partial x} + \frac{1}{\rho}\frac{\partial p}{\partial \rho}\frac{\partial \rho}{\partial x} + \frac{1}{\rho}\frac{\partial p}{\partial S}\frac{\partial S}{\partial x} = 0, \tag{2}$$

$$\frac{\partial S}{\partial t} + u\frac{\partial S}{\partial x} = 0. \tag{3}$$

We shall reduce the system of three quasilinear equations (1)–(3) in the three unknown functions ρ, u and S to characteristic form. According to §2 of Chapter 1, for this it is necessary to compute the roots ξ_1, ξ_2, ξ_3 of the equation

$$\begin{vmatrix} u-\xi & \rho & 0 \\ \frac{1}{\rho}p'_\rho & u-\xi & \frac{1}{\rho}p'_S \\ 0 & 0 & u-\xi \end{vmatrix} = (u-\xi)^3 - p'_\rho(u-\xi) = 0. \tag{4}$$

As mentioned in §1, we assume that

$$\frac{\partial p(\rho, S)}{\partial \rho} > 0. \tag{5}$$

Setting

$$c^2 = c^2(\rho, S) = \frac{\partial p(\rho, S)}{\partial \rho}, \tag{6}$$

we write (4) in the form

$$(u-\xi)[(u-\xi)^2 - c^2] = 0, \tag{7}$$

whence we obtain

$$\xi_1 = u - c; \qquad \xi_2 = u; \qquad \xi_3 = u + c; \qquad \xi_1 < \xi_2 < \xi_3 \quad (c > 0). \tag{8}$$

Thus, if condition (5) is satisfied the system (1)–(3) is a system of hyperbolic type.

In §1 we have seen that $\partial p/\partial \rho = -V^2(\partial p/\partial V) > 0$ is a condition for the stability of the thermodynamic state of the gas. This same condition ensures the hyperbolicity of the equations of gas dynamics and hence the correctness of the Cauchy problem for the equations of gas dynamics. If $\partial p/\partial \rho < 0$, then the Cauchy problem for the system (1)–(3) would not be correct, in general.

We shall reduce this system of equations to characteristic form. For this we multiply (1) by $-c/\rho$, we add it to (2), and to the result we add (3) multiplied by $-(\rho c)^{-1}(\partial p/\partial S)$. We then obtain the equation in characteristic form

$$\frac{\partial u}{\partial t} + (u - c)\frac{\partial u}{\partial x} - \frac{c}{\rho}\left[\frac{\partial \rho}{\partial t} + (u - c)\frac{\partial \rho}{\partial x}\right] - \frac{\partial p}{\partial S}\frac{1}{\rho c}\left[\frac{\partial S}{\partial t} + (u - c)\frac{\partial S}{\partial x}\right] = \frac{\nu c u}{x}. \tag{9}$$

The second characteristic equation corresponding to the eigenvalue $\xi_2 = u$ is (3),

$$\frac{\partial S}{\partial t} + u\frac{\partial S}{\partial x} = 0, \tag{10}$$

while the third is obtained by multiplying (1) by c/ρ, adding the result to (2), and then adding (3) multiplied by $(\rho c)^{-1}(\partial p/\partial S)$. As a result we obtain the third equation in characteristic form:

$$\frac{\partial u}{\partial t} + (u + c)\frac{\partial u}{\partial x} + \frac{c}{\rho}\left[\frac{\partial \rho}{\partial t} + (u + c)\frac{\partial \rho}{\partial x}\right] + \frac{1}{\rho c}\frac{\partial p}{\partial S}\left[\frac{\partial S}{\partial t} + (u + c)\frac{\partial S}{\partial x}\right] = -\frac{\nu c u}{x}. \tag{11}$$

The equations (9)–(11) are thus the characteristic form of the equations of gas dynamics (1)–(3) in Euler coordinates. The quantity

$$c = c(\rho, S) = \sqrt{\frac{\partial p(\rho, S)}{\partial \rho}}$$

is called the *speed of sound*, since small perturbations of the solution propagate along characteristics; the slopes of these characteristics are $u - c$, u, and $u + c$. Therefore, small perturbations propagate relative to the substance with speeds 0 and $\pm c(\rho, S)$. The integral curves of the equations

$$\frac{dx}{dt} = u - c, \qquad \frac{dx}{dt} = u, \qquad \frac{dx}{dt} = u + c \tag{12}$$

are called the characteristics of the system (1)–(3) or (9)–(11); the line $dx/dt = u$ is also called the *trajectory*.

According to (10), the entropy S is constant along the trajectory. We obtain the conclusion we have already noted: the entropy of a particle remains constant as long as the flow remains smooth.

Since $c > 0$, the characteristic $dx/dt = u - c$ of the first family moves to the left relative to the substance with increasing t, while the characteristics of the third family $dx/dt = u + c$ move to the right relative to the substance.

Finally, we note still another form of writing the characteristic system (9)–(11) that is frequently encountered in the literature:

$$\left.\begin{aligned} dx = (u - c)dt, \qquad du - \frac{c}{\rho}\,d\rho - \frac{\partial p}{\partial S}\frac{1}{\rho c}\,dS &= \frac{\nu c u}{x}\,dt; \\ dx = u\,dt, \qquad dS &= 0; \\ dx = (u + c)dt, \qquad du + \frac{c}{\rho}\,d\rho + \frac{\partial p}{\partial S}\frac{1}{\rho c}\,dS &= -\frac{\nu c u}{x}\,dt. \end{aligned}\right\} \tag{13}$$

The lines along which the differential relations among the unknown functions are satisfied are indicated in this notation.

7. ***Isentropic and isothermal flows. The Riemann invariants.*** Suppose that on the line $t = 0$ we are given the initial values of the variables of gas dynamics

$$u = u_0(x), \qquad \rho = \rho_0(x), \qquad S = S_0(x). \tag{1}$$

If we assume that the initial functions u_0, ρ_0 and S_0 possess bounded first derivatives (or are Lipschitz continuous), then the results of Chapter 1 imply the existence of a differential (or Lipschitz continuous) solution of the system (2.6.9)–(2.6.11) in some strip $0 \le t \le t_0$.

We assume that $S_0(x) = S_0 = \text{const}$. From (2.6.10) it then follows that in the entire domain of the variables x and t where a differential solution of (2.6.9)–(2.6.11) exists the entropy remains constant:

$$S(x, t) = S_0(x) = S_0 = \text{const}. \tag{2}$$

Such a flow is called *isentropic*.

The problem of determining an isentropic flow obviously reduces to integrating the system of two quasilinear equations

$$\frac{\partial u}{\partial t} + (u - c)\frac{\partial u}{\partial x} - \frac{c}{\rho}\left[\frac{\partial \rho}{\partial t} + (u - c)\frac{\partial \rho}{\partial x}\right] = \frac{\nu c u}{x}, \tag{3}$$

$$\frac{\partial u}{\partial t} + (u + c)\frac{\partial u}{\partial x} + \frac{c}{\rho}\left[\frac{\partial \rho}{\partial t} + (u + c)\frac{\partial \rho}{\partial x}\right] = -\frac{\nu c u}{x}, \tag{4}$$

where

$$c = c(\rho, S_0) = \sqrt{\frac{\partial p(\rho, S_0)}{\partial \rho}} = c(\rho). \tag{5}$$

Just as any system of two quasilinear equations, the system (3), (4) reduces to Riemann invariants (Chapter 1, §3). Introducing the function

$$\varphi(\rho) = \int_{\rho_0}^{\rho} \frac{c(\rho)d\rho}{\rho} = \int_{\rho_0}^{\rho} \frac{c^2(\rho)\,d\rho}{\rho c(\rho)} = \int_{\rho_0}^{\rho} \frac{dp(\rho, S_0)}{\rho c(\rho)} \tag{6}$$

and the new variables s and r,

$$s = u - \varphi(\rho), \qquad r = u + \varphi(\rho), \tag{7}$$

we can write the system (3), (4) in the form

$$\frac{\partial s}{\partial t} + (u - c)\frac{\partial s}{\partial x} = \frac{\nu c u}{x}, \qquad \frac{\partial r}{\partial t} + (u + c)\frac{\partial r}{\partial x} = -\frac{\nu c u}{x}. \tag{8}$$

The variables r and s are called the *Riemann invariants.*

The functions u and ρ are uniquely determined on the basis of r and s, i.e., the transformation (7) has an inverse. Indeed,

$$u = \frac{r + s}{2} \tag{9}$$

and

$$\varphi(\rho) = \frac{r - s}{2}. \tag{10}$$

Since

$$\varphi'(\rho) = c(\rho)/\rho > 0, \tag{11}$$

ρ is uniquely determined from (10) as a function of $r - s$, i.e., we may assume that

$$\rho = \varphi^{-1}(r - s), \qquad c(\rho) = \psi(r - s) = c(\varphi^{-1}(r - s)), \tag{12}$$

where

$$(\varphi^{-1})' = \frac{\rho}{2c(\rho)} > 0, \qquad \frac{\partial \psi}{\partial r} = -\frac{\partial \psi}{\partial s} = \frac{\rho c'(\rho)}{c(\rho)}. \tag{13}$$

Thus, the system of equations for an isentropic flow can be written in terms of Riemann invariants in the form

$$\frac{\partial s}{\partial t} + \left[\frac{r + s}{2} - \psi(r - s)\right]\frac{\partial s}{\partial x} = \frac{(r + s)\nu\psi(r - s)}{2x}, \tag{14}$$

$$\frac{\partial r}{\partial t} + \left[\frac{r + s}{2} + \psi(r - s)\right]\frac{\partial r}{\partial x} = -\frac{(r + s)\nu\psi(r - s)}{2x}. \tag{15}$$

The equations for an isentropic, plane-symmetric flow are especially simple. In this case $\nu = 0$, and the right sides of (14) and (15) vanish:

$$\left.\begin{aligned} \frac{\partial s}{\partial t} + \left[\frac{r+s}{2} - c\right]\frac{\partial s}{\partial x} = 0, \\ \frac{\partial r}{\partial t} + \left[\frac{r+s}{2} + c\right]\frac{\partial r}{\partial x} = 0, \qquad c = \psi(r-s). \end{aligned}\right\} \tag{16}$$

It follows from (16) that the Riemann invariants r and s preserve constant values along the corresponding characteristics: the invarint s is constant along the characteristic

$$\frac{dx_1}{dt} = u - c = \frac{s+r}{2} - \psi(r-s),$$

while r is constant along the characteristic

$$\frac{dx_2}{dt} = u + c = \frac{r+s}{2} + \psi(r-s).$$

We consider one further special case—the case of an isothermal gas. We assume that the gas has very high thermal conductivity and is connected to a thermostat which maintains a constant temperature T_0. As a consequence of the high thermal conductivity the temperature in the gas equalizes very rapidly, and we may approximately assume that it is constant and equal to T_0. This means that we consider the limiting case of infinite thermal conductivity.

In contrast to the approximation of a locally adiabatic process in which the conservation of the entropy of the particles is violated in regions of abrupt gradients, this model is meaningful and remains consistent even for discontinuous flows. We therefore establish the integral conservation laws which are valid in this case.

It is altogether clear that the laws of conservation of mass and momentum are also valid in the case of this flow model. As concerns the law of conservation of energy, in this case it must be altered, since the gas maintains a constant temperature T_0 by obtaining or releasing energy to the thermostat. The law of conservation of energy is meaningful now only for the closed gas-thermostat system. As concerns the gas itself, application of the law of conservation of energy to any mass of gas reveals only the amount of heat released or obtained from the thermostat. There is an example of this in §4.6.

Writing the equation of state in the form $p = p(\rho, T_0)$, we see that in this case the pressure may be assumed to be a function of the density ρ alone, and $\partial p/\partial \rho > 0$.

From the conservation laws (2.3.13) and (2.3.14) we obtain the differential equations (2.5.1) and (2.5.2) which reduce to the characteristic form (3), (4), where c is now to be understood as the quantity

$$c = c_T(\rho) = \sqrt{\frac{\partial p(\rho, T_0)}{\partial \rho}}; \tag{17}$$

$c = c_T(\rho)$ is the so-called "isothermic speed of sound".

Equations (3) and (4) can, of course, be written in the form (14), (15) also in the case of an isothermic gas; in this case (7) is valid for r and s, (12) holds for $c_T(\rho)$, and (6) holds for $\varphi(\rho)$ provided that $c(\rho)$ is understood as the quantity (17).

8. *The equations in Lagrange coordinates. The case of variable entropy.* We take for the initial equations in Lagrange variables the equations (2.5.4), (2.5.5) and (2.5.6), in which q is the mass coordinate. We write them in the form

$$\frac{\partial V}{\partial t} - x^\nu \frac{\partial u}{\partial q} = \nu x^{\nu-1} u \frac{\partial x}{\partial q} = \frac{\nu u V}{x}, \tag{1}$$

$$\frac{\partial u}{\partial t} + x^\nu \frac{\partial p}{\partial q} = 0, \tag{2}$$

$$\frac{\partial S}{\partial t} = 0, \qquad V = \frac{1}{\rho}; \tag{3}$$

the Euler coordinate $x = x(q, t)$ must be considered as a solution of the differential equation (2.5.7),

$$\frac{\partial x(q, t)}{\partial t} = u(q, t), \tag{4}$$

which satisfies the initial condition that follows from (2.4.9):

$$\int_0^{x(q,0)} \rho_0(r) r^\nu \, dr = q, \tag{5}$$

i.e., it may be assumed that

$$x(q, 0) = x_0(q), \tag{6}$$

where $x_0(q)$ is a monotonically increasing and continuous function of q. As always, to (1)–(4) we adjoin the equation of state, which we assume given in the form

$$p = p(V, S), \qquad p'_V(V, S) < 0. \tag{7}$$

The quantity $x = x(q, t)$ is determined from (4); for $\partial x/\partial q$ we have

$$\frac{\partial x(q, t)}{\partial q} = \frac{1}{x^\nu} V. \tag{8}$$

As usual, the eigenvalues ξ_1, ξ_2, ξ_3 are determined from

$$\begin{vmatrix} -\xi & -x^\nu & 0 \\ x^\nu p'_V & -\xi & x^\nu p'_S \\ 0 & 0 & -\xi \end{vmatrix} = 0, \tag{9}$$

i.e.,

$$\xi[\xi^2 + x^{2\nu} p'_V] = 0. \tag{10}$$

Since

$$p'_V(V, S) = -\rho^2 p'_\rho(\rho, S) = -\rho^2 c^2(\rho, S), \tag{11}$$

it follows that

$$\xi_1 = -\rho c x^\nu = -\frac{c x^\nu}{V}, \qquad \xi_2 = 0, \qquad \xi_3 = \rho c x^\nu = \frac{c x^\nu}{V}.$$

These same formulas for ξ_1, ξ_2, ξ_3 can be obtained from the results of subsection 6 by applying the coordinate transformation from x, t to q, t according to (2.4.5).

The system (1)–(3) is reduced to characteristic form as in subsection 6. We arrive finally at the form

$$\frac{\partial u}{\partial t} - x^\nu \rho c \frac{\partial u}{\partial q} + \rho c\left(\frac{\partial V}{\partial t} - x^\nu \rho c \frac{\partial V}{\partial q}\right) - \frac{p'_S(V, S)}{\rho c}\left(\frac{\partial S}{\partial t} - x^\nu \rho c \frac{\partial S}{\partial q}\right) = \frac{\nu c u}{x}, \tag{12}$$

$$\frac{dS}{dt} = 0, \tag{13}$$

$$\frac{\partial u}{\partial t} + x^\nu \rho c \frac{\partial u}{\partial q} - \rho c\left(\frac{\partial V}{\partial t} + x^\nu \rho c \frac{\partial V}{\partial q}\right) + \frac{p'_S(V, S)}{\rho c}\left(\frac{\partial S}{\partial t} + x^\nu \rho c \frac{\partial S}{\partial q}\right) = -\frac{\nu c u}{x}. \tag{14}$$

To (12)–(14) we adjoin the equation (4) with the initial condition (6).

For an isentropic flow we have $S(q, t) = S_0$. The Riemann invariants

$$s = u - \int_{\rho_0}^{\rho} \frac{c\,d\rho}{\rho}, \qquad r = u + \int_{\rho_0}^{\rho} \frac{c\,d\rho}{\rho}$$

coincide with those introduced in subsection 7 (see (2.7.6) and (2.7.7)). This, by the way, is a general property of the Riemann invariants: they are invariant under change of dependent and independent variables.

Thus, the isentropic flow satisfies the system of two quasilinear equations in invariants

$$\frac{\partial s}{\partial t} - x^\nu \rho c \frac{\partial s}{\partial q} = \frac{\nu c u}{x}, \tag{15}$$

$$\frac{\partial r}{\partial t} + x^\nu \rho c \frac{\partial r}{\partial q} = -\frac{\nu c u}{x}, \tag{16}$$

while the function $x = x(q, t)$ is defined by (4) and (6). The quantity ρc depends only on $\rho = 1/V$ and is a single-valued function of $r - s$:

$$\rho c = \psi(r - s)\varphi^{-1}(r - s) = \xi(r - s).$$

Equations (15) and (16) simplify considerably in the case of plane symmetry, where $\nu = 0$.

In contrast to the case of Euler coordinates, in Lagrange coordinates the problem of determining a smooth flow with variable entropy also reduces to a system of two quasilinear equations and hence admits the introduction of invariants. Indeed, suppose that we know the entropy distribution at time $t = 0$,

$$S(q, 0) = S_0(q), \tag{17}$$

and $S_0(q)$ is a differentiable function. In the region of smoothness of the flow it follows from (13) that

$$S(q, t) = S_0(q), \tag{18}$$

and the problem reduces to the solution of the two equations (12) and (14). By introducing the Riemann invariants as above

$$s = u + \int_{V_0}^{V} \sqrt{-\frac{\partial p(V, S_0(q))}{\partial V}}\, dV, \tag{19}$$

$$r = u - \int_{V_0}^{V} \sqrt{-\frac{\partial p(V, S_0(q))}{\partial V}}\, dV, \tag{20}$$

we can write (12) and (14) in the form

$$\frac{\partial s}{\partial t} - x^\nu \rho c \frac{\partial s}{\partial q} = f_1, \qquad \frac{\partial r}{\partial t} + x^\nu \rho c \frac{\partial r}{\partial t} = f_2, \tag{21}$$

where

$$f_1 = \frac{\nu c u}{x} - \frac{dS_0(q)}{dq}\left[x^\nu p'_S(V, S_0) - x^\nu c \rho \int_{\rho_0}^{\rho} \frac{\partial c(\rho, S_0(q))}{\partial S_0} \frac{d\rho}{\rho}\right], \tag{22}$$

$$f_2 = -\frac{\nu c u}{x} - \frac{dS_0(q)}{dq}\left[x^\nu p'_S(V, S_0) - x^\nu c \rho \int_{\rho_0}^{\rho} \frac{\partial c(\rho, S_0(q))}{\partial S_0} \frac{d\rho}{\rho}\right]. \tag{23}$$

The functions f_1 and f_2 can be represented in the form

$$f_i = f_i\left(x, u, r - s, \frac{\partial S_0(q)}{\partial q}, S_0(q)\right), \tag{24}$$

and the relations (24) are linear in the variables u and $\partial S_0/\partial q$. As an illustration, we write these equations for the case of plane symmetry, i.e., for $\nu = 0$:

$$\frac{\partial s}{\partial t} - \rho c \frac{\partial s}{\partial q} = f_1, \qquad \frac{\partial r}{\partial t} + \rho c \frac{\partial r}{\partial q} = f_2, \tag{25}$$

$$f_1 = f_2 = -\frac{\partial S_0}{\partial q}\left[p'_S - c\rho \int_{\rho_0}^{\rho} \frac{\partial c(\rho, S_0(q))}{\partial S} \frac{d\rho}{\rho}\right]. \tag{26}$$

The quantities ρ and c can be expressed uniquely in terms of r and s.

In the case of an isothermal flow the equations in Lagrange variables again have the form (15), (16), where now the quantity c contained in these equations and in the definition of Riemann invariants r and s is the isothermal speed of sound c_T (see (2.7.17)).

9. ***The equations in invariants for polytropic and isothermic gases.*** For a polytropic gas the pressure p is given by

$$p = \frac{A^2(S)}{\gamma} \rho^\gamma, \qquad \gamma = \frac{c_p}{c_V} > 1, \tag{1}$$

and therefore

$$c^2 = \frac{\partial p}{\partial \rho} = A^2(S)\rho^{\gamma-1}, \qquad c(\rho, S) = A(S)\rho^{(\gamma-1)/2}. \tag{2}$$

For an isentropic flow the Riemann invariants are defined by (2.7.7). Computing $\varphi(\rho)$, we obtain

$$\varphi(p) = \int_{\rho_0}^{\rho} \frac{A(S)\rho^{(\gamma-1)/2}\, d\rho}{\rho} = \frac{2A(S)}{\gamma - 1} [\rho^{(\gamma-1)/2} - \rho_0^{(\gamma-1)/2}],$$

since $\gamma > 1$. For simplicity we set $\rho_0 = 0$; then

$$\varphi(\rho) = \frac{2A(S)}{\gamma - 1} \rho^{(\gamma-1)/2}, \qquad c(\rho) = A(S)\rho^{(\gamma-1)/2} = \frac{\gamma - 1}{2} \varphi(\rho), \tag{3}$$

i.e.,

$$c(\rho) = \frac{\gamma - 1}{4}(r - s). \tag{4}$$

Thus, for a polytropic gas the equations of isentropic flow (2.7.14) and (2.7.15) have the form

$$\frac{\partial s}{\partial t} + (\alpha s + \beta r)\frac{\partial s}{\partial x} = \frac{\nu(\gamma - 1)(r^2 - s^2)}{4x}, \tag{5}$$

$$\frac{\partial r}{\partial t} + (\alpha r + \beta s)\frac{\partial r}{\partial x} = -\frac{\nu(\gamma - 1)(r^2 - s^2)}{4x}, \tag{6}$$

where

$$\alpha = \frac{1}{2} + \frac{\gamma - 1}{4} > \frac{1}{2} > 0, \qquad \beta = \frac{1}{2} - \frac{\gamma - 1}{4}. \tag{7}$$

We write these equations for the case of plane symmetry ($\nu = 0$):

$$\frac{\partial s}{\partial t} + (\alpha s + \beta r)\frac{\partial s}{\partial x} = 0, \qquad \frac{\partial r}{\partial t} + (\alpha r + \beta s)\frac{\partial r}{\partial x} = 0. \tag{8}$$

We also write out (2.7.7) for the invariants r and s

$$s = u - \varphi(\rho) = u - \frac{2}{\gamma - 1}c, \qquad r = u + \varphi(\rho) = u + \frac{2}{\gamma - 1}c, \tag{9}$$

and the inverse formulas

$$u = \frac{r + s}{2}, \qquad c = c(\rho, S_0) = \frac{\gamma - 1}{4}(r - s). \tag{10}$$

In the case of an isothermic, ideal gas we have

$$p = R\rho T_0, \tag{11}$$

and therefore

$$C_T = \sqrt{RT_0} = \text{const} = c_0, \tag{12}$$

$$\varphi(\rho) = \int_{\rho_0}^{\rho} \frac{c_T d\rho}{\rho} = c_0 \ln \frac{\rho}{\rho_0}. \tag{13}$$

Setting $\rho_0 = 1$ to be specific, we have

$$2\varphi(\rho) = (r - s) = 2c_0 \ln \rho, \qquad \rho = \exp\left\{\frac{r - s}{2c_0}\right\}. \tag{14}$$

The equations in invariants for an isothermic gas take the form

$$\frac{\partial s}{\partial t} + \left(\frac{r + s}{2} - c_0\right)\frac{\partial s}{\partial x} = \frac{\nu c_0}{2x}(r + s), \tag{15}$$

$$\frac{\partial r}{\partial t} + \left(\frac{r+s}{2} + c_0\right)\frac{\partial r}{\partial x} = -\frac{\nu c_0}{2x}(r+s). \tag{16}$$

Special cases ($\nu = 0$) of the system (8) are interesting. One of the cases $\gamma = -1$ (corresponding to a so-called Chaplygin gas[5] leads to a weakly nonlinear system:

$$\frac{\partial s}{\partial t} + r\frac{\partial s}{\partial x} = 0, \qquad \frac{\partial r}{\partial t} + s\frac{\partial r}{\partial x} = 0. \tag{17}$$

Another interesting case ($\gamma = 3$) leads to the degenerate system of equations

$$\frac{\partial s}{\partial t} + s\frac{\partial s}{\partial x} = 0, \qquad \frac{\partial r}{\partial t} + r\frac{\partial r}{\partial x} = 0, \tag{18}$$

with characteristics that are straight lines, as we have seen in Chapter 1.

To conclude this section we present the equations in Lagrange variables for the motion of a polytropic gas in the case of plane symmetry ($\nu = 0$) with constant entropy. In this case from (2.8.25) and (2.8.26) we obviously have

$$\frac{\partial s}{\partial t} - B(S_0(q))\,(r-s)^{(\gamma+1)/(\gamma-1)}\frac{\partial s}{\partial q} = f_1, \tag{19}$$

$$\frac{\partial r}{\partial t} + B(S_0(q))\,(r-s)^{(\gamma+1)/(\gamma-1)}\frac{\partial r}{\partial q} = f_2, \tag{20}$$

where

$$B(S_0(q)) = \left[\frac{\gamma-1}{4A(S_0(q))}\right]^{(\gamma+1)/(\gamma-1)} A(S_0(q)), \tag{21}$$

$$f_1 = f_2 = \frac{dS_0(q)}{dq}\left[\frac{1}{\gamma(\gamma-1)}\rho^\gamma\right]\cdot 2A(S_0(q))A'(S_0(q)), \tag{22}$$

while ρ, as always, is determined as a certain function of the difference $r - s$ and the quantity $S_0(q)$.

We note also that in the case of an isentropic flow of a Chaplygin gas ($\gamma = -1$) the equations (19) and (20) become linear:

$$\frac{\partial s}{\partial t} - B\frac{\partial s}{\partial q} = 0, \qquad \frac{\partial r}{\partial t} + B\frac{\partial r}{\partial q} = 0. \tag{23}$$

From this we obtain the general integral

$$s = s_0(q + Bt), \qquad r = r_0(q - Bt),$$

[5] To ascribe physical meaning to the equation of state it should here be assumed that $p = p_0 - A^2(S)/\rho$, where $p_0 = \text{const} > 0$.

where s_0 and r_0 are arbitrary functions of a single argument. From this it follows that if a solution r, s of (23) is smooth at time $t = t_0$ it remains smooth for all t.

The system (23) is the weakly nonlinear system (17) transformed to Lagrange variables. Hence, a solution of (17) also preserves smoothness. This fact may be considered as an illustration of the theorem on weakly nonlinear systems proved in §10 of Chapter 1.

§3. Study of the simplest plane, one-dimensional flows

In this section we shall study the qualitative properties of the simplest, mainly isentropic or isothermal flows in the case of plane symmetry ($\nu = 0$).

1. *General properties. Integration in the case* $\gamma = 3$. In Euler coordinates an isentropic flow is described by a system of two equations which for the case of a polytropic gas has the form

$$\frac{\partial s}{\partial t} + (\alpha s + \beta r)\frac{\partial s}{\partial x} = 0, \qquad \frac{\partial r}{\partial t} + (\alpha r + \beta s)\frac{\partial r}{\partial x} = 0, \tag{1}$$

$$\alpha = \frac{1}{2} + \frac{\gamma - 1}{4}, \qquad \beta = \frac{1}{2} - \frac{\gamma - 1}{4}, \qquad \gamma \neq 1. \tag{2}$$

We shall assume that for the system (1) there are the initial conditions

$$s(x, 0) = s_0(x), \qquad r(x, 0) = r_0(x), \tag{3}$$

and the functions s_0 and r_0 are bounded and possess continuous first derivatives.

From the results of Chapter 1 we now get the existence of a solution of the Cauchy problem (1), (2) in some strip $0 \leq t < t_0$; the quantity t_0 is the time at which the derivatives of the solution become unbounded.

We suppose that a solution $s(x, t)$, $r(x, t)$ is known. We call an integral curve of the equation

$$\frac{dx}{dt} = \xi_1(s, r) = \alpha s(x, t) + \beta r(x, t) = u - c$$

an *s-characteristic*, and, similarly, an integral curve of the equation

$$\frac{dx}{dt} = \xi_2(s, r) = \alpha r(x, t) + \beta s(x, t) = u + c$$

an *r-characteristic*.

Along an s-characteristic $x = x_s(t, x_0)$ the invariant $s = s(x, t)$ is constant, as follows from (1), i.e., $s(x_s(t, x_0), t) \equiv s_0(x_0)$ if $x_s(0, x_0) = x_0$. Similarly,

$$r(x_r(t, x_0), t) \equiv r_0(x_0),$$

if $x = x_r(t, x_0)$ is the equation of the r-characteristic passing through the point $x = x_0$ of the initial axis $t = 0$. For a solution s, r we have the estimates

$$\left.\begin{aligned} \min_x s_0(x) \le s(x, t) \le \max_x s_0(x), \\ \min_x r_0(x) \le r(x, t) \le \max_x r_0(x), \end{aligned}\right\} \tag{4}$$

which are valid only in the region $0 \le t \le t_0$ in which the solution remains continuous.

Since

$$s = u - \frac{2}{\gamma - 1} c, \quad r = u + \frac{2}{\gamma - 1} c, \quad u = \frac{r+s}{2}, \quad c = \frac{\gamma - 1}{4}(r - s), \tag{5}$$

from (4) we obtain estimates for the speed u and the sound speed c (for $\gamma > 1$):

$$\begin{aligned} &\min_x\left[u_0(x) - \frac{2}{\gamma-1}c_0(x)\right] + \min_x\left[u_0(x) + \frac{2}{\gamma-1}c_0(x)\right] \\ &\quad \le 2u(x,t) \le \max_x\left[u_0(x) - \frac{2}{\gamma-1}c_0(x)\right] + \max_x\left[u_0(x) + \frac{2}{\gamma-1}c_0(x)\right], \end{aligned} \tag{6}$$

$$\begin{aligned} &\min_x\left[u_0(x) + \frac{2}{\gamma-1}c_0(x)\right] - \max_x\left[u_0(x) - \frac{2}{\gamma-1}c_0(x)\right] \\ &\quad \le \frac{4}{\gamma-1}c(x,t) \le \max_x\left[u_0(x) + \frac{2}{\gamma-1}c_0(x)\right] - \min_x\left[u_0(x) - \frac{2}{\gamma-1}c_0(x)\right], \end{aligned} \tag{7}$$

where $c_0(x)$ and $u_0(x)$ are the initial values of the sound speed and the speed of the flow.

By making these estimates rougher, we easily find that

$$\begin{aligned} &\min_x u_0(x) - \frac{1}{\gamma-1}\left[\max_x c_0(x) - \min_x c_0(x)\right] \\ &\quad \le u(x,t) \le \max_x u_0(x) + \frac{1}{\gamma-1}\left[\max_x c_0(x) - \min_x c_0(x)\right], \end{aligned} \tag{8}$$

$$\begin{aligned} &\min_x c_0(x) - \frac{\gamma-1}{4}\left[\max_x u_0(x) - \min_x u_0(x)\right] \\ &\quad \le c(x,t) \le \max_x c_0(x) + \frac{\gamma-1}{4}\left[\max_x u_0(x) - \min_x u_0(x)\right]. \end{aligned} \tag{9}$$

Setting

$$\min_x u_0(x) = u_0, \qquad \max_x u_0(x) = U_0, \qquad \min_x c_0(x) = c_0,$$

$$\max_x c_0(x) = C_0, \qquad \Delta u_0 = U_0 - u_0, \qquad \Delta c_0 = C_0 - c_0,$$

we rewrite (8) and (9) more briefly as

$$u_0 - \frac{1}{\gamma - 1}\Delta c_0 \le u(x, t) \le U_0 + \frac{1}{\gamma - 1}\Delta c_0, \tag{10}$$

$$c_0 - \frac{\gamma - 1}{4}\Delta u_0 \le c(x, t) \le C_0 + \frac{\gamma - 1}{4}\Delta u_0. \tag{11}$$

It should be kept in mind that by its very meaning the sound speed $c \ge 0$. Therefore, if $c_0 - 4^{-1}(\gamma - 1)\Delta u_0 < 0$ or

$$\min_x\left[u_0(x) + \frac{2}{\gamma - 1}c_0(x)\right] - \max_x\left[u_0(x) - \frac{2}{\gamma - 1}c_0(x)\right] < 0,$$

then the left side of (11) and (7) should be set equal to zero. From (5) it therefore also follows that

$$\min_x\left[u_0(x) - \frac{2}{\gamma - 1}c_0(x)\right] \le u(x, t) \le \max_x\left[u_0(x) + \frac{2}{\gamma - 1}c_0(x)\right].$$

However, if $c_0 - 2^{-1}(\gamma - 1)\Delta u_0 < 0$, then the estimate $c(x, t) \ge 0$ is also rough. We therefore now show that if the initial functions $u_0(x)$ and $c_0(x)$ possess bounded derivatives, then $c(x, t) \ne 0$ for any finite value $t > 0$. We shall simultaneously establish useful estimates for the derivatives of the solution of the Cauchy problem (1), (3). Since the assertion to be proved has a general character, we shall consider the case of any $\gamma \ge -1$. For $\gamma \ge -1$ we have $\alpha \ge 0$.

THEOREM. *If in an isentropic flow of a polytropic gas* ($\gamma \ge -1$) *no shock waves arise* (*the characteristics of a single family do not intersect*) *and the initial values* $r_0(x)$ *and* $s_0(x)$ *of the Riemann invariants are differentiable, then there exist function* $P(t) < \infty$ *and* $\rho_0(t) > 0$ *such that*

$$\frac{\partial u(x, t)}{\partial x} \le P(t), \qquad \rho(x, t) \ge \rho_0(t).$$

PROOF. We set $\partial s/\partial x = p_1$ and $\partial r/\partial x = p_2$, and we suppose that

$$\frac{ds_0(x)}{dx} = p_1(x, 0) \le P_0, \qquad \frac{dr_0(x)}{dx} = p_2(x, 0) \le P_0, \tag{12}$$

and also

$$0 < \rho_0 \le \rho(x, 0) = \rho_0(x) = \left[\frac{\gamma - 1}{4A}(r_0(x) - s_0(x))\right]^{2/(\gamma-1)} \le R_0. \tag{13}$$

We suppose, first of all, that $r_0(x)$ and $s_0(x)$ are twice continuously differentiable; then the solution of (1) as also twice continuously differentiable.

We set

$$\frac{\partial}{\partial t} + (\alpha s + \beta r)\frac{\partial}{\partial x} = \frac{\partial}{\partial t} + (u - c)\frac{\partial}{\partial x} = \left(\frac{d}{dt}\right)_s,$$

$$\frac{\partial}{\partial t} + (\alpha r + \beta s)\frac{\partial}{\partial x} = \frac{\partial}{\partial t} + (u + c)\frac{\partial}{\partial x} = \left(\frac{d}{dt}\right)_r.$$

Differentiating (1) with respect to x, we obtain

$$\left(\frac{dp_1}{dt}\right)_s = -\alpha p_1^2 - \beta p_1 \cdot p_2 = -p_1[\alpha(p_1 - p_2) + p_2],$$

$$\left(\frac{dp_2}{dt}\right)_r = -\alpha p_2^2 - \beta p_1 \cdot p_2 = -p_2[\alpha(p_2 - p_1) + p_1]. \tag{14}$$

Writing (1) in the form

$$\left(\frac{ds}{dt}\right)_s = \frac{\partial s}{\partial t} + (u - c)p_1 = 0; \qquad \left(\frac{dr}{dt}\right)_r = \frac{\partial r}{\partial t} + (u + c)p_2 = 0,$$

we easily establish that

$$p_1 = \frac{1}{2c}\left(\frac{ds}{dt}\right)_r = -\frac{1}{2c}\left[\frac{d}{dt}(r - s)\right]_r = \frac{2}{\gamma - 1}\frac{1}{c}\left(\frac{dc}{dt}\right)_r = -\frac{1}{\rho}\left(\frac{d\rho}{dt}\right)_r,$$

$$p_2 = -\frac{1}{2c}\left(\frac{dr}{dt}\right)_s = -\frac{1}{2c}\left[\frac{d}{dt}(r - s)\right]_s = -\frac{1}{\rho}\left(\frac{d\rho}{dt}\right)_s. \tag{15}$$

By means of (15) the equations (14) can be brought to the form

$$\left(\frac{d}{dt}\frac{p_1}{\rho}\right)_s = \alpha p_1\left(\frac{p_2}{\rho} - \frac{p_1}{\rho}\right), \qquad \left(\frac{d}{dt}\frac{p_2}{\rho}\right)_r = \alpha p_2\left(\frac{p_1}{\rho} - \frac{p_2}{\rho}\right). \tag{16}$$

Setting

$$p_1/\rho = v_1, \quad p_2/\rho = v_2; \qquad \alpha p_1 = \beta_1, \quad \alpha p_2 = \beta_2,$$

we rewrite (16) as

$$\left(\frac{dv_1}{dt}\right)_s = \beta_1(v_2 - v_1), \qquad \left(\frac{dv_2}{dt}\right)_r = \beta_2(v_1 - v_2). \tag{17}$$

If $\gamma = -1$, then $\alpha = 0$, $\beta_i = 0$; the system (17) can be integrated and yields the estimate

$$\frac{p_i(x, t)}{\rho(x, t)} \le \max_x \left\{\frac{p_i(x, 0)}{\rho(x, 0)}\right\} \le \frac{P_0}{\rho_0}, \qquad i = 1, 2.$$

We thus henceforth assume that $\gamma > -1$ $(\alpha > 0)$.

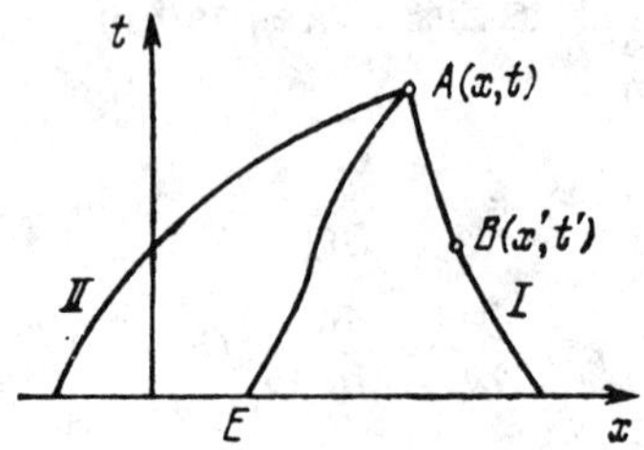

FIGURE 2.10

Let $A(x, t)$ be an arbitrary point of the half plane $t > 0$. Through this point we pass two characteristics: an s-characteristic I and an r-characteristic II (Figure 2.10). We set $v_0 = \max\{v_1(x, t), v_2(x, t)\}$, and we suppose that $v_0 > 0$. We shall prove that on at least one of the characteristics I or II there is a point $B(x', t')$ at which $\max\{v_1(x', t'), v_2(x', t')\} \geq v_0$ and $t' < t$. Suppose, for example, that $v_0 = v_1(x, t)$. Then, integrating the first equation of (17) along the characteristic I from the point $A(x, t)$ to an arbitrary point $B(x', t')$ of the ctaracteristic I, we obtain

$$v_1(x', t') - v_0 = v_1(x', t') - v_1(x, t) = \int_t^{t'} \beta_1(v_2 - v_0) \exp\left(- \int_\tau^{t'} \beta_1 \, d\xi \right) \tag{18}$$

There are two possible cases: $v_2(x, t) < v_0$ and $v_2(x, t) = v_0$. In the first case we have $v_2(x', t') - v_0 < 0$ in some neighborhood of the point (x, t); $\beta_1 > 0$ along the entire characteristic I, since $v_0 > 0$. Therefore, for t' sufficiently close to t and $t' < t$ the integral on the right side of (18) is positive; this means that on the characteristic I there exists a region $0 \leq t - t' < \varepsilon$ in which $v_1(x', t') - v_0 > 0$, i.e., $v_1(x', t') > v_0 = v_1(x, t)$.

In the second case ($v_2(x, t) = v_0$) there are points at which

$$\max\{v_1(x', t'), v_2(x', t')\} \geq v_0$$

on either of the characteristics I or II. We shall show that they exist on the characteristic I. For this we consider an arbitrary point (x', t') on the characteristic I for $t' < t$. If $v_1(x', t') - v_0 \geq 0$, then the required point is (x', t'); if $v_1(x', t') - v_0 < 0$, then we again consider (18). The left side of (18) is negative, the lower limit of the outer integral in (18) is greater than the upper limit,

$\beta > 0$ on the entire characteristic I, and the same goes for the exponential factor. Therefore, the equality (18) is possible only in the case where on the characteristic I there are intervals where $v_2 - v_0 > 0$. This means that on the characteristic I there necessarily exist points at which $\max\{v_1(x', t'), v_2(x', t')\} \geq v_0$ and $t' < t$.

From what has been proved it follows that $\max_x \max\{v_1(x, t), v_2(x, t)\}$ does not increase with increasing t, and therefore

$$\frac{p_i(x, t)}{\rho(x, t)} \leq \max_x \max\left\{\frac{p_1(x, 0)}{\rho(x, 0)}, \frac{p_2(x, 0)}{\rho(x, 0)}\right\} \leq \frac{P_0}{\rho_0}, \tag{19}$$

where P_0 and ρ_0 are given by (12) and (13).

Finally, we observe that (19) does not contain derivatives of the initial functions; it therefore remains valid for initial functions which are differentiable only once. In this case it is necessary to assume that (14)–(17) are satisfied in the "broad sense" (see Chapter 1, §7.1).

Since $u = (r + s)/2$, from (19) there also follows the one-sided estimate

$$\frac{\partial u}{\partial x} \leq \rho(x, t)\frac{P_0}{\rho_0}. \tag{20}$$

We note that the estimates (19) and (20) are sharp: they are achieved in the case $p_1(x, 0) = p_2(x, 0) = P_0$ and $\rho(x, 0) = \rho_0$. Writing the equation of continuity

$$\frac{\partial \rho}{\partial t} + u\frac{\partial \rho}{\partial x} = -\rho\frac{\partial u}{\partial x}$$

in the form

$$\left(\frac{d\rho}{dt}\right)_0 = -\rho\frac{\partial u}{\partial x}, \quad \left(\frac{d}{dt}\right)_0 = \frac{\partial}{\partial t} + u\frac{\partial}{\partial x},$$

we conclude that, according to (20),

$$\left(\frac{d\rho}{dt}\right)_0 \geq -\frac{P_0\rho^2(x, t)}{\rho_0}, \quad \text{i.e.} \left(\frac{d}{dt}\frac{1}{\rho}\right)_0 \leq \frac{P_0}{\rho_0}. \tag{21}$$

Integrating (21) from an arbitrary point $A(x, t)$ to a point E of the initial axis along the trajectory AE (Figure 2.10), we obviously obtain

$$\frac{1}{\rho(x, t)} - \frac{1}{\rho_0(E)} \leq \frac{P_0 t}{\rho_0},$$

i.e.,

$$\rho(x, t) \geq \frac{\rho_0(E)\rho_0}{\rho_0 + \rho_0(E)P_0 t} \geq \frac{\rho_0^2}{\rho_0 + R_0 P_0 t}. \tag{22}$$

Inequality (22) shows that the density $\rho(x, t)$ cannot vanish if the initial values possess a bounded derivative (we must obviously assume that $P_0 \geq 0$). The inequalities (20) and (22) prove the theorem.

Inequality (22) implies that under the indicated conditions for $t \geq 0$ there is a one-to-one correspondence between the Euler and Lagrange coordinates. This circumstance can be physically interpreted as follows. A gas which at time $t = 0$ continuously fills all space with a nonzero density cannot "tear apart" during its motion, i.e., it cannot form a region of vacuum in which $\rho = 0$, provided that for $t = 0$ there are no jumps of the velocity $u_0(x)$ such that $u_0(x - 0) < u_0(x + 0)$. This fact also holds in more general cases. It is valid, in particular, for nonisentropic continuous flows and for nonideal gases as well as for discontinuous solutions (flows with shock waves).

Since $c = A\rho^{(\gamma-1)/2}$, for $\gamma > 1$ we also obtain from (22) a lower bound for $c(x, t)$. Since $c(x, t)$ and hence also $\rho(x, t)$ are bounded above for $\gamma > 1$ by means of (11), from (20) we obtain an estimate of the derivative

$$\frac{\partial u(x, t)}{\partial x} \leq P_0\left[\frac{C_0}{c_0} + \frac{\gamma - 1}{4}\,\frac{\Delta u_0}{c_0}\right]^{2/(\gamma-1)}.$$

We remark that for $\gamma < 1$ the inequalities (11) are changed so that together with (22) they bound $c(x, t)$ above and below.

Above we have established upper bounds for p_1 and p_2. To estimate these quantities from below, we note that it follows from (14) that if $P(t)$ denotes the quantity

$$P(t) = \max_{-\infty<x<\infty} \max\{|p_1(x, t)|, |p_2(x, t)|\},$$

then

$$\frac{dP(t)}{dt} \leq (|\alpha| + |\beta|)P^2(t).$$

Hence

$$\left.\begin{aligned} P(t) &\leq \frac{P_0}{1 - (|\alpha| + |\beta|)\,t\cdot P_0}, \\ p_i(x, t) &\geq \frac{-P_0}{1 - (|\alpha| + |\beta|)\,t\cdot P_0}. \end{aligned}\right\} \tag{23}$$

The estimates (19) and (23) bound the derivatives of the solution above and below. It follows from (23) that for $t < 1/(|\alpha| + |\beta|)P_0$ the derivatives remain bounded both above and below. Thus, the strip $0 \leq t \leq 1/(|\alpha| + |\beta|)P_0$ is a strip in which there clearly exists a classical solution of the Cauchy problem (1), (2).

Of course, the estimate (23) is too rough. Thus, for example, if $ds_0/dx \geq 0$ and $dr_0/dx \geq 0$, then $p_1, p_2 \geq 0$, as follows from (14), and because of (19) a classical solution exists for all $t > 0$.

We shall illustrate our general remarks with the example of the flow of a polytropic gas with exponent $\gamma = 3$. In this case $\alpha = 1$, $\beta = 0$, and the system (1) decomposes into two individual equations:

$$\frac{\partial s}{\partial t} + s\frac{\partial s}{\partial x} = 0, \qquad \frac{\partial r}{\partial t} + r\frac{\partial r}{\partial x} = 0. \tag{24}$$

Since along an s-characteristic the invariant s is constant and the slope of an s-characteristic is equal to s, these characteristics are straight lines.

Along the line $x = x_0 + s_0(x_0)t = x_s(t, x_0)$ we have $s(x, t) = s_0(x_0)$; similarly, $r(x, t) = r_0(x_0)$ along the line $x = x_0 + r_0(x_0)t$. Thus, the solution of the Cauchy problem for (24) is given implicitly by

$$s(\xi + s_0(\xi)t;\, t) = s_0(\xi),$$
$$r(\eta + r_0(\eta)t;\, t) = r_0(\eta).$$

To find the solution $s(x, t)$, $r(x, t)$ explicitly it is necessary to solve

$$x = \xi + s_0(\xi)t, \qquad x = \eta + r_0(\eta)t \tag{25}$$

for ξ and η. Suppose that we have determined them from (25):

$$\xi = \xi(x, t), \qquad \eta = \eta(x, t);$$

then

$$s(x, t) = s_0(\xi(x, t)), \qquad r(x, t) = r_0(\eta(x, t)). \tag{26}$$

The geometric meaning of the quantities $\xi(x, t)$ and $\eta(x, t)$ is clear from Figure 2.11.

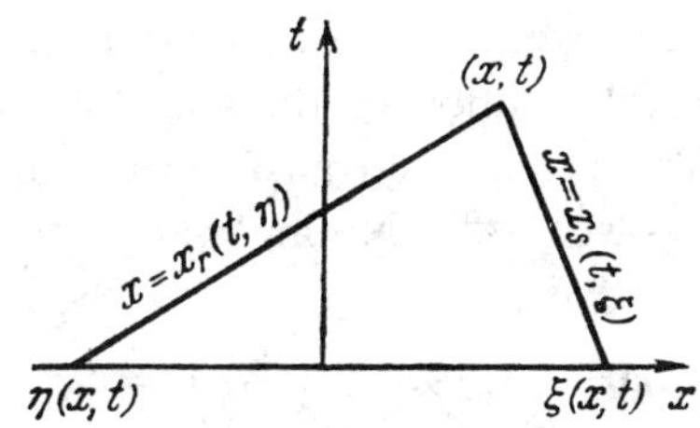

FIGURE 2.11

We shall verify that some of the estimates obtained above are satisfied. Differentiating (25) with respect to x, we obtain

$$\frac{\partial \xi(x, t)}{\partial x} = \frac{1}{1 + t(ds_0(\xi(x, t))/dx)}.$$

Hence, it follows from (26) that

$$\frac{\partial s(x, t)}{\partial x} = \left(\frac{ds_0(x_0)/dx}{1 + t(ds_0(x_0)/dx)}\right)_{x_0=\xi(x,t)},$$

and we see that (23) holds in this case. According to Figure 2.11,

$$r(x, t) - s(x, t) = \frac{\xi(x, t) - \eta(x, t)}{t} = r_0(\eta(x, t)) - s_0(\xi(x, t)). \tag{27}$$

We have

$$r_0(\eta) - s_0(\xi) = r_0(\eta) - s_0(\eta) + s_0(\eta) - s_0(\xi) = 2c_0(\eta) + s_0(\eta) - s_0(\xi). \tag{28}$$

From (27) and (28) it therefore follows that

$$2c(x, t) \geq 2c_0(\eta) - P_0[\xi(x, t) - \eta(x, t)]. \tag{29}$$

Now from (27) we have

$$\xi(x, t) - \eta(x, t) = 2c(x, t)t. \tag{30}$$

Combining (30) with (29), we obtain

$$c(x, t) \geq \frac{c_0(\eta)}{1 + P_0 t} \geq \frac{\rho_0}{1 + P_0 t}. \tag{31}$$

Since $\rho = c$ for $\gamma = 3$, inequality (31) is a minor refinement of (22). All the other estimates made above can similarly be easily verified for the case $\gamma = 3$.

To conclude this subsection we again return to the case of arbitrary $\gamma > -1$, and we shall show that the question of the existence or nonexistence of a global (i.e., for any $t > 0$) classical solution of the Cauchy problem (1), (3) can be resolved unequivocally. To this end we transform the equations (14) by multiplying each of them by $(r - s)^{-\beta/(\alpha-\beta)}$. Noting that

$$\left(\frac{dr}{dt}\right)_s = -(\alpha - \beta)(r - s)p_2; \qquad \left(\frac{ds}{dt}\right)_r = (\alpha - \beta))(r - s)p_1,$$

from (14) we obtain

$$\left(\frac{d}{dt}[p_1(r-s)^{-\beta/(\alpha-\beta)}]\right)_s = -\alpha p_1^2(r-s)^{-\beta/(\alpha-\beta)},$$

$$\left(\frac{d}{dt}[p_2(r-s)^{-\beta/(\alpha-\beta)}]\right)_r = -\alpha p_2^2(r-s)^{-\beta/(\alpha-\beta)}. \tag{32}$$

From (2.9.4) and (2.9.2) we have

$$\rho = B(r-s)^{1/(\alpha-\beta)}, \tag{33}$$

and therefore (32) can be rewritten in the form

$$\left(\frac{d}{dt}\frac{\rho^\beta}{p_1}\right)_s = \alpha\rho^\beta; \qquad \left(\frac{d}{dt}\frac{\rho^\beta}{p_2}\right)_r = \alpha\rho^\beta. \tag{34}$$

We have already mentioned above and it is easy to see from (32) that p_1 and p_2 preserve their sign on the s- and r-characteristics respectively. Therefore, recalling (20) and (22), we conclude that if $p_1(x, 0), p_2(x, 0) \geq 0$, then a classical solution of the Cauchy problem (1), (3) exists globally, i.e., for all $t > 0$.

On the other hand, if $p_1(x, 0)$ or $p_2(x, 0)$ has negative values, then the gradient catastrophe occurs in the solution of the Cauchy problem (1), (3). Indeed, suppose, for example, $p_1(x_0, 0) = p_1^0 < 0$. Integrating the first equation of (34) along the s-characteristic passing through the point $(x_0, 0)$, we obtain

$$\frac{\rho^\beta(t)}{p_1(t)} - \frac{\rho_0^\beta(x_0)}{p_1^0} = \int_0^t \rho^\beta(\tau)\,d\tau$$

or

$$p_1(t) = p_1^0\frac{\rho^\beta(t)}{\rho_0^\beta(x_0) + p_1^0\int_0^t \rho^\beta(\tau)\,d\tau}. \tag{35}$$

Here $\rho(t)$ and $p_1(t)$ are the values of $\rho(x, t)$ and $p_1(x, t)$ along the s-characteristic considered.

The integral in the denominator of (35) tends to ∞ as $t \to \infty$. Indeed, if $0 \leq \beta < 1$ $(3 \geq \gamma > -1)$, then, using the estimate of the density (22), we have

$$\int_0^t \rho^\beta(\tau)\,d\tau \geq \int_0^t \frac{\rho_0^{2\beta}\,d\tau}{(\rho_0 + R_0P_0t)^\beta} \to \infty \quad \text{as } t \to \infty,$$

while if $\beta < 0$ $(\gamma > 3)$, then we use (11), from which it follows that

$$\rho(\tau) \leq B[C_0 + 4^{-1}(\gamma-1)\Delta u_0]^{2/(\gamma-1)} = R.$$

Then for $\beta < 0$

$$\int_0^t \rho^\beta(\tau)\, d\tau \geq \int_0^t R^{-\beta}\, d\tau \to \infty \qquad \text{as } t \to \infty.$$

Therefore, for any $\gamma > -1$ the denominator of the right side of (35) vanishes for a finite value of t (since $p_1^0 < 0$), and the derivative $p_1(x, t)$ becomes infinite.

Thus, in order that for $\gamma > -1$ a classical solution of the Cauchy problem (1), (3) with the conditions (12) and (13) exist for all $t > 0$ it is necessary and sufficient that

$$\frac{\partial s(x, 0)}{\partial x} \geq 0, \qquad \frac{\partial r(x, 0)}{\partial x} \geq 0.$$

2. *Travelling waves (Riemann waves). Compression and rarefaction waves.* Here we consider certain of the simplest isentropic flows in the case of plane symmetry ($\nu = 0$). In studying travelling waves it makes practically no difference whether we use Lagrange or Euler variables. We shall use Lagrange variables. The equations of isentropic flow can then be written in the form

$$\frac{\partial s}{\partial t} - \xi(r - s)\frac{\partial s}{\partial q} = \frac{\partial s}{\partial t} - c\rho\frac{\partial s}{\partial q} = 0, \tag{1}$$

$$\frac{\partial r}{\partial t} + \xi(r - s)\frac{\partial r}{\partial q} = \frac{\partial r}{\partial t} + c\rho\frac{\partial r}{\partial q} = 0, \tag{2}$$

where the Riemann invariants r and s are connected with u and V by the formulas

$$s = u + \int_{\bar{V}}^{V} \frac{c}{V}\, dV = u + \int_{\bar{V}}^{V} \sqrt{-\frac{\partial p}{\partial V}}\, dV,$$

$$r = u - \int_{\bar{V}}^{V} \frac{c}{V}\, dV = u - \int_{\bar{V}}^{V} \sqrt{-\frac{\partial p}{\partial V}}\, dV.$$

For an isentropic flow, p is a function of the single variable V. We shall assume that $p = p(V)$ is an arbitrary differentiable function which satisfies, however, the conditions I and II:

$$\frac{dp}{dV} < 0, \qquad \frac{d^2p}{dV^2} > 0. \tag{3}$$

Denoting

$$\varphi(V) = -\int_{\bar{V}}^{V} \frac{c(V)\, dV}{V} = \frac{r - s}{2},$$

we obtain

$$\varphi'(V) = -\frac{c(V)}{V} < 0$$

and

$$\xi'(r - s) = -\frac{1}{4}\frac{d^2p/dV^2}{dp/dV} > 0.$$

Thus, from (3) it follows that $\xi(r - s)$ is a monotonically increasing function of $r - s$.

An isentropic flow in which one of the Riemann invariants is constant is called a *Riemann wave* or a *travelling wave*. To be specific we set $r = r_0 = \text{const}$; then (2) is satisfied identically, while (1),

$$\frac{\partial s}{\partial t} - \xi(r_0 - s)\frac{\partial s}{\partial q} = 0, \quad (4)$$

serves to determine $s(q, t)$.

The characteristics, which are the integral curves of the equation

$$\frac{dq}{dt} = -\xi(r_0 - s(q, t)),$$

are obviously straight lines in the q, t plane, since $s(q, t)$ is constant along them. It thus follows that along these lines

$$\frac{dq}{dt} = \frac{q - q_0}{t - t_0} = -\xi(r_0 - s(q, t)) \quad (5)$$

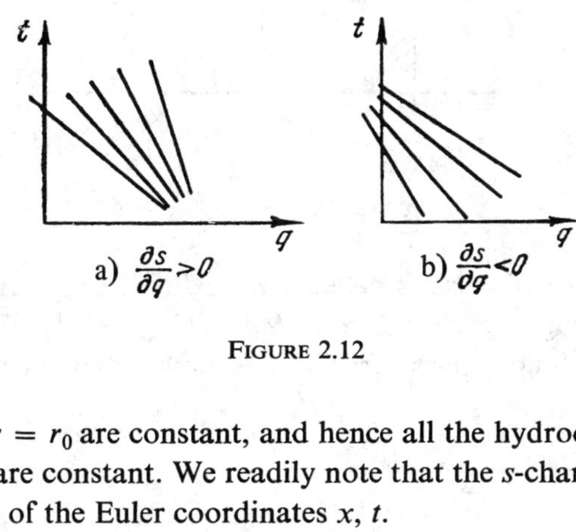

FIGURE 2.12

both $s(q, t)$ and $r = r_0$ are constant, and hence all the hydrodynamic quantities V, p, ρ, c, and u are constant. We readily note that the s-characteristics are also lines in the plane of the Euler coordinates x, t.

If $\partial s/\partial q > 0$ in some region, then the characteristics (5) form a diverging pencil of lines (Figure 2.12, a); if $\partial s/\partial q < 0$, then they form a converging pencil of lines (Figure 2.12, b). Since $u = \frac{1}{2}(r + s) = \frac{1}{2}(r_0 + s)$, $\partial s/\partial q > 0$ implies $\partial u/\partial q > 0$. From the equation of continuity

$$\frac{\partial V}{\partial t} = \frac{\partial u}{\partial q}$$

it therefore follows that for $\partial s/\partial q > 0$ also $\partial V/\partial t > 0$, and the density ρ decreases with increasing time t. Therefore, a Riemann wave in which $\partial s/\partial q > 0$ is called a *rarefaction wave*, while a Riemann wave in which $\partial s/\partial q < 0$ is called a *compression wave*.

The case in which $s = s_0 = \text{const}$ is considered in an altogether analogous manner. In this case also for $\partial r/\partial q < 0$ we have a compression wave, while for $\partial r/\partial q > 0$ we have a rarefaction wave. Thus, the general sign of a compression wave $\partial u/\partial q < 0$ or, equivalently, $\partial u/\partial x < 0$ leads to the condition $\partial s/\partial q < 0$ for a Riemann s-wave ($r = r_0$), and $\partial r/\partial q < 0$ for a Riemann r-wave ($s = s_0$).

For a Chaplygin gas we have $p = AV + p_0$, $d^2p/dV^2 = 0$, and $\xi = \sqrt{-A}$ is a constant quantity. The slope of the characteristics in this case is fixed, and therefore all s-characteristics are parallel to one another, as are the r-characteristics. The condition $\partial s/\partial q > 0$ nevertheless again distinguishes a region of rarefaction in the Riemann wave $r = r_0$.

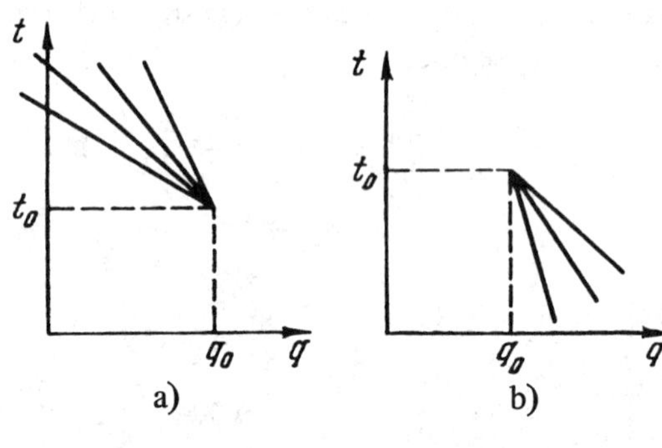

FIGURE 2.13

A Riemann wave ($r = r_0$) is called *centered* if the s-characteristics form a pencil of lines issuing from a single point (q_0, t_0) (Figure 2.13). Since the invariant $s(q, t)$ is constant along s-characteristics, this implies that in a centered Riemann wave

$$s = s\left(\frac{q - q_0}{t - t_0}\right), \qquad r = r_0$$

or

$$r = r\left(\frac{q - q_0}{t - t_0}\right), \qquad s = s_0.$$

Solutions depending only on the variable $y = (q - q_0)/(t - t_0)$ are called *self-similar*. We shall show that the centered Riemann waves give all self-similar solutions of the equations of gas dynamics.

Writing (2.5.4)–(2.5.6) in the form

$$\frac{\partial V}{\partial t} - \frac{\partial u}{\partial q} = 0, \quad \frac{\partial u}{\partial t} + \frac{\partial p}{\partial q} = 0, \quad \frac{\partial S}{\partial t} = 0 \quad (\nu = 0) \tag{6}$$

and assuming that the quantities V, u, p, and S depend only on $y = (q - q_0)/(t - t_0)$, we have

$$\frac{\partial}{\partial t} = -\frac{y}{t - t_0}\frac{d}{dy}, \qquad \frac{\partial}{\partial q} = \frac{1}{t - t_0}\frac{d}{dy},$$

and we arrive at the equations

$$y\frac{dV}{dy} + \frac{du}{dy} = 0, \qquad y\frac{du}{dy} - \frac{dp}{dy} = 0, \qquad \frac{dS}{dy} = 0.$$

From the last equation it follows that $S = S_0 =$ const, and a self-similar flow is hence an isentropic flow.

Passing to Riemann invariants in the two remaining equations, we write them in the form

$$[y + \xi(r - s)]\frac{ds}{dy} = 0, \qquad [y - \xi(r - s)]\frac{dr}{dy} = 0.$$

If $y \neq -\xi(r - s)$, then $ds/dy = 0$ and $s = s_0$. Since $\xi(r - s) > 0$, it follows that for $y = -\xi(r - s)$ we have $r = r_0 =$ const. Thus, for a self-similar solution there are only two possibilities: either

$$s = s_0 = \text{const}, \qquad y = \xi(r - s_0), \tag{7}$$

or

$$r = r_0 = \text{const}, \qquad y = -\xi(r_0 - s), \tag{8}$$

and hence any self-similar solution of (6) is a Riemann wave.

Since $\xi'(r - s) > 0$, by differentiating (7) and (8) with respect to q, we conclude that any self-similar solution is a compression wave for $t < t_0$ and is a rarefaction wave for $t > t_0$. Thus, if we consider the half plane $t \geq 0$, then in this half plane any self-similar solution depending on the variable $y = q/t$ ($q_0 = t_0 = 0$) is a rarefaction Riemann wave.

In Euler variables a centered Riemann wave is given by the conditions

$$s = s_0 = \text{const}, \qquad y = \frac{x - x_0}{t - t_0} = u + c$$

or

$$r = r_0 = \text{const}, \qquad y = \frac{x - x_0}{t - t_0} = u - c.$$

We shall consider some relations which hold for an arbitrary Riemann wave in the case of a polytropic gas. For a polytropic gas

$$s = u - \frac{2}{\gamma - 1} c, \qquad r = u + \frac{2}{\gamma - 1} c.$$

Suppose that in the Riemann wave $r = r_0 = \text{const}$, and let u_0 and c_0 be the values of the velocity and sound speed at some point of the Riemann wave. Then

$$u_0 + \frac{2}{\gamma - 1} c_0 = u + \frac{2}{\gamma - 1} c,$$

or

$$c = c_0\left(1 - \frac{\gamma - 1}{2} \frac{u - u_0}{c_0}\right). \tag{9}$$

Since $p = \gamma^{-1} A^{-2/(\gamma-1)} c^{2\gamma/(\gamma-1)}$, from (9) we have the relation between the pressure p and the velocity u in a Riemann s-wave:

$$p = p_0\left[1 - \frac{\gamma - 1}{2} \frac{u - u_0}{c_0}\right]^{2\gamma/(\gamma-1)} \qquad (r = r_0 = \text{const}). \tag{10}$$

Similarly, in a Riemann r-wave

$$p = p_0\left[1 + \frac{\gamma - 1}{2} \frac{u - u_0}{c_0}\right]^{2\gamma/(\gamma-1)} \qquad (s = s_0 = \text{const}). \tag{11}$$

Finally, we note still another important property of a Riemann wave: any continuous flow next to a zone of constant flow is a Riemann wave. Indeed, suppose that the flow is continuous and constant to the left of the line AB (Figure 2.14). This means that AB is a line through which a solution of the system of equations of gas dynamics cannot be uniquely continued, and hence AB is a characteristic. Since the flow to the left of AB is constant, it follows that

AB is a straight line. Suppose, for example, AB is an s-characteristic; then to the right of AB we have $r = r_0$, and so to the right of AB the flow is a Riemann s-wave.

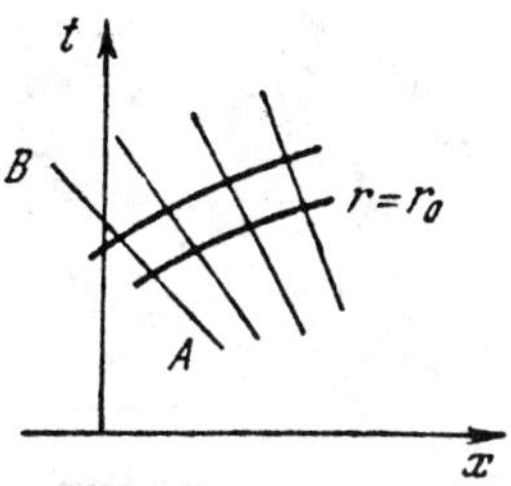

FIGURE 2.14

3. ***Profiles in a Riemann wave. The gradient catastrophe.*** We consider the behavior of the hydrodynamic quantities in travelling compression and rarefaction waves. Suppose, for example, that $r = r_0 = \text{const}$ and $\partial s/\partial q > 0$, i.e., we consider the case of a rarefaction s-wave. As we have seen above, in this case $\partial u/\partial q > 0$ and $\partial V/\partial t > 0$. Since $\xi^2(r - s) = -dp/dV$ and $d^2p/dV^2 > 0$, it follows that as ξ increases V decreases, and hence c increases. Since moreover $\xi'(r - s) > 0$, the conditions $\partial s/\partial q > 0$ and $r = r_0 = \text{const}$ imply that $\partial V/\partial q > 0$, $\partial \rho/\partial q < 0$ and $\partial c/\partial q < 0$. Thus, the profiles of the velocity u and the sound speed c in a rarefaction s-wave ($r = \text{const}$, $\partial s/\partial q > 0$) have the form shown in Figure 2.15, a. In a similar way we obtain the profiles in the case of rarefaction r-waves (Figure 2.15, b).

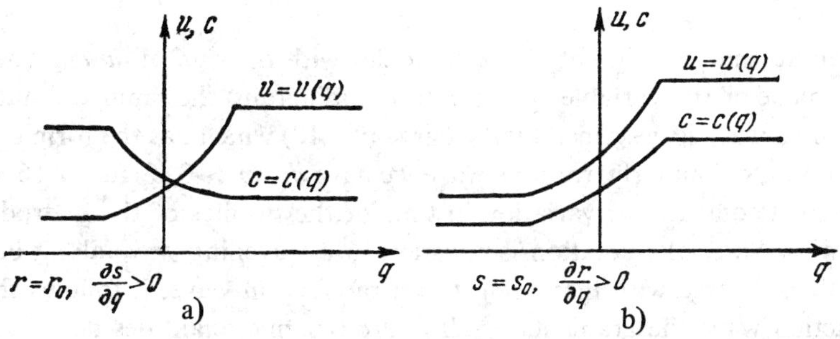

FIGURE 2.15.

Figures 2.16, a and b show the profiles of the hydrodynamic quantities in the case of a compression wave. Thus, $\partial u/\partial q > 0$ in a rarefaction wave, while $\partial u/\partial q < 0$ in a compression wave. The wave $r = \text{const}$ differs from the wave $s = \text{const}$ by the sign of the quantity $(\partial u/\partial q)(\partial c/\partial q)$: for $r = r_0$ this quantity is negative, and for $s = s_0$ it is positive.

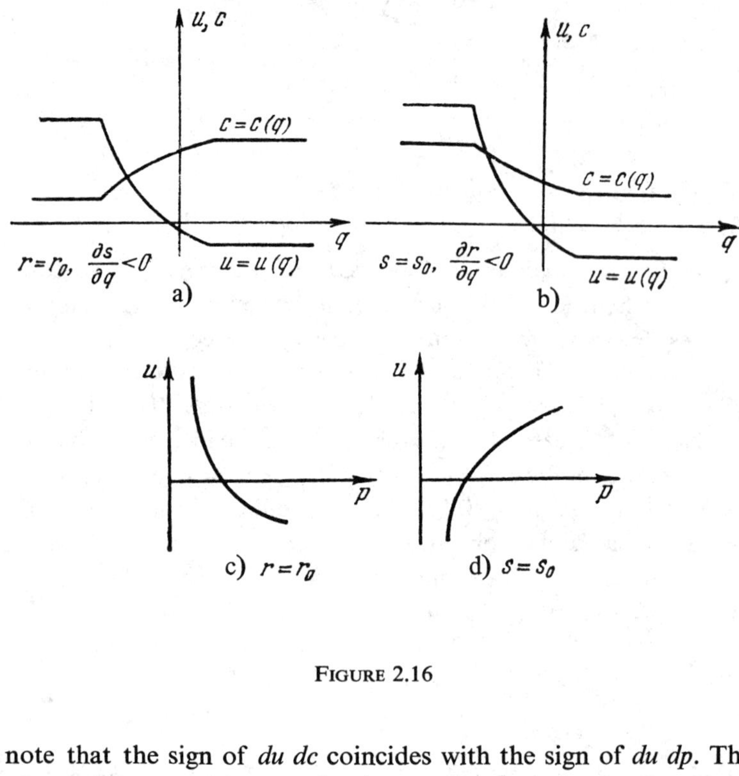

FIGURE 2.16

We note that the sign of *du dc* coincides with the sign of *du dp*. Therefore, in the plane of the variables p and u (a p, u-diagram) the family of states in a Riemann s-wave is described by the curve (3.2.10) which has the form shown in Figure 2.16, c. Similarly, for an r-wave we have $du\,dp > 0$ (Figure 2.16, d).

We shall consider the variation in time of the profiles of the hydrodynamic quantities. Since the solution is constant in a travelling wave along lines and these lines diverge with increasing t in a rarefaction wave, it follows that in a rarefaction wave the gradients of all hydrodynamic quantities decrease in absolute value with increasing time t. On the other hand, in a compression wave the characteristics of the corresponding family converge with increasing time t and the gradients of all hydrodynamic quantities increase in absolute value.

Figure 2.17, a and b show the variation of the profile $s = s(q, t)$ with increasing t in compression and rarefaction waves in the case $r = r_0 = \text{const}$.

The characteristics in a compression wave intersect for some finite value $t = t_0$. At a point of intersection of the characteristics the derivatives of the hydrodynamic quantities become unbounded. It is not hard to compute the time of the formation of unbounded derivatives. Suppose that $r = r_0$ and $s(q, 0) = s_0(q)$. Then, according to (3.2.5), two s-characteristics issuing from the points $q = q_1^0$ and $q = q_2^0$ of the initial axis $t = 0$ intersect at time t_n:

$$t_n = \frac{q_2^0 - q_1^0}{\xi_1(r_0 - s_0(q_1^0)) - \xi(r_0 - s_0(q_2^0))}. \tag{1}$$

If $s_0(q)$ is differentiable we obtain the least value of the time t at which the s-characteristics intersect:

$$t_{\min} = \frac{1}{\max_q\left[-\dfrac{\partial\xi(r_0 - s_0(q))}{\partial q}\right]}. \tag{2}$$

Formula (2) is meaningful only when $\max_q(-\partial\xi/\partial q) > 0$. If $t_{\min} \leq 0$, then the s-characteristics do not intersect for $t > 0$, and the derivatives in the Riemann wave remain bounded. Thus, in any travelling compression wave the gradients increase and become unbounded in finite time. This phenomenon is called the *gradient catastrophe.*

If $\partial^2 p/\partial V^2 > 0$ the gradient catastrophe occurs in a travelling compression wave, while if $\partial^2 p/\partial V^2 < 0$ it occurs in a travelling rarefaction wave. Thus, for $\partial^2 p/\partial V^2 < 0$ a compression shock wave is formed, while for $\partial^2 p/\partial V^2 < 0$ we have a rarefaction shock wave (see §4). For $t > t_{\min} > 0$ there exists no continuous solution: the solution becomes discontinuous.

We shall consider a more complex form of the flow—a flow with local initial data. We say that the initial data are *local* if the initial functions $r_0(q)$ and $s_0(q)$ vary only on a finite segment $a \leq q \leq b$ of the axis $t = 0$, i.e.,

$$r_0(q) = \begin{cases} r^- & \text{for } q < a, \\ r^+ & \text{for } q > b, \end{cases} \qquad s_0(q) = \begin{cases} s^- & \text{for } q < a, \\ s^+ & \text{for } q > b. \end{cases} \tag{3}$$

The flow occurring with the initial conditions (3) is not described by travelling Riemann waves. However, in certain cases for sufficiently large t this flow may consist only of travelling waves. If, for example, $r_0'(q), s_0'(q) \geq 0$, then it follows from the estimates (see subsection 1) that the solution has bounded derivatives

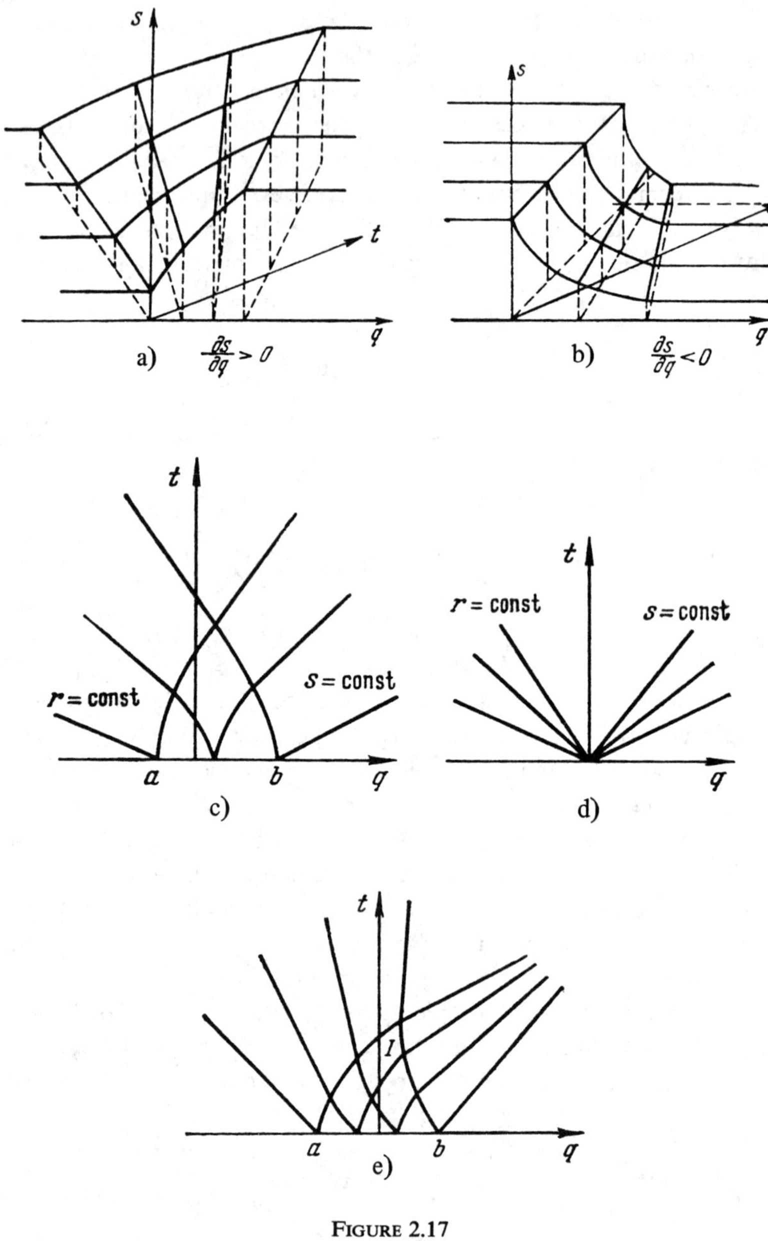

FIGURE 2.17

for all $t > 0$. The values of the invariant s propagate along s-characteristics with velocity $-\rho c < 0$; the values of the invariant r propagate along r-characteristics with velocity $\rho c > 0$. Therefore, for some $t_1 > 0$ the zones of variation of the invariants r and s part, and the solution consists of two travelling rarefaction waves separated by a zone of constant flow (Figure 2.17, c).

If we now let $a \to 0$ and $b \to 0$, i.e., we go over to a problem with piecewise constant initial data, then the solution $r(q, t)$, $s(q, t)$ obviously tends to a self-similar solution consisting of two centered rarefaction waves (Figure 2.17, d).

A Cauchy problem with piecewise constant initial data

$$r_0(q) = \begin{cases} r^-, & q < 0, \\ r^+, & q > 0, \end{cases} \qquad s_0(q) = \begin{cases} s^-, & q < 0, \\ s^+, & q > 0, \end{cases} \tag{4}$$

is called a *problem of decay of an arbitrary discontinuity* and is studied in detail in §6 of this chapter. The above analysis enables us to assert that if $s^- \leq s^+$ and $r^- \leq r^+$, then the decomposition problem has a continuous solution for $t > 0$ consisting of two centered rarefaction waves (Figure 2.17, e).

An analogous approach to the treatment of the problem of decay of a discontinuity as a limit of a problem with local initial data makes it possible to predict the qualitative behavior of the solution without the inequalities $r^- \leq r^+$ and $s^- \leq s^+$. If, for example, $r^- > r^+$, then after the interaction of the travelling waves (zone I) a compression r-wave propagates to the right (Figure 2.17, e). As we have seen above, the gradients in compression waves grow without bound. This leads to discontinuity of the solution. Shock waves appear in the solution. On this basis it can be asserted that if $s^- > s^+$ or $r^- > r^+$, then shock waves necessarily occur in the solution of the decomposition problem.

4. ***The piston problem. The flow of gas into a vacuum.*** Travelling waves find many applications in a number of the simplest problems, and also at times in the qualitative investigation of very complex flows. We shall here consider some simple problems with solutions described by means of travelling waves.

Suppose a gas is located to one side (to the right) of a rigid wall (a piston) which at the initial time $t = 0$ begins to move according to a particular law. We assume that at the initial time the gas is at rest and possesses constant density, pressure, and entropy, i.e., we assume that

$$u(q, 0) = 0, \qquad p(q, 0) = p_0, \qquad \rho(q, 0) = \rho_0, \qquad S(q, 0) = S_0. \tag{1}$$

We assume that an equation of state for the gas is given which satisfies conditions I and II:

$$\frac{\partial p}{\partial V} < 0, \qquad \frac{\partial^2 p}{\partial V^2} > 0, \qquad p(\infty, S) = 0. \tag{2}$$

The piston is located on the boundary of the gas with coordinate q which we take equal to zero. The law of motion of the piston is expressed by

$$u(0, t) = U(t), \tag{3}$$

where $U(t)$ is the speed of the piston given as a function of the time t, and $u(q, t)$ is the velocity of the gas.[6] We assume that $U(t)$ is a continuously differentiable function satisfying the condition

$$U(0) = 0. \tag{4}$$

We first consider the case in which $U'(t) \leq 0$. Construction of the solution to this problem reduces to finding a solution of the equations of isentropic flow which satisfies the initial conditions (1) and the boundary condition (3). In Riemann invariants the problem reduces to finding a solution of the equations

$$\frac{\partial s}{\partial t} - \xi(r - s)\frac{\partial s}{\partial q} = 0, \qquad \frac{\partial r}{\partial t} + \xi(r - s)\frac{\partial r}{\partial q} = 0, \tag{5}$$

which satisfies the initial conditions

$$s(q, 0) = -r(q, 0) = -\varphi(V_0), \qquad V_0 = \frac{1}{\rho_0}, \tag{6}$$

and the boundary condition given on the line $q = 0$

$$s(0, t) + r(0, t) = 2U(t). \tag{7}$$

The zone I (Figure 2.18) to the left of the characteristic OA ($q = c_0\rho_0 t$) is obviously a zone where the flow is constant, i.e., in zone I

$$s = s_0, \qquad r = r_0 = -s_0, \qquad u = 0, \qquad \rho = \rho_0, \qquad c = c_0. \tag{8}$$

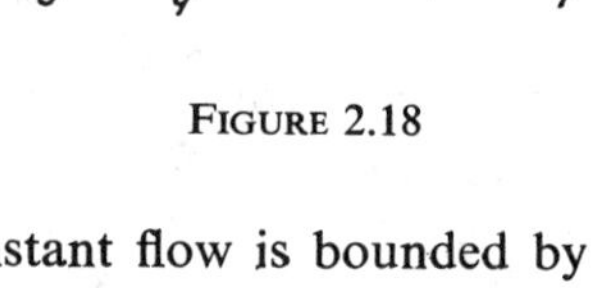

FIGURE 2.18

The zone II of nonconstant flow is bounded by the zone of constant flow along the r-characteristic OA. Hence, the flow in the zone II is a Riemann wave. It is evident from Figure 2.18 that in zone II the Riemann invariant $s(q, t)$

[6] The equality (3) holds only if the gas does not break away from the piston.

is constant with values which propagate in zone II along s-characteristics with initial axis $t = 0$. Thus, in zone II we have $s(q, t) = s_0 = \text{const}$, and it only remains for us to integrate the second equation of (5):

$$\frac{\partial r}{\partial t} + \xi(r - s_0)\frac{\partial r}{\partial q} = 0. \tag{9}$$

If an r-characteristic intersects[7] the axis $q = 0$ at a point $\tau > 0$, then at this point $s(0, \tau) = s_0$, and we can determine $r(0, \tau)$ from (7):

$$r(0, \tau) = 2U(\tau) - s_0 = 2U(\tau) + \varphi(V_0). \tag{10}$$

In the zone II of the Riemann wave the invariant $r(q, t)$ is constant along r-characteristics which are the lines

$$\frac{q}{t - \tau} = \xi(r(0, \tau) - s_0) = \xi(r(q, t) - s_0). \tag{11}$$

Along a line (11) we have $r(q, t) = r(0, \tau)$, where $r(0, \tau)$ is given by (10). The formula (11) defines the solution $r = r(q, t)$ parametrically in terms of the parameter τ (the ordinate of the point of intersection of the characteristic (11) with the axis $q = 0$). It follows from (11) that if $U'(t) < 0$, then in zone II we have $\partial r(q, t)/\partial q > 0$, i.e., the motion in zone II is a rarefaction wave, and the picture of the characteristics is as shown in Figure 2.18. In satisfying the boundary condition (7) it was assumed that an s-characteristic issuing from the ray $q > 0$ of the initial axis intersects the line $q = 0$ for $t > 0$.

According to its physical meaning, the quantity

$$c \geq 0, \tag{12}$$

and therefore (10) is meaningful only if

$$c = \psi(2U(t) - 2s_0) = \psi(2U(t) + 2\varphi(V_0)) \geq 0. \tag{13}$$

According to (2), $\psi(r - s)$ is a monotonically increasing function. Therefore, in general, (13) gives a lower bound for the piston speed $U(t)$ for which the boundary condition (3) can be satisfied, i.e., it gives a lower bound for the speed of the gas on the line $q = 0$.

For $U(t) < 0$ sufficiently small in modulus the inequality (13) is clearly satisfied, since for $U(t) = 0$

$$\psi(2\varphi(V_0)) = \psi(r_0 - s_0) = c_0 > 0.$$

[7] We shall see below that for $U(t) < 0$ sufficiently large in modulus this is not so. i.e., the r-characteristics do not intersect the axis $q = 0$.

Suppose that for some $t = t_1$ inequality (13) becomes an equality, and $U(t) < U(t_1)$ for $t > t_1$. Then for $t > t_1$ the boundary condition (3) becomes meaningless. In this case the piston breaks away from the gas, and between the piston and the gas a region of vacuum is formed. The formula (11) parametrically defines the solution $r(q, t)$ in the entire zone II for the parameter τ in the interval $0 \leq \tau \leq t_1$.

We observe that it follows from the assumptions (2) that if $\sqrt{-\partial p/\partial V} = \xi(r - s) = c\rho = 0$, then $p = 0$, and therefore the boundary $q = 0$ for $t > t_1$ may be considered a free boundary of the gas, i.e., the boundary of the gas with the vaccum in which $p = 0$ and $\rho = 0$. An inconvenience of the Lagrange coordinates emerges in this case, in that in the q, t plane the boundary of the gas $q = 0$ coincides with the position of the piston, since in the region between them $\rho = 1/V = 0$.

The picture becomes clearer if we represent the motion of the gas and piston in Euler coordinates. Figure 2.19 shows the picture of the r-characteristics in Euler coordinates in the case where the gas breaks away from the piston.

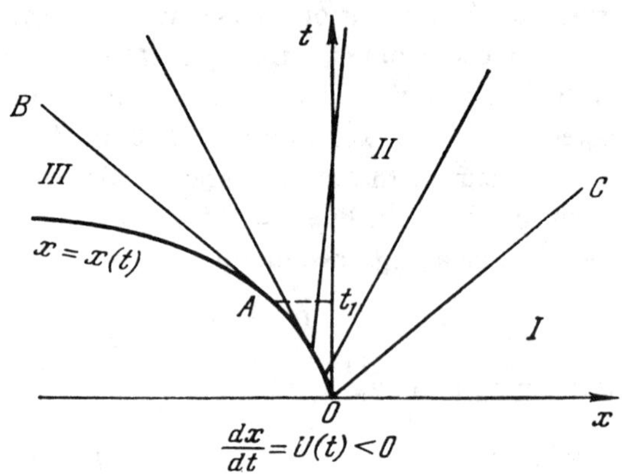

FIGURE 2.19

As before, in the zone I we have a gas at rest; the zone II bounded by the r-characteristics OC and AB is the zone of the rarefaction wave; and the zone III (between the trajectory of the piston $x = x(t)$ and the characteristic AB) is the vacuum zone. At the point A ($t = t_1$) the gas breaks away from the piston. We remark that the line AB forming the boundary between the gas and the vacuum is simultaneously an r- and s-characteristic. For a polytropic gas we have $\varphi(V) = 2(\gamma - 1)^{-1}c$ and $\psi(r - s) = c = 4^{-1}(\gamma - 1)(r - s)$; therefore, for $\gamma > 1$ condition (13) can be written in the form

$$\left[U(t) + \frac{2}{\gamma - 1} c_0\right] \geq 0, \tag{14}$$

i.e.,

$$U(t) \geq -\frac{2}{\gamma - 1} c_0. \tag{15}$$

Thus, for $U(t) < -2(\gamma - 1)^{-1}c_0$ the gas breaks away from the piston. We remark that if an isothermic gas is formally considered as a gas for which $\gamma = 1$, then it follows from (15) that an isothermic gas does not break away from the piston, since $2(\gamma - 1)^{-1}c_0 \to \infty$ as $\gamma \to 1$. This same conclusion can be obtained also from the formula for ρ, since

$$\rho = \exp\left\{\frac{r - s}{2c_T}\right\} > 0, \quad \text{where } c_T = \sqrt{RT} = \text{const.}$$

Another physical piston problem in which the pressure on the piston is given rather than the speed of the piston is solved in almost the same way. This leads to the boundary condition in Lagrange variables

$$p(0, t) = p_0(t). \tag{16}$$

We assume that in the conditions of the preceding problem $U(t)$ decreases monotonically on the segment $0 \leq t \leq t_1$, while $U(t) = U_0 = \text{const}$ for $t \geq t_1$. Figure 2.20, a shows the picture of the characteristics in the q, t plane in this case. We see that if $U_0 \geq -2(\gamma - 1)^{-1}c_0$, then in zone III the flow is constant.

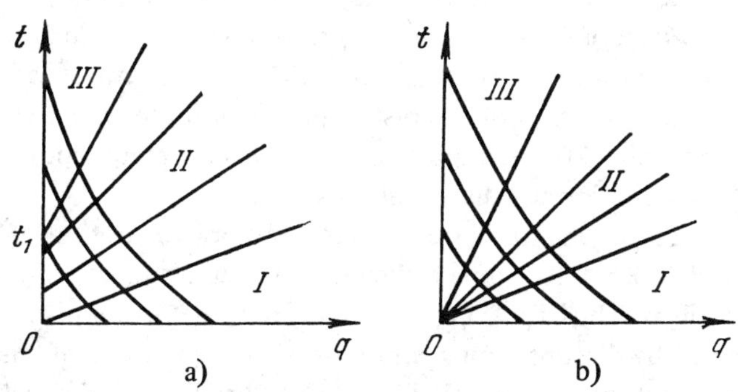

FIGURE 2.20

If we now let t_1 tend to zero, then in the limit not needed we obtain a self-similar solution; the picture of the characterisitcs in this case is shown in Figure 2.20, b. This self-similar solution corresponds to a problem with the piston moving away from the gas with a constant speed $U_0 \geq -2(\gamma - 1)^{-1}c_0$. In the case $U_0 < -2(\gamma - 1)^{-1}c_0$ the zone III vanishes and is replaced by a rarefaction wave, i.e., in this case zone II squeezes out zone III. For $U_0 < -2(\gamma - 1)^{-1}c_0$ the condition $\rho = 0$ is satisfied on the boundary $q = 0$, and the gas is not in contact with the piston, i.e., they are separated by a region of vacuum. This self-similar solution gives the solution of another problem, the problem of the flow of gas into a vacuum. In this problem we may assume that at the initial time $t = 0$ the wall bounding the gas on the left (at $q = 0$) is removed.

We now consider a second case, where

$$U'(t) \geq 0, \qquad U(0) = 0. \tag{17}$$

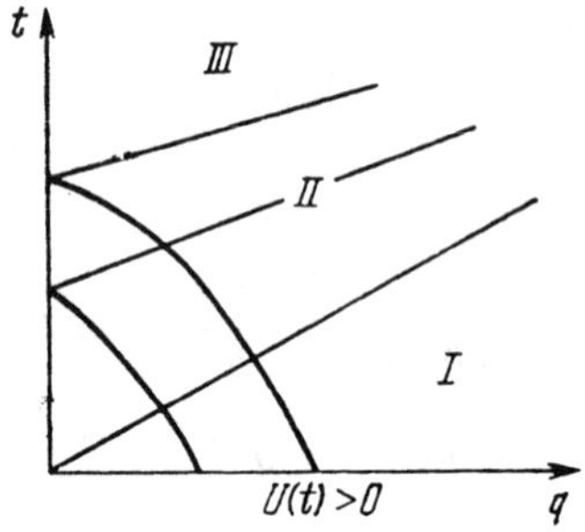

FIGURE 2.21

The boundary condition (7) can now always be satisfied, since for $U(t) > 0$ the inequality (13) is always satisfied. It is easy to see that $\partial r(q, t)/\partial q < 0$ for $U(t) > 0$, and the flow in zone II is hence a compression wave. Figure 2.21 shows the picture of the characteristics in the q, t plane. Since the r-characteristics intersect for $t > 0$, a classical solution of the piston problem in the case $U(t) > 0$ exists for a limited time only. After the intersection of two characteristics the solution of the piston problem becomes discontinuous.

We remark that, in general, if we consider any law of motion of the piston $u = U(t)$ and if $U'(t) > 0$ for some t, then discontinuities must arise in the solution of this problem. Thus, the complete solution of the piston problem in these cases is described by discontinuous solutions of the equations of gas dynamics which contain shock waves. We shall study the properties of shock waves in §4.

We further remark that if it is assumed that the piston begins to move into the gas with a finite speed $U(0) > 0$, then the solution is discontinuous for all $t > 0$. Therefore, in this case it makes no sense to consider the isentropic problem.

5. ***The problem with two pistons. Reflection and refraction of travelling waves on a contact boundary.*** We continue the qualitative study of the simplest flows with the example of these two problems, in which it is necessary to consider the interaction of two Riemann waves.

As we have already noted in subsection 1, in the case of a polytropic gas with an exponent of the adiabat $\gamma = 3$ the characteristics are straight lines in the plane of the Euler variables x, t along which the corresponding Riemann invariant is constant. This circumstance makes it possible to obtain a solution of the Cauchy problem and also of any well-posed boundary value problem in the case where no shock waves form in the solution. Suppose that a polytropic gas with exponent $\gamma = 3$ is in a state of rest ($u = 0$, $\rho = \rho_0$, $p = p_0$, $S = S_0$) at the initial time and is bounded on two sides by pistons having given trajectories of motion: $x = x_1(t)$ and $x = x_2(t)$ (Figure 2.22). We now use the Euler coordinates x, t.

It is obvious that in zone I we have a constant flow which coincides with the initial flow; in zones II and III we have Riemann waves, while zone IV is a zone of interference of the Riemann waves.

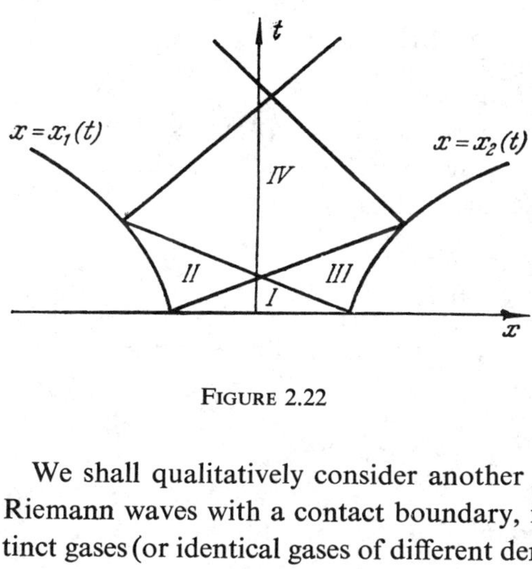

FIGURE 2.22

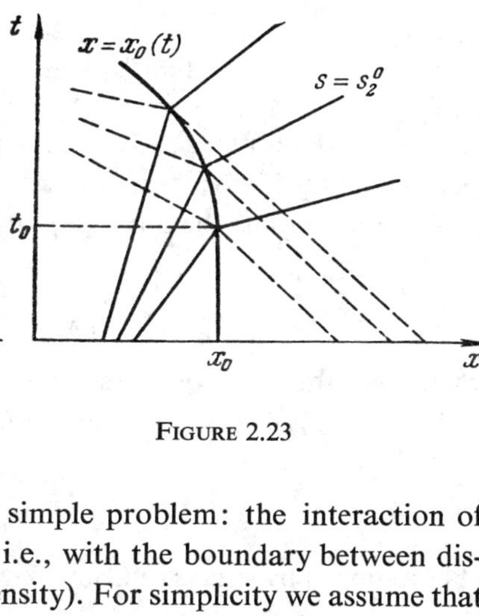

FIGURE 2.23

We shall qualitatively consider another simple problem: the interaction of Riemann waves with a contact boundary, i.e., with the boundary between distinct gases (or identical gases of different density). For simplicity we assume that both gases are polytropic: the gas to the left of the boundary $x = x_0$(Figure 2.23) has an exponent of the adiabat $\gamma = \gamma_1$, while the gas to the right has exponent $\gamma = \gamma_2$, i.e., we suppose that $p = A_1 \rho^{\gamma_1}$ to the left of $x = x_0$, while $p = A_2\rho^{\gamma_2}$ to the right. It is easy to see that for two neighboring gases the boundary between them is a trajectory, and therefore on the boundary of the two gases the conditions

$$u_- = u_+, \qquad p_- = p_+, \tag{1}$$

must be satisfied, where u_-, p_- and u_+, p_+ are the velocity and pressure in the gases to the left and right of the contact boundary. It will be shown in §4 that these equations follow from the laws of conservation of mass, momentum, and energy.

Denoting the Riemann invariants s and r to the left and right of the contact boundary by s_1, r_1 and s_2, r_2 respectively, we conclude that the continuity of the velocity on the contact boundary required by the first of the conditions (1) can be written

$$s_1 + r_1 = s_2 + r_2 = 2U(t), \tag{2}$$

where $U(t)$ is the velocity of the contact boundary. We consider a point $(x_0(t), t)$ on the contact boundary (Figure 2.23) after the arrival there of a Riemann r-wave from the gas to the left. Since the s-characteristics carry on the contact boundary $x = x_0(t)$ in the gas II the constant values of the invariant $s = s_2 = s_2^0$, the motion to the right of the contact boundary is obviously a Riemann r-wave ($s = s_2^0 = \text{const}$). Therefore, according to (3.2.11), the pressure on the contact boundary is determined by the relation

$$p_2 = p_0\left[1 + \frac{\gamma_2 - 1}{2}\frac{U(t)}{c_2^0}\right]^{2\gamma_2/(\gamma_2-1)}, \tag{3}$$

where p_0 is the initial pressure in the gases I and II and c_2^0 is the initial sound speed in the gas II (the initial velocity in gases I and II is equal to zero).

We shall compute the pressure p_1 in the gas I. Since

$$\frac{p_1}{p_0} = \left(\frac{c_1}{c_1^0}\right)^{2\gamma_1/(\gamma_1-1)},$$

where c_1^0 is the value of the sound speed in the gas I at the initial time and $c_1 = 4^{-1}(\gamma_1 - 1)(r_1 - s_1)$, it follows that

$$p_1 = p_0\left[\frac{\gamma_1 - 1}{4}\frac{r_1 - s_1}{c_1^0}\right]^{2\gamma_1/(\gamma_1-1)}. \tag{4}$$

Eliminating s_1 by means of (2), we obtain

$$p_1 = p_0\left[\frac{r_1}{r_1^0} - \frac{\gamma_1 - 1}{2}\frac{U(t)}{c_1^0}\right]^{2\gamma_1/(\gamma_1-1)}, \tag{5}$$

where $r_1^0 = 2(\gamma_1 - 1)^{-1}c_1^0$ is the value of the invariant r in the zone of constant flow of the gas I in front of the Riemann r-wave (we assume here that $U(0) = 0$). Equating the pressures p_1 and p_2, we obtain

$$\left[\frac{r_1}{r_1^0} - \frac{\gamma_1 - 1}{2}\frac{U(t)}{c_1^0}\right]^{2\gamma_1/(\gamma_1-1)} = \left[1 + \frac{\gamma_2 - 1}{2}\frac{U(t)}{c_2^0}\right]^{2\gamma_2/(\gamma_2-1)}. \tag{6}$$

If $r_1 = r_1(t)$ is a known function, then from this equation we could determine $U(t)$. Actually $r_1(t)$ is not known, since the interaction of the two Riemann waves to the left of the contact boundary (Figure 2.23) leads to a distortion of the r-characteristics. However, we can carry out a qualitative investigation of the dependence of $U(t)$ on $r_1(t)$ and qualitatively study the character of the flow.

Suppose, for example, a rarefaction r-wave is incident on the contact boundary. It is then easy to see from Figure 2.23 that $r_1(t)$ is a monotonically decreasing function of the variable t $(dr_1(t)/dt < 0)$. For $t = t_0$ (Figure 2.23) we have $r_1(t_0) = r_1^0$, and it therefore follows from (6) that $U(t_0) = 0$. Since we assume that $\gamma_1, \gamma_2 > 1$, the left side of (6) decreases with increasing $U(t)$, while the right side increases monotonically. This implies that if $r_1(t)$ is a monotonically decreasing function of the variable t, then $U(t) < 0$, and $U(t)$ also decreases monotonically.

The flow in the gas II can thus be considered as the motion caused by the withdrawal of a piston with speed $U(t) < 0$. Therefore, the Riemann r-wave in the gas II is a rarefaction wave. In an entirely similar way we can establish that if the incident wave is a compression wave, then $U'(t) > 0$ and a compression r-wave propagates in the gas II.

It is somewhat more involved to establish the character of an s-wave reflected from the contact boundary, i.e., the sign of $ds_1(t)/dt$. Differentiating (2) with respect to t, we obtain

$$\frac{ds_1(t)}{dt} = 2U'(t) - \frac{dr_1(t)}{dt}. \tag{7}$$

From this we conclude that if $2U'(t) > dr_1(t)/dt$, then the reflected wave is a rarefaction wave; otherwise it is a compression wave. For simplicity we consider only the case of identical gases $(\gamma_1 = \gamma_2 = \gamma)$ with different entropies, i.e., $c_1^0 \neq c_2^0$. Equations (6) can then be written in the form

$$\frac{\gamma - 1}{2} \frac{r_1(t) - U(t)}{c_1^0} = 1 + \frac{\gamma - 1}{2} \frac{U(t)}{c_2^0} \tag{8}$$

and

$$U(t) = \frac{r_1(t) - 2/(\gamma - 1)c_1^0}{1 + c_1^0/c_2^0} \quad \text{and} \quad U'(t) = \frac{1}{1 + c_1^0/c_2^0} r_1'(t).$$

From this we conclude that if $c_1^0/c_2^0 < 1$, then $ds_1(t)/dt$ has the same sign as $dr_1(t)/dt$; if $c_1^0/c_2^0 > 1$, then the signs of $ds_1(t)/dt$ and $dr_1(t)/dt$ are opposite.

For the same gas $(\gamma_1 = \gamma_2 = \gamma)$ the inequality $c_1^0 < c_2^0$ implies that the density ρ_1^0 of the gas I is greater than the density ρ_2^0 of the gas II, since $p = c^2\rho/\gamma$. We can therefore formulate the result obtained in the following way.

From a denser medium ($\rho_2^0 > \rho_1^0$, $c_2^0 < c_1^0$) *a rarefaction* (*compression*) *wave is also reflected as a rarefaction* (*compression*) *wave, and, conversely, from a less dense medium* ($\rho_2^0 < \rho_1^0$, $c_2^0 > c_1^0$) *a rarefaction* (*compression*) *wave is reflected as a compression* (*rarefaction*) *wave.*

We note the special case $c_2^0 = 0$ ($\rho_2^0 = \infty$) in which $U(t) = 0$. This case can be treated as the reflection of a Riemann wave from an infinitely dense gas or as the reflection from a rigid wall ($U(t) = 0$). We conclude that compression (rarefaction) waves are always reflected from a rigid wall as compression (rarefaction) waves.

Formula (6) enabled us to carry out a qualitative investigation of the problem of the refraction of Riemann waves on a contact boundary. A complete solution of this problem is difficult, since the function $r_1(t)$ is actually not known, because the r-characteristics are distorted in the zone of interaction of an incident Riemann wave with one reflected from the contact boundary.

In connection with this we note a special case where the solution of this problem reduces to the solution of an ordinary differential equation. We consider the case in which $\gamma_1 = 3$, while $\gamma_2 > 1$ is arbitrary. The characteristics in the gas I are then straight lines, and we may assume that $r_1(x, t) = f_1(x, t)$, where $f_1(x, t)$ is uniquely determined in the Riemann wave. Now in (6) $r_1(t)$ is the value of the invariant $r_1(x, t)$ on the contact boundary, i.e., at the point $(x_0(t), t)$:

$$r_1(t) = r_1(x_0(t), t) = f_1(x_0(t), t). \tag{9}$$

Because of the meaning of $U(t)$ we have

$$\frac{dx_0(t)}{dt} = U(t), \tag{10}$$

and since by (6) and (9) $U(t)$ can be expressed as a particular function of the variables t and $x_0(t)$, the solution of the entire problem reduces to integrating (10). As the initial condition for (10) we pose the condition (Figure 2.23)

$$x_0(t_0) = x_0.$$

Thus, for example, if $\gamma_1 = \gamma_2 = 3$ and a centered rarefaction wave is incident, then (10) assumes, according to (8), the form

$$\frac{dx_0(t)}{dt} = \frac{x_0(t)/t - c_1^0}{1 + c_0^0/c_2^1}, \qquad x_0(t_0) = c_1^0 t_0, \qquad t_0 > 0, \tag{11}$$

since in a centered rarefaction wave for $\gamma = 3$ we have $r(x, t) = x/t$.

6. ***Remarks on boundary conditions* for the *equations of gas dynamics and an illustration of their solvability for the example of the piston problem.*** In the analysis of the solvability of a problem with boundary conditions for the equations of gas dynamics the conclusions we obtained in considering this question for systems of quasilinear equations of hyperbolic type in §11 of Chapter 1 are fully applicable. Nevertheless, in studying the motion of gases and liquids certain classes of boundary conditions are especially natural and therefore especially important. We shall therefore consider here typical boundary conditions of the equations of gas dynamics.

It is most natural to pose boundary conditions on the boundary of a distinguished volume of gas, i.e., on a trajectory. In Lagrange coordinates, to a trajectory there corresponds a fixed coordinate q; boundary conditions of this type in Lagrange coordinates are therefore posed on lines $q = \text{const}$.

It is possible to distinguish two types of boundary conditions:

1. External boundary conditions or conditions on external boundaries. These are properly boundary conditions which effectively describe the action of an external medium on a given volume of gas. For the one-dimensional flow we consider such conditions are conditions on the left and right boundaries of the region containing the gas.

2. Internal boundary conditions or conditions on internal boundaries. These include the conditions on contact boundaries between gases possessing different properties (different entropies, different equations of state, etc.). Conditions on lines of discontinuity of the solution also belong to the class of internal boundary conditions. These conditions will be discussed in detail in the next section.

As concerns conditions on contact boundaries, in gas dynamics they are always the same and require the continuity of the flow velocity u and the pressure p. An example of the use of these conditions is given in subsection 5, where the problem of the reflection of Riemann waves from a contact boundary is qualitatively studied.

We shall consider external boundary conditions. In the case of the piston problem considered in subsection 4 the boundary condition is posed at the point $q = 0$. It is easy to see that if on the piston the pressure

$$p(0, t) = p(t) \geq 0 \tag{1}$$

is prescribed, then this boundary value problem is uniquely solvable, since the condition (1) implies that $c \geq 0$. In the case where the piston speed is prescribed,

$$u(0, t) = U(t), \tag{2}$$

we see that for $U(t) < 0$ it is possible for the gas to break away from the piston, i.e., the condition (2) is not satisfied and is replaced by the condition $p = 0$.

This situation also occurs for systems of linear equations. If boundary conditions are posed outside the domain of dependence of the initial data, then these conditions generally do not uniquely determine the solution.

In the case of the piston problem the physical formulation of the problem made it possible to obtain a correct solution by replacing, where necessary, the condition (2) by the condition for a "free boundary" $p = 0$.

On the other hand, for $U(t) > 0$ the piston problem with the boundary condition (2) is always solvable, although the solution is discontinuous. This conclusion should be compared with the case of a system of linear equations, for which a boundary value problem is generally not solvable either in the class of smooth or discontinuous solutions if the boundary condition is posed in the domain of determinacy of the solution of the Cauchy problem.

§4. Discontinuities in the one-dimensional flow of compressible gases. Shock waves

1. *The Hugoniot conditions.* In the examples of the simplest flows in §3 we have seen that, as a rule, solutions of the equations of gas dynamics remain continuous only for a limited time before then discontinuities arise in the solution. It is natural that the differential equations lose their meaning for discontinuous flows; however, as we have mentioned above, the integral laws of conservation of mass, momentum, and energy remain meaningful for discontinuous flows. We shall derive the conditions which must be satisfied on lines of discontinuity of solutions of the equations of gas dynamics as a consequence of the integral conservation laws.

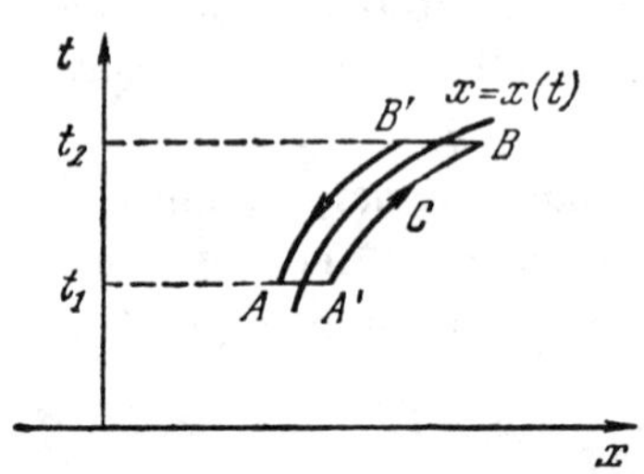

FIGURE 2.24

Suppose that $x = x(t)$ is the equation of one of the lines of discontinuity of the hydrodynamic variables which on the segment $t_1 \leq t \leq t_2$ in question we assume to have a continuous tangent (Figure 2.24). Suppose that $f(x, t)$ suffers a discontinuity on the line $x = x(t)$. We set

$$f_1(t) = f(x(t) - 0, t); \qquad f_2(t) = f(x(t) + 0, t); \qquad [f] = f_2(t) - f_1(t). \qquad (1)$$

The integral conservation laws in Euler coordinates (2.3.13)–(2.3.15) have the form

$$\left.\begin{aligned} &\oint_C \rho x^\nu\,dx - \rho u x^\nu\,dt = 0,\\ &\oint_C \rho u x^\nu\,dx - (p + \rho u^2)x^\nu\,dt = -\iint_{G_C} \nu p x^{\nu-1}\,dx\,dt,\\ &\oint_C \rho\left(\varepsilon + \frac{u^2}{2}\right)x^\nu\,dx - \rho u\left(\varepsilon + \frac{p}{\rho} + \frac{u^2}{2}\right)x^\nu\,dt = 0. \end{aligned}\right\} \tag{2}$$

We write the conservation laws (2) for the contour $AA'BB'$ assuming that the lines $A'B$ and $B'A$ are infinitely close to the line of discontinuity $x(t)$ from the right and left respectively.

In view of the boundedness of all the hydrodynamic variables the integrals over the parts AA' and BB' of the contour C vanish, as does also the double integral $\iint_{G_C} \nu p x^{\nu-1}\,dx\,dt$. Along the line $x = x(t)$ we have $dx = Ddt$, where $D = D(t) = x'(t)$. Therefore, from the first of the equations (2), for example, we obtain

$$\int_{t_1}^{t_2} x^\nu(t)\,\{(\rho_2(t) - \rho_1(t))\,D(t) - (\rho_2(t)u_2(t) - \rho_1(t)u_1(t))\}\,dt = 0. \tag{3}$$

Since the limits of integration in (3) are arbitrary, the integrand must be equal to zero, i.e.,

$$x^\nu(t)\{D(t)[\rho] - [\rho u]\} = 0.$$

Cancelling x^ν, we see that the conditions on the lines of discontinuity are the same for all three symmetries $\nu = 0, 1, 2$. Proceeding analogously with all the conservation laws (2), on the line of discontinuity $x = x(t)$ we obtain the conditions

$$D[\rho] = [\rho u], \tag{4}$$

$$D[\rho u] = [p + \rho u^2], \tag{5}$$

$$D\left[\rho\left(\varepsilon + \frac{u^2}{2}\right)\right] = \left[\rho u\left(\varepsilon + \frac{p}{\rho} + \frac{u^2}{2}\right)\right]. \tag{6}$$

which relate the jumps of the hydrodynamic variables on the line of discontinuity $x = x(t)$ and the speed $D = x'(t)$ of the line of discontinuity. The relations (4)–(6) are called the *conditions for hydrodynamic compatibility of the discontinuity*, or the *Hugoniot conditions* in honor of the French scientist who first obtained them.[8]

According to the notation, $D = D(t)$, and hence $D[f] = [Df]$. Hence, (4) can be rewritten in the form

$$[\rho(u - D)] = 0. \tag{7}$$

Multiplying (7) by D and subtracting it from (5), we obtain

$$[p + \rho(u - D)^2] = 0. \tag{8}$$

Finally, mutiplying (8) by D, subtracting (6), and noting that $[D] = [D^2] = 0$, we obtain

$$\left[\rho(u - D)\left(\varepsilon + \frac{p}{\rho} + \frac{(u - D)^2}{2}\right)\right] = 0. \tag{9}$$

Recalling the notation (1), we can rewrite the Hugoniot conditions (7)–(9) as the equalities

$$\rho_2(u_2 - D) = \rho_1(u_1 - D) = m, \tag{10}$$

$$p_2 + \rho_2(u_2 - D)^2 = p_1 + \rho_1(u_1 - D)^2 = j, \tag{11}$$

$$\begin{aligned}\rho_2(u_2 - D)\left(\varepsilon_2 + \frac{p_2}{\rho_2} + \frac{(u_2 - D)^2}{2}\right) \\ = \rho_1(u_1 - D)\left(\varepsilon_1 + \frac{p_1}{\rho_1} + \frac{(u_1 - D)^2}{2}\right) = f.\end{aligned} \tag{12}$$

According to §2.2 the quantities m, j, and f denote respectively the flux of mass, momentum, and energy computed in a coordinate system moving with speed D relative to the system in which the velocity of the flow u is computed. Therefore, the Hugoniot conditions (10)–(12) require the continuity of the flux of mass, momentum, and energy on the line of discontinuity of the hydrodynamic variables.

We remark, finally, that the integral conservation laws written in Lagrange coordinates lead to the same Hugoniot conditions (4)–(6) if the connection of the Lagrange coordinates with the Euler coordinates is considered.

2. *Various types of discontinuities: shock waves and contact discontinuities. Various forms of the Hugoniot conditions. The Hugoniot adiabat.* We distinguish discontinuities of a solution according to whether $m = 0$ or $m \neq 0$. If $m(t) = m = 0$ we call this type of discontinuity a *contact discontinuity*; if $m(t) \neq 0$ we call the discontinuity a *shock wave*.

[8] In the American literature these conditions are frequently called the Rankine conditions or the Rankine-Hugoniot conditions.

In the case of a contact discontinuity it follows from (4.1.10) that

$$D = u_1 = u_2 = x'(t), \tag{1}$$

i.e., the line of discontinuity coincides with the trajectory of the particle (in Lagrange coordinates a contact discontinuity is therefore represented by a line $q = \text{const}$).

Setting $u_1 = D$ and $u_2 = D$, from (4.1.11) we obtain

$$p_1 = p_2 = j, \tag{2}$$

while condition (4.1.12) for $u_1 = u_2 = D$ is satisfied identically. Thus, on a contact discontinuity the following two conditions are satisfied:

$$u_1 = u_2, \qquad p_1 = p_2, \tag{3}$$

i.e., the pressure and velocity of the flow are continuous. It is easy to demonstrate the converse: if on a discontinuity the conditions (3) are satisfied, then the discontinuity is a contact discontinuity.

The quantities ρ, ε and S may experience an arbitrary jump on a contact discontinuity while satisfying, however, the continuity condition for the pressure (2). A contact discontinuity may, in particular, form the boundary separating two distinct gases satisfying different equations of state. Here the continuity conditions of the velocity and pressure (3) may be considered internal boundary conditions on the boundary separating the different gases.

In the case of a shock wave we have $m \neq 0$, and matter passes across the line of discontinuity $x = x(t)$. In the case $m > 0$ matter crosses the line of discontinuity from left to right; for $m > 0$ we therefore say that the shock wave moves from right to left relative to the matter; conversely, for $m < 0$ we say that the shock wave moves to the right.

We shall consider various representations of the Hugoniot conditions (4.1.10)–(4.1.12) for the case of a shock wave. The condition (4.1.11) of continuity for the flux of momentum can be written as the following equivalent equalities:

$$\left.\begin{aligned} p_2 + m(u_2 - D) &= p_1 + m(u_1 - D), \\ p_2 + V_2 m^2 &= p_1 + V_1 m^2, \\ m^2 &= \frac{p_2 - p_1}{V_1 - V_2}, \end{aligned}\right\} \tag{4}$$

where $V = 1/\rho$. After division by m, (4.1.12) acquires the form

$$\varepsilon_2 + p_2 V_2 + \frac{(u_2 - D)^2}{2} = \varepsilon_1 + p_1 V_1 + \frac{(u_1 - D)^2}{2}. \tag{5}$$

If we denote by the letter $\mathcal{T}$ the quantity $\varepsilon + pV$, which is called the *enthalpy*, then (5) can be written in the form

$$\mathcal{T}_2 + \frac{(u_2 - D)^2}{2} = \mathcal{T}_1 + \frac{(u_1 - D)^2}{2}. \tag{6}$$

From (4.1.10) we have

$$(u_2 - D)(u_1 - D) = \frac{m^2}{\rho_1 \rho_2} = m^2 V_1 V_2. \tag{7}$$

Substituting the third formula of (4), we obtain

$$(u_2 - D)(u_1 - D) = \frac{p_2 - p_1}{V_1 - V_2} V_1 V_2 = \frac{p_2 - p_1}{\rho_2 - \rho_1}. \tag{8}$$

According to (4.1.10),

$$\begin{aligned} (u_1 - D) &= mV_1; \qquad (u_2 - D) = mV_2; \\ (u_1 - D)^2 &= m^2 V_1^2; \qquad (u_2 - D)^2 = m^2 V_2^2. \end{aligned} \tag{9}$$

Substituting the third formula of (4), we obtain

$$(u_1 - D)^2 = \frac{p_2 - p_1}{V_1 - V_2} V_1^2, \qquad (u_2 - D)^2 = \frac{p_2 - p_1}{V_1 - V_2} V_2^2. \tag{10}$$

We note several more useful formulas. From (9) we have

$$u_2 - u_1 = [(u_2 - D) - (u_1 - D)] = m(V_2 - V_1).$$

Therefore,

$$\left.\begin{aligned} (u_2 - u_1)(u_1 - D) &= m^2 V_1 (V_2 - V_1) = (p_1 - p_2) V_1, \\ (u_2 - u_1)(u_2 - D) &= m^2 V_2 (V_2 - V_1) = (p_1 - p_2) V_2, \end{aligned}\right\} \tag{11}$$

and, finally,

$$\left.\begin{aligned} (u_2 - u_1)^2 &= (p_2 - p_1)(V_1 - V_2), \\ (u_2 - D)^2 - (u_1 - D)^2 &= (p_1 - p_2)(V_1 + V_2). \end{aligned}\right\} \tag{12}$$

From these formulas it is possible to draw several conclusions.

1) It follows from (4) that on passing through a shock front the pressure increases or decreases simultaneously with the density.

2) From (7) it follows that the differences $(u_2 - D)$ and $(u_1 - D)$ have the same sign.

3) For finite m the differences $(V_2 - V_1)$ and $(p_2 - p_1)$ have the same order, so that as $V_2 - V_1 \to 0$ also $p_2 - p_1 \to 0$.

The equalities (10) express the relative velocities $u_2 - D$ and $u_1 - D$ in terms of thermodynamic quantities. Therefore, substituting them into (5) and (6), we obtain relations containing only thermodynamic quantities:

$$\varepsilon_2 - \varepsilon_1 = \frac{1}{2}(p_2 + p_1)(V_1 - V_2), \tag{13}$$

$$\mathcal{T}_2 - \mathcal{T}_1 = \frac{1}{2}(p_2 - p_1)(V_2 + V_1). \tag{14}$$

Equality (13) or (14) is called the *equation of the Hugoniot adiabat*.

We introduce the function

$$H(p, V; p_0, V_0) = \varepsilon(p, V) - \varepsilon(p_0, V_0) + (V - V_0)\frac{p + p_0}{2}, \tag{15}$$

which we consider as a function of the two variables p and V which depends parametrically on p_0 and V_0.

Let $M_1 = \{p_1, V_1\}$ and $M_2 = \{p_2, V_2\}$ be points of the p, V plane characterizing the thermodynamic state of a substance on different sides of a line of discontinuity. By (13) we have

$$\begin{aligned} H(M_2, M_1) &= H(p_2, V_2; p_1, V_1) = -H(M_1, M_2) \\ &= -H(p_1, V_1; p_2, V_2) = 0. \end{aligned} \tag{16}$$

Points M_1 and M_2 connected by (16) are called *conjugate points*. The property of conjugacy is not transitive, since $H(M_0, M_1) = 0$ and $H(M_1, M_2) = 0$ do not imply $H(M_0, M_2) = 0$.

We fix a point $M_0(p_0, V_0)$ in the p, V plane, and we consider the set of points $M(p, V)$ conjugate to M_0. They must lie on the curve

$$H(M, M_0) = H(p, V, p_0, V_0) = 0. \tag{17}$$

The curve defined by (17) is called the *Hugoniot adiabat* with center at $M_0(p_0, V_0)$. According to (13), the Hugoniot adiabat is the geometric locus of points (p, V) characterizing the thermodynamic state of a substance on one side of a front of discontinuity (a shock wave) if the state p_0, V_0 on the other side of the front is given.

Suppose that the Hugoniot adiabat $H(M, M_0)$ passes through the point M_1. Then the Hugoniot adiabat $H(M, M_1)$ passes through M_0 but does not coincide with $H(M, M_0)$ (Figure 2.25). This circumstance reflects the fact that the Hugoniot adiabat is not a level line of a function of two variables but is a level line of a function of two variables depending also on two parameters. Therefore, if we choose points (p, V) conjugate to (p_0, V_0) as new centers of

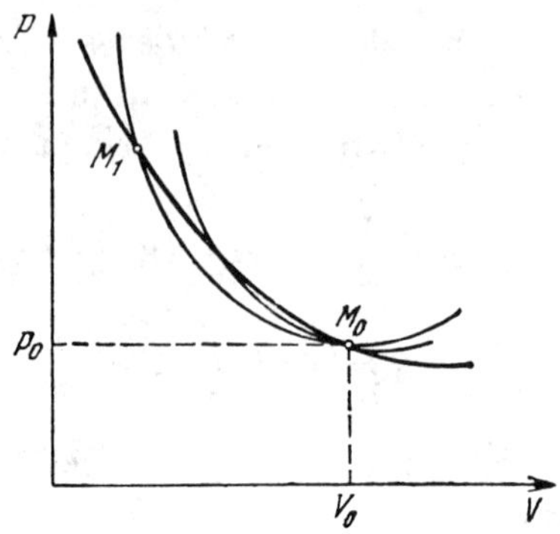

FIGURE 2.25

Hugoniot adiabats, we obtain a one-parameter family of Hugoniot adiabats passing through (p_0, V_0) (a pencil of adiabats).

3. *The Hugoniot adiabat for a normal gas.* Our previous remarks regarding the Hugoniot adiabat pertained to a substance with an arbitrary equation of state. To study the Hugoniot adiabat in more detail we assume that the equations of state of the substance $p = p(V, S)$, $\varepsilon = \varepsilon(V, T)$ satisfy the conditions formulated in §1.4 for a normal gas:

$$\left.\begin{array}{ll} \text{I.} & \dfrac{\partial p(V, S)}{\partial V} < 0, \\ \text{II.} & \dfrac{\partial^2 p(V, S)}{\partial V^2} > 0, \\ \text{III.} & p(V, S) \to \infty \quad \text{as } V \to 0, \\ \text{IV.} & \dfrac{\partial p(V, S)}{\partial S} > 0, \\ \text{V.} & c_V = \dfrac{\partial \varepsilon(V, T)}{\partial T} > 0. \end{array}\right\} \tag{1}$$

Condition VI is the requirement that the domain of the variables p and V in which conditions I–V are satisfied be convex. All our subsequent considerations, except where specially mentioned, pertain to a normal gas.

We consider the total differential dH of the function $H(p, V, p_0, V_0)$ of two variables p and V assuming p_0 and V_0 to be fixed:

$$dH = d\varepsilon + \frac{p + p_0}{2}\, dV + \frac{V - V_0}{2}\, dp. \tag{2}$$

Using the basic thermodynamic relation $d\varepsilon = TdS - pdV$, we obtain the following expression for dH:

$$dH = TdS + \frac{V - V_0}{2}\,dp - \frac{p - p_0}{2}\,dV. \tag{3}$$

Equation (3) can also be written in the form

$$dH = TdS + \frac{(V - V_0)^2}{2}\,dK, \tag{4}$$

where

$$K = \frac{p - p_0}{V - V_0} \tag{5}$$

is the slope of the ray passing through the center $M_0(p_0, V_0)$. According to (4.2.4),

$$K = -m^2, \tag{6}$$

where m is the flux of mass through the shock front separating the states p_0, V_0 and p, V (here $H(p, V, p_0, V_0) = 0$).

We consider the relative positions of the following curves: a) the Poisson adiabat given by the equation $dS = 0$ passing through $M_0(p_0, V_0)$, which we denote by the letter A; b) the Hugoniot adiabat $dH = 0$ passing through $M_0(p_0, V_0)$, which we denote by H. According to (3) and (4), for $V = V_0$ and $p = p_0$ we have $dH = TdS$. Thus, the curves A and H have a common tangent at the point of their intersection $M_0(p_0, V_0)$.

We shall consider the behavior of the differentials dH, dS, and dK on each of these curves.

a) On the adiabat A we have $dS = 0$. According to (4).

$$dH = \frac{(V - V_0)^2}{2}\,dK. \tag{7}$$

According to properties I and II, along A (i.e., for $S = \text{const}$)

$$dK/dV > 0, \tag{8}$$

and therefore on the adiabat A

$$dH/dV > 0. \tag{9}$$

Hence,

$$\left.\begin{aligned} &H < 0 \quad \text{on the upper part of the adiabat } A\ (V < V_0), \\ &H > 0 \quad \text{on the lower part of the adiabat } A\ (V > V_0). \end{aligned}\right\} \tag{10}$$

b) On the ray $dK = 0$

$$dH = TdS. \tag{11}$$

This equality implies that along the ray K the signs of dH and dS coincide, and, in particular, the stationary points of H and S coincide.

It was shown in §1.4 that on the ray $K = \text{const} > 0$ the entropy S has no stationary points and increases monotonically with increasing p; on the ray $K < 0$ the entropy S has a unique stationary point at which S achieves a maximum. The same is therefore valid for the function H as well, so that along the ray $K = \text{const} > 0$ H has no stationary points; on the ray $K < 0$ H has a unique maximum at the same point as S. This implies that on each ray $K = \text{const} < 0$ there exists a unique point M_H of the Hugoniot adiabat H lying between the point M_A of the Poisson adiabat and the center M_0.

Thus, there exists a curve H, the Hugoniot adiabat, passing through M_0. The smoothness of the Hugoniot adiabat follows from the fact that it is an integral curve of the ordinary differential equation (2) passing through M_0.

c) On the adiabat H the relation (4) implies that

$$TdS = -\frac{(V - V_0)^2}{2} dK. \tag{12}$$

This implies that if the point $M(p, V)$ moves along the adiabat H so that the ray $K(M, M_0)$ moves clockwise ($dK < 0$), then the entropy S increases monotonically.

Summarizing our conclusions regarding the behavior of the differentials dH, dS and dK on the curves A, K and H, we conclude that in a neighborhood of M_0 where the adiabats intersect tangentially the curves A and H are positioned as shown in Figure 2.26.

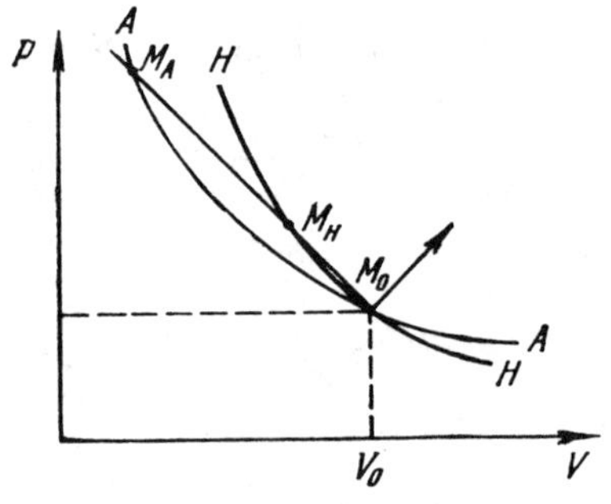

FIGURE 2.26

The relative position of the adiabats A and H in a neighborhood of the center M_0 indicates that the tangency of A and H must have no less than second order. We shall verify this computationally. By (3), along H we have

$$TdS = -\frac{V - V_0}{2}\,dp + \frac{p - p_0}{2}\,dV. \tag{13}$$

Differentiating with respect to V, we get

$$\begin{aligned} dTdS + Td^2S &= -\frac{1}{2}\,dV\,dp - \frac{V - V_0}{2}\,d^2p + \frac{1}{2}\,dp\,dV \\ &= -\frac{V - V_0}{2}\,d^2p, \end{aligned} \tag{14}$$

i.e., $d^2S = 0$ at the point (p_0, V_0), and the tangency of A and H has second order. Finally, differentiating (14) again with respect to V, we find that

$$d^2TdS + 2dTd^2S + Td^3S = -\frac{1}{2}\,dVd^2p - \frac{V - V_0}{2}\,d^3p.$$

Since at M_0 we have $dS = d^2S = 0$, we get

$$Td^3S = -\frac{1}{2}\,dVd^2p \quad \text{and} \quad \frac{d^3S}{dV^3} = -\frac{1}{2T}\,\frac{d^2p}{dV^2} < 0$$

at M_0 according to property II of the equation of state. Thus, at the point M_0 of the curve H only the third derivative d^3S/dV^3 is nonzero.

It is not hard to see that in a neighborhood of the center $M_0(p_0, V_0)$ the following properties of the adiabat H hold.

1) For any point M of H

$$K = \frac{p - p_0}{V - V_0} < 0. \tag{15}$$

2) No ray M_0M_H is tangent to H for $M_H \neq M_0$.

3) Each ray M_0M_H intersects H in no more than one point $M_H \neq M_0$.

4) Each ray intersecting the upper branch of the adiabat A also intersects the upper branch of H. Here on M_0M_H the functions H and S have contours of the form shown in Figure 2.27.

We remark that properties 2) and 3) do not imply the convexity of H, since these properties are valid only for rays issuing from a specific point of the adiabat, namely, its center.

We prove that these properties of H hold not only locally but also globally. Let M_H be any point of H. Since $H(M_H, M_0) = H(M_0, M_0) = 0$, it follows

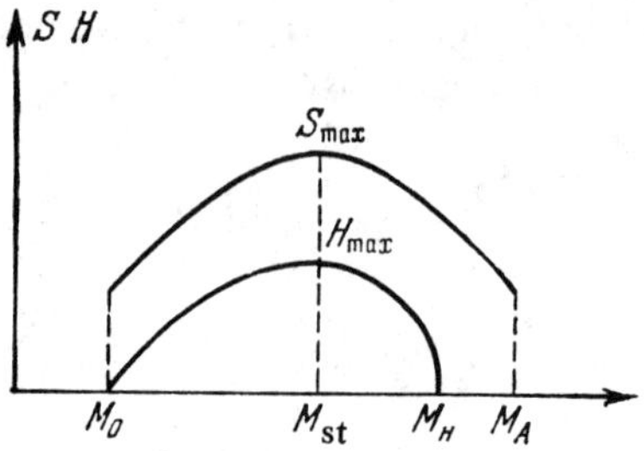

FIGURE 2.27

that on M_0M_H there is a stationary point of the function H (and hence by (11) of S as well). Therefore, $K < 0$, and property 1) of H has been proved.

We prove property 2). The stationary point M_{st} lies strictly inside the interval M_0M_H and is the only stationary point. Tangency of H with the ray K would imply the existence on K of two stationary points of H and hence two stationary points of S, which is impossible. Property 2) is proved.

The existence of a ray intersecting H at two points distinct from M_0 necessarily implies the existence of a ray tangent to H at a point distinct from M_0. This is impossible by property 2), and so property 3) is proved.

Finally, since $S(M_H) > S(M_0)$ for $p_H > p_0$, any ray intersecting the upper part of H first intersects H at M_H and then intersects A at M_A. Property 4) is proved.

In particular, it follows from this that the slope of a ray K intersecting H at the point $M_H \neq M_0$ is within the limits

$$-\infty < K = \frac{p - p_0}{V - V_0} < \frac{\partial p(V_0, S_0)}{\partial V} < 0. \qquad (16)$$

We now consider the following problem. The state on one side of a line of discontinuity (a shock wave) is given. It is characterized by the parameters $u = u_0$, $V = V_0$, $p = p_0$ and $\varepsilon = \varepsilon_0 = \varepsilon(p_0, V_0)$. The flux of mass m across the discontinuity front is also given. It is required to determine the state (u, V, p, ε) on the other side of the line of discontinuity on the basis of the Hugoniot conditions.

We assume that $K = -m^2$ satisfies (16). We shall show that for a normal gas this problem always has a unique solution.

The thermodynamic parameters p, V of the gas are determined by the point of intersection of the Hugoniot adiabat $H(p, V, p_0, V_0)$ with the ray $K = (p - p_0)/(V - V_0) = -m^2$. Since K satisfies condition (16), by properties 1)–4) of the adiabat H the point (p, V) exists and is uniquely determined.

Thus, the thermodynamic parameters p, V and ε are uniquely determined. After this all the remaining parameters of the shock wave are uniquely determined. The differences $u_0 - D = mV_0$ and $u - D = mV$ are determined according to (4.2.9). Since u_0 is known, this determines the speed D of the shock wave. Hence, under the condition (16) the problem posed has a unique solution.

We consider the limit case in which $K = -m^2 = \partial p(V_0, S_0)/\partial V$. For this value of K the ray is tangent to the Hugoniot adiabat (and simultaneously to the Poisson adiabat) at the point $M_0(p_0, V_0)$. The conjugate point $M(p, V)$ characterizing the state of matter on the other side of the front coincides with $M_0(p_0, V_0)$, i.e., the shock wave is infinitely weak. In this case $u = u_0$, and by (4.2.8) we have

$$(u - D)^2 = \frac{\partial p(\rho_0, S_0)}{\partial \rho} = c_0^2, \quad \text{i.e., } u - D = \pm c_0 = \pm c.$$

Thus, an infinitely weak shock wave moves through matter with the speed of sound, i.e., it is a weak discontinuity. For $K = -\infty$ a shock wave is called *infinitely strong.*

4. ***Stable and unstable discontinuities. Conditions for stability and the Zemplén theorem.*** Suppose that the quantities $u_1, p_1, V_1, \varepsilon_1, u_2, p_2, V_2, \varepsilon_2$, and the speed D of the shock wave satisfy the Hugoniot conditions (4.1.10)–(4.1.12). It is easy to see that the Hugoniot conditions do not change whether we consider the state u_1, p_1, V_1, ε_1 as the state of the gas to the left of the front of the shock wave and the state u_2, p_2, V_2, ε_2 as the state to the right or, conversely, u_1, p_1, V_1, ε_1 the state to the right and u_2, p_2, V_2, ε_2 the state to the left. However, as we shall now see, these cases are basically different in their physical interpretation to the extent that one of them should even be conceded impossible.

We begin by supposing that the mass flux $m = \rho_1(u_1 - D) = \rho_2(u_2 - D)$ has a particular sign. Suppose, for example, that $m < 0$. In this case the wave moves to the right relative to the matter, and hence in the process of motion matter intersects the front of the shock wave while moving from right to left relative to the front.

As we mentioned in §1, we conceive of a surface of discontinuity of the flow as a narrow zone of large gradients in which the effect of the dissipative forces of viscosity and thermal conductivity is considerable. It is known from thermodynamics that the effect of these processes leads to an increase in the entropy;

this characterizes the irreversibility of processes with viscosity and thermal conductivity. Since for $m < 0$ the particles of gas in the process of the motion pass from right to left relative to the front of the shock wave, it is obvious that, due to the irreversibility of the processes occuring in the narrow zone which we associate with the shock wave, we should require that $S_r < S_l$, where S_r and S_l are the entropy of the gas to the right and left of the shock wave respectively.

The inequality

$$S_r < S_l \qquad (m < 0) \tag{1}$$

makes it impossible to interchange the states $u_1, p_1, V_1, \varepsilon_1$ and $u_2, p_2, V_2, \varepsilon_2$ and, to the contrary, specifies a particular position of these states relative to the front.

For $m < 0$ we call a shock wave for which condition (1) is satisfied a *stable discontinuity*; if condition (1) is violated, we call such a discontinuity *unstable*.

Altogether analogous arguments lead us to the conclusion that for $m > 0$

$$S_l < S_r \qquad (m > 0). \tag{2}$$

We call (1) and (2) the *stability conditions for a shock wave*.

If we agree to call the state ahead of the front of a shock wave the state to the right of it if $m < 0$ and to the left of it if $m > 0$ and the second state the state behind the front of the shock wave, then inequalities (1) and (2) require that the entropy of the gas behind the front be greater than the entropy of the gas ahead of the wave front.

We shall denote the state ahead of the front by the letters $u_0, p_0, V_0, \varepsilon_0$ and the state behind the front by u, p, V, ε; the stability conditions (1) and (2) can then be given by the single inequality

$$S > S_0. \tag{3}$$

By a shock wave we shall henceforth mean only a stable shock wave, i.e., a discontinuity satisfying the Hugoniot conditions and condition (3).[9]

We return to the problem of determining the state on one side of the front if the state on the other side and the mass flux m are given; this problem was considered in the preceding subsection. If $u_0, p_0, V_0, \varepsilon_0$ is the state ahead of the front of the shock wave, then the problem of determining the state behind the front which satisfies the stability condition (3) is uniquely solved if the following condition is satisfied:

[9] We note that if the equations of state of the gas do not satisfy conditions (4.3.1) for a normal gas, then inequality (3) is insufficient to specify the stability of a shock wave. This question is briefly discussed in subsection 8.

$$-m^2 = K = \frac{p - p_0}{V - V_0} < \frac{\partial p}{\partial V}(V_0, S_0). \tag{4}$$

For a normal gas the stability condition (3) entails the following consequences.

1. To each value of D and each state ahead of the front $u_0, p_0, V_0, \varepsilon_0$ there corresponds one and only one state behind the front, u, p, V, ε if $|D - u_0| > c_0$.

Indeed, $m = \rho_0(u_0 - D)$ and $m^2 > \rho_0^2 c_0^2 = -\partial p(V_0, S_0)/\partial V$. This implies (4.3.16) and the validity of the assertion.

2. As $|D - u_0|$ increases from c_0 to ∞ the entropy behind the front increases monotonically.

3. Shock waves lead only to a compression of matter and an increase in the pressure.

Indeed, on the upper branch of the adiabat H corresponding to the states behind the front, we have $S > S_0, p > p_0$ and $V < V_0$, i.e. $\rho > \rho_0$.

4. A shock wave moves with supersonic speed through the gas ahead of the front and with subsonic speed in the medium behind the front.

This assertion (Zemplén's theorem) can be written as

$$|u_0 - D| > c_0, \qquad |u - D| < c. \tag{5}$$

As we have seen, the first inequality is equivalent to the condition $S > S_0$. In subsection 3 it was shown that

$$-\frac{p - p_0}{V - V_0} < -\frac{\partial p}{\partial V} = \rho^2 c^2 \tag{6}$$

for p and V lying to the right of the Posson adiabat A. Since, according to subsection 3, M_H lies to the right of A, inequality (6) is satisfied. Therefore,

$$\frac{p - p_0}{V - V_0} = -m^2 = -\rho^2(u - D)^2 > -\rho^2 c^2.$$

This implies that $|u - D| < c$, as we wished to prove.

We now observe that it is also possible to consider the problem of determining the state $u_0, p_0, V_0, \varepsilon_0$ ahead of the front on the basis of a given state u, V, p, ε behind the front and the mass flux m (or the speed D) through the front of the discontinuity. In this case the stability condition (3) can be replaced by the equivalent inequality

$$K = \frac{p - p_0}{V - V_0} > \frac{\partial p(V, S)}{\partial V},$$

while the point (p, V) (if it exists) is unique and lies on the lower branch of the Hugoniot adiabat with center at the point $M(p, V)$. Recalling that the speeds ξ_1, ξ_2, ξ_3 of the characteristics are

$$\xi_1 = u - c, \qquad \xi_2 = u, \qquad \xi_3 = u + c, \tag{7}$$

we can also write (5) in the form

$$\xi_{1l} > D > \xi_{1r} \quad \text{for } D < u_r, \tag{8}$$

$$\xi_{3l} > D > \xi_{3r} \quad \text{for } D > u_l, \tag{9}$$

where ξ_l and ξ_r are, respectively, the speeds of the characteristics to the left and right of the discontinuity.

Finally, we note further that the stability conditions for the discontinuity (1), (2), or (3) are equivalent to the inequality

$$u_l > u_r. \tag{10}$$

Indeed, suppose, for example, that $m < 0$. Then according to (4.2.10)

$$u_l - u_r = u - u_0 = m(V - V_0).$$

Since $V > V_0$, this implies (10). The case $m > 0$ is treated in the same way.

The inequalities (8)–(10) make it possible to represent schematically the behavior of the characteristics and flow lines in a neighborhood of a line of discontinuity. Figures 2.28, a and b show the relative positions of the line of discontinuity $x = x(t)$, the flow lines $dx/dt = u$, and the characteristics $dx/dt = \xi_1 = u - c$ and $dx/dt = \xi_3 = u + c$ in the region to the left and right of a shock front in Euler coordinates for the case of a shock wave propagating in the gas to the right ($m < 0$) and to the left ($m > 0$). Figure 2.29 shows the relative positions of the corresponding curves in the plane of the Lagrange coordinates q, t for the case $m < 0$. In this figure $q = q(t)$ is the line of discontinuity, $q =$ const is the flow line, and $dq/dt = \pm\rho c$ are the characteristics of the first and third families.

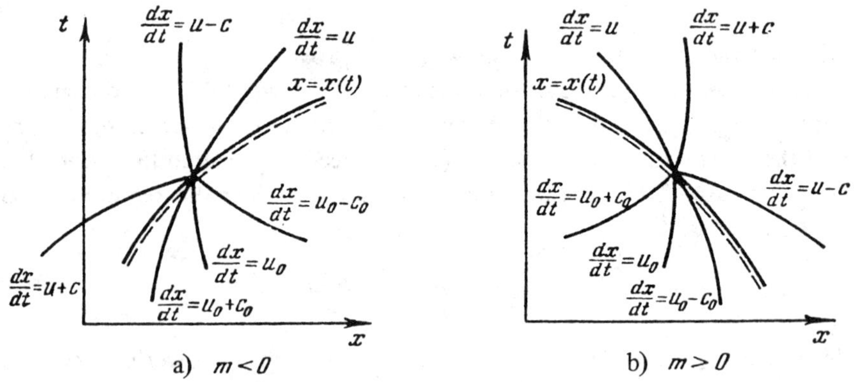

FIGURE 2.28

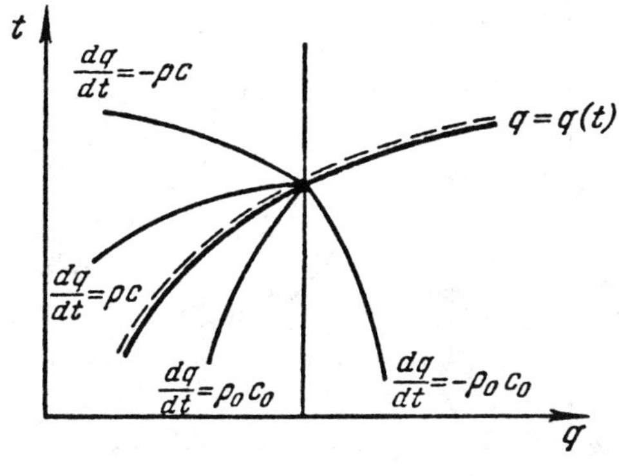

FIGURE 2.29

We note one characteristic circumstance which follows from the picture of the relative positions of the line of discontinuity and the characteristics shown in Figures 2.28 and 2.29. At each point of the line of discontinuity $x = x(t)$ from below (from the region of smaller values of the time t) there arrive four characteristics, while only two characteristics leave it (those defined for larger values of t); moreover, one of the latter is the flow line. In the case $m < 0$ two incoming characteristics of the third family intersect on the line of discontinuity, while for $m > 0$ two characteristics of the first family intersect there. It is said that for $m < 0$ the characteristics of the third family form a "fir tree", while for $m > 0$ the characteristics of the first family form a "fir tree".

This circumstance is connected in an essential way with the stability of a shock wave. In particular, it indicates that the solution in this case is necessarily discontinuous, since on a line of discontinuity incoming characteristics of a single family intersect. These considerations make it possible to formally determine the stability of a discontinuous solution of a system of equations of hyperbolic type as the condition of satisfying the "fir-tree conditions" on lines of discontinuity. This approach to discontinuous solutions is applied and extensively discussed in Chapter 4 (see also subsection 9).

5. ***The Hugoniot conditions for a polytropic gas.*** In the case of a polytropic gas the equation of state has the form

$$\varepsilon = \frac{1}{\gamma - 1}\frac{p}{\rho} = \frac{pV}{\gamma - 1} = \frac{1}{\gamma(\gamma - 1)}c^2, \qquad c^2 = \frac{\partial p(\rho.\ S)}{\partial \rho}. \tag{1}$$

For the enthalpy $\mathcal{T}$ we have the expression

$$\mathcal{T} = \frac{\gamma}{\gamma - 1}\frac{p}{\rho} = \frac{1 - h}{2h}c^2, \qquad 0 < h = \frac{\gamma - 1}{\gamma + 1} < 1, \tag{2}$$

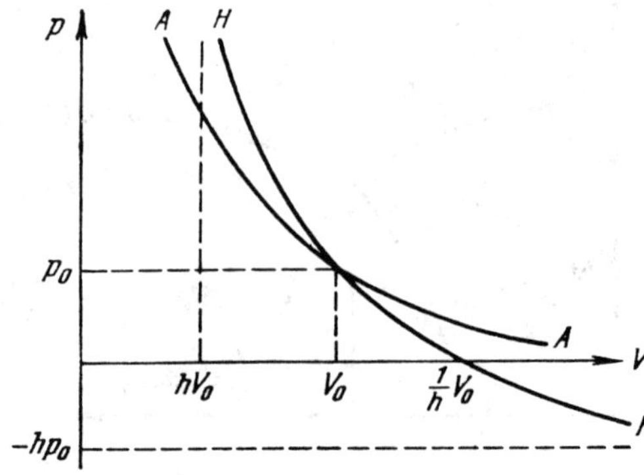

FIGURE 2.30

and the equation of the Hugoniot adiabat with center at the point (p_0, V_0) has the form

$$pV + h(p_0V - pV_0) - p_0V_0 = 0 \tag{3}$$

or

$$(p + hp_0)(V - hV_0) = (1 - h^2)p_0V_0. \tag{4}$$

Figure 2.30 shows the graphs of the Hugoniot adiabat and the Poisson adiabat for the case of a polytropic gas. According to (4), the Hugoniot adiabat is a hyperbola with the asymptotes

$$V = hV_0, \qquad p = -hp_0, \tag{5}$$

while the Poisson adiabat is given by the equation

$$pV^\gamma = a^2(S_0) = \text{const} \tag{6}$$

and has the axes $p = 0$, $V = 0$ as asymptotes.

For a polytropic gas the following properties hold.

1) The pressure p varies along the Hugoniot adiabat from 0 to ∞ as V varies from V_0/h to hV_0; $p = \infty$, $V = hV_0$ corresponds to an infinitely strong shock wave for values p_0, V_0 ahead of the front; $p = 0$, $V = V_0/h$ corresponds to an infinitely strong shock wave for a given state p_0, V_0 behind the front. From Figure 2.30 we conclude that the limiting compression of a polytropic gas by a shock wave is

$$\frac{\rho}{\rho_0} = \frac{V_0}{V} = \frac{1}{h} = \frac{\gamma + 1}{\gamma - 1} > 1. \tag{7}$$

2) Along the Hugoniot adiabat

$$dp/dV < 0. \tag{8}$$

This implies that the pressure p and the density ρ increase monotonically behind the wave front as the quantity $|D - u_0|$ increases.

In subsections 3 and 4 we showed that for a normal gas the state behind the front of a shock wave is uniquely determined from the Hugoniot conditions if the state ahead of the front (u_0, p_0, V_0, ε_0) and a quantity characterizing the strength of the shock wave (the mass flux m or the speed of the shock wave D) are given.

Here we shall indicate the corresponding computational formulas for the case of a polytropic gas. We characterize the strength of the shock wave by the dimensionless quantity

$$M_0 = \frac{|u_0 - D|}{c_0}, \tag{9}$$

which is greater than or equal to 1 by Zemplén's theorem. If M_0 is known, then we find the modulus of the mass flux m through the wave front from the equality

$$|m| = \rho_0 c_0 M_0. \tag{10}$$

We now explicitly express the quantities p, V, c and u in terms of p_0, V_0, c_0, u_0 and M_0. Rewriting the equation of the adiabat (4) in the form

$$\frac{p + hp_0}{V_0} = \frac{p_0 + hp}{V} = \frac{(1 - h)(p - p_0)}{V_0 - V} = (1 - h)m^2, \tag{11}$$

where we have used (4.2.4), we find that

$$p + hp_0 = (1 - h)\rho_0(u_0 - D)^2 = (1 + h)p_0 M_0^2, \tag{12}$$

since $p_0 = c_0^2\rho_0(1 - h)/(1 + h)$. Hence

$$p = p_0[(1 + h)M_0^2 - h], \tag{13}$$

$$\frac{\rho}{\rho_0} = \frac{V_0}{V} = \frac{p + hp_0}{p_0 + hp} = \frac{M_0^2}{(1 - h) + hM_0^2}, \tag{14}$$

$$\frac{c^2}{c_0^2} = \frac{p\rho_0}{p_0\rho} = \frac{[(1 + h)M_0^2 - h]\,[(1 - h) + hM_0^2]}{M_0^2}. \tag{15}$$

The formulas (13)–(15) express the thermodynamic quantities behind the wave front in terms of the known quantities ahead of the front and M_0. To obtain the velocity u behind the front we use (4.2.10), according to which

$$u = u_0 + m(V - V_0) = u_0 \pm \rho_0 c_0 M_0 V_0\left(\frac{V}{V_0} - 1\right)$$
$$= u_0 \pm c_0 M_0\left(\frac{V}{V_0} - 1\right) = u_0 \pm (1 - h)c_0\left(M_0 - \frac{1}{M_0}\right) \tag{16}$$

(the sign $+$ in (16) is chosen for the case of a wave proceeding to the right, i.e., for $m < 0$; if the wave propagates to the left relative to the gas ($m > 0$), then the sign $-$ is chosen).

The equations (13)–(16) explicitly express the state behind the front as rational functions of M_0. It is easily seen that the quantities

$$\frac{p}{p_0}, \qquad \frac{\rho}{\rho_0}, \qquad \frac{c}{c_0}, \qquad \left|\frac{u - u_0}{c_0}\right|$$

are monotonically increasing functions of the parameter M_0.

Because of the symmetry of the Hugoniot conditions the state u_0, p_0, V_0, ε_0 ahead of the front of a shock wave is determined on the basis of the state u, p, V, ε behind the front by the same formulas if in place of M_0 we use the number $M = |u - D|/c$, i.e.,

$$\left.\begin{aligned} \frac{p_0}{p} &= (1 + h)M^2 - h, \qquad \frac{\rho_0}{\rho} = \frac{M^2}{(1 - h) + hM^2}, \\ \frac{c_0}{c} &= \left[(1 + h) - \frac{h}{M^2}\right][(1 - h) + hM^2], \\ u_0 &= u \pm (1 - h)\, c\left[M - \frac{1}{M}\right]; \end{aligned}\right\} \tag{17}$$

here it follows from Zemplén's theorem that $M \leq 1$.

6. *The Hugoniot conditions for an isothermic gas.* An isothermic gas is the limit case of a thermally conducting gas when the coefficient of thermal conductivity tends to infinity, while the temperature in the gas is kept constant by exterior heat sources. In this case the integral conservation laws yield only two conservation laws on the front of a discontinuity—the conservation of mass and momentum:

$$\rho_1(u_1 - D) = \rho_2(u_2 - D) = m, \tag{1}$$

$$p_1 + \rho_1(u_1 - D)^2 = p_2 + \rho_2(u_2 - D) = j, \tag{2}$$

while the pressure p is given by

$$p = p(V, T_0) = F(V). \tag{3}$$

Since in an isothermic gas the temperature is identical ahead of and behind the front of a shock wave, the role of the Hugoniot adiabat H in this case is played by the isotherm (3) with an equation which can be written in the form

$$p - F(V) = p_0 - F(V_0) = 0. \tag{4}$$

In this case two points (p_0, V_0) and (p, V) lying on the isotherm (3) are conjugate.

Figure 2.31 shows the relative positions of the Poisson adiabat A, the Hugoniot adiabat H, and the isotherm T for the case of a normal gas.

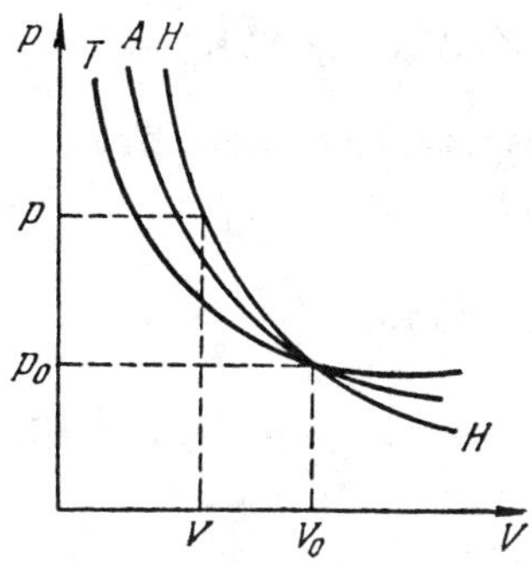

FIGURE 2.31

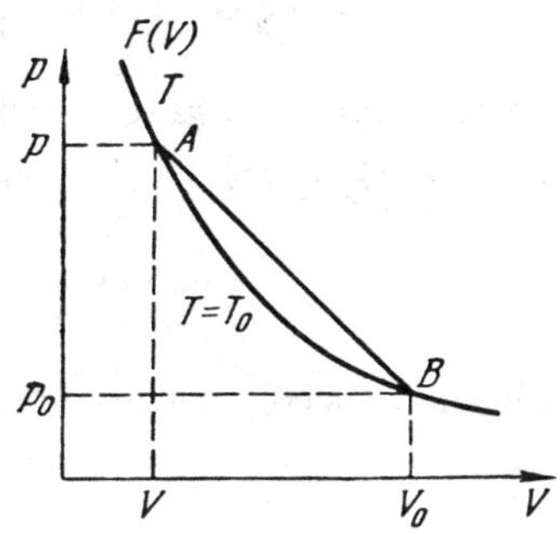

FIGURE 2.32

Under the assumption that the gas possesses properties I–V of subsection 3 the isotherm satisfies the conditions (see §1.4)

$$\frac{\partial p}{\partial V} = \frac{dF}{dV} < 0, \qquad \frac{\partial^2 p}{\partial V^2} = \frac{d^2F}{dV^2} > 0; \quad p \to \infty \text{ as } V \to 0. \tag{5}$$

Hence, the isotherm (3) is a convex curve, and any ray drawn from (p_0, V_0) intersects it in no more than one point. Therefore, the ray $K = (p - p_0)/(V - V_0)$ intersects the upper part of the isotherm (3) in only one point (Figure 2.32) if

$$-\infty < K = -m^2 < \left(\frac{\partial p}{\partial V}\right)_{V_0} = -\left[\frac{c_T(V_0)}{V_0}\right]^2, \tag{6}$$

where c_T is the isothermal speed of sound. For $K = -\infty$ (an infinitely strong shock wave) the compression is also infinite. This same conclusion follows formally from the formulas for the compression of a polytropic gas as $\gamma \to 1$, since $h = (\gamma - 1)/(\gamma + 1) \to 0$ as $\gamma \to 1$.

We shall give computational formulas for a shock wave in an isothermic gas, again taking as the parameter determining the force of the wave the quantity

$$M_0 = \frac{|u_0 - D|}{c_T(V_0)}.$$

From (2) it follows that

$$\frac{F(V) - F(V_0)}{V_0 - V} = m^2 = \rho_0^2 c_T^2(V_0) M_0^2. \tag{7}$$

The volume V is determined from this, after which u is found from the formula

$$u - u_0 = m(V - V_0) = -\rho_0 c_T(V_0) M_0(V - V_0), \tag{8}$$

where to be specific we have taken $m < 0$ (the shock wave moves from left to right relative to the gas). In the case of an ideal gas

$$p = F(V) = \frac{RT_0}{V} = \frac{c_T^2}{V} = c_T^2 \rho, \tag{9}$$

the isothermic speed of sound $c_T = \sqrt{RT_0}$ is constant, and (7) and (8) take the form

$$\frac{V_0}{V} = \frac{\rho}{\rho_0} = M_0^2, \tag{10}$$

$$u - u_0 = c_T M_0\left(1 - \frac{V}{V_0}\right) = c_T M_0\left(1 - \frac{1}{M_0^2}\right) = c_T\left(M_0 - \frac{1}{M_0}\right). \tag{11}$$

Passing to the Riemann invariants (§2.9)

$$s = u - c_T \ln \rho, \qquad r = u + c_T \ln \rho, \tag{12}$$

we give the Hugoniot conditions (10) and (11) the symmetric form

$$s - s_0 = c_T\left[-\ln M_0^2 + \left(M_0 - \frac{1}{M_0}\right)\right] = \psi(M_0), \tag{13}$$

$$r - r_0 = c_T\left[\ln M_0^2 + \left(M_0 - \frac{1}{M_0}\right)\right] = \varphi(M_0), \tag{14}$$

where s_0 and r_0 are the values of the Riemann invariants computed at the point u_0, p_0, V_0.

To resolve the question of the stability of a shock wave in an isothermic gas it suffices to consider the requirement of the second law of thermodynamics for a nonequilibrium isothermal process (see (1.2.9)):

$$\frac{dS}{dt} > \frac{1}{T}\frac{dQ}{dt}. \tag{15}$$

We apply this inequality to a particle of gas which in its motion intersects the shock front. Since per unit time a mass of gas $|\rho_0(u_0 - D)| = |\rho(u - D)| = |m|$ intersects the shock wave, taking this quantity as the mass of the particle indicated, we make the time of its passage from a state ahead of the wave $(u_0, p_0, V_0, \varepsilon_0)$ to a state behind the wave (u, p, V, ε) equal to one. According to the law of conservation of energy (formula (1.1.2)), $\Delta Q = \Delta E + \Delta A$, where ΔE is the increment in the total energy of the particle, and ΔA is the work it does per unit time on the surrounding gas. Since

$$\Delta E = |m|\left(\varepsilon - \varepsilon_0 + \frac{u^2 - u_0^2}{2}\right), \qquad \Delta A = (pu - p_0u_0)\operatorname{sgn} m,$$

inequality (15) gives

$$T_0(S - S_0) + \frac{p_0u_0 - pu}{m} + \varepsilon_0 - \varepsilon + \frac{u_0^2 - u^2}{2} > 0$$

(where $m > 0$ if the wave moves from right to left relative to the gas and $m < 0$ otherwise). Recalling (1) and (2), we obtain

$$T_0(S - S_0) + \frac{p_0}{\rho_0} - \frac{p}{\rho} + \varepsilon_0 - \varepsilon + \frac{(u_0 - D)^2}{2} - \frac{(u - D)^2}{2} > 0. \tag{16}$$

Substituting (4.2.10),

$$(u - D)^2 = \frac{p_0 - p}{V - V_0}V^2, \qquad (u_0 - D)^2 = \frac{p_0 - p}{V - V_0}V_0^2,$$

we bring (16) to the form

$$T_0(S - S_0) - \left[(\varepsilon - \varepsilon_0) + (V - V_0)\frac{p + p_0}{2}\right]$$
$$= T_0(S - S_0) - H(p, V, p_0, V_0) > 0. \tag{17}$$

We shall compute the quantity $T_0(S - S_0)$. For this we integrate the thermodynamic identity $TdS = d\varepsilon + pdV$ along the isotherm T from the point (p_0, V_0) to the point (p, V) (Figure 2.32). We assume that the gas is normal, so that $F = F(V)$ is a curve convex upward. Along the isotherm $p = F(V)$, $T = T_0 =$ const we therefore have

$$T_0(S - S_0) = \varepsilon - \varepsilon_0 + \int_{V_0}^{V} F(V)\,dV. \tag{18}$$

Substituting this into (17), we bring it to the final form

$$\frac{p + p_0}{2}(V_0 - V) - \int_V^{V_0} F(V)\,dV > 0. \tag{19}$$

It is obvious that for a normal gas this condition is equivalent to the condition $V < V_0$, i.e., to states behind the front there corresponds only the upper half of the isotherm T.

As before, from this it follows that a shock wave in the case of an isothermic gas leads to an increase of the pressure and density; the motion ahead of the front is supersonic, while behind the front it is subsonic (the speed of sound there is understood to be the isothermic speed of sound $c_T = V\sqrt{-\partial F/\partial V}$).

The stability condition (19) can be given another form. In the plane of the Lagrange variables (q, t) we consider a portion of gas located between the lines $q = q_1$ and $q = q_2$ $(q_1 < q_2)$. For any successive moments of time t_1 and t_2 $(t_1 < t_2)$ the inequality (15) applied to this portion of gas gives

$$\int_{q_1}^{q_2}\left(\varepsilon + \frac{u^2}{2} - T_0 S\right)_{t=t_1} dq - \int_{q_1}^{q_2}\left(\varepsilon + \frac{u^2}{2} - T_0 S\right)_{t=t_2} dq \geq \Delta A, \tag{20}$$

where ΔA is the work done by this portion of gas on the surrounding gas in time $t_2 - t_1$:

$$\Delta A = \int_{t_1}^{t_2} (pux^\nu)_{q=q_1}\,dt - \int_{t_1}^{t_2} (pux^\nu)_{q=q_2}\,dt \tag{21}$$

(the equality sign in (20) corresponds to the case of a smooth (i.e., quasiequilibrium) flow). From (20) and (21) we easily obtain the following integral condition equivalent to (19):

$$\oint_C \left(\varepsilon + \frac{u^2}{2} - T_0 S\right) dq - pux^\nu\,dx \geq 0 \tag{22}$$

(C is an arbitrary piecewise smooth contour traversed in the positive direction).

In thermodynamics the quantity $\varepsilon + u^2/2 - T_0 S = E - T_0 S$ is called the *free energy* (per unit mass), while (22) expresses a well-known law of thermodynamics: in an isothermic process the work performed by a system is less than the loss of free energy and is equal to it only in the case of a quasiequilibrium process.

In the last case the flow is smooth, and from (22) we obtain the differential relation

$$\frac{\partial}{\partial t}\left(\varepsilon + \frac{u^2}{2} - T_0 S\right) + \frac{\partial}{\partial q}(pux^\nu) = 0. \tag{23}$$

It must be satisfied identically in a smooth part of the flow, i.e., it must be a consequence of the differential equations of motion (2.5.4) and (2.5.5). This is actually the case: from (2.5.4) and (2.5.5) we easily obtain

$$\frac{\partial}{\partial t}\left(\frac{u^2}{2} - \int_1^V F(V)\, dV\right) + \frac{\partial}{\partial q}(pux^\nu) = 0,$$

which by (18) coincides with (23).

We remark that conditions (19) and (22), as explained in §1, can be understood as the requirement of the increase of entropy of the total system consisting of the gas particles and the exterior heat sources (the thermostat). Then (23) implies that the entropy of this system is constant in the case of a smooth flow.

7. ***Strong and weak shock waves. Comparison of a shock wave and a Riemann compression wave.*** For simplicity we assume that the gas ahead of the front of a shock wave is at rest, i.e., $u_0 = 0$. For an infinitely strong shock wave we have $p/p_0 = \infty$. We shall assume that p is finite and $p_0 = 0$.

Passing to the limit as $M_0 \to \infty$, $p_0 \to 0$ and $c_0 \to 0$ in (4.5.13)–(4.5.16), we obtain the Hugoniot conditions for an infinitely strong shock wave in a polytropic gas:

$$p = \lim(1 + h)p_0 M_0^2 = (1 - h)\rho_0(u_0 - D)^2 = \frac{2}{\gamma + 1}\rho_0 D^2, \tag{1}$$

$$\frac{\rho}{\rho_0} = \frac{V_0}{V} = \frac{1}{h} = \frac{\gamma + 1}{\gamma - 1}, \tag{2}$$

$$c = \sqrt{h(1 + h)}\,|D|, \tag{3}$$

$$u = (1 - h)D. \tag{4}$$

We see that the kinetic energy of the gas behind the front of the shock wave is equal to the internal energy, since

$$\varepsilon = \frac{(1 - h)^2}{2h(1 + h)}c^2 = \frac{(1 - h)^2}{2}D^2 = \frac{u^2}{2}.$$

For weak shock waves we assume that $M_0 = |u_0 - D|/c_0$ is close to one. Setting $M_0 = 1 + \varepsilon$, $0 < \varepsilon \ll 1$, we expand (4.5.13)–(4.5.16) in the parameter ε up to terms of order ε^3:

$$\frac{p}{p_0} = 1 + (1 + h)(2\varepsilon + \varepsilon^2), \tag{5}$$

$$\frac{\rho}{\rho_0} = \frac{M_0^2}{(1-h)+hM_0^2} = \frac{1+2\varepsilon+\varepsilon^2}{1+h\varepsilon(2+\varepsilon)}$$
$$\simeq [1-2h\varepsilon - h\varepsilon^2(1-4h)](1+2\varepsilon+\varepsilon^2)$$
$$= 1+2\varepsilon(1-h)+\varepsilon^2[1-5h+4h^2]+O(\varepsilon^3), \tag{6}$$

$$\frac{c}{c_0} = \sqrt{\left(1+h-\frac{h}{M_0^2}\right)(1-h+hM_0^2)} = 1+2h\varepsilon - h\varepsilon^2 + O(\varepsilon^3), \tag{7}$$

$$u-u_0 = (1-h)c_0\left(M_0-\frac{1}{M_0}\right)$$
$$= (1-h)c_0(2\varepsilon-\varepsilon^2)+O(\varepsilon^3) \qquad (m<0). \tag{8}$$

We compute the jumps of the Riemann invariants r and s on the front of a weak shock wave:

$$r-r_0 = u-u_0+\frac{(1-h)}{h}(c-c_0)$$
$$= (1-h)c_0[(2\varepsilon-\varepsilon^2)+(2\varepsilon-\varepsilon^2)]+O(\varepsilon^3)$$
$$= 4(1-h)c_0\varepsilon - 2(1-h)c_0\varepsilon^2 + O(\varepsilon^3) \qquad (m<0), \tag{9}$$

$$s-s_0 = u-u_0-\frac{(1-h)}{h}(c-c_0)$$
$$= (1-h)c_0(2\varepsilon-\varepsilon^2) - c_0\frac{1-h}{h}[2h\varepsilon - h\varepsilon^2]+O(\varepsilon^3)$$
$$= O(\varepsilon^3) \qquad (m<0). \tag{10}$$

As we saw in subsection 3, it follows from general properties of the Hugoniot adiabat H that

$$S-S_0 = O(h^3). \tag{11}$$

Thus, in a weak shock wave propagating to the right through the gas the invariant r has a jump of first order in ε, while the invariant s and the entropy S have jumps of third order. Similarly, in a wave proceeding to the left the invariant r and the entropy S have jumps of third order.

Formulas (10) and (11) show that a weak shock wave behaves like a "short" travelling compression wave. Indeed, as we saw in §3, constancy of the entropy S and one Riemann invariant characterize a Riemann compression wave. For weak shock waves this is violated only in third order, and therefore the relations in a weak shock wave may approximately be considered as the same relations in a travelling compression wave. This makes it possible to approximately replace a weak shock wave by a travelling wave. This is useful in a number of cases, especially in the analysis of the interaction of shock waves with Riemann waves.

Noting that in our approximation $D = c_0(1 + \varepsilon)$ $(m < 0)$, we see that the speed of a weak shock wave is the arithmetic mean of the speeds of the r-characteristics, since the equality

$$D = \frac{u_0 + c_0 + u + c}{2}$$

is satisfied to terms of third order.

8. *Shock transition for media with anomalous thermodynamic properties.* The question of the admissibility of a discontinuity for a gas (or liquid) with equations of state which do not satisfy conditions (4.3.1) for a normal gas becomes considerably more complicated. Globally, i.e., for arbitrary equations of state satisfying only the necessary conditions of thermodynamics, this question has so far not been resolved.

In recent years there has been interest in physical processes described by equations of state with anomalous thermodynamic properties (see Al′tšuler 1965), and Zel′dovič and Raĭzer (1966)). The case in which condition II is violated (see subsection 3) and the quantity $p''_{VV}(V, S)$ may have variable sign is of greatest interest. This case was considered by Galin (1958), Sidorenko (1968), Wendroff (1972), and Liu (1975). It is assumed that the quantity $p''_{VV}(V, S)$ may have variable sign, but the zeros of this function for fixed S are isolated; it is further assumed that

$$p'_s(V, S) > 0, \qquad p'_V(V, S) < 0. \tag{1}$$

Of course, the Hugoniot conditions must be satisfied on any discontinuity for gases with arbitrary equations of state. In particular, therefore, the parameters p_0, V_0 ahead of a front of discontinuity and p, V behind the front must lie on the Hugoniot adiabat, i.e.,

$$H(p, V; p_0, V_0) = 0. \tag{2}$$

In the case where p''_{VV} has variable sign the condition $S \geq S_0$ is not sufficient to distinguish the physically correct discontinuities; generally speaking, it is only a necessary condition.

Sufficient conditions for the admissibility of a discontinuity satisfying condition (2) can be obtained from the following, apparently distinct requirements.

a) The requirement of the stability of the discontinuity relative to its partition into a sequence of several discontinuities moving in the same direction relative to the gas. This condition assumes that in such a partition these discontinuities again combine into the original discontinuity.

b) The condition that the states u_0, p_0, V_0 and u, p, V be limiting states as $x \to \pm\infty$ in a shock transition for a viscous and thermally conducting gas with positive coefficients of viscosity and thermal conductivity (see §5).

If conditions (1) are satisfied these conditions lead to the following criterion for the admissibility of a discontinuity. A discontinuity $\{u, p, V; u_0, p_0, V_0\}$ u_0, p_0 and V_0 are the values of the parameters ahead of the front of the shock wave) is admissible if for any point $(\tilde{p}, \tilde{V})$ of the Hugoniot adiabat (2) which is intermediate relative to its points (p_0, V_0) and (p, V) the following inequality is satisfied:

$$\frac{p - p_0}{V_0 - V} \geq \frac{\tilde{p} - p_0}{V_0 - \tilde{V}}. \tag{3}$$

If the conditions (1) are satisfied, (3) implies that $S > S_0$, i.e., the criterion of admissibility (3) for the case of $p''_{VV}(V, S)$ of variable sign is a stronger condition than the condition that the entropy only increase.

9. *Examples.* We shall consider here two of the simplest problems of a flow with shock waves.

1. *The motion of a piston in a quiescent gas.* In a quiescent, polytropic gas located to the right of a piston and characterized by parameters $u_0 = 0$, p_0, V_0 and c_0 a piston moves with constant speed $U > 0$ for $t > 0$. A shock wave propagates through the gas with constant speed D leaving behind the front a state with parameters u, p, V and c. It is clear that $u = U$. From (4.5.16) we therefore define M_0 as the positive root of the equation

$$M_0^2 - \frac{U}{(1-h)c_0} M_0 - 1 = 0,$$

i.e.,

$$M_0 = \frac{U}{2(1-h)c_0} + \sqrt{\frac{U}{4(1-h)^2c_0^2} + 1}. \tag{1}$$

After this p and V are determined by (4.5.13), (4.5.14) and the equality $D = c_0 M_0$. We note that for $U > 0$ we have $V < V_0$, and the stability condition on the shock wave is satisfied.

2. *The reflection of a shock wave from a rigid wall.* A shock wave propagates though a quiescent gas ($u_0 = 0$, p_0, V_0, c_0) with speed $D > 0$ leaving behind the front the state u, p, V, c computed above (Figure 2.33, a). At time $t = 0$ the shock wave arrives at a rigid wall $x = x_0$ bounding the gas on the right. On the rigid wall the condition $u(x_0, t) = 0$ is prescribed. Therefore, the shock wave is reflected from the wall as a shock wave propagating to the left with velocity $D_1 < 0$. We denote the state behind the front of the reflected shock

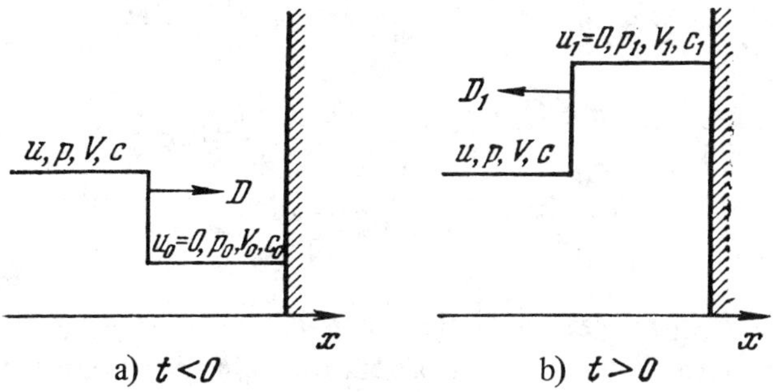

FIGURE 2.33

wave by $u_1 = 0, p_1, V_1, c_1$ (Figure 2.33, b). Thus, the problem reduces to determining just these parameters. We remark that the problem of determining u_1, p_1, V_1 and c_1 on the basis of the known parameters u, p, V and c reduces to the preceding problem, since it is known that $u_1 = 0$.

We introduce the quantities

$$M_0 = \frac{D - u_0}{c_0} = \frac{D}{c_0}, \qquad M = \frac{D - u}{c}, \qquad M_1 = \frac{u - D_1}{c}. \tag{2}$$

The following relations hold:

$$M_0 > 1, \qquad 0 < M < 1, \qquad M_1 > 1, \tag{3}$$

$$u - u_0 = u = (1 - h)c_0\left(M_0 - \frac{1}{M_0}\right), \tag{4}$$

$$u_0 - u = - u = (1 - h)c\left(M - \frac{1}{M}\right), \tag{5}$$

$$u - u_1 = u = (1 - h)c\left(M_1 - \frac{1}{M_1}\right). \tag{6}$$

It follows from (5) and (6) that M_1 and $1/M$ satisfy the same quadratic equation

$$\alpha^2 - \frac{u}{(1 - h)c}\alpha - 1 = 0. \tag{7}$$

Since $M_1 > 1$ and $1/M > 1$, it follows that

$$M_1 = \frac{1}{M}, \quad \text{i.e. } MM_1 = 1. \tag{8}$$

We shall compute the increase of pressure in the reflection of the shock wave. Using (8), we obtain

$$\left.\begin{aligned} \frac{p_0}{p} &= (1+h)M^2 - h, \qquad M^2 = \frac{p_0/p + h}{1+h}, \\ \frac{p_1}{p} &= (1+h)M_1^2 - h = (1+h)\frac{1}{M^2} - h = \frac{(1+2h)p/p_0 - h}{1 + hp/p_0}, \end{aligned}\right\} \tag{9}$$

$$\frac{p_1 - p_0}{p - p_0} = 1 + \frac{(1+h)}{h + p_0/p}. \tag{10}$$

In the case of a weak wave we have $p/p_0 \to 1$ and $(p_1 - p_0)/(p - p_0) \to 2$, which corresponds to an acoustic law of reflection. In the case of a strong wave when $p_0/p \to 0$ we have

$$(p_1 - p_0)/(p - p_0) \to 2 + 1/h = 2 + (\gamma + 1)/(\gamma - 1).$$

For gases with index γ close to 1 we obtain a strong increase in pressure. However, it should not be thought that for an isothermic gas the increase of pressure is inflnite, since the analogy between an isothermic gas and a polytropic gas with $\gamma = 1$ is not applicable here. Indeed, if $p_0 \to 0$ for an isothermic gas, then $\rho_0 \to 0$, since $c_T^2 = RT = \text{const}$. Therefore, there does not exist an isothermic shock wave propagating on a background with $p_0 = 0$ (a vacuum) with finite velocity.

§5. Study of the shock transition. The width of a shock wave

1. ***Formulation of the question for a normal gas.*** We shall assume that the equations of state of the gas satisfy conditions I–VI formulated in §1.4, i.e., the gas is normal.

As we have already mentioned, we consider discontinuous flows as limiting flows of a viscous and thermally conducting fluid as the coefficients of viscosity and thermal conductivity tend to zero. We therefore now consider some simple solutions of the equations of gas dynamics for a gas possessing viscosity and thermal conductivity, and we then obtain discontinuous solutions by passing to the limit. We are thus able to estimate the width of the zone of the shock wave for gases possessing finite viscosity and thermal conductivity.[10]

We shall consider only the case of plane symmetry ($\nu = 0$), since for sufficiently small distances and times any shock wave for $\nu \neq 0$ may approximately be considered a plane shock wave.

We write out the differential equations for a viscous, thermally conducting fluid (2.5.11)–(2.5.13):

[10] Our presentation in subsections 1–4 follows the work of Gilbarg (1951).

$$\frac{\partial \rho}{\partial t} + \frac{\partial \rho u}{\partial x} = 0, \tag{1}$$

$$\frac{\partial \rho u}{\partial t} + \frac{\partial}{\partial x}\left[p - \mu \frac{\partial u}{\partial x} + \rho u^2\right] = 0, \tag{2}$$

$$\frac{\partial}{\partial t}\left(\rho\varepsilon + \rho\frac{u^2}{2}\right) + \frac{\partial}{\partial x}\left[\rho u\left(\varepsilon + \frac{p}{\rho} + \frac{u^2}{2}\right) - \mu u\frac{\partial u}{\partial x} - \kappa\frac{\partial T}{\partial x}\right] = 0. \tag{3}$$

To the equations (1)–(3) we add the equations of state

$$p = p(\rho, T), \qquad \varepsilon = \varepsilon(\rho, T), \tag{4}$$

which we suppose to satisfy conditions I–VI. Multiplying (1) by $u^2/2 - \varepsilon$ and (2) by $-u$, and adding the results to (3), we obtain

$$\begin{aligned}\rho\left(\frac{\partial \varepsilon}{\partial t} + u\frac{\partial \varepsilon}{\partial x}\right) + p\frac{\partial u}{\partial x} &= \frac{\partial}{\partial x}\mu u\frac{\partial u}{\partial x} + \frac{\partial}{\partial x}\left(\kappa\frac{\partial T}{\partial x}\right) - u\frac{\partial}{\partial x}\left(\mu\frac{\partial u}{\partial x}\right)\\ &= \mu\left(\frac{\partial u}{\partial x}\right)^2 + \frac{\partial}{\partial x}\left(\kappa\frac{\partial T}{\partial x}\right).\end{aligned} \tag{5}$$

Solving (1) for $\partial u/\partial x$ and substituting it into (5), we bring (5) to the form

$$\frac{\partial \varepsilon}{\partial t} + u\frac{\partial \varepsilon}{\partial x} + p\left(\frac{\partial V}{\partial t} + u\frac{\partial V}{\partial x}\right) = \frac{\mu}{\rho}\left(\frac{\partial u}{\partial x}\right)^2 + \frac{1}{\rho}\frac{\partial}{\partial x}\left(\kappa\frac{\partial T}{\partial x}\right). \tag{6}$$

Since $TdS = d\varepsilon + pdV$, from (6) it follows that

$$\begin{aligned}\frac{\partial S}{\partial t} + u\frac{\partial S}{\partial x} &= \frac{\mu}{\rho T}\left(\frac{\partial u}{\partial x}\right)^2 + \frac{1}{\rho T}\frac{\partial}{\partial x}\left(\kappa\frac{\partial T}{\partial x}\right)\\ &= \frac{\mu}{\rho T}\left(\frac{\partial u}{\partial x}\right)^2 + \frac{\kappa}{\rho T^2}\left(\frac{\partial T}{\partial x}\right)^2 + \frac{1}{\rho}\frac{\partial}{\partial x}\left(\kappa\frac{\partial \ln T}{\partial x}\right).\end{aligned} \tag{7}$$

Equation (7) shows that the entropy S of a thermally insulated gas increases. Indeed, integrating (7), we obtain

$$\frac{\partial}{\partial t}\int_{q_1}^{q_2} S(q, t)\, dq = \kappa\frac{\partial \ln T}{\partial x}\Big|_{q_1}^{q_2} + \int_{q_1}^{q_2}\left[\frac{\mu}{T}\left(\frac{\partial u}{\partial x}\right)^2 + \frac{\kappa}{T^2}\left(\frac{\partial T}{\partial x}\right)^2\right]dx, \tag{8}$$

where q is the Lagrange coordinate ($dq = \rho dx$), and q_1 and q_2 are the boundaries of the distinguished volume of gas.

If the volume of gas is thermally insulated, then $\kappa(\partial T/\partial x)|_{q_1}^{q_2} = 0$ and from (8) it follows that the total entropy of the gas does not decrease. In particular, (7) implies that in a viscous gas that is not thermally conducting the entropy of each particle does not decrease in time.

We consider the simplest solutions of the system (1)–(3), namely, stationary solutions. Since the system (1)–(3) is invariant under the Galilean transformation $x' = x - Ut$, $u' = u - U$, we consider stationary flows in a coordinate system in which the flow has fixed profiles of all the quantities u, p, ρ and ε. Dropping the primes on the variables x' and u' and setting $\partial\rho/\partial t = \partial u/\partial t = \partial\varepsilon/\partial t = 0$, we obtain the equations determining stationary flows:

$$\frac{d}{dx}\rho u = 0, \tag{9}$$

$$\frac{d}{dx}\left[p + \rho u^2 - \mu\frac{du}{dx}\right] = 0, \tag{10}$$

$$\frac{d}{dx}\left[\rho u\left(\varepsilon + \frac{p}{\rho} + \frac{u^2}{2}\right) - \mu u\frac{du}{dx} - \kappa\frac{dT}{dx}\right] = 0. \tag{11}$$

These equations have the first integrals

$$\rho u = C_1, \qquad p + \rho u^2 - \mu\frac{du}{dx} = C_2,$$
$$C_1\left(\varepsilon + \frac{p}{\rho} + \frac{u^2}{2}\right) - \mu u\frac{du}{dx} - \kappa\frac{dT}{dx} = C_3. \tag{12}$$

The first integral obviously expresses the constancy of the mass flux, the second the constancy of the momentum flux, and the third the constancy of the flux of energy across any arbitrary section $x = \text{const}$. We shall transform (12) to a form which is convenient for what follows. Simple transformations lead to two ordinary differential equations

$$\mu\frac{dV}{dx} = \frac{1}{C_1}\left[C_1^2\left(V - \frac{C_2}{C_1^2}\right) + p(V, T)\right] = \mathscr{M}(V, T), \tag{13}$$

$$\kappa\frac{dT}{dx} = C_1\left[\varepsilon(V, T) - \frac{1}{2}C_1^2\left(V - \frac{C_2}{C_1^2}\right)^2 - \frac{C_3}{C_1} + \frac{1}{2}\frac{C_2^2}{C_1^2}\right] = \mathscr{L}(V, T), \tag{14}$$

which connect the two functions V and T. As a consequence of (13) and (14) we obtain the equation in the V, T plane

$$\frac{\kappa}{\mu}\frac{dT}{dV} = \frac{\mathscr{L}(V, T)}{\mathscr{M}(V, T)}, \tag{15}$$

which plays an important role in the analysis of solutions and conditions for their existence.

We introduce the dimensionless variables

$$\bar{V} = V\frac{C_1^2}{C_2}, \qquad \bar{p} = \frac{p}{C_2}, \qquad \bar{\varepsilon} = \varepsilon\frac{C_1^2}{C_2^2}, \qquad \bar{T} = R\frac{C_1^2}{C_2^2}T. \tag{16}$$

Going over to the dimensionless variables $\bar{V}$, $\bar{p}$, $\bar{\varepsilon}$, $\bar{T}$, we rewrite (13) and (14) in the form

$$\bar{\mu}\frac{d\bar{V}}{dx} = \bar{V} + \bar{p}(\bar{V}, \bar{T}) - 1 = \bar{\mathscr{M}}(\bar{V}, \bar{T}), \tag{17}$$

$$\bar{\kappa}\frac{d\bar{T}}{dx} = \bar{\varepsilon}(\bar{V}, \bar{T}) - \frac{1}{2}(\bar{V} - 1)^2 - \bar{\beta} = \bar{\mathscr{L}}(\bar{V}, \bar{T}), \tag{18}$$

where we have set

$$\left.\begin{aligned} &\bar{\mu} = \frac{\mu}{C_1}, \qquad \bar{\kappa} = \frac{\kappa}{C_1 R}, \qquad \bar{\beta} = \frac{C_3 C_1}{C_2^2} - \frac{1}{2}, \\ &\bar{p}(\bar{V}, \bar{T}) = p(V, T), \qquad \bar{\varepsilon}(\bar{V}, \bar{T}) = \varepsilon(V, T). \end{aligned}\right\} \tag{19}$$

To simplify notation we henceforth drop the bar over all quantities and finally rewrite the system (17), (18) as

$$\mu\frac{dV}{dx} = p(V, T) + V - 1 = \mathscr{M}(V, T), \tag{20}$$

$$\kappa\frac{dT}{dx} = \varepsilon(V, T) - \frac{1}{2}(V - 1)^2 - \beta = \mathscr{L}(V, T). \tag{21}$$

We must, however, remember that the relations (20) and (21) involving $p(V, T)$ and $\varepsilon(V, T)$ are obtained from the equations of state with consideration of (16).

The equations (20) and (21) contain dimensionless quantities $p(V, T)$ and $\varepsilon(V, T)$, where V and T are dimensionless. It is easy to verify that these relations are such that they satisfy conditions I–VI provided that the initial equations of state satisfy these conditions. (Here it should be recalled that $C_2 > 0$ and $C_3 C_1 > 0$.)

For the system (20), (21) we pose the following boundary value problem: find a solution $V(x)$, $T(x)$ of the system (20), (21) which tends to constant values at infinity, i.e., as $x \to +\infty$

$$V(x) \to V_1, \qquad T(x) \to T_1, \tag{22}$$

and as $x \to -\infty$

$$V(x) \to V_2, \qquad T(x) \to T_2. \tag{23}$$

A necessary condition for the existence of a solution is obviously the condition that the points (V_1, T_1) and (V_2, T_2) be stationary points of the system (20), (21), i.e.,

$$\mathscr{M}(V_1, T_1) = \mathscr{L}(V_1, T_1) = \mathscr{M}(V_2, T_2) = \mathscr{L}(V_2, T_2) = 0. \tag{24}$$

In other words, (V_1, T_1) and (V_2, T_2) must be points of intersection of the curves

$$\mathcal{M}(V, T) = 0, \qquad \mathcal{L}(V, T) = 0. \tag{25}$$

Suppose such points (V_1, T_1) and (V_2, T_2) exist. Then at these points, according to (20) and (21), we have

$$\mu \frac{dV}{dx} = \kappa \frac{dT}{dx} = 0;$$

and therefore from the integrals (12) it follows that

$$\rho_1 u_1 = \rho_2 u_2 = C_1, \tag{26}$$

$$p_1 + \rho_1 u_1^2 = p_2 + \rho_2 u_2^2 = C_2, \tag{27}$$

$$\rho_1 u_1\left(\varepsilon_1 + \frac{p_1}{\rho_1} + \frac{u_1^2}{2}\right) = \rho_2 u_2\left(\varepsilon_2 + \frac{p_2}{\rho_2} + \frac{u_2^2}{2}\right) = C_3 \tag{28}$$

(here u, ρ, p and ε are the original functions, not transformed to dimensionless variables).

From this it follows that the states u_1, ρ_1, p_1, ε_1 and u_2, ρ_2, p_2, ε_2 must satisfy the Hugoniot conditions. Indeed, since we are considering a stationary solution of the system (1)–(3) in a moving coordinate system, denoting the velocity U of the system by D, we find that in conditions (26)–(28) the quantities u_1 and u_2 are replaced by $u_1 - D$ and $u_2 - D$ on passage to a stationary coordinate system after which the conditions (26)–(28) take the usual form of the Hugoniot conditions (4.1.10)–(4.1.12). In particular, this implies that the points (p_1, V_1) and (p_2, V_2) in the plane of the variables p and V must lie on the Hugoniot adiabat. Thus, all points of intersection of the curves (25) lie on the Hugoniot adiabat passing through one of them.

2. *Properties of the curves* $\mathcal{M} = 0$ *and* $\mathcal{L} = 0$ *for a normal gas.* We shall show that for a normal gas the functions $\mathcal{L}$ and $\mathcal{M}$ possess the following properties.

1. $\partial\mathcal{L}(V, T)/\partial T > 0$ and $\partial\mathcal{M}(V, T)/\partial T > 0$.
2. The curves $\mathcal{M} = 0$ and $\mathcal{L} = 0$ either have no points of intersection, or they have two and only two common points, (V_0, T_0) and (V_1, T_1). We shall assume that $V_0 > V_1$.
3. $\partial\mathcal{L}(V, T)/\partial V > 0$ on the curve $\mathcal{L} = 0$ for $V_1 < V < V_0$.
4. $\mathcal{L}_V/\mathcal{L}_T < \mathcal{M}_V/\mathcal{M}_T$ at the point (V_0, T_0), and $\mathcal{L}_V/\mathcal{L}_T > \mathcal{M}_V/\mathcal{M}_T$ at (V_1, T_1).

From property 1 it follows that the temperature T is a single-valued function of V along the curves $\mathcal{L} = 0$ and $\mathcal{M} = 0$. We set

$$\left.\begin{aligned} T &= l(V) \quad \text{along the curve } \mathscr{L} = 0, \\ T &= m(V) \quad \text{along the curve } \mathscr{M} = 0. \end{aligned}\right\} \tag{1}$$

From properties 1 and 3 it follows that the curve $\mathscr{L} = 0$ drops monotonically on the segment $[V_1, V_0]$, i.e., the function $l(V)$ decreases monotonically. Indeed, along the curve $\mathscr{L} = 0$ we have

$$\frac{dl(V)}{dV} = \frac{dT}{dV} = -\frac{\partial\mathscr{L}/\partial V}{\partial\mathscr{L}/\partial T} < 0. \tag{2}$$

Finally, property 4 implies that the slope of the curve $\mathscr{M} = 0$ at the point (V_0, T_0) is less than the slope of the curve $\mathscr{L} = 0$, while at (V_1, T_1) it is greater.

Thus, properties 1–4 of the functions $\mathscr{L}$ and $\mathscr{M}$ imply that the curves $\mathscr{L} = 0$ and $\mathscr{M} = 0$ in the V, T plane are approximately as shown in Figure 2.34.

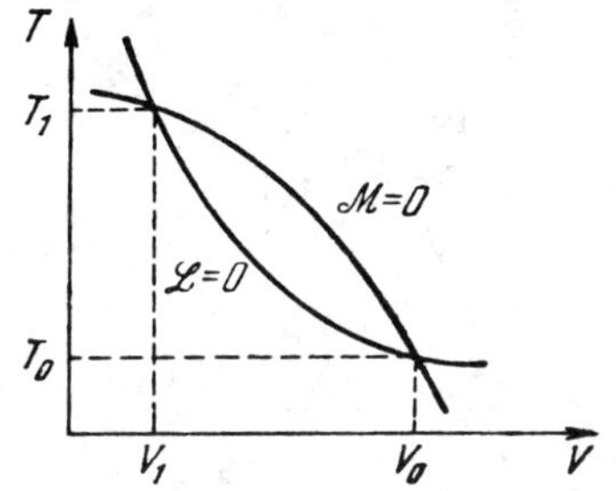

FIGURE 2.34

We now prove that the properties 1–4 of the functions $\mathscr{L}$ and $\mathscr{M}$ follow from properties I–V of the equations of state. Property 1 for $\mathscr{L}$ and $\mathscr{M}$ can be written in the form

$$\frac{\partial p(V, T)}{\partial T} > 0, \qquad \frac{\partial \varepsilon(V, T)}{\partial T} > 0, \tag{3}$$

and, since the dimensionless variables (5.1.16) differ from the original variables only by positive multiples, the inequalities (3) follow from properties IV and V (see also (1.4.23)).

We prove property 2. All points of intersection of the curves $\mathscr{L} = 0$ and $\mathscr{M} = 0$ in the V, T plane must lie on the Hugoniot adiabat passing through one of them. The line $\mathscr{M} = 0$ in the plane of the dimensionless variables p, V is obviously the line

$$p = 1 - V \tag{4}$$

with negative slope. As we saw in §4.3, any line with negative slope intersects the Hugoniot adiabat H in precisely two points. Thus, if the curves $\mathscr{L} = 0$ and $\mathscr{M} = 0$ have one common point, then they have two common points but no more. This proves property 2.

We now prove properties 3 and 4. Using the relation $d\varepsilon = TdS - pdV$ in the expressions (5.1.13) and (5.1.14) for $\mathscr{L}(V, T)$ and $\mathscr{M}(V, T)$ (in dimensional variables), we obtain

$$d\mathscr{L} = C_1[T\,dS - C_1\mathscr{M}dV]. \tag{5}$$

Hence

$$\frac{\partial \mathscr{L}(V, T)}{\partial V} = C_1 T \frac{\partial S(V, T)}{\partial V} - C_1^2 \mathscr{M}. \tag{6}$$

We consider the picture of the intersection of the curves $\mathscr{L} = 0$ and $\mathscr{M} = 0$ in the plane of the dimensional variables V and p (Figure 2.35). From (5) it follows that on the curve $\mathscr{L} = 0$

$$\frac{dS}{dV} = \frac{C_1}{T} M. \tag{7}$$

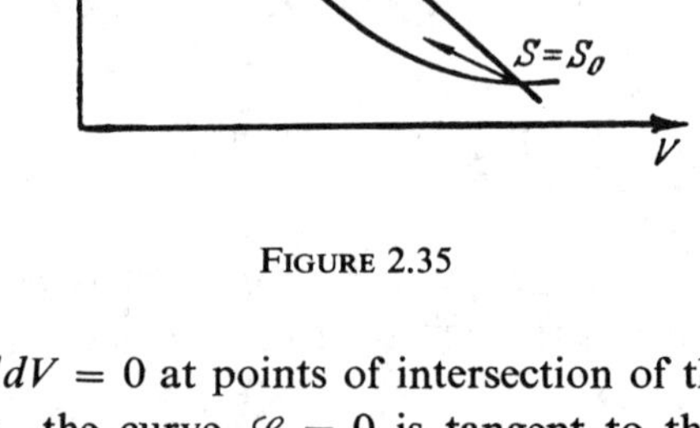

FIGURE 2.35

This implies that $dS/dV = 0$ at points of intersection of the curve $\mathscr{L} = 0$ with the line $\mathscr{M} = 0$, i.e., the curve $\mathscr{L} = 0$ is tangent to the Poisson adiabat A passing through the point of intersection and hence also to the Hugoniot adiabat H. Since for a normal gas the slope of A at the point (p_1, V_1) is greater than the slope of the ray $\mathscr{M} = 0$, while at (p_0, V_0) it is less than the slope of $\mathscr{M} = 0$, this implies that the slope of the curve $\mathscr{L} = 0$ at (p_1, V_1) is greater than the slope of the line $\mathscr{M} = 0$, while at (p_0, V_0) it is less. Under the mapping of the

V, p plane onto the V, T plane, to the lower half plane $\mathcal{M}(V, p) < 0$ there corresponds the region $\mathcal{M}(V, T) < 0$, which lies below the curve $\mathcal{M} = 0$ because of the relation $\partial p(V, T)/\partial T > 0$, which is valid for a normal gas. This implies that the picture of the intersection of the curves $\mathcal{M} = 0$ and $\mathcal{L} = 0$ in the V, p plane is qualitatively the same as in the V, T plane (Figure 2.34). Properties 3 and 4 have thus been proved.[11]

3. *Qualitative investigation of the integral curves of a shock transition.* A solution of the system (5.1.20), (5.1.21) satisfying the boundary conditions (5.1.22), (5.1.23) describes a stationary shock wave in a viscous, thermally conducting gas. We denote by V_0, T_0 the state ahead of the wave front and by V_1, T_1 the state behind the front, and to be specific we assume that $m = C_1 > 0$, i.e., the shock wave propagates through the gas to the left. Then the dimensionless variables $\mu > 0$ and $\kappa > 0$.

A solution $V = V(x)$, $T = T(x)$ of (5.1.20), (5.1.21) can then be considered a parametric definition of an integral curve of the equation

$$\frac{\kappa}{\mu}\frac{dT}{dV} = \frac{\mathcal{L}(V, T)}{\mathcal{M}(V, T)}. \tag{1}$$

Conversely, to each solution $T = T(V)$ of (1) there corresponds a solution $T = T(x)$, $V = V(x)$ of the system (5.1.20), (5.1.21) which is determined up to translation.

To a shock transition there corresponds a solution $T = T(V)$ of (1) passing through the points (V_0, T_0) and (V_1, T_1) of intersection of the curves $\mathcal{L} = 0$ and $\mathcal{M} = 0$. It is clear that these points are singular points of equation (1), and this makes possible the present formulation of the problem.

In the plane of the variables V and T we consider the direction field of vectors $\{\mathcal{M}, \mathcal{L}\}$. The curves $\mathcal{M} = 0$ and $\mathcal{L} = 0$ divide the quadrant $V, T > 0$ into four regions. The number of a region corresponds to the number of the quadrant in which the vector $\{\mathcal{M}, \mathcal{L}\}$ lies (Figure 2.36).

From the picture of the direction field of equation (1) shown in Figure 2.36 it follows that in each of the regions the function $T = T(V)$ is monotone on an integral curve. The integral curves may pass from one region to another in the following order: from region IV to regions I and III; from region II to regions I and III. The following transitions are impossible: from regions I and III to regions II and IV; from region IV to II and, conversely, from II to IV.

This implies that an integral curve joining the points (V_0, T_0) and (V_1, T_1)

[11] We note that the study of the curves $\mathcal{L} = 0$ and $\mathcal{M} = 0$ in the V, p plane is considerably simpler than in the V, T plane. However, for the investigation of an isothermic jump it is more convenient to use the variables V and T.

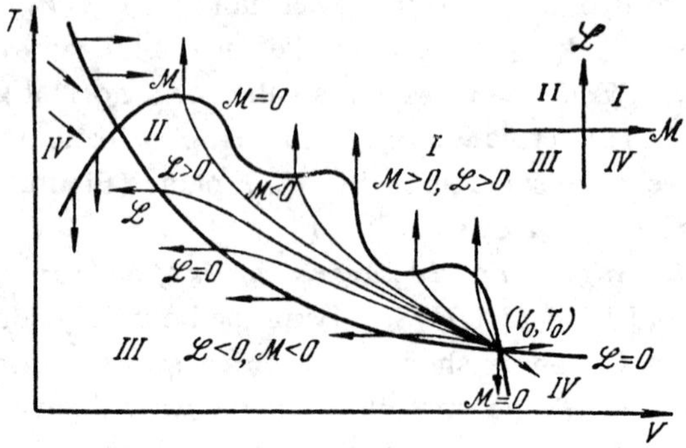

FIGURE 2.36

must lie entirely in region II. Thus, for a shock transition the following relations must hold:

$$\frac{dV}{dx} < 0, \qquad \frac{dT}{dx} > 0 \quad (m = C_1 > 0). \tag{2}$$

We shall prove the existence and uniqueness of the integral curve for a shock transition. To this end we establish the type of the singular points (V_0, T_0) and (V_1, T_1) of (1). It is known that the type of a singular point is determined by the characteristic equation, which has the form

$$\begin{vmatrix} \dfrac{\partial \mathscr{M}}{\partial V} - \lambda & \dfrac{\partial \mathscr{M}}{\partial T} \\ \dfrac{\partial \mathscr{L}}{\partial V} & \dfrac{\partial \mathscr{L}}{\partial T} - \lambda \end{vmatrix} = 0. \tag{3}$$

For the roots λ_1 and λ_2 of the characteristic equation we have the expression

$$\lambda_{1,2} = \frac{1}{2}\left(\frac{\partial \mathscr{M}}{\partial V} + \frac{\partial \mathscr{L}}{\partial T}\right) \pm \sqrt{\frac{1}{4}\left(\frac{\partial \mathscr{M}}{\partial V} + \frac{\partial \mathscr{L}}{\partial T}\right)^2 + \frac{\partial \mathscr{L}}{\partial V}\frac{\partial \mathscr{M}}{\partial T} - \frac{\partial \mathscr{M}}{\partial V}\frac{\partial \mathscr{L}}{\partial T}}.$$

It is easy to see that because of properties 1 and 3 of the functions $\mathscr{L}$ and $\mathscr{M}$ the characteristic roots λ_1 and λ_2 are real and distinct. They have the same sign if

$$\Delta = \begin{vmatrix} \dfrac{\partial \mathscr{M}}{\partial V} & \dfrac{\partial \mathscr{M}}{\partial T} \\ \dfrac{\partial \mathscr{L}}{\partial V} & \dfrac{\partial \mathscr{L}}{\partial T} \end{vmatrix} > 0,$$

and different signs if $\Delta < 0$. Therefore, in view of property 4 of the functions

$\mathscr{L}$ and $\mathscr{M}$ we conclude that, at the point (V_0, T_0), λ_1 and λ_2 have the same (positive) sign, while at the point (V_1, T_1) the signs differ. Thus the point (V_0, T_0) is a node, while (V_1, T_1) is a saddle point for equation (1). We shall show that there exists an integral curve of equation (1) joining (V_0, T_0) and (V_1, T_1). According to Figure 2.36 for the direction field of equation (1), through any point M of the curve $\mathscr{M} = 0$ there passes an integral curve of (1) which simultaneously passes through (V_0, T_0). By letting M tend to (V_1, T_1) along the curve $\mathscr{M} = 0$, we find by continuity that (V_1, T_1) and (V_0, T_0) are joined by an integral curve of (1).

Similarly, through any point L of the curve $\mathscr{L} = 0$ there passes an integral curve of (1) passing through (V_0, T_0). Letting L tend to (V_1, T_1) along the curve $\mathscr{L} = 0$, we find in the limit that there exists an integral curve of (1) joining (V_0, T_0) and (V_1, T_1) which is possibly distinct from the preceding curve.

Since (V_1, T_1) is a saddle point, according to the qualitative theory of differential equations, through it there pass only two integral curves of the equation (1). Therefore, (V_0, T_0) and (V_1, T_1) are joined by one and only one integral curve of (1). Indeed, if these points were joined by two distinct integral curves of (1), then any integral curve drawn through a point lying inside the region bounded by these two integral curves would also pass through these points. This contradicts our conclusion that (V_1, T_1) is a saddle point.

By properties 1 and 3 of the functions $\mathscr{M}$ and $\mathscr{L}$ the curves $\mathscr{M} = 0$ and $\mathscr{L} = 0$ are zeros of first order of these functions. This implies that in a neighborhood of the singular points (for example, (V_0, T_0)) the functions $\mathscr{L}$ and $\mathscr{M}$ can be represented in the form

$$\mathscr{L} = a_{11}(V - V_0) + a_{12}(T - T_0),$$
$$\mathscr{M} = a_{21}(V - V_0) + a_{22}(T - T_0),$$

while along the integral curve joining (V_0, T_0) and (V_1, T_1)

$$\mathscr{M} = b(V - V_0), \qquad b \neq 0.$$

It follows from (5.1.21) that the "width" of the zone of the shock transition is infinite, since the integral

$$\Delta x = \int_{V_0}^{V_1} \frac{\mu\, dV}{\mathscr{M}(V, T)}$$

diverges. Therefore, the graphs of the shock transition have approximately the from shown in Figure 2.37, from which it is evident that the values V_0, T_0 ahead of the front and V_1, T_1 behind the front are achieved asymptotically at infinity.

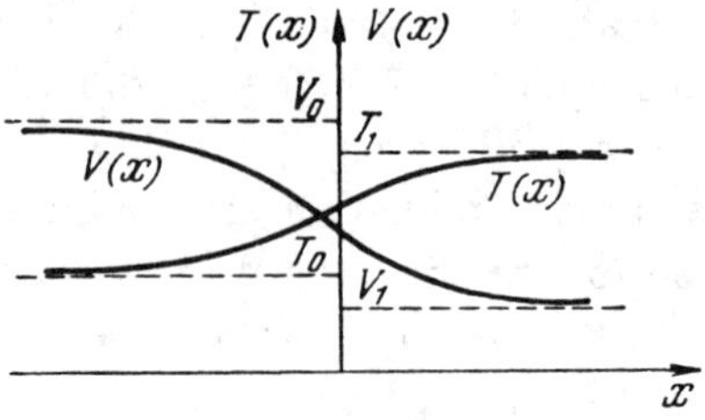

FIGURE 2.37

4. ***Limiting cases. An isothermic jump.*** We shall consider the behavior of the shock transition in two limiting cases: for $\kappa \to 0$ and for $\mu \to 0$.

If we fix $\mu > 0$ and vary κ, then the dependence of shock transitions on κ will be monotone in the sense that to a smaller value of κ there corresponds an integral curve of equation (5.3.1) lying nearer to the curve $\mathscr{L} = 0$. We shall prove that for sufficient small κ the integral curve of (5.3.1) lies in an ε-strip above the curve $\mathscr{L} = 0$. Indeed, at any point of the region II over an ε-strip of the curve $\mathscr{L} = 0$ we have $\mathscr{L} > \delta(\varepsilon)$. By choosing $N > 0$ sufficiently large, it is thus possible to select $\kappa = \kappa(\varepsilon, N)$ so small that at any point over an ε-strip

$$\frac{dT}{dV} = -\frac{\mu}{\kappa}\left|\frac{\mathscr{L}}{\mathscr{M}}\right| < -N,$$

i.e., the slope of the integral curves of (5.3.1) over an ε-strip can be made greater than the maximal slope of the curve bounding the ε-strip from above.[12] If at some point an integral curve leaves the ε-strip, then it does not arrive at the point (V_1, T_1). This means that as $\kappa \to 0$ the shock transition tends to the curve $\mathscr{L} = 0$.

We consider the second limiting case, where $\mu \to 0$ and $\kappa \neq 0$. If the segment of the line $\mathscr{M} = 0$ contained between the singular points is a curve decreasing monotonically as V increases, then it is proved by an analogous argument that as $\mu \to 0$ the shock transition tends to the curve $\mathscr{M} = 0$. In the case where this segment of $\mathscr{M} = 0$ is not a monotone curve the situation is different. For simplicity we consider the case in which the curve $\mathscr{M} = 0$ has a single maximum on the segment $V_1 \leq V \leq V_0$ (Figure 2.38). It is clear that in this case the curve $A_1A_2A_0$ consisting of the chord A_1A_2 parallel to the V axis and the arc A_0A_2 of the curve $\mathscr{M} = 0$ is an upper bound of the integral curves. Arguing as before, it is possible to prove that this boundary is precise, i.e., as $\mu \to 0$ an integral curve tends uniformly to the curve $A_1A_2A_0$. In particular, on the interval

[12] To be specific it may be assumed that the ε-strip is obtained by translating the curve $\mathscr{L} = 0$ an amount ε in the variable T.

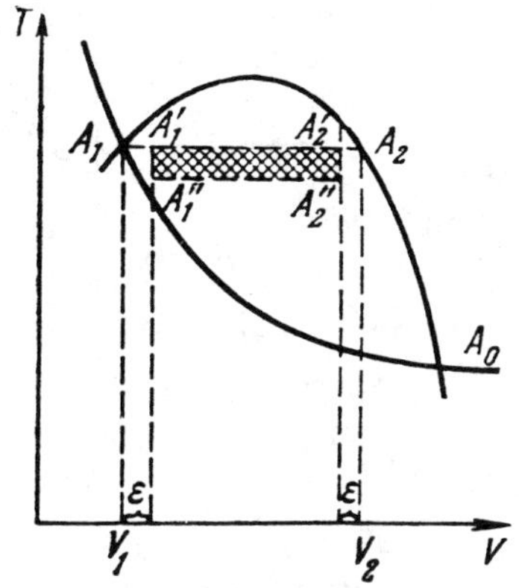

FIGURE 2.38

$[V_1 + \varepsilon,\ V_2 - \varepsilon]$ the shock transition lies in the ε-strip $A_1'A_1''A_2'A_2''$ (the hatched region in Figure 2.38). Since in this strip $|\mathscr{M}(V, T)| > M_0(\varepsilon)$ while $|\mathscr{L}(V, T)| < L_0(\varepsilon)$, the following inequality holds:

$$\frac{\kappa}{\mu}\left|\frac{dT}{dV}\right| < \left|\frac{L_0}{M_0}\right|.$$

This implies that μ can be made so small that on $[V_1 + \varepsilon,\ V_2 - \varepsilon]$ we have

$$\left|\frac{dT}{dV}\right| < \delta_1 = \delta_1(\varepsilon, \mu), \qquad \delta_1 \to 0 \text{ as } \mu \to 0.$$

Hence, on this interval T lies within the limits $T_1 - \alpha(\varepsilon, \mu) \leq T \leq T_1 + \alpha(\varepsilon, \mu)$, where $\alpha(\varepsilon, \mu) \to 0$ as $\mu \to 0$. The segment Δx corresponding to the interval $[V_1 + \varepsilon,\ V_2 - \varepsilon]$ has magnitude

$$\Delta x = \Delta(\varepsilon, \mu) = \int_{V_1+\varepsilon}^{V_2-\varepsilon} \frac{\mu\, dV}{\mathscr{M}(V, T(V))}.$$

Since $\mathscr{M}(V, T) > M_0(\varepsilon) > 0$ on $[V_1 + \varepsilon,\ V_2 - \varepsilon]$, this implies that $\Delta(\varepsilon, \mu) \to 0$ as $\mu \to 0$. The relations obtained are valid for any $\varepsilon \to 0$.

Thus, if the curve $\mathscr{M} = 0$ has a maximum on the segment $V_1 \leq V \leq V_0$ the shock transition in the case $\kappa \neq 0$, $\mu = 0$ consists in smooth variation on the interval $(-\infty, x_0)$ from values V_0, T_0 ahead of the front to values V_2, $T_2 = T_1$ (the point A_2 in Figure 2.38) and in a jump in the specific volume at the point x_0 from a value V_2 at the left to V_1 to the right at a constant temperature $T = T_1$ (Figure 2.39). Thus, in the presence of only thermal conductivity a jump in the density at constant temperature is possible. Such a jump is called *isothermic*.

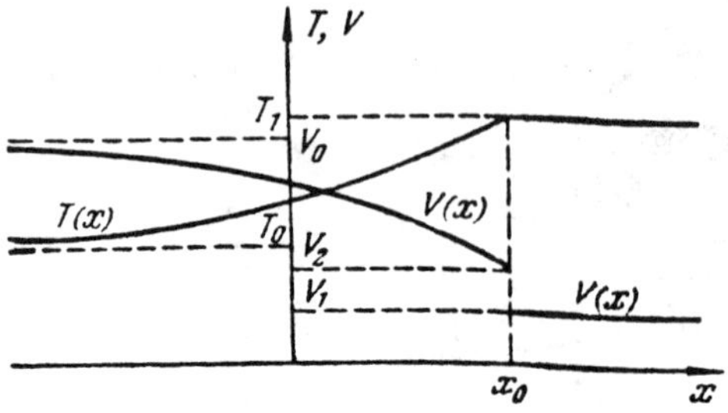

FIGURE 2.39

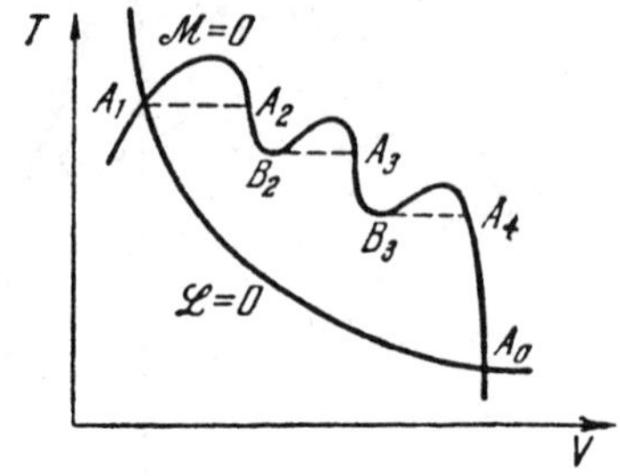

FIGURE 2.40

As a consequence of the positivity of the coefficient of thermal conductivity it follows from this that the specific volume V_2 ahead of the front of an isothermic jump is greater than the specific volume V_1 behind the front of the isothermic jump. We thus find that the limit discontinuities as $\mu \to 0$ satisfy the stability condition for discontinuities in an isothermic gas which we obtained in §4.6.

In the case where the curve $\mathcal{M} = 0$ has several maxima the shock transition will have several isothermic jumps (Figure 2.40) separated by zones of smooth flow. In this case as $\mu \to 0$ the shock transition tends to the curve $A_1A_2B_2A_3B_3A_4A_0$, as is clear from the fact that the integral curve of (5.3.1) always remains in zone II.

We note also that the phenomenon of an isothermic jump depends not only on the equations of state but also on the constants C_1, C_2 and C_3 defining the flow. Generally speaking, an isothermic jump occurs only for sufficiently strong shock waves. We shall demonstrate this below for the case of an ideal gas.

To conclude this subsection we remark that we have obtained stationary flows of a viscous, thermally conducting gas in the form of a "smeared-out" shock wave of the type shown in Figure 2.37. Indeed, the values V_0, T_0, u_0 ahead of the front and V_1, T_1, u_1 behind the front satisfy the Hugoniot conditions and the stability condition $S_1 > S_0$ ($V_1 < V_0$). Therefore, as $\mu \to 0$ and $\kappa \to 0$ we obtain in the limit from a solution $T = T(x)$, $V = V(x)$ of the system (5.1.20) and (5.1.21) a stable discontinuity (a shock wave) satisfying the conditions of dynamic compatibility and the stability condition.

There is the conjecture that any solution of the equations of gas dynamics containing stable discontinuities can be considered as a limit solution of the equations of gas dynamics with viscosity and thermal conductivity as the coefficients of viscosity μ and thermal conductivity κ tend to zero. So far there has been no example to refute this conjecture, although it has not been proved either. The latter is related to the difficulties which arise in a precise consideration of the Cauchy problem for the nonlinear equations describing the flow of a viscous, thermally conducting gas. In the present section we have shown only that this conjecture is true for stationary flows, i.e., for a constant shock wave which exists an indefinitely long time.

5. *Shock transition for the case of an ideal gas* (*Becker's investigation*). The first investigation of a shock transition in a viscous, thermally conducting gas was carried out by Becker (1921). He considered the case of an ideal gas

$$p = \frac{RT}{V} = \frac{R\varepsilon}{c_V V}, \qquad \varepsilon = c_V T. \tag{1}$$

In dimensionless variables (1) takes the form

$$T = pV, \qquad \varepsilon = \frac{1}{\gamma - 1} T = \frac{1}{\gamma - 1} pV, \tag{2}$$

while the functions $\mathscr{L}$ and $\mathscr{M}$ specialize as follows:

$$\mathscr{M}(V, T) = \frac{T}{V} + V - 1, \tag{3}$$

$$\mathscr{L}(V, T) = \frac{1}{\gamma - 1} T - \frac{1}{2}(V - 1)^2 - \beta. \tag{4}$$

Thus, the curve $\mathscr{M}(V, T) = 0$ is the parabola

$$T = V - V^2 = -\left(V - \frac{1}{2}\right)^2 + \frac{1}{4}, \tag{5}$$

which is concave downward with the line $V = 1/2$ as axis and vertex at the point $V = 1/2$, $T = 1/4$ (Figure 2.41). The curve $\mathscr{L}(V, T) = 0$ is also a parabola

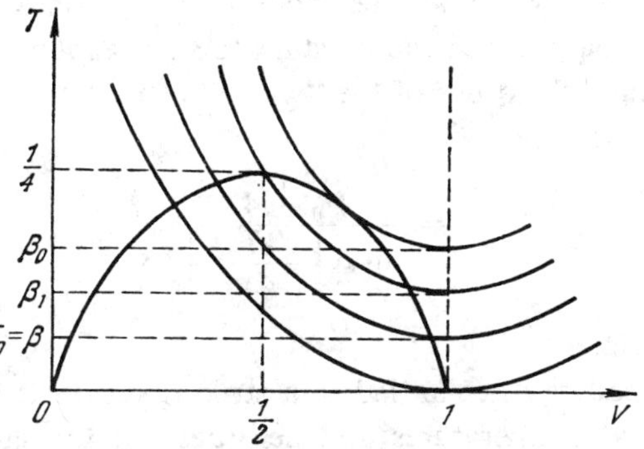

FIGURE 2.41

$$T = \frac{\gamma - 1}{2}(V - 1)^2 + \beta(\gamma - 1), \tag{6}$$

which is convex upward with axis $V = 1$ and vertex at the point $V = 1$, $T = \beta(\gamma - 1)$. Thus, the form of the parabola (6) does not change, while its position depends on β, i.e., on the flow constants.

The parabolas $\mathscr{L} = 0$ and $\mathscr{M} = 0$ intersect at two points for $0 \leq \beta < \beta_0$; the cases $\beta > \beta_0$ and $\beta < 0$ are not physically realizable, since in the first case there are no points of intersection and hence no asymptotic values of V and T, while in the second case $T < 0$. The case $\beta = 0$ corresponds to an infinitely strong shock wave, since $T_1/T_0 = p_1/p_0 = \infty$, while the case $\beta = \beta_0 = 1/2(\gamma^2 - 1)$ corresponds to an infinitely weak shock transition, since in this case the points (V_0, T_0) and (V_1, T_1) coalesce, i.e., $T_1/T_0 = p_1/p_0 = 1$.

In the case $\kappa = 0$, $\mu \neq 0$ we can integrate (5.1.20) and (5.1.21). Indeed,

$$T(x) = \frac{\gamma - 1}{2}[V(x) - 1]^2 + \beta(\gamma - 1), \tag{7}$$

while $V(x)$ satisfies

$$\mu \frac{dV}{dx} = \frac{\gamma + 1}{2} \frac{(V - V_0)(V - V_1)}{V} \quad \text{for } V_1 \leq V \leq V_0. \tag{8}$$

Hence,

$$\frac{2}{\gamma + 1} \int \frac{V\,dV}{(V - V_0)(V - V_1)}$$
$$= \frac{2}{\gamma + 1} \ln\left[(V_0 - V)^{V_0/(V_0 - V_1)}(V - V_1)^{-V/(V_0 - V_1)}\right] = \frac{\kappa}{\mu} + \text{const}. \tag{9}$$

Figure 2.42, a shows the approximate form of the profile $V = V(x)$ defined by (9). The graph of $V(x)$ approaches the constant values $V = V_1$ and $V = V_0$ as $x \to \pm\infty$, so that the "width" of the shock wave is, strictly speaking, infinite. However, this approach to constant values occurs exponentially, i.e., rather rapidly. To determine the order of the "width" of the zone of the shock wave we define it as the quantity

$$l = \frac{V_0 - V}{\max\left|\dfrac{dV(x)}{dx}\right|} \tag{10}$$

(see also Figure 2.42, a).

Computations made by Becker led to a striking result. It was found for the majority of gases at temperatures and densities not too large the quantities μ and κ are such that the width of the zone of a shock transition is of the order

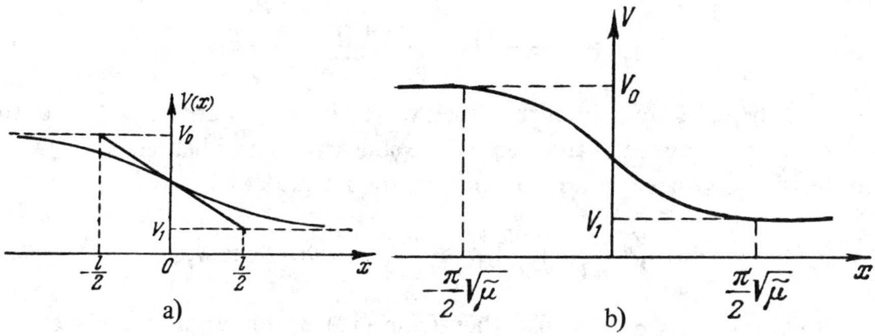

FIGURE 2.42

10^{-4}–10^{-6} cm, i.e., of the order of the mean free path of a molecule of the gas.

If we disregard the nonequilibrium of hydrodynamic flows at distances of the order of the mean free path of molecules this suggests that to a high degree of accuracy a shock transition can be effectively replaced by a travelling discontinuity (a shock wave) with left and right limit values which satisfy the Hugoniot conditions and the condition of increase of entropy (the stability condition). This justifies the generally accepted point of view according to which the flow can be divided into regions of reversible processes where the equations of hydrodynamics without dissipative terms prevail and into regions of irreversible processes which constitute narrow zones and can be effectively described by moving surfaces of discontinuity. We further note that a more precise idea of the zone of a shock transition can be obtained only by using the Boltzmann equation, but the estimate of the order of the width of the zone of the shock wave remains the same.

To conclude this subsection we observe that in the case $\mu = 0$, $\kappa \neq 0$ an isothermic jump exists only for sufficiently strong shock waves where $0 \leq \beta \leq \beta_1$, as is readily apparent from Figure 2.41.

We present the formulas (5.1.13) and (5.1.14), needed below, in Lagrange variables q, t:

$$\bar{\mu}\frac{dV}{dq} = C_1\left(V - \frac{C_2}{C_1^2}\right) + \frac{p(V,T)}{C_1}, \qquad \bar{\mu} = \mu\rho = \frac{\mu}{V}; \tag{11}$$

$$\bar{\kappa}\frac{dT}{dq} = C_1\varepsilon - \frac{1}{2}C_1^3\left(V - \frac{C_2}{C_1^2}\right)^2 - C_3 + \frac{1}{2}\frac{C_2^2}{C_1}, \qquad \bar{\kappa} = \rho\kappa. \tag{12}$$

For an ideal gas, $\varepsilon = (\gamma - 1)^{-1}pV$. We consider the case $\kappa = 0$, $\mu \neq 0$. From (11) and (12) we obtain the equation for V:

$$\bar{\mu}\frac{dV}{dq} = \frac{\gamma + 1}{2} m \frac{(V - V_1)(V - V_0)}{V}, \tag{13}$$

$m = C_1$ is the mass velocity of the shock wave; V_0 and V_1 are the values of the specific volume V ahead of the front of the shock wave and behind it.

In the case μ = const from (13) we have the integral

$$\ln\left(\frac{V_0 - V}{V - V_1}\right)^{1/(V_0 - V)} = \frac{\gamma + 1}{2\mu} mq + \text{const}; \tag{14}$$

if, for example, $\tilde{\mu} = \mu\rho$ = const, then from (13) we obtain a formula closely related to (9):

$$\ln \frac{(V_0 - V)^{V_0/(V_0 - V_1)}}{(V - V_1)^{V/(V_0 - V_1)}} = \frac{\gamma + 1}{2\bar{\mu}} mq + \text{const}. \tag{15}$$

Assuming μ = const and using the definition of the width of the zone of the shock wave according to (10), from (13) we obtain an expression for the width Δq of the zone:

$$\Delta q = \frac{8\mu}{\gamma + 1} \frac{1}{|\Delta u|},$$

where $\Delta u = u_1 - u_0$ is the jump of the velocity on the shock wave.

6. ***Stationary solutions of the equations of gas dynamics with a von Neumann-Richtmyer viscosity.*** We now consider a stationary solution of the equations of gas dynamics for a polytropic gas with thermal conductivity equal to zero and a "viscosity" in (5.1.1)–(5.1.3) of special form:[13]

$$\omega = \frac{\lambda}{V} \frac{\partial u}{\partial x} \left| \frac{\partial u}{\partial x} \right|, \tag{1}$$

i.e., in the preceding treatment we formally set the coefficient μ equal to

$$\mu = \frac{\lambda}{V} \left| \frac{\partial u}{\partial x} \right|. \tag{2}$$

Since according to (5.1.12) and (5.1.16)

$$u = mV, \tag{3}$$

it is easy to see that for a polytropic gas the determination of a stationary solution reduces to integrating the equation

$$\bar{\mu}\frac{dV}{dx}\left|\frac{dV}{dx}\right| = (V - V_0)(V - V_1) \tag{4}$$

[13] A viscosity of this type was first considered by von Neumann and Richtmyer (1950).

(see (5.5.8)) where $\bar{\mu} = 2(\gamma + 1)^{-1}m\lambda$. For the case $m > 0$ we have $\bar{\mu} > 0$ and $dV/dx < 0$ (see Figure 2.42, a), and (4) can therefore be represented in the form

$$\bar{\mu}\left(\frac{dV}{dx}\right)^2 = (V - V_1)(V_0 - V) > 0 \tag{5}$$

and it is easily integrated:

$$\arcsin \frac{V - (V_0 + V_1)/2}{(V_0 - V_1)/2} = \frac{x}{\sqrt{\bar{\mu}}} + C. \tag{6}$$

Figure 2.42, b shows the profile $V = V(x)$ for the solution (6). An essential difference from the preceding case is the finite width $\pi\sqrt{\bar{\mu}}$ of the shock transition. In connection with this we observe that the approach of $V = V(x)$ to the constant values V_0 and V_1 is not analytic, since the second derivative $V''(x)$ at the points $x = \pm 2^{-1}\pi\sqrt{\bar{\mu}}$ suffers a discontinuity; this is easily verified either from (4) or from the formula (6) for the solution.

If in (5.5.13) we similarly set

$$\bar{\mu} = \frac{\lambda}{V}\left|\frac{\partial u}{\partial q}\right|, \tag{7}$$

then we arrive at the equation

$$\bar{\mu}\left|\frac{dV}{dq}\right|\frac{dV}{dq} = (V - V_0)(V - V_1), \qquad \bar{\mu} = \frac{2\lambda}{\gamma + 1}, \tag{8}$$

with the integral

$$\arcsin \frac{V - (V_0 + V_1)/2}{(V_0 - V_1)/2} = \frac{q}{\sqrt{\bar{\mu}}} + C, \tag{9}$$

and the width of the zone of the shock transition is

$$\pi\sqrt{\bar{\mu}} = \pi\sqrt{\frac{2\lambda}{\gamma + 1}}.$$

We remark that in the case of Lagrange coordinates the von Neumann-Richtmyer viscosity leads to a finite width of the shock front which does not depend on the strength of the shock wave.

In some cases a finite width of the shock transition is essential. The von Neumann-Richtmyer "viscosity" (1) is therefore widely used in numerical computations of discontinuous solutions of the equations of gas dynamics. In the equations of gas dynamics for gases free of internal friction and thermal conductivity an artifical "viscosity" of the type (1) with a small coefficient λ is introduced. This makes it possible to spread shock transitions out over a finite region, and this is convenient for numerical computations. This is discussed in more detail in Chapter 3.

7. *Shock transition in a medium with anomalous thermodynamic properties.* We shall briefly consider the question that we raised in §4.8 concerning the admissibility of a shock transition on the basis of the methods developed in this section. Dropping the condition $p''_{VV}(V, S) > 0$ leads to additional requirements on the shock transition which we obtain using the viscosity method. This question was considered by Galin (1958), Sidorenko (1968), Wendroff (1972), and Liu (1975). We call a shock transition (u_0, V_0, p_0), (u_1, V_1, p_1) *admissible* if there exists a solution of the system (5.1.12) for a viscous, thermally conducting gas with positive coefficients of the viscosity $\mu > 0$ and thermal conductivity $\kappa \geq 0$ which joins the points (u_0, V_0, p_0) and (u_1, V_1, p_1). A discontinuity which can be represented as a sequence of several discontinuities admissible in the indicated sense which move relative to the medium with the same velocity is also considered to be admissible.

We note first of all that for any admissible discontinuity and any medium satisfying only the condition $T = \partial\varepsilon(V, S)/\partial S > 0$ the inequality

$$S_1 \geq S_0 \tag{1}$$

is always satisfied, where S_0 is the entropy ahead of the front of the shock wave, and S_1 is the entropy behind the front. Indeed, writing equation (5.1.7) for the entropy in the form

$$\frac{\partial \rho S}{\partial t} + \frac{\partial \rho u S}{\partial x} = \frac{\mu}{T}\left(\frac{\partial u}{\partial x}\right)^2 + \frac{\kappa}{T^2}\left(\frac{\partial T}{\partial x}\right)^2 + \frac{\partial}{\partial x}\left(\kappa \frac{\partial \ln T}{\partial x}\right),$$

we obtain in analogy to (5.1.9)–(5.1.12) an equation for the entropy in the shock transition

$$\frac{d}{dx}\left[\rho u S - \kappa \frac{\partial \ln T}{\partial x}\right] = \frac{\mu}{T}\left(\frac{du}{dx}\right)^2 + \frac{\kappa}{T^2}\left(\frac{dT}{dx}\right)^2. \tag{2}$$

If it is assumed that the shock wave moves to the left relative to the medium ($C_1 > 0$), then S_0 is the value of the entropy as $x \to -\infty$, and S_1 is the entropy as $x \to +\infty$.

Integrating (2) from the point $x = -\infty$ ($S = S_0$, $d(\ln T)/dx = 0$), we obtain

$$\rho u S = \rho_0 u_0 S_0 + \kappa \frac{d \ln T}{dx} + \int_{-\infty}^{x} \left[\frac{\mu}{T}\left(\frac{du}{dx}\right)^2 + \frac{\kappa}{T^2}\left(\frac{dT}{dx}\right)^2\right] dx. \tag{3}$$

Now $\rho u = C_1 > 0$ and $d(\ln T)/dx \to 0$ as $x \to +\infty$; therefore condition (1) is always satisfied, since the integral on the right side of (3) is nonnegative.

We have already indicated in §4.8 that condition (1) is insufficient to distinguish admissible shock transitions in the case of $p''_{VV}(V, S)$ of variable sign. We shall obtain additional conditions by assuming that for the shock transition (u_0, V_0, p_0), (u_1, V_1, p_1) there exists a solution of (5.1.12) for $\mu > 0$ and $\kappa = 0$. From (5.1.14) we then obtain

$$\mathscr{L} = \varepsilon - \frac{1}{2} C_1^2 \left(V - \frac{C_2}{C_1^2} \right)^2 - \frac{C_3}{C_1} + \frac{1}{2} \frac{C_2^2}{C_1^2} = 0, \tag{4}$$

and the problem of determining the curve of the shock transition reduces to integrating the single equation (5.1.13):

$$\mu \frac{dV}{dx} = \frac{1}{C_1} \left[C_1^2 \left(V - \frac{C_2}{C_1^2} \right) + p \right] = \mathscr{M}. \tag{5}$$

We shall consider (4) and (5) in the plane of the variables V and p. As at all points of the shock transition considered ($\kappa = 0$), we have $\mathscr{L} = 0$ at the point (V_0, p_0), and therefore

$$\begin{aligned} \mathscr{L}(V, p) &= \mathscr{L}(V, p) - \mathscr{L}(V_0, p_0) \\ &= \varepsilon - \varepsilon_0 - \frac{1}{2} \cdot C_1^2 \left[V - \frac{C_2}{C_1^2} \right]^2 + \frac{1}{2} C_1^2 \left[V_0 - \frac{C_2}{C_1^2} \right]^2. \end{aligned} \tag{6}$$

At (V_0, p_0) the quantity $\mathscr{M}$ is equal to zero, and we can therefore find C_2 from this:

$$C_2 = p_0 + C_1^2 V_0. \tag{7}$$

We substitute this into formula (6) for $\mathscr{L}(V, p)$:

$$\begin{aligned} \mathscr{L}(V, p) &= \varepsilon - \varepsilon_0 + p_0(V - V_0) - \frac{1}{2} C_1^2 (V - V_0)^2 \\ &= \varepsilon - \varepsilon_0 + p_0(V - V_0) - \frac{1}{2} m^2 (V - V_0)^2. \end{aligned} \tag{8}$$

The condition for the existence of an integral curve of (5) joining the points (V_0, p_0) as $x \to -\infty$ and (V_1, p_1) as $x \to +\infty$ has the form

$$\begin{aligned} C_1(V - V_0)\mathscr{M}(V, p) &= C_1(V - V_0)[\mathscr{M}(V, p) - \mathscr{M}(V_0, p_0)] \\ &= [C_1^2(V - V_0) + (p - p_0)](V_1 - V_0) > 0 \end{aligned} \tag{9}$$

for $(V - V_0)(V - V_1) < 0$.

This condition must be satisfied at all points of the curve $\mathscr{L}(V, p) = 0$, where $\mathscr{L}(V, p)$ is given by (8). Dividing (9) by the positive quantity $(V - V_0)(V_1 - V_0)$, we obtain

$$\frac{p - p_0}{V - V_0} > -C_1^2 = -m^2 = \frac{p_1 - p_0}{V_1 - V_0}; \qquad (V - V_0)(V - V_1) < 0. \qquad (10)$$

Condition (10) along the curve $\mathscr{L}(V, p) = 0$ guarantees the existence of an integral curve for the shock transition. Thus, this is the condition for the admissibility of a shock transition that we were seeking. If on the curve $\mathscr{L} = 0$ between the points (V_0, p_0) and (V_1, p_1) there is a point (V^*, p^*) at which

$$\frac{p^* - p_0}{V^* - V_0} = -m^2; \qquad \mathscr{L}(V^*, p^*) = 0; \qquad (V^* - V_0)(V^* - V_1) < 0, \qquad (11)$$

then obviously there exists no integral curve $V = V(x)$ joining (V_0, p_0) and (V_1, p_1). This is related to the fact that on an integral curve issuing from (V_0, p_0) during approach to a point (V^*, p^*) of the curve $\mathscr{L} = 0$ nearest to it at which (11) is satisfied the coordinate $x \to +\infty$, since the integral

$$\Delta x = \int_V^{V^*} \frac{dV}{\mathscr{M}(V, p)}$$

diverges (we here assume that $\varepsilon(V, p)$ is a differentiable function). Moreover, if on some segment of $\mathscr{L} = 0$ inequality (11) is satisfied, then on this segment equation (5) does not specify the dependence $V = V(x)$ at all, since $\mathscr{M} = 0$ and $V = \text{const}$. A discontinuity (V_0, p_0), (V_1, p_1) for which a generalization of (10) is satisfied,

$$\frac{p - p_0}{V - V_0} \geq -C_1^2 = -m^2 = \frac{p_1 - p_0}{V_1 - V_0};$$
$$\mathscr{L}(V, p) = 0; \qquad (V - V_0)(V - V_1) < 0, \qquad (12)$$

is nevertheless to be considered admissible. This is related to the fact that in the case of isolated points of the curve $\mathscr{L}(V, p) = 0$ at which (11) is satisfied the discontinuity (V_0, p_0), (V_1, p_1) may be considered as a sequence of discontinuities each of which is admissible, and they all move relative to the medium with the same velocity, as follows from (11). Finally, segments of the curve on which (11) is satisfied should be considered to be a set of admissible shock transitions.

In the present case the lack of an integral curve for problem (4), (5) is connected with the lack of a self-similar solution $V = V(x - Ut)$ in the problem of the structure of the shock transition, and not with the failure of the method of viscosity for selecting admissible solutions. The situation is that on an entire segment of the Hugoniot adiabat the Lagrangian velocity of propagation of the shock wave is constant[14] and equal to $m = \sqrt{(p_1 - p_0)/(V_0 - V_1)}$. Due to this,

[14] It is obvious that a segment of the curve $\mathscr{L} = 0$ satisfying condition (10) is simultaneously a segment of the Hugoniot adiabat.

in the problem of the structure of the shock wave there is no free parameter (no velocity of the shock wave), and this problem has no solution.

The situation is similar to the case of linear hyperbolic equations (for example, the equations of acoustics), where the introduction of a viscosity does not lead to a spread-out stationary wave but rather to a progressive increase in the zone of broadening of the shock transition. Thus, for example, the linear equation

$$\frac{\partial u}{\partial t} + a\frac{\partial u}{\partial x} = \mu\frac{\partial^2 u}{\partial x^2} \qquad (a = \text{const}, \mu > 0),$$

has no self-similar solution $u = u(x - Ut)$ assuming different values as $x \to -\infty$ and $x \to +\infty$. Nevertheless, by solving the Cauchy problem for this equation and then letting $\mu \to 0$, we can obtain in the limit an arbitrary discontinuity (shock transition) propagating along the line $x = at + x_0$. Apparently, the situation is analogous in the case where (11) is satisfied on segments of the Hugoniot adiabat $H(V, p; V_0, p_0) = 0$.

Limiting ourselves to these brief remarks regarding the method of viscosity, we consider in more detail the condition of admissibility of a shock transition (12). We consider the transition (V_0, p_0), (V_1, p_1) admissible if conditions (12) are satisfied. We observe immediately that (12) implies that $S_1 \geq S_0$; here $S_1 = S_0$ only in the case where the line (11) is the Hugoniot adiabat, i.e., all points of the ray $(p - p_0)/(V - V_0) = -m^2$ for $V_1 \leq V \leq V_0$ are points of the Hugoniot adiabat. It is evident from this that in the present shock transition we are dealing with a medium with an equation of state given by

$$p = -m^2V + (p_0 + m^2V_0) = -m^2/\rho + C,$$

i.e., with a Chaplygin gas.

We shall now formulate the admissibility conditions (12) so that they appear as conditions on the Hugoniot adiabat $H(V, p; V_0, p_0) = 0$. We shall assume that the medium is such that the following conditions are satisfied:

$$T = \frac{\partial \varepsilon(V, S)}{\partial S} > 0; \qquad p'_V(V, S) < 0, \qquad p'_S(V, S) > 0. \tag{13}$$

To be specific we assume that in some neighborhood of the point (V_0, p_0) the Hugoniot adiabat $H(V, p; V_0, p_0) = 0$ satisfies the condition $S(V, p) > S(V_0, p_0)$ for $V < V_0$, i.e., the shock adiabat begins as a compression shock wave. This occurs, for example, for $p''_{VV}(V_0, S_0) > 0$. On the line

$$\frac{p - p_0}{V - V_0} = -m^2 = \frac{p_1 - p_0}{V_1 - V_0}, \tag{14}$$

it is obvious that $\mathscr{M} = 0$. This implies that all points of intersection of the adiabat $H(V, p; V_0, p_0) = 0$ with the line (14) are simultaneously zeros of the function $\mathscr{L}(V, p)$, i.e., points of intersection of the curve $\mathscr{L}(V, p) = 0$ with the line (14).

Moreover,

$$\frac{\partial \mathscr{L}(V,p,K)}{\partial K} = \frac{\partial}{\partial K}\left[\varepsilon - \varepsilon_0 + p_0(V - V_0) - \frac{1}{2}K(V - V_0)^2\right]$$
$$= -\frac{1}{2}(V - V_0)^2 < 0 \quad \text{for } V \neq V_0, \tag{15}$$

and

$$\frac{\partial \mathscr{L}(V, p, K)}{\partial p} = \frac{\partial \varepsilon(V, p)}{\partial p} = \frac{\varepsilon'_S(V, S)}{p'_S(V, S)} > 0. \tag{16}$$

Consideration of these circumstances leads to the conclusion that the curve $\mathscr{L}(V, p, m^2) = 0$ is situated between the line (14) and the Hugoniot adiabat $H(V, p; V_0, p_0) = 0$, and they have common points of intersection or tangency. Figure 2.43 shows the relative positions of the line $\mathscr{M} = 0$, the Hugoniot adiabat $H = 0$, and the curve $\mathscr{L} = 0$ in the case $p''_{VV}(V_0, S_0) > 0$.

It becomes clear that condition (12) for admissibility of a shock transition can also be formulated as follows if the inequalities (13) are satisfied: a shock transition (V_0, p_0), (V_1, p_1) is admissible if the Hugoniot adiabat

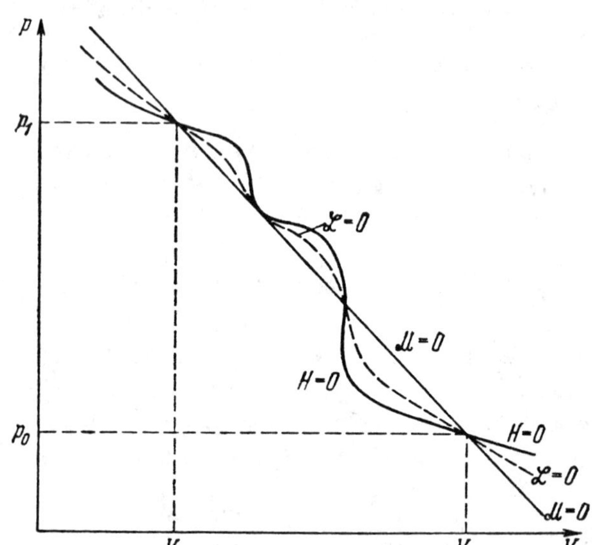

FIGURE 2.43

$H(V, p; V_0, p_0) = 0$ for $V_1 \leq V \leq V_0$ lies in the V, p plane to the right of the ray (14) (see Figure 2.43). This can be expressed by a formula analogous to (12):

$$\frac{p - p_0}{V - V_0} \geq -m^2 = \frac{p_1 - p_0}{V_1 - V_0} \quad \text{for } H(V, p; V_0, p_0) = 0;$$
$$V_1 \leq V \leq V_0. \qquad (17)$$

We now consider the possibility of a shock transition from the point V_0, p_0 to the point V_1, p_1 for $V_1 > V_0$, i.e., the case in which the discontinuity is a rarefaction wave. In this case also the condition for the existence of an integral curve of equation (5) can be written in the form (9) or (10). We remark only that for $V_1 > V_0$ condition (10) requires that the curve $\mathscr{L} = 0$ lie to the right of the ray (14) as shown in Figure 2.44, where AB is tangent to the curve $H = 0$ at the point $A(V_0, p_0)$. According to what has been said above, this simultaneously requires that the Hugoniot adiabat $H(V, p; V_0, p_0) = 0$ for $V_0 < V < V_1$ lie to the right of the line (14).

However, for $p''_{VV}(V_0, S_0) > 0$ this is possible only under the condition that the function $p''_{VV}(V, S)$ have variable sign and the quantity $(V_1 - V_0)$ be sufficiently large. Indeed, according to §4.3, in some neighborhood of (V_0, p_0), while intersecting the line (14) at the two points (V_0, p_0) and (V_1, p_1), the Hugoniot adiabat $H = 0$ remains to the left of it for $(V - V_0)(V - V_1) < 0$.

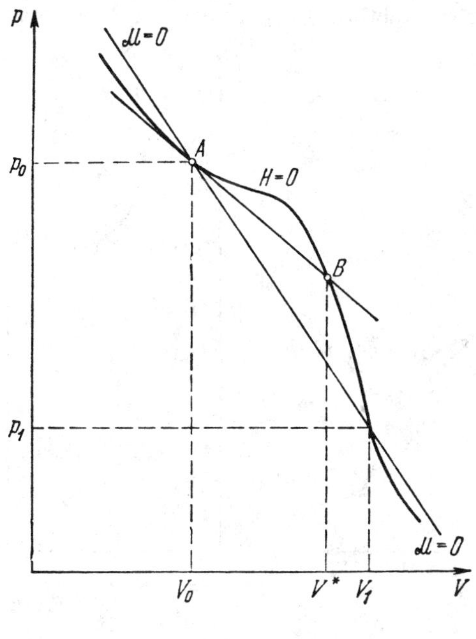

FIGURE 2.44

Therefore, for $p''_{VV}(V_0, S_0) > 0$ a rarefaction shock wave is possible only if $V_1 - V_0 > 0$ is sufficiently large.

We now note that for $p''_{VV}(V_0, S_0) < 0$ conditions (10) and (12), as before, provide conditions for the admissibility of a shock transition. The difference in this case is that for $V_1 > V_0$ rarefaction shock waves of small amplitude are possible, while compression shock waves ($V_1 < V_0$) are possible only for sufficiently large amplitude and $p''_{VV}(V, S)$ of variable sign. The case $p''_{VV}(V_0, S_0) = 0$ occupies an intermediate position and requires special consideration that is not difficult.

We summarize our conclusions in the case of $p''_{VV}(V, S)$ of variable sign. We have established that the set of values (V_1, p_1) behind the front of a shock transition from a point (V_0, p_0) ahead of the front lies on the Hugoniot adiabat $H(V, p; V_0, p_0) = 0$ and must satisfy (12). This implies that for $V_1 < V_0$ the entire segment $V_1 \leq V \leq V_0$ of the adiabat $H(V, p; V_0, p_0) = 0$ must be located to the left of the line (14) or partially coincide with it; for $V_1 > V_0$ the entire segment of the adiabat $H(V, p; V_0, p_0) = 0$ contained between the points (V_0, p_0), (V_1, p_1) must lie to the right of the line (12). Thus, condition (12) distinguishes portions of the Hugoniot adiabat which are physically formed by admissible shock transitions. Figure 2.45 shows the admissible portions of the Hugoniot adiabat for the two cases (they are distinguished by the double lines).

Thus, by applying the method of viscosity, we have obtained the criterion

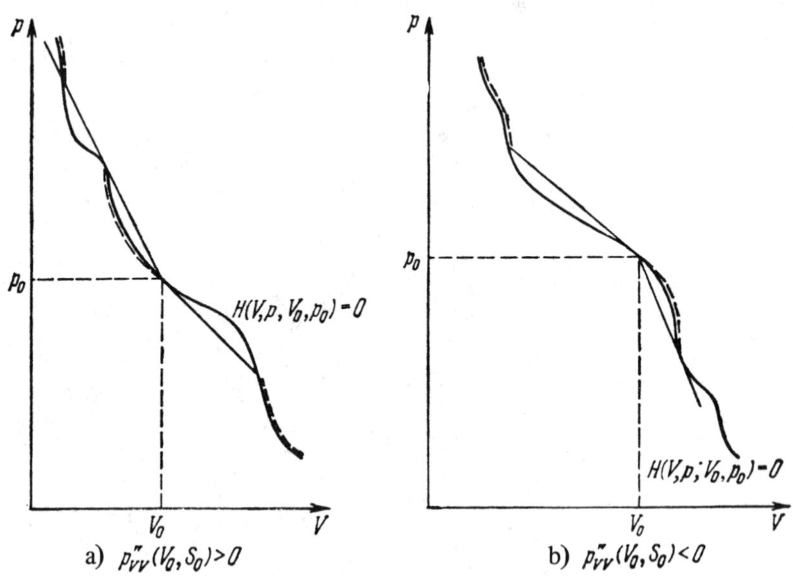

FIGURE 2.45

(12) for the admissiblity of a shock transition. In this investigation we have assumed that the thermal conductivity of the medium is equal to zero. Consideration of a shock transition in a viscous and thermally conducting gas does not alter the criterion (12) for the admissibility of a shock wave.

§6. The problem of the decay of an arbitrary discontinuity

1. ***General properties of the solution of the problem of the decay of a discontinuity.*** In this section we consider in detail the problem of the decay of an initial discontinuity. In the case of an isothermic gas this problem was posed and solved by Riemann (1860). A qualitative investigation of the problem of the decay of a discontinuity for polytropic gases was carried out by Kočin (1926) and for normal gases by Landau and Lifšic (1953).

An *arbitrary discontinuity* is an initial state of two infinite masses of gas characterized by constant parameters $u_1, p_1, V_1, \varepsilon_1, T_1$ and $u_0, p_0, V_0, \varepsilon_0, T_0$ adjoining along the plane $x = 0$ at the initial time $t = 0$. Here the magnitudes of the discontinuity to the left and right are arbitrary and subject only to the equations of state of the gases which may be different for the neighboring gases. The determination of the flow arising for $t > 0$ with these initial conditions is called the *problem of the decay of an arbitrary discontinuity.* Thus, the problem of the decay of the discontinuity is the problem of determining a one-dimensional flow with plane symmetry ($\nu = 0$) satisfying the integral conservation laws

$$\left.\begin{gathered}\oint_C \rho\,dx - \rho u\,dt = 0, \qquad \oint_C \rho u\,dt - [p + \rho u^2]\,dx = 0,\\ \oint_C \rho\left(\varepsilon + \frac{u^2}{2}\right)dt - \left[\rho u\left(\varepsilon + \frac{p}{\rho} + \frac{u^2}{2}\right)\right]dx = 0,\end{gathered}\right\} \tag{1}$$

and the piecewise constant initial conditions: for $t = 0$

$$\left.\begin{aligned} &u = u_1, \quad V = V_1, \quad p = p_1, \quad T = T_1, \quad \varepsilon = \varepsilon_1 \quad \text{for } x < 0;\\ &u = u_0, \quad V = V_0, \quad p = p_0, \quad T = T_0, \quad \varepsilon = \varepsilon_0 \quad \text{for } x > 0.\end{aligned}\right\} \tag{2}$$

In §4 we have seen that conditions of dynamic compatibility (the Hugoniot conditions) must be satisfied on a stable discontinuity. In the case of a shock wave a stability condition (the increase of entropy) is also satisfied on the discontinuity; on a contact discontinuity (the boundary of two gases) the pressure and velocity are continuous. Therefore, if an arbitrary discontinuity is not a contact discontinuity or a shock wave it decomposes by forming some configuration of stable discontinuities and continuous, gas-dynamical flows.

It is easy to see that if we make a similarity transformation of the independent variables $t' = kt$, $x' = kx$ $(k > 0)$, then in the new variables x', t' as well the search for a solution of the decay problem reduces to finding a solution of the conservation laws (1) satisfying the initial conditions (2), provided that by x and t we now understand x' and t'. If we assume the uniqueness of the solution of problem (1), (2), then this implies that

$$z(x, t) = \bar{z}(x', t') = \bar{z}(kx, kt). \tag{3}$$

Here the vector z denotes the collection of hydrodynamical variables $z = \{\rho, u, p, \varepsilon, \ldots\}$, while $\bar{z}$ denotes the same quantities in the variables x', t'.

Setting $k = 1/t > 0$ in (3), we obtain

$$z(x, t) = \bar{z}\left(\frac{x}{t}, 1\right) = z_0\left(\frac{x}{t}\right). \tag{4}$$

Thus, the assumption of the uniqueness of the solution implies the *self-similarity* of the solution of the problem on the decay of an arbitrary discontinuity, i.e., the dependence of all hydrodynamical variables only on the single variable $y = x/t$. In particular, this implies that lines of discontinuity—shock waves and a contact discontinuity—are straight lines in the plane of the variables x and t, i.e., the velocities of shock waves and contact discontinuity are constant.[15]

In §3.2 we have seen that a self-similar flow that is continuous for $t > 0$ is a centered, rarefaction Riemann wave characterized by constancy of the entropy S and one of the Riemann invariants (r or s). Thus, the self-similar solution of the problem on the decay of a discontinuity contains as elements shock waves, rarefaction waves, and a contact discontinuity.

We shall establish some general properties of the self-similar solution of the decay problem which are valid for normal gases.

1. *Not more than one shock wave propagates in each of the gases* "1" (*the left gas*) *and* "0" (*the gas on the right*).

Indeed, by a shock wave we understand only a stable shock wave. As we saw in §4.4, from this we obtain Zemplén's theorem. Let us suppose, for example, that in the gas "0" there propagate two shock waves $x/t = D_1$ and $x/t = D_2$, $D_2 > D_1$ (Figure 2.46), while the flows in zones I, II, and III are constant. Denoting the speed of sound c in zones I, II, and III by c_{I}, c_{II}, and c_{III}, and the velocity u by u_{I}, u_{II}, and u_{III}, we have

$$c_{\mathrm{I}} < c_{\mathrm{II}} < c_{\mathrm{III}}, \qquad u_{\mathrm{I}} < u_{\mathrm{II}} < u_{\mathrm{III}}. \tag{5}$$

However, Zemplén's theorem requires that

[15] It should be kept in mind, however, that general theorems on the uinqueness of a solution of the Cauchy problem for (1) with discontinuous initial conditions have so far not been proved.

$$c_{\mathrm{I}} < D_2 - u_{\mathrm{I}}, \qquad D_2 - u_{\mathrm{II}} < c_{\mathrm{II}}, \tag{6}$$

$$c_{\mathrm{II}} < D_1 - u_{\mathrm{II}}. \tag{7}$$

It is obvious that (7) is not compatible with (6) under the condition $D_1 < D_2$, and this proves our assertion.

2. *In each of the gases there propagates not more than one centered rarefaction wave*; *in the the gas* "0" (*on the right*) $s =$ const *in the rarefaction wave*; *in the gas* "1" (*on the left*) $r =$ const.

The assertion follows easily from the equalities

$$u + c = y = \frac{x}{t} \quad (s = \text{const}), \qquad u - c = y = \frac{x}{t} \quad (r = \text{const}), \tag{8}$$

which are valid in centered rarefaction waves.

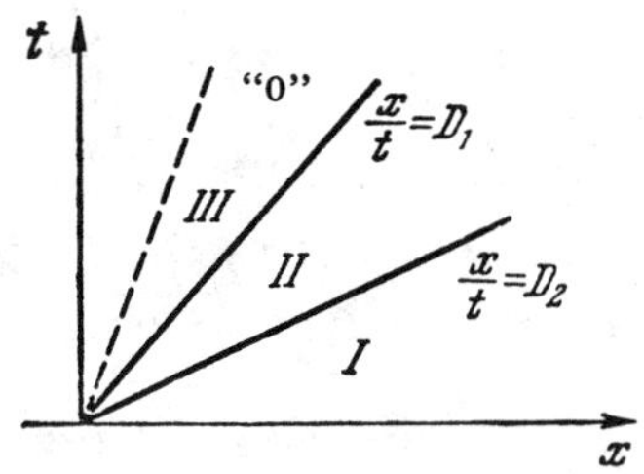

FIGURE 2.46

3. *The presence in one of the gases of a shock wave precludes the possibility of the propagation of a rarefaction wave in this same gas, and, conversely, the propagation in one of the gases of a centered rarefaction wave precludes the possibility of the propagation of a shock wave in it.*

The assertion follows easily from Zemplén's theorem, the equalities $D = y = x/t$, and conditions (8).

As a consequence of these properties, we find that the self-similar solution of the problem of decay of a discontinuity contains a contact discontinuity separating the gases ("0" and "1"); in each of the gases there propagates not more than one (travelling or shock) wave bounded by zones of constant flow. Thus, the construction of a self-similar solution of the decay problem consists in "patching together" elementary solutions (constant flows, centered waves) and the determination of the parameters characterizing these solutions and the discontinuities. Since the discontinuities and elementary solutions are determined by a finite number of parameters, this problem becomes a purely algebraic problem. We shall show below that for normal gases a self-similar solution of the problem on the decay of an arbitrary discontinuity exists and is unique, i.e., it is uniquely determined by the initial data (2).

We begin our treatment with the case of polytropic gases. The exponent of the Poisson adiabat γ for the gas "0" lying to the right of the point $x = 0$ we denote by γ_0, while that for the gas "1" we denote by γ_1. We shall apply the following method of considering the decay problem: on the basis of the specific case of conditions (2) in which the position of the shock waves and rarefaction waves is obvious (the configuration), by continuously varying the parameters of problem (2), we continuously vary the solution, passing through critical values of the parameters which separate one configuration from another.

2. *Configuration* A. Inasmuch as it is clear that a solution of problem (6.1.1), (6.1.2) depends essentially only on the difference $u_1 - u_0$, we shall set $u_0 = 0$. It suffices to consider only the case

$$p_1 \geq p_0. \tag{1}$$

We begin by considering the problem on decay of the discontinuity formed by two quiescent masses of polytropic gases. The problem is posed in the following manner. At the point $x = 0$ there is a partition separating the two masses of gas characterized by the parameters $\gamma_1, S_1, \rho_1, p_1, T_1, u_1 = 0$ on the left and by $\gamma_0, S_0, \rho_0, p_0, T_0, u_0 = 0$ on the right, and the following condition is satisfied:

$$p_1 > p_0. \tag{2}$$

At the time $t = 0$ the partition is removed, and the gases begin to move. Assuming the self-similarity of the motion (or, equivalently, the uniqueness of the solutions of the decay problem), we compute it.

Since no matter flows through the contact boundary (the boundary between the gases "0" and "1"), for each mass of gas the contact boundary may be considered as a piston. By condition (2) the piston begins to move in the direction of the gas "0" and moves away from the gas "1". If a constant (due to self-similarity) velocity U of the piston is prescribed, then the problem can be uniquely solved for each of the gases individually. In order to obtain a solution of the problem of the decay of the discontinuity, we must "patch together" solutions of these two piston problems by requiring that on the contact boundary the pressure p_- from the left is equal to the pressure p_+ from the right. The velocity U of the contact boundary and all the parameters determining the motion are found from this condition.

We know solutions of piston problems (see §§3.4 and 4.8); therefore, the arrangement of the discontinuities has the form shown in Figure 2.47 (configuration A). The four rays Γ_0, Γ_1, Γ_2, and Γ_3 divide the upper half plane into 5 regions. In regions I, II, III, and V we have constant flows, while in region IV we have a centered rarefaction wave. The ray Γ_3 is a shock wave, Γ_2 is a contact discontinuity, and Γ_1 and Γ_0 are lines of weak discontinuity on which the

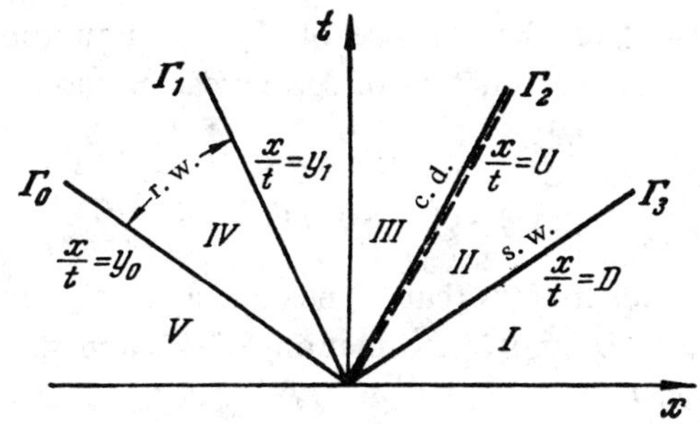

FIGURE 2.47

solution is continuous. On Γ_3 the Hugoniot conditions for a shock wave and the stability condition must be observed; on Γ_2 the pressure and velocity are continuous; on Γ_1 and Γ_0 all hydrodynamic variables are continuous.

In region IV p and u are connected by a relation following from the constancy of the Riemann invariant r (see (3.2.10) and (3.2.11)):

$$p = p_1\left[1 - \frac{\gamma_1 - 1}{2}\,\frac{u - u_1}{c_1}\right]^{2\gamma_1/(\gamma_1-1)} = p_1\left[1 - \frac{\gamma_1 - 1}{2}\,\frac{u}{c_1}\right]^{2\gamma_1/(\gamma_1-1)} \qquad (u_1 = 0). \tag{3}$$

Since zone III is a zone of constant flow, the velocity u on the characteristic Γ_1 is equal to the velocity U of the contact boundary Γ_2. Therefore, if we denote the pressure p in zone III by p_-, then, according to (3),

$$p_- = p_1\left[1 - \frac{\gamma_1 - 1}{2}\,\frac{U - u_1}{c_1}\right]^{2\gamma_1/(\gamma_1-1)} = p_1\left[1 - \frac{\gamma_1 - 1}{2}\,\frac{U}{c_1}\right]^{2\gamma_1/(\gamma_1-1)}. \tag{4}$$

On the shock wave Γ_3 we have (see (4.5.13)–(4.5.16))

$$p = p_0[(1 + h_0)M_0^2 - h_0], \qquad M_0 = \frac{|D - u_0|}{c_0} = \frac{D}{c_0}, \qquad h_0 = \frac{\gamma_0 - 1}{\gamma_0 + 1}, \tag{5}$$

$$u = u_0 + c_0(1 - h_0)\left(M_0 - \frac{1}{M_0}\right) = c_0\left(M_0 - \frac{1}{M_0}\right)(1 - h_0). \tag{6}$$

Considering that zone II is a zone of constancy of the flow and that on the contact boundary Γ_2 the pressure and velocity are continuous, we arrive at the equation

$$\begin{aligned} p_-(M_0) &= p_1\left[1 - \frac{\gamma_1 - 1}{2}\,\frac{(1 - h_0)c_0\,(M_0 - 1/M_0)}{c_1}\right]^{2\gamma_1/(\gamma_1-1)} \\ &= p_+(M_0) = p_0[(1 + h_0)M_0^2 - h_0] \end{aligned} \tag{7}$$

for determining M_0.

The function $p_-(M_0)$ on the left side of (7) is a monotonically decreasing function of M_0; $p_+(M_0)$ is a function of M_0 increasing monotonically to $+\infty$. Since for $M_0 = 1$

$$p_-(1) = p_1 > p_+(1) = p_0, \tag{8}$$

by assumption (2) it thus follows that (7) has one and only one root $M_0 > 1$.

We determine y_0, y_1, U, and D as functions of M_0 and show that the conditions of configuration A

$$y_0 < y_1 < U < D, \tag{9}$$

are satisfied. The inequalities (9) are always satisfied for $U > 0$, $U - u_1 = U > 0$. Indeed, the inequality $U = c_0(1 - h_0)(M_0 - 1/M_0) < D = c_0 M_0$ is obvious. For y_0 and y_1 we have

$$y_0 = u_1 - c_1 = -c_1, \qquad y_1 = U - c_-, \tag{10}$$

where c_- is the speed of sound in zone III, $c_- > 0$. This implies that $y_1 < U$. Finally, $r_1 = (r_1)_-$; therefore,

$$\begin{aligned} y_1 - y_0 &= U - c_- + c_1 = U + \frac{\gamma_1 - 1}{2}(U - u_1) \\ &= U + \frac{\gamma_1 - 1}{2} U = \frac{\gamma_1 + 1}{2} U > 0. \end{aligned} \tag{11}$$

Thus, all the conditions (9) are satisfied, and configuration A is consistent.

We now vary the parameters of the problem. We fix p_1 and p_0 ($p_1 > p_0$), and we vary the velocity u_1 of the gas "1". Setting $u_1 \neq 0$ in (3) and (4), we then arrive at the equation

$$\begin{aligned} p_-(M_0) &= p_1\left[1 - \frac{\gamma_1 - 1}{2}\frac{U - u_1}{c_1}\right]^{2\gamma_1/(\gamma_1-1)} \\ &= p_1\left[1 - \frac{\gamma_1 - 1}{2}\frac{c_0(1 - h_0)(M_0 - 1/M_0) - u_1}{c_1}\right]^{2\gamma_1/(\gamma_1-1)} \\ &= p_+(M_0) = p_0[(1 + h_0)M_0^2 - h_0] \end{aligned} \tag{12}$$

in place of (7). As before, $p_-(M_0)$ and $p_+(M_0)$ are monotone functions of M_0.

We trace the variation of the root M_0 depending on the variation of the parameter u_1. The following assertion holds: *the root M_0 is a monotonically increasing* *function of* u_1. Indeed, $p_-(M_0)$ is a monotone function of u_1. In particular, the value

$$p_-(1) = p_1\left[1 + \frac{\gamma_1 - 1}{2}\frac{u_1}{c_1}\right]^{2\gamma_1/(\gamma_1-1)},$$

to which point B of Figure 2.48 corresponds increases monotonically with increasing u_1. With increasing u_1 the point B and hence the entire curve $p = p_-(M_0, u_1)$ ascend monotonically (Figure 2.48). Since the curve $p = p_+(M_0)$ is fixed, the point C of intersection of the curves $p = p_-(M_0, u_1)$ and $p = p_+(M_0)$ moves to the right with increasing u_1, i.e., the root M_0 of (12) increases with increasing u_1, as required.

We denote by $u_B < 0$ the value of u_1 for which $p_-(1, u_1) = u_+$, i.e.,

$$1 + \frac{\gamma_1 - 1}{2}\frac{u_B}{c_1} = \left(\frac{p_0}{p_1}\right)^{(\gamma_1-1)/2\gamma_1},$$

or

$$u_B = \frac{2}{\gamma_1 - 1}c_1\left[\left(\frac{p_0}{p_1}\right)^{(\gamma_1-1)/2\gamma_1} - 1\right] < 0. \tag{13}$$

In correspondence with Figure 2.48 it is then obvious that for

$$u_B < u_1 \leq 0 \tag{14}$$

equation (12) always has a unique root $M_0 > 1$. For $M_0 > 1$ the shock wave Γ_3 (Figure 2.47) satisfies the stability condition; for $M_0 < 1$ the shock wave Γ_3 is unstable, and hence configuration A is impossible. In order that configuration

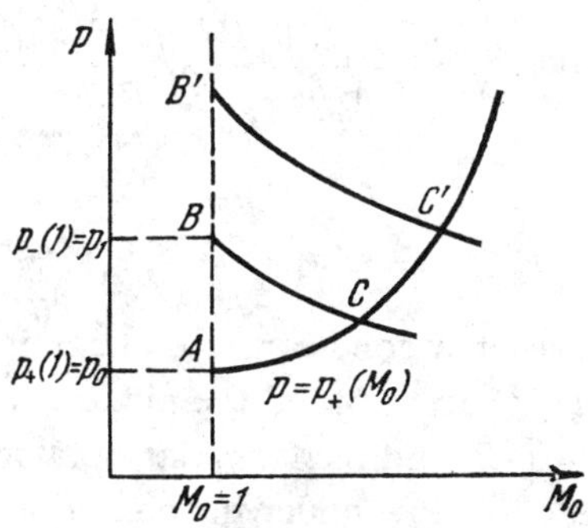

FIGURE 2.48

A be possible (Figure 2.47) it is necessary that the entire chain of inequalitites (9) be satisfied.

We therefore consider these inequalities for $u_1 > 0$. With increasing u_1 the quantities U and D increase to infinity in such a way that the inequality $U < D$ is preserved. The inequality $y_1 < U$ is also preserved. It remains to investigate the difference

$$y_1 - y_0 = (U - c_-) - (u_1 - c_1) = (U - u_1) - (c_- - c_1). \qquad (15)$$

Because of the constancy of the invariant r in zone IV we have

$$r = u_1 + \frac{2}{\gamma_1 - 1} c_1 = U + \frac{2}{\gamma_1 - 1} c_-, \qquad U - u_1 = \frac{2}{\gamma_1 - 1}(c_1 - c_-). \quad (16)$$

Substituting (16) into (15), we obtain

$$y_1 - y_0 = \frac{\gamma_1 + 1}{2}(U - u_1). \qquad (17)$$

We have shown that M_0 increases with increasing u_1. Now from (12) it follows that with increasing M_0 the difference $U - u_1$ decreases. Hence, it follows from (17) that with increasing u_1 the quantity $y_1 - y_0$ decreases. For $U - u_1 > 0$ we have $y_1 - y_0 > 0$. For $U - u_1 = 0$ we have $y_1 - y_0 = 0$. The latter is satified provided that the equalities

$$p_1 = p_0[(1 + h_0)M_{0\,\mathrm{cr}}^2 - h_0], \qquad (18)$$

$$u_1 = U = (1 - h_0)\, c_0\left(M_{0\,\mathrm{cr}} - \frac{1}{M_{0\,\mathrm{cr}}}\right) = u_\mathrm{b}, \qquad (19)$$

containing $M_{0\,\mathrm{cr}}$ as a parameter are simultaneously satisfied.

It is not hard to find u_b explicitly:

$$u_\mathrm{b} = (1 - h_0)c_0\left[\sqrt{\frac{p_1/p_0 + h_0}{1 + h_0}} - \sqrt{\frac{1 + h_0}{p_1/p_0 + h_0}}\right] > 0. \qquad (20)$$

Hence, if the conditions

$$u_B < u_1 < u_\mathrm{b}, \qquad (21)$$

are satisfied, then the consistency conditions (9) of configuration A are satisfied. Therefore, if inequalities (21) are satisfied the picture of the discontinuities has the form shown in Figure 2.47, and the formulas obtained above make it possible to compute completely the flow under the conditions of configuration A.

3. ***Configuration*** B. For $u_1 = u_\mathrm{b}$ we have $y_1 - y_0 = 0$, i.e., the zone of the rarefaction wave vanishes, and the solution consists of one shock wave and a

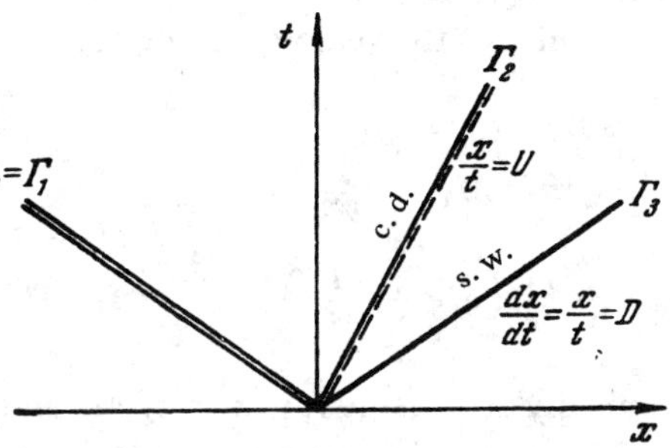

FIGURE 2.49

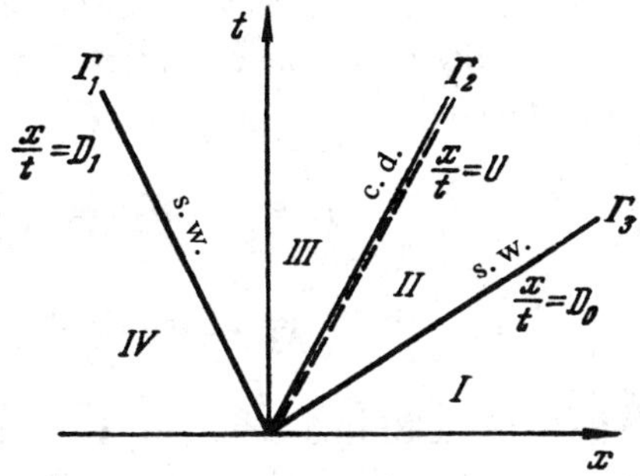

FIGURE 2.50

contact discontinuity (Figure 2.49). With further increase of u_1 the difference $u_1 - U$ becomes negative. Therefore, for $u_1 > u_b$ the contact boundary $dx/dt = U$ is to be considered as a piston moving simultaneously into the gas "0" and the gas "1". Thus, in correspondence with the solution of the piston problem (see §4.8) the solution of the problem of the decay of the discontinuity for $u_1 > u_b$ must be sought in the form of two shock waves, one propagating into the gas "0" and the other into the gas "1" (Figure 2.50). We call this case *configuration* B.

In the case of configuration B we have four zone I–IV of constant flow separated by the shock waves Γ_1 and Γ_3 and the contact discontinuity Γ_2. We shall prove the consistency of configuration B under the condition $u_1 > u_b$. We write out the conditions on the shock waves Γ_1 and Γ_3:

$$p_- = p_1[(1 + h_1)M_1^2 - h_1]; \tag{1}$$

$$u_- = u_1 - c_1(1 - h_1)\left(M_1 - \frac{1}{M_1}\right), \qquad M_1 = \frac{|D_1 - u_1|}{c_1}; \tag{2}$$

$$p_+ = p_0[(1 + h_0)M_0^2 - h_0], \qquad M_0 = \frac{D_0}{c_0}; \tag{3}$$

$$u_+ = c_0(1 - h_0)\left(M_0 - \frac{1}{M_0}\right). \tag{4}$$

Equating the pressures p_- and p_+ on the contact boundary Γ_2, we obtain

$$M_1^2 = \alpha M_0^2 + \beta, \qquad \alpha = \frac{p_0}{p_1}\frac{(1 + h_0)}{(1 + h_1)}, \qquad \beta = \frac{h_1 p_1 - h_0 p_0}{p_1(1 + h_1)}. \tag{5}$$

This implies that M_1 is a monotonically increasing function of M_0. To determine M_0 we write down the condition for the continuity of the velocity on the contact boundary Γ_2:

$$\begin{aligned} u_- = u_-(M_1) &= u_1 - c_1(1 - h_1)\left(M_1 - \frac{1}{M_1}\right) \\ = u_+ = u_+(M_0) &= c_0(1 - h_0)\left(M_0 - \frac{1}{M_0}\right). \end{aligned} \tag{6}$$

The left side of this inequality is a monotonically decreasing function of M_1 and of M_0 by (5); the right side is a function of M_0 increasing monotonically to ∞. We consider the values of $u_-(M_1)$ and $u_+(M_0)$ for $M_0 = M_{0\,\mathrm{cr}}$, where $M_{0\,\mathrm{cr}}$ is determined from (6.2.18) and (6.2.19). Since

$$M_{0\,\mathrm{cr}}^2 = \left(\frac{p_1}{p_0} + h_0\right)\frac{1}{1 + h_0} = \frac{p_1 + p_0 h_0}{p_0(1 + h_0)}, \tag{7}$$

it follows that

$$\left.\begin{aligned} M_1(M_{0\,\mathrm{cr}}) = 1, \qquad u_-(M_1(M_{0\,\mathrm{cr}})) = u_1, \\ u_+(M_{0\,\mathrm{cr}}) = c_0(1 - h_0)\left(M_{0\,\mathrm{cr}} - \frac{1}{M_{0\,\mathrm{cr}}}\right) = u_{\mathrm{b}}. \end{aligned}\right\} \tag{8}$$

Now by hypothesis $u_1 > u_{\mathrm{b}}$; from (8) we therefore have

$$u_-(M_1(M_{0\,\mathrm{cr}})) > u_+(M_{0\,\mathrm{cr}}). \tag{9}$$

Hence, for $u_1 > u_{\mathrm{b}}$ equation (6) always has only one root $M_0 > M_{0\,\mathrm{cr}} > 1$.

After determining M_0 from (5) we find M_1 and all the parameters of the flow in zones II and III. The consistency conditions for configuration B

$$D_1 < u_1 - c_1, \qquad D_1 < U < D_0, \qquad c_0 < D_0, \tag{10}$$

are easily verified and are always satisfied for $u_1 > u_{\mathrm{b}}$, if it is noted that $M_0 > M_{0\,\mathrm{cr}} > 1$ provided that $M_1 > 1$. Thus, for any u_1 satisfying the condition $u_1 > u_{\mathrm{b}}$ we have configuration B.

4. *Configuration* C. For $u_1 = u_B < 0$ (see (6.2.13)) we have $M_0 = 1$ and $U = 0$, i.e., the contact boundary is a stationary piston for the gas "0". For $u_1 = u_B$ we therefore have a solution in which in the gas "0" the shock wave vanishes, and it remains stationary while preserving its initial parameters; in the gas "1" a rarefaction wave $\Gamma_0 O \Gamma_1$ (Figure 2.51) propagates.

As u_1 ($u_1 < u_B < 0$) decreases further the contact boundary begins to move to the left relative to the gas "0" ($U < 0$), so that it can be considered a piston simultaneously moving out of the gases "0" and "1". In correspondence with

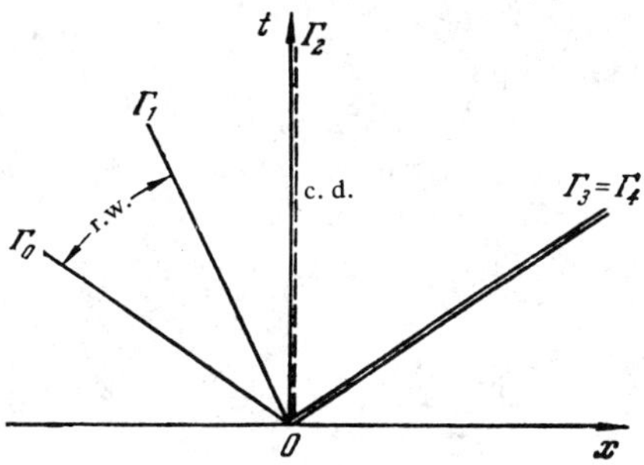

FIGURE 2.51

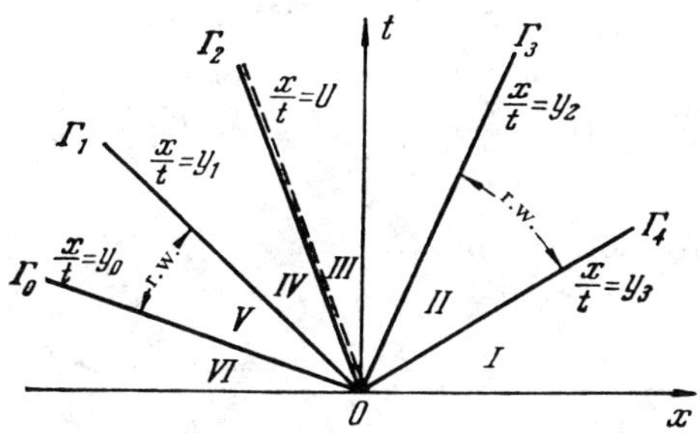

FIGURE 2.52

the solution of the piston problem (§3.4) in this case the solution of the decay problem consists of two centered rarefaction waves propagating into the gases "1" and "0" (Figure 2.52). We call this case *configuration* C. The upper half plane decomposes into 6 regions I–VI separated by four lines of weak discontinuities Γ_0, Γ_1, Γ_3, Γ_4 and a contact boundary Γ_2. Zones I, III, IV, and VI are regions of constant flow, while zones II and V are regions of rarefaction waves; in zone II the invariant s is constant, while in zone V the invariant r is constant.

We shall show that under the condition

$$u_1 < u_B \tag{1}$$

configuration C is consistent. In zones II and V we have

$$p_+ = p_0\left[1 + \frac{\gamma_0 - 1}{2}\,\frac{u - u_0}{c_0}\right]^{2\gamma_0/(\gamma_0-1)}$$

$$= p_0\left[1 + \frac{\gamma_0 - 1}{2}\,\frac{u}{c_0}\right]^{2\gamma_0/(\gamma_0-1)} \qquad (s = \text{const}), \tag{2}$$

$$p_- = p_1\left[1 - \frac{\gamma_1 - 1}{2}\,\frac{u - u_1}{c_1}\right]^{2\gamma_1/(\gamma_1-1)} \qquad (r = \text{const}). \tag{3}$$

In particular, on the contact boundary Γ_2 for $u = u_- = u_+ = U$ we must have $p_- = p_+$, i.e., we arrive at the equation

$$p_- = p_-(U) = p_1\left[1 - \frac{\gamma_1 - 1}{2}\,\frac{U - u_1}{c_1}\right]^{2\gamma_1/(\gamma_1-1)}$$
$$= p_+ = p_+(U) = p_0\left[1 + \frac{\gamma_0 - 1}{2}\,\frac{U}{c_0}\right]^{2\gamma_0/(\gamma_0-1)} \qquad (4)$$

for determining U. Here

$$1 - \frac{\gamma_1 - 1}{2}\,\frac{U - u_1}{c_1} \geq 0, \qquad 1 + \frac{\gamma_0 - 1}{2}\,\frac{U}{c_0} \geq 0. \qquad (5)$$

The condition (5) implies the nonnegativity of the pressure on the boundary. We remark that the equality signs in (5) can only hold simultaneously and correspond to break-away of the gases. As we see, in (4) $p_-(U)$ is a monotonically decreasing function of U, and $p_+(U)$ is a monotonically increasing function of U. For $U = 0$

$$p_-(0) = p_1\left[1 + \frac{\gamma_1 - 1}{2}\,\frac{u_1}{c_1}\right]^{2\gamma_1/(\gamma_1-1)}, \qquad p_+(0) = p_0. \qquad (6)$$

Condition (1) therefore implies that

$$p_-(0) < p_+(0). \qquad (7)$$

As U decreases, $p_-(U)$ increases and $p_+(U)$ decreases. Thus, if $p_+(U)$ does not vanish there exists a unique root $U < 0$ of (4) satisfying the condition

$$U > -\frac{2}{\gamma_0 - 1}\,c_0. \qquad (8)$$

The equations (5) then lead us to

$$u_1 > -\left(\frac{2}{\gamma_1 - 1}\,c_1 + \frac{2}{\gamma_0 - 1}\,c_0\right). \qquad (9)$$

Thus, if (9) is satisfied there exists a root $U < 0$ of (4). We shall show that if $u_1 < u_B$ the configuration C is always consistent, i.e.,

$$y_0 < y_1 \leq U \leq y_2 < y_3. \qquad (10)$$

It is not hard to see that

$$y_3 - y_2 = (u_0 + c_0) - (u_+ + c_+)$$
$$= -U - (c_+ - c_0) = -\frac{\gamma_0 + 1}{2}\,U > 0. \qquad (11)$$

The inequalities

$$\begin{aligned} y_2 - U &= (U + c_+) - U = c_+ \geq 0, \\ U - y_1 &= U - (U - c_-) = c_- \geq 0 \end{aligned} \tag{12}$$

are obvious. Finally,

$$y_1 - y_0 = U - u_1 + (c_1 - c_-) = \frac{\gamma_1 + 1}{2}(U - u_1). \tag{13}$$

Now from (4) and the conditions $p_1 > p_0$ and $U < 0$ it follows that $U - u_1 > 0$. Hence $y_1 - y_0 > 0$. Thus, the configuration C is consistent if $u_2 < u_B$ and condition (9) is satisfied.

If condition (9) is violated, then (4) has no root U. In this case the gases break free of one another, and (4) is replaced by the two equations for a free boundary

$$\left.\begin{aligned} p_- = p_-(U_1) = p_1\left[1 - \frac{\gamma_1 - 1}{2}\frac{U_1 - u_1}{c_1}\right]^{2\gamma_1/(\gamma_1 - 1)} &= 0, \\ \text{i.e. } U_1 = u_1 + \frac{2}{\gamma_1 - 1}c_1, & \end{aligned}\right\} \tag{14}$$

$$\left.\begin{aligned} p_+ = p_+(U_0) = p_0\left[1 + \frac{\gamma_0 - 1}{2}\frac{U_0}{c_0}\right]^{2\gamma_0/(\gamma_0 - 1)} &= 0, \\ \text{i.e. } U_0 = -\frac{2}{\gamma_0 - 1}c_0. & \end{aligned}\right\} \tag{15}$$

Since the nonsatisfaction of (9) implies that $U_1 < U_0$, the gases break free of one another, and the solution has the form shown in Figures 2.53. The regions $\Gamma_0 O \Gamma_1$ and $\Gamma_2 O \Gamma_3$ are regions of rarefaction waves corresponding to $r = \text{const}$ and $s = \text{const}$; the region $\Gamma_1 O \Gamma_2$ is a region of vacuum in which we have $\rho = 0$, $p = 0$ and $c = 0$.

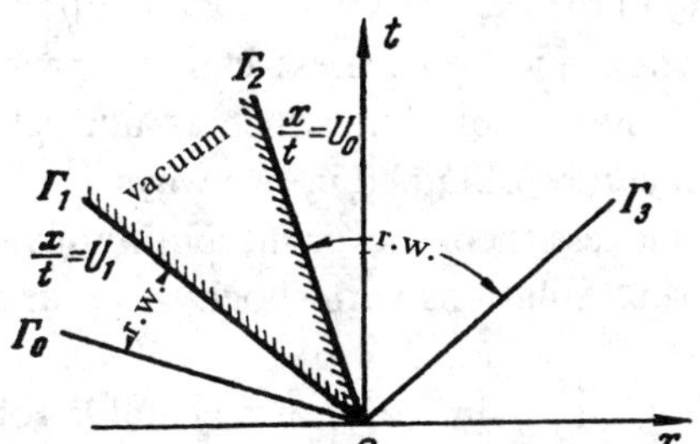

FIGURE 2.53

5. *A survey of configurations. Gases with equal pressure.* We shall now write out conditions for the existence of configurations A, B, and C under the assumption that $p_1 > p_0$ and $u_0 = 0$. Configuration A is possible under the conditions

$$u_B < u_1 < u_b, \tag{1}$$

configuration B is possible for

$$u_1 \geq u_b, \tag{2}$$

and configuration C is possible for

$$u_1 \leq u_B, \tag{3}$$

where

$$u_B = \frac{2}{\gamma_1 - 1} c_1 \left[\left(\frac{p_0}{p_1} \right)^{(\gamma_1 - 1)/2\gamma_1} - 1 \right] < 0. \tag{4}$$

$$u_b = (1 - h_0) c_0 \sqrt{\frac{p_1/p_0 + h_0}{1 + h_0} + \frac{1 + h_0}{p_1/p_0 + h_0} - 2} > 0. \tag{5}$$

In the case $p_1 = p_0$ we have

$$u_B = u_b = 0. \tag{6}$$

Hence, in this case for $u_1 > 0$ we have configuration B, while for $u_1 < 0$ we have configuration C.

We now observe that to any arbitrary discontinuity (collection of quantities $\gamma_1, p_1, \rho_1, c_1, u_1$ and $\gamma_0, p_0, \rho_0, c_0, u_0$) the conditions (1)–(6) assign one and only one configuration with flow parameters which can be computed uniquely, and the stability conditions on the shock waves are satisfied. Therefore, our treatment of the problem of the decay of an arbitrary discontinuity shows that it always has one and only one stable, self-similar solution. We have thus proved an existence and uniqueness theorem for the solution of the problem of the decay of a discontinuity in the class of self-similar solutions for polytropic gases.

However, the following question arises: can the problem on the decay of a discontinuity have a stable solution which is not self-similar? A negative answer to this question can be obtained in two ways:

1) by proving a uniqueness theorem for discontinuous solutions of the equations of gas dynamics, i.e., solutions with shock waves and centered rarefaction waves; or

2) by direct proof of the self-similarity of any stable solution of the problem of the decay of a discontinuity.

As concerns the first method, we must state that at present no sufficiently general uniqueness theorems for discontinuous solutions of the equations of gas dynamics have been obtained, and pursuit of such theorems apparently involves major difficulties, although for polytropic (and normal) gases it seems that no one doubts the uniqueness of the solution. By following the second method, it is actually possible to prove the self-similarity of a solution of the problem of the decay of a discontinuity by using certain specific properties of any stable solution of this problem with piecewise constant initial data. However, we shall not take up this proof here, but as an example we refer the reader to §3.5 of Chapter 4, where a similar problem is solved for a system of two quasilinear equations of rather general form.

6. *The problem of the decay of a discontinuity for an isothermic, ideal gas.* Here we mean by gases "1" and "0" two isothermic, ideal gases with equations of state of the form

$$p_1 = c_1^2\rho_1, \qquad p_0 = c_0^2\rho_0, \qquad c_1^2 = R_1T, \qquad c_0^2 = R_0T. \tag{1}$$

An isothermic, ideal gas can formally be considered as a polytropic gas with exponent $\gamma = 1$. The difference in treatment consists in dropping the equation for the conservation of energy and the third Hugoniot condition while replacing it by the condition $T =$ const. Inasmuch as the analysis of critical configurations involved only the first two Hugoniot conditions, all the results of the preceding subsections can be applied to the isothermic case. In the formulas of subsections 2–4 it is necessary to set $h_0 = h_1 = 0$ and $\gamma_0 = \gamma_1 = 1$ while expanding to eliminate the indeterminacy where required. We shall consider the expressions for u_B and u_b. Expanding the indeterminacy in (6.5.4) as $\gamma_1 \to 1$, we have

$$u_B = c_1 \ln \frac{p_0}{p_1} < 0 \qquad (p_1 > p_0). \tag{2}$$

For u_b we have

$$u_b = c_0\sqrt{\frac{p_1}{p_0} + \frac{p_0}{p_1} - 2} = c_0\left[\sqrt{\frac{p_1}{p_0}} - \sqrt{\frac{p_0}{p_1}}\right]. \tag{3}$$

The conditions for configurations A, B, and C are as before. We also note that for $\gamma_1 = \gamma_0 = 1$ the conditions (6.4.9) are always satisfied, so that for isothermic gases the break-away of the gases and formation of a vacuum are impossible.

The computational formulas after passing to the limit have the following form.

Configuration A:

$$p_-(M_0) = p_1 \exp\left\{\frac{u_1 - c_0(M_0 - 1/M_0)}{c_1}\right\} = p_+(M_0) = p_0M_0^2. \tag{4}$$

This corresponds to (6.2.12).

Configuration B:

$$u_-(M_1) = u_1 - c_1\left(M_1 - \frac{1}{M_1}\right) = u_+(M_0) = c_0\left(M_0 - \frac{1}{M_0}\right) \tag{5}$$

and

$$M_1 = \sqrt{p_0/p_1}\, M_0, \tag{6}$$

so that finally

$$u_1 - c_1\left[\sqrt{\frac{p_0}{p_1}}\, M_0 - \sqrt{\frac{p_1}{p_0}}\, \frac{1}{M_0}\right] = c_0\left[M_0 - \frac{1}{M_0}\right]. \tag{7}$$

This corresponds to (6.3.6) and (6.3.5).

Finally, in the case of configuration C

$$p_1 \exp\left\{-\frac{U - u_1}{c_1}\right\} = p_0 \exp \frac{U}{c_0}. \tag{8}$$

This corresponds to (6.4.4).

We remark that all these computational formulas can be easily obtained by using the Riemann invariants. We leave it to the reader to perform the corresponding computations if desired.

7. *The problem of the decay of a discontinuity for normal gases.* In investigating the problem of the decay of a discontinuity for normal gases, in addition to the usual conditions I–V (see §4.3), we shall require that the following property of the Hugoniot adiabat hold: on its upper branch $(p - p_0)(V_0 - V)$ must increase monotonically to ∞ simultaneously with increase of the entropy S. From the relations (see §4.2)

$$m^2 = \frac{p - p_0}{V_0 - V} > 0, \tag{1}$$

$$(u - u_0)^2 = (p - p_0)(V_0 - V), \tag{2}$$

where u_0, p_0, V_0 denote the parameters of the flow ahead of the front of the shock wave and u, p, V those behind the front, it then follows that p and $|u|$ are monotonically increasing functions[16] of the parameter $|m|$ or, equivalently, of $M_0 = |m|/\rho_0 c_0$. We recall also (see §§3.2 and 2.7) that in travelling waves u and p are connected by the relations

$$S = \text{const}, \qquad u - \Phi(S, p) = \text{const} \quad (s = \text{const}), \tag{3}$$

[16] It is shown in §4.3 that the monotone growth of $p(M_0)$ is a consequence of conditions I–V.

$$S = \text{const}, \qquad u + \Phi(S, p) = \text{const} \quad (r = \text{const}), \tag{4}$$

where $\Phi(S, p)$ is defined by the formula

$$\Phi(S, p) = \int_{\bar{p}}^{p} \frac{dp}{a_m(S, p)}, \tag{5}$$

in which $\bar{p}$ is a fixed limit of integration and

$$a_m(S, p) = \rho c > 0 \tag{6}$$

may be considered a function of the pressure p and entropy S. From (5) and (6) it follows that $\Phi(S, p)$ is a monotonically increasing function of the pressure p. The function $\Phi(S, p)$ depends on the equation of state, and therefore the functions $\Phi_0(S, p)$ and $\Phi_1(S, p)$ for the gases "0" and "1" respectively are, in general, distinct.

We begin our consideration of the decay problem with configuration A, setting

$$u_0 = u_1 = 0, \qquad p_1 > p_0. \tag{7}$$

We assume that the equations of state of the gases "1" and "0" are distinct, and conditions I–V and the condition of monotone growth of $u(M)$ (see (1) and (2)) are satisfied for each of them.

Configuration A (Figure 2.47) follows from the assumption (7). In region IV the invariant r is constant, and therefore

$$u_1 + \Phi_1(S_1, p_1) = u_- + \Phi_1(S_1, p_-). \tag{8}$$

In region II

$$p_+ = p_0(M_0), \qquad u_+ = u_0(M_0), \tag{9}$$

where $p_0(M_0)$, $u_0(M_0)$ denotes the state behind the front of the shock wave in the gas "0" for given M_0, and they increase to ∞ with increasing M_0.

The continuity conditions for the velocity and pressure on the contact boundary Γ_2 lead to an equation for determining M_0:

$$u_-(M_0) = u_1 + \Phi_1(S_1, p_1) - \Phi_1(S_1, p_0(M_0)) = u_+(M_0) = u_0(M_0). \tag{10}$$

The left side of (1) decreases monotonically, while the right side increases monotonically to ∞ with increasing M_0. For $M_0 = 1$, according to (5) and (7), we have

$$\left.\begin{aligned} u_-(1) &= \Phi_1(S_1, p_1) - \Phi_1(S_1, p_0) > 0 \qquad (u_1 = 0), \\ u_+(1) &= u_0(1) = 0. \end{aligned}\right\} \tag{11}$$

As before, from this it follows that (10) has one and only one root $M_0 > 1$. Fixing p_1 and p_0, we study u_1. The function

$$u_- = u_-(M_0, u_1) = u_1 + \Phi_1(S_1, p_1) - \Phi_1(S_1, p_0(M_0))$$

is a monotonically increasing function of u_1. Therefore, the root M_0 of (10) is a monotonically increasing function of u_1.

We shall examine how the inequalities

$$y_0 < y_1 < U < D \tag{12}$$

change as u_1 varies. The inequality $D - U > 0$ is obvious by (4.2.8):

$$(U - D)(u_0 - D) = D(D - U) = \frac{p_0(M_0) - p_0}{\rho_0(M_0) - \rho_0} > 0. \tag{13}$$

The inequality $U - y_1 = U - (U - c_-) = c_- > 0$ is also obvious. We consider the difference

$$y_1 - y_0 = (U - c_-) - (u_1 - c_1) = (U - u_1) - (c_- - c_1). \tag{14}$$

Since $u_- = U$, from (8) it follows that

$$U - u_1 - (c_- - c_1) = \Phi_1(S_1, p_1) - \Phi_1(S_1, p_-) + (c_1 - c_-). \tag{15}$$

With increasing u_1 the quantities $p_- = p_0(M_0)$ and $c_- = c_-(S_1, p_-)$ increase; hence, the differences $\Phi_1(S_1, p_1) - \Phi_1(S_1, p_-)$ and $c_1 - c_-$ decrease. Thus, the difference $y_1 - y_0$ decreases with increasing u_1 and becomes equal to zero for some value $u_1 = u_B$. The region of the rarefaction wave then vanishes and we have configuration AB (Figure 2.49). For $u_1 = u_b$

$$p_- = p_1, \qquad c_- = c_1, \qquad u_- = u_1, \tag{16}$$

and u_b can therefore be found from the equations

$$\left.\begin{aligned} p_1 &= p_0(M_{0\,\mathrm{cr}}), \\ u_b &= u_0(M_{0\,\mathrm{cr}}), \end{aligned}\right\} \tag{17}$$

which contain $M_{0\,\mathrm{cr}}$ as a parameter.

If u_1 decreases we have seen that M_0 decreases and for $u_1 = u_B$ becomes equal to 1. For $u_1 = u_B$ the shock wave vanishes, and the solution adopts the configuration AC (Figure 2.51). The value of u_B can be found from the equation

$$u_B = \Phi_1(S_1, p_0) - \Phi_1(S_1, p_1) < 0, \tag{18}$$

which follows from (10) for $M_0 = 1$. For

$$u_1 > u_b \tag{19}$$

configuration B is always consistent (Figure 2.50).

We write out the conditions for the continuity of the pressure and velocity on the contact boundary Γ_2:

$$p_-(M_1) = p_+(M_0), \tag{20}$$

$$u_-(M_1) = u_+(M_0). \tag{21}$$

The function $p_-(M_1)$ is a monotonically increasing function of the parameter M_1; M_1 is therefore determined from (20) as a monotonically increasing function of M_0, while (21) may be considered an equation for determining M_0.

Noting now that the left side of (21) is a monotonically decreasing function of M_0 while the right is a monotonically increasing function, and that

$$M_1(M_{0\,\mathrm{cr}}) = 1, \qquad u_+(M_{0\,\mathrm{cr}}) = u_b, \qquad u_-(1) = u_1 > u_b,$$

we conclude that (20) and (21) have one and only one root $M_0 > M_{0\,\mathrm{cr}} > 1$; $M_1 > 1$. Thus, the condition for the consistency of configuration B ($M_0 > 1$, $M_1 > 1$) is always satisfied for $u_1 > u_b$. Finally, for

$$u_1 < u_B \tag{22}$$

we have configuration C (Figure 2.52).

The condition for the continuity of the velocity and pressure on the contact boundary Γ_2 has the form

$$\begin{aligned} u_- &= u_1 + \Phi_1(S_1, p_1) - \Phi_1(S_1, p) \\ &= u_+ = \Phi_0(S_0, p) - \Phi_0(S_0, p_0) = U \end{aligned} \tag{23}$$

(p denotes the pressure in zones III and IV, $p_- = p_+ = p$). The quantities p and U are determined from this relation. The left side of (23) decreases monotonically, while the right side increases monotonically with increasing p; for $p = p_0$

$$u_+ = u_+(p_0) = 0;$$

$$u_- = u_-(p_0) = u_1 + \Phi_1(S_1, p_1) - \Phi_1(S_1, p_0) = u_1 - u_B < 0.$$

This implies that (23) has one and only one root $p = p_- = p_+ < p_0 < p_1$. The consistency conditions for configuration C follow from the fact that $0 > U > u_1$. As u_1 decreases further, break-away of the gases is possible.

8. ***The solution of the problem of the decay of a discontinuity in the plane of the variables*** p, u (***the*** p, u-***diagram***). In all cases of configurations A, B, and C the state u_+, p_+, S_+ to the right of the contact boundary was connected with the state $u_0 = 0$, p_0, S_0 in gas "0" at the initial time either by the relation of a rarefaction wave

$$S_+ = S_0, \qquad u_+ - \Phi_0(S_0, p_+) = u_0 - \Phi_0(S_0, p_0) = -\Phi_0(S_0, p_0), \tag{1}$$

or by the Hugoniot relations, from which it follows that

$$p_+ = p_0(S_0, M_0), \qquad u_+ = u_0(S_0, M_0), \tag{2}$$

where $p_0(S_0, M_0)$ and $u_0(S_0, M_0)$ increase monotonically with increasing M_0, and tend to ∞ as $M_0 \to \infty$. This implies that the parameter M_0 can be eliminated from (2) to obtain the new relation

$$u_+ = \Psi_0(S_0, p_+); \tag{3}$$

here $\Psi_0(S_0, p_+)$, just as $\Phi_0(S_0, p_+)$, is a monotonically increasing function of p_+.

A necessary condition for a rarefaction wave ($y_3 - y_2 \geq 0$) is

$$p_+ \leq p_0, \tag{4}$$

from which it follows that to determine u_+ we can use only the half of the curve (1) prescribed by the condition $p_+ \leq p_0$:

$$u_+ = \Phi_0(S_0, p_+) - \Phi_0(S_0, p_0). \tag{5}$$

The condition for the stability of a shock wave ($M_0 \geq 1$), on the other hand, requires that the pressure p_+ be greater than p_0 ($p_+ > p_0$); therefore, to determine u_+ we can use only the half of the curve (3) prescribed by the condition $p_+ \geq p_0$. Therefore, the curve

$$u_+ = \begin{cases} \Phi_0(S_0, p_+) - \Phi_0(S_0, p_0) & \text{for } p_+ \leq p_0, \\ \Psi_0(S_0, p_+) & \text{for } p_+ \geq p_0 \end{cases} \tag{6}$$

determines the velocity u_+ on the contact boundary on the basis of given pressure p_+. The curve

$$u = g_0(S_0, p) = \begin{cases} \Phi_0(S_0, p) - \Phi_0(S_0, p_0) & \text{for } p \leq p_0, \\ \Psi_0(S_0, p) & \text{for } p \geq p_0 \end{cases} \tag{7}$$

in the plane of the variables p and u describes the set of states which can be connected with the state on the right p_0, S_0, $u_0 = 0$ by means of a centered rarefaction wave or a stable ($M_0 \geq 1$) shock wave.

We say that a "state" is determined on the curve (7) although at each of its points only two of the three hydrodynamical parameters are known, namely, p and u. It is easy to see, however, that in a rarefaction wave ($p \leq p_0$) we have $S = S_0$, while for $p \geq p_0$ the entropy S is uniquely determined at each point of this curve by the Hugoniot conditions. More briefly, the curve (7) is the projection onto the plane of the variables p, u of a curve in the space of three variables (p, u, S) describing the set of states (p, u, S) that can be connected with the state on the right p_0, $u_0 = 0$, S_0 by a rarefaction wave or a shock wave.

It is easy to see that the curve (7) passes through the point (p_0, $u_0 = 0$), and from the property of tangency of second order at the point (p_0, S_0) of the Hugoniot adiabat H and the Poisson adiabat A it follows that the curve (7) has a continuously differentiable tangent at the point (p_0, $u_0 = 0$) and hence everywhere. Figure 2.54 shows the approximate form of the curve (7); its upper part corresponds to a shock wave (the relation (3)), while the lower part corresponds to a rarefaction wave (relation (5)).

In an altogther analogous way the curve

$$u = u_1 - g_1(S_1, p), \tag{8}$$

where

$$g_1(S_1, p) = \begin{cases} \Phi_1(S_1, p) - \Phi_1(S_1, p_1), & p \le p_1, \\ \Psi_1(S_1, p), & p \ge p_1, \end{cases} \tag{9}$$

describes in the p, u plane the family of states which can be connected with the state u_1, p_1, S_1 as the state on the left either by a rarefaction wave (r = const) or by a shock wave ($m > 0$, $M_1 \ge 1$). It is easy to see that $g_1(S, p)$ is a monotonically increasing function of p and that the curve (8) passes through the point (p_1, u_1) and has two continuous derivatives. Figure 2.54 shows the approximate form of the curve (8). Figure 2.54 also shows to which parts of the curves

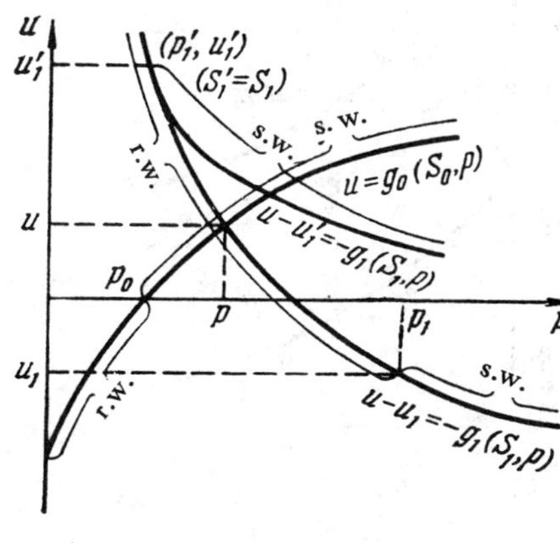

FIGURE 2.54

(7), (8) there correspond rarefaction waves (r. w.) and shock waves (s. w.) in the gases "0" and "1" respectively. Since on a contact discontinuity the continuity of the velocity $u_- = u_+$ and the pressure $p_- = p_+$ is always required (with the exception of the case where the gases break away), the solution of the problem of the decay of a discontinuity reduces to determing points (p, u) $(p = p_- = p_+, u = u_- = u_+)$ of intersection of the curves (7) and (8). If a point of intersection (p, u) of these two curves lies on the upper half $(p > p_0)$ of the curve (7), then a shock wave propagates in the gas "0"; if $p < p_0$, then a rarefaction wave propagates in the gas "0". Similarly, if the point (p, u) of intersection of these two curves lies on the upper half $(p < p_1)$ of the curve (8), then a rarefaction waves propagates in the gas "1"; if $p > p_1$, then a shock wave propagates in the gas "1". In correspondence with this Figure 2.55 shows the possible cases of intersection of these curves. Prescribing, as always, the condition $p_1 > p_0$ and noting that the curves (8) for various u_1 differ only by translation, we present in Figure 2.55 a summary of the possible configurations

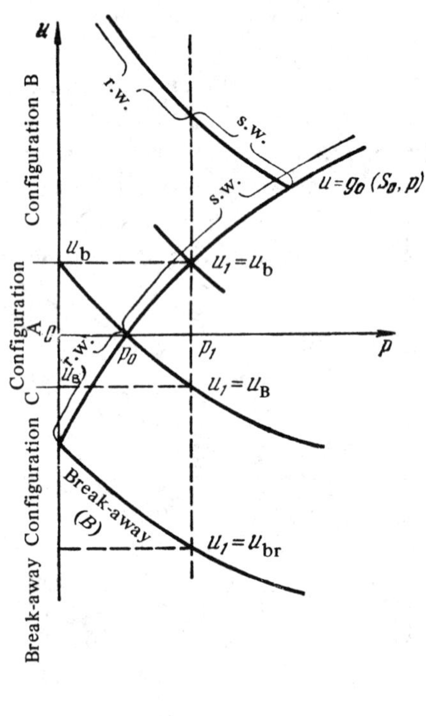

FIGURE 2.55

depending on u_1 and also a graphical method of determining the quantities u_b, u_B, and u_{br}. Figures 2.54 and 2.55 also proved a graphical method of solving the problem of the decay of an arbitrary discontinuity.[17]

An analogous treatment of the problem of the decay of a discontinuity can be carried out also in the projection onto the plane (s, r) of Riemann invariants. In the (s, r) plane parts of the curves (7) and (8) corresponding to rarefaction waves are rays parallel to the coordinate axes; another part of them is a smooth curve which agrees smoothly (with two derivatives) with these rays (Figure 2.56). Application of the (s, r)-diagram is especially convenient in solving the decay problem for identical isothermic gases.

To conclude this subsection we determine the function $g_0(S_0, p)$ for the case of polytropic gases. For a polytropic gas

$$p = \frac{A^2(S)\rho^\gamma}{\gamma}, \tag{10}$$

$$\Phi_0(S_0, p) - \Phi_0(S_0, p_0) = \frac{2}{\gamma_0 - 1}\left[\left(\frac{p}{p_0}\right)^{(\gamma_0-1)/2\gamma_0} - 1\right]c_0(S_0, p_0). \tag{11}$$

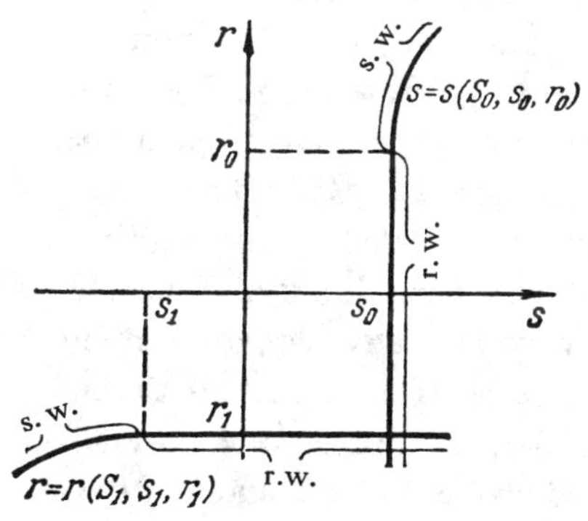

FIGURE 2.56

[17] It should be kept in mind that the curve (8), just as the curve (7), depends parametrically on the entropy S_1 (S_0); therefore, the prescription of only a point (p_1, u_1) still does not determine it. If, however, it is noted that for all u_1 and $p_1 \geq p_0$ the entropy S_1 is fixed, then the curves $u - u_1 = -g_1(S_1, p)$ and $u - u_1' = -g_1(S_1, p)$ coincide for $u_1' - u_1 = g_1(p_1', S_1)$, $p \leq \min(p_1, p_1')$. On the basis of this, Figure 2.55 shows the regions of values p_1, u_1 for $S = S_1 =$ const in which the configuration of the solution of the problem of the decay of a discontinuity is preserved.

In the case of a shock wave ($p > p_0$) we find M_0 from (4.5.13):

$$M_0 = \sqrt{\frac{p + h_0 p_0}{p_0(1 + h_0)}}, \tag{12}$$

after which from (4.5.16) we find $\Psi_0(S_0, p)$:

$$\Psi_0(S_0, p) = (1 - h_0)\, c_0(S_0, p_0)\left[\sqrt{\frac{p + h_0 p_0}{p_0(1 + h_0)}} - \sqrt{\frac{p_0(1 + h_0)}{p + h_0 p_0}}\right]. \tag{13}$$

Thus, for a polytropic gas (7) can be written in the form

$$u = g_0(S_0, p) = \begin{cases} \dfrac{2}{\gamma_0 - 1} c_0(S_0, p_0)\left[\left(\dfrac{p}{p_0}\right)^{(\gamma_0-1)/2\gamma_0} - 1\right] & \text{for } p \le p_0, \\ (1 - h_0) c_0(S_0, p_0)\left[\sqrt{\dfrac{p + h_0 p_0}{p_0(1 + h_0)}} - \sqrt{\dfrac{p_0(1 + h_0)}{p + h_0 p_0}}\right] & \\ \qquad\qquad\qquad\qquad\qquad\qquad\qquad \text{for } p \ge p_0. \end{cases} \tag{14}$$

9. ***The problem of decay of a discontinuity in media with anomalous thermodynamic properties.*** Giving up the condition $p''_{VV}(V, S) > 0$ in the neighboring media in the problem of the decay of a discontinuity considerably complicates this problem and makes the picture of the flow more varied. The question of the uniqueness of a solution to the decay problem becomes more complicated; the uniqueness of the solution is proved under certain restrictions on the equation of state which are not dictated by the requirements of thermodynamics. The conditions for stability (admissibility) of a shock wave (see §§4.8 and 5.7) are formulated in a more complicated manner. A number of questions connected with this problem for media with anomalous thermodynamic properties were considered by Galin (1958), Sidorenko (1968), Wendroff (1972), and Liu (1975).

In the case considered properties 1, 2, and 3 of the decay problem established in subsection 1 are violated; in a medium with anomalous properties several shock transitions as well as centered Riemann waves may propagate following one another in the same direction relative to the gas.

We shall assume that in the media in question conditions (5.7.13)

$$T = \frac{\partial \varepsilon(V, S)}{\partial S} > 0; \qquad p'_V(V, S) < 0, \qquad p'_S(V, S) > 0 \tag{1}$$

are satisfied, and we assume that only those discontinuities are admissible connecting the states (u_0, V_0, p_0) ahead of the front and (u_1, V_1, p_1) behind the front for which the admissibility condition (5.7.17) is satisfied; for any point (V, p) of the Hugoniot adiabat $H(V, p;\, V_0, p_0) = 0$ with center at (V_0, p_0) located between (V_0, p_0) and (V_1, p_1) the following inequality is satisfied:

$$\frac{p - p_0}{V - V_0} \geq -m^2 = \frac{p_1 - p_0}{V_1 - V_0}. \tag{2}$$

It is evident from conditions (2) that for $p''_{VV}(V, S)$ of variable sign discontinuities are possible for which the following equalities are satisfied:

$$\frac{\partial p}{\partial V}(V_1, S_1) = -m^2 = \frac{p_1 - p_0}{V_1 - V_0}, \tag{3}$$

$$\frac{\partial p}{\partial V}(V_0, S_0) = -m^2 = \frac{p_1 - p_0}{V_1 - V_0} \tag{4}$$

and

$$\frac{\partial p}{\partial V}(V_1, S_1) = \frac{\partial p}{\partial V}(V_0, S_0) = -m^2 = \frac{p_1 - p_0}{V_1 - V_0}. \tag{5}$$

The equalities (3) and (4) imply that the speed of the discontinuity D coincides with the speed of the characteristic ξ_1 (for $m > 0$) or ξ_3 (for $m < 0$) on one side of the discontinuity. Therefore, the relative position of the line of discontinuity (the dashed line) and the r-characteristics (solid lines) in a neighborhood of the discontinuity has the form shown in Figure 2.57, a and b for the case $m < 0$. If equalities (3) and (4) are satisfied simultaneously, then to the left and right of the discontinuity the speed of the characteristics (ξ_1 for $m > 0$ or ξ_3 for $m < 0$) coincides with the speed D of the discontinuity, so that the line of discontinuity coincides with the characteristics (Figure 2.57, c). In a number of papers such discontinuities are called contact discontinuities. However, in our opinion this name is not altogether appropriate, since a contact discontinuity in the mechanics of fluids and gases means the boundary between two

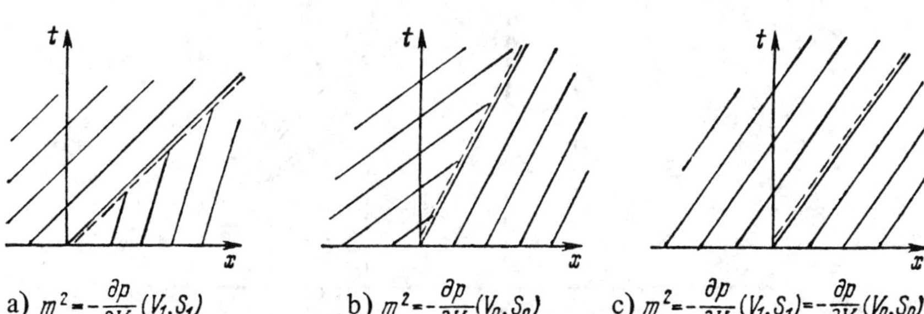

FIGURE 2.57

distinct gases moving with the same velocities $u_0 = u_1 = D$, i.e., with $m = 0$. Therefore, discontinuities satisfying conditions (5) should be called shock waves, as before.

As previously, let u_+, V_+, p_+ denote the flow parameters on the right side of the contact boundary in the solution of the problem of the decay of a discontinuity, and let $u_0 = 0$, V_0, p_0 be the prescribed initial values in the gas "0". We shall study the set $W(V_0, p_0)$ of possible values V_+, p_+, u_+.

It is obvious that the set $W(V_0, p_0)$ contains all physical (admissible) portions of the Hugoniot adiabat $H(V, p;\ V_0, p_0) = 0$ having center at (V_0, p_0) that we introduced in §5.7. We shall consider only the case $p''_{VV}(V_0, S_0) > 0$; then the admissible segments of the Hugoniot adiabat are those indicated by a double line in Figure 2.58. We see that the admissible segments of the Hugoniot adiabat form only part of the set $W(V_0, p_0)$, since by its very meaning this set must be a continuous curve. We henceforth call the curve $V, p \in W(V_0, p_0)$ the *wave adiabat*. The admissible segments of the Hugoniot adiabat give only those values of V_+, p_+ that can be connected with V_0, p_0 by a stable shock wave of compression or rarefaction. In addition to this, we have seen that V_+, p_+ can

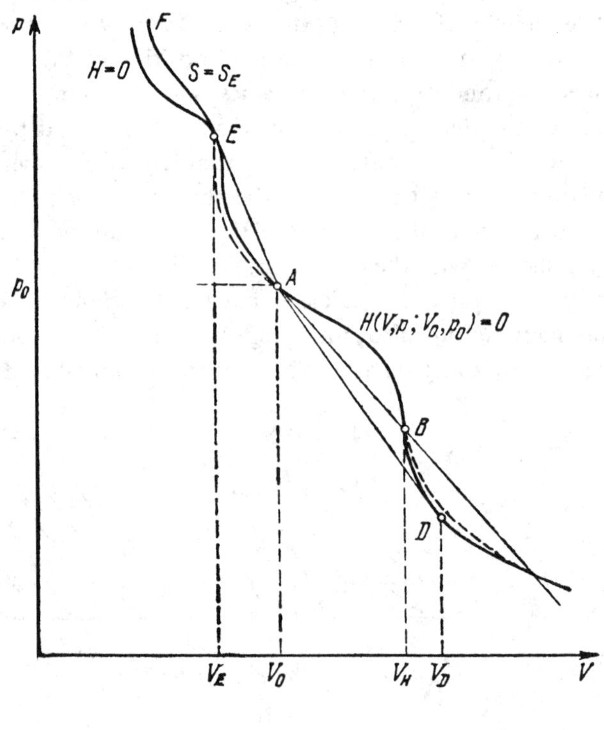

FIGURE 2.58

be connected with V_0, p_0 by a rarefaction Riemann wave. If $p''_{VV}(V_0, S_0) > 0$, then to a rarefaction Riemann wave there corresponds a segment of the Poisson adiabat $S = S_0$ prescribed by the condition $V \geq V_0$ and the condition of monotone decrease of the quantity $\xi = \sqrt{-\partial p(V, S_0)/\partial V}$ with increasing V. The latter condition is a condition for a centered rarefaction wave in which to the transition from the right state u_0, V_0, p_0 to the left state u_+, V_+, p_+ there corresponds a monotone decrease in the speed of the characteristic ξ_3. Therefore, on admissible segments of the Poisson adiabat $S = S_0$ the following condition must be satisfied:

$$\frac{\partial}{\partial V}\left[-\frac{\partial p}{\partial V}(V, S_0)\right] = -\frac{\partial^2 p}{\partial V^2}(V, S_0) < 0. \tag{6}$$

If on the Poisson adiabat there are no segments where condition (6) is violated, then any of its points with $V > V_0$ give possible values of V_+, p_+. However, in the case of interest to us and, in particular, in the case shown in Figure 2.58 there are such segments. Indeed, we saw in §5.7 that on admissible segments of the Hugoniot adiabat we have $S > S_0$, and the value of the entropy at the point B (see Figure 2.58) therefore exceeds S_0. This implies that the Poisson

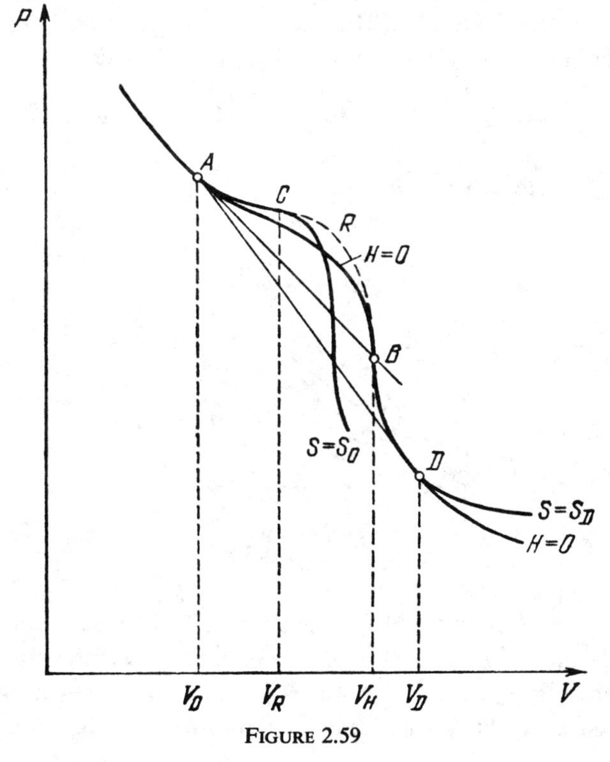

FIGURE 2.59

adiabat $S = S_0$ intersects the Hugoniot adiabat inside the segment $[V_0, V_H]$. On $[V_0, V_H]$ the Hugoniot adiabat changes convexity, and from (1) we conclude that the adiabat $S = S_0$ also changes convexity inside this segment. The relative positions of the Poisson adiabat $S = S_0$ and the Hugoniot adiabat are shown in Figure 2.59. Here the line AB is tangent to the curves $H = 0$ and $S = S_0$ at the point A, AD is tangent to the Hogoniot adiabat $H = 0$ at the point D, and

$$p''_{VV}(V_R, S_0) = 0, \qquad p''_{VV}(V_0, S_0) > 0.$$

We see that the derivative $p''_{VV}(V, S_0)$ changes its sign at the point $V = V_R$ ($p''_{VV}(V, S_0) < 0$ for $V > V_R$). Passage from the point (V_0, p_0) by a rarefaction wave ($S = S_0$) is thus possible only up to the point C, while a shock transition from the point A is possible only for $V > V_H$, i.e., starting at the point B of the Hugoniot adiabat. We have an unfilled portion $[V_R, V_H]$ of the wave adiabat $W(V_0, p_0)$.

Here it is necessary to consider the possibility of the appearance of shock waves of the type shown in Figures 2.57, b where a rarefaction r-wave is completed by an admissible shock transition as shown in Figure 2.60. This is possible if on the Poisson adiabat AC (Figure 2.59) there is a point $(\tilde{V}, \tilde{p})$ contained between the points A and C and a point (V_1, p_1) such that

$$H(V_1, p_1; \tilde{V}, \tilde{p}) = 0; \qquad S(\tilde{V}, \tilde{p}) = S_0,$$
$$\frac{p_1 - \tilde{p}}{V_1 - \tilde{V}} = \frac{\partial p}{\partial V}(\tilde{V}, \tilde{S}), \qquad V_0 \le \tilde{V} \le V_R. \tag{7}$$

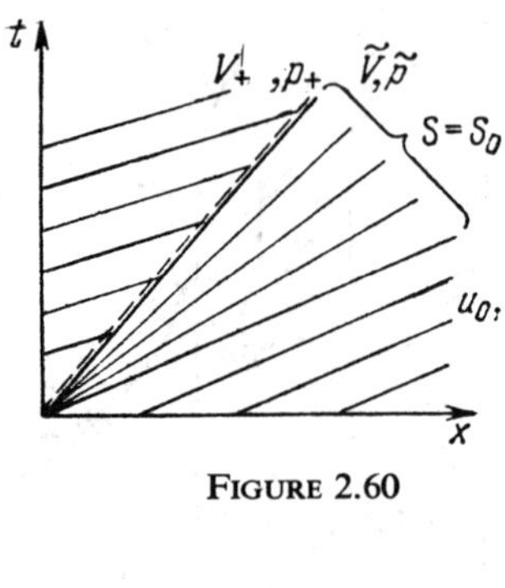

FIGURE 2.60

If here the entire portion of the Hugoniot adiabat $H(V, p; \tilde{V}, \tilde{p})$ for $\tilde{V} < V < V_1$ lies to the right of the line $p - \tilde{p} = m^2(\tilde{V} - V)$, then the transition (7) is admissible: the values V_1, p_1 are possible values of V_+, p_+, and they must be included in the wave adiabat $W(V_0, p_0)$. One such transition is that from point A to point B (Figures 2.59 and 2.58); in this case $\tilde{V} = V_0$, $\tilde{p} = p_0$, and there is no rarefaction wave (Figure 2.60).

A necessary condition for (7) to be satisfied is that $p''_{VV}(V, S)$ change sign on the adiabat $H(V, p; \tilde{V}, \tilde{p})$ for $\tilde{V} \leq V \leq V_1$. We now consider just this case, since $p''_{VV}(V, S_0)$ changes sign at C. This consideration leads to the conclusion that the set of values V_1, p_1 satisfying conditions (7) forms a continuous curve (we shall denote it by R) which joins the points C and B (Figure 2.59). Here to the displacement of the point $(\tilde{V}, \tilde{p})$ along the adiabat $S = S_0$ in the direction from the point C to the point A there corresponds the motion of the point (V_1, p_1) along the curve R from the point C to the point B. We thus conclude that to the right of A the wave adiabat $W(V_0, p_0)$ consists of the segment AC of the Poisson adiabat $S = S_0$, the segment CB of the curve R consisting of values V_1, p_1 satisfying (7), and the part BD of the Hugoniot adiabat $H(V, p; V_0, p_0) = 0$. We have thus constructed a continuous curve $W(V_0, p_0)$—the wave adiabat—to the right of the point A.

We observe that if the admissibility condition is violated on the Hugoniot adiabat BD during its extension to the right, then the wave adiabat will contain new segments consisting of Poisson adiabats $S =$ const and the curves R constructed for them. This corresponds to the picture of r-characteristics shown in Figure 2.61, where several centered rarefaction waves propagate separated by admissible shock transitions.

We briefly consider the question of constructing the wave adiabat $W(V_0, p_0)$ for $V \leq V_0$, keeping in mind that it is basically accomplished via the same considerations as above. As we have seen, in the case we are considering $p''_{VV}(V_0, S_0) > 0$, and the segment AE of the Hugoniot adiabat $H(V, p; V_0, p_0)$ (see Figure 2.58) is admissible; it consists of possible values V_+, p_+ and is thus naturally included in the wave adiabat $W(V_0, p_0)$. A shock transition from the point A to the point E has the picture of r-characteristics shown in Figure 2.57, c. The wave adiabat can be extended further (for decreasing $V = V_+$) by means of a rarefaction r-wave $S = S_E$, where S_E is the entropy at the point E. The

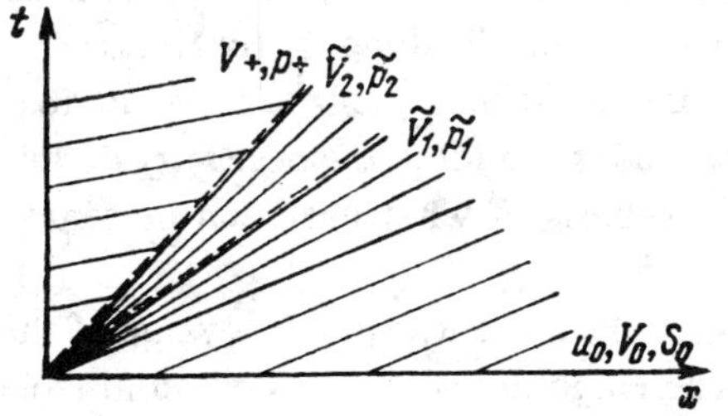

FIGURE 2.61

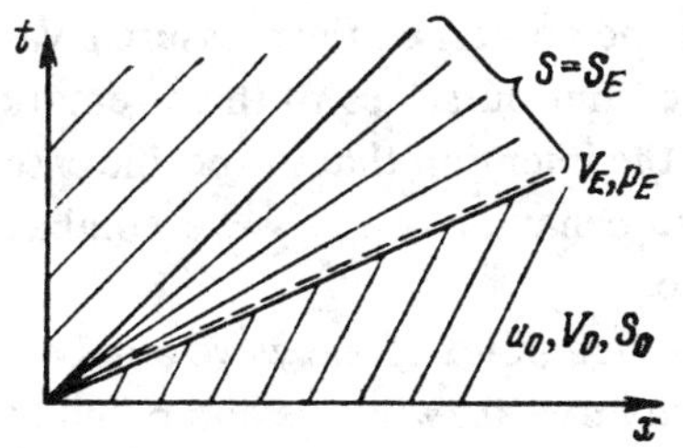

FIGURE 2.62

point E now corresponds to the right state in the rarefaction wave $S = S_E$; therefore, transition to the point F corresponding to values $V_+ < V_E$, p_+ is possible only under the condition that $p''_{VV}(V, S_E) < 0$ for $V_+ < V < V_E$ (this follows from the fact that the speed of the r-characteristics decreases on passing from E to F).

In the case shown in Figure 2.58 this condition ($p''_{VV}(V, S_E) < 0$) is satisfied at $V = V_E$, and thus in some half neighborhood $V < V_E$ the Poisson adiabat $S = S_E$ gives possible values V_+, p_+ and must be included in the wave adiabat. Figure 2.62 shows the picture of r-characteristics for the case where (V_+, p_+) lies on the Poisson adiabat $S = S_E$.

Suppose that at a point F of the adiabat $S = S_E$ the sign of $p''_{VV}(V, S_E)$ changes from minus for $V > V_F$ to plus for $V < V_F$. In this case a transition along the adiabat $S = S_E$ is possible only up to the point F. To continue the wave adiabat it is now necessary to consider shock transitions from points $(\tilde{V}, \tilde{p})$ of the adiabat $S = S_E$ to points (V_1, p_1) under the conditions

$$H(V_1, p_1; \tilde{V}, \tilde{p}) = 0; \qquad S(\tilde{V}, \tilde{p}) = S_E,$$
$$\frac{p_1 - \tilde{p}}{V_1 - \tilde{V}} = \frac{\partial p}{\partial V}(\tilde{V}\,\tilde{S}); \qquad V_F \le \tilde{V} \le V_E, \tag{8}$$

which are analogous to (7).

We again designate the set of points (V_1, p_1) defined by (8) as the curve R. This curve is defined for $p > p_F$; it continuously abuts the segment EF of the Poisson adiabat $S = S_E$ (for $\tilde{V} = V_F$, $\tilde{p} = p_F$), and it terminates on intersecting a new admissible segment of the original Hugoniot adiabat $H(V, p; V_0, p_0) = 0$. This segment of R is also included in the wave adiabat $W(V_0, p_0)$.

The wave adiabat is constructed in this manner. We see that it includes all admissible portions of the Hugoniot adiabat $H(V, p; V_0, p_0) = 0$ and segments of the Poisson adiabat $S = S_i$ and the curves R_i they determine. Here S_i are the values of the entropy at the right end points of admissible segments of the Hugoniot adiabat for $V \ge V_0$ and at the left end points of admissible segments for $V < V_0$. We note that this curve intersects any line $p = \text{const}$ in no more than one point, i.e., the pressure p varies monotonically along this curve.

We limit ourselves to these remarks on the constructing of the wave adiabat with the intention that in specific cases the reader can fill in the necessary details of the behavior of the wave adiabat on the basis of what has been presented above.

To each point of the wave adiabat $W(V_0, p_0)$ there corresponds a value of the velocity $u = u_+$ which is computed from the Hugoniot conditions on admissible segments of the adiabat $H(V, p; V_0, p_0) = 0$ and on the curves R and from the Riemann invariant on segments of the Poisson adiabat $S = S_i$. As a result, we have a continuous curve in the space of the variables u, V, p which gives the set of possible values u_+, V_+, p_+ for fixed u_0, V_0, p_0. The projection of this curve onto the plane $V = 0$ gives the curve

$$u = u_0 + g_0(S_0, p), \tag{9}$$

that we introduced in subsection 8 in solving the problem on the decay of a discontinuity by the method of the p, u-diagram. For the gas "1" the relation

$$u = u_1 - g_1(S_1, p) \tag{10}$$

is constructed in a similar way; here $g_1(S_1, p)$ is the projection onto the plane $V = 0$ of the wave adiabat constructed for the gas "1". The problem of the decay of a discontinuity can be solved as in subsection 8 by determining the point of intersection ($u_+ = u_-$, $p_+ = p_-$) of the curves (9) and (10).

For the uniqueness of the solution on the decay of a discontinuity it is necessary to impose additional restrictions on the equations of state. For example, Wendroff (1972) and Liu (1975) imposed the following added condition:

$$\bar{p}'_V(V, \varepsilon) = \frac{\partial}{\partial V}[p(V, S(V, \varepsilon))] < 0; \tag{11}$$

the uniqueness of the solution of the problem of the decay of a discontinuity is established if this condition is satisfied (for the case where the gases "0" and "1" are described by the same equations of state).

For the existence of a solution of the problem of the decay of an arbitrary discontinuity additional restrictions on the equations of state are required even for normal gases (the increase of $(p - p_0)(V_0 - V)$ to infinity along the Hugoniot adiabat). Naturally, for media with anomalous thermodynamic properties additional restrictions ensuring the existence of a solution of the problem of the decay of an arbitrary discontinuity are required.

10. ***Linearized formulas for the decay of a discontinuity in the case of a polytropic gas.*** It is easy to see that $\Psi_0(S_0, p)$ and $\Phi_0(S_0, p) - \Phi_0(S_0, p_0)$ for $p/p_0 = 1 + \varepsilon$ coincide up to terms of order ε^3. This follows from the analysis of weak shock waves carried out in §4.7. Therefore, expanding (6.8.14) in powers of the small quantity $(p - p_0)/p_0$ and retaining only terms of first order, we obtain

$$u = g_0(S_0, p) = \frac{c_0}{\gamma_0}\frac{p - p_0}{p_0} = \frac{p - p_0}{\rho_0 c_0}. \tag{1}$$

Similarly, for weak waves

$$u_1 - u = g_1(S_1, p) = \frac{c_1}{\gamma_1}\frac{p - p_1}{p_1} = \frac{p - p_1}{\rho_1 c_1}. \tag{2}$$

Therefore, for a decay of a discontinuity with small amplitudes we find that the values of the pressure p and velocity u on a contact discontinuity are expressed by the same formulas regardless of the configuration:

$$p = \frac{u_1 - u_0}{1/(\rho_1 c_1) + 1/(\rho_0 c_0)} + \frac{p_1/(\rho_1 c_1) + p_0/(\rho_0 c_0)}{1/(\rho_1 c_1) + 1/(\rho_0 c_0)}, \tag{3}$$

$$u = \frac{p_1 - p_0}{\rho_0 c_0 + \rho_1 c_1} + \frac{\rho_0 c_0 u_0 + \rho_1 c_1 u_1}{\rho_0 c_0 + \rho_1 c_1}, \tag{4}$$

where we have replaced u_1 by $u_1 - u_0$. The formulas (3) and (4) describe a solution of the problem of the decay of a discontinuity in the acoustic approximation, i.e., for infinitely weak waves.

11. ***Decay of a discontinuity in a channel of variable cross section.*** We consider two semi-infinite cylindrical tubes with cross-sectional areas A_1 and A_2 which are joined in the plane $x = 0$ (Figure 2.63) and are filled with gases characterized at time $t = 0$ by parameters u_1, ρ_1, p_1, S_1 and u_2, ρ_2, p_2, S_2 respectively.

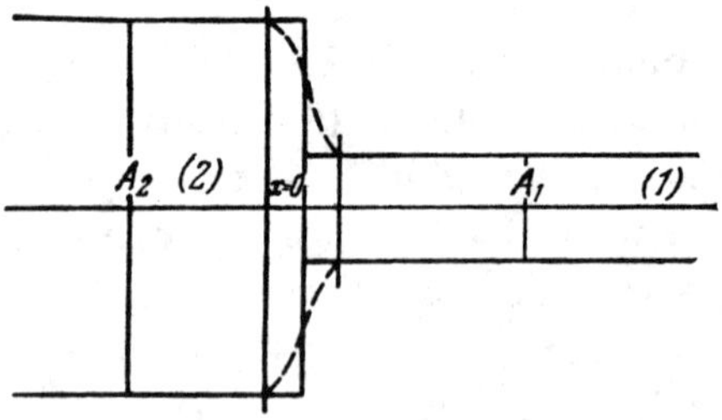

FIGURE 2.63

The motion occurring for $t > 0$ is two-dimensional; however, it may be assumed that the waves propagating in each tube for $t \to \infty$ and $|x| \to \infty$ become one-dimensional. The approximate picture of the flow (we call it a flow by decay of the discontinuity at the jump of the cross section) is based on the assumption that the asymptotic state is established instantaneously, and the flow decomposes into two one-dimensional flows separated by an infinitely thin transition zone contained between the planes $x = -\varepsilon$ and $x = +\varepsilon$. The motion in the transition zone is a stationary flow, so that the quantities $f = (u, \rho, p)$ to the left and the right of the transition zone (we denote them by f^- and f^+ respectively) are connected by the relations

$$A_2(\rho u)^- = A_1(\rho u)^+, \tag{1}$$

$$\left[\varepsilon(p, \rho) + \frac{p}{\rho} + \frac{u^2}{2}\right]^- = \left[\varepsilon(p, \rho) + \frac{p}{\rho} + \frac{u^2}{2}\right]^+. \tag{2}$$

The first relation expresses the law of conservation of mass, and the second expresses Bernoulli's law. To this we adjoin a third relation which in different decay models has a different form. We restrict our attention to the adiabatic model in which it is assumed that the entropy is conserved in the transition zone.[18] For a polytropic gas we thus have

$$\left(\frac{p}{\rho^\gamma}\right)^- = \left(\frac{p}{\rho^\gamma}\right)^+. \tag{3}$$

We consider the simplest case of decay of the discontinuity—the propagation of a shock wave proceeding through the quiescent gas with parameters ρ_1, p_1, u_1 and proceeding from the larger portion of the pipe to the narrower portion. We shall assume that the jump of the cross section is not large, i.e.,

$$\frac{A_2 - A_1}{A_1} = \delta \ll 1. \tag{4}$$

It is then possible to assume that the perturbation of the shock wave is small and to linearize the computational formulas. After the shock wave passes through the jump in the cross section we have the following configuration of discontinuities: a shock wave D_1 propagates to the right followed at the point $x = 0$ by a discontinuity subject to the relations (1)–(3), while a reflected wave propagates to the left; there is a contact boundary between the shock wave which has passed and the section $x = 0$. Figure 2.64 shows the configuration of discontinuities in the x, t plane. The line Γ_4 is the trajectory of the shock wave entering the narrow pipe, Γ_1 is the trajectory of the reflected shock wave, Γ_2 is

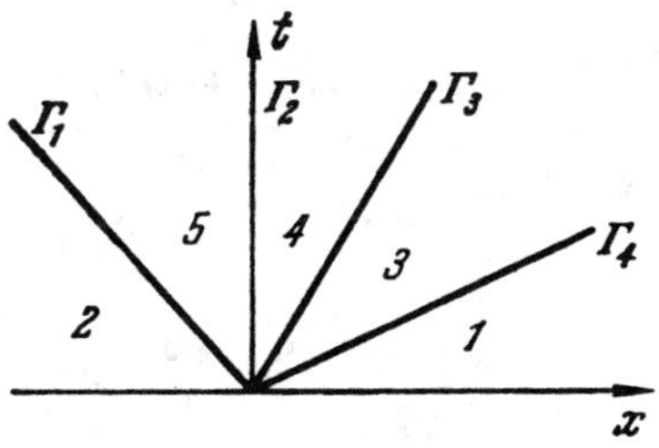

FIGURE 2.64

[18] A complete analysis of the problem of the decay of a discontinuity at a jump of the cross section was given by Dulov (1958) and Jaušev (1967).

the transition zone, and Γ_3 is the trajectory of the contact boundary. On the lines Γ_1, Γ_2 and Γ_3 the flow parameters experience jumps of small amplitude, and the corresponding matching conditions can be linearized. We denote by Δp and Δu the total change of p and u on passing from state (2) to state (3) and by $\Delta_i p$ and $\Delta_i u$ the change of p and u on passing through Γ_i $(i = 1, 2, 3)$. The following linearized relations hold:

$$\left.\begin{aligned}
&\text{a)} && \text{on } \Gamma_1\colon \frac{\Delta_1 p}{p_2} + \frac{1+h}{1-h}\,\frac{\Delta_1 u}{c_2} = 0,\\
&\text{b)} && \text{on } \Gamma_2\colon -\delta + \frac{\Delta_2 u}{u_2} + \frac{\Delta_2 p}{\gamma p_2} = 0,\\
&\text{c)} && \frac{\Delta_2 p}{\rho_2} + u_2 \Delta_2 u = 0,\\
&\text{d)} && \text{on } \Gamma_3\colon \Delta_3 p = \Delta_3 u = 0.
\end{aligned}\right\} \tag{5}$$

The total change Δu, Δp is caused by the change of the force M of the shock wave on passing from the larger part of the tube to the narrower part. Using the Hugoniot conditions for a polytropic gas

$$\left.\begin{aligned}
&\text{a)} && \frac{p_2}{p_1} = (1+h)\,M^2 - h,\\
&\text{b)} && \frac{u_2}{c_1} = (1-h)\left(M - \frac{1}{M}\right),\\
&\text{c)} && \frac{c_2}{c_1} = \sqrt{\left[(1+h) - \frac{h}{M^2}\right][(1-h) + hM^2]},\\
&\text{d)} && \frac{\rho_2}{\rho_1} = \frac{M^2}{1 - h + hM^2},
\end{aligned}\right\} \tag{6}$$

we have

$$\left.\begin{aligned}
&\text{a)} && \frac{\Delta p}{p_1} = \frac{\Delta_1 p + \Delta_2 p}{p_1} = 2(1+h)M\Delta M,\\
&\text{b)} && \frac{\Delta u}{c_1} = \frac{\Delta_1 u + \Delta_2 u}{c_1} = (1-h)\left(1 + \frac{1}{M^2}\right)\Delta M.
\end{aligned}\right\} \tag{7}$$

The relations (6) make it possible to express the quantities with index 2 in terms of quantities with index 1 and known functions of M; the coefficients of (5a), (6b), and (5c) are then expressed in terms of M and quantities with index 1. The equations (5a), (5b), (5c), (7a), and (7b) provide a system of five equations for the four quantities $\Delta_1 u$, $\Delta_2 u$, $\Delta_1 p$ and $\Delta_2 p$. The condition of algebraic compatibility of (7) leads to an equation first obtained by Chester (1950):

$$\frac{dA}{A} = -\frac{2M\,dM}{(M^2-1)k(M)}, \tag{8}$$

where

$$k(M) = 2\left[\left(1 + (1-h)\frac{1-g^2}{g}\right)\left(2g + 1 + \frac{1}{M^2}\right)\right]^{-1}. \tag{9}$$

$$g^2(M) = \frac{(\gamma-1)\,M^2+2}{2\gamma M^2-(\gamma-1)}. \tag{10}$$

Equation (8) provides a relation between the change of the force of the shock wave and the cross section of the tube. Chisnell (1957) proposed using (8) for the decay of a nonstationary shock front moving in a channel with continuously varying cross section. In the theory of Chisnell the channel with continuously varying cross section is approximated by a sequence of cylindrical channels (Figure 2.65) adjoining one another, and the passage of the shock wave from one cylindrical portion to another is described by (8).

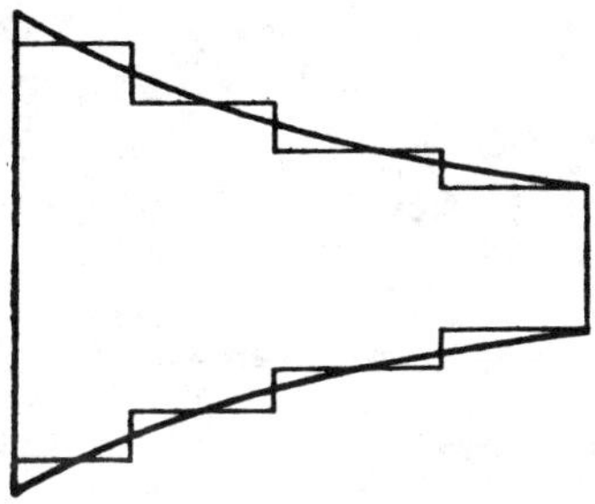

FIGURE 2.65

Integrating (8), we find that

$$Af(M) = \text{const}, \tag{11}$$

where

$$f = z^{1/\gamma}(z-1)(z+h)^{-1/2}\left[\frac{1+R}{1-R}\right]^{\sqrt{\gamma/(2(\gamma-1))}}$$

$$\times\left[\frac{R-(\gamma-1/(2\gamma))^{1/2}}{R+(\gamma-1/(2\gamma))^{1/2}}\right]\exp\left[\left(\frac{2}{\gamma-1}\right)^{1/2}\arctan\left\{\frac{2\gamma^{1/2}R}{\gamma-1}\right\}\right], \tag{12}$$

$$R = \left(1+\frac{1}{hz}\right)^{-1/2}, \tag{13}$$

$$z = \frac{2\gamma}{\gamma+1}M^2 - h. \tag{14}$$

The quantity $k(M)$ is a slowly varying function of M. Thus, for $\gamma = 7/5$ we have $k(1) = 1/2$ and $k(\infty) \simeq 0.394$. If k is assumed constant, then the integral (11) simplifies and reduces to the form

$$A^k(M^2 - 1) = \text{const.} \tag{15}$$

For weak shock waves and arbitrary γ we have $k(M) \to 1/2$, and (15) takes the form

$$M - 1 \sim A^{-1/2}. \tag{16}$$

For strong shock waves

$$\underset{M\to\infty}{k(M)} \to k(\infty) = \frac{2}{(1 + k_1)(1 + k_2)}.$$

$$k_1 = \sqrt{\frac{2}{\gamma(\gamma - 1)}}, \qquad k_2 = \sqrt{\frac{2(\gamma - 1)}{\gamma}}. \tag{17}$$

From (15) we then have

$$M \sim A^{-k_\infty/2}. \tag{18}$$

The estimate (18) was applied by Chisnell to establish the asymptotics of a strong shock wave in the case of cylindrical and spherical symmetry.

From (18) it follows that

$$M \sim x^{-n}, \qquad n = k_\infty/2, \tag{19}$$

for a cylindrical shock wave, and

$$M \sim x^{-n}, \qquad n = k_\infty, \tag{20}$$

for a spherical shock wave (x is the distance to the axis or point of symmetry respectively).

Comparison of (19) and (20) with the self-similar solution of Guderley for a converging shock wave (see §9.5) shows excellent agreement. Thus, for $\gamma = 5/3$ and $7/5$ the comparison of the exponents n is shown in the table.

	Cylindrical wave		Spherical wave	
	Chisnell	Guderley	Chisnell	Guderley
$\gamma = 5/3$	0,2254	0,2260	0,4508	0,4527
$\gamma = 7/5$	0,1971	0,1973	0,3941	0.3944

The theory developed by Chisnell does not take account of the additional effect of secondary waves reflected from the walls of the channel and overtaking the shock wave. In the works of Chisnell (1955, 1957) corrections were made in (18) which, however, were not appreciable.

Whitham (1958) gave a simple interpretation of equation (8). It is known (see Stanjukovič (1955)) that the flow in a channel in the one-dimensional approximation can be described by the equations

$$\left.\begin{aligned} &\frac{\partial\rho}{\partial t} + \frac{1}{A}\frac{\partial}{\partial x}(\rho u A) = 0,\\ &\frac{\partial u}{\partial t} + u\frac{\partial u}{\partial x} + \frac{1}{\rho}\frac{\partial p}{\partial x} = 0,\\ &\frac{\partial}{\partial t}\left(\frac{p}{\rho^\gamma}\right) + u\frac{\partial}{\partial x}\left(\frac{p}{\rho^\gamma}\right) = 0, \end{aligned}\right\} \tag{21}$$

where $A(x)$ is the area of the cross section of the channel. Equations (21) can be rewritten in the characteristic form

$$\left.\begin{aligned} &\text{a)} && \frac{1}{\rho}\frac{dp}{dt_1} - \frac{du}{dt_1} + \frac{cu}{u-c}\frac{d\ln A}{dt_1} = 0,\\ &\text{b)} && \frac{d}{dt_2}\left(\frac{p}{\rho^\gamma}\right) = 0,\\ &\text{c)} && \frac{1}{\rho c}\frac{dp}{dt_3} + \frac{du}{dt_3} + \frac{cu}{u+c}\frac{d\ln A}{dt_3} = 0, \end{aligned}\right\} \tag{22}$$

where

$$\frac{d}{dt_1} = \frac{\partial}{\partial t} + (u-c)\frac{\partial}{\partial x}, \quad \frac{d}{dt_2} = \frac{\partial}{\partial t} + u\frac{\partial}{\partial x}, \quad \frac{d}{dt_3} = \frac{\partial}{\partial t} + (u+c)\frac{\partial}{\partial x}. \tag{23}$$

Assuming, to be specific, that the shock wave moves from left to right and that the difference of slopes of the trajectories of the shock wave and the overtaking characteristic is not large, it may approximately be assumed that the relation

$$d\bar{r} = \frac{1}{\rho c}dp + du + \frac{cu}{u+c}d\ln A = dr + \frac{cu}{u+c}d\ln A = 0 \tag{24}$$

is satisfied not only along the r-characteristic but also on the trajectory of the shock wave. From (6a), (6b) we have

$$dr = du + \frac{1}{\rho_2 c_2}dp = c_1\left[(1-h)\left(1+\frac{1}{M^2}\right) + \frac{2p_1}{\rho_2 c_2 c_1}(1+h)M\right]dM. \tag{25}$$

Substituting (25) into (24), we arrive at the relation

$$\left[(1-h)\left(1+\frac{1}{M^2}\right)+\frac{2\rho_1}{\rho_2 c_2 c_1}(1+h)M\right]dM+\frac{c_2 u_2}{c_1(u_1+c_1)}d\ln A=0, \quad (26)$$

which is equivalent to (8) because of (6). Since the assumption (24) is valid not only on weak waves but also on strong waves passing into the center, this explains the good agreement of the Chisnell theory with the self-similar solution for the problem of a converging shock wave.

For flows possessing the property that the slope of the characteristic overtaking the shock front be close to the slope of the shock front, various approximations have been worked out which well describe the behavior of the flow near the front (the Poincare-Lighthill-Kuo method (see Tsien (1956)), and the method of short waves (see Ryžov and Hristianovič (1958)).

§7. The interaction of strong discontinuities

On the basis of the analysis of the decay of an arbitrary discontinuity in this section we consider several problems regarding the interaction of strong discontinuities such as shock waves and contact boundaries. When strong discontinuities combine, piecewise constant flows arise with parameters which may be taken as initial data for the solution of the problem of the decay of an arbitrary discontinuity. Therefore, the problem of the interaction of strong discontinuities reduces to the problem of the decay of a discontinuity. In this section we consider interactions of strong discontinuities in the following order:

1. Incidence of a shock wave on a contact boundary.
2. Combination of shock waves moving toward one another relative to the gas.
3. Combination of shock waves moving in the same direction in the gas.

These three combinations exhaust all possible interactions of strong discontinuities. The basic problem is to establish the configuration of the flow formed depending on the parameters characterizing the interacting discontinuities. We shall first carry out a more general investigation by the method of the p, u-diagram, and we then consider the case of a polytropic gas.

1. *Incidence of a shock wave on the boundary between two media.* Suppose that on the boundary $x = 0$ of two media characterized by parameters $\bar{p}_0 = p_0$, $\bar{S}_0$, $\bar{u}_0 = 0$ (to the left of $x = 0$) and p_0, S_0, $u_0 = 0$ (to the right) a shock wave is incident at time $t = 0$, propagating in the medium $x < 0$ in the direction from left to right. Thus, at time $t = 0$ an initial discontinuity is formed with parameters p_1, S_1, u_1 (the gas "1" to the left) and p_0, S_0, $u_0 = 0$ (the gas "0" to the right). Here the point (p_1, u_1) must lie on the upper part of the curve (6.8.7) computed for the gas initially found in $x < 0$ and passing through the point $(p_0, 0)$, i.e.,

$$u_1 = \bar{\Psi}_0(\bar{S}_0, p_1), \qquad p_1 > p_0, \ u_1 > 0. \tag{1}$$

Thus, the set of possible states (p_1, u_1) of the gas "1" is depicted in the p, u plane by a curve with the equation (1) (Figure 2.66).

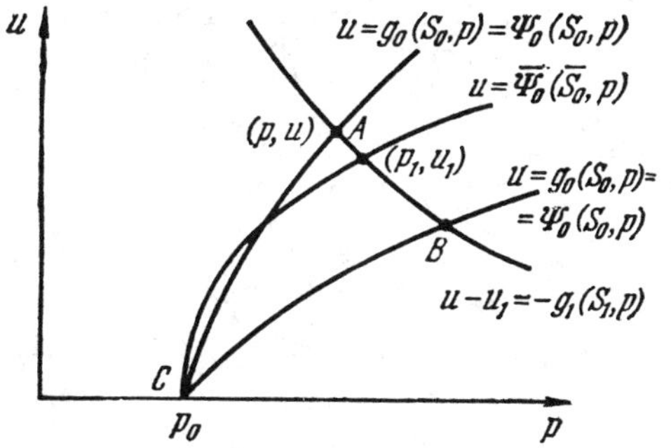

FIGURE 2.66

Suppose that an initial shock wave is given. A point (p_1, u_1) lying on the curve (1) (Figure 2.66) is thus given. In correspondence with §6.8 the solution of the problem on the interaction of a shock wave with a contact boundary reduces to determining the point (p, u) of intersection of the curve (6.8.7)

$$u = g_0(S_0, p) \tag{2}$$

for the gas "0" passing through the point (p_0, u_0) and the curve (6.8.8)

$$u = u_1 - g_1(S_1, p) \tag{3}$$

for the gas "1" passing through the point (p_1, u_1). It is obvious that the curve (2) can intersect (3) only with its upper branch, i.e., for $p > p_0$. Hence, the flow which arises has either configuration A (it corresponds to the point A of intersection of the curve (2) (CA) and the curve (3) (AB) in Figure 2.66) or configuration B (it corresponds to the point B of intersection of the curve (2) (BC) and the curve (3) (AB)).

Thus, in gas "0" a shock wave propagates, while in gas "1" a rarefaction wave or a shock wave propagates depending on the relative positions of the curves (1) and (2).

We shall consider the case of polytropic gases. According to (6.8.13),

$$\bar{\Psi}_0(\bar{S}_0, p) = (1 - h_1)\bar{c}_0\left[\sqrt{\frac{p + h_1 p_0}{p_0(1 + h_1)}} - \sqrt{\frac{p_0(1 + h_1)}{p + h_1 p_0}}\right], \tag{4}$$

$$\Psi_0(S_0, p) = (1 - h_0)\bar{c}_0\left[\sqrt{\frac{p + h_0 p_0}{p_0(1 + h_0)}} - \sqrt{\frac{p_0(1 + h_0)}{p + h_0 p_0}}\right]. \tag{5}$$

We introduce the quantity

$$\begin{aligned} K(\theta) &= \frac{\bar{\Psi}_0^2(\bar{S}_0, p)}{\Psi_0^2(S_0, p)} = \frac{(1 - h_1)^2}{(1 - h_0)^2} \frac{\bar{c}_0^2}{c_0^2} \frac{1 + h_0}{1 + h_1} \frac{\theta + h_0}{\theta + h_1} \frac{(\theta - 1)^2}{(\theta - 1)^2} \\ &= \frac{(1 - h_1)^2}{(1 - h_0)^2} \frac{(1 + h_0)}{(1 + h_1)} \frac{\theta + h_0}{\theta + h_1} \left(\frac{\bar{c}_0}{c_0}\right)^2, \end{aligned} \tag{6}$$

where

$$\theta = p/p_0. \tag{7}$$

From this we get

$$K(1) = \frac{(1 + h_0)^2}{(1 - h_0)^2} \frac{(1 - h_1)^2}{(1 + h_1)^2} \left(\frac{\bar{c}_0}{c_0}\right)^2 = \left(\frac{\gamma_0}{\gamma_1}\right)^2 \left(\frac{\bar{c}_0}{c_0}\right)^2, \tag{8}$$

$$K(\infty) = \frac{(1 - h_1)^2}{(1 - h_0)^2} \frac{(1 + h_0)}{(1 + h_1)} \left(\frac{\bar{c}_0}{c_0}\right)^2 = \frac{1 + h_1}{1 + h_0} K(1). \tag{9}$$

If $h_0 < h_1$ ($\gamma_0 < \gamma_1$), then $K(\infty) > K(1)$, and $K(\theta)$ is a monotonically increasing function of θ. If $h_0 > h_1$ ($\gamma_0 > \gamma_1$), then $K(\infty) < K(1)$, and $K(\theta)$ is a monotonically decreasing function for $0 < \theta < \infty$.

We consider the following cases:

1. $K(1) \geq 1$ and $K(\infty) \geq 1$. Then $K(\theta) \geq 1$ and $\bar{\Psi}_0(\bar{S}_0, p) \geq \Psi_0(S_0, p)$ for all $p \geq p_0$. In this case the curve $u = \Psi_0(S_0, p)$ lies below $u = \bar{\Psi}_0(\bar{S}_0, p)$ (Figure 2.66) and for any strength of the shock wave configuration B arises as a result of the interaction with the contact boundary, i.e., for $t > 0$ two shock waves propagate in different directions away from the contact boundary (Figure 2.67).

2. $K(1) \leq 1$ and $K(\infty) \leq 1$. Then $\bar{\Psi}_0(\bar{S}_0, p) \leq \Psi_0(S_0, p)$, and for any strength of the shock wave the flow for $t > 0$ has configuration A, i.e., a shock wave propagates in the gas "0" (as always), while a centered rarefaction wave propagates in the gas "1".

3. If $K(1) > 1$ and $K(\infty) < 1$ ($\gamma_0 > \gamma_1$), then for some $p = P$ the curves $u = \Psi_0(S_0, p)$ and $u = \bar{\Psi}_0(\bar{S}_0, p)$ intersect. For $p_0 \leq p_1 \leq P$ we have configuration B; for $p_1 > P$, i.e., for sufficiently strong shock waves, we have configuration A.

4. $K(1) < 1$ and $K(\infty) > 1$ ($\gamma_0 < \gamma_1$). Then the curves $u = \bar{\Psi}_0(\bar{S}_0, p)$ and $u = \Psi_0(S_0, p)$ intersect for $p = P$; in the region $p_0 \leq p_1 \leq P$ the flow has configuration A, while for $p_1 > P$ it has configuration B.

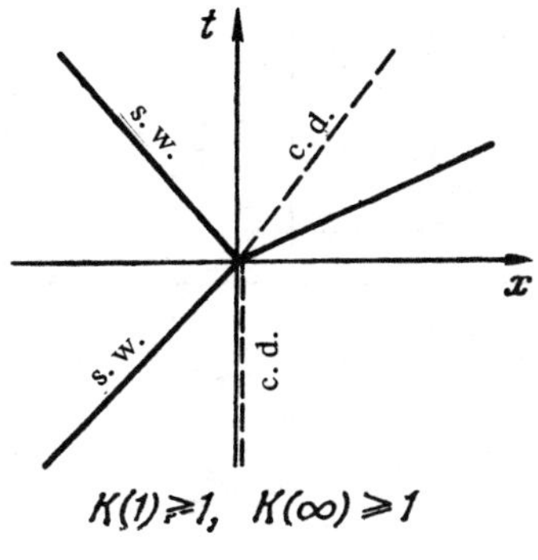

FIGURE 2.67

We consider the special case of identical gases ($\gamma_0 = \gamma_1 = \gamma$). Then

$$K(1) = K(\infty) = \left(\frac{\bar{c}_0}{c_0}\right)^2. \tag{10}$$

Considering the relation $c^2 = \gamma p/\rho$, we write (10) in the form

$$K(1) = K(\infty) = \rho_0/\bar{\rho}_0. \tag{11}$$

This implies that when a shock wave goes from a less dense gas to a denser gas ($K > 1$), a shock wave is reflected into the less dense gas (we have configuration B). If the shock wave goes from a denser gas to a less dense gas ($K < 1$), then a rarefaction wave is reflected (configuration A).

This result can be compared with the conclusions of §3.5, where the refraction of a travelling wave on a contact boundary was studied. There we found that from a denser medium a wave of the same type as the incident wave is reflected. Since a weak shock wave may be considered as a weak Riemann compression wave, we see that qualitatively the refraction of shock waves and Riemann waves on contact boundaries have the same character.

2. *The encounter of two shock waves.* Two shock waves propagate toward one another through a medium with parameters $\bar{u}_0 = 0$, $\bar{p}_0$, $\bar{S}_0$ leaving behind states $u_1 = \overline{\Psi}_0(\bar{S}_0, p_1), p_1, S_1$ behind the left wave and $u_0 = -\overline{\Psi}_0(\bar{S}_0, p_0)$, p_0, S_0 behind the right wave (Figure 2.68). At time $t = 0$ the shock waves meet at the point $x = 0$ forming an initial discontinuity with parameters $u_1 = \overline{\Psi}_0(\bar{S}_0, p_1)$, p_1, S_1 (on the left) and $u_0 = -\overline{\Psi}_0(\bar{S}_0, p_0)$, p_0, S_0 (on the right). To be specific,

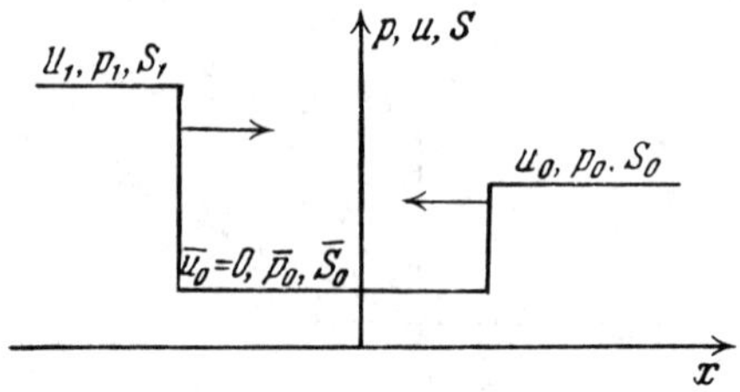

FIGURE 2.68

we assume that

$$p_1 \geq p_0. \tag{1}$$

There are two possible cases:

a) The curve $u - u_0 = \Psi_0(S_0, p)$ is always below $u = \overline{\Psi}_0(\bar{S}_0, p)$ (Figure 2.69, a).

b) The curve $u - u_0 = \Psi_0(S_0, p)$ intersects $u = \overline{\Psi}_0(\bar{S}_0, p)$ at some point (P, U) (Figure 2.69, b).

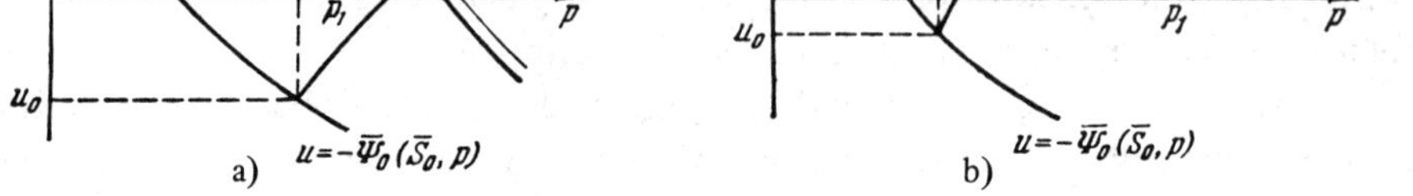

FIGURE 2.69

In the first case only configuration B is possible; in the second case both B and A are possible, but in any case a shock wave propagates in the gas "0".

We consider in more detail the case of a polytropic gas for which the intersection of the curves $u = \overline{\Psi}_0(\bar{S}_0, p)$ and $u = u_0 + \Psi(S_0, p) = \Psi_0(S_0, p) - \overline{\Psi}_0(\bar{S}_0, p)$ is impossible. We must thus prove the inequality

$$\overline{\Psi}_0(\bar{S}_0, p) + \overline{\Psi}_0(\bar{S}_0, p_0) > \Psi_0(S_0, p) \quad \text{for } p \geq p_0 > \bar{p}_0. \tag{2}$$

Considering (6.8.13) for $\Psi_0(S_0, p)$ and (6.8.12), we reduce (2) to the form

$$\frac{M_1^2 - 1}{M_1} + \frac{M_0^2 - 1}{M_0} > \frac{c_0}{\bar{c}_0} \frac{M^2 - 1}{M}, \tag{3}$$

where

$$\left. \begin{aligned} M_0 &= -\frac{D_0}{\bar{c}_0} = \sqrt{\frac{p_0 + h\bar{p}_0}{\bar{p}_0(1+h)}} > 1 \qquad (p_0 > \bar{p}_0), \\ M_1 &= \frac{D_1}{\bar{c}_0} = \sqrt{\frac{p + h\bar{p}_0}{\bar{p}_0(1+h)}} \geq M_0 > 1 \qquad (p \geq p_0 > \bar{p}_0), \\ M &= \frac{|D - u_0|}{c_0} = \sqrt{\frac{p + hp_0}{p_0(1 + h)}} > 1 \qquad (p > p_0). \end{aligned} \right\} \tag{4}$$

This implies the inequality

$$(1 + h)M^2 - h = \frac{(1 + h)M_1^2 - h}{(1 + h)M_0^2 - h}. \tag{5}$$

Recalling (4.5.15)

$$\frac{c_0^2}{\bar{c}_0^2} = \frac{[(1 + h)M_0^2 - h][(1 - h) + hM_0^2]}{M_0^2}, \tag{6}$$

and also the consequence of (5)

$$M^2 - 1 = \frac{M_1^2 - M_0^2}{(1 + h)\, M_0^2 - h}, \tag{7}$$

we transform (3) to the form

$$\frac{(M_1 M_0 - 1)^2}{M_1^2} > \frac{1}{M^2} \frac{(M_1 - M_0)^2 (1 - h + hM_0^2)}{(1 + h)M_0^2 - h}. \tag{8}$$

Eliminating M from (8) by means of (5), we get

$$\frac{(M_1 M_0 - 1)^2}{M_1^2} > \frac{(M_1 - M_0)^2 [1 - h + hM_0^2]}{M_1^2 - h + hM_0^2}. \tag{9}$$

For $M_1 = M_0$ the inequality (9) is obviously satisfied (this case corresponds to the encounter of shock waves of the same force). We shall show that (9) is satisfied for any $M_1 \geq M_0$. For this we transform it to the form

$$(M_1^2 - 1)(M_0^2 - 1)[(1 - h)M_1^2 + 2hM_0 M_1 - h] > 0,$$

which easily implies that it is always satisfied for $M_1 \geq M_0 > 1$. Thus, in a polytropic gas during the encounter of two shock waves only configuration B occurs, i.e., during the encounter of two shock waves each of them "passes through" the other, as it were.

3. ***Combining of shock waves propagating in the same direction.*** In a gas with parameters p_0, S_0, $u_0 = 0$ a shock wave propagates with constant velocity $D_0 > 0$ from left to right leaving behind a state $\bar{u}_0$, $\bar{p}_0$, $\bar{S}_0$ on the curve

$$u = \Psi_0(S_0, p). \tag{1}$$

Behind the first shock wave a second shock wave propagates with constant velocity D_1, $D_1 > D_0 > \bar{u}_0$ leaving behind a state u_1, p_1, S_1 located on the curve

$$u - \bar{u}_0 = \overline{\Psi}_0(\bar{S}_0, p), \tag{2}$$

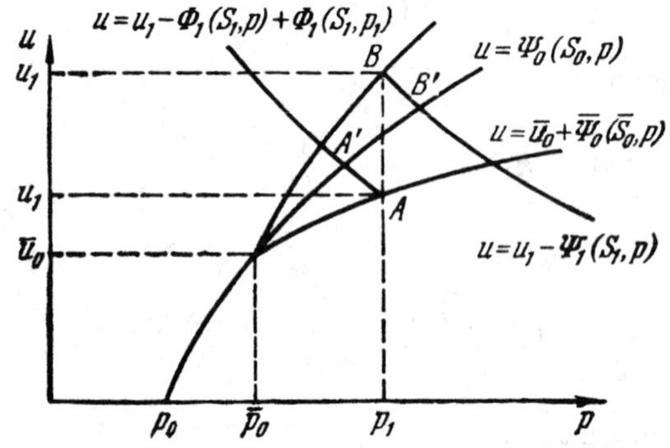

FIGURE 2.70

passing through the point $(\bar{p}_0, \Psi_0(S_0, \bar{p}_0))$ (Figure 2.70). At the moment of combination of the shock waves an initial discontinuity arises which in decomposing produces a flow with configuration B if the point $(p_1, \bar{u}_0 + \overline{\Psi}_0(\bar{S}_0, p_1))$ lies above the curve (1) (this case is shown by the point B in Figure 2.70) and a flow with configuration A if the point $(p_1, \bar{u}_0 + \overline{\Psi}_0(\bar{S}_0, p_1))$ lies below the curve (1) (the corresponding solutions are labelled by the points B' and A' in the figure). However, a shock wave always propagates in the direction of the gas "0".

We again consider the case of a polytropic gas in more detail. The relative positions of the curves (1) and (2) are determined by the sign of the difference

$$\Delta = \Psi_0(S_0, \bar{p}_0) + \overline{\Psi}_0(\bar{S}_0, p) - \Psi_0(S_0, p) \tag{3}$$

with the given restrictions

$$p_0 < \bar{p}_0 < p. \tag{4}$$

We introduce the notation

$$\frac{D_0}{c_0} = M_0, \qquad \frac{D_1 - \bar{u}_0}{\bar{c}_0} = \bar{M}, \qquad \frac{D}{c_0} = M. \tag{5}$$

Here M_0 corresponds to the first shock wave propagating on the background $(0, p_0)$ and leaving behind the state $(\bar{u}_0, \bar{p}_0)$; $\bar{M}$ corresponds to the second wave propagating on the background $(\bar{u}_0, \bar{p}_0)$ and leaving behind the state (u_1, p_1); M corresponds to a possible shock wave propagating on the background $(0, p_0)$ and leaving behind the state (u, p). The difference (3) can then be represented in the form

$$\Delta = (1-h)c_0\frac{M_0^2 - 1}{M_0} + (1-h)\bar{c}_0\frac{\bar{M}^2 - 1}{\bar{M}} - (1-h)c_0\frac{M^2 - 1}{M}. \tag{6}$$

From the relations

$$\frac{\bar{p}_0}{p_0} = (1+h)M_0^2 - h, \qquad \frac{p_1}{\bar{p}_0} = (1+h)\bar{M}^2 - h, \qquad \frac{p_1}{p_0} = (1+h)M^2 - h \tag{7}$$

it follows that

$$(1+h)\bar{M}^2 - h = \frac{(1+h)M^2 - h}{(1+h)M_0^2 - h}, \quad \text{or} \quad \bar{M}^2 = \frac{M^2 + hM_0^2 - h}{(1+h)M_0^2 - h}. \tag{8}$$

Substituting (8) into (6), we obtain

$$\begin{aligned}\Delta &= (1-h)c_0\left[\frac{M_0^2 - 1}{M_0} - \frac{M^2 - 1}{M} + \frac{\bar{c}_0}{c_0}\frac{M^2 - M_0^2}{(1+h)M_0^2 - h}\frac{1}{\bar{M}}\right] \\ &= (1-h)c_0\left[\frac{(M_0 - M)(MM_0 + 1)}{MM_0} + \frac{\bar{c}_0}{c_0}\frac{M^2 - M_0^2}{(1+h)M_0^2 - h}\frac{1}{\bar{M}}\right].\end{aligned} \tag{9}$$

Applying (4.5.15) (see also (7.2.6)), which in the present case can be written in the form

$$\frac{\bar{c}_0^2}{c_0^2} = \frac{[(1+h)M_0^2 - h][(1-h) + hM_0^2]}{M_0^2}, \tag{10}$$

we obtain from (9)

$$\begin{aligned}\Delta &= (1-h)c_0\left[\frac{(M_0 - M)(MM_0 + 1)}{MM_0} + \frac{\sqrt{(1-h) + hM_0^2}(M^2 - M_0^2)}{M_0\sqrt{(1+h)M_0^2 - h}}\frac{1}{\bar{M}}\right] \\ &= (1-h)c_0(M - M_0)\left[\frac{\sqrt{1 - h + hM_0^2}(M + M_0)}{M_0\bar{M}\sqrt{(1+h)M_0^2 - h}} - \frac{(MM_0 + 1)}{MM_0}\right].\end{aligned} \tag{11}$$

Since, according to (4), $M > M_0 > 1$, we see that the sign of Δ coincides with the sign of the expression

$$\frac{(1 - h + hM_0^2)(M + M_0)^2}{M_0^2 \bar{M}^2[(1 + h)M_0^2 - h]} - \frac{(MM_0 + 1)^2}{M^2 M_0^2}. \tag{12}$$

Finally, substituting $\bar{M}^2$ from (8), we obtain

$$\begin{aligned}\operatorname{sgn} \Delta &= \operatorname{sgn}\left\{\frac{[1 - h + hM_0^2]}{M^2 + hM_0^2 - h} \frac{(M + M_0)^2}{M_0^2} - \frac{(MM_0 + 1)^2}{M^2 M_0^2}\right\} \\ &= \operatorname{sgn}\{M^2(M + M_0)^2[1 + h(M_0^2 - 1)] \\ &\quad - (MM_0 + 1)^2[M^2 + h(M_0^2 - 1)]\} \\ &= -\operatorname{sgn}(M^2 - 1)(M_0^2 - 1)[(1 - h)M^2 - 2hMM_0 - h]. \end{aligned} \tag{13}$$

Thus, if

$$(1 - h)M^2 - 2hMM_0 - h > 0, \tag{14}$$

it follows that $\Delta < 0$, and we have configuration A; if

$$(1 - h)M^2 - 2hMM_0 - h < 0, \tag{15}$$

then configuration B holds.

It is not hard to see that when the second shock wave is sufficiently strong ($\bar{M} \gg 1$, $M \gg 1$) inequality (14) is satisfied, and we have configuration A. In order that the equation

$$(1 - h)M^2 - 2hMM_0 - h = 0 \tag{16}$$

have a root M ($M > M_0 > 1$) it is necessary and sufficient that

$$1 < M_0^2 < \frac{h}{1 - 3h} \tag{17}$$

or $h \geq 1/3$, i.e. $\gamma \geq 2$. If

$$(1 - 3h) \geq h, \quad \text{i.e. } h \leq \tfrac{1}{4}, \gamma \leq \tfrac{5}{3}, \tag{18}$$

then (15) cannot be satisfied for any $M_0 > 1$, and we always have configuration A. If $\gamma > 5/3$, then configuration B can also occur. More precisely, for $M \leq M_{cr}$, where M_{cr} is greater than M_0, the root of (16), or, equivalently, for

$$\bar{M}^2 \leq \frac{M_{cr}^2 + hM_0^2 - h}{(1 + h)M_0^2 - h}$$

configuration B occurs; otherwise we have configuration A.

4. ***The interaction of strong discontinuities in an isothermic gas.*** In the case of an isothermic gas the analysis of the interaction of discontinuities simplifies. Indeed, an isothermic gas may be considered as a polytropic gas with $\gamma = 1$. For an isothermic gas we have

$$\Psi_0(p) = c_0\left[\sqrt{\frac{p}{p_0}} - \sqrt{\frac{p_0}{p}}\right], \tag{1}$$

$$\Phi_0(p) = c_0 \ln \frac{p}{p_0}. \tag{2}$$

We shall consider the problem of the incidence of a shock wave on a contact discontinuity (see Figure 2.66). As we saw in subsection 1, the choice between configurations A and B is determined by the sign of the difference $\Delta = \bar{\Psi}_0(p) - \Psi_0(p)$ under the condition $p > p_0$. For $\Delta < 0$ we have configuration A, while for $\Delta > 0$ we have configuration B. Recalling (1), we have

$$\Delta = (\bar{c}_0 - c_0)[\sqrt{p/p_0} - \sqrt{p_0/p}]. \tag{3}$$

Since $p/p_0 > 1$, configuration A occurs for $\bar{c}_0 < c_0$, and configuration B occurs for $\bar{c}_0 > c_0$. Recalling the relation

$$p_0 = c_0^2\rho_0 = \bar{c}_0^2\bar{\rho}_0, \tag{4}$$

we find that configuration A occurs for $\bar{\rho}_0 < \rho_0$ and configuration B for $\bar{\rho}_0 > \rho_0$. This result also follows directly from the results of subsection 1.

In solving the problem on the interaction of shock waves it is necessary to note that $\bar{c}_0 = c_0$. We first consider the interaction of shock waves moving toward one another (see Figures 2.68 and 2.69). Using the notation of subsection 2 and setting $h = 0$, we have

$$\frac{p_0}{\bar{p}_0} = M_0^2, \qquad \frac{p_1}{\bar{p}_0} = M_1^2, \qquad \frac{p_1}{p_0} = M^2, \tag{5}$$

$$M^2 = \frac{M_1^2}{M_0^2}, \qquad M = \frac{M_1}{M_0}. \tag{6}$$

The condition for configuration B has the form (see (7.2.3))

$$\frac{M_1^2 - 1}{M_1} + \frac{M_0^2 - 1}{M_0} > \frac{(M_1/M_0)^2 - 1}{(M_1/M_0)} = \frac{M_1}{M_0} - \frac{M_0}{M_1}, \tag{7}$$

whence we have

$$(M_1 + M_0)(M_1 + 1)(M_0 - 1) > 0. \tag{8}$$

Thus, we always have configuration B.

From the relations

$$\frac{p_1}{\bar{p}_0} = M_1^2, \qquad \frac{p_0}{\bar{p}_0} = M_0^2, \qquad \frac{p}{p_1} = \bar{M}_0^2, \qquad \frac{p}{p_0} = \bar{M}_1^2 \tag{9}$$

we have

$$M_1\bar{M}_0 = M_0\bar{M}_1, \qquad \frac{\bar{M}_0}{M_0} = \frac{\bar{M}_1}{M_1} = K. \tag{10}$$

From the relations

$$u = \bar{c}\left[M_1 - \frac{1}{M_1} - \left(\bar{M}_0 - \frac{1}{\bar{M}_0}\right)\right], \tag{11}$$

$$u = \bar{c}\left[\bar{M}_1 - \frac{c}{\bar{M}_1} - \left(M_0 - \frac{1}{M_0}\right)\right]. \tag{12}$$

On recalling (10), we have

$$(M_0 + M_1)(1 - K)\left[1 + \frac{1}{KM_0M_1}\right] = 0. \tag{13}$$

From this is follows that

$$K = 1. \tag{14}$$

This means that after encounter of the waves each of them retains its force:

$$\bar{M}_0 = M, \qquad \bar{M}_1 = M_1; \tag{15}$$

the relative velocities of the shock waves are correspondingly preserved:

$$D_1 = D_r - \bar{u}_0 = D + |u_0|, \qquad D_0 = D_l - u_1, \tag{16}$$

where D_r is the velocity of the shock wave proceeding to the right after the encounter, and D_l is the velocity of the one proceeding to the left.

Thus, the computational formulas for the interaction of colliding waves have the form

$$p = M_0^2M_1^2\bar{p}_0, \qquad u = \bar{c}_0\left[\left(M_1 - \frac{1}{M_1}\right) - \left(M_0 - \frac{1}{M_0}\right)\right], \tag{17}$$

where p and u are the pressure and velocity between the diverging wave fronts. Their velocities are determined by (16).

In conclusion we consider the problem of the combination of waves proceeding in the same direction. Condition (7.3.14) is satisfied for $h = 0$. Hence, configuration A always occurs in the case of an isothermic gas during the combination of shock waves proceeding in the same direction.

§8. The interaction of shock waves with travelling waves

If a shock wave moving with constant velocity over a constant background runs into a travelling wave, then the strength does not change. In the case of a polytropic gas the entropy behind the wave front becomes variable, which complicates the analytic investigation. Therefore, we restrict consideration to barotropic, polytropic gases and, in particular, to isothermic gases. We first make a remark which is valid for barotropic, polytropic gases. Suppose that a shock wave moves from left to right with constant velocity D_0 on a constant background (p_0, ρ_0, u_0) leaving behind a constant background (p_1, ρ_1, u_1) connected with (p_0, ρ_0, u_0) by the Hugoniot conditions. At some time t_0 the shock wave enters a region of perturbed motion which may be either a travelling wave or a region of interference of travelling waves. The motion which ensues behind the shock wave is then a travelling wave until a new shock wave forms there (Figure 2.71).

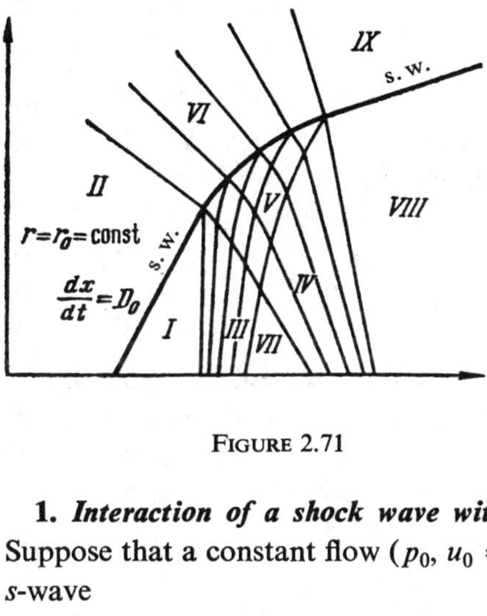

FIGURE 2.71

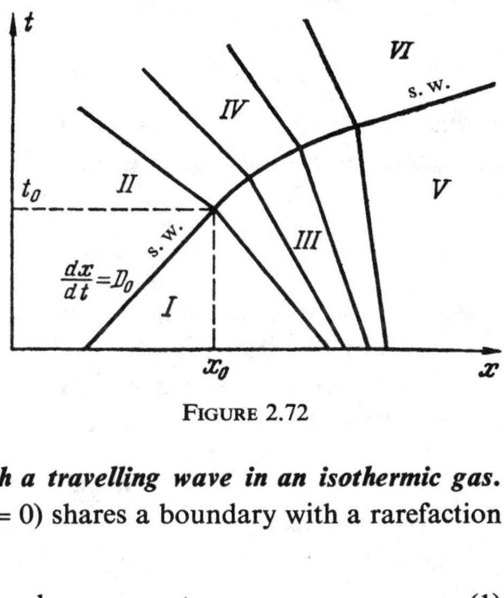

FIGURE 2.72

1. ***Interaction of a shock wave with a travelling wave in an isothermic gas.*** Suppose that a constant flow $(p_0, u_0 = 0)$ shares a boundary with a rarefaction s-wave

$$r = r_0 = u + c \ln \rho = \text{const.} \tag{1}$$

A shock wave moves with velocity D_0 over a constant background $(p_0, u_0 = 0)$ from left to right leaving behind the state

$$\left.\begin{aligned} p_1 &= c^2\rho_1, \\ \rho_1 &= \rho_0 M_0^2, \\ M_0 &= D_0/c. \end{aligned}\right\} \tag{2}$$

At time $t = t_0$ the shock wave enters region III of the travelling wave (Figure 2.72) where the pressure and velocity grow in opposite directions, i.e.,

$$\frac{\partial u}{\partial x}\frac{\partial p}{\partial x} < 0.$$

In region III of the travelling wave the Hugoniot conditions have the form (see (4.6.13) and (4.6.14))

$$r^- - r^+ = \varphi(M), \qquad s^- - s^+ = \psi(M), \tag{3}$$

$$M = \frac{D - u^+}{c}, \qquad \varphi(x) = c\left(x - \frac{1}{x} + \ln x^2\right),$$

$$\psi(x) = c\left(x - \frac{1}{x} - \ln x^2\right), \tag{4}$$

where D is the velocity of the shock wave; quantities with the index $+$ denote quantities in the wave III ahead of the front of the shock wave; the quantities with the index $-$ denotes those behind the front of the shock wave (region IV).

In the present case of an s-wave

$$r^- = r_1, \qquad r^+ = r_0. \tag{5}$$

From (3) and (5) it follows that M is constant and equal to M_0. Thus, the amplitude of the shock wave is unchanged, and the quantities

$$\frac{p_+}{p_-}, \qquad \frac{\rho_+}{\rho_-}, \qquad u^+ - u^- \tag{6}$$

preserve a constant value on the wave front.

In region IV we have a rarefaction (compression) s-wave if in region III the Riemann waves is a rarefaction (compression) wave. Actually, since $u^- - u^+$ is constant on the wave front, it follows that

$$\frac{\partial u^-}{\partial t} + D\frac{\partial u^-}{\partial x} = \frac{\partial u^+}{\partial t} + D\frac{\partial u^+}{\partial x}, \tag{7}$$

or

$$\left(1 - \frac{u^- - c}{D}\right)\frac{\partial u^-}{\partial x} = \left(1 - \frac{u^+ - c}{D}\right)\frac{\partial u^+}{\partial x},$$

so that the derivatives $\partial u^-/\partial x$ and $\partial u^+/\partial x$ (in regions III and IV) have the same sign. Since in the s-wave with a constant invariant r (in region III) the Riemann relation

$$x - (u - c)t = f(u) \tag{8}$$

holds, where $f(u)$ is some function of u, and on the shock wave

$$\frac{dx}{dt} = D = u + cM_0 = u^+ + D_0, \tag{9}$$

on combining (8) and (9), we obtain a differential equation for the trajectory of the shock wave:

$$x - \left[\frac{dx}{dt} - c(M_0 + 1)\right]t = f\left(\frac{dx}{dt} - cM_0\right). \tag{10}$$

In the special case of a centered rarefaction wave when

$$\frac{x - x_1}{t - t_1} = u - c \tag{11}$$

where (x_1, t_1) is the center of the wave, we obtain the easily integrable equation

$$\frac{dx}{dt} = \frac{x - x_1}{t - t_1} + D_0 + c. \tag{12}$$

It is clear that an analogous analysis holds also for the case of a shock wave proceeding from right to left and incident on a Riemann r-wave.

In the case of a rarefaction wave ($\partial u/\partial x > 0$, $\partial p/\partial x < 0$) the shock wave goes "downhill" while accelerating; in the case of a compression wave ($\partial u/\partial x < 0$, $\partial p/\partial x > 0$) the shock waves goes "uphill" decelerating. This picture of the interaction can be realized in the problem with two pistons; the right piston first moves out of the gas forming a rarefaction wave (r = const), and then the left piston moves into the gas forming a shock wave travelling into the rarefaction wave.

We now consider the problem of the interaction of a shock wave moving to the right with a Riemann r-wave ($s = s_0$ = const) (Figure 2.73). In this case the pressure and velocity ahead of the shock wave increase in the same direction ($(\partial u/\partial x)(\partial p/\partial x) > 0$). In (3) it is necessary to set

$$r^- = r = \text{const}, \qquad s^+ = s_0. \tag{13}$$

In the case of a rarefaction (compression) wave in region III we have

$$\left.\begin{aligned} \frac{\partial u^+}{\partial x} > 0 \quad \left(\frac{\partial u^+}{\partial x} < 0\right), \\ \frac{\partial r^+}{\partial x} > 0 \quad \left(\frac{\partial r^+}{\partial x} < 0\right), \end{aligned}\right\} \tag{14}$$

and from (3) we find that

$$\left.\begin{aligned} \frac{\partial M}{\partial x} &< 0 \quad \left(\frac{\partial M}{\partial x} > 0\right), \\ \frac{\partial s^-}{\partial x} &< 0 \quad \left(\frac{\partial s^-}{\partial x} > 0\right), \\ \frac{\partial u^-}{\partial x} &< 0 \quad \left(\frac{\partial u^-}{\partial x} > 0\right). \end{aligned}\right\} \tag{15}$$

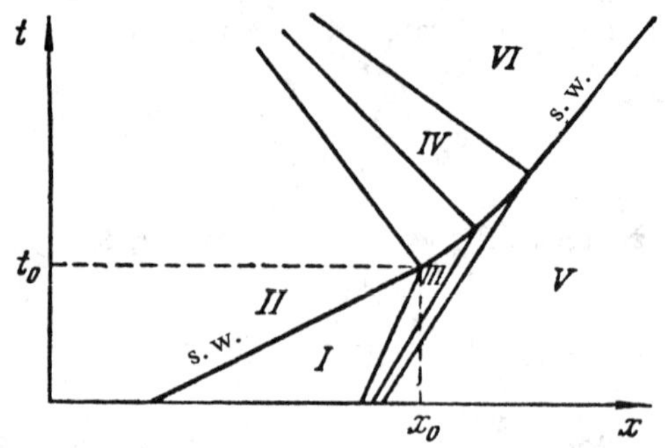

FIGURE 2.73

This implies that for a rarefaction wave III the refracted wave IV is a compression wave; for a compression wave III the wave IV is a rarefaction wave. In the first case the strength M of the wave decreases, while in the second it increases.

The trajectory of the shock wave can be found in analogy with the foregoing. The picture can be realized in the problem with a single piston: while moving away, the piston first forms a rarefaction wave, and then, moving into the gas, it forms a compression wave (a shock wave).

2. *Asymptotics of the interaction of a shock wave and a centered rarefaction wave.* As a result of the interaction of a shock wave with a rarefaction wave III, after it passes out of the latter onto the constant background V the shock wave acquires a velocity D_1 in place of the initial velocity D_0. It is clear that the jump in the velocity $D_1 - D_0$ of the shock wave does not depend on the point of entry (x_1, t_1) if a smooth flow prevails in region IV. Suppose that the point of entry (x_1, t_1) approaches the center (x_0, t_0) of the wave. The difference $D_1 - D_0$ then remains unchanged. If $x_1 = x_0$ and $t_1 = t_0$, then it is possible to speak of an instantaneous interaction of a shock wave with a concentrated rarefaction wave. In this case we have an arbitrary discontinuity characterized by the states (II, V). The following question can be posed: will the velocity D_1 of the shock wave in the problem of the decay of a discontinuity and the velocity

D_1 after the shock wave emerges from the rarefaction wave III coincide? We first consider the case of a centered rarefaction s-wave (Figure 2.74, a).

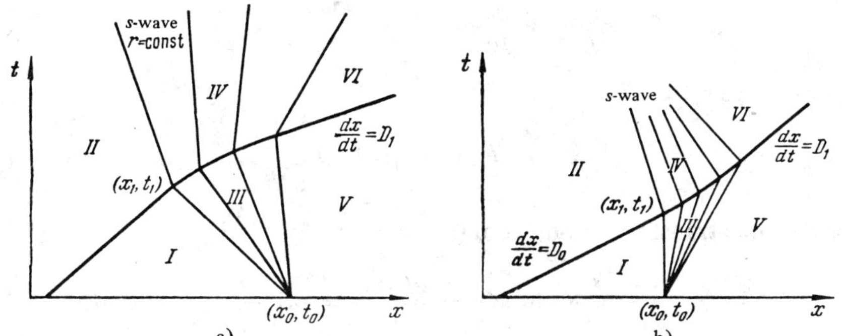

FIGURE 2.74

The states II and V are determined by the formulas

$$\left.\begin{aligned} u_2 = u_1 + c\left(M_0 - \frac{1}{M_0}\right), \qquad p_2 = p_1 M_0^2, \\ u_2 - u_1 = c[\sqrt{p_2/p_1} - \sqrt{p_1/p_2}], \end{aligned}\right\} \tag{1}$$

$$u_5 = u_1 + c \ln \frac{p_1}{p_5}. \tag{2}$$

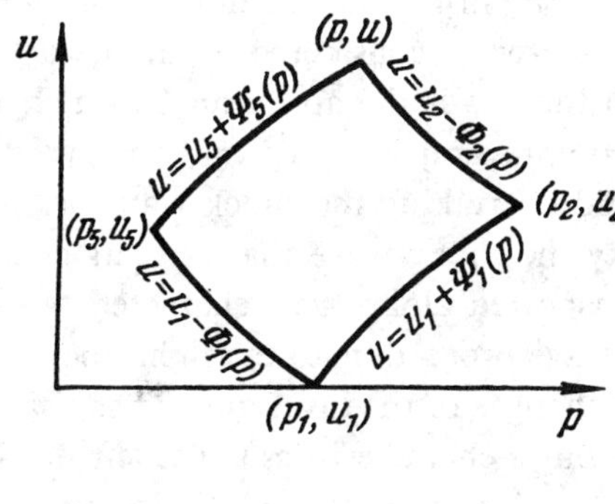

FIGURE 2.75

We compute the decay of the discontinuity (p_2, u_2), (p_5, u_5) by using the (p, u)-diagram (Figure 2.75). In the present case we have configuration A. The quantities u and p are found by solving the equation

$$M - \frac{1}{M} - \ln \frac{p_5}{p_1} = M_0 - \frac{1}{M_0} - \ln \frac{p}{p_2}, \tag{3}$$

where we have set

$$M^2 = \frac{p}{p_5}, \qquad M_0^2 = \frac{p_2}{p_1}. \tag{4}$$

After a simple transformation, from (3) we have

$$\frac{\varphi(M)}{c} = M - \frac{1}{M} + \ln M^2 = \frac{\varphi(M_0)}{c} = M_0 - \frac{1}{M_0} + \ln M_0^2. \tag{5}$$

Hence

$$M = M_0. \tag{6}$$

Thus, after the interaction of the shock wave proceeding to the right with the s-wave it acquires the same velocity as it has as a result of the instantaneous interaction of the shock wave and a concentrated rarefaction wave computed on the basis of the formulas for the decay of a discontinuity. This agreement is occasioned by the fact that the refracted wave IV is a rarefaction wave, and as $t \to \infty$ no singularities arise in the motion behind the wave front. A similar picture holds for the interaction of a shock wave proceeding to the left with a rarefaction r-wave.

We now consider the interaction of a shock wave proceeding to the right with an r-wave (Figure 2.74, b) (Arsenin and Janenko (1956)). In this case the shock wave, in passing through region III, acquires a different velocity than that acquired as the result of the instantaneous interaction of a shock wave and a concentrated rarefaction r-wave considered as the decay of an arbitrary discontinuity. The reason for this is the appearance of a shock wave proceeding to the left from the compression wave IV and then a rarefaction wave proceeding to the right and overtaking the shock wave propagating to the right. In changing its velocity the shock wave that appears changes the invariant r. These changes are transported along the r-characteristics, and, on arriving at the front of the shock wave proceeding to the right, they change the velocity of the latter thus causing changes of the invariant s. The changes of s on the right wave are transported along s-characteristics to the left shock wave, thus causing change in the invariant r. A certain asymptotic regime emerges as a result of the interaction of the trailing fronts of the right and left shock waves. The limit configuration and flow coincide with the configuration and flow obtained as a result of the decay of the discontinuity II, V (Figure 2.76).

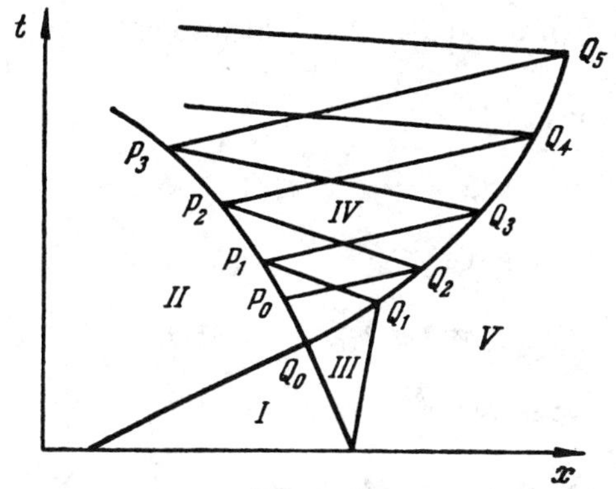

FIGURE 2.76

We shall consider the interaction of the fronts in detail. We introduce the following notation:

$r_{\mathrm{I}}, s_{\mathrm{I}}$ are the values of the invariants in I;

$r_{\mathrm{III}}, s_{\mathrm{III}}$ are the values of the invariants on the leading front of the wave in III;

$r_{\mathrm{V}}, s_{\mathrm{V}}$ are the values of the invariants in V;

$r_{\mathrm{II}}, s_{\mathrm{II}}$ are the values of the invariants in II;

r, s are the values of the invariants on the trailing front of the right wave;

R, S are the values of the invariants on the trailing front of the left wave;

D is the velocity of the wave proceeding to the right;

$\bar{D}$ is the velocity of the wave proceeding to the left;

$M = (D - u_0)/c_0 > 1$;

$\bar{M} = |(\bar{D} - \bar{u})/\bar{c}| > 1$;

u_0, c_0 are the quantities ahead of the front of the right wave;

$\bar{u}$, $\bar{c}$ are the quantities ahead of the front of the left wave;

r_i, s_i, M_i are the values of r, s, M at the points Q_i; and

R_i, S_i, $\bar{M}_i$ are the values of R, S, $\bar{M}$ at the points P_i.

On the fronts $(Q_0, Q_1, Q_2, \ldots)$ and $(P_1, P_2, P_3, P_4, \ldots)$ we have the relations

$$s - s_{\mathrm{V}} = \psi'(M), \qquad r - r_{\mathrm{III}} = \varphi(M),$$
$$s - s_{\mathrm{III}} = \psi(M), \qquad r - r_{\mathrm{V}} = \varphi(M), \tag{7}$$

and, respectively,

$$R - r_{\mathrm{II}} = -\psi(\bar{M}), \qquad S - s_{\mathrm{II}} = -\varphi(\bar{M}).$$

The arcs $(P_1, P_2), (P_3, P_4), \ldots, (P_{2i-1}, P_{2i})$ and $(Q_1, Q_2), (Q_3, Q_4), \ldots, (Q_{2i-1}, Q_{2i})$ correspond to segments of the trajectories of the shock waves when they move with constant velocity; the arcs $(P_0, P_1), (P_2, P_3), \ldots, (P_{2i}, P_{2i+1})$ and $(Q_0, Q_1), (Q_2, Q_3), \ldots, (Q_{2i}, Q_{2i+1})$ correspond to segments of variable velocity.

We have the relations

$$S_i = s_i, \qquad r_{i+2} = R_i. \tag{8}$$

On the segment Q_0Q_1 we have

$$M_1 < M_0, \qquad s_1 < s_0. \tag{9}$$

Considering (7)–(9), we have

$$S_1 < s_0, \qquad \bar{M}_1 > \bar{M}_0, \qquad R_1 < R_0. \tag{10}$$

Thus, on P_0P_1 the left wave intensifies, while R and S abate. By (8) on the segment Q_2Q_3 we have

$$r_3 < r_2, \qquad M_3 < M_2, \qquad s_3 < s_2, \tag{11}$$

i.e., on Q_2Q_3 the right wave becomes weaker. The picture is subsequently repeated: the wave proceeding to the left intensifies, the wave going to the right becomes weaker, M_i decreases monotonically, and $\bar{M}_i$ increases monotonically.

We denote by M_∞ and $\bar{M}_\infty$ the limits of M_i and $\bar{M}_i$ as $i \to \infty$. By (7) and (8) the following relations then hold:

$$\left.\begin{aligned} r - r_V &= \varphi(M_\infty), & s - s_V &= \psi(M_\infty), \\ R - r_{II} &= -\varphi(\bar{M}_\infty), & S - s_{II} &= -\psi(M_\infty). \end{aligned}\right\} \tag{12}$$

These are the formulas for the decay of the discontinuity for the states (II, V). We have thus proved that the asymptotic regime of the interaction of a shock wave with a rarefaction wave and the flow obtained from the decay of the discontinuity (II, V) coincide.

We remark that if the amplitude of the rarefaction wave III is sufficiently large, then the shock wave entering it may become a rarefaction wave, and configuration B is replaced by configuration A.

3. ***The interaction of shock waves with travelling waves in barotropic, polytropic gases.*** In the case of barotropic, polytropic gases the third Hugoniot condition is replaced by the constancy of the entropy.

The Hugoniot conditions for a shock wave proceeding to the right take the form

$$u_1 - u_0 = (1 - h)c_0\left(M_0 - \frac{1}{M_0}\right), \qquad M_0 = \frac{D - u_0}{c_0}, \tag{1}$$

$$\frac{p_1}{p_0} = (1 + h)M_0^2 - h, \tag{2}$$

$$\frac{\rho_1^\gamma}{\rho_0^\gamma} = \frac{p_1}{p_0} = (1 + h)M_0^2 - h, \tag{3}$$

$$\frac{c_1}{c_0} = \left(\frac{p_1}{\rho_1}\frac{\rho_0}{p_0}\right)^{1/2} = \left(\frac{\rho_1}{\rho_0}\right)^{(\gamma-1)/2} = [(1 + h)M_0^2 - h]^{(\gamma-1)/2\gamma}. \tag{4}$$

We first consider the incidence of a shock wave proceeding from left to right on a rarefaction s-wave. From conditions (1)–(4) we have

$$r_1 - r_0 = c_0\left[(1-h)\left(M_0 - \frac{1}{M_0}\right) + \frac{2}{\gamma-1}\left(\frac{c_1}{c_0} - 1\right)\right] = c_0F(M_0), \tag{5}$$

$$s_1 - s_0 = c_0\left[(1-h)\left(M_0 - \frac{1}{M_0}\right) - \frac{2}{\gamma-1}\left(\frac{c_1}{c_0} - 1\right)\right] = c_0G(M_0). \tag{6}$$

Since $F(M_0)$ in (5) is a monotonically increasing function of M_0, while c_0 satisfies the condition

$$\frac{\partial c_0}{\partial x} = -\frac{\gamma - 1}{2}\frac{\partial u_0}{\partial x} < 0 \qquad (r = r_0 = \text{const}) \tag{7}$$

and

$$F(M_0)\frac{\partial c_0}{\partial x} + F'(M_0)c_0\frac{\partial M_0}{\partial x} = 0,$$

it follows that

$$\frac{\partial M_0}{\partial x} > 0, \tag{8}$$

i.e., the strength of the shock wave increases.

It is clear that behind the front of the shock wave we have a Riemann s-wave ($r = r_0$). We shall establish the sign of $\partial u/\partial x$ or, equivalently, of $\partial s/\partial x$ behind the front of the shock wave. From (1), (5), and (7) we have

$$\frac{\partial u_1}{\partial x} = \left[1 + h\left(1 + \frac{1}{M_0^2}\right)\frac{F(M_0)}{F'(M_0)} - h\left(M_0 - \frac{1}{M_0}\right)\right]\frac{\partial u_0}{\partial x}. \tag{9}$$

Since the approximation of a barotropic gas is valid for M_0 close to 1, the expression in square brackets is close to 1. From this we have

$$\frac{\partial u_1}{\partial x} > 0. \tag{10}$$

Thus, the wave IV is a rarefaction wave (Figure 2.74, a).

We now consider the incidence of a shock wave on a rarefaction r-wave. In (5) and (6) we then have $r_1 = \text{const}$ and $s_0 = \text{const}$. From (5) we have

$$r_1 = r_0 + c_0 F(M_0) = s_0 + c_0\left[\frac{4}{\gamma - 1} + F(M_0)\right]. \tag{11}$$

Since $\partial c_0/\partial x > 0$, it follows from (11) that

$$\frac{\partial M_0}{\partial x} < 0, \tag{12}$$

i.e., the amplitude of the shock wave decreases. We shall determine the sign of $\partial u_1/\partial x = \frac{1}{2}(\partial s_1/\partial x)$ behind the front of the shock wave. For sufficiently small $|M_0 - 1|$ the quantity $G(M_0)$ is a monotonically increasing function of M_0. From (11) it follows that

$$\frac{\partial s_1}{\partial x} < 0,$$

i.e., the refracted wave IV is a compression wave.

We thus see that in the barotropic approximation, i.e., for sufficiently small $|M_0 - 1|$, the qualitative picture of the interaction is the same as for an isothermic gas.

§9. Analytic solutions of one-dimensional gas dynamics

1. ***The general integral of isentropic, one-dimensional plane flow.*** One-dimensional, isentropic plane flow is described, as we know, by the equations in invariants (see §2.7)

$$\left.\begin{aligned} \frac{\partial r}{\partial t} + \left[\frac{r+s}{2} + K(r-s)\right]\frac{\partial r}{\partial x} = 0, \\ \frac{\partial s}{\partial t} + \left[\frac{r+s}{2} - K(r-s)\right]\frac{\partial s}{\partial x} = 0, \end{aligned}\right\} \tag{1}$$

where

$$r = u + \int c\, d\ln\rho, \qquad s = u - \int c\, d\ln\rho, \tag{2}$$

and x is the Euler coordinate.

The function $K(r - s) = c$ is connected with the equation of state

$$p = p(S, \rho) = F(\rho) \tag{3}$$

by the relation

$$K(r - s) = g[h^{-1}(r - s)], \tag{4}$$

where

$$g = \sqrt{F'(\rho)}, \qquad h = 2\int \sqrt{F'(\rho)}\,\frac{d\rho}{\rho}. \tag{5}$$

The system reduces by a hodograph transformation to a linear system:

$$\left.\begin{aligned} \frac{\partial x}{\partial s} - \left[\frac{r+s}{2} + K(r-s)\right]\frac{\partial t}{\partial s} &= 0, \\ \frac{\partial x}{\partial r} - \left[\frac{r+s}{2} - K(r-s)\right]\frac{\partial t}{\partial r} &= 0. \end{aligned}\right\} \tag{6}$$

We rewrite (6) in the form

$$\left.\begin{aligned} \frac{\partial}{\partial s}\left[x - \left(\frac{r+s}{2} + K\right)t\right] &= -t\left(\frac{1}{2} - K'\right), \\ \frac{\partial}{\partial r}\left[x - \left(\frac{r+s}{2} - K\right)t\right] &= -t\left(\frac{1}{2} - K'\right). \end{aligned}\right\} \tag{7}$$

From (7) it follows that the expression

$$dW = \left[x - \left(\frac{r+s}{2} + K\right)t\right]dr + \left[x - \left(\frac{r+s}{2} - K\right)t\right]ds \tag{8}$$

is the total differential of some function $W(r, s)$, which we call a *potential.* From the expression for the total differential we have

$$\left.\begin{aligned} x - \left(\frac{r+s}{2} + K\right)t &= \frac{\partial W}{\partial r}, \\ x - \left(\frac{r+s}{2} - K\right)t &= \frac{\partial W}{\partial s}. \end{aligned}\right\} \tag{9}$$

By (9), x and t can be expressed in terms of W_r and W_s:

$$\begin{aligned} t &= -\frac{\partial W/\partial r - \partial W/\partial s}{2K(r-s)}, \\ x &= -\frac{((r+s)/2 - K)\partial W/\partial r - ((r+s)/2 + K)\partial W/\partial s}{2K(r-s)}. \end{aligned} \tag{10}$$

Because of (7) and (10) W satisfies the second order equation

$$\frac{\partial^2 W}{\partial r \partial s} = \frac{(\frac{1}{2} - K')}{2K(r-s)}\left(\frac{\partial W}{\partial r} - \frac{\partial W}{\partial s}\right). \tag{11}$$

Thus, the potential function satisfies the special Darboux equation

$$\frac{\partial^2 W}{\partial x_1 \partial x_2} - f(x_1 + x_2)\left(\frac{\partial W}{\partial x_1} + \frac{\partial W}{\partial x_2}\right) = 0, \tag{12}$$

where we have set

$$x_1 = r, \quad x_2 = -s, \quad f = \frac{K' - 1/2}{2K}. \tag{13}$$

It is not hard to see that by the substitution

$$W = \varphi v, \qquad \varphi = ce^{\int f(\theta)d\theta}, \qquad \theta = x_1 + x_2, \tag{14}$$

(12) can be brought to the form

$$\frac{\partial^2 v}{\partial x_1 \partial x_2} = F(x_1 + x_2)v, \qquad F = f^2 - f'. \tag{15}$$

We now consider those equations of state for which the general integral and the Riemann function can be represented in closed form. We begin with polytropic gases for which p is a power function of ρ:

$$p = a^2\rho^\gamma, \qquad \gamma = \frac{c_p}{c_V} > 1. \tag{16}$$

In this case

$$K(r-s) = c = \frac{\gamma - 1}{4}(r - s). \tag{17}$$

Hence

$$\begin{aligned} f &= \frac{-1/2 + (\gamma - 1)/4}{2((\gamma - 1)/4)(r - s)} = \frac{-(1 - (\gamma - 1)/2)/(\gamma - 1)}{r - s} \\ &= \frac{\gamma - 3}{2(\gamma - 1)} \frac{1}{x_1 + x_2} = -\frac{m}{x_1 + x_2}, \end{aligned} \tag{18}$$

For (12) with the function $f = -m/(x_1 + x_2)$ we have the following reduction property which was established by Darboux and makes it possible to pass from one value of m to another. If W is a solution of (12) for $f = -m/(x_1 + x_2)$, then

$$W' = LW, \qquad L = \frac{1}{x_1 + x_2}\left(\frac{\partial}{\partial x_1} + \frac{\partial}{\partial x_2}\right), \tag{19}$$

is a solution of (12) for

$$f = -\frac{m'}{x_1 + x_2}, \qquad m' = m + 1. \tag{20}$$

For $\gamma = +3$ we have $m = 0$, and W satisfies the equation

$$\frac{\partial^2 W}{\partial x_1 \partial x_2} = 0, \tag{21}$$

which has the well-known general integral (the d'Alembert integral)

$$W = F(x_1) + G(x_2). \tag{22}$$

To positive integral values of m there correspond the following values of γ:

$$\gamma = \frac{2m + 3}{2m + 1}; \tag{23}$$

for $m = 1$ we have $\gamma = 5/3$ (a monatomic gas); for $m = 2$ we have $\gamma = 7/5$ (a diatomic gas), etc. Because of the reduction property, to such values of m and γ there corresponds the general integral

$$W(x_1, x_2) = L^m[F(x_1) + G(x_2)], \tag{24}$$

where the operator L is given by (19).

We shall show that the expression (24) for the general integral can be transformed to the form

$$W(x_1, x_2) = \frac{\partial^{m-1}}{\partial x_1^{m-1}} \frac{F'(x_1)}{(x_1 + x_2)^m} + \frac{\partial^{m-1}}{\partial x_2^{m-1}} \frac{G'(x_2)}{(x_1 + x_2)^m}. \tag{25}$$

We henceforth set

$$F'(x_1) = \Phi(x_1), \qquad G'(x_2) = \Psi(x_2). \tag{26}$$

For $m = 1$ the representation (25) obviously follows from (24). We shall prove the equivalence of (24) and (25) for any m by induction. Suppose that

$$L^m F(x_1) = \frac{\partial^{m-1}}{\partial x_1^{m-1}} \frac{\Phi(x_1)}{(x_1 + x_2)^m}. \tag{27}$$

We shall prove that

$$L^{m+1} F(x_1) = \frac{\partial^m}{\partial x_1^m} \frac{\Phi(x_1)}{(x_1 + x_2)^{m+1}}. \tag{28}$$

Using (27), we have

$$L^{m+1}F(x_1) = LL^mF(x_1) = L\left[\frac{\partial^{m-1}}{\partial x_1^{m-1}} \frac{\Phi(x_1)}{(x_1+x_2)^m}\right]$$
$$= \frac{1}{x_1+x_2}\frac{\partial^m}{\partial x_1^m}\frac{\Phi(x_1)}{(x_1+x_2)^m} - \frac{m}{x_1+x_2}\frac{\partial^{m-1}}{\partial x_1^{m-1}}\frac{\Phi(x_1)}{(x_1+x_2)^{m+1}}. \tag{29}$$

After this it is not hard to see that (28) is equivalent to

$$(\theta H)^{(m)} = \theta H^{(m)} + mH^{(m-1)}, \tag{30}$$

where for fixed x_2 we have set

$$\theta = x_1 + x_2, \qquad \frac{\partial}{\partial x_1} = \frac{d}{d\theta}, \qquad f^{(m)} = \frac{d^m}{d\theta^m} f, \qquad H = \frac{\Phi}{\theta^{m+1}}. \tag{31}$$

We have thus proved that the general integral of (12) for $f = -m/(x_1 + x_2)$ has the form

$$W(x_1 + x_2) = \frac{\partial^{m-1}}{\partial x_1^{m-1}}\left[\frac{\Phi(x_1)}{(x_1+x_2)^m}\right] + \frac{\partial^{m-1}}{\partial x_2^{m-1}}\left[\frac{\Psi(x_2)}{(x_1+x_2)^m}\right]. \tag{32}$$

If x_1 and x_2 are assumed to be complex numbers and W, Φ, and Ψ are analytic functions of their arguments, then by applying the well-known Cauchy representation for the derivative of an analytic function formula (32) can be written (see Copson (1953))

$$W(x_1, x_2) = D_1^{m-1}\frac{\Phi(x_1)}{(x_1+x_2)^m} + D_2^{m-1}\frac{\Psi(x_2)}{(x_1+x_2)^m}$$
$$= \frac{(m-1)!}{2\pi i}\left[\oint_{C_1}\frac{\Phi(t)}{(t+x_2)^m(t-x_1)^m}dt + \oint_{C_2}\frac{\Psi(t)}{(x_1+t)^m(t-x_2)^m}dt\right]. \tag{33}$$

Here $D_i = \partial/\partial x_i$; the contour C_1 is taken in the complex plane of the variable x_1 with x_2 considered as a complex parameter; the contour C_2 is taken in the x_2 plane with x_1 considered as a parameter.

Formula (33) is also meaningful for fractional m if $(m - 1)!$ is replaced by $\Gamma(m)$; D_1^{m-1} and D_2^{m-1} then have the meaning of the fractional derivatives first introduced by Riemann and Liouville.

We now proceed to the problem of determining the Riemann function for equation (12) for $f = -m/(x_1 + x_2)$. We shall use (15), which for this f takes the form

$$\frac{\partial^2 v}{\partial x_1 \partial x_2} = \frac{a}{\theta^2}v, \qquad \theta = x_1 + x_2, \qquad a = m(m-1). \tag{34}$$

For the Riemann function $R(\xi_1, \xi_2; x_1, x_2)$ of (34) we have the representation (see Chapter 1, §13.4)

$$R(\xi_1, \xi_2; x_1, x_2) = (1 - \xi)^m F(m, m, 1, \xi), \tag{35}$$

where

$$\xi = \frac{(x_1 - \xi_1)(x_2 + \xi_2)}{(x_1 - \xi_2)(x_2 + \xi_1)} \tag{36}$$

and

$$\begin{aligned} F(\alpha, \beta, \gamma, x) = 1 &+ \frac{\alpha\beta}{\gamma} x + \cdots \\ &+ \frac{\alpha(\alpha+1)\cdots(\alpha+k-1)\beta(\beta+1)\cdots(\beta+k-1)}{k!\gamma(\gamma+1)\cdots(\gamma+k-1)} x^k + \cdots \end{aligned} \tag{37}$$

is the well-known hypergeometric series of Gauss.

We remark that for integral m the series (37) becomes a polynomial. It can be shown that for integral m (34) admits a differential constraint of mth order, and, conversely, if an equation of the form (15) admits a differential constraint of mth order, then $F(x_1 + x_2)$ satisfies an equation which contains the functions $m(m-1)/(x_1 + x_2)^2$ among its solutions. We shall illustrate this assertion with an example. We consider equation (15), admitting a differential constraint of second order. Then, as shown in Chapter 1, §12, $F(x_1 + x_2)$ satisfies the equation

$$\frac{\partial^2 \ln F}{\partial x_1 \partial x_2} = F, \tag{38}$$

which in our case takes the form

$$FF'' - F'^2 = F^3, \tag{39}$$

where the prime denotes the derivative with respect to $\theta = x_1 + x_2$.

The function

$$F(\theta) = \frac{m(m-1)}{\theta^2}$$

satisfies (39) for $m = 2$. The general integral of (38) has the form

$$\theta + C_2 = \begin{cases} \dfrac{1}{\sqrt{C_1}} \ln \dfrac{\sqrt{C_1 + 2F} - \sqrt{C_1}}{\sqrt{C_1 + 2F} + \sqrt{C_1}}, & C_1 > 0, \\ \dfrac{2}{\sqrt{-C_1}} \arctan \sqrt{\dfrac{C_1 + 2F}{-C_1}}, & C_1 < 0. \end{cases} \tag{40}$$

Using the arbitrary constants C_1 and C_2, the functions $F(\theta)$ can be approximated by means of the function (40), and good approximations can be obtained. Dombrovskiĭ (1964) obtained approximations of the Darboux equation in this way.

In the next subsection we shall show how the knowledge of the general integral of the Darboux equation makes it possible to solve certain problems of gas dynamics leading to the interference of simple waves.

2. *Problems of the interaction of elementary solutions.* We shall consider two problems:

a) the interaction of two Riemann waves;

b) the incidence of a Riemann wave on the boundary separating two media.

Suppose that at time $t = t_0$ a centered r-wave begins to propagate from the point $x = x_0$, while at time $t = t_1$ a centered s-wave begins to propagate from $x = x_1$. We may assume that the centered waves are obtained as a result of the withdrawal of pistons from the gas with constant speed (Figure 2.77).

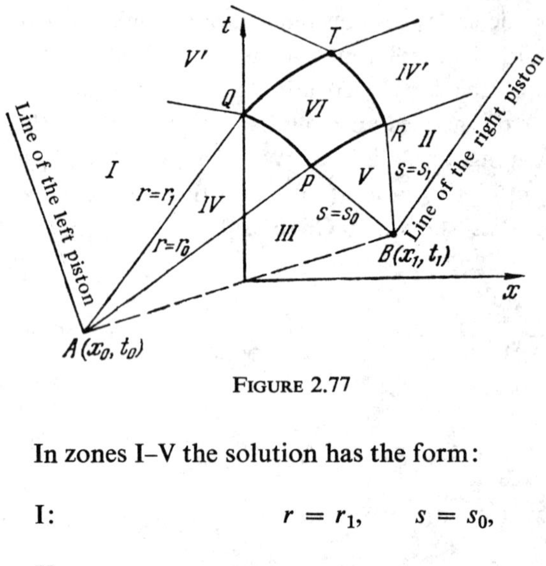

FIGURE 2.77

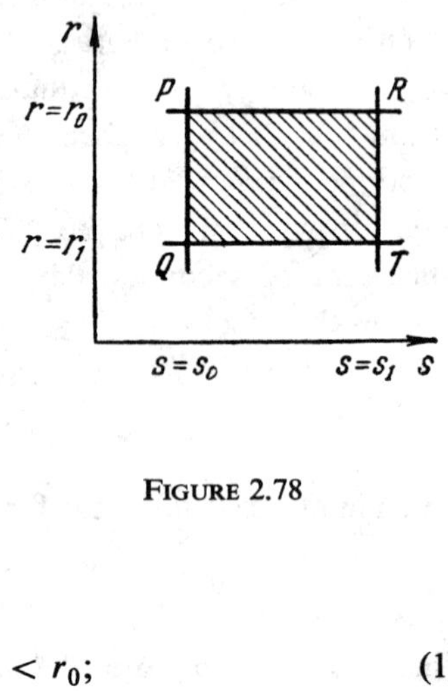

FIGURE 2.78

In zones I–V the solution has the form:

I: $$r = r_1, \quad s = s_0, \quad r_1 < r_0; \tag{1}$$

II: $$r = r_0, \quad s = s_1, \quad s_1 > s_0; \tag{2}$$

III: $$u=u_0=0, \quad c=c_0, \quad r=r_0=\frac{2}{\gamma-1}c_0, \quad s=s_0=-\frac{2}{\gamma-1}c_0; \tag{3}$$

IV: $$\frac{dx}{dt} = \frac{x-x_0}{t-t_0} = \frac{x+h}{t+\tau} = \alpha r + \beta s_0; \tag{4}$$

V: $$\frac{dx}{dt} = \frac{x-x_1}{t-t_1} = \frac{x-h}{t-\tau} = \alpha s + \beta r_0. \tag{5}$$

In zone VI, which is an interference zone and maps into the rectangle $PQTR$ in the r, s plane (Figure 2.78), we have the representation

$$\left.\begin{aligned} x - (\alpha r + \beta s)t &= \frac{\partial W}{\partial r}, \\ x - (\alpha s + \beta r)t &= \frac{\partial W}{\partial s}, \end{aligned}\right\} \tag{6}$$

where W satisfies the equation

$$\frac{\partial^2 W}{\partial r\,\partial s} = \frac{m}{r - s}\left(\frac{\partial W}{\partial r} - \frac{\partial W}{\partial s}\right), \qquad m = \frac{3 - \gamma}{2(\gamma - 1)}. \tag{7}$$

Using (4) and (5), on the sides PR and PQ of the rectangle $PQTR$ it is possible to pose boundary conditions constituting a Goursat problem for W:

$$\left.\begin{aligned} PR\colon\quad r = r_0, \qquad \frac{\partial W}{\partial s} &= h - \tau(\alpha s + \beta r_0); \\ PQ\colon\quad s = s_0, \qquad \frac{\partial W}{\partial r} &= -h + \tau(\alpha r + \beta s_0). \end{aligned}\right\} \tag{8}$$

We shall solve this problem for $\gamma = 5/3$. The general integral of (7) has the form

$$W = \frac{f(r) + g(s)}{r - s}. \tag{9}$$

The boundary conditions (8) lead to the relations

$$\left.\begin{aligned} \frac{(r_0 - s)g'(s) + g(s) + f(r_0)}{(r_0 - s)^2} &= h - \tau(\alpha s + \beta r_0), \\ \frac{(r - s_0)f'(r) - f(r) - g(s_0)}{(r - s_0)^2} &= -h + \tau(\alpha r + \beta s_0). \end{aligned}\right\} \tag{10}$$

For simplicity we assume that $\tau = 0$, i.e., withdrawal of the pistons occurs simultaneously. The relations (10) then take the form

$$\left.\begin{aligned} (r_0 - s)g'(s) + g(s) + f(r_0) &= h(r_0 - s)^2, \\ (r - s_0)f'(r) - f(r) - g(s_0) &= -h(r - s_0)^2, \end{aligned}\right\} \tag{11}$$

i.e., $f(r)$ and $g(s)$ are polynomials of second degree in their variables. Thus, noting that W is determined up to an additive constant, it is possible to set

$$W = \frac{Ar^2 + Bs^2 + C(r + s) + D}{r - s}. \tag{12}$$

By satisfying the boundary conditions (8), we find A, B, C, and D and obtain for W the final expression

$$W = \frac{-h(r^2 + s^2) + 18hc_0^2}{r - s}. \tag{13}$$

Formulas (6) and (13) determine the solution in zone VI. In zones IV′ and V′ we have travelling Riemann waves which are no longer centered and are not hard to compute.

We now consider the interaction of a travelling wave with a contact boundary. A centered Riemann s-wave propagates from the point $x = 0$, $t = 0$ and is incident on a contact boundary located at the point $x = -h$ for $t = 0$ (Figure 2.79). We again consider the case where both media are polytropic gases with $\gamma = 5/3$. The picture of the flow is described by

region I: $r = r_0$, $s = s_0 = -r_0$, $c = c_0$, $S = S_0$,

$$u = u_0 = 0; \tag{14}$$

region II: $r = r_0$, $\alpha s + \beta r_0 = \xi = x/t$; (15)

region III: $r = r_0$, $s = s_2$; (16)

region VII: $r = r_1$, $s = s_1 = -r_1$, $c = c_1$, $S = S_1$. (17)

Region IV is the region of interference of the incident and reflected waves; regions II and V correspond to travelling waves, and regions VI, VII, and VIII are regions of constant flow.

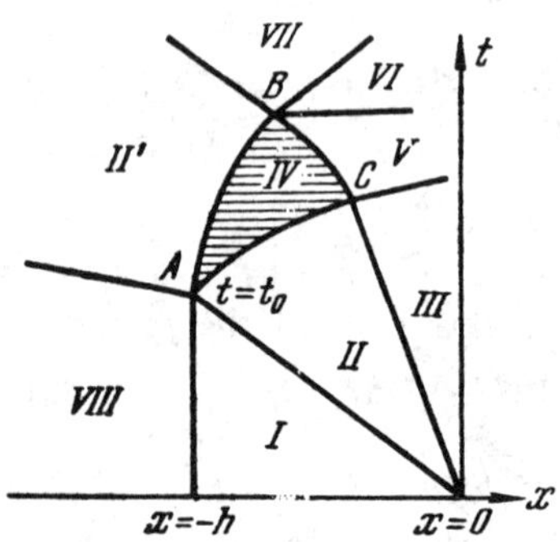

FIGURE 2.79

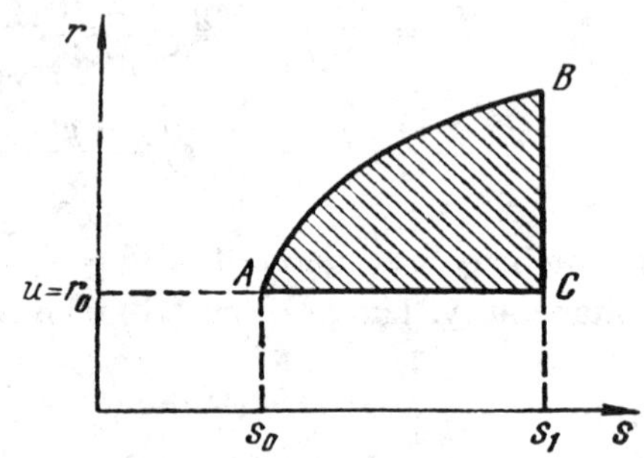

FIGURE 2.80

We first determine the flow in region IV (Figure 2.80). Using (6), we can pose boundary conditions for $W(r, s)$ on the line AC:

$$r = r_0, \qquad x - (\alpha s + \beta r_0)t = W_s(r, s) = 0. \tag{18}$$

Using the formula for the general integral, we obtain

$$g'(s)(r_0 - s) + f(r_0) + g(s) = 0. \tag{19}$$

It follows that $g(s)$ is a linear function of s. Since W is determined up to an additive constant, the sum $f(r) + g(s)$ may contain a term $C(r - s)$ with an arbitrary constant C. By choice of C we can set $g(s) = \text{const} = B$. Combining B into $f(r)$, we finally obtain

$$W(r, s) = \frac{f(r)}{r - s}, \qquad f(r_0) = 0. \tag{20}$$

We now consider the boundary condition on the contact boundary AB. We first determine the line AB in the plane of the hodograph. Noting that the invariant $r = r_1$ is constant to the left of the boundary, from the conditions on the contact boundary we find that the quantities r and s on the right side of the boundary are connected by the linear relation

$$s = \frac{1 + \kappa}{1 - \kappa} r - \frac{2\kappa}{1 - \kappa} r_1, \qquad \kappa = \left(\frac{A_0}{A_1}\right)^{1/\gamma}, \tag{21}$$

where A_0 and A_1 are the entropy constants of the gases to the right and left of the boundary respectively.

The boundary condition on AB in the x, t plane has the form

$$\frac{dx}{dt} = \frac{r + s}{2}. \tag{22}$$

Using (6) and (21), we obtain

$$\begin{aligned} &[(\beta - \alpha)(r - s)(W_{rr} + W_{rs}) + W_r - W_s]\,dr \\ &\qquad + [(\beta - \alpha)(r - s)(W_{rs} + W_{ss}) + W_r - W_s]\,ds = 0, \end{aligned} \tag{23}$$

$$ds = \frac{1 + \kappa}{1 - \kappa}\,dr. \tag{24}$$

Observing that (21) implies that

$$r - s = \frac{2\kappa}{\kappa - 1}(r - r_1), \tag{25}$$

from this we obtain an equation for $f(r)$:

$$\kappa^2(r - r_1)^2 f'' + \kappa(3 - \kappa)(r - r_1)f' + 3(1 - \kappa)f = 0, \tag{26}$$

with a general integral of the form

$$f(r) = C_1(r - r_1)^{\lambda_1} + C_2(r - r_1)^{\lambda_2}, \tag{27}$$

where C_1 and C_2 are arbitrary constants, and λ_1 and λ_2 are roots of the characteristic equation

$$\kappa^2\lambda^2 + \kappa(3 - 2\kappa)\lambda + 3(1 - \kappa) = 0. \tag{28}$$

The constants C_1 and C_2 are determined from the conditions

$$f(r_0) = C_1(r_0 - r_1)^{\lambda_1} + C_2(r_0 - r_1)^{\lambda_2} = 0, \tag{29}$$

$$f'(r_0) = C_1\lambda_1(r_0 - r_1)^{\lambda_1 - 1} + C_2\lambda_2(r_0 - r_1)^{\lambda_2 - 1} = -2r_0(h_0 + c_0t_0). \tag{30}$$

We remark in conclusion that the transmitted wave II′ is a travelling rarefaction wave, while the wave V may be either a rarefaction wave or a compression wave. In the case $r_B < r_C$ we have a rarefaction wave, while in the case $r_B > r_C$ we have a compression wave.

The investigations of §3.5 show that the reflected wave V is a rarefaction wave for $\kappa > 1$ and a compression wave for $\kappa < 1$. In the case $\kappa > 1$ the regions II′ and V constitute two rarefaction waves, and the solution tends as $h \to 0$ to the solution of the corresponding problem on the decay of a discontinuity (see §8.2).

The solution presented above was indicated to us by N. N. Anučina. A general solution of the problem of the incidence of a Riemann wave on the boundary separating two media was given by Taub (1946).

3. *Plane one-dimensional flows with variable entropy. Martin's method.* M. H. Martin (1953b) succeeded in reducing the equations of gas dynamics to the Monge-Ampère equation and then applied the method of the intermediate integral (see Chapter 1, §12) to obtain a generalization of the Riemann invariants to the case of a flow with variable entropy. Subsequent investigations of Martin (1953a), Ludford (1955), and Zav'jalov (1955) made it possible to completely determine a class of equations of state and distribution functions for the entropy to which the method of the intermediate integral is applicable.

We start from the equations in Lagrange coordinates

$$\frac{\partial u}{\partial t} + \frac{\partial p}{\partial q} = 0, \qquad \frac{\partial V}{\partial t} - \frac{\partial u}{\partial q} = 0, \qquad \frac{\partial S}{\partial t} = 0. \tag{1}$$

To the first two equations of (1) there correspond the potentials φ_1 and φ_2 (see Chapter 1, §5.3):

$$V\,dq + u\,dt = d\varphi_1, \tag{2}$$

$$u\,dq - p\,dt = d\varphi_2. \tag{3}$$

Martin introduced the potential function ξ by means of the equality

$$d\xi = u\,dq + t\,dp = d\varphi_2 + d(pt). \tag{4}$$

From (4) we have

$$\frac{\partial \xi}{\partial q} = u, \qquad \frac{\partial \xi}{\partial p} = t; \tag{5}$$

$$du = \xi_{qp}\,dp + \xi_{qq}\,dq, \qquad dt = \xi_{pp}dp + \xi_{pq}dq; \tag{6}$$

$$\frac{\partial u}{\partial p} = \xi_{pq}, \qquad \frac{\partial u}{\partial q} = \xi_{qq}; \tag{7}$$

$$\frac{\partial t}{\partial p} = \xi_{pp}, \qquad \frac{\partial t}{\partial q} = \xi_{pq}. \tag{8}$$

Substituting (6) into (2), we find that

$$d\varphi_1 = (V + u\xi_{pq})dq + u\xi_{pp}dp. \tag{9}$$

The condition that (9) be a total differential with (7) leads to the Monge-Ampère equation for the function ξ:

$$\xi_{qq}\xi_{pp} - \xi_{pq}^2 = \frac{\partial V(p, S)}{\partial p}. \tag{10}$$

To determine the equation of state $V = V(p, S)$ and the prescribed entropy distribution $S = S(q)$ on the right side of (10) we have a specific function

$$-f^2(p, q) = \frac{\partial V[p, S(q)]}{\partial p}, \tag{11}$$

and (10) takes the form

$$\xi_{pp}\xi_{qq} - \xi_{pq}^2 + f^2(p, q) = 0, \tag{12}$$

an equation we studied in Chapter 1, §12.

We remark that the choice of p and q as the independent variables implies that p is not a function of q or, equivalently, of S, i.e., the thermodynamic parameters of the flow form a two-dimensional manifold. We shall consider later the exceptional case in which $p = p(S)$.

Martin's method consists in seeking a family of solutions of (12) with single-function arbitrariness which satisfy the additional relation

$$\varphi(p, q, \xi, \xi_p, \xi_q) = \varphi(p, q, \xi, t, u) = \text{const.} \tag{13}$$

The relation (13) is an intermediate integral of the Monge-Ampère equation (12).

It is clear that (12) and (13) are compatible only for a specific function $f(p, q)$. The complete analysis of compatibility carried out by Martin, Ludford, and (independently) Zav'jalov showed that an intermediate integral exists for the following functions $f(p, q)$:

$$f(p, q) = \sqrt{F'(\alpha_1 p + \alpha_2 q)}, \tag{14}$$

$$f(p, q) = \frac{1}{(q+\alpha_2)^2}\sqrt{F'\left(\frac{p+\alpha_1}{q+\alpha_2}\right)}. \tag{15}$$

In the first case

$$\varphi = \alpha_2 t - \alpha_1 u \pm g(\alpha_1 p + \alpha_2 q). \tag{16}$$

In the second case

$$\varphi = \xi - t(p+\alpha_1) - u(q+\alpha_2) \pm g\left(\frac{p+\alpha_1}{q+\alpha_2}\right). \tag{17}$$

In both cases $g(\theta)$ is connected with $F(\theta)$ by the relation

$$g'(\theta) = \sqrt{F'(\theta)}. \tag{18}$$

We consider the case of a polytropic gas, where

$$V = A^{2/\gamma} p^{-1/\gamma}, \qquad A^2 = A^2(S), \tag{19}$$

$$f = \sqrt{-\left(\frac{\partial V}{\partial p}\right)_S} = A^{1/\gamma}\frac{1}{\sqrt{\gamma}} p^{-(\gamma+1)/2\gamma}. \tag{20}$$

If the entropy S is constant, then f is a function of p only, and we can use the representation (14) with $\alpha_2 = 0$. Comparing (14) and (20), we find that

$$\alpha_1 = 1, \qquad F = A^{2/\gamma} p^{-1/\gamma},$$

$$g = \frac{2}{\gamma - 1}\sqrt{\gamma}\, A^{1/\gamma} p^{(\gamma-1)/2\gamma} = \frac{2}{\gamma-1} c. \tag{21}$$

We then have

$$\varphi = -u \pm \frac{2}{\gamma - 1} c, \tag{22}$$

i.e., the intermediate integral is none other than the Riemann invariant. This justifies the term "generalized Riemann invariant" introduced by Martin and Ludford for the intermediate integral (13) of equation (10).

We now consider the case of variable entropy. We shall show that among flows with a constant Riemann invariant there are flows adjoining the quiet region across the shock wave. We use for this the representation (15), (17). We shall assume that the shock wave moves on a quiet background ($\rho_0 = 1, u_0 = 1$) and is strong, i.e., for the quantities behind the wave front formulas (4.7.1)–(4.7.4) hold:

$$\left.\begin{aligned} p &= \frac{2}{\gamma+1}\rho_0 D^2 = \frac{2}{\gamma+1}D^2, \\ \rho_1 &= \frac{\gamma+1}{\gamma-1}\rho_0 = \frac{\gamma+1}{\gamma-1}, \\ u &= \frac{2}{\gamma+1}D, \end{aligned}\right\} \tag{23}$$

where

$$D = \frac{dx}{dt} = \frac{dq}{dt}. \tag{24}$$

Using (18) and (19), for $A(S)$ and $g(\theta)$ we obtain the expressions

$$A^{2/\gamma} = c^2 \gamma q^{(1-3\gamma)/\gamma}, \tag{25}$$

$$g = \frac{2\gamma}{\gamma-1} c \left(\frac{p}{q}\right)^{(\gamma-1)/2\gamma}, \tag{26}$$

where c is a constant, and α_1 and α_2 have been set equal to zero. Suppose that at time $t = 0$ a shock wave begins to move in the quiet gas according to the power law

$$q = Ct^{\alpha}. \tag{27}$$

We shall find the exponent α to which there corresponds a contant value of the function (17) on the wave front. The function φ will then be identically constant behind the wave front and will also have a constant generalized Riemann invariant.

Considering (3), from (17) we find

$$d\varphi_2 = u\,dq - p\,dt = d(uq) \mp dg\left(\frac{p}{q}\right), \tag{28}$$

whence it follows that

$$q\,du + p\,dt \mp dg\left(\frac{p}{q}\right) = 0. \tag{29}$$

By the assumption (27) the relation (29) on the trailing wave front after division by dt takes the form

$$\left(\frac{2}{\gamma+1}\alpha(\alpha-1)C^2+\frac{2}{\gamma+1}C^2\alpha^2\right)t^{2\alpha-2}+Bt^{((\gamma-1)/2\gamma)(\alpha-2)-1}=0, \quad (30)$$

where

$$B=\mp c\left(\frac{2}{\gamma+1}\right)^{(\gamma-1)/2\gamma}(C\alpha^2)^{(\gamma-1)/2\gamma}(\alpha-2). \quad (31)$$

This implies that

$$\alpha=\frac{2}{3\gamma+1}. \quad (32)$$

Generalized Riemann waves can be considered as solutions with a differential constraint (Chapter 1, §12)

$$\Phi(q,\,t,\,\varphi_1,\,\varphi_2,\,u,\,p,\,S(q))=0, \quad (33)$$

where φ_1 and φ_2 are potentials of the system (1). As has been shown, determination of Φ reduces to integrating a linear homogeneous system. Since in this case q and t are independent variables, the differential constraint (33) for (1) also yields the case excluded from Martin's treatment in which $p=p(S)$. This case will be considered in the next subsection.

4. ***The equation of the hydrodynamical surface.*** In the case of an ideal gas (9.3.1) takes the form

$$\frac{\partial u}{\partial t}+\frac{\partial p}{\partial q}=0, \qquad \psi p^{-\kappa}\frac{\partial p}{\partial t}+\frac{\partial u}{\partial q}=0, \qquad \frac{\partial \psi}{\partial t}=0, \quad (1)$$

$$\kappa=\frac{\gamma+1}{\gamma}, \qquad \psi=\frac{1}{\gamma}A^{2/\gamma}. \quad (2)$$

For this system we shall obtain an equation in the space of the hodograph u, p, ψ. Applying the algorithm indicated in §12 of Chapter 1 for the surface

$$\psi=\psi(p,u), \quad (3)$$

we arrive at the following equations:

$$\left.\begin{aligned}\frac{\partial q}{\partial p}&=\frac{\partial \psi}{\partial p}\mu, & \frac{\partial q}{\partial u}&=\frac{\partial \psi}{\partial u}\mu,\\ \frac{\partial t}{\partial p}&=-\psi p^{-\kappa}\frac{\partial \psi}{\partial u}\mu, & \frac{\partial t}{\partial u}&=-\frac{\partial \psi}{\partial p}\mu,\end{aligned}\right\} \quad (4)$$

where μ in an undetermined factor.

The integrability conditions for (4) lead to equations for μ:

$$\left.\begin{aligned}\psi p^{-\kappa}\frac{\partial\psi}{\partial u}\frac{\partial\ln\mu}{\partial u}-\frac{\partial\psi}{\partial p}\frac{\partial\ln\mu}{\partial p}=\frac{\partial^2\psi}{\partial p^2}-\frac{\partial(\psi p^{-\kappa}\,\partial\psi/\partial u)}{\partial u}=-\Phi,\\ \frac{\partial\psi}{\partial p}\frac{\partial\ln\mu}{\partial u}-\frac{\partial\psi}{\partial u}\frac{\partial\ln\mu}{\partial p}=0.\end{aligned}\right\}\qquad(5)$$

Solving (5) for $\partial\ln\mu/\partial u$ and $\partial\ln\mu/\partial p$, we find that

$$\frac{\partial\ln\mu}{\partial u}=\frac{\partial\psi}{\partial u}\varphi,\qquad\frac{\partial\ln\mu}{\partial p}=\frac{\partial\psi}{\partial p}\varphi,\qquad(6)$$

where

$$\varphi=\frac{\Phi}{\Delta},\qquad\Delta=\left(\frac{\partial\psi}{\partial p}\right)^2-\psi p^{-\kappa}\left(\frac{\partial\psi}{\partial u}\right)^2.\qquad(7)$$

Finally, the integrability conditions for (6) give the following equation of third order:

$$\begin{vmatrix}\dfrac{\partial\psi}{\partial u} & \dfrac{\partial\psi}{\partial p}\\[2mm] \dfrac{\partial\varphi}{\partial u} & \dfrac{\partial\varphi}{\partial p}\end{vmatrix}=0.\qquad(8)$$

Integrating this we arrive at the second order equation

$$\varphi=\frac{\partial(\psi p^{-\kappa}\partial\psi/\partial u)/\partial u-\partial^2\psi/\partial p^2}{(\partial\psi/\partial p)^2-\psi p^{-\kappa}(\partial\psi/\partial u)^2}=f(\psi),\qquad(9)$$

where $f(\psi)$ is an arbitrary function of ψ.

It is not hard to see that (9) admits the following integrals:

$$\psi=g(p),\qquad(10)$$

$$u=h(\psi),\qquad(11)$$

where g and h are arbitrary functions of their arguments. The solutions corresponding to them have the form

$$u=Ct+C_1,\qquad p=-Cq+C_2,\qquad\psi=g(p),\qquad(12)$$

$$\left.\begin{aligned}\int\frac{1}{\psi}h'(\psi)\,d\psi=Cq+C_1,\\ \frac{1}{1-\kappa}p^{1-\kappa}=-Ct+C_2,\qquad u=h(\psi).\end{aligned}\right\}\qquad(13)$$

The relation (10) gives precisely the exceptional case when transformation to the Monge-Ampère equation is not possible.

Solutions of the form (12), (13) were obtained by Stanjukovič (1955). If $\psi = \psi(u, p)$ is a solution of (9), then a solution of the original equation is obtained by successive integration of the completely integrable systems (6) and (4).

The formulas delivering the solution have the form

$$\ln \mu = \int f(\psi)\, d\psi, \qquad \mu = e^{\int f(\psi) d\psi} = F(\psi), \tag{14}$$

$$q = \int \mu\left(\frac{\partial \psi}{\partial p} dp + \frac{\partial \psi}{\partial u} du\right) = \int \mu\, d\psi = \int F(\psi)\, d\psi, \tag{15}$$

$$t = -\int F(\psi)\left[\left(\psi p^{-\kappa} \frac{\partial \psi}{\partial u}\right) dp + \frac{\partial \psi}{\partial p}\, du\right]. \tag{16}$$

To the given surface $\Phi(\psi, p, u) = 0$ there corresponds via (14)–(16) a solution of (1) which is unique up to a constant. Uniqueness is violated for the case when $\psi = \text{const}$, to which there corresponds a family of solutions depending on two arbitrary functions of a single argument.

5. ***Solutions of the equations of gas dynamics characterized by differential contraints.*** We write (9.3.1) in the form

$$\frac{\partial u}{\partial t} + \frac{\partial p}{\partial q} = 0, \qquad \frac{\partial V}{\partial t} - \frac{\partial u}{\partial q} = 0, \qquad \frac{\partial p}{\partial t} + M(p, V)\frac{\partial u}{\partial q} = 0,$$
$$M(p, V) = \left(\frac{\partial S}{\partial V}\right)\left(\frac{\partial S}{\partial p}\right)^{-1}. \tag{1}$$

It can be shown that, in general, differential constraints of first order compatible with the system (1) are quasilinear. As a result of a complete analysis, we establish that in the case of one differential constraint adjoined to (1) the overdetermined system is compatible with two-function arbitrariness in the solution if the constraint has one of the two forms (Raspopov, Šapeev, and Janenko (1977)):

a) $$\frac{\partial V}{\partial q} + M^{-1}\frac{\partial p}{\partial q} = f(V, p, q), \tag{2}$$

b) $$\frac{\partial u}{\partial q} \mp M^{-1/2}\frac{\partial p}{\partial q} = f(u, p, q, t). \tag{3}$$

In the first case f is determined from the equation

$$\frac{\partial f}{\partial V} - M\frac{\partial f}{\partial p} + M^{-1}\frac{\partial M}{\partial V} f = 0.$$

In the second case f is determined from a system of two equations, and M can depend only on p. If $f \not\equiv 0$, then the function M is uniquely determined; otherwise $M = M(p)$ is an arbitrary function.

The differential constraints we have found make it possible to construct solutions of (1). For example, in case b) we set $f \equiv 0$ and $M = a^2p^n$; then the general solution of the system (1), (3) has the form

$$u = \pm \frac{2p^{1-n/2}}{(2-n)a} + c_1, \qquad V = \frac{p^{1-n}}{(n-1)a^2} + g(q),$$
$$G(p, \mp q + ap^{n/2}t) = 0, \tag{4}$$

where g and G are arbitrary functions, and c_1, n and a are arbitrary constants.

The system (1) is compatible with two differential constraints of first order in either of the following cases:

a) The constraints have the form

$$\frac{\partial V}{\partial q} + M^{-1}\frac{\partial p}{\partial q} = f_1(u, p, V, q, t),$$
$$\frac{\partial u}{\partial q} \pm M^{-1/2}\frac{\partial p}{\partial q} = f_2(u, p, V, q, t); \tag{5}$$

here f_1 and f_2 are determined from a system of four quasilinear equations which are compatible only for certain forms of the equation of state (for example, in the case of a polytropic gas $M = \gamma pV^{-1}$, where γ, the exponent of the polytrope, can only be equal to 3).

b) The constraints have the form

$$\frac{\partial u}{\partial q} = \frac{f_1(u, p, q, t)}{M(p, V)}, \qquad \frac{\partial p}{\partial q} = f_2(u, p, q, t), \tag{6}$$

where f_1 and f_2 are determined from the equations

$$\frac{1}{M}\frac{\partial f_1}{\partial t} - \frac{f_2}{M}\frac{\partial f_2}{\partial u} - \frac{f_1}{M}\frac{\partial f_1}{\partial p} + \frac{f_1}{M}\frac{\partial f_2}{\partial u} + f_2\frac{\partial f_2}{\partial p} + \frac{\partial f_2}{\partial q}$$
$$+ \left(\frac{f_1}{M}\right)^2\frac{\partial M}{\partial p} - \frac{f_1^2}{M^3}\frac{\partial M}{\partial V} = 0,$$
$$\frac{\partial f_1}{\partial q} + \frac{f_1}{M}\frac{\partial f_1}{\partial u} + f_2\frac{\partial f_1}{\partial p} - f_2\frac{\partial f_2}{\partial u} - f_1\frac{\partial f_2}{\partial p} + \frac{\partial f_2}{\partial t} = 0. \tag{7}$$

The system (7) is compatible, and the arbitrariness in its solution depends on the form of the equation of state.

For a polytropic gas ($M = \gamma pV^{-1}$) there are the following cases:

b$_1$) $$f_1 = \mathscr{F}(t)p, \qquad f_2 = -(u + c_1)\left(\frac{\mathscr{F}(t)}{\gamma} + \frac{1}{\mathscr{F}(t)}\frac{d\mathscr{F}(t)}{dt}\right), \tag{8}$$

where c_1 is an arbitrary constant, and $\mathscr{F}(t)$ is found from the equation

$$\mathscr{F}'' + \frac{\gamma+3}{\gamma}\mathscr{F}\mathscr{F}' + \frac{\gamma+1}{\gamma^2}\mathscr{F}^3 = 0. \tag{9}$$

Among solutions of (1) characterized by the differential constraints (6), (8), and (9) there is L. I. Sedov's solution with functional arbitrariness describing the flow of a gas with a linear velocity profile.

b$_2$) The functions $f_1 = f_1(p, q, t)$, $f_2 = f_2(p, q, t)$ are determined implicitly from the relations

$$G_1(f_1 p^{-(\gamma+1)/\gamma}, \gamma p f_1^{-1} - t, q) = 0, \qquad G_2(f_2, q f_2 - p, t) = 0,$$

in such a way that

$$\frac{\partial f_1}{\partial q} + f_2 \frac{\partial f_1}{\partial p} - f_1 \frac{\partial f_2}{\partial p} + \frac{\partial f_2}{\partial t} = 0. \tag{10}$$

For example, (10) is valid if

$$f_1 = \frac{\gamma p}{t + c_1}, \qquad f_2 = \frac{c_2(t+c_1)^{-\gamma} + p}{q};$$

$$f_1 = \frac{\gamma p}{t + c_1}, \qquad f_2 = c_2(t+c_1)^{-\gamma};$$

$$f_1 = c_1(q + c_2)^{-1/\gamma} p^{\gamma+1/\gamma}, \qquad f_2 = \frac{p}{q + c_2}. \tag{11}$$

The solution of the system (1), (6), (11) can be written in the explicit form

$$u = \frac{c_3}{\gamma - 1}(t + c_1)^{1-\gamma} + \beta(q), \qquad p = (c_3 q + c_2)(t + c_1)^{-\gamma},$$
$$V = \beta'(q)(t + c_1).$$

Since we are considering a polytropic gas $p = \Psi(S(q))V^{-\gamma} = \phi(q)V^{-\gamma}$, here the relation $(\beta'(q))^\gamma(c_3 q + c_2) = \phi(q)$ must be satisfied.

Finally, in resolving the question of the existence of solutions of (1) characterized by three differential constraints

$$\frac{\partial V}{\partial q} = g_1(u, p, V, q, t), \qquad \frac{\partial u}{\partial q} = g_2(u, p, V, q, t),$$

$$\frac{\partial p}{\partial q} = g_3(u, p, V, q, t), \tag{12}$$

we find that they exist for any equation of state; g_1, g_2, and g_3 are determined from a certain Cauchy-Kovalevskaja system, and the system (1), (12) is completely integrable.

Other solutions of the equations of gas dynamics characterized by differential constraints of first order are presented by Raspopov, Šapeev, and Janenko (1974). They showed that the class of solutions of the system (1) characterized by differential constraints contains the solutions of Martin, Ludford, and Zav'jalov, since the present method is more general than the method of the intermediate integral. In the same paper the solutions are applied to the problem of the motion of a shock wave. This exact solution has been found a convenient test for numerical methods (Vorožcov (1977)).

We consider the following problem: a contact zone $u = 0$, $p = p_0 = \text{const}$, $S = S_1(q)$ is bounded on the left by a piston $q = q_0$ which moves according to the law $u = U(t)$, $U(0) = 0$. We wish to determine the parameters of the moving gas. We write the solution (4) in the form

$$u = \frac{2p_0^{1-n/2}}{a(n-2)} - \frac{2p^{1-n/2}}{a(n-2)}, \tag{13}$$

$$V = \frac{p^{1-n}}{a^2(n-1)} + g_1(q), \tag{14}$$

$$ap^{n/2}t - q = g_2(p). \tag{15}$$

From the condition of the problem and (13) we find that on the piston

$$t = U^{-1}\left(\frac{2p_0^{1-n/2}}{a(n-2)} - \frac{2p^{1-n/2}}{a(n-2)}\right),$$

where U^{-1} denotes the function inverse to U. According to (15), the pressure p is determined as a function of q and t from the relation

$$ap^{n/2}\,t - q = ap^{n/2}U^{-1}\left(\frac{2p_0^{1-n/2}}{a(n-2)} - \frac{2p^{1-n/2}}{a(n-2)}\right) - q_0.$$

To find $u(q, t)$ we substitute $p(q, t)$ into (13). The function $V(q, t)$, where $g_1(q) = S_1(q)$, is found from (14). The boundary condition on the moving piston is satisfied by construction. The quiet region is separated from the region of motion by the rectilinear characteristic $ap_0^{n/2}t - q = -q_0$. The problem is solved.

6. ***A solution of the one-dimensional equations of gas dynamics with constant arbitrariness.*** We use the general equations of one-dimensional flows

$$\left.\begin{aligned} \rho\left(\frac{\partial u}{\partial t} + u\frac{\partial u}{\partial x}\right) + \frac{\partial p}{\partial x} = 0, \\ \frac{\partial \rho}{\partial t} + u\frac{\partial \rho}{\partial x} + \rho\frac{\partial u}{\partial x} + \frac{\nu\rho u}{x} = 0, \\ \frac{\partial S}{\partial t} + u\frac{\partial S}{\partial x} = 0, \end{aligned}\right\} \tag{1}$$

where the parameter ν is equal to 0 in the plane case, to 1 in the cylindrical case, and to 2 in the spherical case.

We henceforth assume that the equation of state has the form

$$p = a^2(S)\rho^\gamma. \tag{2}$$

In this case the system (1) can be transformed to the form

$$\left.\begin{aligned} \frac{\partial u}{\partial t} + u\frac{\partial u}{\partial x} + \frac{p}{\rho}\frac{\partial \ln p}{\partial x} = 0, \\ \frac{\partial \ln \rho}{\partial t} + u\frac{\partial \ln \rho}{\partial x} + \frac{\partial u}{\partial x} + \frac{\nu u}{x} = 0, \\ \frac{\partial \ln p}{\partial t} + u\frac{\partial \ln p}{\partial x} + \gamma\left[\frac{\partial u}{\partial x} + \frac{\nu u}{x}\right] = 0. \end{aligned}\right\} \tag{3}$$

In the plane case ($\nu = 0$) the systems (1) and (3) have coefficients depending only on the unknown functions and admit transformations of similarity and translation in the independent variables x and t. In the cylindrical and spherical cases only a similarity transformation in x and t is possible. Going over to logarithmic scales in x and t and dimensionless velocities

$$\tau = \ln t, \qquad \lambda = \ln x, \qquad u = \frac{x}{t}U, \qquad c = \frac{x}{t}C, \tag{4}$$

we can write (1) in the form[19]

$$\frac{\partial U}{\partial \tau} + U\frac{\partial U}{\partial \lambda} + C^2\frac{\partial \varphi}{\partial \lambda} + C^2\frac{\partial \psi}{\partial \lambda} + U^2 - U = 0, \tag{5}$$

$$\frac{\partial \varphi}{\partial \tau} + U\frac{\partial \varphi}{\partial \lambda} + \frac{\partial U}{\partial \lambda} + (\nu + 1)\,U = 0, \tag{6}$$

$$\frac{\partial \psi}{\partial \tau} + U\frac{\partial \psi}{\partial \lambda} = 0, \tag{7}$$

where $\varphi = \ln \rho$ and $\psi = \gamma^{-1}\ln a^2$. In the system (5)–(7) for any ν the coefficients

[19] The equations of gas dynamics were considered in this form by Stanjukovič (1955).

do not depend on the arguments τ and λ, and hence, as the analysis in Chapter 1, §12, shows, it has the simple waves

$$U = U(\theta), \qquad \varphi = \varphi(\theta), \qquad \psi = \psi(\theta),$$
$$\theta = \theta_0 + a_1\lambda + a_2\tau. \tag{8}$$

Since the coefficients of the system (5)–(7) do not depend on φ or ψ, solutions of the form

$$U = U(\theta), \qquad \varphi = \varphi(\theta) + l\tau, \qquad \psi = \psi(\theta) - \frac{\gamma - 1}{\gamma} l\tau, \tag{9}$$

are possible where l is an arbitrary constant.

Going over to the variables u, ρ, p we obtain the representations

$$u = \frac{x}{t} U(\xi) = t^{-(a_2/a_1+1)} \left(\frac{\xi}{\xi_0}\right)^{1/a_1} U(\xi),$$
$$p = \rho_0 \left(\frac{x}{t}\right)^2 t^l P(\xi) = \rho_0 t^{l-2(a_2/a_1+1)} \left(\frac{\xi}{\xi_0}\right)^{2/a_1} P(\xi), \qquad \rho = \rho_0 t^l R(\xi), \tag{10}$$

where

$$\xi = e^\theta = \xi_0 x^{a_1} t^{a_2}, \qquad \xi_0 = e^{\theta_0}. \tag{11}$$

Solutions of the form (10) are called *self-similar*.[20] It is clear that the exponents a_1 and a_2 are determined up to a factor, so that the essential parameter is only their ratio $a_1/a_2 = -k$.

The system of equations for $U(\xi)$, $R(\xi)$ and $P(\xi)$ corresponding to (3) has the form

$$\left.\begin{aligned}
&\text{a)} \quad (a_1U + a_2)\frac{dU}{d\theta} + \frac{C^2}{\gamma} a_1 \frac{d\ln P}{d\theta} = U - U^2 - 2\frac{C^2}{\gamma},\\
&\text{b)} \quad (a_1U + a_2)\frac{d\ln R}{d\theta} + a_1\frac{dU}{d\theta} = -(\nu + 1)U - l,\\
&\text{c)} \quad (a_1U + a_2)\frac{d\ln P}{d\theta} + \gamma a_1\frac{dU}{d\theta} = -[(\nu + 1)\gamma + 2]U - l + 2.
\end{aligned}\right\} \tag{12}$$

A linear combination of (12b) and (12c) leads to the integral (the so-called adiabatic integral)

[20] Self-similar solutions of one-dimensional gas dynamics as solutions invariant under certain groups of transformations were first considered by Bechert (1941).

$$\ln P - \kappa \ln R + (\gamma - \kappa)\ln\left(\frac{1}{k} - U\right) + [(\nu + 1)(\gamma - \kappa) + 2]\frac{\theta}{a_1} = \ln C_1,$$

$$\kappa = \frac{(\nu + 1)\gamma + 2 + k(l - 2)}{kl + (\nu + 1)}, \tag{13}$$

which makes it possible to reduce the number of equations (12) to two for any k and l.

We shall show that for certain k and l the system (12) admits a further integral. As a conservation law, the equations (1) have the relation

$$\frac{\partial}{\partial t}\left[\rho x^\nu\left(\frac{1}{\gamma - 1}\frac{p}{\rho} + \frac{1}{2}u^2\right)\right] + \frac{\partial}{\partial x}\left[\rho x^\nu u\left(\frac{\gamma}{\gamma - 1}\frac{p}{\rho} + \frac{1}{2}u^2\right)\right] = 0, \tag{14}$$

which expresses the law of conservation of energy. To it there corresponds the self-similar equation

$$\frac{dA}{d\xi} + [\nu + 3 + k(l - 3)]\,B - k\xi\frac{dB}{d\xi} = 0, \tag{15}$$

where

$$\left.\begin{aligned} A &= \xi^{l-2}\left[\frac{P}{\gamma - 1} + \frac{PU^2}{2}\right], \\ B &= \xi^{l-3}U\left[\frac{\gamma}{\gamma - 1}P + \frac{PU^2}{2}\right], \qquad \xi = \xi_0 t x^{-k}. \end{aligned}\right\} \tag{16}$$

Under the condition

$$k = \frac{\nu + 3}{2 - l}, \tag{17}$$

(15) admits the integral (the energy integral)

$$A - k(\xi B) = \text{const.} \tag{18}$$

With the energy integral, (12) reduces to a single equation.

We shall give a brief survey of some problems leading to self-similar solutions. Basically these solutions describe flows adjoining a quiet region across a shock wave or weak discontinuity. Here the characteristic circumstance is that the pressure in the quiet region is equal to zero. The shock wave can then be considered infinitely strong, and the Hugoniot conditions acquire a homogeneous form which makes it possible to seek a solution in self-similar form.

It is known (see §4.7) that the Hugoniot conditions on a strong shock wave have the form

$$\frac{\rho}{\rho_0} = \frac{\gamma + 1}{\gamma - 1}, \qquad u = \frac{2}{\gamma + 1} D, \qquad p = \frac{2}{\gamma + 1} \rho_0 D^2, \tag{19}$$

where $D = dx/dt$ is the velocity of the shock wave. We shall assume that the trajectory of the shock wave is a ξ-line (i.e., a line with the equation $\xi = \text{const}$). From (11) it then follows that

$$D = \frac{dx}{dt} = -\frac{a_2}{a_1}\frac{x}{t} = \frac{1}{k}\frac{x}{t}. \tag{20}$$

The Hugoniot conditions (19) for dimensionless quantities U, R, P take the form

$$U(\xi) = \frac{2}{\gamma + 1}\frac{1}{k}, \qquad R(\xi) = \frac{\gamma + 1}{\gamma - 1} t^{-l}, \qquad P(\xi) = \frac{2}{\gamma + 1}\frac{1}{k^2} t^{-l}. \tag{21}$$

This implies that $l = 0$. After these preliminary remarks we consider a number of self-similar solutions.

1. *A converging shock wave.* The problem was posed and solved by Guderley (1942) and independently by Stanjukovič for the case $\gamma = 7/5$, and it was investigated for all γ by a group of Soviet mathematicians (see Brušlinskiĭ and Każdan (1963)). The problem can be formulated in the following manner. A spherical ($\nu = 2$) shock wave moves while accelerating toward the point $x = 0$ in a quiet gas with parameters p_0, ρ_0 and $u_0 = 0$. At time $t = 0$ it arrives with infinite velocity at the center, which is a singular point of the system (1) (Figure 2.81). The analytic character of the singularity of the solution at the point $x = 0$, $t = 0$ is very complicated. It is conjectured that the solution in a neighborhood of the center can be represented by analytic functions of an argument $\xi = \xi_0 x^{a_1} t^{a_2}$ in fractional powers, and hence it has self-similar character. For convenience in formulating the boundary conditions in (11) we set

$$a_2 = 1, \qquad a_1 = -k. \tag{22}$$

Since the shock wave is infinitely strong, conditions (19) and (21) hold. We choose ξ_0 so that $\xi = 1$ on the shock wave. The conditions (21) then take the form

$$U(1) = \frac{2}{\gamma + 1}\frac{1}{k}, \qquad R(1) = \frac{\gamma + 1}{\gamma - 1}, \qquad P(1) = \frac{2}{\gamma + 1}\frac{1}{k^2}. \tag{23}$$

The constant C_1 in (13) is determined from (23). Using (13) and introducing the new variables

$$y = 1 - kU, \qquad z = \frac{k^2 C^2}{1 - kU} = \frac{k^2 \gamma P}{R(1 - kU)}, \tag{24}$$

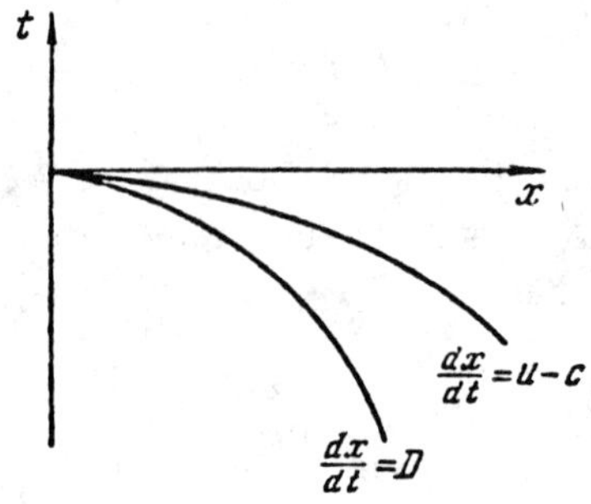

FIGURE 2.81

we reduce the system (12) to two equations:

$$\left.\begin{aligned} &\text{a)} \qquad k\xi \frac{dy}{d\xi} = \frac{q(y, z)}{z - y}, \\ &\text{b)} \qquad k\xi \frac{dz}{d\xi} = \frac{z}{y} \frac{p(y, z)}{z - y}, \end{aligned}\right\} \tag{25}$$

where

$$\left.\begin{aligned} p(y, z) &= (3 - y)z - (2\gamma - 1)y^2 \\ &\quad + [2\gamma - 3 + (k - 1)(\gamma - 2)]y - \gamma(k - 1), \\ q(y, z) &= (1 - y)(y - 3z) - (1 - k)\left(\frac{2}{\gamma} z + 1 - y\right). \end{aligned}\right\} \tag{26}$$

From (23) we find that

$$y(1) = \frac{\gamma - 1}{\gamma + 1}, \qquad z(1) = \frac{2\gamma}{\gamma + 1}. \tag{27}$$

For $t = 0$ the solution $u(x, t)$, $\rho(x, t)$, $p(x, t)$ must be bounded. Hence

$$U(0) = 0, \qquad P(0) = 0, \tag{28}$$

which in the variables y, z implies that

$$y(0) = 1, \qquad z(0) = 0. \tag{29}$$

We have thus reduced the problem of finding a self-similar solution to the boundary value problem (27), (29) for the system (25).

Dividing (25b) by (25a), we find that

$$\frac{dz}{dy} = \frac{z}{y}\frac{p}{q}. \tag{30}$$

We have arrived at an equivalent formulation: find a solution of (30) passing through the points $M_1((\gamma - 1)/(\gamma + 1),\ 2\gamma/(\gamma + 1))$ and $M_2(1, 0)$ in the y, z plane.

The problem is overdetermined and is solvable only for certain values of k. It is not hard to see that the characteristic

$$\frac{dx}{dt} = u - c, \tag{31}$$

going into the center is a ξ-line and is represented in the y, z plane by the line (Figure 2.82)

$$y - z = 0. \tag{32}$$

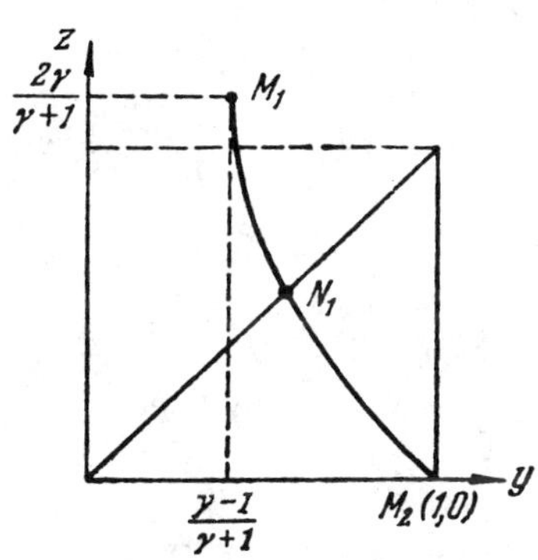

FIGURE 2.82

The points M_1 and M_2 lie on different sides of the line $z - y = 0$, and hence the desired integral curve $y = y(\xi)$, $z = z(\xi)$ must intersect this line at some point M corresponding to a certain value of the parameter ξ_M.

If as the curve $y(\xi)$, $z(\xi)$ passes through the point M: $\{y = y(\xi_M), z = z(\xi_M)\}$ the functions $p(y, z)$ and $q(y, z)$ do not change sign, then the right sides of (25) change sign, and the functions $y(\xi)$ and $z(\xi)$ become multiple valued, which is impossible. This implies that at the point M: $\{y(\xi_M), z(\xi_M)\}$ the quantities $p(y, z)$ and $q(y, z)$ also vanish, and it is a singular point. The conditions $y - z = 0$ and $p = 0$ lead to the equation

$$y^2 - \left[1 - \frac{\gamma - 2}{2\gamma}(1 - k)\right]y - \frac{1 - k}{2} = 0. \tag{33}$$

Thus, there are two singular points N_1, N_2 on the line $y - z = 0$. The position of these points and also the character of the singularities depend on the two parameters γ and k. In the case $\gamma = 7/5$ (the investigation of Guderley) the point N_1 is a saddle point, and there exists a unique integral curve joining M_1 and M_2. Moreover, it passes through N_1 (Figure 2.82). We mention only that there are ranges of the parameters γ and k for which the problem is uniquely solvable.

2. *The problem of filling a spherical cavity* is investigated in a similar way. A mass of isentropic gas escapes to a vacuum, so that the boundary of the gas with the vacuum moves with acceleration toward the center and at time $t = 0$ arrives at the center $x = 0$ with infinite velocity.

For the dimensionless invariants

$$r = k\left(u + \frac{2}{\gamma - 1}c\right), \qquad s = k\left(u - \frac{2}{\gamma - 1}c\right), \tag{34}$$

we obtain the system of equations

$$k\xi\frac{dr}{d\xi} = \frac{p(s, r)}{q(r, s)}, \qquad k\xi\frac{ds}{d\xi} = \frac{p(r, s)}{q(s, r)}, \tag{35}$$

where

$$\left.\begin{aligned} p(r, s) &= \frac{4}{\gamma - 1}ks - 2\left(\frac{1}{\gamma - 1} + 1\right)s^2 - \left(\frac{2}{\gamma - 1} - 1\right)rs + r^2, \\ q(r, s) &= \frac{4}{\gamma - 1} - \left(\frac{2}{\gamma - 1} + 1\right)r - \left(\frac{2}{\gamma - 1} - 1\right)s. \end{aligned}\right\} \tag{36}$$

The boundary with the vacuum is a ξ-line. The quantity ξ_0 is chosen so that $\xi = 1$ on the front. The boundary conditions for (35) then have the form

$$r(1) = s(1) = 1, \tag{37}$$

$$r(0) = s(0) = 0. \tag{38}$$

In the r, s plane we again arrive at a boundary value problem: find an integral curve of the equation

$$\frac{ds}{dr} = \frac{p(r, s)q(r, s)}{p(s, r)q(s, r)}, \tag{39}$$

passing through $M_1(1, 1)$ and $M_2(0, 0)$.

It can be proved similarly that the desired integral curve must pass through singular points of (39). A detailed analysis reveals the lack of a unique analytic solution.

3. *The problem of point explosion* was solved by Sedov (1946) and G. I. Taylor (see, for example, Courant and Friedrichs (1950)). At the point $x = 0$ at time $t = 0$ a finite amount of energy E is released instantaneously which goes into energy of motion of a gas separated from the quiescent medium by a shock wave. Since for small t the concentration of energy and pressure are large, the pressure p_0 of the quiescent gas may be neglected, and this makes possible a self-similar approximation. For convenience in formulating the boundary conditions in (11) we set $a_1 = 1$ and $a_2 = -1/k$, and we choose ξ_0 so that $\xi = 1$ on the shock wave. The flow arising as a result of the point explosion is then described by a solution of (12) that satisfies (23) and, in addition, the symmetry condition

$$\lim \xi U(\xi) \to 0, \qquad \xi \to 0, \tag{40}$$

which implies that $u(0, t) = 0$.

We shall see that there is also an energy integral in addition to the adiabatic integral (13). Because of the assumption $p_0 = 0$ the flux of energy through the shock wave is equal to zero, and the energy of the moving gas is identically equal to E:

$$E = \int_0^x \rho\left(\varepsilon + \frac{u^2}{2}\right) x^\nu \, dx. \tag{41}$$

Going over to the quantities P, R, U, we obtain

$$E = \frac{\rho_0}{\xi_0^{\nu+3}} t^{(\nu+3)/k-2} \int_0^1 \left(\frac{1}{\gamma - 1} P + \frac{1}{2} RU^2\right) \xi^{\nu+2} \, d\xi. \tag{42}$$

This implies that $k = (\nu + 3)/2$. Under the condition $l = 0$ this is necessary and sufficient for the existence of an energy integral. From the conditions on the shock wave it follows that the constant of integration in (18) is equal to zero, and the energy integral takes the form

$$kU\left(\frac{\gamma}{\gamma - 1} P + \frac{1}{2} RU^2\right) = \frac{1}{\gamma - 1} P + \frac{1}{2} RU^2. \tag{43}$$

Noting that $l = 0$, we can write (13) in the form

$$\ln P - (\gamma - 1) \ln R + \ln\left(\frac{1}{k} - U\right) + (\nu + 3)\,\theta = \ln C_1. \tag{44}$$

The constant C_1 is determined from the conditions on the shock wave:

$$C_1 = \frac{2a^3(\gamma - 1)^\gamma}{(\gamma + 1)^{\gamma+1}}, \qquad a = \frac{1}{k}. \tag{45}$$

We transform (43) to the form

$$\ln P - \ln R + \ln\left(\gamma U - \frac{1}{k}\right) - \ln\left(\frac{1}{k} - U\right) - 2\ln U = \ln \frac{\gamma - 1}{2}. \tag{46}$$

Differentiating (44) and (46) with respect to θ and eliminating $d(\ln P)/d\theta$ and $d(\ln R)/d\theta$ from (44), (46), and (12b), we arrive at the equation for U

$$\frac{dU}{d\theta} = U \frac{(2 - bU)(\gamma U - a)}{\gamma(1 + \gamma) U^2 - 2a(\gamma + 1) U + 2a^2}, \tag{47}$$

where

$$b = (\nu + 3) - (2 - \gamma)(\nu + 1) > 0. \tag{48}$$

Integrating (47), we find that

$$C\xi = U^{-\alpha}(\alpha_1 - U)^{-\beta_1}(U - \alpha_2)^{\beta_2}, \tag{49}$$

where

$$\left.\begin{aligned} \alpha_1 &= \frac{2}{b}, \qquad \alpha_2 = \frac{a}{\gamma}, \\ \beta_1 &= \frac{a^2b^2 + 2(\gamma + 1)(\gamma - ab)}{b(2\gamma - ab)}, \qquad \beta_2 = \frac{(\gamma - 1)a}{2\gamma - ab}. \end{aligned}\right\} \tag{50}$$

We note that α_1, α_2, $\alpha_1 - \alpha_2$ and β_2 are positive for any γ and ν, and β_1 is positive for all $\gamma > 1$ and all ν.

The constant C in (49) is determined from the conditions on the shock wave ($\xi = 1$) and is equal to

$$C = \left(\frac{2}{\gamma + 1} a\right)^{-\alpha}\left(\alpha_1 - \frac{2}{\gamma + 1} a\right)^{-\beta_1}\left(\frac{2}{\gamma + 1} a - \alpha_2\right)^{\beta_2}. \tag{51}$$

If we now use the energy integral (43) and the integral (44), (45), we obtain the following expressions for R and P:

$$R = \left[4a^3 \frac{(\gamma - 1)^{\gamma-1}}{(\gamma + 1)^{\gamma+1}}\right]^{1/(2-\gamma)} \left[\frac{\gamma U - a}{(a - U)^2 U^2}\right]^{1/(2-\gamma)} \xi^{(\nu+3)/(\gamma-2)}, \tag{52}$$

$$P = \frac{\gamma - 1}{2} R U^2 \frac{a - U}{\gamma U - a}. \tag{53}$$

If follows from (49) that $U \to \alpha_2$ as $\xi \to 0$, and so the symmetry condition (40) is satisfied.

A detailed analysis of the formulas for this solution shows that U achieves the value α_2 as $\xi \to 0$ for $\gamma < 7$ if $\nu = 2$ and for all γ if $\nu = 0$ or 1. For $\gamma > 7$ and $\nu = 2$ the point explosion is accompanied by the formation of a cavity with a boundary in the x, t plane which is represented by a ξ-line.

We have considered the problem of point explosion in the adiabatic approximation without considering thermal conductivity. It is clear that for sufficiently small t the concentration of energy is high, and the process should be considered thermally conducting. The point explosion problem with consideration of thermal conductivity has been considered by Korobeĭnikov (1961) and Neuvažaev (1962).

A problem closely related in its formulation to the point explosion problem is the problem of the motion arising from the action of a piston having a spherical (cylindrical) surface. If it is assumed that the velocity of the piston varies according to the power law

$$u_p = \text{const } t^{\alpha}, \tag{54}$$

then the ensuing flow is described by the self-similar equations (12). In contrast to the problem of point explosion, here there is only the adiabatic integral (13), and there is no energy integral. Therefore, the system (12), from which p is eliminated by means of (13), reduces to two equations for $u(\xi)$ and $R(\xi)$.

The flow in question is bounded by two ξ-lines: the line of the shock wave and the line of the piston (Figure 2.83). This implies that

$$\alpha = \frac{1}{k}, \qquad l = 0. \tag{55}$$

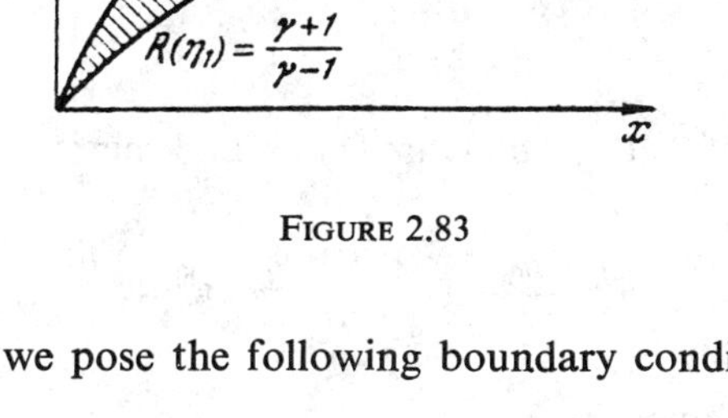

FIGURE 2.83

For $u(\xi)$ and $R(\xi)$ we pose the following boundary conditions: on the shock wave where $\xi = \eta_1$

$$u(\eta_1) = \frac{2}{\gamma + 1}\frac{1}{k}, \qquad R(\eta_1) = \frac{\gamma + 1}{\gamma - 1}; \tag{56}$$

on the line of the piston where $\xi = \eta_0$

$$u(\eta_0) = \frac{1}{k}. \tag{57}$$

Integration of the system for $u(\xi)$ and $R(\xi)$ proceeds from the value $\xi = \eta_1$ to the value $\xi = \eta_0$ for which condition (57) is satisfied. Since there is no energy integral, the exponent k is arbitrary.

The piston problem has been investigated in detail by N. L. Krašeninnikov, N. S. Mel′nikova, and N. N. Kočina. We refer the reader to the monograph of Korobeĭnkov (1961) in which a detailed analysis, accompanied by references, is given for the piston problem and the problem of point explosion.

There is a very interesting solution of still another self-similar problem—the problem of an instantaneous percussion—which was given by Zel′dovič (1956) and was further investigated by Adamskiĭ (1956) and Žukov and Každan (1956).

7. *Self-similar solutions in Lagrange coordinates.* After going over to the Lagrange mass coordinate

$$q(x, t) = \int_0^x \rho(x, t)\, x^\nu\, dx \tag{1}$$

(9.6.1) take the form

$$\left.\begin{aligned}
&\text{a)} && \frac{\partial u}{\partial t} + x^\nu \frac{\partial p}{\partial q} = 0,\\
&\text{b)} && \frac{\partial V}{\partial t} - \frac{\partial (u x^\nu)}{\partial q} = 0,\\
&\text{c)} && \frac{\partial S}{\partial t} = 0,\\
&\text{d)} && \frac{\partial x}{\partial t} = u.
\end{aligned}\right\} \tag{2}$$

Using the representation (9.6.10), from (1) we find that

$$q = \frac{\rho_0 t^{l+(\nu+1)/k}}{\xi_0^{\nu+1}} \int_0^\xi R(\xi)\xi^\nu\, d\xi. \tag{3}$$

This implies that the quantity

$$\eta = \eta_0 q t^{-[l+(\nu+1)/k]} = \frac{\xi_0^{\nu+1}}{\rho_0} q t^{-[l+(\nu+1)/k]} \tag{4}$$

is a function of ξ. Using (9.6.10), we find that f can be represented in the form

$$f = f_0 t^\beta F(\eta), \tag{5}$$

which proves self-similarity in the Lagrange coordinates.

8. ***Flows with a linear velocity profile.*** Sedov (1954) considered a very interesting class of solutions possessing functional arbitrariness. This class is completely determined by the presence of a linear velocity profile[21]

$$u = A(t)x. \tag{1}$$

Differentiating (1) with respect to t and considering (9.7.2d), we find that

$$\frac{\partial u}{\partial t} = \left(\frac{\dot{A}}{A} + A\right)u, \qquad \dot{A} = \frac{dA}{dt}. \tag{2}$$

Integrating (2) with respect to t, we obtain

$$u = B(t)U(q), \qquad B(t) = e^{\int(\dot{A}/A+A)dt}, \tag{3}$$

where $U(q)$ is an arbitrary function of q.

Integrating (9.7.2d) with respect to t, we have

$$x = C(t)U(q), \qquad C(t) = \int B(t)\,dt. \tag{4}$$

The equalities (1), (3), and (4) are not contradictory, since it is not hard to see that $B = AC$. The quantity V is determined from (9.7.2b):

$$V = \frac{C^{\nu+1}(t)}{\nu+1} \cdot \frac{dU^{\nu+1}(q)}{dq}. \tag{5}$$

Using the equation of state (9.6.2), from this we obtain an expression for p,

$$p = \psi(q)V^{-\gamma} = (\nu+1)^\gamma C(t)^{-(\nu+1)\gamma}\psi(q)\left[\frac{dU^{\nu+1}(q)}{dq}\right]^{-\gamma}, \tag{6}$$

where

$$\psi(q) = \psi[S(q)]. \tag{7}$$

Equation (9.7.2a) makes it possible to determine $C(t)$ and coordinate the

---

[21] The case of an inhomogeneous constraint $u = A(t)x + B(t)$ contradicts the symmetry condition $u(0, t) = 0$. Flows of this type were also considered by A. I. Žukov (private communication), Häfele (1954), von Hoerner (1955), and Keller (1956).

arbitrary functions $U(q)$ and $\psi(q)$. Substituting (3) and (6) into (9.7.2a), we find after separation of variables that

$$C^{\gamma(\nu+1)-\nu}\ddot{C}(t) = -(\nu+1)\gamma\mu, \tag{8}$$

$$U^{\nu-1}\frac{d}{dq}\left\{\psi(q)\left[\frac{dU^{\nu+1}}{dq}\right]^{-\gamma}\right\} = \mu, \tag{9}$$

where μ is a constant.

If $C(t)$ satisfies (8) and the functions $U(q)$ and $\psi(q)$ are coordinated by means of (9), then (3)–(6) determine a solution depending on a single arbitrary function. The solution (3)–(6) can be realized as the escape to a vacuum of a spherical volume of gas and the flow of gas behind a converging spherical wave. A detailed investigation of these flows is given in the monograph of Sedov (1957).

CHAPTER 3

DIFFERENCE METHODS FOR SOLVING THE EQUATIONS OF GAS DYNAMICS

In this chapter we present the basic concepts and facts from the theory of difference schemes and methods of constructing computational algorithms which find application in the numerical solution of problems of gas dynamics. Here we consider only the basic questions of this theory while referring the reader to monographs and journal articles for details. In recent years several monographs and textbooks devoted to the questions we consider have appeared (see, for example, Godunov and Rjaben'kiĭ (1973), Marčuk (1973), Richtmyer and Morton (1967), Samarskiĭ (1971), Samarskiĭ and Gulin (1973), and others).

We point out the principal distinguishing feature connected with the numerical solution of problems of gas dynamics. It consists in the fact that the majority of the problems considered here are nonlinear, while the theory of difference methods was developed primarily for linear problems. Therefore, if assertions and constructions which are rigorous only for linear problems are also applied to problems of gas dynamics, it must be kept in mind that this transfer to nonlinear problems of methods applicable to linear problems is not always justified, although it is necessary.

The theory of difference methods has two basic aspects:

1) methods of constructing difference schemes;

2) justification of the difference scheme chosen, i.e., investigation of the convergence of the computational algorithm or, better, estimation of the precision of solving the problem posed.

From a practical point of view investigation of the efficiency of the algorithm with regard to the expenditure of machine time required to achieve the necessary accuracy is also very important.

We first recall a number of facts from the theory of partial differential equations and functional analysis which are needed for the subsequent exposition.

§1. The Cauchy problem in a Banach space for systems of linear differential equations

1. ***Linear operators in normed spaces.*** A finite-dimensional unitary space U_N is a complex space X_N of vectors x in which each element $x = (x_1, \ldots, x_N) \in X_N$ has a norm $\|x\|$ defined by

$$\|x\|^2 = \sum_{i=1}^{N} x_i \bar{x}_i, \tag{1}$$

where the x_i are the components of x, and $\bar{x}_i$ is the complex conjugate of x_i.

Let A be a linear operator in U_N. Its *Hermitian norm* $\|A\|$ is the supremum of the quantity $\|Ax\|/\|x\|$, where $x \neq 0$ is an element of U_N, and $\|x\|$ and $\|Ax\|$ are understood in the sense of (1).

A linear function space $X = \{u\}$ is called *normed* if for each element (function) $u \in X$ a nonnegative number $\|u\|$ is defined, called the norm of u, such that the following conditions are satisfied:

1) $\|u\| > 0$ for any nonzero element $u \in X$; the norm of the zero element is equal to zero;

2) $\|u_1 + u_2\| \leq \|u_1\| + \|u_2\|$ (the triangle inequality);

3) $\|cu\| = |c| \cdot \|u\|$.

The introduction of a norm makes it possible to define a limit process in the space X. By definition, $u = \lim_{i \to \infty} u_i$ if $\|u - u_i\| \to 0$ as $i \to \infty$, u_i, $u \in X$. A sequence $\{u_i\}$ is called a *fundamental sequence* if for $i, j \geq N$

$$\|u_i - u_j\| \leq \varepsilon(N)$$

and $\varepsilon(N) \to 0$ as $N \to \infty$. A normed space X is called *complete* or a *Banach space* if every fundamental sequence $\{u_i\}$ converges to some element $u \in X$. Henceforth if no specific norm is indicated we denote complete normed spaces by the letter B.

Let $U \subset B$ be some class of functions. We form the set $\bar{U}$ (the *closure* of U) in the following manner: $u \in \bar{U}$ if u is the limit of a sequence $\{u_i\}$, $u_i \in U$. It is clear that $U \subseteq \bar{U}$, and $\bar{U}$ can be defined as the *completion* of U by limit elements. We call a class $U \subset B$ *dense* in B if $\bar{U} = B$.

We shall consider some examples of function spaces. The linear space of functions $u(x)$ continuous on the segment $[a, b]$ together with all derivatives through order p becomes a Banach space by introducing the norm

$$\|u\| = \max_x \max\{|u(x)|, |u'(x)|, \ldots, |u^{(p)}(x)|\}, \qquad x \in [a, b].$$

We denote this space by $C_p(a, b)$. In particular, the space of continuous functions $u(x)$ with norm $\|u\| = \max_x |u(x)|$ is denoted by $C_0(a, b)$ or simply $C(a, b)$.

The space of square-summable functions on $[a, b]$ with the norm

$$\|u\|^2 = \int_a^b |u(x)|^2\,dx,$$

is denoted by $L_2(a, b)$. For the norm in L_2 we have the relation (the Bunjakovskiĭ-Schwarz inequality)

$$\|\omega\|^2 \le \|u\|\ \|v\|,$$

where $w(x) = |u(x)v(x)|^{1/2}$.

The collection of polynomials $P_n(x) = a_\alpha x^\alpha$ ($\alpha = 0, 1, \ldots, n$) is dense in $C(a, b)$ (the Weierstrass theorem). The collection of trigonometric polynomials $T_n(x) = a_\alpha e^{i\alpha x}$ ($\alpha = -n, \ldots, 0, \ldots, n$) is dense in the space $C(a, b)$ of continuous periodic functions for $b - a \le 2\pi$. The space $C(a, b)$ is dense in $L_2(a, b)$.

Let A be a linear operator defined on some dense class $U \subset B$ and taking a function $u \in U$ into a function $v \in B$. The *norm* $\|A\|$ of A is the quantity

$$\|A\| = \sup \frac{\|Au\|}{\|u\|}, \qquad u \in U,\ \|u\| \neq 0.$$

An operator A is called *bounded* if $\|A\| < \infty$, and *unbounded* otherwise. A bounded operator A possesses the following properties:

a) $Au = \lim_{i\to\infty} Au_i$, if $u = \lim_{i\to\infty} u_i$, $u, u_i \in U$.

b) If $\{u_i\}$ is a fundamental sequence, so is $\{Au_i\}$.

If A is a bounded operator, then its domain U can be extended to the entire space B (the extension of A).[1] We denote the extended operator by $\bar{A}$. By definition,

$$\bar{A}u = \lim_{i\to\infty} \bar{A}u_i, \qquad u = \lim_{i\to\infty} u_i.$$

It is not hard to show that $\bar{A}u = Au$ for $u \in U$, and $\|\bar{A}\| = \|A\|$. An example of the extension of a nonlinear operator was presented in §8 of Chapter 1. We consider the space B of Lipschitz continuous functions $u(x)$ with norm

$$\|u\| = \sup_{x, x'} \max\left\{|u(x)|, \frac{|u(x) - u(x')|}{|x - x'|}\right\}.$$

The differentiable functions $u \in C_1$ form a dense class in this space. In §8 of Chapter 1 we defined the operator S solving the Cauchy problem for quasilinear

[1] Regarding the extension of an operator, see, for example, Smirnov (1959), Ljusternik and Sobolev (1951), or Kantorovič and Akilov (1959).

equations on C_1, while the solution $u = Su_0$ is bounded in the norm of B. Therefore, the operator S can be extended to the class B of Lipschitz continuous functions. The extended operator $\bar{S}$ assigns to each $u_0 \in B$ a solution $u = \bar{S}u_0$ which is bounded in B and is a generalized solution of the system of equations.

A collection of operators A defined in B forms a linear space X_A. This space becomes a normed space if as the norm of an operator A considered as an element of X_A we take its norm as an operator in B (the *induced norm*). It is then possible to define the closeness of bounded operators, and we say that a family of bounded operators $A(\tau)$ depending on a parameter τ converges to A in the sense of the *uniform topology* if $\|A(\tau) - A\| \to 0$ as $\tau \to 0$. The operators $A(\tau)$ converge to A in the sense of the *strong topology* if $\|[A(\tau) - A]u\| \to 0$ as $\tau \to 0$ for any $u \in B$.

Finally, if a family of operators $A(\tau)$ is unbounded, then closeness of the operators $A(\tau)$ can be estimated on some function class $U \subset B$. In this case if $\|[A(\tau) - A]u\| \to 0$ as $\tau \to 0$ for any $u \in U$, then we say that the family $A(\tau)$ *approximates* the operator A on the class U, and we write $A(\tau) \sim A$. We henceforth say for brevity that the operator $A(\tau)$ depending on a parameter τ *approximates* the operator A.

We now consider some spaces connected with difference schemes. Briefly, difference methods for integrating systems of differential, integro-differential and integral equations of mathematical physics consist in going over from derivatives to difference quotients and from integrals to sums. At a practical level this means going over from an infinite-dimensional space of functions of a continuous argument to a finite-dimensional space of mesh functions and the reduction of equations for functions of a continuous argument to algebraic relations. While convenient in practice, this procedure occasions difficulties in the proof of convergence, since the mesh function and the function of continuous argument it approximates are defined in different spaces with different norms. Moreover, the norm of a mesh function depends on the parameters of the mesh and varies together with these parameters.

In a theoretical investigation it is therefore more convenient to consider difference operators in the same function space as the operators they approximate. In this procedure we assume that the difference equations are satisfied by functions of continuous argument at each point of the region in question. As we shall see, this approach is not always possible.

We illustrate both methods with a simple example. For the heat equation

$$\frac{\partial u}{\partial t} = a^2 \frac{\partial^2 u}{\partial x^2}, \qquad a = \text{const}, \tag{2}$$

we pose the mixed Cauchy problem

$$u(x, 0) = u_0(x), \qquad 0 < x < l; \tag{3}$$

$$u(0, t) = u(l, t) = 0, \qquad 0 \le t \le t_0. \tag{4}$$

We introduce a mesh in the x, t plane by setting

$$\left.\begin{aligned} x_i &= ih \quad (i = 0, 1, \ldots, N+1), \qquad N+1 = 1/h, \\ t_m &= m\tau \quad (m = 0, 1, \ldots, M), \qquad M = t_0/\tau. \end{aligned}\right\} \tag{5}$$

We define the mesh function u_i^m at points of the mesh (5), denoting by u_i^m the value of the function at the point $x_i = ih$, $t_m = m\tau$. We replace (2)–(4) by the algebraic relations

$$\frac{u_i^{m+1} - u_i^m}{\tau} = a^2 \frac{u_{i-1}^m - 2u_i^m + u_{i+1}^m}{h_2},$$

$$u_i^0 = u_0(ih), \qquad u_0^m = 0, \qquad u_{N+1}^m = 0.$$

For fixed m we take the quantities u_i^m $(i = 1, \ldots, N)$ as the components of an N-dimensional vector u^m for which it is possible to define some norm, for example,

$$\|u^m\| = \max_i |u_i^m| \qquad (i = 1, \ldots, N)$$

or

$$\|u^m\| = \sqrt{h \textstyle\sum_{i=1}^N (u_i^m)^2}.$$

Usually the mesh function u_i^m is extended by some means, for example, by interpolation, to the entire region $0 \le x \le l$, $0 \le t \le t_0$. A proof of convergence of u_i^m to $u(x, t)$ can be obtained by proving the convergence of the function $\tilde{u}(x, t)$ obtained by interpolation to $u(x, t)$ or by proving that $\{u_i^m\} \to \{u(ih, m\tau)\}$ for all $0 \le i \le l/h$, $0 \le m \le t_0/\tau$ and $\tau, h \to 0$.

It is easy to see that the solution of the mixed problem (2)–(4) reduces to solving the Cauchy problem for (2) with the initial condition

$$u(x, 0) = u_0(x), \qquad -\infty < x < \infty, \tag{6}$$

posed on the entire line $t = 0$ provided that $u_0(x)$ is periodic with period $2l$ and is odd on the segment $[-l, l]$. Indeed, in this case the initial function $u_0(x)$ and the solution $u(x, t)$ of problem (2), (6) can be represented by the series

$$u_0(x) = \sum_{k=1}^{\infty} a_k \sin \frac{k\pi x}{l}, \qquad u(x, t) = \sum_{k=1}^{\infty} a_k(t) \sin \frac{k\pi x}{l}, \qquad a_k(0) = a_k,$$

and hence the initial data (3) and the boundary conditions (4) are satisfied. Here the solution $u(x, t)$ of problem (2), (6) is a periodic function of x with period $2l$ which in the strip $0 \leq x \leq l$ coincides with the solution of the mixed problem (2)–(4). The Cauchy problem (2), (6) with a periodic function $u_0(x)$ we replaced by the difference problem

$$\frac{\tilde{u}(x, t+\tau) - \tilde{u}(x, t)}{\tau} = a^2 \frac{\tilde{u}(x+h, t) - 2\tilde{u}(x, t) + \tilde{u}(x-h, t)}{h^2}, \qquad (7)$$

$$\tilde{u}(x, 0) = u_0(x), \qquad -\infty < x < \infty, \qquad (8)$$

for the function $\tilde{u}(x, t)$.

In this approach the boundary conditions have vanished, and the initial condition (8) in combination with the difference equation (7) makes it possible to determine $\tilde{u}(x, t)$ on the lines $t = k\tau$ $(k = 1, 2, \ldots)$. Thus, for fixed $t = k\tau$ the functions $u(x, t)$ and $\tilde{u}(x, t)$ are defined on the entire line $-\infty < x < \infty$. The proof of the convergence of $\tilde{u}(x, t)$ to $u(x, t)$ reduces to the proof of convergence of functions of a single variable on the lines $t =$ const. It is clear that this approach is not possible for arbitrary boundary conditions (4).

In those cases where we must consider questions connected with boundary conditions we shall go over to mesh functions. In following subsections we shall give a comparative analysis of Cauchy problems for difference and differential equations.

To conclude this subsection we agree on the following notation. If a function (x, t) as a function of x for any fixed t belongs to a Banach space B, then we consider it a one-parameter family of elements of this space and write $u(t) \in B$. In particular, $u(t) \in C_q(a, b)$ means that, for fixed t, $u(x, t)$ has q continuous derivatives with respect to x on $[a, b]$; $u(t) \in L_2(-l, l)$ means that

$$\int_{-l}^{l} u^2(x, t)dx < \infty.$$

Sometimes for brevity we omit the domain of functions and write simply C_q, L_2, etc.

2. ***Correctness of the Cauchy problem in a Banach space for systems of linear differential equations.*** The theory of generalized solutions of differential equations has developed relatively recently—since the 1940s. Referring the reader wishing to become familiar with this question more deeply to the monographs of Sobolev (1950), Schwartz (1950), Gel'fand and Šilov (1958), and Hörmander (1963), we briefly present a more special theory which does not require the concepts of a generalized function and a generalized derivative.[2]

[2] See Hille (1948) and Richtmyer and Morton (1967).

In the strip G: $|x| < \infty$, $0 \leq t \leq \bar{t}$ we consider the system

$$\frac{\partial u}{\partial t} = L(D)u, \tag{1}$$

where $u = \{u_1(x, t), \ldots, u_n(x, t)\}$ is a vector-valued function of x and t; $L(D)$ is a matrix differential operator with coefficients that depend on x and t:

$$L(D) = A_\alpha(x, t)\, D^\alpha, \qquad \alpha = 1, \ldots, p, D = \frac{\partial}{\partial x}, \tag{2}$$

$$A_\alpha(x, t) = \|a_{ij}^\alpha(x, t)\| \qquad (i, j = 1, \ldots, n; \alpha = 1, \ldots, p). \tag{3}$$

A solution of (1) is a function $u(x, t)$ having the continuous derivatives figuring in (1) and satisfying (1). It is thus required that a solution $u(x, t)$ be differentiable with respect to t and have a continuous derivative $\partial^p u(x, t)/\partial x^p$ for any t, i.e., $u(t) \in C_p$. For the system (1) following initial conditions can be posed:

$$u(x, t_0) = u_0(x), \qquad 0 \leq t_0 \leq \bar{t}. \tag{4}$$

We assume the periodicity of the coefficients of (1) and of the initial function $u_0(x)$ in x with period $2l$. With regard to (1) we further assume that for any $t_0 \in [0, \bar{t}]$ and any function $u_0(x) \in C_q$ $(q \geq p)$ there exists a unique solution $u(x, t)$ defined in the strip $t_0 \leq t \leq \bar{t}$.

It is obvious that this solution $u(x, t)$ is a periodic function of x with the same period $2l$. Thus, consideration of a problem (1), (4) periodic in x enables us at the first stage of studying this problem to avoid complex questions of the correct formulation of the boundary conditions. More precisely, we are considering the altogether definite boundary condition given by the periodicity condition $u(x + 2l, t) \equiv u(x, t)$, for which the correctness of the formulation of problem (1), (4) can be studied rather simply, in particular, by the Fourier method.

We apply an analogous method in studying the difference analogue of problem (1), (4) in §2. By imposing the condition of periodicity in the variable x, we again postpone the consideration of the question of boundary conditions of other types, which for a difference boundary value problem is also very complicated.

The correspondence $u(t_0) \to u(t)$ $(t_0 \leq t \leq \bar{t})$, which can be written in the form

$$u(t) = S(t, t_0)u_0(x) = S(t, t_0)\, u(t_0), \tag{5}$$

defines a *transition operator* $S(t, t_0)$. If for any $u_0(x) \in C_q$ we have $u(t) \in C_q$,

$t_0 \le t \le \bar{t}$, then we say of the system (1) that it possesses the *extension property* in C_q. In this case the family of operators $S(t, t_0)$ possesses the *composition property* in C_q, i.e.,

$$S(t_2, t_0) = S(t_2, t_1)S(t_1, t_0), \qquad 0 \le t_0 \le t_1 \le t_2 \le \bar{t}, \tag{6}$$

and also the property of continuity

$$\|[S(t + \tau, t) - E]u(t)\|_B \to 0, \qquad \tau \to 0, \tag{6a}$$

The equality (6) implies that repeated application of the transition operator does not take $u(t)$ out of the space C_q. If here

$$\|S(t_2, t_1)\|_{C_q} \le M(\bar{t}), \qquad 0 \le t_1 \le t_2 \le \bar{t}, \tag{7}$$

then we call the Cauchy problem (1), (4) *correct* in C_q. If

$$\|S(t + \tau, t)\|_{C_q} \le 1 + C(\bar{t})\tau, \qquad 0 \le t < t + \tau \le \bar{t}, \tag{8}$$

then we call problem (1), (4) uniformly correct in C_q. The operator $S(t_2, t_1)$ we shall also call the *solution operator* of the Cauchy problem, and the operator $S(t + \tau, t)$ the *step operator*.

If the initial data do not belong to C_q or the system (1) does not possess the continuation property in C_q, we are faced with the necessity of defining a generalized solution, i.e., a solution belonging to a larger space than C_q.

We shall assume that there exists a Banach space B containing C_q as a dense class and such that the operator $S(t_2, t_1)$ is bounded in the norm of B on the class $u(t) \in C_q$. The operator $S(t_2, t_1)$ can then be extended to B with preservation of norm. Equality (5), in which by $S(t, t_0)$ we understand the extended operator and by $u(t_0)$ and $u(t)$ functions in B, defines a *generalized solution* $u(x, t)$ of the Cauchy problem (1), (4) which is *correct* in B if (7) is satisfied and *uniformly correct* in B if (8) is satisfied. Here, of course, the norm of C_q in (7) and (9) is replaced by the norm of B.

If the Cauchy problem admits the introduction of a norm in which the transition operator $S(t_2, t_1)$ satisfies properties (6), (6a) and (7) or (8), we say that the collection of transition operators $S(t_2, t_1)$ forms a continuous semigroup.

The semigroup property is very important. Indeed, in practice the solution $u(t)$ at the time t can be obtained from the initial data $u(t_0)$ not all at once but in steps by means of the transitions $u(t_0) \to u(t_1) \to \cdots \to u(t_m) = u(t)$. In this case we must be assured that this sequence of transitions is equivalent to the single transition $u(t_0) \to u(t)$ independently of the choice of intermediate times.

The semigroup property is especially important for difference solutions of the Cauchy problem which are obtained by stepwise advance in time with the unavoidable introduction of round-off errors at each step. The semigroup property implies the possibility of a recurrent method of obtaining the solution. The basic tendency in modern numerical algorithms is to reduce any algorithm to a collection of recurrent algorithms.

We shall illustrate the concepts we have introduced. We define the shift operator $T(h)$ by the equality

$$T(h)u(x) = u(x + h). \tag{9}$$

It is easy to see that in the space C_q and L_p of periodic functions

$$\|T(h)\| = 1. \tag{10}$$

We consider the equation

$$\frac{\partial u}{\partial t} + a\frac{\partial u}{\partial x} = 0, \qquad a = \text{const} > 0, \tag{11}$$

with the initial data

$$u(x, 0) = u_0(x). \tag{12}$$

If $u_0 \in C_1$, then the Cauchy problem (11), (12) has a solution. The transition operator can be represented in the form

$$S(t_2, t_1) = T[-a(t_2 - t_1)]. \tag{13}$$

Suppose $u_0 \in L_2$ but $u_0 \notin C_1$. Then a solution of problem (11), (12) does not exist in C_1, but the equality

$$u(t) = T(-at)u_0 \tag{14}$$

is meaningful and defines a generalized solution of the Cauchy problem (11), (12) in L_2. An analogous approach is possible also for the acoustic equations

$$\frac{\partial u}{\partial t} - a^2\frac{\partial v}{\partial x} = 0, \qquad \frac{\partial v}{\partial t} - \frac{\partial u}{\partial x} = 0, \qquad a = \text{const} > 0, \tag{15}$$

$$u(x, 0) = u_0(x), \qquad v(x, 0) = v_0(x). \tag{16}$$

If $u_0 \in C_1$ and $v_0 \in C_1$, then $u(t) \in C_1$, $v(t) \in C_1$, and the equations (15) are equivalent to the system in invariants

$$\frac{\partial r}{\partial t} + a\frac{\partial r}{\partial x} = 0, \qquad \frac{\partial s}{\partial t} - a\frac{\partial s}{\partial x} = 0, \tag{17}$$

where $r = u - av$ and $s = u + av$ are the Riemann invariants. If $u_0 \in L_2$ and $v_0 \in L_2$, then $r_0 \in L_2$ and $s_0 \in L_2$, and the equalities

$$r(t) = T(-at)r_0, \qquad s(t) = T(at)s_0 \tag{18}$$

define a generalized solution in L_2. Moreover, in the norm of L_2 we have

$$\|r(t)\| = \|r_0\|, \qquad \|s(t)\| = \|s_0\|, \tag{19}$$

if the initial functions r_0 and s_0 (u_0 and v_0) are periodic with period $2l$. If we define the norm of the vector-valued function $f = \{u, v\}$ by means of the equality

$$\|f\|^2 = \int_{-l}^{l} (u^2 + a^2v^2)\, dx, \tag{20}$$

then from (18) and (19) it follows that

$$\begin{aligned}\|f(t)\|^2 &= \frac{1}{2}\int_{-l}^{l} [r^2(x, t) + s^2(x, t)]\, dx \\ &= \frac{1}{2}\int_{-l}^{l} [r^2(x, 0) + s^2(x, 0)]\, dx = \|f(0)\|^2,\end{aligned} \tag{21}$$

i.e., the generalized solution is extendable in the norm (20). It follows from (18) and (19) that the Cauchy problem (15), (16) is uniformly correct for $0 < t < \infty$ in C_1, and from (21) that it is also uniformly correct in L_2.

It is not always possible to introduce (or even to find) a norm in which the transition operators of the Cauchy problem form a semigroup. In this case it is necessary to introduce a broader concept of correctness and define a generalized solution correspondingly.

Suppose that the operator $S(t_2, t_1)$ possesses the property that for $u(t_1) \in B_1$ the relation

$$u(t_2) = S(t_2, t_1)u(t_1), \qquad u(t_2) \in B_2,$$

holds, and

$$\|S(t_2, t_1)\| = \sup \frac{\|u(t_2)\|_{B_2}}{\|u(t_1)\|_{B_1}} = C(t_2, t_1) \leq C(\bar{t}) < \infty.$$

We then say that the Cauchy problem is correct in the sense of I. G. Petrovskiĭ. This concept will be considered in more detail in the next subsection.

3. *The Fourier method.* For a system (1.2.1) with coefficients $A_\alpha(t)$ depending only on time it is possible to effectively construct a solution operator by the Fourier method.[3] Suppose that $u_0 \in C_q$ is a periodic vector-valued function with period 2π. Then $u_0(x)$ can be represented by the Fourier series

[3] Regarding the theory of Fourier series, see the monographs of Bari (1961), Zygmund (1959), and Hardy and Rogosinski (1950).

$$u_0(x) = \sum_{k=-\infty}^{\infty} C_0(k)e^{ikx}, \tag{1}$$

where

$$C_0(k) = \frac{1}{2\pi} \int_{-\pi}^{\pi} u_0(x)e^{-ikx}\, dx, \tag{2}$$

which for $q \geq 1$ converges to $u_0(x)$ absolutely and uniformly. Assuming that q is sufficiently large, we shall seek a solution of problem (1.2.1), (1.2.4) in the form of a Fourier series

$$u(x, t) = \sum_{k=-\infty}^{\infty} C(k, t)e^{ikx}, \tag{3}$$

i.e., as a superposition of functions

$$v(x, t) = C(k, t)e^{ikx}. \tag{4}$$

These functions we shall call *harmonics*.

We shall establish a condition under which a harmonic is a solution of (1.2.1). Substituting (4) into (1.2.1), we find

$$\left[\frac{\partial}{\partial t} - L(D)\right] v = \left[\frac{\partial C}{\partial t} - L(ik, t)C\right] e^{ikx}, \tag{5}$$

where $L(ik, t)$ is the matrix

$$L(ik, t) = A_\alpha(t)(ik)^\alpha, \qquad \alpha = 1, \ldots, p. \tag{6}$$

It follows from (5) that the harmonic (4) is a solution of (1.2.1) if the vector $C(k, t)$ satisfies the differential equation

$$\frac{\partial C(k, t)}{\partial t} - L(ik, t)\, C(k, t) = 0. \tag{7}$$

Hence, in order that (3) represent a solution of the Cauchy problem (1.2.1), (1.2.4) it is necessary that $C(k, t)$ satisfy (7) for $k = 0, \pm 1, \pm 2, \ldots$ and the initial conditions

$$C(k, t_0) = C_0(k), \qquad k = 0, \pm 1, \pm 2, \ldots, \tag{8}$$

since for $t = t_0$ the series (3) must go over into (1). If (3) belongs to C_r, $r \geq p$, then it represents a solution of the Cauchy problem (1.2.1), (1.2.4).

We shall estimate the smoothness of (3) as it depends on the smoothness of the initial data. It is known that there exist no necessary and sufficient conditions that the Fourier series (3) belong to C_q, and this complicates the analysis of the extendability of a solution in C_q. It is therefore more convenient to use the concept of a generalized derivative, which we consider in the special case of Fourier series in a single variable x. If

$$u = u(x) = \sum_{k=-\infty}^{\infty} \lambda_k e^{ikx}, \qquad \lambda_{-k} = \lambda_k^* \tag{9}$$

and the condition

$$\sum_{k=-\infty}^{\infty} |\lambda_k|^2 k^{2q} < \infty \tag{10}$$

is satisfied, then, as is known, $u \in W_2^q(-\pi, \pi)$, and we say that the series $u(x)$ has a generalized derivative of order q,

$$u^{(q)}(x) = \sum_{k=-\infty}^{\infty} \lambda_k (ik)^q \, e^{ikx},$$

which is obtained by formal differentiation of (9). Here $u^{(q)} \in W_2^0(-\pi, \pi)$. For a function $u \in W_2^q$ it is possible to define a norm in the following manner:

$$\|u\|_q^2 = \frac{1}{2\pi} \int_{-\pi}^{\pi} [u^2 + (u^{(1)})^2 + \cdots + (u^{(q)})^2] \, dx$$

$$= \sum_{k=-\infty}^{\infty} |\lambda_k|^2 [1 + k^2 + \cdots + k^{2q}] \leq \text{const} \cdot \sum_{k=-\infty}^{\infty} |\lambda_k|^2 (1 + |k|^q)^2 < \infty.$$

Supose that $S(k, t_2, t_1)$ is the transition operator of the system (7) corresponding to any fixed k. We call the operator $S(k, t_2, t_1)$ the *spectral image* or the *Fourier image* of $S(t_2, t_1)$. By the definition of $S(k, t_2, t_1)$, we have

$$C(k, t_2) = S(k, t_2, t_1) C(k, t_1). \tag{11}$$

We shall assume that in the interval $[0, \bar{t}]$ the system (7) of ordinary differential equations is uniformly stable in k, i.e.,

$$\sup_k \|S(k, t_2, t_1)\| = N(t_2, t_1) < \infty, \tag{12}$$

where $\|S(k, t_2, t_1)\|$ is the Hermitian norm of $S(k, t_2, t_1)$ in the space U_n of vectors $C(k, t)$. If $u(t_0) \in W_2^q$, then by (11) and (12) we have

$$|C(k, t)| \leq N \, |C(k, 0)|,$$

from which it follows that $u(t) \in W_2^q$. Hence, there exists a generalized solution of problem (1.2.1), (1.2.4) in W_2^q for any q. In particular, for $q = 0$ we have that $W_2^0 = L_2$ and

$$\|u(t)\|^2 = 2\pi \sum_{k=-\infty}^{\infty} |C(k, t)|^2.$$

Considering (11), we now see that

$$\|S(t_2, t_1)\|_{L_2} = \sup_k \|S(k, t_2, t_1)\|_{U_n} = N(t_2, t_1) < \infty. \tag{13}$$

Hence, in L_2 there exists a generalized solution. Thus (13) is a condition for the correctness of the Cauchy problem (1.2.1), (1.2.4) in $L_2(-\pi, \pi)$. We shall give an explicit expression for the transition operator $S(t_2, t_1)$ in terms of $S(k, t_2, t_1)$. From (2) and (3) we have

$$\begin{aligned} u(x, t_2) &= \sum_{k=-\infty}^{\infty} S(k, t_2. t_1)C(k, t_1)e^{ikx} \\ &= \frac{1}{2\pi} \sum_{k=-\infty}^{\infty} S(k, t_2, t_1)\left(\int_{-\pi}^{\pi} u(s, t_1)e^{-iks}\, ds\right)e^{ikx} \\ &= \frac{1}{2\pi}\int_{-\pi}^{\pi} K(t_2, t_1, x - s)u(s, t_1)\, ds, \end{aligned} \tag{14}$$

where

$$K(t_2, t_1, x - s) = \sum_{k=-\infty}^{\infty} S(k, t_2, t_1)e^{ik(x-s)}. \tag{15}$$

In interchanging the order of integration and summation we have used the theorem on the possibility of termwise integration of a Fourier series.

Thus, in the present case the transition operator $S(t_2, t_1)$ is an integral operator of *convolution type*. We consider the special case where (1.2.1) is an equation with constant coefficients. Then the matrix $L(ik, t) = L(ik)$ does not depend on t, and the solution of problem (7), (8) has the form

$$C(k, t) = e^{L(ik)(t-t_0)}\, C_0(k). \tag{16}$$

Here $e^{L(ik)t}$ is the (exponential) matrix which can be represented by the series

$$e^{L(ik)t} = E + \sum_{m=1}^{\infty} \frac{L^m(ik)t^m}{m!}.$$

From (16) it follows that

$$S(k, t_2, t_1) = e^{L(ik)(t_2-t_1)}, \qquad S(k, t_2, t_1) = S(k, t_2 - t_1, 0).$$

For the transition operator $S(t_2, t_1)$ we obtain the estimate

$$\|S(t_2, t_1)\| = \sup_k \|e^{L(ik)(t_2-t_1)}\|_{U_n} \le \sup_k \exp(\|L(ik)(t_2 - t_1)\|_{U_n}).$$

Thus, the Cauchy problem with constant coefficients is correct in L_2 if

$$\sup_k \|e^{L(ik)(t_2-t_1)}\|_{U_n} \le N(\bar{t}), \qquad t_1 \le t_2 \le \bar{t}. \tag{17}$$

We can interpret condition (17) in the following manner. We consider a solution of (1.2.1) of the form

$$u(x, t) = u_0 e^{\omega t + ikx}, \tag{18}$$

where u_0 is a constant vector. If (18) is a solution of (1.2.1), then

$$\text{Det } \|\omega E - L(ik)\| = 0. \tag{19}$$

It is clear that for the correctness of the Cauchy problem for (1.2.1) it is necessary and sufficient that

$$\text{Re } \omega \le \mu, \tag{20}$$

where μ is a constant not depending on k. The condition (20) implies that any harmonic solution grows in amplitude no faster than $e^{\mu t}$.

We shall now define correctness in the sense of Petrovskiĭ, restricting attention to the special case in which the spaces B are the spaces W_2^q. Suppose that in place of (12) there is the weaker estimate

$$\sup_k \|S(k, t_2, t_1)\| \le N(T)(1 + |k|^q) \tag{21}$$

and q is the least integral exponent possessing this property. Then

$$|C(k, t_2)| \le N(T)(1 + |k|^q)|C(k, t_1)|. \tag{22}$$

Let

$$u(t_1) \in W_2^q. \tag{23}$$

From (22) it follows that

$$\|u(t_2)\|_{L_2} \le N(T) < \infty, \qquad u(t_2) \in W_2^0. \tag{24}$$

Conversely, (23) and (24) imply (21) and (22). Thus, (21) is a necessary and sufficient condition that the system (1.2.1) be correct in the sense of Petrovskiĭ with the mapping $B_1 \to B_2$, where $B_1 = W_2^q$ and $B_2 = W_2^0$.

We consider the system

$$\frac{\partial u}{\partial t} + A\frac{\partial u}{\partial x} = \mu \frac{\partial^2 u}{\partial x^2}, \qquad \mu \ge 0, \tag{25}$$

with a constant matrix A. If the system (25) for $\mu = 0$ is hyperbolic, then by a linear transformation in the space of the components $u_1, \ldots, u_n$ it can be brought to the canonical form

$$\frac{\partial r_j}{\partial t} + \xi_j \frac{\partial r_j}{\partial x} = \mu \frac{\partial^2 r_j}{\partial x^2}, \qquad j = 1, \ldots, n. \tag{26}$$

Fourier transformation of (26) leads to a system of equations for the Fourier coefficients $C_j(k, t)$ of the functions $r_j(x, t)$:

$$\frac{\partial C_j}{\partial t} + (i\xi_j k + \mu k^2) C_j = 0 \qquad (j = 1, \ldots, n). \tag{27}$$

Hence

$$C_j(k, t) = e^{-(\mu k^2 + i\xi_j k)t} C_j(k, 0), \tag{28}$$

$$S(k, t_2, t_1) = \|e^{-(\mu k^2 + i\xi_j k)(t_2 - t_1)} \delta_{jl}\| \qquad (j, l = 1, \ldots, n), \tag{29}$$

$$\|S(t_2, t_1)\|_{L_2} = \sup_k \|S(k, t_2, t_1)\| = \sup_k |e^{-(\mu k^2 + i\xi_j k)(t_2 - t_1)}| = 1. \tag{30}$$

The system (25) is correct in L_2 for $\mu \geq 0$.

The Fourier method is an effective method of estimating the norms of operators and the analysis of the properties of solutions in L_2. It carries over directly to the theory of difference equations with coefficients depending on t. For equations with coefficients depending on x and t the investigation of correctness becomes much more complicated.

With the example of the acoustic equations with a variable speed of sound

$$\frac{\partial u}{\partial t} - a^2 \frac{\partial v}{\partial x} = 0, \quad \frac{\partial v}{\partial t} - \frac{\partial u}{\partial x} = 0, \quad 0 < a_0 \leq a(x, t) \leq a_1 < \infty, \tag{31}$$

we shall illustrate the application of another method of studying correctness—the method of *energy inequalities* or the method of the energy integral. Multiplying the first equation of (31) by $2u$, the second by $2a^2 v$, and adding, we obtain

$$\frac{\partial u^2}{\partial t} + \frac{\partial (a^2 v^2)}{\partial t} = 2 \frac{\partial}{\partial x} (a^2 uv) + \frac{\partial a^2}{\partial t} v^2 - 2 \frac{\partial a^2}{\partial x} uv. \tag{32}$$

From this we get the inequality

$$\frac{\partial}{\partial t} (u^2 + a^2 v^2) \leq 2 \frac{\partial}{\partial x} (a^2 uv) + b(u^2 + a^2 v^2), \tag{33}$$

where

$$b = \max \left| \frac{\partial \ln a^2}{\partial t} \right| + \max \left| \frac{1}{a} \frac{\partial a^2}{\partial x} \right|. \tag{34}$$

Suppose that the functions $a(x, t)$, $u(x, t)$, and $v(x, t)$ are periodic in x with period 2π. Integrating (33) with respect to x over $[-\pi, \pi]$, we then obtain

$$\frac{\partial}{\partial t} \|f\|^2 \leq b \|f\|^2, \tag{35}$$

where

$$f(t) = \{u(t), v(t)\}, \qquad \|f\|^2 = \int_{-\pi}^{\pi} (u^2 + a^2v^2)\, dx. \tag{36}$$

From (35) it follows that

$$\|f(t)\|^2 \leq e^{bt}\|f(0)\|^2. \tag{37}$$

Thus, the Cauchy problem for (31) is correct in L_2.

The dependence of the norm (36) on t is not essential: a solution $f(t) = \{u(t), v(t)\}$ of (31) with a variable coefficient $a(x, t)$ is correct in the norm (36) with a constant a, $a_0 \leq a \leq a_1$.

To conclude this subsection we consider a *local analysis* of the correctness of equations with variable coefficients, which is also called the method of "freezing" the coefficients. With each point $P = (\tilde{x}, \tilde{t})$ we associated a system with constant coefficients

$$\frac{\partial u}{\partial t} = L(P, D)u, \qquad L(P, D) = A_\alpha(\tilde{x}, \tilde{t})D^\alpha \quad (\alpha = 1, \ldots, p), \tag{38}$$

where $\tilde{x}$ and $\tilde{t}$ are considered as parameters.

The harmonic solution (18) of the system (38) corresponding to some point $P = (\tilde{x}, \tilde{t})$ for sufficiently large k is a strongly oscillating function of x. In a sufficiently small region G_ε: $\tilde{x} - \varepsilon \leq x \leq \tilde{x} + \varepsilon$, $\tilde{t} - \varepsilon \leq t \leq \tilde{t} + \varepsilon$ the coefficients $A_i(x, t)$ of (1.2.1), which we assume to be continuous and smooth, may be regarded as approximately constant, while the harmonic solution (18) varies strongly. It is clear that the harmonic (18), as a solution of (38) with constant coefficients, is at the same time to a high degree of accuracy a solution of (1.2.1) in G_ε. Therefore, the behavior of (18) in G_ε gives some idea of the correctness properties of the system (1.2.1).

There is a conjecture, valid for many equations, that a necessary and sufficient condition for the correctness of (1.2.1) in L_2 is the correctness of the local system (38) in L_2 for any point $P = (\tilde{x}, \tilde{t})$ (the *local correctness conjecture*).

This treatment becomes more precise the larger k (high-frequency harmonics); therefore we also call the present criterion of correctenss the *asymptotic correctness criterion*.

A local analysis of the correctness of the Cauchy problem can also be applied to systems of nonlinear differential equations. However, for nonlinear equations the conjectures mentioned above are still more problematical. In view of the inapplicability in this case of the superposition principle and the absence gererally of a sufficiently broad class of special solutions from which it is possible to construct more general solutions, the investigation of correctness of various problems for systems of nonlinear equations is considerably more complicated. Establishing correctness of the Cauchy problem for systems of nonlinear differential equations reduces essentially to the proofs of the basic theorems: uniqueness and existence of a solution and its continuous dependence on the initial data of the problem (in particular, on the initial values). In Chapter 1 we studied the Cauchy problem for systems of quasilinear equations of hyperbolic type and established that correctness of this problem in C_1 holds only in a narrow strip $0 \leq t \leq t^*$, while the magnitude of t^* depends on $\|u_0'(x)\|$ and $t^* \to 0$ if $\|u_0'(x)\| \to \infty$. We thus have global incorrectness of the Cauchy problem for systems of quasilinear equations of hyperbolic type in the class C_1, i.e., for any $t > 0$.

Incorrectness of the problem in C_1 is related to the fact that it does not possess the property of extendability in C_1: for $t \geq t^*$ the solution $u(x, t) \notin C_1$. For some types of systems of quasilinear equations there is a correctness class for the Cauchy problem which is a broader class—the generalized solutions of these systems. The formulation and study of the Cauchy problem in this class is discussed in detail in Chapter 4.

Thus, the investigation of correctness on the basis of the principle of "freezing" the coefficientis and the conjecture of "local correctness" may lead, and frequently does lead, to incorrect conclusions in the case of nonlinear differential equations. For example, for the equation

$$\frac{\partial u}{\partial t} + u\frac{\partial u}{\partial x} = 0$$

"freezing" the coefficients and the study of local stability leads to an incorrect conclusion regarding the correctness of the Cauchy problem for this equation in C_q $(q \geq 1)$.

A method of investigating the stability of solutions of systems of nonlinear differential equations under small perturbations is nevertheless broadly applied in practice. We shall illustrate this method with the example of the system of quasilinear equations

$$\frac{\partial u}{\partial t} + A(u)\frac{\partial u}{\partial x} = 0. \tag{39}$$

Suppose that $u(x, t)$ is a solution of (39) and that $\bar{u}(x, t)$ is another solution which differs little from $u(x, t)$ at time t_0. Their difference $v = \bar{u} - u$ satisfies the equations

$$\frac{\partial v}{\partial t} + A(u)\frac{\partial v}{\partial x} + [A(\bar{u}) - A(u)]\frac{\partial u}{\partial x} + [A(\bar{u}) - A(u)]\frac{\partial v}{\partial x} = 0, \tag{40}$$

which we represent in the form

$$\frac{\partial v}{\partial t} + A(u)\frac{\partial v}{\partial x} + \left(B(u, \bar{u})\frac{\partial u}{\partial x}\right)v + (B(u, \bar{u})v)\frac{\partial v}{\partial x} = 0. \tag{41}$$

Here $B(u, \bar{u})$ is a matrix with three-index components which satisfies the condition

$$A(\bar{u}) - A(u) = B(u, \bar{u})(\bar{u} - u),$$

or, in terms of indices,

$$a_{ij}(\bar{u}) - a_{ij}(u) = b^{\alpha}_{ij}(\bar{u}_\alpha - u_\alpha), \qquad i, j = 1, \ldots, n;\ \alpha = 1, \ldots, n.$$

It is known that if the matrix $A(u)$ is continuously differentiable, then

$$B(u, \bar{u}) \to \dot{A}(u) = \left(\left(\frac{\partial a_{ij}(u)}{\partial u_\alpha}\right)\right) = ((b^{\alpha}_{ij}(u, u)))$$

as $|\bar{u} - u| \to 0$. Thus, for small $\|v\|$ we have for the difference $v = \bar{u} - u$ the system

$$\frac{\partial v}{\partial t} + A(u)\frac{\partial v}{\partial x} + \left(\dot{A}(u)\frac{\partial u}{\partial x}\right)v + \left(\dot{A}(u)\frac{\partial v}{\partial x}\right)v = 0, \tag{42}$$

if we neglect terms of order v^2.

Considering small perturbations such that $\|\partial v/\partial x\| \ll \|\partial u/\partial x\|$ and dropping the nonlinear terms in (42), we obtain for v a linear *system of variational equations*

$$\frac{\partial v}{\partial t} + A(u)\frac{\partial v}{\partial x} + \left(\dot{A}(u)\frac{\partial u}{\partial x}\right)v = 0. \tag{43}$$

For sufficiently small $v(x, t_0)$ and $v'_x(x, t_0)$ the system (43) approximately describes the evolution of perturbations in time in a strip $t_0 \leq t \leq t_0 + \delta$. By studying the linear system (43) we can get an idea of the stability of the solution $u(x, t)$ of the original nonlinear system (39). We note again that (43) describes the evolution of rapidly oscillating small perturbations poorly or altogether incorrectly, since in (42) we have dropped terms of order $|v| \cdot |\partial v/\partial x|$. Therefore, if, for example,

$$v(x, t_0) = \varepsilon v_0 \sin kx,$$

then even for $\varepsilon v_0 \ll |u|$ the system (43) is applicable only for $|k| \ll |\partial u/\partial x| \cdot 1/\varepsilon v_0$. If $|k\varepsilon v_0| \sim 1$, then we are again dealing with the nonlinear system (42), and there is essentially no advantage in replacing (39) for $u(x, t)$ by the approximate system (42) for $v(x, t)$,

In spite of all these imperfections in the method of linearization of nonlinear equations, it is almost the only algorithmic method affording an idea of the correctness properties of nonlinear problems and the stability of their solutions. For this reason it is widely applied in nonlinear problems, for example, in the study of the onset of turbulence in flows of a viscous fluid.

There is still another area where the study of stability by the method of linearization is completely justified and necessary. In solving nonlinear problems on a computer very small perturbations constantly enter due to round-off error; the frequency of these pulsations usually satisfies the requirement $|k\varepsilon v_0| \ll 1$. Therefore, the evolution of these perturbations can be described by linear systems of the type (43) (or their difference analogues). If the system (43) indicates no growth or weak growth of the perturbations, then this may bear witness to the stability of the method applied to compute the problem.

§2. Basic concepts of the theory of difference schemes

1. *The difference Cauchy problem.* With the Cauchy problem (1.2.1), (1.2.4) considered in §1.

$$\frac{\partial u}{\partial t} = L(D)u, \tag{1}$$

$$u(x, 0) = u_0(x); \qquad 0 \le t \le \bar{t}, \tag{2}$$

where $u = \{u_1, \ldots, u_n\}$, $D = \partial/\partial x$ and

$$L(D) = A_\alpha D^\alpha, \qquad \alpha = 1, \ldots, p,$$
$$A_\alpha = \|a_{ij}^\alpha(x, t)\| \quad (i, j = 1, \ldots, n, \alpha = 1, \ldots, p),$$

we associate the difference Cauchy problem

$$\frac{u^{m+1}(x) - u^m(x)}{\tau_{m+1}} = \Lambda_1 u^{m+1}(x) + \Lambda_0 u^m(x), \tag{3}$$

$$u^0(x) = u_0(x). \tag{4}$$

Here Λ_1 and Λ_0 are operators defined in B depending on $t_m = \tau_1 + \cdots + \tau_m$,

$\tau_{m+1} > 0$, and also on other parameters. The system (3) relating the functions u^{m+1} and u^m at the two time levels $t = t_{m+1}$ and $t = t_m$ is usually called a *two-level difference scheme*. Just as in §1, we assume that the operator $L(D)$ and the vector-valued function $u_0(x)$ are periodic in x with period $2l$. We require that the operators Λ_1 and Λ_0 be periodic in x with period $2l$; we can then consider periodic solutions of problem (3), (4) for which

$$u^m(x + 2l) \equiv u^m(x), \qquad m = 0, 1 \ldots .$$

We shall assume that (3) is uniquely solvable in B for $u^{m+1}(x)$, i.e., that the operator $B_m = E - \tau_{m+1}\Lambda_1$ is invertible in B for all values of the parameters considered. The system (3) can then be written in the form

$$u^{m+1} = C_{m+1}u^m, \qquad C_{m+1} = B_m^{-1}(E + \tau_{m+1}\Lambda_0). \tag{5}$$

From (5) it follows that

$$u^m = C_{m,k}u^k, \tag{6}$$

We call $C_{m,k}$ the *transition operator*, $C_m = C_{m,m-1}$ the *step operator*, and $C_{m,0}$ the *solution operator* of problem (3), (4).

DEFINITION. The problem (3), (4) is *correct* in B if there is a number $M(\bar{t}) > 0$ such that

$$\|C_{m,k}\|_B \leq M(\bar{t}) \tag{7}$$

for all $0 \leq k \leq m - 1$, $t_m \leq \bar{t}$ and any sufficiently small $\{\tau_k\}$; it is called *uniformly correct* in B if there is an $N > 0$ such that

$$\|C_m\|_B = \|C_{m,m-1}\|_B \leq 1 + N\tau_m \tag{8}$$

for all m and sufficiently small τ_m.

If the Cauchy problem (3), (4) is correct we call the difference scheme (3) *stable*.

DEFINITION. Problem (3), (4) *approximates* the problem (1), (2) in the space B if

$$\|(S_m - C_m)u(t_{m-1})\|_B = \tau_m\varepsilon_m(\tau_m), \tag{9}$$

where $S_m = S(t_m, t_{m-1})$ is the transition operator of the system (1), $u_0(x)$ is an arbitrary element of B, $u(t) = S(t, 0)u_0(x)$ is a generalized solution of problem (1), (2), and $\varepsilon_m(\tau) \to 0$ as $\tau \to 0$ uniformly with respect to m, i.e., $\varepsilon_m(\tau) \leq \varepsilon(\tau)$ and $\varepsilon(\tau) \to 0$ as $\tau \to 0$.

DEFINITION. We say that the difference scheme (3), (4) *converges* in B to the Cauchy problem (1), (2) if for any initial function $u_0 \in B$ the corresponding solution $u^m(x)$ of problem (3), (4) converges to the solution $u(x, t)$ of the Cauchy problem (1), (2) in the norm of B, i.e.,

$$\max_m \|u^m(x) - u(x, t_m)\|_B = \max_m \|[C_{m,0} - S(t_m, 0)]u_0(x)\|_B \to 0, \qquad 0 \leq t_m \leq \bar{t}, \quad (10)$$

for $\tau = \max \tau_m \to 0$.

We emphasize that the convergence (10) implies that for any t, $0 \leq t \leq \bar{t}$,

$$\|\tilde{u}(x, t) - u(x, t)\|_B \to 0$$

as $\tau \to 0$. Here $\tilde{u}(x, t) = u^m(x)$, where t_m is the point of the sequence $\{t_k\}$ nearest t.

First Convergence Theorem. *If* 1) *problems* (1), (2) *and* (3), (4) *are correct in B, and* 2) *problem* (3), (4) *approximates problem* (1), (2), *then the solution* $u^m(x)$ *of problem* (3), (4) *converges in B to a solution* $u(x, t)$ *of problem* (1), (2).

Proof. To estimate the quantity (10) we use the representation

$$S(t_m, 0) = S_m S_{m-1} \ldots S_1, \qquad C_{m0} = C_m C_{m-1} \ldots C_1. \quad (11)$$

This implies that

$$\begin{aligned} C_{m0} - S(t_m, 0) &= \sum_{k=1}^{m} C_m \cdots C_{k+1}(C_k - S_k)S_{k-1} \cdots S_1 \\ &= \sum_{k=1}^{m} C_{m,k}(C_k - S_k)S_{k-1,0}. \end{aligned} \quad (12)$$

Using (12), we obtain

$$\begin{aligned} \|u^m - u(t_m)\|_B &= \|[C_{m,0} - S(t_m, 0)]u_0\|_B \\ &\leq \sum_{k=1}^{m} \|C_{m,k}\|_B \|(C_k - S_k)S(t_{k-1}, 0)u_0\|_B \\ &= \sum_{k=1}^{m} \|C_{m,k}\|_B \|(C_k - S_k)u(t_{k-1})\|_B. \end{aligned} \quad (13)$$

Applying the correctness estimate (7) and the approximation estimate (9), we find

$$\|u^m - u(t_m)\|_B \leq M(\bar{t}) \sum_{k=1}^{m} \tau_k \varepsilon_k(\tau_k) \leq M(\bar{t}) t_m \max_k \varepsilon_k(\tau_k). \quad (14)$$

This implies the convergence

$$\|u^m - u(t_m)\|_B \to 0 \quad \text{as } \tau \to 0, \quad (15)$$

and the theorem is proved.

Convergence theorems of this type were formulated by Rjaben'kiĭ (1952), Meĭman (1954), and Lax and Richtmyer (1956). The scheme of proof we have presented is due to Lax and Richtmyer.

We defined approximation in terms of bounded operators S_m and C_m. In practice it is more convenient to define approximation in terms of the unbounded operators L, Λ_1 and Λ_0.[4] In this case closeness is estimated not on generalized solutions $u(t) \in B$ but on solutions $u(t) \in C_1$ of problem (1), (2).

DEFINITION. Problem (3), (4) *approximates* problem (1), (2) on the class C_q in the norm of B if for any solution $u(t) \in C_q$ of problem (1), (2)

$$\max_m \|R_{m+1}u(t_m)\|_B = \max \left\| \frac{u(t_{m+1}) - u(t_m)}{\tau_{m+1}} - \Lambda_1 u(t_{m+1}) - \Lambda_0 u(t_m) \right\|_B \to 0 \tag{16}$$

as $\tau = \max \tau_m \to 0$. The operator

$$R_{m+1} = \frac{T_0(\tau_{m+1}) - E}{\tau_{m+1}} - \Lambda_1 T_0(\tau_{m+1}) - \Lambda_0,$$

where $T_0(\tau_{m+1})u(t_m) = u(t_m + \tau_{m+1}) = u(t_{m+1})$ and its value on the solution of problem (1), (2) are called the *remainder* of the difference scheme (3).

SECOND CONVERGENCE THEOREM. *Let the following conditions hold*:

1) *The problems* (1), (2) *and* (3), (4) *are correct in B*.

2) $\|B_m^{-1}\|_B = \|(E - \tau_{m+1}\Lambda_1)^{-1}\|_B \leq N(\bar{t})$.

3) *The difference scheme* (3), (4) *approximates the problem* (1), (2) *in the sense* (16).

Then the solution of problem (3), (4) *converges to the solution* $u(t)$ *of the Cauchy problem* (1), (2) *in the norm of B if the latter belongs to the class* C_q.

PROOF. The difference $v^m = u^m - u(t_m)$ satisfies the difference equation

$$\frac{v^{m+1} - v^m}{\tau_{m+1}} = \Lambda_1 v^{m+1} + \Lambda_0 v^m - R_{m+1}u(t_m)$$

and the equation

$$v^{m+1} = C_{m+1}v^m + f_{m+1},$$

by the invertibility of B_m; here

$$f_{m+1} = -\tau_{m+1} B_m^{-1} R_{m+1} u(t_m).$$

[4] As a rule, the operator L is unbounded, while Λ_1 and Λ_0 are bounded for fixed τ and h but become unbounded as $\tau, h \to 0$.

The solution of this difference equation has the form

$$v^{m+1} = C_{m+1,0}\, v^0 + C_{m+1,\alpha} f_\alpha \qquad (\alpha = 1, 2, \dots, m+1). \tag{17}$$

This can be verified without difficulty by induction. Using condition 2) of the theorem and the approximation condition (16), we obtain a uniform estimate for f_{m+1}:

$$\| f_{m+1} \|_B \le \tau_{m+1} N(\bar{t}) \beta(\tau), \tag{18}$$

where $\beta(\tau) \to 0$ as $\tau = \max \tau_m \to 0$.

From (17), (18), the correctness condition (7) for the difference problem (3), (4), and the equality $v^0 = 0$ we now obtain an estimate for v^{m+1}:

$$\begin{aligned} \| v^{m+1} \|_B = \| C_{m+1,\alpha} f_\alpha \|_B &\le \left(\sum_{k=1}^{m+1} \tau_k \right) M(\bar{t}) N(\bar{t}) \beta(\tau) \\ &= t_{m+1} M(\bar{t}) N(\bar{t}) \beta(\tau). \end{aligned} \tag{19}$$

This estimate implies the convergence of the solution of the difference problem to the solution of the Cauchy problem (1), (2), since $\| v^{m+1} \|_B \to 0$ for any m $(t_m \le \bar{t})$ and $\tau \to 0$. It simultaneously implies the convergence of the difference scheme (3), (4) to the Cauchy problem (1), (2). The theorem is proved.

We now consider in more detail the structure of the remainder term of the difference scheme (3). Recalling that the solution $u(t) \in C_q$ of problem (1), (2) is differentiable with respect to t, we write

$$\begin{aligned} R_{m+1} u(t_m) = &\left[\frac{u(t_{m+1}) - u(t_m)}{\tau_{m+1}} - \frac{\partial u(t_m)}{\partial t} \right] \\ &+ [L(D) u(t_m) - \Lambda_1 u(t_{m+1}) - \Lambda_0 u(t_m)]. \end{aligned} \tag{20}$$

In view of the obvious estimates

$$\left.\begin{aligned} \left\| \frac{u(t_{m+1}) - u(t_m)}{\tau_{m+1}} - \frac{\partial u(t_m)}{\partial t} \right\|_B &= \alpha_1(\tau_{m+1}), \\ \| \Lambda_1 u(t_{m+1}) - \Lambda_1 u(t_m) \|_B &= \alpha_2(\tau_{m+1}), \end{aligned}\right\} \tag{21}$$

where $\alpha_i(\tau) \to 0$ as $\tau \to 0$, we see that (16) can be written in the form

$$\| (L(D) - \Lambda_1 - \Lambda_0) u(t_m) \|_B \to 0 \tag{22}$$

as $\tau_{m+1} \to 0$.

If (22) is satisfied for any function $u(t) \in C_q$, we say that the operator $\Lambda_1 + \Lambda_0$ *approximates* $L(D)$ *in the class* C_q in the norm of B as $\tau = \max \tau_m \to 0$. We denote the approximation of operators by the symbol

$$\Lambda_1 + \Lambda_0 \underset{B}{\sim} L(D). \tag{23}$$

We have thus arrived at the following theorem.

THIRD CONVERGENCE THEOREM. *Let the following conditions hold:*

1) *Problems* (1), (2) *and* (3), (4) *are correct in* B,

2) $\|B_m^{-1}\|_B = \|(E - \tau_{m+1}\Lambda_1)^{-1}\|_B \leq N(\bar{t})$.

3) $\Lambda_1 + \Lambda_0 \sim_B L(D)$.

Then the solution of the difference problem (3), (4) *converges in the norm of* B *to the solution* $u(x, t) \in C_q$ *of problem* (1), (2).

We shall now compare these three definitions of approximation. As we have just seen, the third definition (23) is equivalent to the second definition (16), and for this reason the third convergence theorem needs no separate proof.

Under the condition that the operator B_m is invertible in B the second definition of approximation (16) implies the first definition (9) in the class $C_q \subset B$ in which (16) holds. Indeed, suppose that $\|B_m^{-1}\|_B \leq N(t)$; from (16) we then have

$$\begin{aligned}\tau_{m+1}R_{m+1}u(t_m) &= B_m[u(t_{m+1}) - B_m^{-1}(E + \tau_{m+1}\Lambda_0)u(t_m)]\\ &= B_m[(S_{m+1} - C_{m+1})u(t_m)]\end{aligned}$$

for $u(t) \in C_q$. Hence

$$\begin{gathered}(S_{m+1} - C_{m+1})u(t_m) = \tau_{m+1}B_m^{-1}R_{m+1}u(t_m),\\ \|(S_{m+1} - C_{m+1})u(t_m)\|_B \leq \tau_{m+1}\, N(\bar{t})\|R_{m+1}u(t_m)\|_B\end{gathered}$$

and (9) holds, since $\|R_{m+1}u(t_m)\|_B \to 0$ uniformly with respect to m as $\tau = \max \tau_{m+1} \to 0$.

We see that when B_m is invertible in B all three definitions of approximation by the difference scheme (3) of the system of differential equations (1) are sufficient for convergence, but the simplest and most formalized verification of approximation is according to definitions (16) and (23), since it requires verification of the local approximation of the operator $L(D)$.

However, the approximation of $L(D)$ according to definitions (16) and (23) acquires practical value only after the boundedness of the family of operators B_m^{-1} is established, since without this the second and third convergence theorems do not hold. Therefore, all three definitions of approximation are equally useful in the basic problem in the theory of difference methods of solving differential equations—establishing the convergence of the difference scheme and estimating its precision.

Nevertheless, in practical work the investigation of approximation often concludes only with establishing it in the sense of the second definition (16) or, equivalently, the third definition (23). This is related to the fact that the verification of the invertibility of B_m and, more important, the uniform estimate of $\|B_m^{-1}\|_B$ constitute a considerably more difficult problem. However, if this is not done there cannot be complete confidence in the convergence of the difference scheme.

In this sense a large number of problems are solved by difference methods without complete certainty of the convergence of the difference scheme. This shortcoming is partially compensated by carrying out several computations with decreased step size τ and by comparison of results and sometimes by comparison with experiment.

Above we have considered the case of the difference scheme (3) with the variable step size τ_{m+1} in time. Although such schemes are used in practical computations, for simplicity difference schemes with constant step size τ are investigated. We shall henceforth consider difference schemes with constant step size τ almost exclusively.

Until now we have considered operators Λ_1 and Λ_0 of arbitrary structure defined in B, and the convergence theorems are essentially theorems of functional analysis. We now consider the theoretically more special case, which is in practice perhaps the most important, where Λ_1 and Λ_0 are finite-difference operators.

We introduce the linear space B_1 of functions periodic in x with period $2l$ and defined for $0 \leq t \leq \bar{t}$ which for any $t \in [0, \bar{t}]$ belong to the space B, i.e., $u(t) = u(x, t) \in B$. In this space we define the B_1-norm by

$$\|u(x, t)\|_B = \sup_t \|u(t)\|_B. \tag{24}$$

In B_1 we define the shift operator $T(h, \tau)$. The equality

$$T(h, \tau)u(x, t) = u(x + h, t + \tau) \tag{25}$$

defines $T(h, \tau)$ for $0 \leq t \leq \bar{t} - \tau$ and any x. For brevity we set

$$\begin{aligned} T(h, 0) &= T_1, \qquad T(-h, 0) = T_{-1} = T_1^{-1}, \\ T(0, \tau) &= T_0, \qquad T(0, -\tau) = T_{-0} = T_0^{-1}, \quad h, \tau > 0. \end{aligned} \tag{26}$$

The operator

$$\begin{aligned} \Lambda(\tau, h, T) = B_{\beta_0\beta_1} T_0^{\beta_0} T_1^{\beta_1}, \qquad \beta_0 &= -q_0, -q_0 + 1, \dots, q_0; \\ \beta_1 = -q_1, -q_1 + 1, \dots, q_1, & \end{aligned} \tag{27}$$

where $B_{\beta_0\beta_1}$ is a matrix in the n-dimensional space of components of the vector-valued function $u(x, t) = \{u_1(x, t), \dots, u_n(x, t)\}$, we call a *difference operator*. It is obvious that Λ is defined in B_1 for $q_0\tau \leq t \leq \bar{t} - q_0\tau$ and any x.

The family (27) of operators Λ depending on τ and h is called a *family of finitary operators* if q_0, $q_1 < Q$, where Q does not depend on τ, h. It is clear that the sum and product of such operators is again a finitary operator. However an operator inverse to a finitary operator is not in general finitary.

Let

$$\Omega = A_{\alpha_0\alpha_1} D_0^{\alpha_0} D_1^{\alpha_1}, \qquad \alpha_0 = 0, 1, \ldots, p_0; \alpha_1 = 0, 1, \ldots, p_1, \\ D_0 = \partial/\partial t, \qquad D_1 = \partial/\partial x \tag{28}$$

be a matrix differential operator. We shall define an approximation $\Lambda \sim \Omega$ of the operator (28) by the operator (27).

We consider the function

$$R(\tau, h)u = [\Lambda(\tau, h, T) - \Omega(D)]u(x, t), \tag{29}$$

where $u(x, t) \in C_{q_0 q_1} \subset B_1$ and the operators Λ, Ω and R are considered in B_1. Here $C_{q_0 q_1}$ is the space of functions $u(x, t)$ which are differentiable q_0 times with respect to t and q_1 times with respect to x, $q_0 \geq p_0$, $q_1 \geq p_1$. For brevity we shall write C_q instead of $C_{q_0 q_1}$.

If for any function $u(x, t) \in C_q \subset B_1$ as $\tau, h \to 0$

$$\|R(\tau, h)u(x, t)\|_{B_1} = O(\tau^\alpha) + O(h^\beta), \qquad \alpha, \beta > 0, \tag{30}$$

then we say that the operator Λ *absolutely approximates* the operator Ω in the norm of B_1 on the class C_q with *orders of approximation* α in τ and β in h. It follows from (30) that for a sufficiently small neighborhood $0 < \tau^2 + h^2 < \rho_0^2$, $\tau, h > 0$, of the point $\tau = 0$, $h = 0$ and any function $u(x, t) \in C_q$ there is a $C > 0$ such that

$$\|R(\tau, h)u(x, t)\|_{B_1} \leq C(\tau^\alpha + h^\beta). \tag{31}$$

If, however, (30) and (31) do not hold for arbitrary τ and h in a neighborhood of the point $\tau = 0$, $h = 0$ but for some limiting procedure $h = h(\tau)$

$$\|R(\tau, h)u(x, t)\|_{B_1} = O(\tau^\alpha), \qquad \alpha > 0, \tag{32}$$

then we say that Λ *conditionally approximates* the operator Ω with *order* of conditional approximation α in τ for the limiting procedure $h = h(\tau)$. We remark that the order of conditional approximation α depends on the limiting procedure $h = h(\tau)$. As a rule, for a conditionally approximating operator there are estimates of the form

$$\|R(\tau, h)u(x, t)\|_{B_1} = O(\tau^\alpha) + O(h^\beta) + O(\tau^\gamma h^\delta),$$

where one of the quantities γ or δ may be negative.

For analytic functions $u(x, t)$ we have the equalities

$$\left.\begin{aligned} u(x+h,t) &= T_1u(x,t) = \sum_{k=0}^{\infty} \frac{\partial u^k}{\partial x^k}(x,t)\frac{h^k}{k!} = e^{hD}u(x,t), \\ u(x,t+\tau) &= T_0u(x,t) = \sum_{k=0}^{\infty} \frac{\partial^k u}{\partial t^k}(x,t)\frac{\tau}{k!} = e^{\tau D_0}u(x,t), \end{aligned}\right\} \tag{33}$$

i.e.,

$$T_1 = e^{hD_1}, \quad T_0 = e^{\tau D_0}, \quad T_1^{\beta_1} = e^{\beta_1 hD_1}, \quad T_0^{\beta_0} = e^{\beta_0\tau D_0}, \quad T(h,\tau) = e^{hD_1+\tau D_0}. \tag{34}$$

We shall also use (33) and (34) for nonanalytic functions; in this case the series (33) are broken off after an appropriate number of terms and the remainder is estimated. The approximation $\Lambda \sim \Omega$ can then be verified in the following manner.

Formulas (34) are substituted into the expression

$$R = \Lambda - \Omega = B_{\beta_0\beta}T_0^{\beta_0}T_1^{\beta} - A_{\alpha_0\alpha_1}D_0^{\alpha_0}D_1^{\alpha_1}, \tag{35}$$

after which the exponentials in (34) are expanded in power series in the parameters τ and h. We obtain a power series in τ and h which begins with terms of order τ^α and h^β, while all other terms are of higher order of smallness in a neighborhood of $\tau = 0$, $h = 0$. It is obvious that in this case α and β are integers; they determine the order of approximation $\Lambda \sim \Omega$. Approximation holds for the class $C_{q_0q_1}$ in which the function $u(x, t)$ has the derivatives found in the coefficients of the powers τ^α and h^β and also all derivatives contained in the coefficients of lower powers of τ and h in (35) before reducing like terms.

If the limiting procedure $h = C\tau^\gamma$ ($\gamma > 0$) is known, then this expression for h is substituted into (35), after which R takes the form

$$R = \Lambda - \Omega = P(\tau, D_0, D_1). \tag{36}$$

Here $P(\tau, D_0, D_1)$ is a matrix series with fractional powers in the parameter τ and integral powers in the operators D_0 and D_1. The least power α to which τ is contained in (36) is the order of conditional approximation $\Lambda \sim \Omega$ for the limiting procedure $h = C\tau^\gamma$. We note that the order of approximation $\Lambda \sim \Omega$ depends on the class C_q of comparison functions. If $C_{\bar{q}} \subset C_q$, then the order of approximation in $C_{\bar{q}}$ is less than in C_q.

We consider a number of examples which illustrate the concepts of order of approximation and absolute and conditional approximation. We shall determine the order of approximation $\Lambda \sim \Omega$ for the operators

$$\Omega = D_0 + aD_1, \qquad \Lambda = \frac{T_0 - E}{\tau} + a\frac{E - T_{-1}}{h}, \quad a = \text{const}. \tag{37}$$

Following the general rule outlined above, we find that

$$R = \Lambda - \Omega = \frac{1}{2!}\tau D_0^2 + \frac{1}{3!}\tau^2 D_0^3 + \cdots - \frac{a}{2!}hD_1^2 + \frac{a}{3!}h^2 D_1^3 - \cdots. \tag{38}$$

It follows from (38) that on the class $C_{q_1 q_2}$ $(q_1, q_2 \geq 2)$ the approximation is absolute and has first order in both τ and h.

We consider the approximation $\Lambda \sim \Omega$ on the class U of analytic solutions of the equation $\Omega u = 0$. Here in (38) we must formally set $D_0 = -aD_1$, after where we have

$$R = \Lambda - \Omega = ah\Big[\frac{D_1^2}{2!}(\kappa - 1) - h\frac{D_1^3}{3!}(\kappa^2 - 1) + \cdots \\ \cdots + (-1)^m h^{m-2}\frac{D_1^m}{m!}(\kappa^{m-1} - 1) + \cdots\Big], \tag{39}$$

where $\kappa = a\tau/h$. For $\kappa = 1$ $(h = a\tau)$ the conditional approximation $\Lambda \sim \Omega$ on analytic functions of the equation $\Omega u = 0$ has infinite order. We remark that this is possible only for $a \geq 0$, since otherwise we obtain $h < 0$, while we have always assumed that $h > 0$. In the present case, by the way, this means only that the conditional approximation of infinite order for $a < 0$ is given by a different difference operator, namely,

$$\Lambda = \frac{T_0 - E}{\tau} + a\frac{T_1 - E}{h}, \qquad a < 0$$

for $\kappa = -a\tau/h = 1$.

For equations with constant coefficients approximation is conveniently established in the L_2 norm on trigonometric polynomials

$$P_N = \sum_{k=-N}^{N} C_k e^{\omega t + ikx}. \tag{40}$$

In this case it is convenient to go over to the *spectral images* $\bar{\Lambda}$ and $\bar{\Omega}$ of Λ and Ω, which are defined by the equalities

$$\left.\begin{aligned} \Lambda(C_k e^{\omega t + ikx}) &= (\bar{\Lambda} C_k) e^{\omega t + ikx}, \\ \Omega(C_k e^{\omega t + ikx}) &= (\bar{\Omega} C_k) e^{\omega t + ikx}. \end{aligned}\right\} \tag{41}$$

Like Λ and Ω, the matrices $\bar{\Lambda}$ and $\bar{\Omega}$ are defined in the n-dimensional space of components of $C_k = \{C_k^1, \ldots, C_k^n\}$ and are obtained from the matrix operators $\Lambda(\tau, h, T)$ and $\Omega(D_0, D_1)$ by the formal replacement

$$T_0 \to e^{\omega\tau}, \qquad T_1 \to e^{ikh}, \qquad D_0 \to \omega, \qquad D_1 \to ik. \tag{42}$$

To (35) there then corresponds the equality

$$\bar{R}(\tau, h, \omega, k) = \bar{\Lambda}(\tau, h, \omega, k) - \bar{\Omega}(\omega, k) = \Lambda(\tau, h, e^{\omega\tau}, e^{ikh}) - \Omega(\omega, ik)$$
$$= B_{\beta_0\beta_1} e^{\beta_0\omega\tau + i\beta_1 kh} - A_{\alpha_0\alpha} \omega^{\alpha_0}(ik)^{\alpha_1}. \tag{43}$$

Equation (43) is applicable to a class of functions which can be expanded in Fourier series.

Returning to the example (37), we see that infinite order of approximation $\Lambda \sim \Omega$ occurs for $\kappa = 1$ also in a class of discontinuous solutions of the equation $\Omega = 0$ which can be represented by Fourier series. It is not hard to see that this is related to the coincidence of the solution operators for the equations $\Lambda u = 0$ and $\Omega u = 0$. Indeed, for $\kappa = 1$

$$S(t + \tau, t) = C_{m+1,m} = T_1(-a\tau), \qquad t = m\tau.$$

We now present an example of a scheme with conditional approximation. With the equation

$$\Omega u = \frac{\partial u}{\partial t} + a \frac{\partial u}{\partial x} = (D_0 + aD_1)u = 0, \qquad a = \text{const}, \tag{44}$$

we associate the difference equation (the *Lax scheme*)

$$\Lambda u = \left[\frac{2T_0 - (T_1 + T_{-1})}{2\tau} + a \frac{T_1 - T_{-1}}{2h} \right] u = 0. \tag{45}$$

We represent Λ in the form

$$\Lambda = \frac{T_0 - E}{\tau} + a \frac{T_1 - T_{-1}}{2h} - \frac{h^2}{2\tau} \frac{T_1 - 2E + T_{-1}}{h^2}. \tag{46}$$

In verifying the approximation $\Lambda \sim \Omega$, we find that (45) approximates (44) for the limiting procedure $h = \text{const}\,\tau$, while for the limiting procedure $h = \mu\sqrt{2\tau}$ it approximates the parabolic equation

$$\frac{\partial u}{\partial t} + a \frac{\partial u}{\partial x} = \mu^2 \frac{\partial^2 u}{\partial x^2}, \qquad \mu = \text{const}. \tag{47}$$

This example shows that in the case of conditional approximation the difference operator may approximate different differential operators for different limiting procedures.

Figure 3.1 shows the approximation properties of the Lax scheme. When $(\tau, h) \to (0, 0)$ while moving along the lower parabola $\tau = h^2/2\mu_2^2$, the scheme (46) approximates the heat equation (47) with $\mu = \mu_2$. When (τ, h) moves along the upper parabola, the scheme (46) approximates (47) with $\mu = \mu_1$. If $(\tau, h) \to (0, 0)$ while moving along a curve alternately tangent to these parabolas, the scheme (46) does not approximate any differential equation. On the other hand,

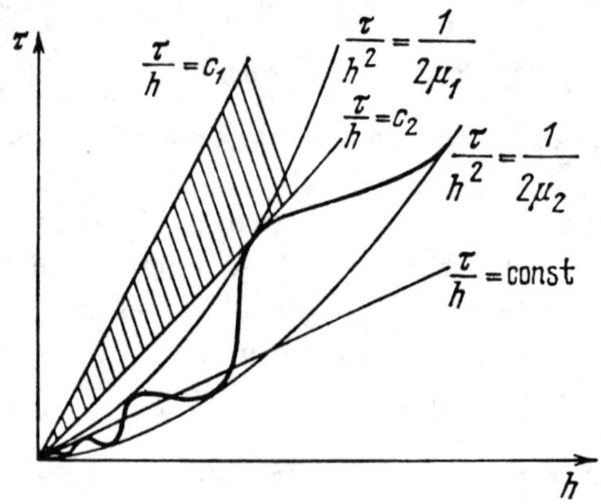

FIGURE 3.1

if (τ, h) tends to $(0, 0)$ in the sector bounded by the lines $\tau/h = C_1$ and $\tau/h = C_2$, $0 < C_1 < C_2 < \infty$, the scheme (46) approximates the hyperbolic equation (44).

In the case of finite finite-difference operators Λ_1 and Λ_0 which are represented by matrix polynomials in the shift operators T_1 and T_{-1} it is possible to introduce the concept of explicit and implicit schemes. A two-level difference scheme (3) is called *explicit* if the step operator C_{m+1} defined in (5) is finite, and *implicit* otherwise. Explicit and implicit schemes differ essentially in the methods of their realization and also in their limit properties as $\tau, h \to 0$.

2. *Dispersion analysis of a difference scheme.* In the preceding subsection for the differential operator

$$\Omega = A_{\alpha_0\alpha_1} D_0^{\alpha_0} D_1^{\alpha_1}, \qquad \alpha_0 = 0, 1, \ldots.\, p_0,\ \alpha_1 = 0, 1, \ldots, p_1, \tag{1}$$

and the difference operator

$$\Lambda = B_{\beta_0\beta_1} T_0^{\beta_0} T_1^{\beta_1}, \qquad \beta_0 = -q_0 \ldots, q_0,\ \beta_1 = -q_1, \ldots, q_1, \tag{2}$$

with constant coefficient matrices $A_{\alpha_0\alpha_1}$ and $B_{\beta_0\beta_1}$ we introduced their spectral images—the matrices $\bar{\Omega}$ and $\bar{\Lambda}$ defined by

$$\bar{\Omega} = A_{\alpha_0\alpha_1} \omega^{\alpha_0} (ik)^{\alpha_1}, \qquad \bar{\Lambda} = B_{\beta_0\beta_1} e^{\beta_0 \omega\tau + ik\beta_1 h}. \tag{3}$$

In order that a harmonic $u = u_0 e^{\omega t + ikx}$ be a solution of the equation

$$\Omega u = 0 \tag{4}$$

or

$$\Lambda u = 0, \tag{5}$$

the quantities ω and k must satisfy the condition

$$\mathrm{Det}\,\|\bar{\Omega}(\omega, k)\| = 0 \tag{6}$$

or

$$\mathrm{Det}\,\|\bar{\Lambda}(\tau, h, \omega, k)\| = 0 \tag{7}$$

respectively. Equations (6) and (7) are called the *dispersion relations* of equations (4) and (5) respectively; solutions $\omega = \omega(k)$ and $\omega = \omega(\tau, h, k)$ of these equations are called *dispersion relations.* The relations $\omega = \omega(k)$ and $\omega = \omega(\tau, h, k)$ provide complete information on the properties of periodic solutions of (4) and (5) respectively.

Thus, the condition of the approximation $\Lambda \sim \Omega$ of a differential operator by a difference operator has the form

$$\omega(\tau, h, k) \to \omega(k) \tag{8}$$

as $\tau, h \to 0$ for any fixed value of k. If the limit (8) occurs as τ and h tend to zero independently, then the approximation is absolute, while if it requires a relation $h = h(\tau)$, then it is conditional.[5]

Correctness questions for equations (4) and (5) depend, of course, on the formulation of additional conditions which uniquely determine solutions of them. For example, these may be initial conditions for $t = 0$ consistent with the operators (1) and (2). If these complex questions concerning correct formulation of initial conditions are not considered and it is simply assumed that everything has been done correctly, then it is only possible to assert that for the stable behavior of solutions of (4) and (5) it is necessary to require that

$$\mathrm{Re}\,\omega(k) \le \mu_1, \qquad \mathrm{Re}\,\omega(\tau, h, k) \le \mu_2, \tag{9}$$

where the constants μ_1 and μ_2 do not depend on k, τ or h.

Therefore, it is expedient to consider the question of applying spectral images of operators to the investigation of correctness and stability of differential and difference problems for the more special case of the Cauchy problems (2.1.1), (2.1.2) and (2.1.3), (2.1.4); in this case

$$\Omega = ED_0 - L(D_1) = ED_0 - A_\alpha D_1^\alpha, \tag{10}$$

[5] The convergence (8) must, of course, occur for all roots of (6). Therefore, a correspondence between the roots of (6) and (7) must be established. In particular, the question of "extraneous" roots of (7) arises here. It is necessary to require that for "extraneous" roots $\mathrm{Re}\,\omega(\tau, h, k) \to -\infty$ as $\tau, h \to 0$.

$$\begin{aligned}\tau\Lambda &= [ET_0 - \tau T_0\Lambda_1(T_1)] - [E + \tau\Lambda_0(T_1)] \\ &= [ET_0 - \tau B^1_\beta T_0 T^\beta_1] - [E + \tau B^0_\beta T^\beta_1],\end{aligned} \tag{11}$$

where E is the identity matrix, A_α, B^1_β and B^0_β are constant $n \times n$ matrices, and $u = \{u_1, \ldots, u_n\}$.

In this connection we remark that many well-posed problems of mathematical physics for equations (4) and (5) with operators of the type (1) and (2) reduce to the Cauchy problem for operators of the type (10) and (11) by raising the dimension n of the space of solutions. We have in mind the reduction of systems of equations with derivatives in t of higher order to a system of equations of first order in t by introducing new variables. For example, by introducing the new variables $\bar{u} = \{\bar{u}_1, \bar{u}_2\} = \{\partial u/\partial x, \partial u/\partial t\}$ the wave equation $u_{tt} = u_{xx}$ can be reduced to the system of equations

$$\frac{\partial \bar{u}_1}{\partial t} = \frac{\partial \bar{u}_2}{\partial x}, \qquad \frac{\partial \bar{u}_2}{\partial t} = \frac{\partial \bar{u}_1}{\partial x},$$

of first order in the variable t.

In the case of the operators (10) and (11) we consider the Cauchy problem and use the definitions of correctness and stability presented in subsection 1 of this section. With this specialization of the operators Ω and Λ there is no question of "extraneous" roots $\omega(\tau, h, k)$: the orders of the characteristic equations (6) and (7) for ω and $\exp(\omega\tau)$ are the same and equal to n.

A necessary condition for correctness of the Cauchy problem for (4) with an operator Ω of the form (10) is that the first condition of (9) be satisfied for all k; in the same way, a necessary condition for the correctness (stability) of the difference problem (5) with an operator Λ of the form (11) is that the second condition of (9) be satisfied for all k with a constant μ_2 not depending on τ or h in the case of unconditional stability and depending on the limiting procedure $h = h(\tau)$ but bounded as $\tau \to 0$ in the case of conditional convergence.

We consider the question of applying the dispersion relations to the investigation of the stability of the difference Cauchy problem for equation (5) under condition (11) in more detail. To the step operator $(E - \tau\Lambda_1)^{-1}(E + \tau\Lambda_0)$ of the difference equation (5) the Fourier transform assigns a step operator $(E - \tau\bar{\Lambda}_1)^{-1}(E + \tau\bar{\Lambda}_0)$ according to the difference scheme for the Fourier coefficients $\bar{u}^{m+1}(k)$:

$$\frac{\bar{u}^{m+1}(k) - \bar{u}^m(k)}{\tau} = \bar{\Lambda}_1\bar{u}^{m+1}(k) + \bar{\Lambda}_0\bar{u}^m(k). \tag{12}$$

Here $\bar{\Lambda}_1$ and $\bar{\Lambda}_0$ are the spectral images of the operators Λ_1 and Λ_0:

$$\bar{\Lambda}_1(\tau, h, k) = B^1_\beta e^{ik\beta h}, \qquad \bar{\Lambda}_0(\tau, h, k) = B^0_\beta e^{ik\beta h}; \tag{13}$$

B_β^1 and B_β^0 may depend on τ and h, while $\bar{u}^{m+1}$ is the Fourier coefficient of the solution $u^{m+1}(x)$ which is periodic in x:

$$u^{m+1}(x) = \sum_{k=-\infty}^{+\infty} \bar{u}^{m+1}(x)e^{ikx}. \tag{14}$$

In analogy to (1.3.13) we have

$$\|C_{m,l}\|_{L_2} = \sup_k \|\bar{C}_{m,l}(k)\|_{U_n}. \tag{15}$$

Here $C_{m,l}$ is the transition operator from the level l ($t = t_l$) to the level m ($t = t_m$) of the difference Cauchy problem, and $\bar{C}_{m,l}$, is the transition operator for the same levels of the system of difference equations (12).

If the step operator $\bar{C}_{m+1,m}(k)$ of the system (12) admits the estimate

$$\sup_k \|\bar{C}_{m+1,m}(k)\|_{U_n} \leq 1 + C(\bar{t})\tau, \tag{16}$$

then by (15) the difference scheme (5), (11) is stable in L_2. Thus, (15) reduces the test of stability of the difference scheme considered to the purely algebraic problem of determining the norm of the matrix $\bar{C}_{m+1,m}(k)$.

It is known that the spectral radius of a matrix, i.e., the maximum of the modulus of its characteristic roots, does not exceed the norm of the matrix. From this we obtain a necessary criterion for stability (the *von Neumann criterion*) of the difference scheme (5), (11): in order that the scheme (5), (11) be uniformly stable it is *necessary* that the spectral radius R_λ of the transition matrix $\bar{C}_{m+1,m}(k)$ of the system (12) admit the estimate

$$R_\lambda(\bar{C}_{m+1,m}(k)) \leq 1 + M(\bar{t})\tau, \tag{17}$$

which is valid for all k, $\tau \leq \tau_0$ and other parameters (for example, h) of the scheme.

In the case where $\bar{C}_{m+1,m}(k)$ is a normal matrix, i.e., it commutes with its adjoint, its norm coincides with the spectral radius, and the criterion (17) expresses the necessary and sufficient condition for the uniform stability of the scheme (5), (11). A detailed analysis of different estimates of matrix norms can be found in the book Richtmyer and Morton (1967).

If, however, $\bar{C}_{m+1,m}(k)$ is not a normal matrix, then condition (17) is not sufficient, and there are well-known examples (see, for example, Godunov and Rjaben'kiĭ (1973)) where condition (17) is satisfied, but the scheme is instable.

We shall consider the application of the dispersion relations in the case where the matrix $\bar{C}_{m+1,m}(k)$ is normal. We consider only the more customary difference schemes of the type

$$\frac{u^{m+1}(x) - u^m(x)}{\tau} = \gamma\Lambda_0 u^{m+1}(x) + (1 - \gamma)\Lambda_0 u^m(x), \tag{18}$$

where

$$\Lambda_0 = B^0_\beta T^\beta_1 \tag{19}$$

is the operator approximating the difference operator $L(D_1) = A_\alpha D^\alpha$. In this case it is obvious that $\Lambda(\tau, h) \sim \Omega$. For the scheme (18) the dispersion relation gives

$$e^{\omega_i \tau} = \frac{1 + (1 - \gamma)\tau\lambda_i(h, kh)}{1 - \gamma\tau\lambda_i(h, kh)}, \qquad i = 1, 2, \ldots, n, \tag{20}$$

where $\omega_i = \omega_i(\tau, h, k)$, and $\lambda_i = \lambda_i(h, kh)$ are the eigenvalues of the matrix

$$\bar{B}(h, kh) = B^0_\beta e^{i\beta kh} = \sum_{\beta=-q_1}^{q_1} B^0_\beta e^{i\beta kh}. \tag{21}$$

(For simplicity we assume that the matrices B^0_β depend only on h and not on τ.)

For well-posed problems of mathematical physics, so-called *evolution* problems, the operator $L(D_1)$ is such that the eigenvalues $\lambda_i(k)$ of its spectral matrix $\bar{L}(ik)$ satisfy the estimate

$$\operatorname{Re} \lambda_i(k) \leq \rho < \infty, \tag{22}$$

for $i = 1, \ldots, n$ and any real k. It can be shown that (22) is valid not only for systems which are correct in B but also for systems correct in the sense of Petrovskiĭ.

The estimate (22) implies that all harmonics (any k) have bounded growth in time; even high-frequency harmonics where $k \to \infty$ do not grow strongly (in the majority of problems they are strongly damped, i.e., $\operatorname{Re}\lambda_i(k) \to -\infty$ as $|k| \to \infty$).

Since $\gamma\Lambda_0 + (1 - \gamma)\Lambda_0 \sim L(D_0)$, it follows, in particular, that (8) is satisfied. Therefore, it is natural to require that the operator Λ_0 also have an analogous property: the eigenvalues $\lambda_i(h, kh)$ of the matrix $\bar{\Lambda}_0(h, kh) = \bar{B}(h, kh)$ corresponding to it (the latter is given by (21)) should have real parts that are bounded above:

$$\operatorname{Re} \lambda_i(h, kh) \leq \rho_1 < \infty. \tag{23}$$

Under our assumptions the matrix

$$\bar{C}_{m+1,m}(\tau, h, k) = [E - \gamma\tau\bar{B}(h, kh)]^{-1}[E + (1 - \gamma)\tau\bar{B}(h, kh)] \tag{24}$$

has the quantities $e^{\omega_i\tau}$ given by (20) as its eigenvalues. Therefore, if the matrix

$\bar{B}(h, kh)$ is normal, then a necessary and sufficient condition for the uniform stability of the difference scheme (18) can be written in the form

$$\sup_k \max_i \left| \frac{1 + (1 - \gamma)\tau\lambda_i(h, kh)}{1 - \gamma\tau\lambda_i(h, kh)} \right| \leq 1 + M(\bar{t})\tau, \tag{25}$$

where the constant $M(\bar{t})$ does not depend on τ. If this requires a particular limiting procedure $h = h(\tau)$, then the uniform stability of (18) has conditional character; otherwise, the stability is unconditional.

If we consider the special but frequently encountered case where $\lambda_i(h, kh) \leq 0$, then it follows from (25) that the implicit scheme is absolutely stable for $\gamma \geq 1/2$, which cannot be asserted, in general, in the case $\gamma = 0$ (the explicit scheme). In subsection 3 the approximation and stability properties of implicit and explicit schemes will be compared.

We remark that the wave number k is contained in the matrix $\bar{B}(h, kh)$ only as integral powers of the exponential e^{ikh}. This implies that $\bar{B}(h, kh)$ is periodic in the parameter k with period $2\pi/h$. For this reason it suffices to determine the supremum in (25) on the segment $|k| \leq \pi/h$ of the wave numbers k.

The periodicity of $\bar{B}(h, kh)$ has a very simple explanation: the difference scheme (18) with an operator Λ_0 of the form (19) "does not distinguish" harmonics with a wave length $\lambda = 2\pi/k < h$. On a mesh with step size h in the variable x the function e^{ikx} for $|k| > 2\pi/h$ coincides with the function $e^{i\bar{k}x}$, where $|\bar{k}| = |k \pm 2\pi n/h| \leq \pi/h$, n an integer. It is therefore said that the difference operator Λ_0 "allows" the maximal wave number $2k_0 = 2\pi/h$. In the case of real matrices B_β^0, which we consider the basic case, $\bar{B}(h, -kh) = [\bar{B}(h, kh)]^*$. Therefore, $\lambda(h, -kh) = \lambda^*(h, kh)$ and $\omega(\tau, h, -k) = \omega^*(\tau, h, k)$. Hence, the behavior of the harmonics $u_0 e^{\omega t + ikx}$ for any k is completely characterized in this case by their behavior for $k \in [0, k_0]$, $k_0 = \pi/h$. We call the segment $[0, k_0]$ of wave numbers the *spectrum of the difference operator*.

We remark that here if for the differential and difference problems we pose the periodicity condition $u(x - l) \equiv u(x + l)$, then the *spectrum of the difference boundary value problem* consists of points $k_n = \pi n/l \in [0, k_0]$, $n = 0, \pm 1, \ldots$. If we consider the difference Cauchy problem on the infinite line $-\infty < x < \infty$ with the condition of boundedness of the solution at infinity, then the harmonic $u_0 e^{\omega t + ikx}$ is a solution for any real k. Therefore, the spectrum of this difference Cauchy problem coincides with the spectrum of Λ_0; it is continuous and consists of the segment $[0, k_0]$ of wave numbers.

We remark also that in speaking of the spectrum of a difference problem not only the admissible set of wave numbers k but also the dispersion relation $\omega = \omega(\tau, h, k)$ is often implied.

The segment $[0, k_0]$ of the spectrum of the difference operator has three characteristic portions: the segment of small k: $0 \leq k \leq k_1$, the segment of intermediate k: $k_1 \leq k \leq k_2$, and the *end of the spectrum*—the portion $k_2 \leq k \leq k_0$. Although the description has qualitative character, it is possible to imagine that $k_1 = k_0/3$ and $k_2 = 2k_0/3$.

The segment of small k corresponds to small derivatives of the harmonics $e^{\omega t + ikx}$ with respect to x, and it correspondingly described properties of the difference scheme on solutions with small derivatives u_x, u_{xx}, u_{xxx}, etc. On this segment the spectral image $\bar{R}(\tau, h, k)$ of the remainder $\Lambda_0 - L(D_1)$ is usually small, which bears witness to good approximation of the differential operator $L(D_1)$ by the difference operator Λ_0 on slowly varying functions. Namely, in this region we have the convergence.[6]

$$\lambda_i(h, kh) \to \lambda_i(k) \tag{26}$$

as $h \to 0$; more precisely, in this part of the spectrum the relations (26) are violated with acceptable accuracy. For these reasons it is natural to call this portion of the spectrum the *approximation portion.*

In the region $k_1 \leq k \leq k_2$ of intermediate wave numbers the derivatives of the harmonics (and correspondingly the solutions of the difference and differential equations they describe) become considerable. For this reason the remainder $\Lambda_0 - LD$ is not small, and it is not possible to speak of a good approximation $\Lambda_0 \sim L(D_1)$ for these types of oscillations. The limit relation (26) is not satisfied by a wide margin. However, most frequently in cases of practical importance the spectral function $\lambda_i = \lambda_i(h, kh)$ of the difference problem still qualitatively follows the behavior of the spectral function $\lambda_i(k)$ of the differential problem, for example, the decay of Re $\lambda_i(k)$ with increasing k.

Finally, on the end of the spectrum $[k_2, k_0]$ the derivatives of the solutions described by this portion are so large that it is not possible to speak of any approximation $\Lambda_0 \sim L(D_1)$, and (26) does not hold here. In principle this portion of the spectrum describes a difference operator Λ_0 which here has nothing in common with the differential operator $L(D_1)$.

It is interesting to note that on the end of the spectrum even the basic qualitative features of the behavior of the spectral function $\lambda_i(k)$ of the differential problem. For example, it is possible that Re $\lambda_i(h, kh)$ in this part of the spectrum may become an increasing function of k and surpass Re $\lambda_i(h, kh)$ in the approximation portion, i.e., for $k \sim 0$; the quantity Re $\lambda_i(h, kh)$ may become positive for $k \sim k_0$ and may even tend to $+\infty$ as $h \to 0$. In the latter case the difference scheme (18) becomes absolutely (for arbitrarily small step size τ) unstable and

---

[6] Strictly speaking, the convergence (26) does not occur in any finite part $[0, k_0]$ of the spectrum. The convergence (26) holds in any fixed segment $[0, \bar{k}]$, where $\bar{k}$ does not depend on h.

is unsuitable even for rough computations; this follows easily from a careful consideration of (20).

In this subsection we shall present the simplest example of a difference operator approximating the differential operator D_1^2 which has an anomalous behavior on the end of the spectrum. An example of anomalous behavior of the spectrum in more real cases is presented in the work of Gary and Helgason (1970); roots $\lambda_i = \lambda_i(h, kh)$ for which Re $\lambda_i \to +\infty$ as $h \to 0$ are called there "spurious" roots. Roždestvenskiĭ, Ermakova and Priĭmak (1977) presented new examples of such behavior of the spectrum of a difference operator, and constructed the elements of a theory of spurious roots.

It must be noted that such anomalous behavior of the spectrum of a difference problem in which rough estimates are not admitted does not always occur. Until now it has been observed only in nonselfadjoint problems with variable coefficients; sometimes the reason for these anomalies is a difference approximation of boundary conditions of high order.

We thus see that the condition (23) ensuring for $\gamma \geq 0.5$ the uniform stability of the difference scheme (18) does not follow from the approximation $\Lambda_0 \sim L(D)$, but must be provided by the choice of the appropriate operator Λ_0 from the family of operators approximating $L(D_1)$.

The computer usually strives to raise the accuracy of the approximation on the segment $[0, k_1]$ of the spectrum and to preserve the qualitative behavior of the spectral function $\lambda_i(h, kh)$ on the remainder of the spectrum. In a number of simple problems it is possible to do this, but for the involved cases indicated above these two desiderata are contradictory. In raising the accuracy of the approximation for small k, we apply a multipoint spatial approximation for which the behavior of the spectrum at the end becomes more complicated and may become undesirable in the sense indicated.

We remark that the detection of anomalies at the end of the spectrum is complicated by the fact that for large $k \sim 1/h$ it is not possible to use the estimates of the eigenvalues of the matrix $\bar{B}$ that are usually valid for small k. Therefore, knowledge of the entire spectrum of the difference operator is required to establish the estimate (23), and we must solve the *complete eigenvalue problem* for the spectral matrix $\bar{B}(h, kh)$.

We shall illustrate the concepts introduced with the simple example of a difference approximation for the heat equation. In this case

$$L(D_1) = D_1^2, \qquad \Omega = D_0 - D_1^2. \tag{27}$$

We consider the simplest approximation

$$\Lambda_0(h) = [T_1(-h) - 2E + T_1(h)]\frac{1}{h^2} \sim D_1^2. \tag{28}$$

This corresponds to the difference scheme

$$\frac{u^{m+1} - u^m}{\tau} = \gamma \frac{u^{m+1}(x-h) - 2u^{m+1}(x) + u^{m+1}(x+h)}{h^2} + (1-\gamma)\frac{u^m(x-h) - 2u^m(x) + u^m(x+h)}{h^2}. \tag{29}$$

The operator $\Lambda_0(h)$ has the spectral function

$$\lambda(h, kh) = -\frac{4}{h^2}\sin^2\frac{kh}{2}, \tag{30}$$

which should be compared with the spectral function $\lambda(k) = -k^2$ of the operator D_1^2. Expanding $\lambda(h, kh)$ in a series of powers of h, we find that

$$\lambda(h, kh) = -k^2 + O(k^4h^2) = \lambda(k) + O(k^4h^2).$$

For small k in the approximation region we see that $\lambda(h, kh)$ and $\lambda(k)$ are close to one another, and there is close behavior in time of the harmonics $e^{\omega t + ikx}$. For $k = \pi/3h = k_0/3 = k_1$ we have

$$\lambda(h, k_1h) = -\frac{1}{h^2}, \qquad \lambda(k_1) = -\frac{\pi^2}{9h^2} \approx -1.0966\frac{1}{h^2}.$$

This agreement is not bad at all.

However, the function $\lambda(h, kh)$ decreases monotonically with increasing k, is periodic in k with period $2\pi/h$, and preserves the main property of $\lambda(k)$—nonnegativity and monotonic decrease with increasing k. At the end of the spectrum for $k = k_0 = \pi/h$ we have

$$\lambda(h, k_0h) = -\frac{4}{h^2}, \qquad \lambda(k_0) = -\frac{\pi^2}{h^2} \simeq \frac{-9{,}78}{h^2};$$

that is, $|\lambda(h, k_0h)|$ is almost 2.5 times less than $|\lambda(k_0)|$.

We see that for the approximation (29) there are no spurious roots; for the selfadjoint operator $L(D_1)$ we have constructed a selfadjoint difference operator[7] Λ_0 which reflects well the qualitative features of the entire spectrum of the operator $L(D_1)$ and approximates the spectrum of $L(D_1)$ on the approximation portion very well (to second order in h).

Since $\lambda(h, kh) < 0$, $0 \leq \gamma \leq 1$, from (20) we conclude that $0 < e^{\omega\tau} \leq 1$ for $\gamma \geq 1/2$ and any τ and h, while $-1 \leq e^{\omega\tau} \leq 1$ for $\gamma = 0$ if

[7] Selfadjointness of the operators $L(D_1)$ and Λ_0 under the condition of periodicity in x is intended.

$$\frac{2\tau}{h^2} = \frac{\tau}{2}|\lambda(h, k_0h)| \leq 1.$$

Since for the approximation (29) the matrix $\bar{B}(h, kh)$ is normal, from our estimates we get estimates for the norm of the transition matrix $C_{m+1,m}$ of the difference scheme (29):

$$\|C_{m,m+1}\|_{L_2} \leq 1 \quad \text{for } \gamma \geq 0.5 \text{ and any } \tau, h;$$
$$\|C_{m,m+1}\|_{L_2} \leq 1 \quad \text{for } \gamma = 0 \text{ and } \tau \leq h^2/2.$$

In the first case we have unconditional stability of the implicit ($\gamma \geq 1/2$) difference scheme (29); in the second case we have conditional stability for $\tau \leq h^2/2$ of the explicit scheme (29).

In a similar way from (20) we derive estimates of stability of the difference scheme in more general cases if the matrix $\bar{B}(h, kh)$ is normal.

We now consider another approximation of the operator D_1^2:

$$\Lambda_0(h) = [E - 2T_1(h) + T_1(2h)]\frac{1}{h^2} \sim D_1^2,$$

or

$$\Lambda_0(h)u(x) = \frac{u(x) - 2u(x+h) + u(x+2h)}{h^2}, \tag{31}$$

For this operator

$$\begin{aligned}\lambda(h, kh) &= -\frac{4e^{ikh}}{h^2}\sin^2\frac{kh}{2}\\ &= -\frac{4}{h^2}\left[\cos kh \sin^2\frac{kh}{2} + i\sin kh \sin^2\frac{kh}{2}\right]\\ &= -k^2 - ik^3h + O(k^4h^2).\end{aligned}$$

It is easy to see that the difference operator (31) has anomalous behavior at the end of the spectrum; in particular, $\operatorname{Re}\lambda(h, kh) > 0$ for $kh > \pi/2$ and $\operatorname{Re}\lambda(h, k_0h) = \operatorname{Re}\lambda(h, \pi) = 4/h^2 \to +\infty$ as $h \to 0$ (Figure 3.2).

It is clear, of course, that the approximation (31) is not good; it is unsymmetric, has only first order of accuracy in h, and is inconvenient for the formulation of the boundary condition. It is therefore not surprising that the difference scheme (18) using the operator (31) is unstable.

We have presented this example only to demonstrate in the simplest form the anomalies at the end of the spectrum of a difference operator. In complicated problems with variable coefficients and involved boundary conditions anomalies at the end of the spectrum may arise even with the apparently most natural approximations of the problem. This naturalness may lead to a situation where the person doing the computation does not even suspect that he is solving a

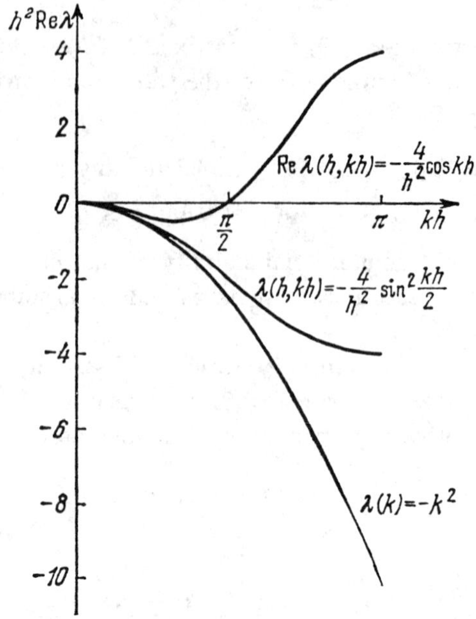

FIGURE 3.2

problem on the basis of an absolutely unstable difference scheme (in this regard see Roždestvenskiĭ, Ermakova and Priĭmak (1977), and Hačaturov (1977)).

We shall consider the application of dispersion relations to the study of the properties of a three-level explicit difference scheme of Du Fort-Frankel type (see Richtmyer and Morton (1967)). This example affords us the possibility of illustrating the effect of "extraneous" roots of the dispersion equation and also phenomena at the end of the spectrum reminiscent of the effect of spurious roots.

For the heat equation we consider the Du Fort-Frankel scheme

$$\begin{aligned}\frac{u^{m+1}(x)-u^{m-1}(x)}{2\tau} &= \frac{u^m(x+h)-u^{m+1}(x)-u^{m-1}(x)+u^m(x-h)}{h^2}\\ &= \frac{u^m(x+h)-2u^m(x)+u^m(x-h)}{h^2}\\ &\quad -\frac{\tau^2}{h^2}\,\frac{u^{m+1}(x)-2u^m(x)+u^{m-1}(x)}{\tau^2},\end{aligned} \tag{32}$$

or

$$\frac{u^{m+1}(x)-u^{m-1}(x)}{2\tau}=\Lambda_0(h)u^m(x)-\frac{\tau^2}{h^2}\frac{u^{m+1}(x)-2u^m(x)+u^{m-1}(x)}{\tau^2}. \tag{33}$$

where $\Lambda_0(h)$ is the operator (28) approximating the operator D_1^2.

The dispersion equation of the scheme (33) has the form

$$\frac{\rho^2-1}{2\tau}=\lambda_0(h,\,kh)\,\rho-\frac{(\rho-1)^2}{h^2},\qquad \rho=e^{\tau\omega(\tau,\,h,\,k)}, \tag{34}$$

where $\lambda_0(h,\,kh)$ is the spectral function of $\Lambda_0(h)$. The equation (34) is quadratic in ρ; it has two roots ρ_1 and ρ_2 and correspondingly two dispersion relations $\omega=\omega_1(\tau,\,h,\,k)$ and $\omega=\omega_2(\tau,\,h,\,k)$. We remark that for the differential equation $D_0u=D_1^2u$ we have only the one root $\omega=-k^2$.[8]

We write out the formulas for ρ_1 and ρ_2:

$$\rho_{1,2}=\frac{\kappa+\tau\lambda_0\pm\sqrt{1+2\kappa\tau\lambda_0+\tau^2\lambda_0^2}}{1+\kappa}, \tag{35}$$

$$\kappa=2\tau/h^2,\qquad \lambda_0=\lambda_0(h,\,kh). \tag{36}$$

From (35), (36) and (30) it follows that the Du Fort-Frankel scheme (33) provides conditional approximation of the heat equation, since a limiting procedure for which

$$\kappa\tau=\frac{2\tau^2}{h^2}\to 0 \tag{37}$$

as $\tau,\,h\to 0$ is required for the approximation. From

$$\rho_{1,2}=\frac{\kappa\cos kh\pm\sqrt{1-\kappa^2\sin kh}}{1+\kappa} \tag{38}$$

it follows that $|\rho_1|\cdot|\rho_2|\le 1$ for any k and $\kappa>0$, which implies the absolute stability of the difference scheme (32). If kh is small and (37) is satisfied, then

$$\rho_2=\frac{\kappa+\tau\lambda_0-\sqrt{1+2\kappa\tau\lambda_0+\tau^2\lambda_0^2}}{1+\kappa}\simeq-\frac{(1-\kappa)}{1+\kappa}(1+\tau\lambda_0).$$

We see that this root has nothing in common with $e^{\lambda\tau}$, but $|\rho_2|<1-\varepsilon$, $\varepsilon<0$, for sufficiently small τ, and therefore

$$\operatorname{Re}\,\omega_2(\tau,\,h,\,k)\to-\infty \tag{39}$$

as $\tau,\,h\to 0$ while observing (37).

[8] We see that the scheme (32) has an "extraneous" root of the dispersion equation. The question of extraneous roots obviously always arises when the order $2q_0$ of the difference equation in the variable t is greater than the order in t of the differential equation it approximates.

The asymptotic behavior (39) of the extraneous root ω_2 bears witness to the fact that the harmonics corresponding to it are damped very rapidly in time for small step size τ and may be neglected, since precisely these harmonics correspond to the root ω_1, and they are moderately damped. This example shows that for extraneous roots of the dispersion equation it suffices to require that a condition of the type (39) be satisfied.

However, we note one unpleasant circumstance that reduces the value of stability of the scheme (33) for any κ. This is that for any κ and $kh = \pi$ according to (38) we have $\rho_2 = -1$. This means that on the end of the spectrum for $k = k_0 = \pi/h$ we have Re $\omega_2 = 0$, i.e., the harmonics $e^{i\pi x/h}$ are not damped in time but rather only change sign at each step. It is especially unpleasant that this circumstance does not depend on the step size τ; reduction of τ does not eliminate these pulsations. This means that the Du Fort-Frankel scheme (33) for $k \sim k_0$, i.e., at the end of the spectrum, is at the limit of stability, and small perturbations related, for example, to boundary conditions or variable coefficients suffice to make it absolutely unstable.

Indeed, if the operator Λ_0 for some $k \in [0, k_0]$ has an eigenvalue such that

$$\lambda_0(h, kh) < -4/h^2, \tag{40}$$

then by (35) we have

$$\rho_2(\tau, h, k) = \frac{\tau[2/h^2 + \lambda_0] - \sqrt{1 + \tau^2\lambda_0(4/h^2 + \lambda_0)}}{1 + \kappa} < -1.$$

Indeed, $\tau(2/h^2 + \lambda_0) < -2\tau/h^2 < -\kappa$, $\lambda_0(4/h^2 + \lambda_0) > 0$, $\sqrt{1 + \tau^2\lambda_0(4/h^2 + \lambda_0)} > 1$, etc. This is also evident from the expansion of ρ_2 for small τ.

$$\rho_2(\tau, h, k) \simeq \frac{-1 - \tau\left[-\lambda_0 - \frac{2}{h^2}\right] - \frac{\tau^2\lambda_0}{2}\left(\frac{4}{h^2} + \lambda_0\right)}{1 + \kappa},$$

from which it follows that

$$\omega_2(\tau, h, k) \simeq \frac{i\pi}{\tau} + \left[-\lambda_0(h, kh) - \frac{4}{h^2}\right], \tag{41}$$

i.e., Re $\omega_2(\tau, h, k) > 0$.

Thus, if the conditions (40) are satisfied, harmonics $e^{ikx+\omega_2 t}$ arise which increase at the rate $e^{\alpha t}$, where

$$\alpha = -[\lambda_0(h, kh) + 4/h^2] > 0,$$

and these change sign at each step in time.[9] Sometimes the quantity α may be of order $1/h$, i.e., very large, and the scheme becomes absolutely unstable.

We arrive at the following conclusion: if the operator Λ_0 becomes extremely dissipative (condition (40) is satisfied), then the Du Fort-Frankel scheme becomes absolutely unstable.

It is not hard to construct an example (see Hačaturov (1977)) where an approximation of a correct boundary value problem for the heat equation leads to the condition (40) being satisfied at the end of the spectrum of Λ_0.

3. ***Approximation viscosity and the first differential approximation of a difference scheme.*** The dispersion relations for difference schemes assist in understanding the properties of these schemes and the specific effects they contribute. It is possible to classify difference schemes on the basis of dispersion relations.

We consider a system of linear equations of hyperbolic type

$$\frac{\partial u}{\partial t} + A\frac{\partial u}{\partial x} = 0 \tag{1}$$

with a real, constant matrix A having distinct, real eigenvalues $\xi_1, \ldots, \xi_n$. The dispersion equation of the system (1) has the form

$$(\omega + i\xi_1 k)(\omega + i\xi_2 k) \cdots (\omega + i\xi_n k) = 0, \tag{2}$$

whence we conclude that the harmonics $u_0 e^{\omega t + ikx}$ corresponding to the roots $\xi_1, \ldots, \xi_n$ are undamped in time, since

$$\operatorname{Re} \omega_i(k) = 0, \qquad |\rho_i| = 1, \qquad \rho_i = e^{\omega_i \tau}, \quad i = 1, \ldots, n. \tag{3}$$

For (1) we consider the difference scheme

$$\Lambda u = \left(\frac{T_0 - E}{\tau} + \Lambda_1 T_0 + \Lambda_0\right) u = 0. \tag{4}$$

We say that the scheme (4) possesses *approximation viscosity* if

$$\left.\begin{array}{ll} |\rho| < 1 & \text{for } k \neq 0, \\ |\rho| = 1 & \text{for } k = 0,\ \rho = e^{\omega(\tau, h, k)\tau}, \end{array}\right\} \tag{5}$$

where $\omega(\tau, h, k)$ is the solution of the dispersion equation for the scheme (4) introduced in the preceding subsection.

We see that for schemes possessing approximation viscosity the amplitude of the harmonics is damped in time. Solutions of differential equations of

[9] The same occurs also in the case of complex $\lambda_0(h, kh)$ if the generalization $\operatorname{Re} \lambda_0(h, kh) < -4/h^2$ of (40) is satisfied.

parabolic type—for example, the heat and diffusion equations—possess an analogous property.

Further classification of difference schemes for systems of equations of hyperbolic type based on dispersion relations is possible. Thus, for example, the difference scheme (4) is called *dissipative of order* $2d$ if for all $k \in [0, k_0]$ and sufficiently small $\tau \leq \tau_0$ there is a $\delta < 0$ such that $|\rho| < 1 - \delta\,|kh|^{2d}$.

For the single linear equation

$$\frac{\partial u}{\partial t} + a\frac{\partial u}{\partial x} = 0, \qquad a > 0, \tag{6}$$

we consider the *running-account difference scheme*

$$\frac{u^{m+1} - u^m}{\tau} + a\frac{\Delta_{-1}u^m}{h} = 0, \tag{7}$$

where $\Delta_{-1} = E - T_{-1}$ and $\Delta_1 = T_1 - E$.

The dispersion equation of the scheme (7) has the form

$$\rho - 1 + \frac{a\tau}{h}(1 - e^{-ikh}) = 0. \tag{8}$$

If the *Courant condition* $\kappa = a\tau/h \leq 1$ is satisfied, it follows from (8) that $|\rho| \leq 1$ for all k. Thus the scheme (7) is conditionally stable for $\kappa \leq 1$. For $\kappa < 1$ conditions (5) are satisfied, and hence the scheme (7) possesses approximation viscosity. However, for $\kappa = 1$ it follows from (8) that $\rho = e^{-ikh}$, $\omega = -ika$, and hence the scheme (7) in this case does not possess approximation viscosity. We recall (see subsection 1) that the scheme (7) for $\kappa = 1$ possesses infinite order of approximation. This, by the way, also follows from the dispersion equation, since $\omega(\tau, h, k) = -ika$ (for $|k| \leq \pi/h$).

The scheme (7) possesses the property of monotonicity (see Godunov (1959)) for $\kappa \leq 1$, which implies that a monotone profile $u^m(x)$ again goes over into a monotone profile $u^{m+1}(x)$. Indeed, if $\Delta_{-1}u^m(x) \leq 0$ for all x, then from (7) we have

$$u^{m+1}(x) = (1 - \kappa)u^m(x) + \kappa u^m(x - h),$$
$$\Delta_{-1}u^{m+1}(x) = (1 - \kappa)\Delta_{-1}u^m(x) + \kappa\Delta_{-1}u^m(x - h) \leq 0.$$

The presence of approximation viscosity for the scheme (7) is expressed in the fact that the profile of $u^m(x)$ smooths out with increasing time (increasing m), in contrast to the profile of the solution $u(x, t) = u_0(x - at)$ of the differential equation (6) which maintains its form unchanged while advancing to the right. This is explained by the fact that the high-frequency harmonics corresponding to large values of k are damped out faster than the low-frequency harmonics for $\kappa < 1$.

It is possible to observe how the scheme (7) "smears out" the discontinuity in the initial data

$$u^0(x) = \begin{cases} 1, & x \le 0, \\ 0, & x > 0. \end{cases} \tag{9}$$

Problem (7), (9) can be solved analytically. From (7) we have

$$\begin{aligned} u^{m+1} &= Cu^m, \qquad C = \alpha E + \beta T_{-1}, \qquad \alpha = 1 - \kappa, \qquad \beta = \kappa, \\ u^m &= C^m u^0 = (\alpha E + \beta T_{-1})^m u^0 = \sum_{k=0}^{m} \binom{m}{k} \alpha^{m-k} \beta^k T_{-k} u^0. \end{aligned} \tag{10}$$

It follows that $u^m(x) = 1$ for $x \le 0$, $1 \ge u^m(x) \ge 0$ for $0 \le x \le mh$, and $u^m(x) = 0$ for $x \ge mh$. An interesting feature is that $u^m(x)$ is a discontinuous function of x; namely, $u^m(x)$ is constant for $(k-1)h \le x \le kh$, and each point $x = kh$ for $k \ge 0$ is a point of discontinuity of $u^m(x)$. This is connected with the fact that we have considered the Cauchy problem (7), (9) for the difference scheme (7) for all x, and the initial function is discontinuous. It is possible to consider the problem (7), (9) only for the points $x = kh$; we then have a discrete collection of values u_k^m which provide a picture of the smooth variation of u_k^m. Figure 3.3 shows the graph of $u^m(x)$ (fine line) and the smoothed profile of u_k^m (heavy line). Among other things, the fact that $u^m(x)$ has disconti-

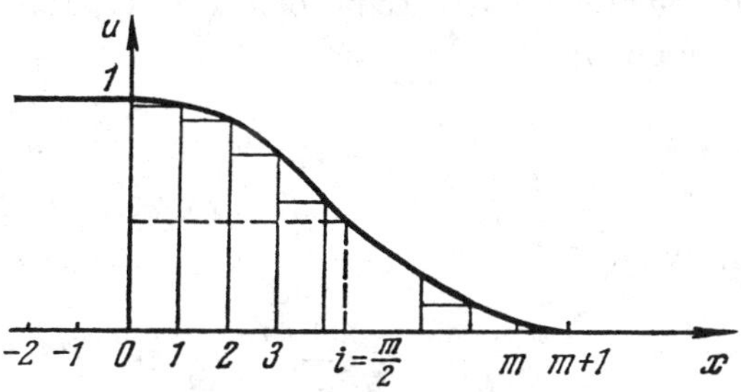

FIGURE 3.3

nuities (although they decrease in amplitude with increasing m) implies that the concept of approximation viscosity reflects only general features of the behavior of solutions of difference equations.

We note, for example, that for $\kappa = 1/2$, $\alpha = \beta = 1/2$ and for an even number of steps $m = 2n$ the mean value $u = 1/2$ is located at the point $x = mh/2$, and the profile is symmetric relative to the point $x = mh/2$, $u = 1/2$. We see from Figure 3.3 that the mean value $u = 1/2$ moves over the mesh with speed

a, while the profile is smoothed symmetrically relative to the mean point. We have here an analogy with the smoothing of an initial discontinuity in the Cauchy problem for the equation

$$\frac{\partial u}{\partial t} + a\frac{\partial u}{\partial x} = \mu\frac{\partial^2 u}{\partial x^2}.$$

We obtain a still more illustrative example of the effect of the approximation viscosity of the scheme (7) by considering for (7) the Cauchy problem

$$u^0(x) = \begin{cases} \frac{1}{h} & \text{for } |x| \le \frac{h}{2}, \\ 0 & \text{for } |x| > \frac{h}{2} \end{cases} \tag{11}$$

(see Godunov (1959) and Collatz (1951), where this problem is used to analyze the propagation of errors in solutions of difference equations).

We compute the value of the solution $u_i^m = u^m(ih)$ at the points $x = ih$ by (10). We have

$$u_i^m = \frac{1}{h}\sum_{k=0}^{m}\binom{m}{k}\alpha^{m-k}\beta^k\delta_{i-k}^0 = \frac{1}{h}\binom{m}{i}\alpha^{m-i}\beta^i. \tag{12}$$

The profile of u_i^m has the form of the binomal distribution law, and the maximum of the profile is displaced with speed a.

Using the analogy with probability theory,[10] it is possible to study the limit behavior of the solution of the difference problem (7), (11) as $m \to \infty$. Applying Stirling's formula, we obtain

$$\binom{m}{i}\alpha^i\beta^{m-i} = \frac{1}{\sqrt{2\pi m\alpha\beta}}\exp\left(-\frac{1}{2}\left(\frac{i - m\alpha}{\sqrt{m\,\alpha\beta}}\right)\right)^2 + \theta\frac{C_1}{m}, \tag{13}$$

where C_1 is a quantity depending on α but not on i, and $|\theta| < 1$. For large m

$$\begin{aligned} u_i^m &\approx \frac{1}{h\sqrt{2\pi m\alpha\beta}}\exp\left(-\frac{1}{2}\left(\frac{m-i-m\alpha}{\sqrt{m\alpha\beta}}\right)\right)^2 \\ &= \frac{1}{h\sqrt{2\pi}}\frac{1}{\sqrt{m\alpha\beta}}\exp\left(-\frac{1}{2}\left(\frac{m\beta - i}{\sqrt{m\alpha\beta}}\right)\right)^2. \end{aligned} \tag{14}$$

We introduce the quantities

$$U_i^m = u_i^m\sqrt{m\alpha\beta}, \qquad y = \frac{i - m\alpha}{\sqrt{m\alpha\beta}},$$

[10] A. I. Žukov (1959) pointed out the analogy of the asymptotic properties of difference solutions with the limit theorems of probability theory. Our analysis mainly follows his work.

Then $U^{(m)}(y) \to e^{-y^2/2}/\sqrt{2\pi h}$ if $m \to \infty$, and i varies in such a way that y is finite.

We shall show that the asymptotic behavior of u^m is determined by the spectral properties of the step operator $C = \alpha E + \beta T_{-1}$ of the difference scheme (7). We denote by $S = T_1(-a\tau)$ the step operator for the Cauchy problem (6), (11), by $u^m(x)$ the solution (10) of the Cauchy problem (7), (11), and by $\bar{u}^m(x) = \bar{u}(m\tau, x)$ the solution of problem (6), (11). Then

$$u^m = C^m u^0, \qquad \bar{u}^m = S^m u^0,$$
$$u^m = C^m S^{-m} \bar{u}^m = (CS^{-1})^m \bar{u}^m.$$

The operator characterizes the deviation at one step of the difference solution from the exact solution.

We go over to the spectral images of the operators C, S and CS^{-1}. The coefficients of multiplication $\rho = e^{\omega\tau}$ corresponding to these operators have the form

$$\rho_C = (1 - \kappa) + \kappa e^{-ikh}, \qquad \rho_S = e^{-ia\tau k},$$
$$\rho_{CS^{-1}} = \rho_C \rho_S^{-1} = (1 - \kappa + \kappa e^{-ikh}) e^{ia\tau k}$$
$$= 1 - \frac{1-\kappa}{\kappa} \frac{a^2\tau^2}{2} k^2 + O(k^3\tau^3). \tag{15}$$

We now consider the asymptotic properties of the deviation operators $(CS^{-1})^m$. If k is fixed, then as $m \to \infty$ and $\tau \to 0$, $m\tau = t$, we have

$$\rho_{(CS^{-1})^m} = \left[1 - \frac{1-\kappa}{\kappa} \frac{a^2\tau^2}{2} k^2 + O(\tau^3 k^3)\right]^m \to e^{-b^2k^2t},$$
$$b^2 = \frac{1-\kappa}{\kappa} \frac{a^2\tau}{2}. \tag{16}$$

Thus, the solution operator $\Sigma(t)$ of the difference equation (7) has the asymptotic representation $\Sigma(t) = \Omega(t)S(t)$, where $S(t)$ is the solution operator of (6) and is obviously equal to $T_1(-at)$, and $\Omega(t)$ is the deviation operator with spectral image corresponding to (16), i.e.,

$$e^{\omega(k)t} = e^{-b^2k^2t} \quad \text{and} \quad \omega(k) = -b^2k^2. \tag{17}$$

This implies that for Σ the spectral function has the form

$$\omega_\Sigma(k) = -iak - b^2k^2. \tag{18}$$

It is evident from (18) that $\Sigma(t)$ coincides with the solution operator of the parabolic equation

$$\frac{\partial u}{\partial t} + a \frac{\partial u}{\partial x} = b^2 \frac{\partial^2 u}{\partial x^2}. \tag{19}$$

Indeed, the solution of (19) with the initial condition (11) is given by

$$u(x, t) \approx \frac{1}{2\sqrt{\pi b^2 t}} e^{-(x-at)^2/(4b^2t)}. \tag{20}$$

As $h \to 0$, (20) becomes exact.

Setting $i = x/h$ and $m = t/\tau$, we see that (14) and (20) coincide. Thus, we see again that (20) describes the asymptotic behavior of solutions of (7) as $\tau, h \to 0$, $\kappa \leq 1$.

We shall indicate a formal method of obtaining (19) from (7). Expanding the difference equation (7)

$$\left(\frac{T_0 - E}{\tau} + a\,\frac{E - T_{-1}}{h}\right)u = \left(\frac{e^{\tau D_0} - E}{\tau} + a\,\frac{E - e^{-hD_1}}{h}\right)u = 0$$

in a series in powers of the parameters τ and h, we obtain

$$\left(D_0 + aD_1 + \frac{1}{2}\tau D_0^2 + \frac{1}{2}ahD_1^2 + \cdots\right)u = 0. \tag{21}$$

Dropping higher order terms in (21) and then setting $D_0^2 = a^2 D_1^2$, we obtain

$$\frac{\partial u}{\partial t} + a\,\frac{\partial u}{\partial x} = \frac{1}{2}ah(1 - \kappa)\frac{\partial^2 u}{\partial x^2}, \tag{22}$$

which coincides with (19). We call (22) the *first differential approximation* of the difference equation (7).

It is easy to see that (7) approximates (22) on solutions $u \in C_3$ up to quantities of order τ^2 and h^2. Therefore, up to quantities of second order it is possible to say that the difference scheme (7) "adds" to (6) the viscosity $b^2(\partial^2 u/\partial x^2)$. This technique of obtaining the first differential approximation of a difference scheme is due to A. I. Žukov.

In recent years the concept of the first differential approximation of a difference scheme has been studied in detail, and it has found a number of applications in the study of properties of difference schemes mainly for the case of hyperbolic equations (see, for example, Janenko and Šokin (1968), Šokin (1973, 1976), and Kuznecov (1972)).

Here we shall briefly consider methods of studying and applying the first and subsequent differential approximations of the explicit difference scheme

$$\frac{u^{m+1}(x) - u^m(x)}{\tau} = \Lambda_0 u^m(x). \tag{23}$$

Suppose that Λ_0 does not depend on x and is an analytic function of $\tau > 0$, $\Lambda_0 = \Lambda_0(\tau)$. For the system (23) we pose the initial condition

$$u^0(x) = u_0(x). \tag{24}$$

The solution of the Cauchy problem (23), (24) has the form

$$u^m(x) = \tilde{u}(t, x) = (E + \tau\Lambda_0)^{t/\tau} u_0(x); \tag{25}$$

here $t = m\tau$. We extend the representation of the difference solution $\tilde{u}(t, x)$ in the form (25) to any t and $\tau > 0$; this means that we apply it also for t that are not multiples of τ. Differentiating (25) with respect to t, we obtain

$$\frac{d\tilde{u}(t, x)}{dt} = \frac{\ln[E + \tau\Lambda_0]}{\tau} \tilde{u}(t, x) \tag{26}$$

which is a system of differential equations that describe the behavior of the solution of (23) in terms of the function $\tilde{u}(t, x)$ of the continuous argument t. We remark that $\tilde{u}(t, x)$, defined as a solution of (26) with the initial condition $u(0, x) = u_0(x)$ following from (24), coincides with $u^m(x)$ at the points $t = m\tau$, i.e., the solutions $u_\tau^m(x)$ of (23), (24) and $\tilde{u}_\tau(t, x)$ of (26) are connected by the conditions

$$u_\tau^m(x) = \tilde{u}_\tau(m\tau, x). \tag{27}$$

We note that formally writing (26) in the form

$$\frac{d\tilde{u}(t, x)}{dt} = \left[\Lambda_0 + \sum_{n=1}^{\infty}(-1)^n \frac{\tau^n}{n+1} \Lambda_0^{n+1}\right]\tilde{u}, \tag{28}$$

by expanding the right side of (26) in powers of τ, we arrive at one of the possible formal representations of the difference scheme (23). Equation (27) or (38) is sometimes called the Π-form of the differential representation of the difference scheme (23).

By leaving several of the leading terms on the right side of (28) as $\tau \to 0$, we obtain differential approximations of the difference scheme (23) of various orders. Here we must consider the dependence of Λ_0 on the parameter τ. In order that these differential approximations describe the behavior of the difference solution $\tilde{u}(t, x)$ as $\tau \to 0$ it is necessary that the series on the right side of (28) converge in some neighborhood of $\tau = 0$.

We shall illustrate obtaining differential approximations with several examples. Suppose that Λ_0 is bounded and does not depend on τ. Then the series (28) converges for $\tau\|\Lambda_0\| < 1$ and diverges for $\tau\|\Lambda_0\| > 1$. Here the norm is the Hermitian norm of Λ_0 in the space of components of the vector $u = \{u_1, \ldots, u_n\}$ at the points $\{x_k\}$ which are related by the difference scheme (23). In this case (23) approximates a system of ordinary differential equations

$$\frac{d\bar{u}(t, x)}{dt} = \Lambda_0 \bar{u}(t, x), \tag{29}$$

which in the general case represents an infinite system of equations. In the case of periodic boundary conditions it reduces to a finite system.

The equations (29) constitute a differential approximation of the difference scheme of order zero; leaving on the right side of (28) all terms of kth order in τ, we obtain the differential approximation of the difference scheme (23) of order k.

It is obvious that for any τ such that $\tau\|\Lambda_0\| < 1$ the differential approximations describe the behavior of solutions of the difference scheme more precisely the higher their order. For example, in the case $\Lambda_0 = \alpha E$ there is the first differential approximation

$$\frac{d\bar{u}}{dt} = \alpha\left(1 - \frac{\alpha\tau}{2}\right)\bar{u},$$

from which it is evident that the approximation (23) of the equation $u_t = \alpha u$ reduces the rate of growth of $|u|$ for $\alpha > 0$ and raises the rate of decrease of $|u|$ for $\alpha < 0$.

Suppose now that Λ_0 depends on τ. For convenience of interpretation we write this dependence in the form $\Lambda_0 = \Lambda_0(h\tau))$. If

$$\tau\|\Lambda_0(h(\tau))\| < 1 \tag{30}$$

then what was said above holds regarding the description of difference solutions by means of solutions of differential approximations of kth order

$$\frac{\partial\bar{u}}{\partial t} = \left[\Lambda_0 + \sum_{n=1}^{k}(-1)^n \frac{\tau^n}{n+1} \Lambda_0^{n+1}(h(\tau))\right]\bar{u}, \tag{31}$$

i.e., the higher the order k of the differential approximation, the more precisely its solutions describe the solutions of the difference scheme (23). We point out that we have not expanded $\Lambda_0(h(\tau))$ in the parameter τ.

We observe that (30) is similar to the usual conditions for stability of the difference scheme (23) but nevertheless does not coincide with them. For example, for the scheme (23) with $\Lambda_0(h) = h^{-2}[T_1(h) - 2E + T_{-1}(h)]$ approximating the heat equation (30) leads to a restriction to a step size $\tau < h^2/4$, while the stability condition of the explicit scheme admits twice the step size, $\tau < h^2/2$.

There is, however, no paradox in all this. It is simply that for $h^2/2 \geq \tau > h^2/4$ there exist solutions of the difference equation (23) with behavior which cannot be described by any differential equation of finite order. These solutions correspond to pulsating harmonics at each step in t which are damped out with increasing time t.

We consider still another example, the difference

$$\frac{u^{m+1}(x) - u^m(x)}{\tau} = a\frac{T_{-1}(h) - E}{h}u^m, \qquad a > 0. \tag{32}$$

Here

$$\Lambda_0 = a\frac{T_{-1}(h) - E}{h}, \qquad \|\Lambda_0\| = \frac{2a}{h}$$

and condition (30) means that $2\tau a/h = 2\kappa < 1$. This is again twice as strong a restriction on the step size τ than required by the stability condition ($\kappa \leq 1$), and this is again explained by the fact that the equation of first order (31) cannot describe the solutions of variable sign at each step in t of the difference scheme (32), and it has such solutions for $\kappa > 1/2$.

The above examples show that the differential approximations (31) of the difference scheme may not describe the behavior of all solutions of the difference scheme (23), and for this reason they may create a wrong impression of the stability properties of the difference scheme—properties inherent in the difference scheme itself and related to the entire spectrum of the difference scheme. We note, however, that solutions of (31) may satisfactorily reflect the behavior of solutions of the difference scheme (23) even if condition (30) is violated, provided that $\tau\|\Lambda_0(h(\tau))u_0(x)\| < 1$ and the difference scheme (23) is stable.

In a similar way we construct differential approximations for the implicit difference scheme

$$\frac{u^{m+1}(x) - u^m(x)}{\tau} = \Lambda_0 u^{m+1}(x).$$

In this case (26) is replaced by

$$\frac{d\tilde{u}(t, x)}{dt} = -\frac{\ln(E - \tau\Lambda_0)}{\tau}\tilde{u}(t, x)$$

and (28) is replaced by

$$\frac{d\tilde{u}(t, x)}{dt} = \left(\Lambda_0 + \sum_{n=1}^{\infty}\frac{\tau^n}{n+1}\Lambda_0^{u+1}\right)\tilde{u}.$$

It is further possible to use the approximation property $\Lambda_0(h) \sim L(D)$ to transform (28). If

$$\Lambda_0(h) = L(D) + \sum_{n=1}^{\infty} h^n L_n(D), \tag{33}$$

then we can substitute this representation of $\Lambda_0(h)$ into the right side of (28) and obtain a new representation of the difference scheme (23) in the form of a differential equation (28) with a right side containing an operator series in powers of τ and h.

This equation is a formal version of the scheme (23) which is, however, valid if (30) is satisfied. The condition (30) actually restricts the limiting procedure, since from it we obtain the restriction $\tau < 1/\|\Lambda_0(h)\|$, a condition under which the difference scheme (23) is asymptotically equivalent to the equation of infinite order (in x) (28) as $\tau \to 0$. The condition (30) indicates which terms in (33) should be considered when substituting (33) into (31) if we restrict attention to the powers τ^k. Thus, we obtain the first differential approximation as a partial differential equation if in (31) we set $k = 1$ and consider the corresponding terms in h according to (33). For the scheme (32) we have the first differential approximation

$$\frac{\partial u}{\partial t} + a\frac{\partial u}{\partial x} = \frac{1}{2}ah(1-\kappa)\frac{\partial^2 u}{\partial x^2}, \qquad \kappa = \frac{a\tau}{h}, \tag{34}$$

which coincides with (22).

It is clear from what has been said that differential approximations of a difference scheme cannot, in general, determine its stability or instability conditions. Nevertheless, for a number of schemes approximating equations of hyperbolic type it has been shown that the stability conditions of the difference scheme coincide with the correctness condition in the Cauchy problem for its first differential approximation (see Šokin (1973)).

Thus, for the system of equations (1) with a symmetric matrix A we consider the simple difference scheme

$$u^{m+1}(x) = \sum_{\alpha=1}^{2} B_\alpha u^m(x + \tau\lambda_\alpha), \tag{35}$$

with a first differential approximation having the form

$$\frac{\partial u}{\partial t} + A\frac{\partial u}{\partial x} = C_2\frac{\partial^2 u}{\partial x_2}, \tag{36}$$

where

$$C_2 = \frac{\tau}{2}\left[\sum_{\alpha=1}^{2} \lambda_\alpha^2 B_\alpha - A^2\right],$$

$$\sum_{\alpha=1}^{2} B_\alpha = E, \qquad \sum_{\alpha=1}^{2} \lambda_\alpha B_\alpha = -A.$$

THEOREM. *For the stability of the difference scheme* (35) *it is necessary and sufficient that the Cauchy problem for its first differential approximation be correct.*

PROOF. If the difference scheme is stable, then the von Neumann condition is satisfied, i.e.,

$$|\rho_j|^2 = 1 - 4\mu_j(1 - \mu_j)\sin^2\frac{k\tau}{2}(\lambda_1 - \lambda_2) \leq 1 \qquad (j = 1, \ldots, n), \tag{37}$$

where $\rho_1, \ldots, \rho_n$ are the eigenvalues of the transition matrix of the scheme

$$G = B_1 e^{ik\tau\lambda_1} + (E - B_1)e^{ik\tau\lambda_2},$$

and μ_j are the eigenvalues of the matrix B_1 ($j = 1, \ldots, n$). From (37) it then follows that $0 \leq \mu_j \leq 1$, and hence the eigenvalues of the matrix $B_2 = E - B_1$ are between 0 and 1. In the present case

$$C_2 = \frac{\tau}{2}(\lambda_1 - \lambda_2)^2 B_1 B_2 = \frac{\tau}{2}(\lambda_1 - \lambda_2)^2 B_1(E - B_1). \tag{38}$$

Therefore, the eigenvalues of the matrix C_2 are nonnegative, and the Cauchy problem for the system (36) is correct in L_2. Conversely, if $C_2 \geq 0$, it follows from (38) that $B_\alpha \geq 0$ ($\alpha = 1, 2$), and the scheme (35) is stable by the theorem of Friedrichs (1954) (see §3).

For the system (1) we consider the majorant scheme

$$u^{m+1}(x) = \sum_{\alpha=-1}^{1} B_\alpha u^m(x + \alpha h), \tag{39}$$

where

$$B_1 = +\kappa A^+, \qquad B_{-1} = -\kappa A^-, \qquad A^+ + A^- = -A,$$
$$A^+ \geq 0, \qquad A^- \leq 0, \qquad B_0 = E - \kappa(A^+ - A^-).$$

Its first differential approximation has the form (36), where

$$C_2 = \frac{h^2}{2\tau}(E - B_0)B_0.$$

In this case there is an analogous theorem.

THEOREM. *For the stability of the majorant scheme* (39) *it is necessary and sufficient that the Cauchy problem for its first differential approximation be correct.*

The proof is similar to the proof of the preceding theorem.

In the general case of two-level difference schene for equation (6) the following assertion holds: *stability of a two-level difference scheme for* (6) *implies the correctness of the Cauchy problem for its first differential approximation in the case of odd order of approximation and for its second differential approximation in the case of even order of approximation.*

The converse does not hold, in general, as the example of the following scheme shows:

$$\frac{u^{m+1}(x) - u^m(x)}{\tau} = \Lambda_0 u^m(x) = \frac{u^m(x-2h) - u^m(x-h)}{h}.$$

The correctness condition of the Cauchy problem for the first differential approximation implies that $\kappa = \tau/h \leq 3$, while this scheme is unstable for any $\kappa > 0$.

We now consider the application of dispersion analysis to the investigation of stability of explicit and implicit difference schemes. The next two theorems (Il'in (1965)) establish the existence of a conditionally stable explicit scheme and an absolutely stable implicit scheme for correct systems of equations with constant coefficients.

THEOREM 1. *Suppose that the system of equations*

$$\frac{\partial u}{\partial t} = L(D)u = \sum_{\alpha=0}^{p} A_\alpha D^\alpha u \tag{40}$$

has constant coefficients (i.e., matrices A_α) and is correct in the sense of Petrovskiĭ. Then the explicit scheme

$$\frac{u^{m+1} - u^m}{\tau} = L\left(\frac{\Delta}{h}\right)u^m, \qquad \Delta = \frac{T_1 - T_{-1}}{2}, \tag{41}$$

absolutely approximates the system (40) *and is stable in the sense of Petrovskiĭ under the condition*

$$\tau/h^{2p} \leq c, \tag{42}$$

where p is the order of the system (40) *and c is a positive constant.*

PROOF. The spectral image of the solution operator of the difference scheme (41) has the form

$$\bar{C}_{m,0}(is) = \bar{C}_{m,0}(k, \tau) = [E + \tau L(is)]^m,$$

where $s = \sin kh/h$, while the spectral image of the solution operator of (40) is $\bar{S}(k, t, 0) = e^{tL(ik)}$. Since (40) is Petrovskiĭ-correct, we have

$$\|e^{tL(i\lambda)}\| \leq \text{const} \cdot (1 + |\lambda|^r) \tag{43}$$

for some integer r and all real λ. We estimate the norm of the operator

$$\Phi_m(s) = [E + \tau L(is)]^m e^{-tL(is)}.$$

Since

$$\|L(is)\| \leq \text{const} \cdot h^{-p}, \tag{44}$$

where the constant does not depend on k, it follows in view of (42) that for sufficiently small h we have $\|\tau L(is)\| < 1/2$, and then

$$\begin{aligned}\Phi_m(s) &= \exp[m \ln (E + \tau L(is)) - tL(is)] \\ &= \exp\left\{m\left[\tau L(is) - \frac{\tau^2}{2} L^2(is)(1 + O(\tau\|L(is)\|))\right] - tL(is)\right\} \\ &= \exp\left\{- m\frac{\tau^2}{2} L^2(is)(1 + O(\tau\|L(is)\|))\right\}.\end{aligned}$$

From (42) and (44) it follows that $\|\Phi_m(s)\| \leq$ const uniformly for all k, m, τ with $0 \leq m\tau = t \leq \bar{t}$. Considering (43), we now obtain

$$\|\bar{C}_{m,0}(is)\| \leq \|\Phi_m(s)\| \, \|e^{tL(is)}\| \leq \text{const}(1 + |s|^r) \leq \text{const}(1 + |k|^r),$$

i.e., the scheme (41) is Petrovskiĭ-stable.

THEOREM 2. *Suppose that the system* (40) *is Petrovskiĭ-correct. Then the implicit difference scheme*

$$\frac{u^{m+1} - u^m}{\tau} = L\left(\frac{\Delta}{h}\right)u^{m+1} \tag{45}$$

absolutely approximates (40) *and is absolutely Petrovskiĭ-stable.*

PROOF. The spectral image of the step operator of the scheme (45) has the form

$$\bar{C}(is) = \bar{C}(k, \tau) = [E - \tau L(is)]^{-1},$$

and the spectral radius is correspondingly

$$\rho_\lambda = \max_j \left\| \frac{1}{1 - \tau\lambda_j(is)} \right\|,$$

where $\lambda_j(is)$ are the characteristic roots of the matrix $L(is)$. It is known that for systems which are Petrovskiĭ-correct the eigenvalues $\lambda_j(is)$ have finite real part, Re $\lambda_j(is) \leq \mu < \infty$, and hence

$$|\rho_\lambda| \leq 1 + \text{const} \cdot \tau. \tag{46}$$

We observe that if the matrix $L(is)$ is normal, then this immediately implies the correctness of the scheme in L_2. In the general case we apply more precise estimates of the matrix $\bar{C}(is)$.

Since any function of a matrix $f(A)$ is completely determined by the values of f on the spectrum of A, we can represent $f(A)$ as a Newton interpolation polynomial

$$f(A) = b_0E + b_1(A - \lambda_1E) \\ + b_2(A-\lambda_1E)(A-\lambda_2E) + \cdots + b_{n-1}(A-\lambda_1E)\cdots(A-\lambda_{n-1}E),$$

where $|b_k| \leq \max_{\lambda\in\Pi}|f^{(j)}(\lambda)|$, and Π is the smallest convex polygon in the complex plane containing all eigenvalues $\lambda_1, \ldots, \lambda_n$ of A. From this we obtain

$$\|f(A)\| \leq C_1 \sum_{\alpha=0}^{n-1} \|A^\alpha\| \max_{z\in\Pi} |f^{(\alpha)}(z)|, \tag{47}$$

where C_1 depends only on n (n is the order of the matrix A). We set $A = L(is)$ and $f(A) = (E - \tau A)^{-m} = (E - \tau L(is))^{-m}$. Then (47) takes the form

$$\begin{aligned} \|[E-\tau L(is)^{-m}\| &\leq C_1 \sum_{\alpha=0}^{n-1} \tau^\alpha \|L(is)\|^\alpha \max_{z\in\Pi} \frac{m(m+1)\cdots(m+\alpha-1)}{|1 - z|^{m+\alpha}} \\ &\leq \text{const} \cdot (1+|k|^q) \max_{z\in\Pi} \frac{1}{|1-z|^{m+\alpha}}, \end{aligned} \tag{48}$$

where $q = p(n - 1)$, the constant does not depend on k, m or τ, and we have used the relation

$$\|L(is)\| \leq \text{const}(1 + |k|^p)$$

for the correctness of the system (40) in the sense of Petrovskiĭ (see (1.3.21)). Now

$$\text{Re}\ \lambda(A) = \text{Re}\ \lambda(\tau L(is)) \leq \tau\mu$$

and for τ so small that $\tau\mu < 1$ we have

$$\begin{aligned} |1 - z|^{-m-\alpha} &\leq |1 - \text{Re}\ z|^{-m-\alpha} \leq |1 - \mu\tau|^{-m-\alpha} \\ &= |1 - \mu\tau|^{-t/\tau-\alpha} \leq |1 - \mu\tau|^{-\bar{t}/\tau-n+1} \leq \text{const}. \end{aligned} \tag{49}$$

Substituting (49) into (48), we have

$$\|[E - \tau L(is)]^{-m}\| \leq \text{const}(1 + |k|^q),$$

i.e., the difference scheme is Petrovskiĭ-correct with $q = p(n - 1)$.

The next theorem establishes a connection between the type of the difference approximation and the stability of the corresponding difference scheme (Janenko and Demidov (1971).

THEOREM. *Suppose that the system* (40) *with constant coefficients has an operator* $L(D)$ *with unbounded spectrum and the two-level finite difference operator* $(T_0 - E)/\tau - \Lambda_0(T_1)$ *absolutely approximates the differential operator* $\partial/\partial t - L(D)$. *Then the explicit scheme*

$$\frac{u^{m+1} - u^m}{\tau} - \Lambda_0(T_1)u^m = 0 \tag{50}$$

cannot be absolutely stable.

PROOF. The difference operator $\Lambda_0(T_1)$ absolutely approximating $L(D)$ can be represented in the form

$$\Lambda_0(T_1) = a_{\alpha\beta}\left(\frac{\Delta_1}{h}\right)^\alpha T^\beta_{-1},$$

where

$$\sum_\beta a_{\alpha\beta} = a_\alpha = \{a^i_{j\alpha}\}, \qquad L(D) = \{a^i_{j\alpha}D^\alpha\}.$$

The spectral image of the step operator of the scheme has the form

$$\bar{C}(k, \tau) = E + \tau a_{\alpha\beta}(ik)^\alpha e^{-ik\beta h} = E + \tau A.$$

We suppose that the scheme (50) is absolutely stable. We must then have $\|A\| \leq \text{const}/\tau$ for all h, where the constant does not depend on τ, k or h. Fixing τ here and letting h tend to zero, we find that

$$|a_\alpha(ik)^\alpha| \leq \text{const}/\tau$$

for all k. This contradicts the assumption that the spectrum of $L(D)$ is unbounded. The proof of the theorem is complete.

We mention also a result of Kreiss (1958) which for a system (40) with constant coefficients that is correct in L_2 establishes the existence of an approximating explicit scheme which is stable under the condition $\tau/h^p \leq \text{const}$ (p is the order of the system), i.e., under a less stringent condition than the corresponding theorem of Il'in.

§3. Investigation of the stability of difference schemes

The theorems formulated in §2 make it possible to prove the convergence of a difference scheme as soon as approximation and stability have been established.

Criteria of approximation are relatively simple, have local character, and for the most part require only the expansion of the scheme in a Taylor series. The investigation of stability is considerably more involved, since it requires establishing uniform boundedness in the parameters $\tau \to 0$ and $h \to 0$ of the norm of the solution operator of the problem. In many cases the properties of the solution operator of are unknown for all practical purposes, and it is therefore necessary to resort to various types of stability criteria based on a number of simplifying hypotheses and on experience with practical computations.

In this section we give a brief survey of the methods of investigation, criteria, and characteristics of the stability of difference schemes. We mention that in recent years monographs and textbooks have appeared in which questions of the stability of difference schemes are considered rather completely. We mention the thorough monograph of Samarskiĭ and Gulin (1973) and the books of Godunov and Rjaben'kiĭ (1973), Richmyer and Morton (1967), and Rjaben'kiĭ and Filippov (1956).

1. ***The spectral method of investigation.*** If a difference scheme is uniformly stable, then, of course, it is also stable. In most cases this makes it possible to reduce the investigation of stability to an estimate of the norm of the step operator of the difference scheme.

In §2.2 we considered the application of the Fourier transform to obtain dispersion relations and investigate the stability of the simplest difference scheme

$$\frac{u^{m+1}(x) - u^m(x)}{\tau} = \Lambda_1 u^{m+1}(x) + \Lambda_0 u^m(x) \tag{1}$$

for the case where the operators Λ_1 and Λ_0 do not depend on x and t, and the solution $u^m(x)$ satisfy a periodicity condition in x. Under the condition of periodicity of $u^m(x)$ the case of a difference scheme (1) in which the operators Λ_1 and Λ_0 depend on t but not on x can be investigated similarly.

The Fourier transform

$$u^m(x) = \sum_{k=-\infty}^{\infty} \bar{u}^m(k) e^{ikx} \tag{2}$$

reduces the difference scheme (1) to a system of equations for the Fourier coefficients $\bar{u}^m(k)$:

$$\frac{\bar{u}^{m+1}(k) - \bar{u}^m(k)}{\tau} = \bar{\Lambda}_1^{m+1} \bar{u}^{m+1}(k) + \bar{\Lambda}_0^m \bar{u}^m(k). \tag{3}$$

Here the matrices $\bar{\Lambda}_1^{m+1}$ and $\bar{\Lambda}_0^m$ are the spectral images of the operators Λ_1^{m+1} and Λ_0^m, and depend on t. The step operator of the system (3) is the matrix

$$\bar{C}_{m+1,m}(k) = (E - \tau\bar{\Lambda}_1^{m+1})^{-1}(E + \tau\bar{\Lambda}_0^m). \tag{4}$$

Because of the relation

$$\|C_{m+1,m}\|_{L_2} = \sup_k \|\bar{C}_{m+1,m}(k)\|_{U_n}$$

we see that the estimate of the norm of the step operator $C_{m+1,m}$ of the scheme (1) reduces to an estimate of the norm of the matrix (4). If the estimate

$$\|\bar{C}_{m+1,m}(k)\| \leq 1 + C(\bar{t})\tau$$

holds with a constant $C(\bar{t})$ not depending on τ, h, k or $t \leq \bar{t}$, then the difference scheme (1) is uniformly stable.

If the operators Λ_1^{m+1} and Λ_0^m depend smoothly on t, then so do the matrices $\bar{\Lambda}_1^{m+1}$ and $\bar{\Lambda}_0^m$. Therefore, in this case as well a necessary condition for the uniform stability of (1) can be written in the form

$$R_\lambda(\bar{C}_{m+1,m}(k)) \leq 1 + M(\bar{t})\tau, \tag{5}$$

where $R_\lambda(\bar{C})$ is the spectral radius of the matrix $\bar{C}$, and the constant $M(\bar{t})$ does not depend on τ, m or k. If the matrix $\bar{C}_{m+1,m}(k)$ is normal, condition (5) is also sufficient for the uniform stability of (1).

Some authors have studied necessary and sufficient conditions for the boundedness of the norm of the solution operator $\bar{C}_{m,0} = C_{m,m-1}\bar{C}_{m-1,m-2} \cdots \bar{C}_{1,0}$ of the system (3) without assuming that the matrices $\bar{C}_{m,m-1}$ are normal. There is a survey of the work in this direction in the book of Richtmyer and Morton (1967) (see also the survey of works on the stability of difference schemes in Samarskiĭ and Gulin (1973)).

We now consider the case of a difference scheme with coefficients depending on x but not on t. In this case it is also possible to consider difference boundary conditions of rather general type. We denote by $C = C_{m+1,m}$ the step operator for this difference boundary value problem. The matrix C depends on τ and the mesh size in the variable x (it depends on h). The formula

$$u^{m+1} = C^{m+1}u^0 \tag{6}$$

expresses the values u^{m+1} at points of the mesh in terms of the value u^0. We see that stability of the difference scheme (6) requires the uniform boundedness of the norm of powers of C:

$$\|C^m\| \leq N, \qquad \tau m = t \leq \bar{t}, \tag{7}$$

where the constant N does not depend on τ, h or $t < \bar{t}$.

If the spectrum of C is known, i.e., the collection of eigenvalues $\{q_k\}$ and the corresponding eigenfunctions u_k of the matrix,

$$Cu_k = q_k u_k, \tag{8}$$

then a necessary condition for the stability of (6) can be written in the form

$$|q_k| \le 1 + M\tau, \tag{9}$$

where M does not depend on τ or k. This follows from the fact that

$$\|C\| \ge \max |q_k|.$$

Thus, a necessary condition for stability is that all the eigenvalues (multipliers) q_k of the transition matrix $C = C_{m+1,m}$ lie inside a disk of radius $1 + M\tau$ in the complex plane. However, this condition is far from sufficient and for nonself-adjoint matrices C may exceed the region of stability of the difference boundary value problem.

S. K. Godunov and V. S. Rjaben'kiĭ introduced the concept of the spectrum of a family of operators $C(\tau, h)$. This concept can be introduced for a one-parameter family of transition operators C which we obtain, for example, by setting $h = h(\tau)$; then $C = C(\tau) = C_\tau$. A complex number λ is called a point of the *spectrum of a family of difference operators* $\{C_\tau\}$ if for any $\tau_0 > 0$ and $\varepsilon > 0$ there is a $\tau < \tau_0$ such that the inequality

$$\|C_\tau u - \lambda u\| \le \varepsilon \|u\|$$

has a nontrivial solution. Godunov and Rjaben'kiĭ showed that for the stability of a difference boundary value problem it is necessary that the entire spectrum of the operators $\{C_\tau\}$ lie inside or on the boundary of the unit disk $|\lambda| \le 1$. They also showed that this condition is not only necessary but in a certain sense is also close to a sufficient condition for stability.

We thus see that the spectral method makes it possible to investigate stability of a difference boundary value problem in a simple way only for the case of a normal matrix C; if this is not the case, then in general it is necessary to verify that condition (7) is satisfied. This condition is a direct consequence of the definition of stability.

We make some general remarks concerning the spectral method of investigating the stability of a difference boundary value problem. In general it requires the determination of the entire spectrum of the transition matrix C, which has nN rows (and columns), where n is the dimension of the vector $u = \{u_1, \ldots, u_n\}$ and $N(h)$ is the number of points in the variable x connected by the difference scheme. We see that in many cases of practical importance nN is a very large number, and the determination of the entire spectrum of C is a very difficult problem even using modern computers. It is therefore necessary to resort to various types of simplifying hypotheses which make it possible to estimate the location of the spectrum of the difference operator.

In a number of problems the inhomogeneity of the coefficients of the difference scheme is introduced only by the approximation of the boundary conditions. In this case it is expedient to investigate the correctness of a difference Cauchy problem with constant coefficients, which is a comparatively simple problem. After this it is possible to study separately the effect of difference boundary conditions by considering two difference problems on the semi-infinite line and requiring stability of these problems. Stability of the difference boundary value problem is expected if the conditions ensuring stability of the difference problems indicated are combined. This approach was proposed by K.I. Babenko and I.M. Gel'fand (see the Babenko-Gel'fand criterion in Godunov and Rjaben'kiĭ (1973)). It is based on the simple circumstance that for many evolution problems the effect of the boundary conditions is damped out far from the boundaries of the region, and therefore for a sufficiently fine step size h the same occurs for the difference boundary value problem as well.

Other simplifying hypotheses are connected with the principle of "freezing" the coefficients and requiring correctness of the Cauchy problem or boundary value problem for the difference operators with "frozen" coefficients.

2. *The principle of frozen coefficients and the local algebraic method.* In studying the properties of stability of difference schemes the method of freezing variable coefficients is widely applied; this method is analogous to that applied to differential equations in §1.

In a systematic way this technique consists in fixing the values of the variables $x = x_0$ and $t = t_0$ in the difference scheme

$$u^{m+1} = C(x, t, \tau)u^m. \tag{1}$$

After this the constant transition operator $C(x_0, t_0, \tau)$ can be investigated, in particular, by the spectral method.

If conditions are satisfied under which

$$\|C(x_0, t_0, \tau)\| \leq 1 + M(\bar{t})\tau \tag{2}$$

for any values x_0 and t_0 considered with a constant $M(\bar{t})$ not depending on x_0, t_0 or τ, then there is reason to expect that the difference scheme (1) is stable. There are, nevertheless, known examples in which condition (2) is neither necessary nor sufficient for the stability of the Cauchy problem (1) with variable coefficients. However, for some classes of difference problems it can be shown that condition (2) is sufficient for the stability of the difference scheme (1).

We consider the justification of the local algebraic method for the example of the simplest difference scheme (the "cross" scheme) for the acoustic equations

$$\frac{\partial u}{\partial t} = a^2 \frac{\partial v}{\partial x}, \qquad \frac{\partial v}{\partial t} = \frac{\partial u}{\partial x} \tag{3}$$

with a variable speed of sound $a = a(x)$. On the mesh $x = jh$ $(j = 0, 1, \ldots, N + 1)$, $t_m = m\tau$ $(m = 0, 1, \ldots, M)$, $(N + 1)h = 1$, $M\tau = \bar{t}$ this scheme can be written in the form

$$\begin{gathered} \frac{u_j^{m+1} - u_j^m}{\tau} = a_{j-1/2}^2 \frac{v_j^m - v_{j-1}^m}{h}, \\ \frac{v_j^{m+1} - v_j^m}{\tau} = \frac{u_{j+1}^{m+1} - u_j^{m+1}}{h}, \qquad a_{j-1/2}^2 = a^2(x_{j-1/2}). \end{gathered} \tag{4}$$

Eliminating u^m and u^{m+1} from (4), we arrive at the three-level scheme for v

$$\frac{v_j^{m+1} - 2v_j^m + a_j^{m-1}}{\tau^2} = \frac{a_{j+1/2}^2(v_{j+1}^m - v_j^m) - a_{j-1/2}^2(v_j^m - v_{j-1}^m)}{h^2}. \tag{5}$$

For (5) we pose the boundary conditions

$$v_0^m = v_{N+1}^m = 0, \qquad m = 0, 1, \ldots. \tag{6}$$

Denoting by w^m the vector

$$w^m = \{0, v_1^m, \ldots, v_N^m, 0\}, \tag{7}$$

we see that the general solution of the difference system (5), (6) is

$$w^m = \sum_{k=1}^{N} \rho_k^m \overset{k}{w}, \tag{8}$$

where $\overset{k}{w} = \{0, \overset{k}{v}_1, \ldots, \overset{k}{v}_N, 0\}$ is an eigenfunction of the difference boundary value problem

$$\begin{gathered} a_{j+1/2}^2(\overset{k}{v}_{j+1} - \overset{k}{v}_j) - a_{j-1/2}^2(\overset{k}{v}_j - \overset{k}{v}_{j-1}) = \lambda_k h^2 \overset{k}{v}_j, \\ j = 1, 2, \ldots, N, \ \overset{k}{v}_0 = \overset{k}{v}_{N+1} = 0, \end{gathered} \tag{9}$$

and λ_k is the corresponding eigenvalue.

The eigenvalue problem (9) is a difference analogue of a Sturm-Liouville problem for a selfadjoint differential equation of second order. This is one of the difference eigenvalue problems that we considered in the preceding subsection.

The matrix of the system (9) is a symmetric, tridiagonal, Jacobi matrix. It is known (see, for example, Gantmaher (1967)) that such a matrix has a simple structure, and for $a(x) > 0$ its eigenvalues λ_k are real and negative. The system

(9) has N distinct eigenvalues corresponding to the N eigenvectors $\overset{1}{w}, \ldots, \overset{N}{w}$ which form an orthonormal basis in E_N. From this and (8) it follows immediately that the norm of the solution operator C^m of the scheme (5) or (4) does not exceed (and coincides with) $\max_k |\rho_k^m|$.

From (5) and (9), for ρ_k^m we have

$$\frac{\rho_k^{m+1} - 2\rho_k^m + \rho_k^{m-1}}{\tau^2} = \lambda_k \rho_k^m,$$

whence it follows that

$$\rho_k^m = A_k(z_{k1})^m + B_k(z_{k2})^m, \tag{10}$$

where the multipliers z_{k1} and z_{k2} are roots of the characteristic equation

$$\frac{z_k^2 - 2z_k + 1}{\tau^2} = \lambda_k z_k. \tag{11}$$

From (10) it follows that the criterion for uniform stability of the scheme in question has the form

$$|z_{ki}| \leq 1 + M\tau \qquad (k = 1, 2, \ldots, N; i = 1, 2),$$

where M is a constant not depending on τ, h or k.[11]

However, in this form the criterion is not sufficiently effective, since for an arbitrary $a(x)$ the eigenvalues λ_k are unknown. We now formulate and later prove the following effective criterion for stability.

Let $\lambda_k(x)$ *be the eigenvalues of the local difference operator* (5) *corresponding to the values* $a_{j+1/2} = a(x)$, $j = 0, 1, \ldots, N$, $x \in [0, 1]$, *and let* $z_{ki}(x)$ *be the multipliers corresponding to* $\lambda_k(x)$ *according to formula* (11). *If the estimates*

$$|z_{ki}(x)| \leq 1 + M\tau \qquad (k = 1, 2, \ldots, N; i = 1, 2) \tag{12}$$

hold for any $x \in [0, 1]$, *and* M *does not depend on* x, τ, h or k, *then the difference scheme* (5) (*or* (4)) *is uniformly correct.*

This local criterion reduces the problem of determining correctness of a difference scheme with variable coefficients to the considerably simpler algebraic problem of determining the stability of a difference scheme with constant coefficients. For constant $a_{j+1/2} = a(x)$ the system (9) has solutions

$$\overset{k}{v}_j = \sin \pi k x_j = \sin \pi k j h, \qquad \lambda_k = -\frac{4a^2}{h^2} \sin^2 \frac{\pi k h}{2}. \tag{13}$$

[11] Stability is investigated in a similar way by Rjaben'kiĭ and Filipoov (1956) and Douglis (1956).

From (11) we see that $z_{k1}z_{k2} = 1$, and hence if $\tau^2\lambda_k \geq -4$, it follows that $|z_{k1}| = |z_{k2}| = 1$. Therefore, if the condition

$$\kappa(x) = a(x)\frac{\tau}{h} \leq 1 \tag{14}$$

is satisfied for any $x \in [0, 1]$, then $\tau^2\lambda_k(x) \geq -4$, and the local stability criterion (12) is satisfied. Condition (14) is the local Courant condition.[12]

Following Bojarincev (1966), we shall prove the sufficiency of the criterion (12) for the uniform stability of the difference scheme (4) or (5). The proof is based on the property of monotonicity of the characteristic numbers of Jacobi matrices.

We consider the symmetric, tridiagonal Jacobi matrices $B_s = (b_{ij}^s)$, $s = 1, 2$. The matrix B_1 is constructed from the function $a_1(x)$ and B_2 from $a_2(x)$ according to the rule

$$b_{ij}^s = (a_{i-1/2}^s)^2\,\delta_{i-1}^j - [(a_{i-1/2}^s)^2 + (a_{i+1/2}^s)^2]\delta_i^j + (a_{i+1/2}^s)^2\,\delta_{i+1}^j,$$
$$i, j = 1, 2, \ldots, N;\ a_1^2(x), a_2^2(x) > 0,$$

We consider the pencil of matrices $B(\alpha) = B_1 + \alpha(B_2 - B_1)$. We denote by $\omega(\alpha)$ and $\lambda(\alpha)$ a normalized eigenvector and an eigenvalue of $B(\alpha)$; then

$$\begin{aligned} B(\alpha)\omega(\alpha) &= \lambda(\alpha)\omega(\alpha), \qquad B'(\alpha) = B_2 - B_1, \\ B'(\alpha)\omega(\alpha) + B(\alpha)\omega'(\alpha) &= \lambda'(\alpha)\omega(\alpha) + \lambda\omega'(\alpha). \end{aligned} \tag{15}$$

Taking the inner product of (15) with $\omega(\alpha)$ and using the orthogonality of $\omega(\alpha)$ and $\omega'(\alpha)$ (because of the condition $|\omega(\alpha)| = 1$), we obtain

$$([B_2 - B_1]\omega(\alpha), \omega(\alpha)) = \lambda'(\alpha). \tag{16}$$

If $a_2^2(x) \geq a_1^2(x)$, then $([B_2 - B_1]\omega, \omega) \leq 0$, and it follows from (16) that

$$\lambda'(\alpha) \leq 0, \qquad \overset{2}{\lambda}_k \leq \overset{1}{\lambda}_k < 0, \qquad |\overset{2}{\lambda}_k| \geq |\overset{1}{\lambda}_k|,$$

where the $\overset{s}{\lambda}_k$ are the eigenvalues of B_s. We have thus proved the monotone dependence of the roots λ_k on $a^2(x)$.

We further consider the dependence of the roots z_k of (11) on the λ_k. If the roots z_{k1} and z_{k2} of (11) are conjugate, then they are both equal to 1 in modulus. If they are real, then

[12] The criterion (14) as a condition for convergence of a "cross" difference scheme was first formulated for the wave equation with constant speed of sound a in the work of Courant, Friedrichs, and Lewy (1928).

$$\max\{|z_{k1}|, |z_{k2}|\} = \frac{\tau^2}{2}|\lambda_k|(1 + \sqrt{1 - 4/\tau^2|\lambda_k|}) - 1. \tag{17}$$

This implies that nondecrease of $\max\{|z_{k1}|, |z_{k2}|\}$ corresponds to an increase of $|\lambda_k|$.

These monotonicity properites of λ_k and $\max_s |z_{ks}|$ make it possible to prove the local stability criterion. Indeed, if all the local schemes are stable, then, in particular, the scheme with $a_0^2 = \max_x a^2(x)$ is stable. But then by the monotonicity property for the scheme with $a^2 = a^2(x) \leq a_0^2$, λ_k (for $a^2 = a^2(x)$) is smaller in modulus than the eigenvalue of the scheme with $a^2 = a_0^2$. By the monotonicity property $\max|z_{ks}|$ with $a^2 = a^2(x)$ is less than the corresponding quantity with $a^2 = a_0^2$. This implies the stability of the scheme with the variable coefficient $a^2(x)$, which we were required to prove.

Bojarincev (1966) proved a local stability criterion for the implicit scheme

$$\frac{v^{m+1} - 2v^m + v^{m-1}}{\tau^2} = \frac{\Delta_{-1}}{h}\sum_{\alpha=-1}^{1} \gamma_\alpha (a^2)^{m+\alpha} \frac{\Delta_1}{h} v^{m+\alpha} \tag{18}$$

with weights $\gamma_{-1}, \gamma_0, \gamma_1$ ($\gamma_{-1} + \gamma_0 + \gamma_1 = 1$) and an arbitrary smooth function $a^2 = a^2(x, t) > 0$. The local difference operators Λ then depend on x and t, and they are noncommutative. The local theorem in this case is based on the following comparison theorem.

If the difference scheme (18) *is stable for* $a^2(x, t) = a_2^2(x, t)$, *then it is also stable for* $a^2(x, t) = a_1^2(x, t) \leq a_2^2(x, t)$. *If the difference scheme* (18) *is unstable for* $a^2 = a_1^2(x, t)$, *then it is also unstable for* $a^2 = a_2^2(x, t) \geq a_1^2(x, t)$.

This immediately implies a local stability criterion for the scheme (18): *if all the local schemes* (18) *are stable* (*unstable*), *then the scheme* (18) *is also stable* (*unstable*).

Lax (1960) proved the validity of the local von Neumann criterion for explicit difference schemes approximating a system of equations of hyperbolic type

$$\frac{\partial u}{\partial t} + A\frac{\partial u}{\partial x} = 0 \tag{19}$$

with a variable matrix A.

We present one more theorem (Kreiss (1964)) justifying the criterion of local stability for the explicit scheme

$$u^{m+1} = C(\tau, x)u^m, \tag{20}$$

approximating the hyperbolic system (19) with matrix $A = A(x)$.

Suppose that the matrices $A(x)$ *and* $C(\tau, x)$ *are Hermitian, uniformly bounded, and uniformly Lipschitz continuous in* x. *Suppose the scheme* (20) *is dissipative of order* $2r$ (*see* §1.3) *and approximates the system* (19) *with order of approximation* $2r - 1$ (*r is a natural number*). *Then the difference scheme* (20) *is stable.*

In this theorem the dissipative property is assumed to be *uniform with respect to* x; for the spectral image $\bar{C}(\tau, x, k)$ of the local operator $C(\tau, x)$ (x is fixed) and its eigenvalues $q = q_i(\tau, x, k)$ for $|k| \leq \pi/h$ the conditions

$$|q_i(\tau, x, k)| \leq 1 - \delta|kh|^{2r}$$

are satisfied, where δ does not depend on $i = 1, \ldots, n$, x or k.

In conclusion we note that there are papers in which a criterion of local stability is justified for difference schemes approximating equations and systems of equations of parabolic type as well (see, for example, Janenko and Bojarincev (1961), Konoval′cev (1968), and Strang (1966).

3. ***The method of majorants or a priori estimates.*** This method of investigating the stability of difference schemes is broadly applied in practice in the numerical solution of problems of mathematical physics. For many problems with variable coefficients it is practically the only method which makes it possible to estimate the time step size τ for which stability obtains. Still another merit of this method is that in many physical problems it has a direct analogy with the so-called *method of energy inequalities* or the *method of the energy integral* in the theory of differential equations which, as a rule, has a simple physical interpretation—the *law of conservation of energy* is satisfied.

The simplest majorant estimate is the estimate for schemes with positive coefficients. For the equation

$$\frac{\partial r}{\partial t} + \xi(x, t)\frac{\partial r}{\partial x} = g(x, t)r, \qquad \xi(x, t) \geq 0, \tag{1}$$

we consider the difference scheme

$$\frac{r^{m+1}(x) - r^m(x)}{\tau} + \xi^m(x)\frac{r^m(x) - r^m(x-h)}{h} = g^m(x)r^m(x) \tag{2}$$

or

$$r^{m+1}(x) = [1 - \kappa^m(x)]r^m(x) + \kappa^m(x)r^m(x-h) + \tau g^m(x)r^m(x), \tag{3}$$

where

$$\kappa^m(x) = \xi^m(x)\tau/h, \quad \xi^m(x) = \xi(x, m\tau).$$

Under the condition

$$0 \leq 1 - \kappa^m(x) \leq 1 \tag{4}$$

the following estimate holds:

$$|r^{m+1}(x)| \leq [1 - \kappa^m(x)]|r^m(x)| + \kappa^m(x)|r^m(x-h)| + \tau|g^m(x)|\,|r^m(x)|. \tag{5}$$

Choosing as a norm of the solution $r^m(x)$ the quantity $\|r^m\| = \max_x|r^m(x)|$, from (5) we find that

$$\|r^{m+1}\| \leq (1 + g\tau)\|r^m\|, \tag{6}$$

where $g = \max_{x,m}|g^m(x)| \leq \max_{x,t}|g(x, t)|$. This implies the uniform stability of the scheme (2) in the space $C(-l, l)$ (for any l) if the stability condition (4) is satisfied:

$$0 \leq \kappa^m(x) = \xi^m(x)\frac{\tau}{h} \leq 1. \tag{7}$$

We remark that for $\xi^m(x) < 0$ the scheme is unstable.

We shall demonstrate the uniform stability of the scheme (2) with periodic boundary conditions in $L_2(-l, l)$ assuming periodicity of the coefficients $\xi(x, t)$ and $g(x, t)$ and the Lipschitz continuity of $\xi(x, t)$ in x. Multiplying (3) by $r^{m+1}(x)$ and recalling (4), we have

$$\begin{aligned}
[r^{m+1}(x)]^2 &= [1-\kappa^m(x)]r^m(x)r^{m+1}(x)+\kappa^m(x)r^m(x-h)r^{m+1}(x) \\
&\quad + \tau g^m(x)r^m(x)r^{m+1}(x) \leq \frac{1-\kappa^m(x)}{2}[r^m(x)]^2 \\
&\quad + \frac{1-\kappa^m(x)}{2}[r^{m+1}(x)]^2 + \frac{\kappa^m(x)}{2}[r^m(x-h)]^2 + \frac{\kappa^m(x)}{2}[r^{m+1}(x)]^2 \\
&\quad + \tau g\frac{[r^m(x)]^2+[r^{m+1}(x)]^2}{2} \\
&= \frac{1}{2}[r^{m+1}(x)]^2+\frac{1-\kappa^m(x)}{2}[r^m(x)]^2 \\
&\quad + \frac{\kappa^m(x)}{2}[r^m(x-h)]^2+\tau g\frac{[r^m(x)]^2+[r^{m+1}(x])^2}{2}.
\end{aligned}$$

Hence

$$\begin{aligned}
&\left(\frac{1}{2} - g\frac{\tau}{2}\right)[r^{m+1}(x)]^2 \\
&\quad \leq \frac{1-\kappa^m(x)}{2}[r^m(x)]^2 + \frac{\kappa^m(x)}{2}[r^m(x-h)]^2 + \frac{g\tau}{2}[r^m(x)]^2 \\
&\quad = \frac{1-\kappa^m(x)}{2}[r^m(x)]^2 + \frac{\kappa^m(x-h)}{2}[r^m(x-h)]^2 \\
&\quad\quad + \frac{g\tau}{2}[r^m(x)]^2 + \frac{\kappa^m(x)-\kappa^m(x-h)}{2}[r^m(x-h)]^2.
\end{aligned}$$

Integrating with respect to x from $-l$ to l, we have

$$\|r^{m+1}\| \leq \frac{1 + g\tau + C\tau}{1 - g\tau}\|r^m\|^2 = [1 + O(\tau)]\|r^m\|^2, \tag{8}$$

where we have set

$$C = \sup_{h, x, m} \left| \frac{\xi^m(x) - \xi^m(x-h)}{h} \right|, \qquad \|r^m\|^2 = \int_{-l}^{l} |r^m(x)|^2 \, dx.$$

This implies the uniform stability of the scheme (3) in L_2 if (7) is satisfied and $\xi^m(x)$ is Lipschitz continuous.

These estimates carry over without major changes to schemes approximating systems of equations in invariants

$$\frac{\partial r_k}{\partial t} + \xi_k \frac{\partial r_k}{\partial x} = g_\alpha^k r_\alpha, \qquad \alpha, k = 1, \ldots, n.$$

Friedrichs (1954) introduced the general concept of positive schemes—difference schemes with positive matrices—and established a sufficient criterion of correctness in L_2 for them. We shall formulate Friedrichs' criterion in the case of one space variable.

Suppose the linear system

$$\frac{\partial u}{\partial t} = A(x, t) \frac{\partial u}{\partial x} \tag{9}$$

is approximated by the explicit difference scheme

$$u^{m+1}(x) = \sum_{\alpha=-q_1}^{q_1} B_\alpha(x, t, \tau, h) u^m(x + \alpha h). \tag{10}$$

Here $A = \|a_{ij}\|$ and $B_\alpha = \|b_{ij}^\alpha\|$ are real matrices on E_n. Suppose that the scheme (10) satisfies the condition

$$\sum_{\alpha=-q_1}^{q_1} B_\alpha = E, \tag{11}$$

which means that the constant vector $u^m(x) \equiv \text{const}$ is a solution of (10). Then the scheme (10) approximates (9) under the condition

$$\frac{\tau}{h} \sum_{\alpha=-q_1}^{q_1} \alpha B_\alpha = A + O(\tau). \tag{12}$$

The Friedrichs criterion can be formulated as follows.

The scheme (10) *is correct in* L_2 *if the matrices* B *are symmetric, positive, and Lipschitz continuous in* x, *so that*

$$\frac{h}{\tau} \left\| \frac{B_\alpha(x+h) - B_\alpha(x)}{h} \right\| \leq b. \tag{13}$$

Taking the inner product[13] of (10) with u^{m+1} and using (11) and the inequality

$$(B_\alpha u, v) \le \sqrt{(B_\alpha u, u)}\ \sqrt{(B_\alpha v, v)} \le \frac{(B_\alpha u, u) + (B_\alpha v, v)}{2},$$

we have

$$\begin{aligned}(u^{m+1}, u^{m+1}) &= \sum_{\alpha=-q_1}^{q_1} (B_\alpha(x)u^m(x+\alpha h), u^{m+1}(x)) \\ &\le \frac{1}{2}\Big(\sum_{\alpha=-q_1}^{q} B_\alpha(x)u^{m+1}(x), u^{m+1}(x)\Big) \\ &\quad + \frac{1}{2}\sum_{\alpha=-q_1}^{q_1} (B_\alpha(x)u^m(x+\alpha h), u^m(x+\alpha h)) \\ &= \frac{1}{2}(u^{m+1}(x), u^{m+1}(x)) + \frac{1}{2}\sum_{\alpha=-q_1}^{q_1} (B_\alpha(x+\alpha h)u^m(x+\alpha h), u^m(x+\alpha h)) \\ &\quad - \frac{1}{2}\sum_{\alpha=-q_1}^{q_1} ([B_\alpha(x+\alpha h) - B_\alpha(x)]u^m(x+\alpha h), u^m(x+\alpha h)). \qquad (14)\end{aligned}$$

Integrating (14) from $-l$ to l, we find that

$$\|u^{m+1}\|^2 \le \frac{1}{2}\|u^{m+1}\|^2 + \frac{1}{2}b\tau\|u^m\|^2 + \frac{1}{2}\int_{-l}^{l}\sum_{\alpha=-q_1}^{q_1} (B_\alpha(x+\alpha h)u^m(x+\alpha h), u^m(x+\alpha h))\,dx,$$

where we have set

$$\|u\|^2 = \int_{-l}^{l} (u(x), u(x))\,dx.$$

Using the periodicity of $u(x)$ and (11), we make the following transformation:

$$\begin{aligned}&\int_{-l}^{l}\sum_{\alpha=-q_1}^{q_1} (B_\alpha(x+\alpha h)u^m(x+\alpha h), u^m(x+\alpha h))\,dx \\ &\quad = \sum_{\alpha=-q_1}^{q_1}\int_{-l}^{l} (B_\alpha(x)u^m(x), u^m(x))\,dx = \int_{-l}^{l}\sum_{\alpha=-q_1}^{q_1} (B_\alpha u^m(x), u^m(x))\,dx \\ &\quad = \int_{-l}^{l}\Big(\Big(\sum_{\alpha=-q_1}^{q_1} B_\alpha\Big)u^m(x), u^m(x)\Big)\,dx = \|u^m\|^2.\end{aligned}$$

[13] The inner product in the space of the real vectors $u = \{u_1, \dots, u_n\}$, $v = \{v_1, \dots, v_n\}$ is intended: $(u, v) = \sum_1^n u_\alpha v_\alpha$.

We have finally

$$\|u^{m+1}\|^2 \leq (1 + b\tau)\|u^m\|^2. \tag{15}$$

The assertion is proved.

We mention that for positive difference schemes (10) a local stability criterion is also valid.

Schemes with positive coefficients and matrices constitute a restricted but very important class of difference schemes. As a rule, these are schemes of first-order accuracy in which the derivatives are approximated by one-sided difference quotients. In approximations of higher order accuracy where central differences are used, as a rule, we do not obtain positive coefficients. In this case the majorant estimates of stability become more complicated. Estimates of this type are called *a priori estimates.*

The method of a priori estimates for difference schemes is analogous to the corresponding method for differential equations, but in the difference case its realization meets with greater difficulties. This is related, of course, to the special nature of difference analysis in which many relations of ordinary analysis either do not hold or assume a more involved form. We consider the method of a priori estimates for the example of the acoustic equations (3.2.3), and for their integration we use the implicit scheme

$$\frac{u^{m+1} - u^m}{\tau} = (a^2)^{m+1}\frac{\Delta_1}{h}v^{m+1}, \tag{16}$$

$$\begin{gathered}\frac{v^{m+1} - v^m}{\tau} = \frac{\Delta_{-1}}{h}u^{m+1},\\ \Delta_1 = T_1 - E, \qquad \Delta_{-1} = E - R_{-1}.\end{gathered} \tag{17}$$

The a priori estimate for this system is analogous to the energy inequality for the system (3.2.3) established in subsection 3. Multiplying (16) by $2u^{m+1}$, (17) by $2(a^2)^{m+1}v^{m+1}$, and adding, after a simple transformation we obtain

$$\begin{aligned}&[(u^{m+1})^2 + (a^{m+1}v^{m+1})^2] - [(u^m)^2 + (a^m v^m)^2]\\ &= -(u^{m+1} - u^m)^2 - [a^{m+1}(v^{m+1} - v^m)]^2 + [(a^{m+1})^2 - (a^m)^2](v^m)^2\\ &+ \frac{2\tau}{h}(a^{m+1})^2(u^{m+1}\Delta_1 v^{m+1} + v^{m+1}\Delta_{-1}u^{m+1}).\end{aligned} \tag{18}$$

Using the formulas for the difference differentiation of a product

$$\begin{aligned}\Delta_1(fg) &= (\Delta_1 f)g + (T_1 f)\Delta_1 g = (\Delta_1 f)T_1 g + f\Delta_1 g,\\ \Delta_{-1}(fg) &= (\Delta_{-1} f)g + (T_{-1} f)\Delta_{-1} g = (\Delta_{-1} f)T_{-1} g + f\Delta_{-1} g,\end{aligned}$$

we transform the expression

$$(a^{m+1})^2(u^{m+1}\Delta_1 v^{m+1} + v^{m+1}\Delta_{-1}u^{m+1}) = d(u^{m+1}, v^{m+1})$$

to the form

$$d(u^{m+1}, v^{m+1}) = \Delta_1[(a^{m+1})^2 v^{m+1} T_{-1}u^{m+1}] - u^{m+1}T_1 v^{m+1}\Lambda_1(a^{m+1})^2.$$

From (18) it then follows that

$$\begin{aligned}&[(u^{m+1})^2 + (a^{m+1}v^{m+1})^2] - [(u^m)^2 + (a^m v^m)^2]\\ &\quad\le \frac{2\tau}{h}\Delta_1[(a^{m+1})^2 v^{m+1}T_{-1}u^{m+1}] + b_1\tau[(u^m)^2 + (a^m v^m)^2]\\ &\quad+ b_2\tau[(u^{m+1})^2 + T_1(a^{m+1}T_1 v^{m+1})^2],\end{aligned} \tag{19}$$

where

$$b_1 = \sup_{x,m,\tau}\left|\frac{(a^{m+1})^2 - (a^m)^2}{\tau(a^m)^2}\right|, \quad b_2 = \sup_{x,m,h}\left|\frac{1}{\sqrt{T_1(a^{m+1})^2}}\,\frac{\Delta_1(a^{m+1})^2}{h}\right|. \tag{20}$$

Integrating (19) with respect to x, we obtain

$$\|\Phi^{m+1}\|^2 - \|\Phi^m\|^2 \le b_1\tau\|\Phi^m\|^2 + b_2\tau\|\Phi^{m+1}\|^2,$$

where $\|\Phi\|^2 = \int_{-l}^{l}(u^2 + a^2v^2)\,dx$. From this we obtain

$$\|\Phi^{m+1}\|^2 \le \frac{1 + b_1\tau}{1 - b_2\tau}\|\Phi^m\|^2,$$

which proves the unconditional uniform correctness of the scheme (16), (17) in the case of a Lipschitz continuous function $a(x, t)$.

Analogous estimates can be established for a scheme in which the quantities on the upper level enter with weight α and those on the lower with weight β:

$$\begin{aligned}\frac{u^{m+1} - u^m}{\tau} &= \alpha_1(a^{m+1})^2\frac{\Delta_1}{h}v^{m+1} + \beta_1(a^m)^2\frac{\Delta_1}{h}v^m,\\ \frac{v^{m+1} - v^m}{\tau} &= \frac{\Delta_{-1}}{h}[a_2u^{m+1} + \beta_2u^m],\\ \alpha_i + \beta_i = 1, \quad &\alpha_i \ge 0, \quad \beta_i \ge 0, \quad i = 1, 2.\end{aligned} \tag{21}$$

To conclude this subsection we mention that the method of a priori estimates and energy inequalities is frequently applied to estimate the stability of difference schemes with variable coefficients including those with variable step size h_i in the space coordinate x. A considerable number of difference schemes to which this method is applicable are presented in the monograph of Samarskiĭ and Gulin (1973).

4. ***A practical approach to the problem of computational stability. Application of the Picard method to suppress instability.*** The concept of stability and instability of a difference scheme discussed in the preceding subsections has asymptotic character, i.e., instability of a scheme becomes manifest as $\tau \to 0$ or as $t \to \infty$ ($\tau = \text{const}$). Instability of a difference scheme nevertheless appears in practical computations with a finite step size τ, and also very rapidly. It becomes apparent in the rapid growth of solutions which are rapidly oscillating in space and time and correspond to the end of the spectrum, and it leads to overfilling and "abend" of the computer.

It therefore makes sense to discuss the question of practical methods of solving differential equations with finite step size τ and $h > 0$ which avoid the effects of instability. We consider them for the example of the evolution problem for the system

$$\frac{\partial u}{\partial t} = L(x)u, \tag{1}$$

$$u(x, 0) = u_0(x) \tag{2}$$

with an initial function $u_0(x)$ and a matrix operator $L(x)$ which are periodic in x with period $2l$. The operator $L(x)$ may be taken to be a differential operator, although the treatment holds for the case of integral and integro-differential operators as well.

We suppose that on $[0, 2l]$ we have introduced a mesh $x_j = jh, j = 0, 1, \ldots, N, h = 2l/N$, which we assume to be sufficiently fine for an acceptable description of $u(x, t)$ in terms of its values at the points $u(x_j, t)$.

In approximating $L(x)$ on this mesh, we replace it by a matrix operator $\Lambda(h)$. We obtain an idea of the error of this approximation by estimating $\|\Lambda(h) - L(x)u\|$ on solutions (frequently only on sufficiently smooth functions) $u(x, t)$ of problem (1), (2). Comparison of the spectra of the difference operator $\Lambda(h)$ and the original operator $L(x)$ provides a convenient notion of this *error of spatial approximation*.

After we have chosen an approximation $\Lambda(h) \sim L(x)$, it remains to choose a difference scheme for the numerical solution of problem (1), (2) in time. Here we discuss and compare three schemes for the approximate solution of problem (1), (2):

$$\frac{u_i^{m+1} - u_i^m}{\tau} = \Lambda(h)u_i^m, \qquad u_i^0 = u_0(x_i), \tag{3}$$

$$\frac{u_i^{m+1} - u_i^m}{\tau} = \Lambda(h)u_i^{m+1}, \qquad u_i^0 = u_0(x_i) \tag{4}$$

and

$$\frac{du_i}{dt} = \Lambda(h)u_i, \qquad u_i(0) = u_0(x_i), \tag{5}$$

The scheme (3) is obviously an explicit scheme of first-order accuracy in τ; (4) is an explicit scheme of the same order accuracy in τ; finally, (5) is a system of ordinary differential equations which may be considered the limit of (3) or (4) as $\tau \to 0$ (or it may be considered a scheme of infinite-order accuracy in τ).

In order that the discussion be objective, we assume that the operator $L(x)$ has eigenvalues $\{\lambda_k\}$ for which

$$\operatorname{Re} \lambda_k \leq a, \qquad \operatorname{Re} \lambda_{k+1} \leq \operatorname{Re} \lambda_k, \qquad \operatorname{Re} \lambda_k \to -\infty \quad \text{as } k \to \infty. \tag{6}$$

The conditions (6) are typical for many problems of mathematical physics.

We know from §2 that the spectrum of the difference operator $\Lambda(h)$ is finite; it consists of $nN = \bar{N}$ eigenvalues $\lambda_k(h)$ and the eigenvectors corresponding to them. We shall assume that the eigenvalues $\lambda_k(h)$ are also subject to the conditions (6) (for large $\bar{N}$) and that they satisfactorily approximate the original eigenvalues λ_k and retain the qualitative behavior of λ_k on the remainder of the spectrum. There are no specious roots $\lambda_k(h)$ at the end of the spectrum (for $k \sim \bar{N}$), for otherwise all three schemes are unacceptable for the computation of problem (1), (2).

A restriction on the step size for the explicit scheme (3) follows from the dispersion relation

$$\max_k |\rho_k| = \max_k |e^{\omega_k \tau}| + \max_k |1 + \lambda_k(h)\tau| < 1. \tag{7}$$

If (6) is satisfied there are usually no restrictions on the step size (for example, for $a = 0$) for the implicit scheme; a restriction may arise, however, if there are no good methods for solving the system of linear equations (4) for u_i^{m+1}, and the method of successive approximations is applied, for example,

$$\overset{(s+1)}{u_i^{m+1}} = u_i^m + \tau\Lambda(h)\overset{(s)}{u_i^{m+1}}. \tag{8}$$

In this case for the convergence as $s \to \infty$ of the iterations (8) it is necessary to require that $\tau\|\Lambda(h)\| < 1$ and

$$\tau \max_k |\lambda_k(h)| < 1. \tag{9}$$

Finally, if (6) is satisfied the system of ordinary differential equations (5) is a "stable" system on any finite interval $[0, \bar{t}]$.

The question of how to solve the system (5) and what step size τ will make this not too laborious has many aspects. Of course, it is possible to use ready-made pıograms for integrating systems of ordinary differential equations, for example, those using the Runga-Kutta method of high-order accuracy with automatic choice of the step size, etc. An increase in computer time may occur here, since these programs do not take account of the special features of systems of the type (5) which are characterisic of equations of mathematical physics. In order to be able to discuss the scheme (5), we propose a specific method of solving this system—Picard's method of successive approximations. It is known that this method is applicable to any systems of ordinary differential equations including those which are nonlinear and depend on t (see, for example, Hartman (1964)). It becomes an especially convenient computational process for linear systems (5) with constant (not depending on t) matrices $\Lambda(h)$. Obviously this process can be generalized to the case of variable matrices $\Lambda(h)$ and nonlinear systems (5); it is also abvious that certain difficulties and complications arise here.

Roždestvenskiĭ (1974) described some questions of the application of Picard's method to the numerical solution of problems of mathematical physics are described. Here we briefly consider only the basic features.

The solution of (5) at one step from t_m to $t_{m+1} = t_m + \tau$ is given by

$$u^{m+1} = e^{\tau\Lambda(h)}u^m = \sum_{s=0}^{\infty} \frac{\tau^s}{s!} \Lambda^s(h)u^m, \tag{10}$$

where $u^m = \{u_j^m\} = \{u_j(t_m)\}$. It is known that *the series* (10) *converges for any* τ. This circumstance is very important if we are interested in computing with a large step size τ. A second important feature is that as the step size τ increases no loss of accuracy of the solution of problem (1), (2) occurs, since we have the possibility of computing the sum of (10) to prescribed accuracy.

Thus, in solving (5) we do not introduce additional error, the *error of time approximation*, which is inherent in the schemes (3) and (4). It should be kept in mind, however, that sometimes the error of time approximation reduces or even eliminates the error of spatial approximation, as in the case of an explicit scheme for the equation $u_t + au_x = 0$ for $\kappa = a\tau/h = 1$.

The computation of a finite sum of the series (10) can be organized as a convenient computational process in which the computation of each successive term of the series is equivalent in volume of work to a single step of the explicit scheme (3); each successive term of (10) increases by one the order of accuracy of the scheme (5) in τ. Thus, in principle it is possible to prescribe such high accuracy in the computation of the sum of (10) by varying the number of terms of the series that the relation (10)

$$u^{m+1} = e^{\tau\Lambda(h)}u^m$$

is satisfied to prescribed accuracy. This implies that the norm of the step operator of the scheme (5) is bounded:

$$\|e^{\tau\Lambda(h)}\| \le e^{a\tau}$$

(in the absence of degeneracy).

However, in the practical computation of the sum of (10) on a computer there is an upper bound on τ. It is easy to understand this restriction. Let us imagine that we compute $e^{-\lambda\tau}$, $\lambda > 0$, by means of (10), where $\lambda\tau$ has considerable magnitude. Then $e^{-\lambda\tau}$ is close to zero, but many terms of (10) are large, and to compute the sum of (10) to a particular accuracy, say, to five decimal digits, it is necessary to compute the terms and finite sums of (10) with a considerably larger numer of signs. However, computations are carried out on a computer with a large but restricted number of signs, and this poses a restriction on the magnitude of τ. A specific bound for τ depends on the representation of numbers in the computer on which the computations are done: the length of the mantissa of a number, the length of the order of a number, and some other details. For domestic computers with a 48-digit binary representation of a number this bound is given by the inequality

$$\tau\|\Lambda(h)\| \le c \sim 20 \div 30. \tag{11}$$

If we recall that for an explicit scheme the restriction on the step size has the form

$$\tau\|\Lambda(h)\| \le d \simeq 2,$$

conclude that Picard's method makes it possible to carry out a stable computation according to the scheme (5) with a step size τ considerably greater than that permitted by the explicit scheme. This is possible because we are solving the evolution problem (5) quite accurately, and for evolution problems the high-frequency harmonics are rapidly damped out. By prescribing a particular accuracy of computing (10) we automatically suppress these high-frequency harmonics to which there corresponds large damping by increasing the number of terms of the series (and by more accurate computation of $e^{\tau\Lambda}$).

In order to understand more fully the Picard method it is necssary to represent the system (5) in a basis of eigenvectors of the matrix $\Lambda(h)$. The system (5) then has the same form but with a diagonal matrix $\Lambda'(h) = \operatorname{diag}\{\lambda_k h\}$. It thus becomes clear that computation according to (10) provides a high accuracy of solving (5), and the higher the accuracy of solving (5), the higher the stability of the results of the solution, since the system (5) possesses the evolution property.

The question of the expediency of applying Picard's methods in place of the explicit scheme (3) or the implicit scheme (4) depends on several factors and, first of all, on the relation of the error of the time and spatial approximations introduced by the schemes (3) and (4). If the operator $\Lambda(h)$ approximates $L(x)$ to high accuracy, then it is expedient to apply schemes of high-order accuracy in τ, and this is automatically provided by application of the Picard method. Here it is possible to use a step size τ approximately 10 times greater than that permitted by the Courant condition with no loss of stability. The loss of stability in the Picard method has the special character we have mentioned: either the memory unit of the computer gets overfilled, or the results become very inaccurate and have a sporadic character.

On the other hand, for difference schemes (3) and (4) in which the stability condition (or convergence of the iterations) permits a step size τ considerably greater than is required from considerations of the accuracy of the time approximation as compared with the accuracy of the spatial approximation it is altogether possible that in this case the explicit scheme is more expedient than the Picard method.

We mention also that for linear problems (1), (2) it is possible to achieve very high efficiency by computing the fundamental matrix $e^{\tau\Lambda(h)}$ of this system by means of a series in τ rather than a single solution of (5). For evolution problems it is then possible to obtain a solution of (5) practically with no loss of accuracy for very large $t > 0$ by using the *doubling formulas*

$$e^{2\tau\Lambda(h)} = [e^{\tau\Lambda(h)}]^2, \qquad e^{4\tau\Lambda(h)} = [e^{2\tau\Lambda(h)}]^2 \ldots .$$

Several computations of various problems of mathematical physics have been carried out using the Picard method which demonstrate its effectiveness (see, for example, Roždestvenskiĭ (1974), and Pinskiĭ and Ruzanov (1976)).

We make one more remark concerning the possibilities of applying methods for the numerical integration of systems of ordinary differential equations to solving problem (1), (2) by means of the scheme (5). Recently, efficient methods have been developed for the numerical integration of so-called "stiff systems" of ordinary differential equations which are characterized by a large spread of the characteristic exponents λ_k of the system (5) (in general, nonlinear). In particular, in these methods it is also assumed that the properties (6) are satisfied and, moreover, that $-\mathrm{Re}\lambda_k$ is very large for large k. The idea of the method developed by Gear and Nordsieck consists in applying integration schemes of high-order accuracy which in the range of λ_k of moderate modulus ensure high accuracy, while for $\mathrm{Re}\,\lambda_k < -b$, where b is sufficiently large, they only ensure stability of the computation. The latter means, for example, that for a linear system (5) in the approximation region $[0, k_1]$ of the spectrum good accuracy of reproducing λ_k in the time computation is provided; the intermediate part and end of the spectrum considerably distort the integration scheme in time, but damping of the eigenfunctions corresponding to this part of the spectrum is provided, i.e., here only stability is ensured. Methods of integrating "stiff systems" are described, for example, by Gear (1967, 1969) and Nordsieck (1962); see also Pavlov and Povzner (1973).

Of course, these methods of integration can also be applied to solve the Cauchy problem for the system (5). However, for the most characteristic problems of mathematical physics of the type (1), (2), as a rule, they are found to be less efficient than the Picard method. The reason for this is that in the majority of problems of mathematical physics one strives to make the spatial approximation $\Lambda(h) \sim L(x)$ comparatively good even on the end of the spectrum $k \sim k_0$. For example, we have seen in §2.5 that in the approximation region for the approximation of the operator D^2 by the operator $\Lambda_1\Lambda_{-1}/h^2$ for $k \sim k_0/3$ we have $\lambda(h, kh) = -1/h^2$, while at the end of the spectrum for $k = k_0 = \pi/h$ we have $\lambda_0(h, \pi) = -4/h^2$. The difference of $\lambda_0(h, kh)$ in the approximation region and at the end is thus not great. In such situations it is said that the "stiffness" index of the system is not large.

It is just for this reason that methods for stiff systems intended for a difference of $\mathrm{Re}\lambda_0(h, kh)$ in the approximation region and at the end of the spectrum of several decimal orders are not always effective even as compared with ordinary difference schemes. Nevertheless, for a number of problems of mathematical physics characterizied by several processes occurring at considerably different rates the methods of integrating "stiff" systems may also be effective for solving these problems by means of a scheme of the type (5).

§4. Analysis of the simplest difference schemes

In this section we consider a number of the simplest difference schemes for linear equations of hyperbolic type (a single equation and the system of acoustic equations), and we use the concepts introduced above to study them.

1. ***Schemes for the single equation.*** $u_t + au_x = 0$. For this equation

$$\frac{\partial u}{\partial t} + a(x, t)\frac{\partial u}{\partial x} = 0, \tag{1}$$

where $a(x, t) > 0$, $a(x, t) \in C_1$, we consider the Cauchy problem

$$u(x, 0) = u_0(x), \qquad |x| < \infty \tag{2}$$

or the mixed problem

$$u(x, 0) = u_0(x), \qquad x \geq 0,\ u(0, t) = \varphi(t) \tag{3}$$

with a boundary condition at $x = 0$.

The simplest scheme for (1) has the form

$$\frac{u^{m+1} - u^m}{\tau} + a^m \frac{\Delta_{-1} u^m}{h} = \frac{u_i^{m+1} - u_i^m}{\tau} + a_i^m \frac{u_i^m - u_{i-1}^m}{h} = 0,$$
$$a^m = a_i^m = a(ih, m\tau). \tag{4}$$

Equation (4) can be solved for u_i^{m+1}:

$$u_i^{m+1} = (1 - \kappa_i^m) u_i^m + \kappa_i^m u_{i-1}^m, \qquad \kappa_i^m = a_i^m \frac{\tau}{h} \tag{5}$$

and (5) has the following geometric interpretation. According to (5), $u_i^{m+1} = u^{m+1}(x_i) = u^m(x_i^*)$, where the point $x = x_i^*$ on the line $t = m\tau$ is the point of its intersection with the line

$$x - ih = a_i^m[t - (m+1)\tau], \tag{6}$$

i.e., $x_i^* = ih - \kappa_i^m h$, and the value of $u^m(x_i^*)$ at the point x_i^* is obtained by linear interpolation ($\kappa_i^m \leq 1$) or extrapolation ($\kappa_i^m > 1$) from the values u_i^m and u_{i-1}^m (Figure 3.4).

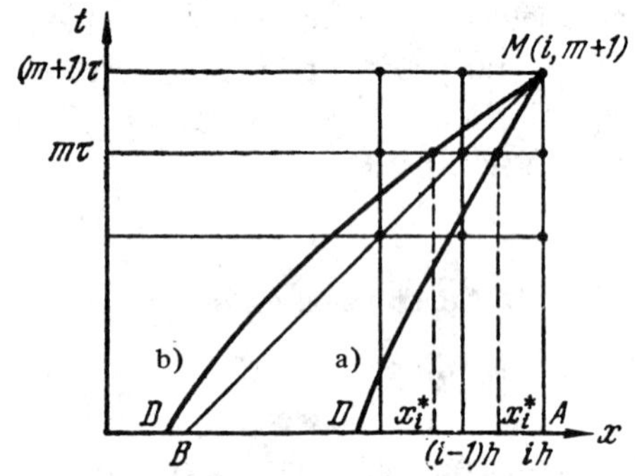

a) $\varkappa_i^m \leqslant 1$; b) $\varkappa_i^m > 1$

FIGURE 3.4

Up to quantities of order τ^2 the segment of the line (6) for $m\tau \leq t \leq (m+1)\tau$ approximately coincides with the characteristic of (1) drawn through the point $(ih, (m+1)\tau)$. Therefore, the scheme (5) can be interpreted as interpolation (extrapolation for $\kappa_i^m > 1$) of the function $u^m(x)$ and subsequent transfer of the values to the next level $t = (m+1)\tau$.

If $\kappa_i^m \leq 1$ for all i and m then the scheme satisfies a maximum principle. Indeed, according to (5)

$$\max_i |u_i^{m+1}| \leq \max_i \{(1 - \kappa_i^m)|u_i^m| + \kappa_i^m |u_{i-1}^m|\} \leq \max_i |u_i^m|. \tag{7}$$

The maximum principle (7) implies stability of the scheme (5) in $C(-\infty, \infty)$ for the Cauchy problem and in $C(0, \infty)$ for the mixed problem. If $\kappa_i^m > 1$ for some i and n, then the maximum principle does not hold, since from (5) in this case we obtain only the estimate

$$\max_i |u_i^{m+1}| \leq \max_i (2\kappa_i^m - 1) \max_i |u_i^m|, \tag{8}$$

where $\max_i (2\kappa_i^m - 1) > 1$.

Moreover, it can be shown that if in some finite region of the variables x and t the quantities κ_i^m remain greater than 1 as $\tau \to 0$ and $h \to 0$, then the scheme (5) is unstable in L_2 and in C. It suffices to demonstrate this in the case $a(x, t) = a = \text{const}$. Then $\kappa_i^m = \kappa > 1$. The dispersion equation of the scheme (5) has the form

$$\rho = 1 - \kappa(1 - e^{ikh}).$$

Hence

$$|\rho| \leq 1, \quad \text{if } \kappa \leq 1,$$

$$|\rho| \geq 1 + \varepsilon(\kappa, kh), \quad \text{if } \kappa > 1,$$

where $\varepsilon(\kappa, kh)$ is some positive function for $k \neq 0$. This implies the instability of (5) for $\kappa > 1$.

Thus, the scheme (4) (or (5)) approximates (1) and is stable if for all i and m

$$\kappa_i^m \leq 1. \tag{9}$$

It is easy to verify that u_i^m is determined only from the values $u_{i-m}^0, u_{i-m+1}^0, \ldots, u_i^0$ of the initial function, i.e., the domain of dependence of the solution of the Cauchy problem for the difference equation (5) consists of points of the mesh located between the points A and B (Figure 3.4), where MA is a vertical line and MB a diagonal line of the mesh.

If $\kappa_i^m > 1$ in some region of (x, t), then the characteristic MD of (1) drawn from the point M lies to the left of the line MB. Thus, for $\kappa_i^m > 1$ the domain of dependence of the solution of the Cauchy problem for the differential equation (1) (which, by the way, consists of the single point D) lies outside the domain of dependence for the difference equation. This will be valid for any refinement of the mesh, provided that

$$\kappa_i^m = \frac{a_i^m \tau}{h} \geq \kappa > 1.$$

This implies that if $\kappa_i^m > 1$ in some region of (x, t), then the solution of (4) does not converge to the solution of (1). Indeed, if the initial function $u_0(x)$ is changed at the point D and in a neighborhood of it, then the solution $u(x, t)$ of (1) varies correspondingly at M. At the same time the solution u_i^m of the difference problem at the same point M does not notice this change. This implies that the solution $u^m(x)$ does not converge to $u(x, t)$ either in the norm of C or in the norm of L_2.

On the other hand, if the Courant condition (9) is satisfied, then the point D lies on the segment AB, and variations of the initial function in a neighborhood of D are transmitted both to the exact solution and to the difference solution.

In correspondence with this analysis it is clear that if $a(x, t) < 0$, then for (1) the approximation

$$\frac{u^{m+1} - u^m}{\tau} + a^m \frac{\Delta_1}{h} u^m = 0$$

is applicable, and the approximation (4) is not applicable.

For (1) with $a(x, t) > 0$ we now consider the implicit approximation

$$\frac{u_i^{m+1} - u_i^m}{\tau} + a_i^m \frac{u_i^{m+1} - u_{i-1}^{m+1}}{h} = 0. \tag{10}$$

Equation (10) can be rewritten in the form

$$u_i^{m+1} = \frac{1}{1 + \kappa_i^m} u_i^m + \frac{\kappa_i^m}{1 + \kappa_i^m} u_{i-1}^{m+1} = u^*, \tag{11}$$

where $\kappa_i^m = a_i^m \tau / h$. Equations (10) and (11) absolutely approximate (1) with first order approximation both in τ and h. From (11) we obtain

$$\min\{u_i^m, u_{i-1}^{m+1}\} \le u_i^{m+1} \le \max\{u_i^m, u_{i-1}^{m+1}\}. \tag{12}$$

This implies that the difference solution u_i^m assumes its minimum and maximum

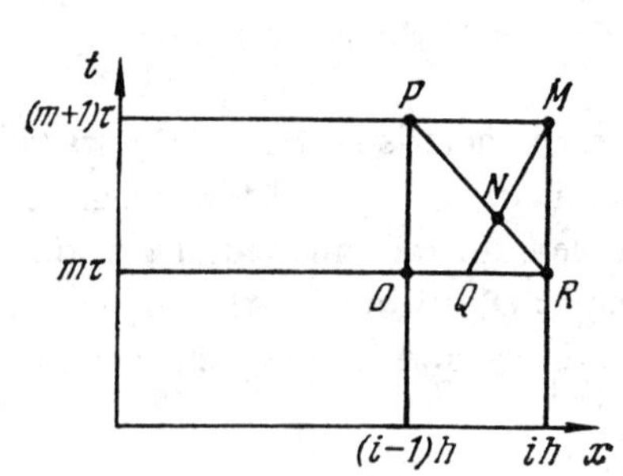

FIGURE 3.5

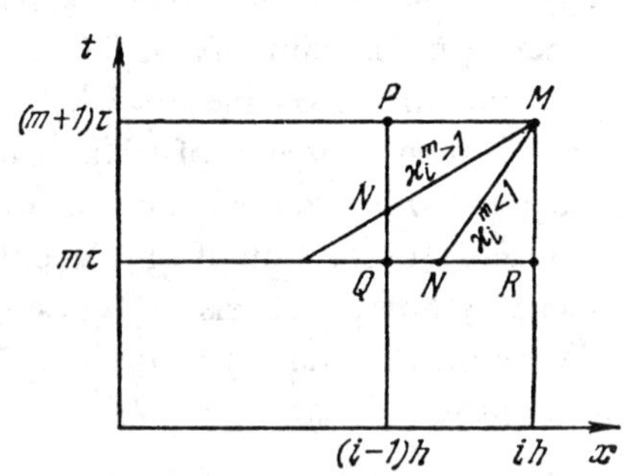

FIGURE 3.6

values on the boundary $t = 0$, $x \geq 0$, $x = 0$, $t \geq 0$, which implies the stability of the scheme (11) in $C(0, \infty)$ for any values of τ and h.

According to (11), $u_i^{m+1} = u^*(N)$, where $u^*(N)$ is the value of u at the point N of the diagonal PR (Figure 3.5) interpolated from the values u_{i-1}^{m+1} and u_i^m. Here N is determined as the intersection of the line (6) (an approximate characteristic MQ of (1)) with the diagonal PR. This implies that the domain of dependence of the difference solution of the mixed problem (1), (3) at the point $x = x_i$, $t = t_m$ is the part of the boundary

$$x = 0, \quad 0 \leq t \leq t_m; \qquad t = 0, \quad 0 \leq x \leq x_i$$

and it always contains the domain of dependence of the exact solution of problem (1), (3).

There is still another version of the implicit scheme for (1):

$$\frac{u_{i-1}^{m+1} - u_{i-1}^m}{\tau} + a_i^m \frac{u_i^{m+1} - u_{i-1}^{m+1}}{h} = 0$$

or in a form solved for u_i^{m+1}

$$u_i^{m+1} = \frac{\kappa_i^m - 1}{\kappa_1^m} u_{i-1}^{m+1} + \frac{1}{\kappa_i^m} u_{i-1}^m = u_i^* \tag{13}$$

which can be interpreted geometrically as the transfer to the point M of the value u_i^* of the difference solution at the point N of the vertical line PQ, where N is the intersection of the characteristic (6) with PQ, and the value at this point is interpolated linearly (for $\kappa_i^m \geq 1$) from the values u_{i-1}^{m+1} and u_{i-1}^m (Figure 3.6). The scheme (13) is conditionally stable in C for $\kappa_i^m \geq 1$. For $\kappa_i^m < 1$ the point N leaves the segment PQ; the interpolation is replaced by extrapolation, and this leads to instability.

It follows from (9) that the explicit scheme for (1) is stable if

$$\tau \leq \frac{h}{\max a(x, t)}, \tag{14a}$$

and is unstable if

$$\tau > \frac{h}{\min a(x, t)}. \tag{14b}$$

The case in which $(\max a(x, t))/(\min a(x, t)) \gg 1$ and the stability condition $\kappa_i^m < 1$ is violated at a few points is especially unpleasant for the explicit scheme. The condition (14a) requires an unjustifiably small step size τ.

In this case it is possible to apply the implicit scheme (10) or (11), which is absolutely stable for any τ and h. However, application of implicit schemes greatly enlarges the domain of dependence of the difference solution (which becomes infinite in the case of the Cauchy problem). In a practical situation this circumstance emerges as increased smoothing of the difference profiles $u^m(x)$ and a loss of characteristic features of the solution. Therefore, if the criterion of stability is violated at a few points, then it makes sense to apply the explicit scheme at points where the stability criterion is satisfied, and the implicit scheme at the remaining points. This objective is served by the explicit-implicit scheme of Carlson

$$\frac{u_i^{m+1} - u_i^m}{\tau} + a_i^m \frac{u_i^m - u_{i-1}^m}{h} = 0, \tag{15a}$$

$$\frac{u_{i-1}^{m+1} - u_{i-1}^m}{\tau} + a_i^m \frac{u_i^{m+1} - u_{i-1}^{m+1}}{h} = 0, \tag{15b}$$

in which (15a) is used for $\kappa_i^m \leq 1$ and (15b) for $\kappa_i^m > 1$. Geometrically the scheme (15) is interpreted as the interpolation of u_i^{m+1} from the values u_i^m and u_{i-1}^m if the line (6) intersects the segment QR ($\kappa_i^m \leq 1$) and from the values u_{i-1}^{m+1} and u_{i-1}^m if (6) intersects PQ ($\kappa_i^m > 1$) (see Figure 3.6).

It is clear that in the case of a negative coefficient $a(x, t)$ the left difference should be replaced by a right difference in all the approximations applied. Thus, (10) now takes the form

$$\frac{u_i^{m+1} - u_i^m}{\tau} + a_i^m \frac{u_{i+1}^{m+1} - u_i^{m+1}}{h} = 0. \tag{16}$$

In this case, of course, the boundary condition should be posed on the right rather than the left boundary.

We now consider the case where $a(x, t)$ changes sign, and for (1) we pose the mixed problem

$$u(-l, t) = f_1(t), \qquad u(l, t) = f_2(t), \qquad u(x, 0) = u_0(x). \tag{17}$$

It is clear that in this case we must have

$$a(-l, t) \geq 0, \qquad a(l, t) \leq 0.$$

Here logical difficulties appear in the case of an implicit scheme. Indeed, the recurrent formula (10) for $a_i^m \geq 0$ is stable in computing from left to right and unstable in computing from right to left; on the other hand, the recurrent formula (16) for $a_i^m < 0$ is stable in computing from right to left and unstable in the opposite direction. Thus, if $a(x, t)$ changes sign we must change not only the choice of (10) and (16) but also simultaneously the direction of recurrent computation. This means that for each $t = t_m$ we must partition $[-l, l]$ into segments

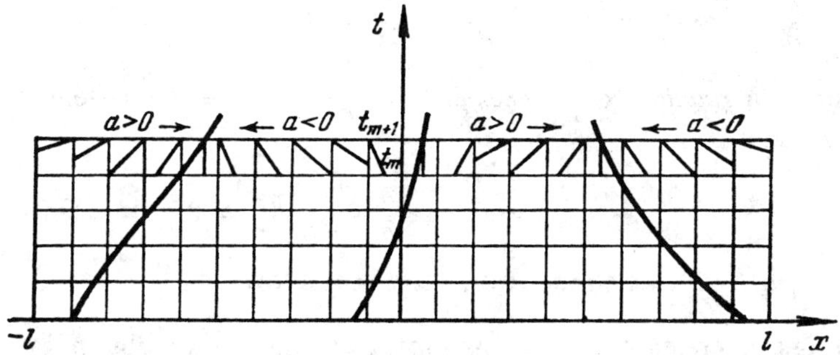

FIGURE 3.7

of constant sign of the function $a^m(x)$, and for each such segment we must solve a one-sided boundary value problem (Figure 3.7). Here the boundary conditions are given only on those lines $a(x, t) = 0$ to the left of which $a < 0$ and to the right of which $a > 0$. The values $u^{m+1}(\xi)$ on these lines are determined by means of the explicit scheme which can be applied, since the Courant condition is satisfied in a neighborhood of the line $a(x, t) = 0$. In this algorithm on the lines $a = 0$, in general, left and right values are obtained which do not coincide but differ by a quantity of order h^2. If we apply the explicit-implicit scheme with consideration of the sign of a, then the values u^{m+1} will be uniquely determined at all points.

There is another method of solving the mixed problem (17) for (1) by means of an implicit scheme. We apply the implicit central approximation

$$\frac{u_i^{m+1} - u_i^m}{\tau} + a_i^m \frac{u_{i+1}^{m+1} - u_{i-1}^{m+1}}{2h} = 0. \tag{18}$$

The relations (18)

$$-\frac{\tau a_i^m}{2h} u_{i-1}^{m+1} + u_i^{m+1} + \frac{\tau a_i^m}{2h} u_{i+1}^{m+1} = u_i^m$$

together with the boundary conditions

$$u_{-N}^{m+1} = f_1[(m+1)\tau], \qquad u_N^{m+1} = f_2[(m+1)\tau]$$

form a closed system of equations for u_i^{m+1} which can be solved by the well-known double sweep method (see, for example, Godunov and Rjaben'kiĭ (1973) and Samarskiĭ (1971)). Here there is a restriction on τ connected with the stability of the double sweep method (pointed out by G. I. Marčuk).

2. ***The "running-count" scheme for the equations of acoustics.*** For the two acoustic equations

$$\frac{\partial u}{\partial t} - a^2 \frac{\partial v}{\partial x} = 0, \qquad \frac{\partial v}{\partial t} - \frac{\partial u}{\partial x} = 0, \quad a = \text{const} > 0 \tag{1}$$

we construct a *running-count scheme*. For this we write the system (1) in invariants r and s:

$$\frac{\partial r}{\partial t} + a\frac{\partial r}{\partial x} = 0, \qquad \frac{\partial s}{\partial t} - a\frac{\partial s}{\partial x} = 0, \tag{2}$$

$$r = u - av, \qquad s = u + av.$$

Applying explicit stable approximations to each of (2), we obtain

$$\frac{r_i^{m+1} - r_i^m}{\tau} + a\frac{r_i^m - r_{i-1}^m}{h} = 0, \qquad \frac{s_i^{m+1} - s_i^m}{\tau} - a\frac{s_{i+1}^m - s_i^m}{h} = 0. \tag{3}$$

The running-count scheme (3) was proposed by Courant, Isaacson, and Rees (1952). This scheme is an explicit scheme and is stable for $\kappa = a\tau/h \leq 1$.

Implicit running-count schemes are constructed similarly:

$$\frac{r_i^{m+1} - r_i^m}{\tau} + a\frac{r_i^{m+1} - r_{i-1}^{m+1}}{h} = 0, \qquad \frac{s_i^{m+1} - s_i^m}{\tau} - a\frac{s_{i+1}^{m+1} - s_i^{m+1}}{h} = 0. \tag{4}$$

The difference schemes (3) and (4) obviously generalize to the case of a variable speed of sound $a = a(x, t)$; in this case right sides linear in r and s obviously appear in (2) and (3). In the case of variable sound speed (or variable step size h_i of the mesh) a combined explicit-implicit scheme is sometimes applied in which an explicit scheme similar to (3) is used at points where $\kappa_i^m = a_i^m\tau/h_i \leq 1$ and an implicit scheme similar to (4) is used at the remaining points of the mesh.

Returning to the variables u, v, we write the explicit scheme (3) in the form

$$\left.\begin{aligned} \frac{u^{m+1} - u^m}{\tau} - a^2\frac{\Delta_1 + \Delta_{-1}}{2h} v^m &= \frac{ah}{2}\frac{\Delta_1\Delta_{-1}}{h^2} u^m, \\ \frac{v^{m+1} - v^m}{\tau} - \frac{\Delta_1 + \Delta_{-1}}{2h} u^m &= \frac{ah}{2}\frac{\Delta_1\Delta_{-1}}{h^2} v^m, \end{aligned}\right\} \tag{5}$$

and the implicit scheme (4) in the form

$$\left.\begin{aligned} \frac{u^{m+1} - u^m}{\tau} - a^2\frac{\Delta_1 + \Delta_{-1}}{2h} v^{m+1} &= \frac{ah}{2}\frac{\Delta_1\Delta_{-1}}{h^2} u^{m+1}, \\ \frac{v^{m+1} - v^m}{\tau} - \frac{\Delta_1 + \Delta_{-1}}{2h} u^{m+1} &= \frac{ah}{2}\frac{\Delta_1\Delta_{-1}}{h^2} v^{m+1}. \end{aligned}\right\} \tag{6}$$

The scheme (5) is the acoustic analogue of Godunov's scheme (the scheme of decay of discontinuities). We consider this in more detail in §8.4.

3. *The "cross" scheme and the implicit scheme with weights.* We have already considered the "cross" scheme

$$\left.\begin{aligned} \frac{u_i^{m+1} - u_i^m}{\tau} &= a_{i-1/2}^2 \frac{v_i^m - v_{i-1}^m}{h}, \\ \frac{v_i^{m+1} - v_i^m}{\tau} &= \frac{u_{i+1}^{m+1} - u_i^{m+1}}{h} \end{aligned}\right\} \tag{1}$$

for the equations of acoustics in considerable detail in §3.2 in the case of variable sound speed $a = a(x)$. It was found there that this scheme is conditionally stable under the condition

$$\kappa(x) = \frac{a(x)\tau}{h} \leq 1. \tag{2}$$

We continue to consider this scheme, assuming for simplicity that $a(x) = \text{const}$.

Although (1) appears to be an implicit scheme (its right side contains u^{m+1}), by first computing u^{m+1} from the first equation and then v^{m+1} from the second, we see that it is equivalent to an explicit scheme. From (1) we obtain the equations ($a = \text{const}$)

$$\frac{u_i^{m+1} - 2u_i^m + u_i^{m-1}}{\tau^2} = a^2 \frac{u_{i+1}^m - 2u_i^m + u_{i-1}^m}{h^2}, \tag{3}$$

$$\frac{v_i^{m+1} - 2v_i^m + v_i^{m-1}}{\tau^2} = a^2 \frac{v_{i+1}^m - 2v_i^m + v_{i-1}^m}{h^2}, \tag{4}$$

from which it becomes obvious that the "cross" scheme (1) has second order of approximation of the acoustic equations both in τ and h.

From (3) and (4) the reason for the name "cross" scheme becomes clear: the points of the mesh $x = (i + \alpha)h$, $t = m\tau$ and $x = ih$, $t = (m + \alpha)\tau$ for $\alpha = -1, 0, 1$ contained in (3) and (4) form a cross with three points on each of the horizontal and vertical lines passing through the point $x = ih$, $t = m\tau$ (Figure 3.8).

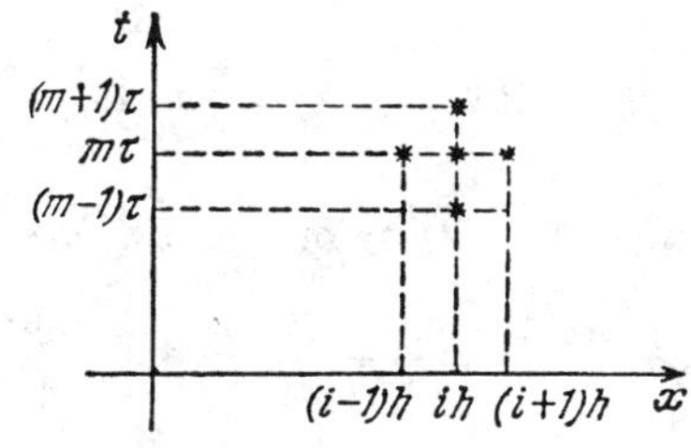

FIGURE 3.8

The dispersion equation for the scheme (1), or, equivalently, for (3) and (4), has the form

$$\rho^2 - 2\left[1 - 2\kappa^2 \sin^2 \frac{kh}{2}\right]\rho + 1 = 0, \qquad \kappa = a\tau/h, \tag{5}$$

from which it follows that $|\rho_{1,2}| = 1$ for $\kappa \leq 1$, while for $\kappa > 1$ there exists kh so large that $\max\{|\rho_1|, |\rho_2|\} > 1$ and the scheme is unstable.

We consider the more general scheme with parameters α_1 and α_2:

$$\left.\begin{aligned} \frac{u^{m+1} - u^m}{\tau} - a^2 \frac{\Delta_{-1}}{h}[\alpha_1 v^{m+1} + (1 - \alpha_1)v^m] &= 0, \\ \frac{v^{m+1} - v^m}{\tau} - \frac{\Delta_1}{h}[\alpha_2 u^{m+1} + (1 - \alpha_2)u^m] &= 0. \end{aligned}\right\} \tag{6}$$

For $\alpha_1 = 0$ and $\alpha_2 = 1$ we have the "cross" scheme. For $\alpha_1 = \alpha_2 = 1/2$ we have a scheme of second-order accuracy which is absolutely stable. This is easily seen by going over to the equivalent equation for u:

$$\frac{u^{m+1} - 2u^m + u^{m-1}}{\tau^2} - a^2 \frac{\Delta_1\Delta_{-1}}{h^2}\left[\frac{u^{m-1} + 2u^m + u^{m+1}}{4}\right] = 0. \tag{7}$$

The dispersion equation

$$\rho^2 - 2\rho \frac{1 - \kappa^2 \sin^2 (kh/2)}{1 + \kappa^2 \sin^2 (kh/2)} + 1 = 0, \qquad \kappa = \frac{a\tau}{h},$$

of the scheme (7) for any κ has complex conjugate roots equal to 1 in modulus.

4. ***The Lax scheme.*** Lax (1954) proposed a scheme which in the case of the acoustic equations takes the form

$$\frac{u_i^{m+1} - \bar{u}_i^m}{\tau} - a^2 \frac{v_{i+1}^m - v_{i-1}^m}{2h} = 0, \tag{1}$$

$$\frac{v_i^{m+1} - \bar{v}_i^m}{\tau} - \frac{u_{i+1}^m - u_{i-1}^m}{2h} = 0, \tag{2}$$

where

$$\bar{u}_i^m = \frac{u_{i-1}^m + u_{i+1}^m}{2}, \qquad \bar{v}_i^m = \frac{v_{i-1}^m + v_{i+1}^m}{2}. \tag{3}$$

Eliminating $\bar{u}_i$ and $\bar{v}_i$ from (1) and (2) by means of (3), we arrive at the scheme

$$\left.\begin{aligned} \frac{u_i^{m+1} - u_i^m}{\tau} - a^2 \frac{v_{i+1}^m - v_{i-1}^m}{2h} &= \frac{h^2}{2\tau}\frac{u_{i-1}^m - 2u_i^m + u_{i+1}^m}{h^2}, \\ \frac{v_i^{m+1} - u_i^m}{\tau} - \frac{u_{i+1}^m - u_{i-1}^m}{2h} &= \frac{h^2}{2\tau}\frac{v_{i-1}^m - 2v_i^m + v_{i+1}^m}{h^2}. \end{aligned}\right\} \tag{4}$$

The dispersion equation has the form

$$\left(\frac{\rho - 1}{\tau} + \frac{2}{\tau}\sin^2\frac{kh}{2}\right)^2 + a^2\frac{\sin^2 kh}{h^2} = 0,$$

whence

$$|\rho_1|^2 = |\rho_2|^2 = 1 - (1 - \kappa^2)\sin^2 kh.$$

The scheme (4) is thus stable under the condition $\kappa \leq 1$. The scheme (4) approximates the acoustic equations (4.2.1) conditionally: for the limiting procedure $\tau/h = \text{const}$ the scheme (4) approximates (4.2.1) with accuracy $O(\tau)$; for the limiting procedure $\tau/h^2 = \text{const}$ the scheme (4) approximates the parabolic system

$$\frac{\partial u}{\partial t} - a^2\frac{\partial v}{\partial x} = \mu\frac{\partial^2 u}{\partial x^2},$$

$$\frac{\partial v}{\partial t} - \frac{\partial u}{\partial x} = \mu\frac{\partial^2 v}{\partial x^2}, \qquad \mu = \frac{h^2}{2\tau}.$$

5. ***A symmetric scheme of second-order accuracy (the predictor-corrector scheme).*** The scheme (4.4.1), (4.4.2) can be used as a component of a scheme of second-order accuracy in which the integration proceeds in two steps: at the first (predictor) step the auxiliary quantities are computed according to the Lax scheme, and at the second (corrector) step refinement occurs according to a scheme of "cross" type.

We consider the scheme

$$\left.\begin{aligned} &\frac{u_i^{m+1} - u_i^m}{\tau} - a^2\frac{v_{i+1/2}^{m+1/2} - v_{i-1/2}^{m+1/2}}{h} = 0,\\ &\frac{v_i^{m+1} - v_i^m}{\tau} - \frac{u_{i+1/2}^{m+1/2} - u_{i-1/2}^{m+1/2}}{h} = 0, \end{aligned}\right\} \tag{1}$$

where the quantities $u_{i+1/2}^{m+1/2}$ and $v_{i+1/2}^{m+1/2}$ are computed by the Lax scheme (4.4.1), (4.4.2) with step size $\tau/2$:

$$\left.\begin{aligned} &\frac{u_{i+1/2}^{m+1/2} - \bar{u}_{i+1/2}^m}{\tau/2} - a^2\frac{v_{i+1}^m - v_i^m}{h} = 0,\\ &\frac{v_{i+1/2}^{m+1/2} - \bar{v}_{i+1/2}^m}{\tau/2} - \frac{u_{i+1}^m - u_i^m}{h} = 0, \end{aligned}\right\} \tag{2}$$

where

$$\bar{u}_{i+1/2}^m = \frac{u_i^m + u_{i+1}^m}{2}, \qquad \bar{v}_{i+1/2}^m = \frac{v_i^m + v_{i+1}^m}{2}.$$

After eliminating $u^{m+1/2}$ and $v^{m+1/2}$ from (1) and (2), we arrive at a homogeneous, symmetric scheme having second-order accuracy in τ and h:

$$\left.\begin{aligned} \frac{u_i^{m+1} - u_i^m}{\tau} - a^2 \frac{v_{i+1}^m - v_{i-1}^m}{/2h} &= \mu \frac{u_{i+1}^m - 2u_i^m + u_{i-1}^m}{h^2}, \\ \frac{v_i^{m+1} - v_i^m}{\tau} - \frac{u_{i+1}^m - u_{i-1}^m}{2h} &= \mu \frac{v_{i+1}^m - 2v_i^m + v_{i-1}^m}{h^2}, \end{aligned}\right\} \tag{3}$$

where $\mu = a^2\tau/2$. The dispersion equation has the form

$$\left(\rho - 1 + 2\kappa^2 \sin^2 \frac{kh}{2}\right)^2 + \kappa^2 \sin^2 kh = 0.$$

For $\kappa = a\tau/h \leq 1$ we have $|\rho_1| = |\rho_2| \leq 1$, and the scheme is stable.

§5. Methods of constructing difference schemes for the equations of gas dynamics

1. *General remarks.* In preceding sections we have presented elements of the theory of difference schemes applied exclusively to linear problems of mathematical physics. In proceeding to a discusion of the methods of solving problems of gas dynamics, we note that this theory provides an understanding of the basic regular and singular features of the numerical solution not only of linear problems but of nonlinear problems as well; problems of gas dynamics are almost always nonlinear.

In this connection it must be acknowledged that almost all the basic concepts and conclusions of the theory of difference schemes developed for linear problems are inapplicable in a precise sense to the nonlinear difference problems arising in gas dynamics. The basic concepts of the theory of difference schemes—the concepts of approximation, stability, and convergence—require revision and generalization, and this revision is at present far from complete.

Since motion in media without viscosity and thermal conductivity (and this model is used rather extensively) is described by discontinuous functions, in general, differential equations are not applicable for its description. As we know from Chapter 2, the motion in such media can be described by systems of integral or integro-differential equations—the conservation laws of mass, momentum, and energy. Thus, in gas dynamics we deal, in general, with discontinuous solutions of systems of integro-differential equations. Therefore, approximation should be understood as the approximation of integral conservation laws in a class of discontinuous solutions. In particular, it is not possible to require of the exact solution that it have bounded derivatives, since they do not exist.

Therefore, the concept of order of approximation must be obtained by other means; that is, by other norms of the closeness of solutions.

In the same way, the concept of convergence of a difference scheme depends in an essential way on the class of solutions of the conservation laws. It is clear that the norm of closeness of two generalized (discontinuous) solutions is weak. In Chapter 4 we discuss a norm of closeness of two generalized solutions; the most suitable is the norm of the space L_1.

Similarly, the concept of stability of a nonlinear difference scheme carried over from the linear theory has very limited value due to the lack of a principle for the superposition of solutions. Briefly, the question of approximation with prescribed accuracy of a solution of systems of nonlinear conservation laws by a solution of a system of nonlinear difference equations does not break down into simple, separate requirements but, as a rule, is solved globally and in a particular way for each restricted class of problems.

The importance of the linear theory of difference schemes should not, however, be underestimated, especially since it is at present the only tool for investigating nonlinear schemes. In addition to lines of discontinuity, a solution of a system of conservation laws has regions in which it is a classical solution of the differential equations of gas dynamics. Therefore, in these regions one can apply the concepts of approximation, accuracy of approximation, and other concepts of the linear theory. On a given background (that is, by fixing some solution of the equations of gas dynamics) one can study the development of small deviations of the solution by means of the concept of stability from the linear theory.

Thus, our general conclusion is as follows. It is necessary to apply all the concepts and methods of the linear theory of difference schemes, but for the case of nonlinear schemes it is equally necessary to remember that they have not been justified and may lead to false conclusions.

2. ***Methods of describing gas-dynamical flows and constructing difference schemes.*** The character of the schemes applied to integrate the equations of gas dynamics depends essentially on the method of describing the flow. In the preceding chapters we have used the following three methods of describing the flow.

FIRST METHOD. The region G of the x, t plane in which the motion is considered is decomposed by strong and weak discontinuities into regions G_i of smooth flow in which the equations of gas dynamics are satisfied, while compatibility conditions are satisfied on the discontinuities. In this treatment a generalized solution is a collection of smooth solutions defined in regions G_i adjoining one another across lines of discontinuities with observation of the compatibility conditions. In this description it is necessary to integrate the equations of gas dynamics numerically in the regions G_i, and to satisfy the matching conditions on the lines of discontinuity.

The best-known difference method corresponding to the first method of description is the method of characteristics. Indeed, among the separating lines we have weak discontinuities and contact boundaries which are characteristics, and this makes the characteristic difference scheme convenient. The full details of the description of the flow, which constitute a positive feature of the method of characteristics, at the same time impedes its realization on a computer due to the complicated logic of computing singularities and the construction of the computational front. Of course, the method of characteristics is not the only difference method that can be applied within the framework of a detailed description of the flow.

SECOND METHOD. The integral conservation laws in Euler or Lagrange coordinates constitute a generalized solution. This provides a unified description, since both the equations of gas dynamics and the compatibility conditions are consequences of the conservation laws. Difference schemes corresponding to the second method of description are obtained by a unified approximation of the conservation laws independent of the character of the flow and are therefore called *homogeneous schemes* or *schemes of continuous count*.[14]

THIRD METHOD. A generalized solution is defined as the limit of classical solutions of a certain system of quasilinear parabolic equations with a small parameter in the highest-order derivatives. If

$$\frac{\partial u}{\partial t} + \frac{\partial \varphi(u)}{\partial x} = f(u) \tag{1}$$

is the original system of equations of gas dynamics written as conservation laws, then the corresponding parabolic system has the form

$$\frac{\partial u}{\partial t} + \frac{\partial \varphi(u)}{\partial x} = f(u) + \frac{\partial}{\partial x}\left(\mu B(u) \frac{\partial u}{\partial x}\right). \tag{2}$$

Here $u = u(x, t)$ is a vector-valued function describing the flow, $f(u)$ and $\varphi(u)$ are vector-valued functions of a vector argument u, $B(u)$ is a square matrix, and μ is a small parameter.[15] The matrix $B(u)$ must be chosen so that the solution $u(x, t)$ of (2) possesses sufficient smoothness and as $\mu \to 0$ tends in some sense to the solution of (1).

Defference schemes based on the third method also have the character of

[14] The concept of homogeneous schemes was introduced and studied by Tihonov and Samarskiĭ (see, for example, their 1961 paper).

[15] Some schemes with an artificial viscosity μ contain a function of $\partial u/\partial x$.

schemes of continuous count. In some cases the second and third approaches lead to the same schemes.

3. ***Boundary conditions in problems of gas dynamics.*** We first consider boundary conditions for the case of a gas with no viscosity or thermal conductivity, and we then comment on treating these dissipative processes in the boundary conditions.

The formulation of boundary conditions naturally depends on the problem considered, but their form depends also on the method of describing the flow and, first of all, on the system of coordinates adopted. In most gas-dynamical problems the boundary conditions are most simply and naturally written in Lagrange coordinates. This is also reflected in the difference methods: difference schemes using Lagrange coordinates account most accurately with inner and outer boundaries. On an inner boundary $q = \text{const}$ (q is the Lagrange coordinate) conditions for the continuity of the pressure p and velocity u are posed:

$$p(q + 0, t) = p(q - 0, t), \qquad u(q + 0, t) = u(q - 0, t). \tag{1}$$

Condition (1) is the compatibility condition for a *contact discontinuity*. The boundary $q = \text{const}$ separating two different gases is a contact discontinuity. In order that a difference scheme admit a contact discontinuity (without smearing it out) it is necessary that the representation of this difference scheme in Lagrange coordinates not contain any (independent) spatial gradients other than the gradients of pressure and velocity. Thus, for example, a difference approximation of the conservation law

$$\frac{\partial v}{\partial t} - \frac{\partial u}{\partial q} = 0, \qquad \frac{\partial u}{\partial t} + \frac{\partial p}{\partial q} = 0, \qquad \frac{\partial}{\partial t}\left(\varepsilon + \frac{\mu^2}{2}\right) + \frac{\partial pu}{\partial q} = 0$$

will satisfy this requirement, since only quantities which are continuous on the contact boundary are differentiated with respect to the space variable. When viscosity and thermal conductivity are present, conditions for the continuity of the force of friction and the flow of heat are usually posed on contact boundaries.

The formulation of boundary conditions on outer bounaries has some diversity, corresponding to the relatively large number of problems of practical interest. We note that it is very convenient to consider problems with fixed masses of the gases in Lagrange coordinates. On the other hand, exterior boundary conditions posed on cross sections through which gases flow are not so conveniently described in Lagrange coordinates, and sometimes it is more natural to formulate and approximate them in Euler coordinates.

It is very inconvenient to approximate internal boundary conditions (1) on contact boundaries on meshes with constant Euler coordinates (Euler meshes), since the contact boundaries move relative to these meshes in passing from one computational interval to another. This leads to the necessity of selecting special difference schemes in Euler coordinates which possess the property that they do not smear out or only weakly smear out the contact boundaries (the *K*-property of a difference scheme).

The method of characteristics is perhaps the most universal in the sense of the approximation on boundary conditions. However, it has its own drawbacks: it is applicable only to equations of hyperbolic type, which in the case of gas dynamics implies the absence of dissipative processes, and it has variable space step size depending on the solution—which is not always convenient.

Just the enumeration of the advantages and drawbacks of approximation of only the boundary conditions in different coordinate systems shows that there is no universals or best method for the difference solution of problems of gas dynamics, and that in real computations it is necessary to apply various methods.

Finally, we mention also that shock waves in gases can also be considered as mobile inner boundary conditions. These conditions enter explicitly only in the method of characteristics; they are missing in other difference methods, since in these schemes shock waves are considered in another sense as a narrow but finite zone of smooth shock transition.

§6. The method of characteristics

The method of characteristics is one of the most popular methods of integrating systems of hyperbolic equations. Its characteristic feature as compared with other difference methods is the minimum use of interpolation operators and the accompanying maximum closeness of the domain of dependence of the difference scheme and the domain of dependence of the system of differential equations. The smoothing of profiles which is characteristic of difference schemes with a fixed mesh is a minimum in the method of characteristics, since the mesh the latter employs is constructed with consideration of the domain of dependence of the system.

The method of characteristics is expounded in detail in the monographs of Panov (1957) and Žukov (1960), and we shall therefore limit ourselves to a brief exposition of the features of this method and the elements of the difference algorithm.

The method of characteristics starts from an approximation of the system of characteristic equations of gas dynamics in a characteristic mesh. Here it is possible to use either Euler or Lagrange coordinates.

1. ***The method of characteristics for smooth flows.*** We first consider the method of characteristics in application to a system of quasilinear equations in invariants.

$$\left.\begin{aligned}\left(\frac{\partial s}{\partial t}\right)_1 &= \frac{\partial s}{\partial t} + \xi_1(r, s)\frac{\partial s}{\partial x} = F_1(r, s, x, t),\\ \left(\frac{\partial r}{\partial t}\right)_2 &= \frac{\partial r}{\partial t} + \xi_2(r, s)\frac{\partial r}{\partial x} = F_2(r, s, x, t).\end{aligned}\right\} \tag{1}$$

As is known, a system of equations of the type (1) represents the equations of gas dynamics in the case of plane, cylindrical, and spherical symmetry and constant entropy (Euler coordinates) and in the case of plane symmetry and variable entropy (Lagrange coordinates) (see Chapter 2).

Suppose that for the system (1) we pose the problem with initial data

$$r(x, 0) = r_0(x), \qquad s(x, 0) = s_0(x), \qquad a \le x \le b, \tag{2}$$

which has a smooth solution in some region G containing the segment $[a, b]$ of the x axis (Figure 3.9).

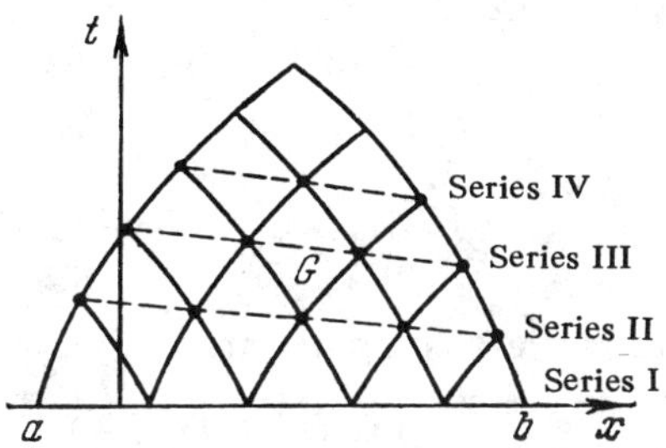

FIGURE 3.9

We decompose $[a, b]$ into intervals $[x_i, x_{i+1}]$, where $x_0 = a$ and $x_{N+1} = b$. The points $(x_i, 0)$ form the first series of calculating points (series I). The next series of calculating points (series II) contains the points of intersection of the r- and s-characteristics issuing from points of series I. If the mth series of calculating points (x_i^m, t_i^m) has been determined, then the next series (x_i^{m+1}, t_i^{m+1}) is determined from the formulas (first approximation)

$$\left.\begin{aligned}\frac{x_i^{m+1} - x_i^m}{t_i^{m+1} - t_i^m} &= \xi_{2i}^m,\\ \frac{x_i^{m+1} - x_{i+1}^m}{t_i^{m+1} - t_{i+1}^m} &= \xi_{1\,i+1}^m,\end{aligned}\right\} \tag{3}$$

where

$$\xi_{\alpha i}^m = \xi_\alpha(r_i^m, s_i^m, x_i^m, t_i^m), \qquad \alpha = 1, 2,$$

and the r_i^m and s_i^m are the values of the invariants at the points x_i^m and t_i^m.

After this the invariants r_i^m and s_{i+1}^m are carried to the point (x_i^{m+1}, t_i^{m+1}) with the corresponding increments

$$\left.\begin{aligned} r_i^{m+1} &= r_i^m + \Delta r_i^m = r_i^m + F_{2i}^m(t_i^{m+1} - t_i^m), \\ s_i^{m+1} &= s_{i+1}^m + \Delta s_{i+1}^m = s_{i+1}^m + F_{1\,i+1}^m(t_i^{m+1} - t_{i+1}^m), \\ F_{\alpha i}^m &= F_\alpha(r_i^m, s_i^m, x_i^m, t_i^m), \qquad \alpha = 1, 2. \end{aligned}\right\} \tag{4}$$

The scheme of first-order accuracy involves the replacement of characteristics passing through points of the lower series x_i^m, t_i^m by the lines (3) and the approximate integration of (1) along the characteristics by the Euler method.

To refine (3) and (4), both the points of the series (x_i^{m+1}, t_i^{m+1}) and the values r_i^{m+1} and s_i^{m+1} are recalculated. In (3) the right sides (the slope of the lines approximating the characteristics) are replaced by averages

$$\frac{x_i^{m+1} - x_i^m}{t_i^{m+1} - t_i^m} = \frac{\xi_{2i}^m + \xi_{2i}^{m+1}}{2}; \qquad \frac{x_i^{m+1} - x_{i+1}^m}{t_i^{m+1} - t_{i+1}^m} = \frac{\xi_{1\,i+1}^m + \xi_{1i}^{m+1}}{2}, \tag{5}$$

where

$$\xi_{\alpha i}^{m+1} = \xi_\alpha(r_i^{m+1}, s_i^{m+1}, x_i^{m+1}, t_i^{m+1}), \tag{6}$$

and x_i^{m+1}, t_i^{m+1}, r_i^{m+1} and s_i^{m+1} are computed from (3) and (4).

The invariants r_i^m and s_i^m are carried to the points (x_i^{m+1}, t_i^{m+1}) in the manner indicated with increments refined by the trapezoidal formula:

$$\left.\begin{aligned} r_i^{m+1} &= r_i^m + \Delta r_i^m = r_i^m + \frac{F_{2i}^m + F_{2i}^{m+1}}{2}(t_i^{m+1} - t_i^m), \\ s_i^{m+1} &= s_{i+1}^m + \Delta s_{i+1}^m = s_{i+1}^m + \frac{F_{1\,i+1}^m + F_{1i}^{m+1}}{2}(t_i^{m+1} - t_{i+1}^m). \end{aligned}\right\} \tag{7}$$

Here F_{1i}^m and F_{2i}^m preserve the previous values from (4), while F_{1i}^{m+1} and F_{2i}^{m+1} are determined from

$$F_{\alpha i}^{m+1} = F_\alpha(r_i^{m+1}, s_i^{m+1}, x_i^{m+1}, t_i^{m+1}) \qquad (\alpha = 1, 2), \tag{8}$$

where r_i^{m+1} and s_i^{m+1} are computed from (4), and x_i^{m+1} and t_i^{m+1} from (5). Recalculation according to (5)–(8) raises the order of accuracy of the scheme, but further recalculation does not lead to further increase in the order of accuracy, and thus one recalculation suffices.

For a polytropic gas with $\gamma = 3$, plane symmetry, and constant entropy equations (1) take the form (see Chapter 2, §2.9)

$$\frac{\partial r}{\partial t} + r\frac{\partial r}{\partial x} = 0, \qquad \frac{\partial s}{\partial t} + s\frac{\partial s}{\partial x} = 0, \tag{9}$$

In this case (3) and (4) without recalculation already give the exact solution of the Cauchy problem for (9), since the characteristics of (9) are lines. The scheme (3), (4), which has first-order accuracy on smooth functions, in the present case possesses an infinitely large order of accuracy on the class of smooth solutions of (9). On the other hand, a scheme of any order accuracy with a constant mesh gives only an approximate solution of (9). This example illustrates well the advantages of the characteristic mesh which minimizes the difference of the domains of dependence of the scheme and the equation and thus the magnitude of the remainder term.

The construction of the computational front can also be carried out in the regular case in another way, not along spacelike lines as shown in Figure 3.8 but along characteristic lines. This algorithm of computation carries over to more general systems of equations. Suppose that

$$l_\alpha^k\left(\frac{du_\alpha}{dt}\right)_k = l_\alpha^k\left(\frac{\partial u_\alpha}{\partial t} + \xi_k\frac{\partial u_\alpha}{\partial x}\right) = f_k \qquad (k, \alpha = 1, \ldots, n) \tag{10}$$

is a hyperbolic system in characteristic form for which there exist invariants.

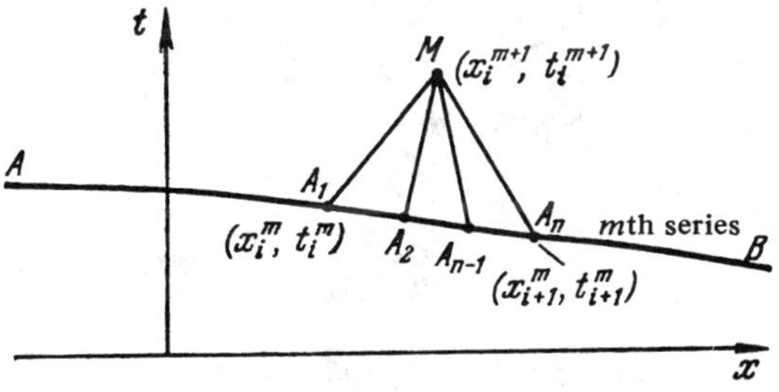

FIGURE 3.10

The equations $dx/dt = \xi_k(x, t, u)$ define n one-parameter families of characteristics. In the general case any pair of families forms a characteristic mesh which does not coincide with the mesh corresponding to another pair. To be specific, suppose we have chosen a pair of characteristics corresponding to the indices $k = 1$ and $k = n$, let M be a calculating point of the mesh (Figure 3.10). and let A_k be the base of the kth characteristic dropped from M to the line AB. The points A_1 and A_n are nodes of the mesh, the points $A_2, \ldots, A_{n-1}$ are located between them, and to determine u at $A_2, \ldots, A_{n-1}$ it is required to interpolate from $u(A_1)$ and $u(A_n)$. Thus, in addition to transport along the characteristic, an interpolation operator arises which leads to smoothing effects inherent in the usual difference methods.

In the general case the equations (10) do not have invariants, and it is necessary to go over to the more complicated prolonged system (see Chapter 1, §5). Neverthless, the method of characteristics preserves the domain of dependence of the hyperbolic system with great accuracy in this case as well.

The regular algorithm indicated for constructing a characteristic difference mesh and the approximate integration of equations in invariants is possible in the region of a smooth solution. If in a neighborhood of a calculating point there is a singularity (a shock wave, a contact boundary, any discontinuity, the site of origin of a shock wave, or a centered rarefaction wave), then the formulas are altered in correspondence with the nature of the singularity and the configuration of the mesh. We shall consider a number of typical configurations.

2. *The method of characteristics in a neighborhood of a contact boundary.* We shall illustrate singularities of the algorithm with the example of a plane-symmetric piecewise isentropic flow and a Lagrange coordinate system.

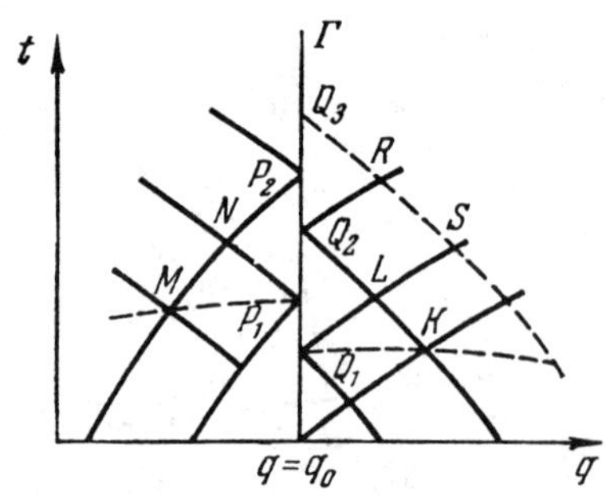

FIGURE 3.11

The calculating points must now be placed not only on the r- and s-characteristics but also on the contact boundary which is the coordinate line q = cost (Figure 3.11). The inhomogeneity connected with the boundary leads to two computational fronts.

Let P_1 and Q_1 be the lowest calculating points on the contact boundary Γ located respectively on the left and right side of the boundary; let M and K be points of the computational front next to them. The points N and L are computed in the regular manner. In first approximation, proceeding from N, it is possible to compute the point P_2 of intersection of the r-characteristic MN

with the boundary Γ and to determine the value of the invariant r_l at this point. Similarly, proceeding from L, it is possible to determine Q_2 and the value of the invariant s_r at this point. On the basis of P_1 and P_2 the value of r_l at Q_2 is interpolated.

The conditions of continuity of P and u at the contact boundary taken at Q_2 lead to the relations

$$\left.\begin{aligned} p &= P_l(S_l, r_l - s_l) = P_r(S_r, r_r - s_r), \\ u &= \frac{r_l + s_l}{2} = \frac{r_r + s_r}{2}. \end{aligned}\right\} \qquad (1)$$

The quantities r_r and s_l at Q_2 are determined from (1). Thus, the computational front on the right moves one step, and we can compute R, Q_3 and the invariant s_r at the latter point in first approximation. In analogy to the foregoing this makes it possible to calculate P_2, after which in first approximation the cycle of computations is completed. It is also possible to construct the formulas of the second approximation, but these are rather complicated.

3. ***The method of characteristics in a neighborhood of a shock wave.*** The presence of a discontinuity again leads to two computational fronts, but, in contrast to the preceding case, the line of discontinuity is now not a timelike line. It follows from Zemplén's theorem that the line of a shock wave is spacelike on the leading front and timelike on the trailing front (Figure 3.12).

The line of the shock wave "cuts" the mesh of characteristics in front of it, and all values of the quantities on the leading front are known, since they are delivered by characteristics from below.

On the other hand, on the trailing front only the invariant r is carried along a characteristic (in the case of a wave proceeding to the right), while the invariant s is carried away from the line of the shock wave and is determined on it from the Hugoniot conditions. For simplicity we consider the case of an isothermic gas where the Hugoniot conditions are written in invariants (see Chapter 2, §4.6):

$$r_l - r_r = a\varphi(M), \qquad (1)$$

$$s_l - s_r = a\psi(M), \qquad (2)$$

$$\left.\begin{aligned} M = \frac{D-u_r}{a}, \qquad \varphi(M) &= M - \frac{1}{M} + \ln M^2, \\ \psi(M) &= M - \frac{1}{M} - \ln M^2, \end{aligned}\right\} \qquad (3)$$

where a denotes the isothermic speed of sound.

Knowing r_l, r_r and u_r, we determine the speed of the shock wave D from (1)

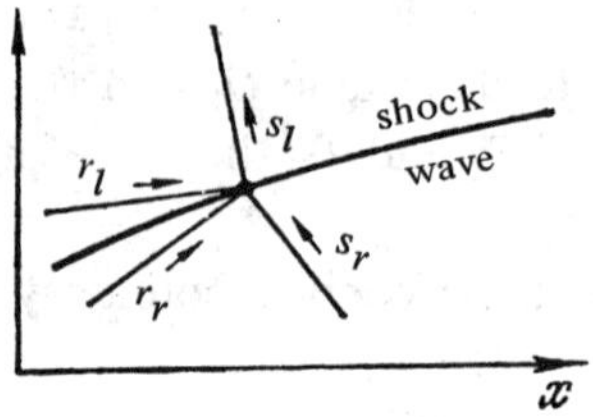

FIGURE 3.12

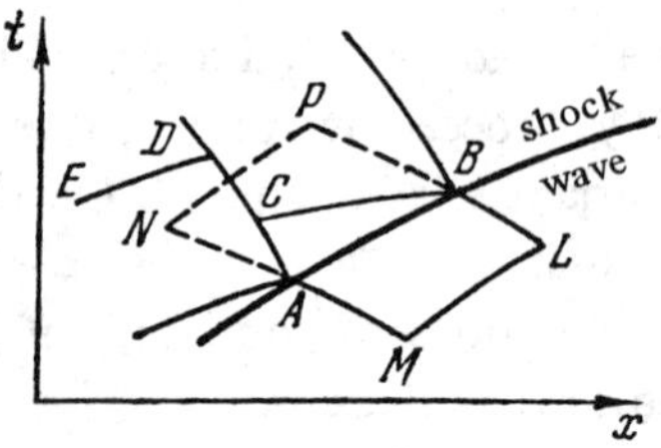

FIGURE 3.13

and (3), and from (2) we determine s_l. Suppose that at the points A and B the line of the shock wave (Figure 3.13) intersects an element (cell) $LMNP$ of the characteristic mesh already computed, and suppose that at A the quantities r_l, s_l, r_r, s_r and hence D are known. Then the characteristic triangle EAD to the left of the line of the shock wave can be completely computed. From A we draw a segment of the line AB

$$\frac{x_B - x_A}{t_B - t_A} = D_A,$$

approximating the trajectory of the shock wave in a neighborhood of A. On the characteristic PL it is possible to interpolate the values r_r and s_r at the point B on the basis of their values at P and L. Since the segment AD has been completely computed, on this segment one can find a point C such that in the first approximation the r-characteristic passing through it passes through the point B. This problem is solved by linear interpolation of the invariants r_D, s_D, r_A and s_A at C under the condition that a segment of the line

$$\frac{x - x_C}{t - t_C} = \frac{r_C + s_C}{2} + a$$

passes through B. If we denote the parameter of linear interpolation along AD by θ, then the computational formulas for determining θ have the form

$$\frac{x_B - x_C}{t_B - t_C} = \frac{x_B - [x_A + \theta(x_D - x_A)]}{t_B - [t_A + \theta(t_D - t_A)]} = (u_A)_l + \theta[u_D - (u_A)_l] + a.$$

After this all quantities are determined. Having determined $(r_B)_l$, one can compute D_B, after which the point B is recalculated by the trapezoidal formula

$$\frac{x_B - x_A}{t_B - t_A} = \frac{D_A + D_B}{2}.$$

This makes it possible to recalculate C and also the invariants $(r_B)_l$, $(s_B)_l$ and $(r_B)_r$, after which the point B is determined, and the computational front to the left of the shock wave can move forward another step.

The computation can be carried out in a similar way, with some complication, in the case of a flow with variable entropy; here it is necessary to consider the trajectory.

Of course, other computational formulas and other configurations of the relative positions of the line of the shock wave and the characteristic mesh are possible.

Analogous difficulties arise in considering singularities of the following types:

3) a centered rarefaction wave;

4) decay of a discontinuity;

5) intersection of characteristics of the same family with subsequent formation of a shock wave; and

6) a boundary with the vacuum in which degeneration of the element of the mesh occurs.

In each specific situation the problem of moving the computational front one step in a neighborhood of the given singularity reduces to interpolation and solution of problems of analytic geometry.

Difficulties arise in the method of characteristics in constructing the computational front in the presence of a large number of singularities of different type. The computation then becomes irregular, and the determination of the possible are added difficulties in allocating memory caused by the shock wave cutting the characteristic mesh.

In spite of the major logical difficulties of realizing the method of characteristics on a computer, in the USSR programs have been created which make it possible to compute with great accuracy flows containing a large number of singularities (see Žukov (1960)).

§7. Explicit running-count schemes

The running-count schemes are closest to the method of characteristics. Just as the method of characteristics, they start from equations in invariants or in characteristic form, but the differene mesh is not a characteristic mesh.

We begin by considering explicit running-count schemes for a one-dimensional, plane, piecewise isentropic flow of gas not containing shock waves. It is known that such a flow is described in Lagrange coordinates by the system of equations in invariants (see Chapter 2, §3).

$$\frac{\partial r}{\partial t} + a(S, r - s)\frac{\partial r}{\partial q} = 0, \qquad \frac{\partial s}{\partial t} - a(S, r - s)\frac{\partial s}{\partial q} = 0, \tag{1}$$

where $a = c\rho$ is the mass speed of sound, and the entropy S is a piecewise constant function with discontinuities which are contact boundaries.

For the system (1) we pose the problem with initial and boundary conditions

$$r(q, 0) = r_0(q), \qquad a(q, 0) = s_0(q), \quad 0 \le q \le Q, \tag{2}$$

$$u(0, t) = f(t), \qquad p(Q, t) = g(t). \tag{3}$$

The boundary conditions (3) mean that the velocity on the left boundary and the pressure on the right boundary are given. Of course, it is possible to choose another combination of boundary conditions (for example, the velocity or pressure on both boundaries). The equation of state is given by the formula[16]

$$p = P(S, r - s). \tag{4}$$

In the absence of shock waves in the flow the entropy is constant on each line $q = \text{const}$, including the boundaries $q = 0$ and $q = Q$. Therefore, the boundary conditions (3) can be written in the form

$$\begin{aligned} &r(0, t) + s(0, t) = 2f(t), \\ &r(Q, t) - s(Q, t) = k(t) = P^{-1}(p, S) = P^{-1}(g(t), S). \end{aligned} \tag{5}$$

On the inner contact boundaries $q = q_i$ we pose the usual conditions

$$p_i^+ = p_i^-, \qquad u_i^+ = u_i^-, \tag{6}$$

which in terms of invariants have the form

$$\left.\begin{aligned} P(S_i^+, r_i^+ - s_i^+) &= P(S_i^-, r_i^- - s_i^-), \\ r_i^+ + s_i^+ &= r_i^- + s_i^-. \end{aligned}\right\} \tag{7}$$

Here the + sign denotes quantities to the right and the − sign quantities to the left of the ith boundary.

In the case of a polytropic gas where

$$P(S, r - s) = \mathscr{F}(S)(r - s)^\alpha, \tag{8}$$

with $\alpha = 2\gamma/(\gamma - 1)$, the conditions (7) become linear:

$$r_i^+ - s_i^+ = \kappa_i(r_i^- - s_i^-), \qquad \kappa_i = \left(\frac{\mathscr{F}_i^-}{\mathscr{F}_i^+}\right)^{1/\alpha}. \tag{9}$$

Passing to difference equations and assuming initially that there are no contact boundaries, we construct a mesh which is uniform with respect to the

[16] We assume for simplicity that the contact boundaries separate gases with the same equation of state (4) but with different values of the entropy. The problem is considerably more complicated if the gases are assumed distinct.

mass, so that

$$\left.\begin{aligned} h = \Delta q_i = q_{i+1} - q_i = \text{const} = \frac{Q}{N+1}, \\ q_0 = 0, \qquad q_1 = h, \ldots, q_{N+1} = (N+1)h = Q. \end{aligned}\right\} \tag{10}$$

To solve the problem it is possible to used an explicit running-count scheme

$$r_i^{m+1} = (1 - \kappa_i^m) r_i^m + \kappa_i^m r_{i-1}^m, \tag{11}$$

$$s_i^{m+1} = (1 - \kappa_i^m) s_i^m + \kappa_i^m s_{i+1}^m \qquad (i = 1, \ldots, N),$$

$$\kappa_i^m = a_i^m \cdot \frac{\tau}{h}, \tag{12}$$

to which there are added the initial data

$$r_i^0 = r_{0i}, \qquad s_i^0 = s_{0i}$$

and the boundary conditions

$$r_0^m + s_0^m = 2f^m, \tag{13}$$

$$r_{N+1}^m - s_{N+1}^m = k^m. \tag{14}$$

By means of the relations (11) the r_i^{m+1} are determined successively from $i = 1$ to $i = N + 1$, and s_{N+1}^{m+1} is determined from (14). Similarly, the s_i^{m+1} are determined from (12) ($i = N, \ldots, 0$), and r_0^{m+1} is determined from (13). After this the $(m + 1)$st level has been computed, and to determine the $(m + 2)$nd level the procedure is repeated entirely. Under the condition

$$\kappa_i^m \le 1 \tag{15}$$

the scheme (11), (12) is a scheme with positive coefficients, which ensures its stability as was shown in §3.3. We shall prove the convergence of the solution of problem (11)–(14) to a solution of problem (1), (3).

We suppose that problem (1)–(3) has a solution $r(q, t)$, $s(q, t)$ possessing continuous second derivatives with respect to q and t. We denote by δr_i^m and δs_i^m the differences

$$\delta r_i^m = r_i^m - r(ih, m\tau), \qquad \delta s_i^m = s_i^m - s(ih, m\tau),$$

where $r(ih, m\tau)$, $s(ih, m\tau)$ is the exact solution of problem (1)–(3) at the mesh points $q = ih$, $t = m\tau$. The quantities δr_i^m and δs_i^m satisfy the difference equations

$$\left.\begin{aligned}\delta r_i^{m+1} &= (1 - \kappa_i^m)\,\delta r_i^m + \kappa_i^m \delta r_{i-1}^m - \tau\{a(r_i^m - s_i^m)\\ &\quad - a[r(ih, m\tau) - s(ih, m\tau)]\}\frac{r(ih, m\tau) - r((i-1)h, m\tau)}{h} + R_{1i}^m,\\ \delta s_i^{m+1} &= (1 - \kappa_i^m)\delta s_i^m + \kappa_i^m\,\delta s_{i+1}^m + \tau\{a(r_i^m - s_i^m)\\ &\quad - a[r(ih, m\tau) - s(ih, m\tau)]\}\frac{s((i+1)h, m\tau) - s(ih, m\tau)}{h} + R_{2i}^m\end{aligned}\right\} \tag{16}$$

with the initial and boundary conditions

$$\left.\begin{aligned}\delta r_i^0 &= 0, & \delta s_i^0 &= 0,\\ \delta r_0^m + \delta s_0^m &= 0, & \delta r_{N+1}^m - \delta s_{N+1}^m &= 0.\end{aligned}\right\} \tag{17}$$

The remainder terms R_{1i}^m, R_{2i}^m have order $O(\tau^2)$.

Considering the smoothness of the solution $r(q, t)$, $s(q, t)$, the system (16) can be rewritten in the form

$$\left.\begin{aligned}\delta r_i^{m+1} &= (1 - \kappa_i^m)\delta r_i^m + \kappa_i^m\,\delta r_{i-1}^m - \tau\alpha_i^m(\delta r_i^m - \delta s_i^m) + R_{1i}^m,\\ \delta s_i^{m+1} &= (1 - \kappa_i^m)\delta s_i^m + \kappa_i^m\,\delta s_{i+1}^m + \tau\beta_i^m(\delta r_i^m - \delta s_i^m) + R_{2i}^m,\end{aligned}\right\} \tag{18}$$

where

$$\alpha_i^m = a'(\theta^*)\,\frac{r(ih, m\tau) - r((i-1)h, m\tau)}{h},$$

$$\beta_i^m = a'(\theta^*)\,\frac{s((i+1)\,h, m\tau) - s(ih, m\tau)}{h}$$

are bounded quantities, $\theta = r - s$, and θ^* is the intermediate value of θ in the formula for finite increments.

It is not hard to see that under condition (15) for the step operator C_{m+1} of problem (17), (18) there is the estimate

$$\|C_{m+1}\| \le 1 + 2A\tau,$$

where $A = \max_{i,m} \max\{|\alpha_i^m|, |\beta_i^m|\}$, and the norm of the vector $\{\delta r^m, \delta s^m\} = \{\delta r_i^m, \delta s_i^m\}$ is defined by

$$\max_i \max\{|\delta r_i^m|, |\delta s_i^m|\}.$$

Hence, we have the estimate (see §2.1)

$$|\delta r_i^m|, |\delta s_i^m| \le Ce^{2At} \max_{i,m} \max\left\{\left|\frac{R_{1i}^m}{\tau}\right|, \left|\frac{R_{2i}^m}{\tau}\right|\right\} = O(\tau),$$

which implies the convergence

$$|\delta r_i^m| \to 0, \qquad |\delta s_i^m| \to 0 \quad \text{as } \tau \to 0$$

uniformly with respect to i and m in the domain of existence of the solution $r(q, t)$, $s(q, t)$.

If there are contact boundaries, then the relations (7) on the boundaries also enter the difference equations. For simplicity we assume that there is one contact boundary situated at the point $q = jh$ of the mesh. The invariants r and s are then discontinuous at the point $q = jh$. For the left values $(r_j^{m+1})_-$ and right values $(s_j^{m+1})_+$ we obtain (11) and (12), which assume the form

$$(r_j^{m+1})_- = (1 - \kappa_j^m)_-(r_j^m)_- + (\kappa_j^m)_- r_{j-1}^m,$$
$$(s_j^{m+1})_+ = (1 - \kappa_j^m)_+(s_j^m)_+ + (\kappa_j^m)_+ s_{j+1}^m.$$

To these there are added the relations (7):

$$P[(S_j)_-, (r_j^{m+1})_- - (s_j^{m+1})_-] = P[(S_j)_+, (r_j^{m+1})_+ - (s_j^{m+1})_+],$$
$$(r_j^{m+1})_+ + (s_j^{m+1})_+ = (r_j^{m+1})_- + (s_j^{m+1})_-,$$

which for known $(r_j^{m+1})_-$ and $(s_j^{m+1})_+$ make it possible to determine $(r_j^{m+1})_+$ and $(s_j^{m+1})_-$, after which the computation continues as usual. In the presence of several contact boundaries each of them is accounted for according to the algorithm indicated.

If there are no shock waves in the flow, but the entropy within each layer is variable, then the equations describing the motion become inhomogeneous:

$$\frac{\partial s}{\partial t} - a(q, r - s)\frac{\partial s}{\partial q} = F_1(q, r - s),$$
$$\frac{\partial r}{\partial t} + a(q, r - s)\frac{\partial r}{\partial q} = F_2(q, r - s).$$

Then the left sides of the equations can be approximated as in the homogeneous case; the right sides are taken from the mth level, and convergence is proved similarly under the same condition (15).

In the general case a system of quasilinear equations including the equation of hydrodynamics does not reduce to invariants. However, in this case as well a running-count scheme is possible. The scheme was proposed by Courant, Isaacson, and Rees (1952).

Suppose for the hyperbolic system

$$l_\alpha^k(u_1, \ldots, u_n)\left[\frac{\partial u_\alpha}{\partial t} + \xi_k(u_1, \ldots, u_n)\frac{\partial u_\alpha}{\partial x}\right] = f_k(u_1, \ldots, u_n)$$
$$(k, \alpha = 1, \ldots, n) \qquad (19)$$

we pose the initial data

$$u_\alpha(x, 0) = u_{\alpha 0}(x). \tag{20}$$

We approximate the Cauchy problem (19), (20) by the difference Cauchy problem

$$l_\alpha^k(u_1^m \ldots, u_n^m)\left[\frac{u_\alpha^{m+1} - u_\alpha^m}{\tau} + \xi_k(u_1^m, \ldots, u_n^m)\frac{\Delta u_\alpha^m}{h}\right] = f_k(u_1^m, \ldots, u_n^m), \tag{21}$$

$$u_\alpha^2(x) = u_{\alpha 0}(x), \tag{22}$$

where $u_\alpha^m(x)$ is a solution of (21), (22) defined at time $t = m\tau$, and

$$\begin{aligned} \Delta &= \Delta_{-1} \quad \text{for } \xi_k(u_1^m, \ldots, u_n^m) \geq 0, \\ \Delta &= \Delta_1 \quad \text{for } \xi_k(u_1^m, \ldots, u_n^m) < 0. \end{aligned} \tag{23}$$

The difference scheme (21) can be rewritten in the form

$$\left.\begin{aligned} l_\alpha^{km} u_\alpha^{m+1} &= l_\alpha^{km}[(1 - \kappa_k^m)E + \kappa_k^m T_{-1}]u_\alpha^m + f_k^m \tau, \quad \kappa_k^m \geq 0, \\ l_\alpha^{km} u_\alpha^{m+1} &= l_\alpha^{km}[(1 + \kappa_k^m)E - \kappa_k^m T_1]u_\alpha^m + f_k^m \tau, \quad \kappa_k^m < 0, \end{aligned}\right\} \tag{24}$$

where we have set

$$l_\alpha^{km} = l_\alpha^k(u_1^m, \ldots, u_n^m), \quad \kappa_k^m = \frac{\xi_k^m \tau}{h}, \quad \xi_k^m = \xi_k(u_1^m, \ldots, u_n^m). \tag{25}$$

For $|\kappa_k^m| \leq 1$ the difference operators

$$(1 - \kappa_k^m)E + \kappa_k^m T_{-1} \quad (\kappa_k^m \geq 0), \quad (1 + \kappa_k^m)E - \kappa_k^m T_1 \quad (\kappa_k^m < 0)$$

become positive, and the scheme (24) is analogous to schemes positive in the Friedrichs sense. In the paper mentioned above it is shown that under the condition $|\kappa_k^m| \leq 1$ the solution $u^m(x)$ of the difference problem (21), (22) converges in C to a solution of the problem (19), (20).

We have considered two versions of the running-count scheme: a) in invariants and b) for a characteristic system. With regard to formulation of the boundary conditions and simplicity of the algorithm, computation in invariants is preferable. In the general case reduction to invariants is possible only for the prolonged system (see Chapter 1, §4.3), after which the running-count method becomes applicable.

§8. Homogenous difference schemes. Schemes with pseudoviscosity

The characteristic and running-count schemes we have considered are distinguished by the fact that regular and unified computation with them is possible only in regions of smooth flow. The presence of discontinuities leads to sharp complication of these techniques. Therefore, the need arises for a unified scheme with formulas of the same type at different points of the mesh regardless of the presence and character of singularities of the solution in a neighborhood of a point. Such computational schemes have been called *homogeneous.*

This formulation of the problem naturally makes added demands on the computational algorithm, since it must now, at least in principle, "equally well" describe both smooth and discontinuous flows. We therefore briefly discuss these requirements.

1. ***Methods for the uniform description of gas-dynamical flows.*** If we consider flows of compressible gases and fluids in the absence of viscous friction and thermal conductivity, then it follows from the content of Chapters 1 and 2 that they are describe by discontinuous quantities. Therefore, there is no system of differential equations applicable to the description of discontinuous flow for the simple reason that the parameters of discontinuous flows are not differentiable.

The singularities arising in the flow parameters are as follows: a weak discontinuity (discontinuity of the derivatives), a contact discontinuity (a boundary separating gases with different thermodynamic parameters), and, finally, a strong disontinuity (a shock wave). The set of all flows with the singularities indicated can be described in a unified way—the laws of conservation of mass, momentum, and energy must be satisfied for any distinguished portion of the gas or space. As we have seen in §2 of Chapter 2, this reduces to satisfying the integral conservation laws, which we again write out in the case of one-dimensional flows with various symmetries ($\nu = 0, 1, 2$). In Euler coordinates they have the form

$$\oint_C \rho x^\nu \, dx - \rho u x^\nu \, dt = 0, \tag{1}$$

$$\oint_C \rho u x^\nu \, dx - (p + \rho u^2) x^\nu \, dt = - \iint_{G_C} \nu p x^{\nu - 1} \, dx \, dt, \tag{2}$$

$$\oint_C \rho\left(\varepsilon + \frac{u^2}{2}\right) x^\nu \, dx - \rho u \left(\varepsilon + \frac{p}{2} + \frac{u^2}{2}\right) x^\nu \, dt = 0 \tag{3}$$

(see formulas (2.3.13)–(2.3.15) of Chapter 2).

In Lagrange coordinates (q is the mass coordinate) these conservation laws have the form

$$\oint_C V\,dq + x^\nu u\,dt = 0, \tag{4}$$

$$\oint_C u\,dq - px^\nu\,dt = -\iint_{G_C} \frac{\nu p V}{x}\,dq\,dt, \tag{5}$$

$$\oint_C \left(\varepsilon + \frac{u^2}{2}\right) dq - upx^\nu\,dt = 0, \tag{6}$$

where $x = x(q, t)$ is the Euler coordinate of the point (q, t) given by

$$\frac{\partial x(q, t)}{\partial t} = u(q, t) \tag{7}$$

(see (2.4.10)–(2.4.12) of Chapter 2).

In these formulas C is any piecewise smooth, closed contour, and G_C is the region bounded by it in the plane of the variables x, t and q, t respectively.

In general, these conservation laws are not sufficient for a complete description of the flow: they must be augmented by requiring a nondecrease of entropy of any fixed mass of gas. This requirement excludes the appearance of unstable discontinuites. The integral conservation laws (1)–(3) or (4)–(6) with the condition of nondecrease of the entropy form the *conservative system of equations of gas dynamics* applicable to any flows including those with discontinuous parameters.

The system of integral conservation laws is extremely inconvenient to approximate by differences on a fixed mesh in the case of the presence of discontinuities of the solution, since it requires explicitly distinguishing the lines of discontinuity, approximating the differential equations away from the lines of discontinuity, and satisfying the integral conservation laws on the lines of discontinuity (the Hugoniot conditions). We conclude that the method of characteristics describes a flow in just this manner. It is clear, therefore, that the method of characteristics is one of the most precise methods of solving the integral conservation laws (1), (2), (3) also in the case of discontinuous solutions provided, of course, it also considers with sufficient accuracy all the singularities of the solution which arise.

Ignoring the discontinuites that arise in the flow and approximating the integral conservation laws directly on fixed cellular meshes in the same way as for smooth flows lead, as a rule, to an unstable computational procedure in which a shock wave is replaced by oscillations of large amplitude. Direct approximation of integral conservation laws can lead to a stable approximate description of the flow parameters only in the case where the difference scheme used here contains dissipation or "approximation viscosity". In approximating nondissipative terms of the equations, the difference approximation contributes small—and for discontinuous solutions large—increments which may destabilize or stabilize the numerical solution. A good difference approximation of the conservation laws should contain "positive" viscosity which is later minimized. The Lax, Lax-Wendroff, and Godunov schemes are stable schemes approximating the integral conservation laws without the explicit introduction of pseudoviscosity.

A characteristic feature of the direct approximation of integral conservation laws is the so-called *conservative* or *divergence* property of the difference schemes obtained by it.[17] This property consists in the fact that the equations of the difference scheme can be interpreted as the integral conservation laws (1), (2), (3) or (4), (5), (6) for a cell of the mesh formed by the intersection of the lines $t = m\tau$ and $t = (m + 1)\tau$ with the lines $x = x_i$ and $x = x_{i+1}$ ($q = q_i$ and $q = q_{i+1}$) with some approximation (or interpolation) of the quantities contained in the conservation laws on the boundaries of the cell. This approximation remains constant on the given boundary of the cell, i.e., it is the same for the conservation laws in neighboring cells.

This ensures the property of *additivity of a conservation difference scheme* which consists in the fact that in summing the difference equations over neighboring cells we obtain a new equation which can be considered as the conservation laws for the outer boundary of the region formed from the union of these cells. More briefly, the difference equations possess the same properties as the contour integrals (1), (2), and (3).

This property provides the following advantage for conservative difference schemes. For a conservative difference scheme it is possible to apply arbitrary approximations (interpolations) of the quantities on the boundaries of the cells. If the difference scheme then converges, i.e., the family of difference solutions has a limit as τ, $h(\tau) \to 0$ (in some weak norm), then this limit satisfies precisely the required conservation laws (1), (2) and (3), and no others. By making proper use of the arbitrariness in interpolating quantities on the boundaries of the cell, it is possible to obtain additional advantages of conservative difference schemes.

This property of conservative difference schemes is similar to the property of a system of differential equations of parabolic type

$$\frac{\partial u_\mu}{\partial t} + \frac{\partial \varphi(u_\mu, x, t)}{\partial x} = \mu \frac{\partial}{\partial x}\left(B(u_\mu, x, t)\frac{\partial u_\mu}{\partial x}\right), \tag{8}$$

in which the "viscosity"—the right side of (8)—is contained in conservative or

[17] Tihonov and Samarskiĭ (see, for example, their paper (1959)) indicated the necessity of applying such schemes for problems with discontinuous coefficients and solutions back in 1951.

divergence form as a derivative of the expression $B(\partial u_\mu/\partial x)$, where B is a rather arbitrary matrix. For the system (8) there is a similar property. If as $\mu \to 0$ the solution $u_\mu(x, t)$ has a limit $u(x, t)$, then this limit satisfies the conservation law

$$\oint_C u\,dx - \varphi(u, x, t)\,dt = 0, \tag{9}$$

i.e., it is a generalized solution of the system

$$\frac{\partial u}{\partial t} + \frac{\partial \varphi(u, x, t)}{\partial x} = 0. \tag{10}$$

(For more details on this, see Chapter 4.) This similarity of stable approximations of integral conservation laws and the "method of vanishing viscosity" for constructing discontinuous solutions of systems of quasilinear equations has not only a superficial but also a deep-seated character. We have already mentioned above that to ensure stability of a difference method in the case of discontinuous solutions the scheme must carry "approximation viscosity".

We now proceed to the second and most widespread method of unified description of gas-dynamical flows, the viscosity or "pseudoviscostiy" method. In this method we avoid a detailed consideration of shock waves[18] and consider the flow of gases possessing some viscosity (and thermal conductivity) which is sometimes quite unrelated to the physical viscosity. In this case it is called "pseudoviscosity". By introducing this viscosity into the equations of gas dynamics, we approximately describe shock waves as a smooth shock transition which was considered in considerable detail in Chapter 2, §5.

The introduction of viscosity into the conservation laws of gas dynamics is achieved by replacing the pressure p in (1)–(3) or (5)–(6) by a quantity $\bar{p}$:

$$\bar{p} = p + \omega, \qquad \omega = -\mu\,\partial u/\partial x = -\mu\rho\,\partial u/\partial q = -\bar{\mu}\,\partial u/\partial q, \tag{11}$$

where μ is the coefficient of viscosity. We observe that the introduction of viscosity raises the order of the differential equations in x (q); this requires the formulation of additional conditions on the inner and outer boundaries. On the inner (contact) boundaries this condition follows from the requirement of continuity of the vectors of momentum and energy flux

$$[\omega] = \omega(x + 0, t) - \omega(x - 0, t) = -\left[\mu\frac{\partial u}{\partial x}\right] = -\left[\mu\rho\frac{\partial u}{\partial q}\right] = 0. \tag{12}$$

On the outer boundaries the condition $\omega = 0$ is usually imposed. The "physical"

[18] The same goes, by the way, for the majority of other methods. The method of characteristics in which all discontinuities are explicitly distinguished is an exception.

viscosity (11) admits contact discontinuities; it does not "smear them out". This merit of the physical viscosity one always strives to maintain in introducing other types of "viscosity"—the so-called "pseudoviscosity".

It was shown in Chapter 2, §5.6, that in the case of a constant coefficient of viscosity μ and a polytropic gas for the effctive width l of a shock transition in Lagrange coordinates there is the expression

$$l = \Delta q = \frac{8\mu}{(\gamma + 1)\,|\Delta u|}, \tag{13}$$

where Δu is the jump of velocity on the shock wave.

For real gases the coefficient of viscosity is rather small; the width l of a shock wave corresponding to a real physical viscosity has the order of the length of the mean free path of a molecule. Such a small real viscosity is insufficient for numerical computation, and it is therefore usually assumed that

$$\mu = \mu_0 h, \tag{14}$$

where μ_0 is chosen so that the width l computed by (13) for the typical condition $|\Delta u| \sim c$ (c is the speed of sound) is acceptable. Formula (13) correctly conveys the width of a shock wave in difference calculation only under the condition that $l \gg h$; in practice cases with $l \sim 5h \div 10h$ are achieved.

A drawback of the viscosity (14) is that it acts throughout the flow, so that strong smoothing of a shock wave corresponding to large μ_0 is always connected with a reduction of accuracy of the computation in regions of smooth flow. Proceeding from these considerations, von Neumann and Richmyer (1950) proposed the nonlinear viscosity

$$\mu = \mu_0 h^2 \rho \left|\frac{\partial u}{\partial q}\right|. \tag{15}$$

Stationary solutions of the equations of gas dynamics with the viscosity (15) were studied in Chapter 2, §5.6. According to this investigation, the width of the shock transition for the viscosity (15) is equal to $l = \pi\sqrt{2\mu_0/(\gamma + 1)}h$, i.e., is has order $O(h)$ and does not depend on the strength of the shock wave. The viscous term ω in a smooth part of the flow has order $O(h^2)$ and hence does not strongly affect the accuracy of the computation. These are the advantages of the artificial viscosity (pseudoviscosity) introduced by von Neumann and Richtmyer.

We note another special feature of the von Neumann-Richtmyer viscosity. In contrast to the linear viscosity in which the approach of the limiting profile $v(q)$ to the asymptotic values v_0 and v_1 occurs analytically, in the case of the von Neumann-Richtmyer viscosity this approach does not occur analytically:

at conjugate points q_0 and q_1 the second derivative has a discontinuity. Since in the zone of a shock wave the gradients are large, this discontinuity of the derivatives leads to a constant source of perturbations which cause strong oscillations of the hydrodynamical quantities in a neighborhood of the front. The profile of a shock wave in the difference scheme has approximately the form shown in Figure 3.14. The amplitude of the oscillations decreases as the coefficient of viscosity decreases. The strong dependence of the profile of the shock wave on μ_0 constitutes a characteristic feature of the von Neumann-Richtmyer scheme which in some cases hampers interpretation of the results.

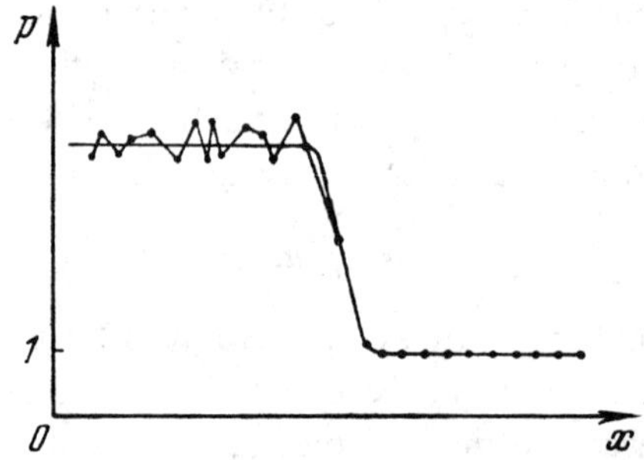

FIGURE 3.14

2. *The "cross" difference scheme for a system of equations with viscosity.* In the presence of viscosity the only type of discontinuities admitted by the conservation laws are contact discontinuities. Considering our remarks on the Lagrange method of description, the flow is therefore described by the system of differential equations ($\nu = 0$)

$$\frac{\partial u}{\partial t} + \frac{\partial \bar{p}}{\partial q} = 0, \qquad \frac{\partial v}{\partial t} - \frac{\partial u}{\partial q} = 0,$$

$$\frac{\partial}{\partial t}\left(\varepsilon + \frac{u^2}{2}\right) + \frac{\partial \bar{p} u}{\partial q} = 0,$$

$$\bar{p} = p + \omega, \qquad \omega = -\mu\rho \frac{\partial u}{\partial q},$$

where μ is given by (8.1.14) or (8.1.15) depending on the choice of the viscosity.

In the case of flows without shock waves the merits and advantages of conservative difference schemes become less apparent. If it is noted, however, that a shock transition "is smeared out" only over several computing intervals, then it can be understood that this property remains useful.

The first difference scheme to be published using pseudovicosity was the von Neumann-Richtmyer (1950) scheme already cited. This is a scheme of "cross" type. The "cross" scheme is contructed on an approximation of the first two conservation laws (8.1.4), (8.1.5) on rectangular cells of the difference mesh in the plane of the Lagrange variables q, t. Here to achieve second-order accuracy and to avoid interpolations the thermodynamic quantities p, $\bar{p}$, v and the velocity u are taken at different points of the mesh—at half integral and integral points respectively. The third equation in used in nondivergence form

$$\frac{\partial \varepsilon}{\partial t} + \bar{p}\,\frac{\partial v}{\partial t} = 0, \tag{2}$$

and relates only the thermodynamic quantities.

The difference scheme has the form

$$\frac{u_i^{m+1/2} - u_i^{m-1/2}}{\tau} + \frac{\bar{p}_{i+1/2}^m - \bar{p}_{i-1/2}^m}{h} = 0, \tag{3}$$

$$\frac{v_{i+1/2}^{m+1} - v_{i+1/2}^m}{u} - \frac{u_{i+1}^{m+1/2} - u_i^{m+1/2}}{h} = 0, \tag{4}$$

$$\frac{\varepsilon_{i+1/2}^{m+1} - \varepsilon_{i+1/2}^m}{\tau} + \frac{\bar{p}_{i+1/2}^{m+1} + \bar{p}_{i+1/2}^m}{2} \cdot \frac{v_{i+1/2}^{m+1} - v_{i+1/2}^m}{\tau} = 0, \tag{5}$$

where

$$\bar{p}_{i+1/2}^m = p_{i+1/2}^m + \omega_{i+1/2}^m, \tag{6}$$

$$\omega_{i+1/2}^m = -\mu_0 h^2 \rho_{i+1/2}^m \frac{|u_{i+1}^{m-1/2} - u_i^{m-1/2}|}{h} \cdot \frac{u_{i+1}^{m-1/2} - u_i^{m-1/2}}{h}, \tag{7}$$

$$\varepsilon_{i+1/2}^m = \varepsilon(p_{i+1/2}^m, v_{i+1/2}^m). \tag{8}$$

In the region of smooth flow the scheme has second-order accuracy, since (3) and (4) approximate the conservation laws by an integration formula with

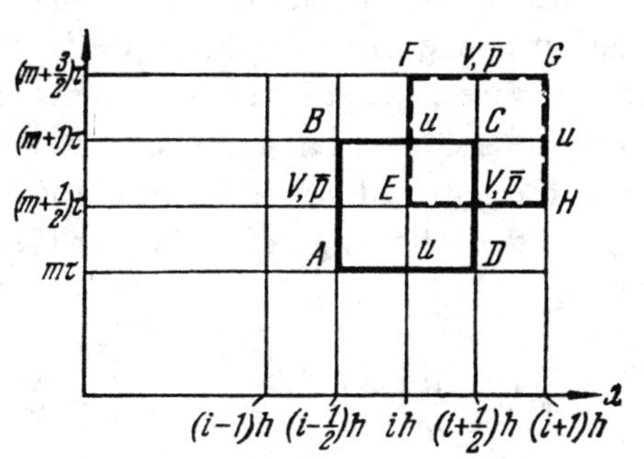

FIGURE 3.15

centered points, and the viscous term has order $O(h^2)$. Formula (5) also has second-order accuracy. Figure 3.15 show the integration cells for the conservation laws.

In a practical computation the notation of fractional steps in time is avoided by a translation of the time index:

$$u_i^{m+1/2} \to u_i^{m+1}. \tag{9}$$

Then (3), (4), and (7) take the form

$$\frac{u_i^{m+1} - u_i^m}{\tau} + \frac{\bar{p}_{i+1/2}^m - \bar{p}_{i-1/2}^m}{h} = 0, \tag{10}$$

$$\frac{v_{i+1/2}^{m+1} - v_{i+1/2}^m}{\tau} - \frac{u_{i+1}^{m+1} - u_i^{m+1}}{h} = 0, \tag{11}$$

$$\omega_{i+1/2}^m = -\mu_0 h^2 \rho_{i+1/2}^m \frac{|u_{i+1}^m - u_i^m|}{h} \frac{u_{i+1}^m - u_i^m}{h}. \tag{12}$$

For an arbitrary equation of state (8) formula (5) requires iteration to determine $\bar{p}_{i+1/2}^{m+1}$. In the case of an ideal gas, (5) admits an explicit solution for $\bar{p}_{i+1/2}^{m+1}$. If the law of conservation of energy

$$\oint \left(\varepsilon + \frac{u^2}{2}\right) dq - (\bar{p}u)\, dt = 0,$$

rather than (2) is approximated, then it is necessary to interpolate u while preserving the position of the points of the mesh at which p, v, ε and the velocity u are computed.[19]

We shall consider a number of other schemes with viscosity. Latter (1955) suggested the following modification of the von Neumann-Richtmyer method. Since the viscosity is introduced to smooth existing shock waves and those arising from compression waves while in rarefaction waves the gradients decrease even in the absence of viscosity, it follows that in difference calculations to increase the accuracy it is expedient to eliminate the effect of viscosity in the region of rarefaction waves, i.e., to "nullify" the coefficient of viscosity. In the planar case in compression waves and shock waves the inequality (see Chapter 2, §4) $\Delta u/\Delta q < 0$ is satisfied, while for rarefaction waves $\Delta u/\Delta q \geq 0$. Therefore, Latter suggests the following expression for the viscous term:

$$\omega = \begin{cases} 0, & \Delta u \geq 0, \\ -\mu_0 h^2 \rho \left|\dfrac{\Delta u}{\Delta q}\right| \dfrac{\Delta u}{\Delta q} = \mu_0 \rho (\Delta u)^2, & \Delta u < 0. \end{cases}$$

[19] Schemes of this type were used by I. M. Gel'fand.

This technique becomes especially effective if the linear viscosity (8.1.14) is used in the difference calculation. The profile of the shock wave is analytic, oscillation effects become considerably less, and at the same time the accuracy in the region of rarefaction waves is satisfactory. The schemes (10), (11) with the condition $\omega = 0$ on rarefaction waves and with a linear coefficient of viscosity $\omega = -\mu_0 ah\,\partial u/\partial q$ (a is the speed of sound) were investigated by Samarskiĭ and Arsenin (1961).

Stability of solutions of the system (1) of equations of gas dynamics with viscosity is verified in correspondence with the conjecture of local stability by the method of "freezing" the coefficients. In this approximation we give a comparative dispersion analysis of the original system (1) and the difference scheme (3)–(5). This enables us to clarify a number of the characteristic features of the von Neumann-Richtmyer scheme. We first rewrite (2) in the form

$$\frac{\partial p}{\partial t} + \bar{a}^2 \frac{\partial v}{\partial t} = \frac{\partial p}{\partial t} + \bar{a}^2 \frac{\partial u}{\partial q} = 0.$$

where we have set

$$\bar{a}^2 = \frac{\partial \varepsilon/\partial v + \bar{p}}{\partial \varepsilon/\partial p} > 0.$$

If $\mu = 0$, then $\bar{p} = p$ and $\bar{a} = a$, where a is the mass speed of sound. If $\mu \neq 0$, then on the shock wave where $\partial u/\partial q < 0$ we have $\bar{p} > p$ and $\bar{a}^2 > a^2$. We shall investigate the stability of solutions of the system of equations of hydrodynamics

$$\left.\begin{aligned} \frac{\partial u}{\partial t} + \frac{\partial p}{\partial q} &= \frac{\partial}{\partial q}\left(\bar{\mu}\frac{\partial u}{\partial q}\right), \\ \frac{\partial v}{\partial t} - \frac{\partial u}{\partial q} &= 0, \\ \frac{\partial p}{\partial t} + \bar{a}^2 \frac{\partial u}{\partial q} &= 0 \end{aligned}\right\} \tag{13}$$

in the approximation of "frozen" coefficients by setting $\bar{\mu} = \text{const}$ and $\bar{a}^2 = \text{const}$. In this approximation the equations for the variation of the solution $(\delta u, \delta v, \delta p)$ coincide with (13). Representing the variation $(\delta u, \delta v, \delta p)$ as a harmonic

$$\delta u = \delta u_0 e^{\omega t + ikq}, \qquad \delta v = \delta v_0 e^{\omega t + ikq}, \qquad \delta p = \delta p_0 e^{\omega t + ikq} \tag{14}$$

and substituting (14) into (13), we obtain equations for δu_0, δv_0 and δp_0:

$$(\omega + \bar{\mu}k^2)\delta u_0 + 0\delta v_0 + ik\delta p_0 = 0,$$
$$- ik\delta u_0 + \omega\delta v_0 + 0\delta p_0 = 0,$$
$$\bar{a}^2 ik\,\delta u_0 + 0\,\delta v_0 + \omega\,\delta p_0 = 0.$$

The characteristic equation has the form

$$\begin{vmatrix} \omega + \bar{\mu}k^2 & 0 & ik \\ -ik & \omega & 0 \\ \bar{a}^2 ik & 0 & \omega \end{vmatrix} = 0. \tag{15}$$

From this we obtain expressions for the roots:

$$\omega_1 = 0, \qquad \omega_{2,3} = -\frac{\bar{\mu}k^2}{2} \pm \sqrt{(\bar{\mu}k^2/2)^2 - \bar{a}^2k^2}. \tag{16}$$

We see that the local system (13) satisfies the correctness condition Re $\omega \leq 0$ for all k.

The nature of dispersion of the harmonics is evident from (16). For large k both roots ω_2 and ω_3 are real and negative; for k sufficiently small ω_2 and ω_3 are complex conjugates; the harmonics die out with oscillation.

An analysis of stability for the "cross' difference scheme of the approximating system (13) is carried out in similar fashion. Here we make the shift of indices $i + 1/2 \to i$, $m + 1/2 \to m + 1$. The "cross" scheme then takes the form

$$\left.\begin{aligned} \frac{u_i^{m+1} - u_i^m}{\tau} + \frac{p_i^m - p_{i-1}^m}{h} &= \bar{\mu}\,\frac{u_{i-1}^m - 2_i^m + u_{i+1}^m}{h^2}, \\ \frac{v_i^{m+1} - v_i^m}{\tau} - \frac{u_{i+1}^{m+1} - u_i^{m+1}}{h} &= 0, \\ \frac{p_i^{m+1} - p_i^m}{\tau} + \bar{a}^2\,\frac{u_{i+1}^{m+1} - u_i^{m+1}}{h} &= 0. \end{aligned}\right\} \tag{17}$$

Assuming that the coefficients $\bar{\mu}$ and $\bar{a}^2$ are constant, we carry out a harmonic analysis of the difference scheme (17). For this we set

$$\delta u_j^m = \delta u_0 \rho^m \xi^j, \qquad \delta v_j^m = \delta v_0 \rho^m \xi^l, \qquad \delta p_j^m = \delta p_0 \rho^m \xi^l, \tag{18}$$

where $\xi = e^{ikh}$, $\rho = e^{\omega\tau}$ and k is real. On substituting (18) into (17), we obtain equations for δu_0, δv_0 and δp_0:

$$\left.\begin{aligned} \left(\frac{\rho-1}{\tau} + \frac{4\bar{\mu}}{h^2}\sin^2\frac{kh}{2}\right)\delta u_0 + 0\delta v_0 + \frac{1-e^{-ikh}}{h}\,\delta p_0 &= 0, \\ -\frac{e^{ikh} - 1}{h}\,\rho\delta u_0 + \frac{\rho - 1}{\tau}\,\delta v_0 + 0\delta p_0 &= 0, \\ \bar{a}^2\,\frac{e^{ikh} - 1}{h}\,\rho\delta u_0 + 0\,\delta v_0 + \frac{\rho - 1}{\tau}\,\delta p_0 &= 0. \end{aligned}\right\} \tag{19}$$

We thus arrive at the characteristic equation

$$\frac{\rho-1}{\tau}\left[\left(\frac{\rho-1}{\tau}\right)^2+\frac{4\bar{\mu}}{h^2}\sin^2\frac{kh}{2}\frac{\rho-1}{\tau}+4\bar{a}^2\frac{\sin^2 kh/2}{h^2}\rho\right]=0, \tag{20}$$

with roots

$$\left.\begin{aligned}&\rho_1=1,\quad \rho_{2,3}=1-2b(r+\bar{\kappa}^2)\pm 2\sqrt{b[b(r+\bar{\kappa}^2)^2-\bar{\kappa}^2]},\\ &b=\sin^2\frac{kh}{2},\qquad r=\frac{\bar{\mu}\tau}{h^2},\qquad \bar{\kappa}=\frac{\bar{a}\tau}{h}.\end{aligned}\right\} \tag{21}$$

From (21) we draw the following conclusion: *if*

$$\max_k b(r+\overline{\kappa^2})^2=(r+\overline{\kappa^2})^2\le\overline{\kappa^2}, \tag{22}$$

then the roots $\rho_{2,3}$ are complex conjugate, and the "cross" scheme with viscosity is stable in the approximation of frozen coefficients.

For $\mu=0$ the Courant condition $\bar{\kappa}\le 1$ is a necessary and sufficient condition for stability. For $\mu<0$ the stability condition is more restrictive for τ. Indeed, instability can occur only for real roots. A simple analysis then shows that a necessary condition for stability is $\bar{\kappa}^2\le 1-Zr$. From this we find that

$$\kappa=\frac{a\tau}{h}\le\sqrt{(1-2r)a^2/\bar{a}^2}=\sqrt{(1-2r)p/\bar{p}}. \tag{23}$$

On a shock wave we have $p/\bar{p}<1$, and the condition (23) sometimes leads to a strong reduction in the step size as compared with the usual Courant condition.

We now consider the dispersion of harmonics in the difference scheme. In the case $\mu=0$ if the correctness condition is satisfied we have $|\rho_i|=1$, i.e., all the roots ρ_i $(i=1,2,3)$ of the characteristic equation lie on the unit circle. Although dispersion occurs, the amplitude of any harmonic does not decrease: oscillations, once they arise, are not damped out.

If $\mu>0$, then if the stability condition (22) is satisfied the amplitude of any harmonic corresponding to the roots ρ_2 and ρ_3 is damped out. Indeed, in this case the roots are complex conjugate with modulus equal to $\sqrt{1-4br}<1$. Damping also occurs for real roots.

It is clear that the larger r, the stronger the damping. We shall consider the linear and quadratic viscosities from this point of view. In the case of linear viscosity we have

$$r=\frac{\bar{\mu}\tau}{h^2}=\mu_0\kappa.$$

Thus, r is finite, and harmonics with large k are strongly damped. In the case of quadratic viscosity we have

$$r = \frac{\mu_0|\Delta u/\Delta q|h^2}{h^2}\tau = \mu_0\left|\frac{\Delta u}{\Delta q}\right|\tau.$$

On the front of a shock wave where $|\Delta u/\Delta q|h$ is finite the damping is strong; away from the shock wave r has order $Q(\tau)$ and damping is weaker.

The harmonics corresponding to the roots ρ_2 and ρ_3 are travelling waves propagating in the q, t plane with speed

$$\frac{dq}{dt} = \frac{\omega}{ik}.$$

The harmonics corresponding to the root $\rho_1 = 1$ are standing waves. The damping of travelling harmonic waves indicates the effect of approximation viscosity on the shock fronts, which leads to smoothing of the latter. The lack of damping of standing harmonic waves indicates that the approximation viscosity does not affect contact boundaries; the contact boundary is not smoothed, and nonuniformities of the pressure and density that arise near the boundary are preserved (the *entropy trace*). Indeed, in a standing wave we have $\rho = \rho_1 = 1$, and from (19) it follows that $\delta u_0 = \delta p_0 = 0$. This means that the variations δu and δp of the pressure and velocity in a standing harmonic wave are equal to zero.

The analysis of stability assuming that the coefficients $\bar{\mu}$ and $\bar{a}^2$ are frozen gives only a qualitative estimate of stability. This analysis essentially involves the linearization of the equations in a neighborhood of a flow with constant parameters or in any case with small gradients.

If the flow gradients are considered, the analysis becomes much more complicated. Nevertheless, the qualitative conclusions obtained from the simplified analysis (reduction of step size in the region of a shock wave, the presence of entropy traces) are corroborated by practical computations.

The "cross" scheme considered above is not a conservative difference scheme, since it approximates one of the equations of gas dynamics (equation (2)) in nondivergence form. In the presence of viscosity this, of course, is inadmissible. However, it is possible to construct conservative difference schemes approximating the conservation laws for a viscous gas both in Lagrange and Euler coordinates. One advantage of such schemes is that they convey the integral characteristics of the flow for rather rough meshes more precisely than nonconservative schemes. Popov and Samarskiĭ (1969, 1970) considered an implicit difference scheme with weights which we write out in the case of Lagrange variables:

$$\frac{u_i^{m+1}-u_i^m}{\tau}+\frac{1}{h}[(\alpha_1\bar{p}_{i+1/2}^{m+1}+(1-\alpha_1)\bar{p}_{i+1/2}^m)-(\alpha_1\bar{p}_{i-1/2}^{m+1}+(1-\alpha_1)\bar{p}_{i-1/2}^m)]=0, \quad (24)$$

$$\frac{x_i^{m+1} - x_i^m}{\tau} = \alpha_2 u_i^{m+1} + (1 - \alpha_2) u_i^m, \tag{25}$$

$$\frac{v_{i+1/2}^{m+1} - v_{i+1/2}^m}{\tau} - \frac{1}{h}[(\alpha_3 u_{i+1}^{m+1} + (1 - \alpha_3) u_{i+1}^m) - (\alpha_3 u_i^{m+1} + (1 - \alpha_3) u_i^m)] = 0, \tag{26}$$

$$\frac{\varepsilon_{i+1/2}^{m+1} - \varepsilon_{i+1/2}^m}{\tau} + \frac{1}{h}[\alpha_5 \bar{p}_{i+1/2}^{m+1} + (1 - \alpha_5)\, \bar{p}_{i+1/2}^m] \times [\alpha_4 (u_{i+1}^{m+1} - u_i^{m+1}) + (1 - \alpha_4)(u_{i+1}^m - u_i^m)] = 0. \tag{27}$$

Equations (24) and (25) may be considered an approximation of the conservation laws of momentum and volume on a difference cell, and they are therefore conservative. By making use of the parameters α_i ($i = 1, \ldots, 5$), one can arrange that (27) in conjunction with (24)–(26) is equivalent to an approximation on a cell to a difference scheme for the integral law of conservation of energy. This is achieved for

$$\alpha_1 = \alpha_4 = \alpha, \qquad \alpha_5 = 0.5, \tag{28}$$

where α is a free parameter. By utilizing the free parameters, one can alter the character of interpolation in time. For $\alpha_3 = \alpha_5 = 0.5$ equation (27) can be transformed to a form so that it has the appearance of an approximation of the equation $\varepsilon_t + \bar{p} v_t = 0$. If these conditions are satisfied ($\alpha_1 = \alpha_4 = \alpha$; $\alpha_3 = \alpha_5 = 0.5$) the authors call the scheme (24)–(27) *completely conservative*; here it is understood that, in addition to the divergence approximation of the divergence equation of conservation of total energy, the scheme approximates with good accuracy the other (nondivergence) forms of the equations of gas dynamics as well.

For $\alpha_k = 0.5$ ($k = 1, \ldots, 5$) the scheme (24)–(27) has order of approximation $O(\tau^2) + O(h^2)$. All the other schemes of this class have order of approximation $O(\tau) + O(h^2)$. For $\alpha \geq 0.5$ the difference scheme is absolutely stable (in the sense of local stability); for $\alpha < 0.5$ the scheme is conditionally stable.

Starting from other considerations, Gol'din, Ionkin, and Kalitkin (1969) also constructed a completely conservative difference scheme with approximation $O(\tau^2) + O(h^2)$ for the equations in Lagrange coordinates, taking account of viscosity and thermal conductivity. Troščiev (1970) constructed a completely conservative scheme with approximation of second order which is explicit. In the case of plane symmetry ($\nu = 0$) Troščiev's scheme coincides with the "cross" scheme (3)–(5).

The von Neumann-Richtmyer pseudoviscosity was introduced into the equations and difference scheme in analogy with the physical viscosity in the equations of gas dynamics. The following question arises: Is it possible to construct a scheme not based on the analogy with physical viscosity but providing damping of the harmonics and the accompanying smoothing of the profiles? An example of such a scheme was given by Lax (1954).

3. ***The Lax, Lax-Wendroff, and predictor-corrector difference schemes.*** For the system (8.2.1), which we write in the form (8.2.13) and set $\bar{\mu} = 0$, $\bar{p} = p$ and $\bar{a} = a$, the Lax approximation has the form

$$\left.\begin{aligned} \frac{u_i^{m+1} - \bar{u}_i^m}{\tau} + \frac{p_{i+1}^m - p_{i-1}^m}{2h} &= 0, \\ \frac{v_i^{m+1} - \bar{v}^m}{\tau} - \frac{u_{i+1}^m - u_{i-1}^m}{2h} &= 0, \\ \frac{p_i^{m+1} - \bar{p}_i^m}{\tau} + (a^2)_i^m \frac{u_{i+1}^m - u_{i-1}^m}{2h} &= 0, \end{aligned}\right\} \tag{1}$$

where the $\bar{f}_i^m$ denote the quantities $(f_{i+1}^m + f_{i-1}^m)/2$. The equations (1) can be represented in the form

$$\left.\begin{aligned} \frac{\Delta_0 u^m}{\tau} + \frac{\Delta_1 + \Delta_{-1}}{2h} p^m &= \mu \frac{\Delta_1 \Delta_{-1}}{h^2} u^m, \\ \frac{\Delta_0 v^m}{\tau} - \frac{\Delta_1 + \Delta_{-1}}{2h} u^m &= \mu \frac{\Delta_1 \Delta_{-1}}{h^2} v^m, \\ \frac{\Delta_0 p^m}{\tau} + (a^2)^m \frac{\Delta_1 + \Delta_{-1}}{2h} u^m &= \mu \frac{\Delta_1 \Delta_{-1}}{h^2} p^m, \end{aligned}\right\} \tag{2}$$

where $\mu = h^2/2\tau$. The terms on the right sides approximate expressions of the form $\mu \partial^2 f/\partial q^2$, and the scheme (1), (2) therefore absolutely approximates the system of equations

$$\left.\begin{aligned} &\frac{\partial u}{\partial t} + \frac{\partial p}{\partial q} = \mu \frac{\partial^2 u}{\partial q^2}, \qquad \frac{\partial v}{\partial t} - \frac{\partial u}{\partial q} = \mu \frac{\partial^2 v}{\partial q^2}, \\ &\frac{\partial p}{\partial t} + a^2 \frac{\partial u}{\partial q} = \mu \frac{\partial^2 p}{\partial q^2}. \end{aligned}\right\} \tag{3}$$

This implies that the approximation (1) of the system (8.2.13) is conditional: for $h^2/2\tau = \mu = \text{const}$ the scheme (1) approximates the parabolic system (3), while for $\tau/h = \text{const}$ the scheme (1) approximates the original hyperbolic system (8.2.13). The limiting procedure $\tau/h = \text{const}$ is typical for the equations of gas dynamics. However, for $a_{\max}^2/a_{\min}^2 \gg 1$ in practical computation the Courant criterion $(\tau/h)a_{\max} \leq 1$ at points for which $a \ll a_{\max}$ diverts the approximation (1) from (8.2.13) to (3).

We shall establish the stability condition for the Lax scheme. We write out the dispersion equation for the scheme (2) by "freezing" the quantity a^2. The roots $\rho_{1,2,3}$ of this equation are

$$\left.\begin{aligned} &\rho_1 = 1 - 4br, \qquad \rho_{2,3} = 1 - 4br \pm i\kappa \sin kh, \\ &r = \frac{\mu\tau}{h^2}, \qquad b = \sin^2\frac{kh}{2}, \qquad \kappa = \frac{a\tau}{h}. \end{aligned}\right\} \tag{4}$$

The scheme is stable for $\kappa \leq 1$. It follows from (4) that all harmonics die out.

We observe that the scheme (2) contains viscous terms in all three equations and, in particular, in the equation for v. This implies that the difference scheme (2) does not admit a contact boundary, in contrast to the system of equations of gas dynamics and in contrast to the von Neumann-Richtmyer scheme. This indicates that a discontinuity of the density will ne "smeared out". The smearing-out effect becomes still more noticeable for small τ, since the coefficient μ of approximation viscosity increases with decreasing τ.

Lax and Wendroff (1960) constructed a scheme in which the approximation viscosity does not lead to smearing-out of the contact boundaries. The scheme they constructed is analogous to a symmetric scheme of second-order accuracy with the added introduction of an artificial viscosity.

Let

$$\frac{\partial u}{\partial t} - \frac{\partial \varphi(u)}{\partial x} = 0 \tag{5}$$

be a conservative system of quasilinear equations to which there corresponds the nondivergence form

$$\frac{\partial u}{\partial t} - A\frac{\partial u}{\partial x} = 0, \qquad A = \left\| \frac{\partial \varphi}{\partial u} \right\|. \tag{6}$$

We shall apply to (5) an approximation of predictor-corrector type. The formulas for the auxiliary (predictor) step have the form

$$\frac{u_i^* - u_i^m}{\tau^*} = A(\bar{u}_i^m)\frac{u_{i+1/2}^m - u_{i-1/2}^m}{h}, \quad \bar{u}_i^m = \frac{u_{i-1/2}^m + u_{i+1/2}^m}{2}. \tag{7}$$

The auxiliary quantities u_i^* pertain to the time $t^* = t_m + \tau^*$. The next (corrector) step

$$\frac{u_{i+1/2}^{m+1} - u_{i+1/2}^m}{\tau} - \frac{\varphi(u_{i+1}^*) - \varphi(u_i^*)}{h} = 0 \tag{8}$$

is an approximation of the conservation laws for (5) for the cell of the mesh

$$\left(i - \frac{1}{2}\right)h \leq x \leq \left(i + \frac{1}{2}\right)h, \qquad m\tau \leq t \leq (m+1)\tau.$$

It is not hard to see that for $\tau^* = \tau/2$ the scheme (7), (8) is equivalent up to terms of second order of smallness to the scheme

$$\frac{u_{i+1/2}^{m+1} - u_{i+1/2}^{m}}{\tau} = \frac{g_{i+1}^{m} - g_i^m}{h}, \tag{9}$$

where the vector g_i^m is given by

$$g_i^m = \frac{\varphi(u_{i-1/2}^m) + \varphi(u_{i+1/2}^m)}{2} + \frac{\tau}{2h} A^2(u_i^m)(u_{i+1/2}^m - u_{i-1/2}^m). \tag{10}$$

The scheme (8), (10) possesses accuracy of order $O(\tau^2) + O(h^2)$, the roots of its local dispersion equation lie on the unit circle, and harmonic solutions of the local systems are not damped.

To damp the harmonics and smooth oscillations and additional term of the form

$$\omega = h\, \frac{B_{i+1/2}^m - B_{i-1/2}^m}{h}\, \frac{u_{i+1/2}^m - u_{i-1/2}^m}{h} \tag{11}$$

is introduced in the function g_i^m; here $B_{i+1/2}^m = B(u_{i+1}^m, u_i^m)$ is a nonnegative, symmetric matrix satisfying the condition $B(u, u) = 0$. The quantity ω of (11) is a pseudoviscous term and specifies the approximation viscosity of the scheme (9), (10).

In application to the equations of hydrodynamics in Lagrange coordinates with $\mu = 0$, when the set of quantities $\{u, v, E = \varepsilon + u^2/2\}$ is taken as the unknown vector u, we have

$$A = \left\| \begin{matrix} u\dfrac{\partial p}{\partial \varepsilon} & -\dfrac{\partial p}{\partial v} & -\dfrac{\partial p}{\partial \varepsilon} \\ 1 & 0 & 0 \\ u^2\dfrac{\partial p}{\partial \varepsilon} - p & -u\dfrac{\partial p}{\partial v} & -u\dfrac{\partial p}{\partial \varepsilon} \end{matrix} \right\|. \tag{12}$$

The matrix B is defined by

$$B(\bar{u}_1, \bar{u}_2) = \mu\, \frac{|u_1 - u_2|}{a^2}\, A^2,$$

where a is the mass speed of sound and μ is the coefficient of viscosity.

We remark that this same effect of damping the harmonics indicating the presence of approximation viscosity can be obtained in another manner without introducing an additional term but by changing the position of the intermediate level t^*. We shall carry out the corresponding analysis in the acoustic approximation. Suppose that

$$\frac{\partial u}{\partial t} = A\frac{\partial u}{\partial x} = \frac{\partial v}{\partial x}, \qquad v = Au, \quad u = \{u_1, \ldots, u_n\}, \quad v = \{v_1, \ldots, v_n\} \tag{13}$$

is a hyperbolic system with the constant matrix $A = \{a_{ij}\}$, $i, j = 1, \ldots, n$. We write the predictor formulas in the form

$$\begin{aligned} \frac{u^*(x) - \bar{u}^m(x)}{\tau^*} &= \frac{v^m(x + h/2) - v^m(x - h/2)}{h} \\ &= A\frac{u^m(x + h/2) - u^m(x - h/2)}{h}, \\ \bar{u}^m(x) &= \frac{u^m(x - h/2) + u^m(x + h/2)}{2}. \end{aligned} \tag{14}$$

The corrector formulas take the form

$$\begin{aligned} \frac{u^{m+1}(x) - u^m(x)}{\tau} &= \frac{v^*(x + h/2) - v^*(x - h/2)}{h} \\ &= A\frac{u^*(x + h/2) - u^*(x - h/2)}{h}. \end{aligned} \tag{15}$$

Substituting into (15) the expressions $u^*(x + h/2)$ and $u^*(x - h/2)$ obtained from (14) for $x = x + h/2$ and $x = x - h/2$, we find that

$$\begin{aligned} \frac{u^{m+1}(x) - u^m(x)}{\tau} &= \frac{\bar{v}^m(x + h/2) - \bar{v}^m(x - h/2)}{h} \\ &\quad + \tau^* A\frac{v^m(x + h) - 2v^m(x) + v^m(x - h)}{h^2} \\ &= A\frac{u^m(x + h) - u^m(x - h)}{2h} \\ &\quad + \tau^* A^2\frac{u^m(x + h) - 2u^m(x) + u^m(x - h)}{h^2}. \end{aligned} \tag{16}$$

We now determine the structure of the approximation viscosity in the first differential approximation (see §2.3).

The first differential approximation of the scheme (16) has the form

$$\frac{\partial u}{\partial t} = A\frac{\partial u}{\partial x} + \left(\tau^* - \frac{1}{2}\tau\right)A^2\frac{\partial^2 u}{\partial x^2}. \tag{17}$$

The system (17) is of parabolic type for $\tau^* > \frac{1}{2}\tau$. By comparing (16), (11), and (13), it is not hard to see that in the acoustic approximation the Lax-Wendroff scheme is analogous to the scheme (16).

We now compare the predictor-corrector scheme (16) with the Lax scheme

$$\frac{u^{m+1}(x) - u^m(x)}{\tau} - A\frac{u^m(x+h) - u^m(x-h)}{2h} = \mu\frac{u^m(x+h) - 2u^m(x) + u^m(x-h)}{h^2}, \qquad \mu = \frac{h^2}{2\tau}. \tag{18}$$

The first differential approximation of the Lax scheme has the form

$$\frac{\partial u}{\partial t} = A\frac{\partial u}{\partial x} + \left(\mu E - \frac{1}{2}\tau A^2\right)\frac{\partial^2 u}{\partial x^2}, \qquad \mu = \frac{h^2}{2\tau}. \tag{19}$$

We shall clarify the difference in the dispersion of the harmonics in the Lax scheme and the predictor-corrector scheme. For simplicity we carry out this analysis in the first differential approximation. It is not hard to see that if λ_i are the eigenvalues of the matrix A and r_i are the Riemann invariants of the system (6) with a constant matrix A, then after passing to the invariants r_i the equations (17) and (19) take the respective forms

$$\frac{\partial r_i}{\partial t} - \lambda_i\frac{\partial r_i}{\partial x} = \left(\tau^* - \frac{1}{2}\tau\right)\lambda_i^2\frac{\partial^2 r_i}{\partial x^2}, \tag{20}$$

and

$$\frac{\partial r_i}{\partial t} - \lambda_i\frac{\partial r_i}{\partial x} = \left(\mu - \frac{1}{2}\tau\lambda_i^2\right)\frac{\partial^2 r_i}{\partial x^2}, \qquad \mu = \frac{h^2}{2\tau}. \tag{21}$$

A considerable difference is now evident between the Lax scheme and the predictor-corrector scheme. If $\lambda_i = 0$ for some, then in the predictor-corrector scheme the corresponding Riemann invariant is not subject to the effect of viscosity and undergoes no change (a standing wave), while in the Lax scheme all invariants are subject to the effect of viscosity. On passing to the equations of hydrodynamics in Lagrange coordinates this means that in the predictor-corrector scheme the entropy (an invariant preserved along a contact characteristic—a contact boundary) is not subject to the effect of viscosity; the contact discontinuity is not smoothed. On the other hand, it has been shown that the Lax viscosity smooths contact discontinuities. The Lax-Wendroff viscosity possesses the analogous property of preserving the contact boundary.

MacCormack (1969) proposed a two-step scheme of second-order accuracy which in application to the system (5) has the form

$$\tilde{u}^{m+1}(x) = u^m(x) + \frac{\tau}{h}[\varphi(u(x+h)) - \varphi(u(x))], \tag{22}$$

$$u^{m+1}(x) = \frac{1}{2}\left\{u^m(x) + \tilde{u}^{m+1}(x) + \frac{\tau}{h}[\varphi(\tilde{u}^{m+1}(x)) - \varphi(\tilde{u}^{m+1}(x-h))]\right\}. \tag{23}$$

Another version of this scheme is obtained from (22) and (23) if on the right sides of (22) and (23) the difference formulas for differentiation forward and backward are interchanged. Anderson (1974) considered this scheme as a noncentral version of the Lax-Wendroff scheme. Kutler and Lomax (1971) and Kutler, Sakell and Aiello (1975) suggested that a term approximating $\delta h^4(\partial^4 u/\partial x^4)$, $\delta = \text{const}$, be added to the right side of (23) with the objective of smoothing the numerical solution. We mention that the MacCormack scheme is widely used in aerodynamical computations.

The question of schemes of the form

$$u^{m+1}(x) = \sum_{\alpha} B_\alpha u^m(x + \alpha h), \tag{24}$$

approximating the hyperbolic system (13) with a constant matrix A and admitting contact discontinuities (property K) was considered by Janenko and Šokin (1968) and Šokin (1973). The matrices B_α are constant and satisfy the conditions

$$\sum_{\alpha} B_\alpha = E, \qquad \sum_{\alpha} \alpha B_\alpha = \kappa A, \quad \kappa = \frac{\tau}{h}.$$

The treatment is based on the first differential approximation of the scheme (24), which has the form

$$\frac{\partial u}{\partial t} = A \frac{\partial u}{\partial x} + C \frac{\partial^2 u}{\partial x^2},$$

where

$$C = \frac{h^2}{2\tau}\left[\sum_{\alpha} \alpha^2 B_\alpha - \kappa^2 A^2\right].$$

Suppose that the matrix A has distinct eigenvalues $\xi_1, \ldots, \xi_n$. We fix one of them, say $\xi_j = \xi$, and determine the corresponding left eigenvector l: $lA = \xi l$. We say that the difference scheme (24) possesses property K if $lC = 0$. Janenko and Šokin (1968) and Šokin (1973) found necessary and sufficient conditions for a number of classes of difference schemes to possess property K. As an example, we formulate one such assertion.

THEOREM. *In order that a three-point difference scheme of the form* (24) ($\alpha = -1, 0, 1$) *possess property K it is necessary and sufficient that* $lB_0 = (1 - \kappa^2\xi^2)l$.

In particular, this assertion implies that the Lax difference scheme ($C = (h^2/2\tau)(1 - \kappa^2 A^2)$, $lC = (h^2/2\tau)(1 - \kappa^2\xi^2)l$) does not possess property K (with the exception of the case $\kappa\xi = 1$) and that the majorant scheme ($\alpha = -1$, 0, 1, $B_1 = \kappa A^+$, $B_{-1} = -\kappa A^-$, $A^+ \geq 0$, $A^- \leq 0$, $A = A^+ + A^-$) does possess property K if $\xi = 0$ (an analogue of the system of equations of gas dynamics in Lagrange coordinates) or $\xi\kappa = 1$.

It is not hard to see that if a difference scheme for the equations of gas dynamics possesses property K for the eigenvalue $\xi = 0$ in the case of Lagrange coordinates and for $\xi = -u$ in the case of Euler coordinates, then the approximation viscosity has no effect on the entropy in the system of equations of the first differential approximation, and hence contact boundaries are not smeared out.

4. *The schemes of Godunov and Kuropatenko.* The schemes considered in the preceding subsection proceeded from an approximation of the equations of gas dynamics with an artificial viscosity which provides smoothing of a shock front. In the present subsection we consider schemes based on another approach which can be formulated as follows.

Computational formulas should basically be of unified character and approximate the conservation laws in such a way that the scheme obtained approximates the generalized solutions both in the region of smooth flow, where it approximates the equations of gas dynamics, and on the discontinuities, where it approximates the Hugoniot conditions. Because of the unified character of the computational formulas the flow in the smooth region is computed in the same way as the flow in a neighborhood of a discontinuity, and this leads to the appearance of the approximation viscosity.

Thus, the scheme presumes the presence of discontinuities in each calculation interval and the construction of computational formulas which take account of possible discontinuities. The first such scheme was the *scheme of decay of a discontinuity* proposed by Godunov (1959).

At each time t_m the functions $\rho^m(q)$, $p^m(q)$, $\varepsilon^m(q)$ and $u^m(q)$ describing the flow are approximated by piecewise constant functions which characterize the mean values of ρ^m, p^m, ε^m and u^m in the calculation interval. The interaction of masses of gas contained in the calculation intervals and characterized by the averaged parameters ρ^m, p^m, ε^m and u^m is described by means of formulas for the decay of a discontinuity on each boundary of two intervals. Assuming the velocity of the contact boundary between the intervals (q_{i-1}, q_i) and (q_i, q_{i+1}) to be constant over the time interval (t_m, t_{m+1}), we can compute the state of the gas in the intervals at the next moment of time t_{m+1} and average the parameters over the intervals, and then repeat the computation.

We approximate the conservation laws

$$\left.\begin{aligned} &\oint u\,dq - p\,dt = 0, \qquad \int v\,dq + u\,dt = 0, \\ &\oint E\,dq - pu\,dt = = 0, \qquad E = \varepsilon + \frac{u^2}{2} \end{aligned}\right\} \tag{1}$$

by the formulas

$$\left.\begin{aligned}
&\frac{u_{i+1/2}^{m+1} - u_{i+1/2}^{m}}{\tau} + \frac{P_{i+1}^{m} - P_{i}^{m}}{h} = 0,\\
&\frac{v_{i+1/2}^{m+1} - v_{i+1/2}^{m}}{\tau} - \frac{U_{i+1}^{m} - U_{i}^{m}}{h} = 0,\\
&\frac{E_{i+1/2}^{m+1} - u_{i+1/2}^{m}}{\tau} + \frac{P_{i+1}^{m}U_{i+1}^{m} - P_{i}^{m}U_{i}^{m}}{h} = 0,
\end{aligned}\right\} \qquad (2)$$

where $u_{i+1/2}^{m}$, $v_{i+1/2}^{m}$, $P_{i+1/2}^{m}$ and $E_{i+1/2}^{m}$ are the average values of the corresponding functions on the interval (q_i, q_{i+1}) at time t_m.

Since the interaction of the intervals (q_{i-1}, q_i) and (q_i, q_{i+1}) is considered on the basis of decay of a discontinuity, it follows that U_i^m and P_i^m are the quantities of velocity and pressure formed on the boundary $q = q_i$ due to the decay of a discontinuity characterized by the quantities

$$u_{i-1/2}^{m}, \qquad v_{i-1/2}^{m}, \qquad E_{i-1/2}^{m}, \qquad q < q_i,$$

$$u_{i+1/2}^{m}, \qquad v_{i+1/2}^{m}, \qquad E_{i+1/2}^{m}, \qquad q > q_i.$$

The known formulas for the decay of a discontinuity (see Chapter 2, §6.10) may be used to determine U_i^m and P_i^m.

If the discontinuities of u and p are not large, for example, relative to the amplitudes

$$\frac{u_{i+1/2} - u_{i-1/2}}{u_{i+1/2}}, \qquad \frac{p_{i+1/2} - p_{i-1/2}}{p_{i+1/2}},$$

then it is possible to apply the linearized decay formulas.

For a polytropic gas the linearized formulas for the decay of a discontinuity have the form (see Chapter 2, §6.10)

$$\left.\begin{aligned}
u_1 - U &= \frac{P - p_1}{\rho_1 c_1} = \frac{c_1}{\gamma_1}\frac{P - p_1}{p_1},\\
U - u_0 &= \frac{P - p_0}{\rho_0 c_0} = \frac{c_0}{\gamma_0}\frac{P - p_0}{p_0},
\end{aligned}\right\} \qquad (3)$$

where U and P denote the velocity and pressure formed on the contact boundary of two media of which the left medium is denoted by the index 1 and the right medium by the index 0; c_1 and c_0 are the speed of sound in media 1 and 0 respectively. From this we obtain the expressions for P and U

$$
\left.\begin{aligned}
P = p_k &= \frac{u_1 - u_0}{1/\rho_1 c_1 + 1/\rho_0 c_0} + \frac{p_1/\rho_1 c_1 + p_0/\rho_0 c_0}{1/\rho_1 c_1 + 1/\rho_0 c_0}, \\
U = u_k &= \frac{p_1 - p_0}{\rho_0 c_0 + \rho_1 c_1} + \frac{\rho_0 c_0 u_0 + \rho_1 c_1 u_1}{\rho_0 c_0 + \rho_1 c_1},
\end{aligned}\right\} \tag{4}
$$

which can be used in the computational formulas of the scheme for the decomposition of a discontinuity when Δu and ΔP are small.

We consider, for example, an isothermic gas. Then $c_1 = c_0 = c$, $\gamma_0 = \gamma_1 = 1$ and $pv = c^2$. Moreover, in (3), where $u_1 - u_0$, $P - p_1$, and $P - p_0$ are small of first order, in the denominators it is possible to set $\rho_1 = \rho_0 = \rho$ and $p_1 = p_0 = p$. Formulas (3) then take the form

$$
u_1 - U = \frac{P - p_1}{\rho c} = \frac{c(P - p_1)}{p}; \qquad U - u_0 = \frac{P - p_0}{\rho c} = \frac{c(P - p_0)}{p}.
$$

From this we have

$$
\left.\begin{aligned}
P &= \frac{p}{2c}(u_1 - u_0) + \frac{p_1 + p_0}{2}, \\
U &= \frac{c}{2p}(p_1 - p_0) + \frac{u_1 + u_0}{2}.
\end{aligned}\right\} \tag{5}
$$

By virtue of the relation $\Delta p/p = -\Delta v/v$, which holds to second order of smallness, (5) can be transformed to the form

$$
\left.\begin{aligned}
P &= \frac{c}{2v}(u_1 - u_0) + \frac{p_1 + p_0}{2}, \\
U &= -\frac{c}{2v}(v_1 - v_0) + \frac{u_1 + u_0}{2}.
\end{aligned}\right\} \tag{6}
$$

Applying (6), we obtain

$$
\left.\begin{aligned}
P_i^m &= \frac{c}{2v_{i+1/2}^m}(u_{i-1/2}^m - u_{i+1/2}^m) + \frac{p_{i-1/2}^m + p_{i+1/2}^m}{2}, \\
U_i^m &= -\frac{c}{2v_{i+1/2}^m}(v_{i-1/2}^m - v_{i+1/2}^m) + \frac{u_{i-1/2}^m + u_{i+1/2}^m}{2}.
\end{aligned}\right\} \tag{7}
$$

Substituting (7) into the first two equations of the difference scheme (2), we obtain the difference scheme

$$\left.\begin{aligned}
&\frac{u_{i+1/2}^{m+1}-u_{i+1/2}^{m}}{\tau}+\frac{p_{i+3/2}^{m}-p_{i-1/2}^{m}}{2h}\\
&\qquad=\frac{ch}{2}\frac{1}{h^2}\left[\frac{u_{i+3/2}^{m}-u_{i+1/2}^{m}}{v_{i+3/2}^{m}}-\frac{u_{i+1/2}^{m}-u_{i-1/2}^{m}}{v_{i+1/2}^{m}}\right],\\
&\frac{v_{i+1/2}^{m+1}-v_{i+1/2}^{m}}{\tau}-\frac{u_{i+3/2}^{m}-u_{i-1/2}^{m}}{2h}\\
&\qquad=\frac{ch}{2}\frac{1}{h^2}\left[\frac{v_{i+3/2}^{m}-v_{i+1/2}^{m}}{v_{i+3/2}^{m}}-\frac{v_{i+1/2}^{m}-v_{i-1/2}^{m}}{v_{i+1/2}^{m}}\right].
\end{aligned}\right\} \tag{8}$$

For the linearized equation of state $p = p_0 - a^2 v$ (a is the mass speed of sound) formulas (7) and (8) take the respective forms

$$\left.\begin{aligned}
P_i^m &= a\frac{u_{i-1/2}^{m}-u_{i+1/2}^{m}}{2}+\frac{p_{i-1/2}^{m}+p_{i+1/2}^{m}}{2},\\
U_i^m &= \frac{p_{i-1/2}^{m}-p_{i+1/2}^{m}}{2}+\frac{u_{i+1/2}^{m}+u_{i-1/2}^{m}}{2},
\end{aligned}\right\} \tag{9}$$

and

$$\left.\begin{aligned}
&\frac{u_{i+1/2}^{m+1}-u_{i+1/2}^{m}}{\tau}+\frac{p_{i+3/2}^{m}-p_{i-1/2}^{m}}{2h}=\mu\frac{u_{i+3/2}^{m}-2u_{i+1/2}^{m}+u_{i-1/2}^{m}}{h^2},\\
&\frac{v_{i+1/2}^{m+1}-v_{i+1/2}^{m}}{\tau}-\frac{u_{i+3/2}^{m}-u_{i-1/2}^{m}}{2h}=\mu\frac{v_{i+3/2}^{m}-2v_{i+1/2}^{m}+v_{i-1/2}^{m}}{h^2},\\
&\mu=\frac{ah}{2}.
\end{aligned}\right\} \tag{10}$$

If we consider μ in (10) as an aribtrary parameter, then for $\mu = ah/2$ we see that (10) gives the scheme for the decay of a discontinuity and the running-count scheme, for $\mu = h^2/2\tau$ it gives the Lax scheme, and for $\mu = a^2\tau/2$ the symmetric scheme of second-order accuracy. For $\mu = a\tau/h = 1$ all the schemes coincide.

The scheme proposed by Kuropatenko (1962, 1966) is based on the approximation of a compression wave or shock wave of finite intensity by a sequence ("train") of shock waves of lesser intensity. In contrast to the scheme of decay of a discontinuity, in this scheme a different approximation is used for compression waves and shock waves on the one hand and for rarefaction waves on the other. As previously, to distinguish compression waves and rarefaction waves there is the condition

$$\Delta u < 0 \quad \text{(compression wave)}, \qquad \Delta u \geq 0 \quad \text{(rarefaction wave)}. \tag{11}$$

Various realizations of a scheme of this type are possible. We consider only one scheme, in which the allocation of quantities in the mesh is analogous to the von Neumann-Richtmyer scheme: thermodynamic quantities are referred to half-integral points and the velocity to integral points. The system of difference equations has the form

$$\frac{u_i^{m+1} - u_i^m}{\tau} + \frac{\bar{p}_{i+1/2}^m - \bar{p}_{i-1/2}^m}{h} = 0, \tag{12}$$

$$\frac{v_{i+1/2}^{m+1} - v_{i+1/2}^m}{\tau} - \frac{u_{i+1}^m - u_i^m}{h} = 0, \tag{13}$$

$$\frac{\varepsilon_{i+1/2}^{m+1} - \varepsilon_{i+1/2}^m}{\tau} + \frac{\bar{p}_{i+1/2}^m + \bar{p}_{i+1/2}^{m+1}}{2\tau}(v_{i+1/2}^{m+1} - v_{i+1/2}^m) = 0, \tag{14}$$

where

$$\bar{p}_{i+1/2}^{m+1} = p_{i+1/2}^{m+1} + \omega_{i+1/2}^m, \tag{15}$$

$$\varepsilon_{i+1/2}^m = \varepsilon(p_{i+1/2}^m, v_{i+1/2}^m). \tag{16}$$

The "viscous" increment $\omega_{i+1/2}^m$ is equal to 0 in the case of a rarefaction wave and is computed by special formulas in the case of a compression wave (shock wave and is computed by special formulas in the case of a compression wave (shock wave). Thus, on a rarefaction wave the scheme (12)–(16) is analogous to the "cross" scheme and has second-order accuracy.

Equation (14) requires iteration because of the nonlinearity of the equation of state (16). The order of computation in this case is as follows: $v_{i+1/2}^{m+1}$ is determined from (13), $p_{i+1/2}^{m+1} = \bar{p}_{i+1/2}^{m+1}$ is determined from (14) by iteration using (15), and then u_i^{m+1} is determined from (12).

In the case of a compression wave (shock wave) $\omega \neq 0$. The term ω in (15) accounts with the presence of a shock wave in the interval (q_i, q_{i+1}) (Figure 3.16).

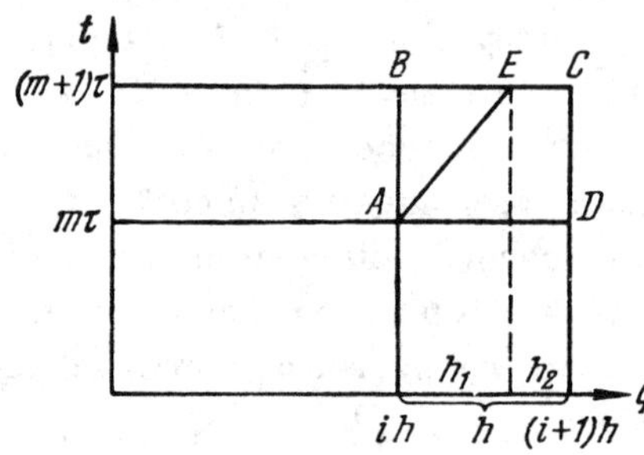

FIGURE 3.16

The equations

$$(\bar{p}_{i+1/2}^{m+1} - p_{i+1/2}^m)(\bar{v}_{i+1/2}^{m+1} - v_{i+1/2}^m) = -\Delta u^2 = -(u_{i+1}^m - u_i^m)^2, \tag{17}$$

$$\bar{\varepsilon}^{m+1}_{i+1/2} = \varepsilon^m_{i+1/2} + \frac{1}{2}\Delta u^2 - p^m_{i+1/2}(\bar{v}^{m+1}_{i+1/2} - v^m_{i+1/2}), \tag{18}$$

are invoked to determine $\bar{p}^{m+1}_{i+1/2}$. Here ε^{m+1}, p^{m+1}, and v^{m+1} are connected by (16), as before. The quantity $\bar{p}^{m+1}_{i+1/2}$ is found from (16)–(18), after which $\varepsilon^{m+1}_{i+1/2}$ is determined from (14).

The relations (17) and (18) are the Hugoniot conditions (see Chapter 2, §4.1) if u^m_{i+1}, $p^m_{i+1/2}$ and $v^m_{i+1/2}$ are taken to characterize the flow on one side of the front and u^m_i, $\bar{p}^m_{i+1/2}$ and $\bar{v}^m_{i+1/2}$ on the other side of the front.

We now determine the viscous increment $\omega^{m+1}_{i+1/2} = \bar{p}^{m+1}_{i+1/2} - p^{m+1}_{i+1/2}$. For this we find $p^{m+1}_{i+1/2}$ from the averaging formula (see Figure 3.16)

$$p^{m+1}_{i+1/2} = \frac{\bar{p}^{m+1}_{i+1/2}h_1 + p^m_{i+1/2}h_2}{h_1 + h_2} = \frac{\bar{p}^{m+1}_{i+1/2}D\tau + (h - D\tau)p^m_{i+1/2}}{h}, \tag{19}$$

where D is the mass velocity of the shock wave computed by the formula

$$D = \frac{\bar{p}^m_{i+1/2} - p^m_{i+1/2}}{u^m_{i+1} - u^m_i} = \frac{u^m_{i+1} - u^m_i}{\bar{v}^{m+1}_{i+1/2} - v^m_{i+1/2}}. \tag{20}$$

Hence

$$\omega^{m+1}_{i+1/2} = \bar{p}^{m+1}_{i+1/2} - p^{m+1}_{i+1/2} = \left(1 - D\frac{\tau}{h}\right)(\bar{p}^{m+1}_{i+1/2} - p^m_{i+1/2}). \tag{21}$$

In the case of an ideal gas, from (17) and (18) we have

$$\omega^{m+1}_{i+1/2} = \left(1 - D\frac{\tau}{h}\right)\left[\frac{\gamma+1}{4v^m_{i+1/2}}\Delta u^2 + \sqrt{\left(\frac{\gamma+1}{4v^m_{i+1/2}}\Delta u^2\right)^2 + \frac{\gamma p^m_{i+1/2}}{v^m_{i+1/2}}\Delta u^2}\right]. \tag{22}$$

It follows from (22) that the viscosity ω is linear,

$$\omega^{m+1}_{i+1/2} = \left(1 - D\frac{\tau}{h}\right)\sqrt{\frac{\gamma p^m_{i+1/2}}{v^m_{i+1/2}}}\,h\frac{\Delta u}{h},$$

if

$$\left(\frac{\gamma+1}{4v^m_{i+1/2}}\Delta u\right)^2 \ll \frac{\gamma p^m_{i+1/2}}{v^m_{i+1/2}},$$

and is quadratic,

$$\omega^{m+1}_{i+1/2} = \left(1 - D\frac{\tau}{h}\right)\frac{\gamma+1}{2v^m_{i+1/2}}\Delta u^2,$$

if

$$\left(\frac{\gamma + 1}{4v_{i+1/2}^m}\Delta u\right)^2 \gg \frac{\gamma p_{i+1/2}^m}{v_{i+1/2}^m}.$$

Thus, the viscous term ω is linear in the case of weak waves and quadratic in the case of strong waves.

This property of the viscosity ensures sufficient monotonicity of the profile of the difference shock wave, since the oscillations are strongly damped both on the shock wave itself and away from it. The property of monotonicity of the difference shock front and the damping of the oscillations make it possible to apply the so-called "differential analyzer" proposed by Kuropatenko (private communication).

Since the effect of the viscous term ω is stronger at the center of the front of the shock wave and abates away from it, the point where ω achieves a maximum is conveniently taken as the center of the difference shock wave. As the boundary of the difference transition, it is possible to take, for example, points closest to the center, where $\omega = 0$ (transition of a compression wave into a rarefaction wave). On the basis of values at the extreme points of the shock transition it is then possible to determine all the parameters of the shock wave: its velocity and the jumps of the thermodynamic quantities.

The differential analyzer has been used successfully in computations and has made it possible to localize shock waves rather precisely. Recently Janenko, Vorožcov, and Fomin (1976 a, b, c) constructed a theory of differential analyzers of shock waves that is based on the concept of the *center of smoothed shock waves* obtained in the numerical solution of problems of gas dynamics. The center is taken to be the coordinate $x(h)$ of the point at which the difference solution u, v, p depend weakly (or not at all) on τ and h. To determine the center of a wave it is therefore necessary to carry out two computations with different step size τ and h. In spite of the fact that this interpretation of the center of a shock wave does not at all imply that is exists and is uniquely determined, the theory developed on the basis of the first differential approximation corroborates the existence and uniqueness of a center. Many calculations of problems with shock waves have also confirmed the effectiveness of this concept of a center. A practical criterion for determining the center of a shock wave consists in finding max $|\omega|$, as above, and for a number of schemes the center coincides with the point where $|\partial u/\partial x|$ attains a maximum. Determination of the center of a shock wave makes it possible to considerably raise the accuracy of the difference solution and also facilitates the interpretation of results.

Figure 3.17 shows the graphs of the density and velocity in a neighborhood of a shock wave obtained on the basis of a two-step Lax-Wendroff scheme with an artificial viscosity $q = 3h^2\rho[\min(\partial u/\partial x, 0)]^2$ introduced additively in the pressure; it shows three different meshes, $h_1 = 1/40$, $h_2 = 2h_1$, $h_3 = 0.5h_1$, when the shock wave has already passed a distance $\approx 23h_1$. It is evident from Figure 3.17 that a center of a finite-difference shock wave exists and is invariant under h in spite of the presence in the numerical solution of parasitic oscillations behind the front of the smeared-out shock wave. The abscissa of the point of intersection of the curves $u(x, t)$ or $\rho(x, t)$ for different h coincides with the exact position of the shock front.

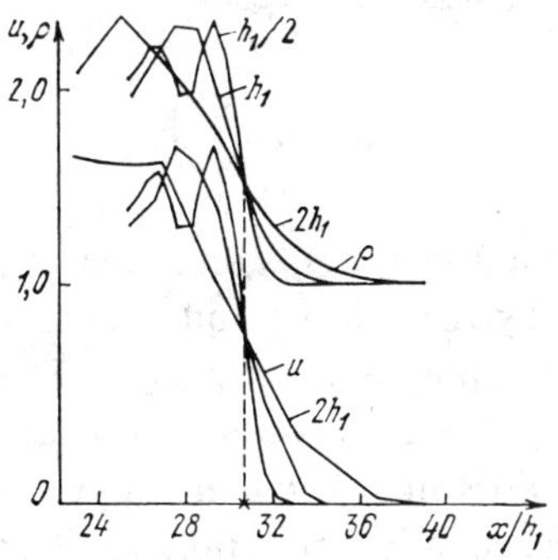

FIGURE 3.17

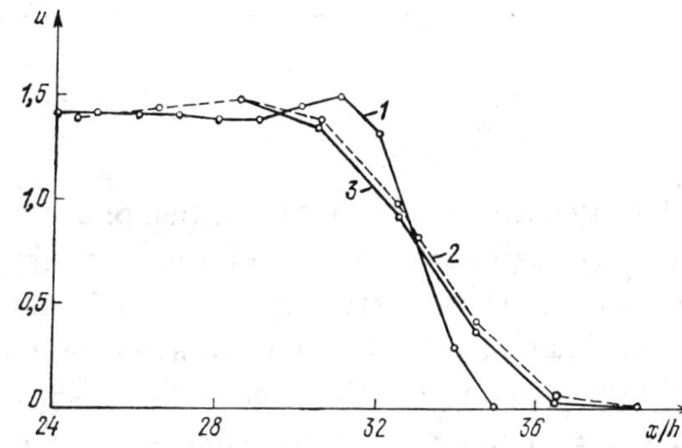

FIGURE 3.18

Figure 3.18 shows the results of computations according to a modified method of particles in boxes (see Harlow (1967), and Perrenko and Vorožcov (1973)) for various values of the Courant number. These results corroborate the presence of a center of the finite-difference shock wave for different $\kappa = \tau/h$ (τ is the step in time). The value $h = 0.1$ and the Courant number $\kappa = 0.5$ correspond to curve 1; for curve 2 we have $h = 0.2$ and $\kappa = 0.25$; for curve 3 we have $h = 0.2$ and $\kappa = 0.375$. Figure 3.19 shows the velocity profiles obtained by the method indicated at the same moment of time as in Figure 3.18 for fixed κ on three different meshes ($h_1 = 1/10$).

Schemes obtained by the method of elementary solutions also possess approximation viscosity, just like the schemes with aritificial viscosity considered in subsections 2 and 3. They possess a number of special features as compared with the latter.

The coefficient of artificial viscosity in the von Neumann-Richtmyer scheme contains an arbitrary constant μ_0 which determines the width of the front and at the same time strongly affects the qualitative character of the solution. Therefore, interpretation of the results depends on the normalization of the

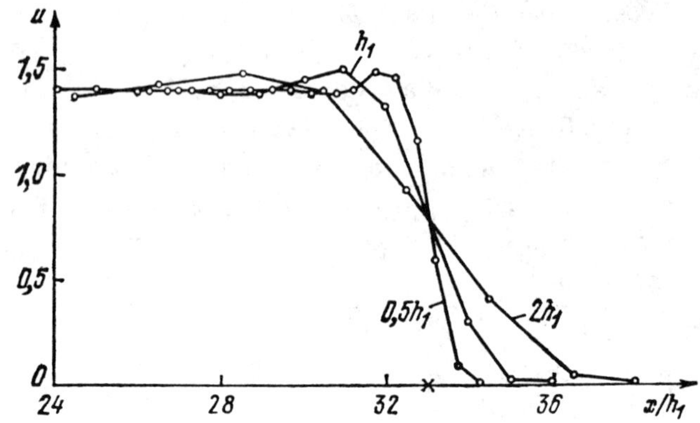

FIGURE 3.19

arbitrary coefficient μ_0. In the method of decay of a discontinuity and the wave train the approximation viscosity is normalized by the scheme and does not contain arbitrary parameters; this facilitates interpretation of results. The approximation viscosity in the Lax scheme for a fixed grid also does not contain arbitrary parameters. However, the Lax scheme does not absolutely approximate the equations and introduces viscosity in the equation of continuity. For small τ this strongly smears out the contact boundaries.

In a region where the flow contains shock waves the schemes based on the method of elementary solutions are apparently preferable, since they provide better approximation of the Hugoniot conditions.

The scheme of decay of a discontinuity, which accounts especially well with the interaction of discontinuities, possesses accuracy of first order in all equations including the energy equation, and this leads to an artificial change of the entropy—artificial heating or cooling of the gas. The scheme of the wave train possesses accuracy of second order on rarefaction waves and at the same time smooths the shock fronts quite monotonically.

5. ***Schemes of higher accuracy.*** The reader has undoubtedly already noted that the concept of accuracy of a difference scheme for computations of discontinuous solutions of gas dynamics has not, in fact, been defined. Indeed, a reasonable introduction of the concept of accuracy in this class of flows encounters major difficulties. The accuracy of a difference scheme is usually understood to mean in this case (of discontinuous flows) the accuracy that it has on smooth solutions of the equations of gas dynamics. It is certainly not obvious and probably not even true that an increase of accuracy of the difference scheme in this sense leads to an increase of accuracy in the computation of the parameters of discontinuous flows. It is thus the more interesting that a number of schemes of higher accuracy have shown good results when applied to computations of discontinuous flows.

The first difference scheme for the equations of gas dynamics of third-order accuracy was constructed by Rusanov (1968). This method was subsequently applied by Burstein and Mirin (1970) to construct a one-parameter family of difference schemes of third-order accuracy. In the work of V. B. Balakin (1970) this method was extended to obtain schemes of higher orders of accuracy than third. In particular, in this work a new difference scheme of third-order accuracy was constructed which does not belong to the one-parameter family indicated above.

The algorithm for constructing difference schemes of higher order accuracy is analogous to the Runge-Kutta iteration method used for systems of ordinary differential equations. Writing the system of equations of gas dynamics in Lagrange coordinates in the form

$$\frac{\partial w}{\partial t} = \frac{\partial F(w)}{\partial x}, \quad \omega = \left\{u,\ v,\ E = \varepsilon + \frac{u^2}{2}\right\}, \quad F(w) = \{-p, u, -up\}, \tag{1}$$

we seek a solution w^{m+1} by means of the representation

$$w^{m+1} = w^m + \tau \sum_{s=0}^{r-1} \alpha_{rs} \frac{\partial F}{\partial x}(w^{m+s/r}). \tag{2}$$

The quantities $w^{m+k/r}$ for $k = 1, \ldots, r+1$ are determined successively from the formulas

$$w^{m+k/r} = w^m + \tau \sum_{s=0}^{k-1} \alpha_{rs} \frac{\partial F}{\partial x}(w^{m+s/r}). \tag{3}$$

The coefficients α_{rs} are determined from the requirement that the expansion of w^{m+1} in powers of τ coincide with the expansion

$$w^{m+1} = w^m + \tau \frac{\partial w^m}{\partial t} + \frac{\tau^2}{2!} \frac{\partial^2 w^m}{\partial t^2} + \cdots + O(\tau^{p+1}), \tag{4}$$

where p is the desired order of accuracy.

Rusanov's scheme and the family of difference schemes of Burstein and Mirin are constructed in this manner for $r = 2$ and $p = 3$. These schemes are constructed for the usual rectangular mesh; the spatial approximation connects five points at each level. Between the two levels t^m and t^{m+1} two auxiliary levels are introduced with checkered disposition of the nodes relative to one another. It is possible to introduce a parameter α which is the distance between the level $t = t^m$ and the first auxiliary level $t = t^{m+1/2}$ with $\tau/3 \leq \alpha \leq 2\tau/3$. For

$\alpha = \tau/3$ we obtain Rusanov's scheme. In solving problems with discontinuities an artificial viscosity is introduced in the right side of (1).

Schemes of higher order accuracy were developed further by Warming, Kutler and Lomax (1973) and Abarbanel and Gottlieb (1973). In the case where the system of equations (1) is linear and has constant coefficients this method leads to the schemes of Strang (1962)—to difference schemes of maximal order of approximation for a given number of points connected by the spatial and time approximation.

In the numerical solution of the equations of gas dynamics, due to their nonlinearity, particular difficulties arise which lead to the necessity of introducing special approximations. The situation is that the coefficients of viscosity considered in gas dynamics are small parameters, and in the solutions of the equations there are singularities such as shock transitions with width of order ν and boundary layers with width of order $\sqrt{\nu}$, where ν is the coefficient of viscosity. The entire flow decomposes into a region in which the gradients are finite and small regions where they are proportional to $1/\nu$ or $1/\sqrt{\nu}$. These singularities require special approximation, since in the difference scheme there are also the small parameters τ and h in addition to the small physical parameters; the difficulty in solving these equations consists in the fact that it is necessary to analyze and compare the orders of smallness of these quantities, and the estimates usually contain rational expressions in these parameters; we thus have no uniform estimate of convergence.

We shall illustrate what we have said with the example of solving a model problem in which there is a boundary layer and an approximation of special type is required. The latter circumstance was pointed out by A. M. Il'in, although in integrating the Navier-Stokes equations such approximations had also been used earlier (Allen and Southwell (1955)). However, the idea is expressed most clearly by Il'in (1969).

For the boundary value problem

$$\nu u'' + u' = 0, \qquad u(0) = 0, \qquad u(1) = 1,$$

where $\nu > 0$ is a small parameter, the exact solution has the form

$$u(x) = \frac{1 - e^{-x/\nu}}{1 - e^{-1/\nu}}.$$

As $x \to 0$ the derivative $u'(x)$ for small ν has order $1/\nu$, and as $\nu \to 0$ the region of large gradients contracts. Thus, we have an illustration of the phenomenon of a boundary layer which arises in flow problems.

For this problem we consider the difference scheme

$$\nu \frac{T_1 - 2E + T_{-1}}{h^2} u + \frac{T_1 - T_{-1}}{2h} u = 0. \tag{5}$$

The exact solution of this difference problem is

$$u(x) = u(kh) = \frac{1 - \left(\dfrac{1 - h/2\nu}{1 + h/2\nu}\right)^k}{1 - \left(\dfrac{1 - h/2\nu}{1 + h/2\nu}\right)^N},$$

where $k = 1, \ldots, N$, $Nh = 1$; it is not hard to see that for $\nu \ll h$ it has nothing in common with the solution of the original problem (while for $\nu = h/2$ it doesn't even exist). We must therefore either strongly refine the mesh or apply other approximations. The scheme with one-sided difference

$$\nu \frac{T_{-1} - 2E + T_1}{h^2} u + \frac{T_1 - E}{h} u = 0$$

has an approximation viscosity of order $O(h)$, i.e., essentially, another differential equation is being solved with a coefficient of the leading derivative equal to $\nu + O(h)$. The solution of the difference equation does not tend to a solution of the original problem as $h \to 0$ uniformly with respect to ν.

A. M. Il'in proposed altering the coefficients of the difference scheme (5) so that exact solutions of the differential equation would also be solutions of the difference equation. This scheme has the form

$$\lambda \frac{T_{-1} - 2E + T}{h^2} u + \frac{T_1 - T_{-1}}{2h} u = 0,$$

where $\lambda = (h/2)\coth(h/2\nu)$. It is not hard to see that for $h \ll \nu$ the condition of approximation of the differential equation is satisfied (since in this case λ is close to ν). It was proved that the solution of this difference scheme converges as $h \to 0$ to the solution of the original problem uniformly with respect to ν.

In the papers of Il'in (1969), Emel'janov (1970, 1973) and Titov and Šiškin (1976) this method of obtaining schemes of uniform approximation is extended to multidimensional elliptic equations with variable coefficients, to parabolic equations, and also to the case where several small parameters are present.

It should be mentioned that Marčuk (1961) used an equivalent approach in constructing difference schemes for diffusion problems. This is actually a method of eliminating the convective term.

§9. Schemes in Euler coordinates and implicit schemes

1. ***Schemes in Euler coordinates.*** The choice of a coordinate system (Lagrange or Euler) for the computation of gas flow is determined by the formulation of the problem. If we are interested in the parameters of the flow in a prescribed region (for example, the flow of gas in a pipeline or water in a channel), then it is natural to choose Euler coordinates. If we must investigate in detail hydrodynamic processes in some fixed amount of material, then it is expedient to use Lagrange coordinates.

We consider, for example, the flow of a certain mass of gas into a vacuum. The quiescent gas is maintained in a cylindrical pipe by partitions at the sections of the pipe $x = x_1$ and $x = x_2$. At time $t = 0$ one of the partitions, say $x = x_2$, is removed, and the gas begins to flow out to the right into the vacuum. If an Euler mesh is used for the computation, then the right boundary will include al lnew points, and the number of points to be computed will grow; this is not connected with the required accuracy of the computation (Figure 3.20, a).

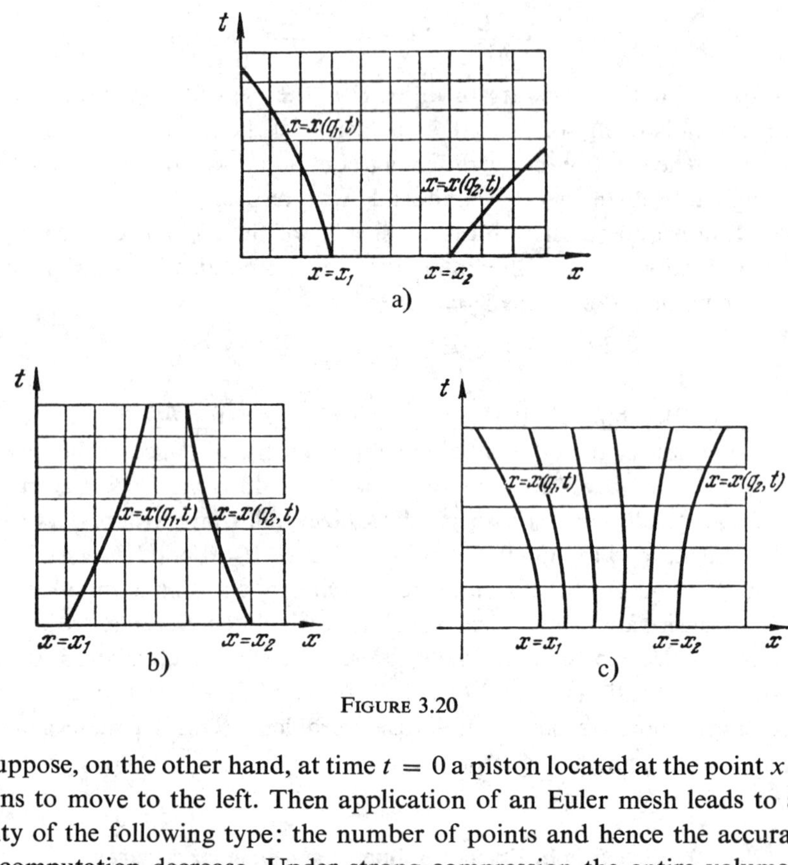

FIGURE 3.20

Suppose, on the other hand, at time $t = 0$ a piston located at the point $x = x_2$ begins to move to the left. Then application of an Euler mesh leads to a difficulty of the following type: the number of points and hence the accuracy of the computation decrease. Under strong compression the entire volume considered may fall into one calculation interval, which implies a total loss of accuracy (Figure 3.20, b). In both this case and the preceding case it is expedient to use a Lagrange mesh (Figure 3.20, c).

The precision of computing a fixed amount of material can be preserved also in Euler coordinates if they are made mobile (this was suggested by S. K. Godunov). In the case of a mobile Euler mesh the points of the mesh connected with contact boundaries move together with the latter. Intermediate points of the mesh are obtained by any law while preserving a particular minimum or maximum of points. In the simplest case the number of points between the contact boundaries remains unchanged. The choice of some version of a mobile mesh is determined by the computational accuracy required. Figure 3.20 shows possible version of a mesh for the case of moving contact boundaries $x = x(q_1, t)$ and $x = x(q_2, t)$.

Aside from the special features of the mesh, the character of approximation of the equations of gas dynamics changes on going over from a Lagrange coordinate system to an Euler system. This is related basically to the fact that $\partial/\partial t$ in Lagrange coordinates is replaced by the two-term expression $\partial/\partial t + u(\partial/\partial x)$ in Euler coordinates.

We consider the equations of isentropic flow in the form

$$\left.\begin{aligned} &\frac{\partial \varphi}{\partial t} + u\frac{\partial \varphi}{\partial x} + \frac{\partial u}{\partial x} = 0, \qquad \varphi = \ln \rho, \\ &\frac{\partial u}{\partial t} + u\frac{\partial u}{\partial x} + c^2\frac{\partial \varphi}{\partial x} = 0, \qquad c^2 = \left(\frac{\partial p}{\partial \rho}\right)_S. \end{aligned}\right\} \tag{1}$$

An explicit approximation of $u(\partial u/\partial x)$ and $u(\partial \varphi/\partial x)$ by means of central differences leads to unstable schemes. Approximation of the spatial gradients which consider the sign of u must therefore be applied. An example of a scheme of this type is the Lelevier scheme

$$\left.\begin{aligned} &\frac{\varphi^{m+1} - \varphi^m}{\tau} + u^m\frac{\Delta}{h}\varphi^m + \frac{\Delta_1 + \Delta_{-1}}{2h}u^m = 0, \\ &\frac{u^{m+1} - u^m}{\tau} + u^m\frac{\Delta}{h}u^m + (c^m)^2\frac{\Delta_1 + \Delta_{-1}}{2h}\varphi^{m+1} = 0, \end{aligned}\right\} \tag{2}$$

where $\Delta = \Delta_{-1}$ for $u > 0$ and $\Delta = \Delta_1$ for $u < 0$. A "cross" type approximation scheme is impossible here.

If we go over from (1) to the divergence system

$$\frac{\partial \rho}{\partial t} + \frac{\partial(\rho u)}{\partial x} = 0, \qquad \frac{\partial \rho u}{\partial t} + \frac{\partial}{\partial x}(p + \rho u^2) = 0, \tag{3}$$

then a complete analogue of the "cross" scheme is impossible, since the quantities p, ρ and u account be allocated to different points as was the case in

Lagrange coordinates. By referring p, ρ and u to the same points, we obtain the symmetric scheme of second order

$$\left.\begin{aligned} \frac{\rho^{m+1}-\rho^{m-1}}{2\tau}+\frac{\Delta_1+\Delta_{-1}}{2h}\rho^m u^m=0,\\ \frac{\rho^{m+1}u^{m+1}-\rho^{m-1}u^{m-1}}{2\tau}+\frac{\Delta_1+\Delta_{-1}}{2h}[p^m+\rho^m(u^m)^2]=0, \end{aligned}\right\} \tag{4}$$

which is unstable. Apparently, stable schemes of second-order accuracy in Euler coordinates can be obtained only by a more involved approximation, for example, by means of an implicit approximation or the predictor-corrector method.

In addition to the indicated schemes, there is also an admissible computational scheme which makes it possible to actually do the computation in Lagrange coordinates with subsequent interpolation to Euler coordinates (this was proposed by V. F. Kuropatenko). Suppose that $\{x_i\}$ is an Euler mesh. Over the step $t_m \le t \le t_m + \tau$ we consider it to be a Lagrange mesh, i.e., we follow the motion of the material particles contained at time $t = t_m$ in the intervals (x_i, x_{i+1}) (Figure 3.21). At time $t = t_m$ we introduce the mass Lagrange coordinate by setting

$$q_{i+1}-q_i=(x_{i+1}-x_i)\,\rho^m_{i+1/2}. \tag{5}$$

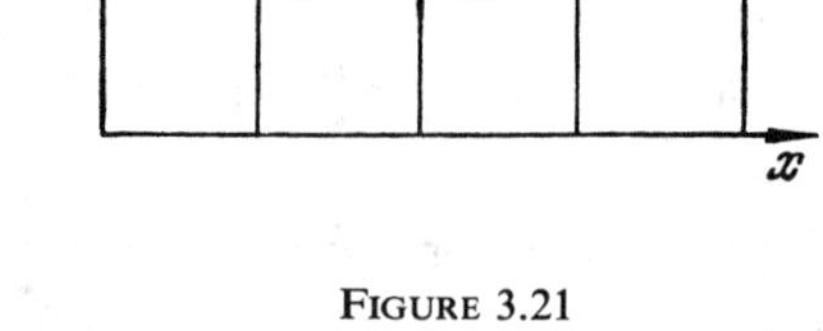

FIGURE 3.21

Applying a "cross" scheme to the computation of the material elements (q_i, q_{i+1}) and $(q_{i-1/2}, q_{i+1/2})$, we obtain

$$\frac{u_i^{m+1}-u_i^m}{\tau}+\frac{p^m_{i+1/2}-p^m_{i-1/2}}{h_i}=0,\quad \frac{v^{m+1}_{i+1/2}-v^m_{i+1/2}}{\tau}-\frac{u^{m+1}_{i+1}-u_i^{m+1}}{h_i}=0, \tag{6}$$

where we have set

$$h_i = q_{i+1} - q_i, \quad \bar{h}_i = q_{i+1/2} - q_{i-1/2}, \quad q_{i+1/2} = \frac{q_i + q_{i+1}}{2}. \tag{7}$$

The quantities u_i^{m+1} are referred to the points x_i', the quantities $v_{i+1/2}^m$ to the centers of the intervals (x_i', x_{i+1}'), and the line AB is the trajectory of the point $x = x(q_i, t)$.

All quantities pertaining to the intervals (x_i', x_{i+1}') are computed in this manner. Interpolation to the mesh $\{x_i\}$ is then made while observing the conservation laws.

Another version of the combination of Lagrange and Euler coordinates in a difference algorithm was realized in the original method of "particles in boxes" of Harlow (1967).[20] We briefly describe this method, considering it as a variant of the method of separating by means of physical processes.

We consider the motion of a polytropic gas with an equation of state $p = p(\rho)$ described by (3). In Harlow's method additional "fractional" steps are introduced. At the first fractional step we solve the system of equations

$$\frac{1}{2}\frac{\partial \rho}{\partial t} = 0, \qquad \frac{1}{2}\rho\frac{\partial u}{\partial t} + \frac{\partial p}{\partial x} = 0, \tag{8}$$

and at the second fractional step we solve the system

$$\frac{1}{2}\frac{\partial \rho}{\partial t} + \frac{\partial \rho u}{\partial x} = 0, \qquad \frac{\partial \rho u}{\partial t} + \frac{\partial \rho u^2}{\partial x} = 0. \tag{9}$$

The system (8) corresponds to the model of slow flow of incompressible gas when inertial forces and compressibility are neglected. It can be integrated at the first half-step by a symmetric difference scheme in an Euler mesh. The system (9) corresponds to the model of the free motion of interially noninteracting fluid particles; it can be integrated by appling a Lagrange mesh. These two processes are combined into a full step by averaging the density and pressure over intervals of the Euler mesh. The presence of fluctuations and self-oscillations are characteristic for the scheme of the method of particles in boxes.

We remark that implicit, absolutely stable schemes in an Euler mesh with separation of the type (8), (9) have been applied in problems of meteorology (see, for example, Marčuk (1967)). The method of particles in boxes is especially effective in the numerical solution of problems of gas dynamics with free surfaces, slippage surfaces, and regions of strong rarefaction.

Harlow's method, like some other schemes, is not invariant under the full

[20]Harlow's method is most often applied in two-dimensional and three-dimensional problems of gas dynamics. We consider it here in the one-dimensional version in order to demonstrate that the idea of "separating" processes has general character.

group of point transformations in the space of the independent and dependent variables admitted by the system of equations of gas dynamics (see Ovsjannikov (1962)). Any difference scheme is realized on a specific mesh; this in itself carries noninvariance into the computational algorithm. This noninvariance may be manifest, for example, in computations of singularities of the flow which move at various angles to the lines of the mesh. Thus, for example, the noninvariance of Harlow's method relative to Galilean transformation leads to the appearance of self-oscillations in the computations. Passage to a difference scheme hinders the group analysis, since difference operators possess different group properties than differential operators. It has been found expedient to carry out the group classification of difference schemes on the basis of their first differential approximations. Invariance properties of difference schemes of gas dynamics and group properties of difference schemes have been investigated by means of the first differential approximation by Janenko and Šokin (1973) and Fedotova and Šokin (1975).

We say that a difference scheme is invariant under a group of transformations if its first differential approximation is invariant under this group of transformations. In correspondence with this all difference schemes can be subdivided into two classes: schemes which preserve the group properties and schemes which do not. Janenko and Šokin (1973) formulated conditions under which the system of equations of the first differential approximation admits the groups of transformations which are admitted by the original system of equations of gas dynamics. We formulate these conditions for the difference scheme

$$\begin{aligned}\frac{W^{m+1}(x) - W^m(x)}{\tau} &= \frac{f^m(x+h) - f^m(x-h)}{2h} \\ &\quad + \frac{1}{h^2}\Big[\Omega\Big(x+\frac{h}{2}\Big)(W^m(x+h) - W^m(x)) \\ &\quad - \Omega\Big(x-\frac{h}{2}\Big)(W^m(x) - W^m(x-h))\Big], \end{aligned} \tag{10}$$

which approximates the system of equations of gas dynamics in Euler coordinates

$$\frac{\partial W}{\partial t} = \frac{\partial f}{\partial x}. \tag{11}$$

Here $\Omega = \|\Omega_{ij}\|_1^3$, $\Omega_{ij} = O(\tau)$, and

$$W = \begin{pmatrix} \rho u \\ \rho \\ \rho E \end{pmatrix}, \quad f = \begin{pmatrix} -p - \rho u^2 \\ -\rho u \\ -\rho u E - up \end{pmatrix}, \quad E = \varepsilon + \frac{1}{2}u^2, \quad \varepsilon = \varepsilon(p, \rho).$$

The first differential approximation of the difference scheme (10) has the form

$$\frac{\partial W}{\partial t} = \frac{\partial f}{\partial x} + \frac{\partial}{\partial x}\left(C\,\frac{\partial W}{\partial x}\right),$$

where $C = \Omega - \tau A^2/2$ and $A = \partial f/\partial W$.

THEOREM. *If in the difference scheme* (10) *the elements of the matrix* Ω *satisfy the conditions*

$$\frac{\partial \Omega_{ij}}{\partial t} = \frac{\partial \Omega_{ij}}{\partial x} = 0 \qquad (i, j = 1, 2, 3),$$

$$\frac{\partial N_{1x}}{\partial u} = N_{2x}, \qquad \frac{\partial N_{2x}}{\partial u} = 0, \qquad \frac{\partial N_{3x}}{\partial u} = 0,$$

$$\left(u_x\frac{\partial}{\partial u_x} + \rho_x\frac{\partial}{\partial \rho_x} + p_x\frac{\partial}{\partial p_x} + 2u_{xx}\frac{\partial}{\partial u_{xx}} + 2\rho_{xx}\frac{\partial}{\partial p_{xx}} + 2p_{xx}\frac{\partial}{\partial p_{xx}}\right)N_{ix} = N_{ix},$$

where

$$N = \begin{pmatrix} N_1 \\ N_2 \\ N_3 \end{pmatrix} = CW_x,$$

then the difference scheme (10) *admits the same group of transformations as the system* (11).

We remark that for $N_2 = 0$ the law of conservation of mass is satisfied in the system of equations of the first differential approximation of the difference scheme.

Fedotova and Šokin (1975) have shown that if $C = \alpha_0 E + \alpha_1 A + \alpha_2 A^2$, then the difference scheme (10) is invariant if the matrix C can be represented in the form

$$C = \beta_0 E + \beta_1(A - uE) + \beta_2(A - uE)^2,$$

where

$$\frac{\partial \beta_i}{\partial t} = \frac{\partial \beta_i}{\partial x} = \frac{\partial \beta_i}{\partial u} = 0 \qquad (i = 0, 1, 2).$$

We note that the case in which C (and hence Ω) depends on higher powers of A can be reduced to the preceding case, since

$$(A - uE)^{2k+1} = c^{2k}(A - uE), \quad (A - uE)^{2k+2} = c^{2k}(A - uE)^2$$
$$(k = 0, 1, 2, \ldots,) \qquad (12)$$

where c is the speed of sound. In the case of Lagrange coordinates the relations (12) have the form (a is the mass speed of sound)

$$A^{2k+1} = a^{2k}A, \qquad A^{2k+2} = a^{2k}A^2 \quad (k = 0, 1, 2, \ldots).$$

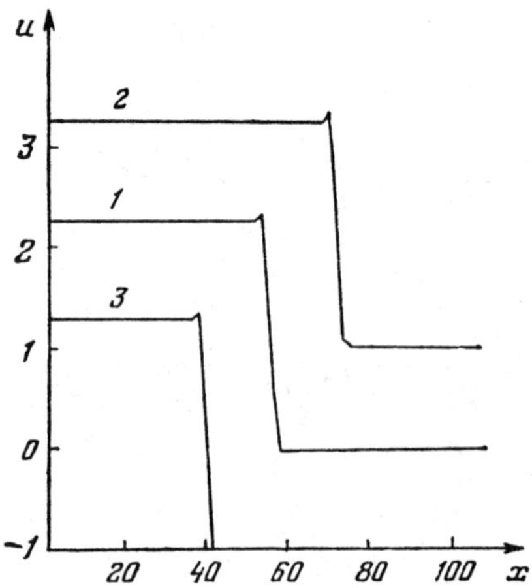

FIGURE 3.22

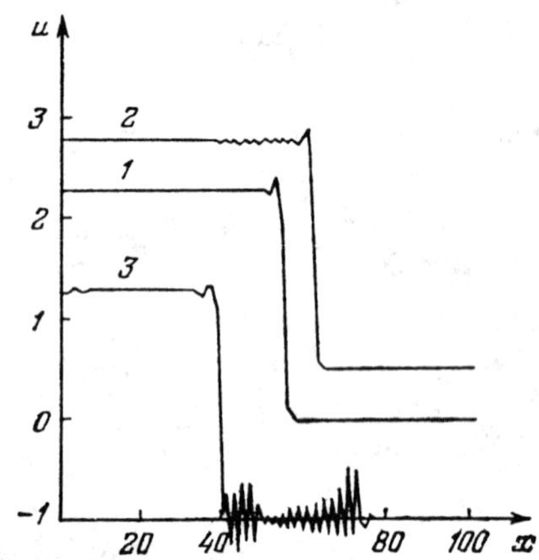

FIGURE 3.23

Figures 3.22 and 3.23 show the velocity profiles of a stationary shock wave in three different Euler coordinate systems obtained one from the other by means of Galilean transformation. The curves in Figure 3.22 are obtained by computing in an explicit, invariant difference scheme; those in Figure 3.23 are obtain on the basis of the Lax-Wendroff scheme, which is not invariant.

2. *Implicit schemes.* It is rather obvious that, other things being equal, the difference solution reproduces the solution of a differential equation the more precisely the closer the domain of dependence of the difference equation and the domain of dependence of the differential equation. From this general consideration it would seem to follow that it is necessary to apply explicit schemes for hyperbolic equations and implicit schemes for parabolic equations. The first schemes for integrating hyperbolic equations were actually explicit schemes, while for the parabolic diffusion equation the implicit double sweep scheme was immediately recognized. However, a more concrete and detailed investigation leads to the conclusion that implicit schemes are expedient in some problems of gas dynamics.

We present a number of arguments in favor of implicit schemes.

1. For large inhomogeneity of the hydrodynamic flow the local speed of sound c at a number of points may greatly exceed the mean value of the speed of sound characteristic of the given flow. In this case the local stability criterion

$$c\tau/h \leq 1, \tag{1}$$

which holds for explicit schemes, may lead to an unjustifiably small step size whereby the domain of dependence of the difference equation becomes considerably greater than the domain of dependence of the differential equation. This circumstance reduces the accuracy of the computation and leads to smoothing of the profiles.

2. For the precise computation of a shock wave it is desirable that the wave front pass at each step from a point of the mesh to the nearest point. By Zemplén's theorem (see Chapter 2, §3.4) this means that in the region behind the wave the stability criterion (1) is violated. Hence, in this case an implicit scheme is needed.

3. The magnitude of the time step size of the scheme is determined by two requirements: accuracy and stability. The accuracy requirement is determined basically by the magnitude of the gradients: the larger the gradients, the faster the hydrodynamic process occurs, and the smaller the admissible step size; and conversely. The requirement of stability of the scheme relative to local perturbations is determined because of the criterion (1) by the magnitude of the speed of sound, i.e., it is not related to the gradients of the flow at all. Therefore, in the case of hydrodynamic flows with small gradients (long-wave profiles) the step size admissible for accuracy is much greater than the step size admissible for stability (stability of the scheme relative to short-wave perturbations). Such flows are, for example, problems of meteorology, channel flows, gas lines, etc. It is clear that in this case an implicit scheme is needed. We remark that explicit and implicit schemes are not opposite to one another. The requirement of maximal closeness of the domains of dependence of the difference and differential equations indicates that explicit and implicit methods of computation should be included as elements in the computational algorithm. It is possible to apply an explicit scheme in some zones and an implicit scheme in others.

We proceed now to the consideration of specific implicit schemes. The first theretical investigation of implicit schemes for the wave equation with variable coefficients by the method of a priori estimates is due to Ladyženskaja (1952). The first implicit scheme for integrating the characteristic equations of hydrodynamics was published by Landau, Meĭman, and Halatnikov (1958).[21]

For the equations of isentropic flow

[21]The scheme was suggested and applied in computations in 1951.

$$\left.\begin{aligned} \frac{\partial s}{\partial t} - a\frac{\partial s}{\partial q} &= F_1, \\ \frac{\partial r}{\partial t} + a\frac{\partial r}{\partial q} &= F_2 \end{aligned}\right\} \tag{2}$$

we pose the mixed Cauchy problem

$$u = \frac{r+s}{2} = f(t), \qquad q = 0, \qquad r - s = g(t), \qquad q = q_0, \tag{3}$$

$$r(q, 0) = r_0(q), \qquad s(q, 0) = s_0(q), \qquad 0 \le q \le q_0. \tag{4}$$

The simplest implicit scheme for solving problem (2)–(4) has the form

$$\left.\begin{aligned} \frac{s^{m+1} - s_i^m}{\tau} - a_i^m \frac{s_{i+1}^{m+1} - s_i^{m+1}}{h} &= F_{1i}^m, \\ \frac{r_i^{m+1} - r_i^m}{\tau} + a_i^m \frac{r_i^{m+1} - r_{i-1}^{m+1}}{h} &= F_{2i}^m, \end{aligned}\right\} \tag{5}$$

$$r_0^m + s_0^m = 2f^m, \tag{6}$$

$$r_{N+1}^m - s_{N+1}^m = g^m, \tag{7}$$

$$r_i^0 = r_{0i}, \qquad s_i^0 = s_{0i}, \qquad (N+1)h = q_0. \tag{8}$$

The formulas (5) make it possible to consider r_i^{m+1} in a stable manner from left to right, running successively through the indices $i = 1, 2, \ldots, N+1$. Similarly, for s_i^{m+1} the formulas (5) provide stable computation from right to left, $i = N, \ldots, 0$.

The boundary conditions are considered in the following manner. The quantity s_0^{m+1} is determined from the formula of explicit calculation

$$s_0^{m+1} = (1 - \kappa_0^m)s_0^m + \kappa_0^m s_1^m + F_{10}^m\tau, \tag{9}$$

after which r_0^{m+1} is found by means of (6).

The quantities r_i^{m+1} are determined recurrently from left to right by (5). The quantity s_{N+1}^{m+1} is determined from (7), after which the s_i^{m+1} are determined recurrently from right to left by (5).

The value of s_0^{m+1} arriving at the left boundary does not coincide with the value determined from (9); it is therefore necessary to refine the value of s_0^{m+1}. Since (5) determines r_{N+1}^{m+1} as a linear function of r_0^{m+1}, and s_0^{m+1} as a linear function of s_{N+1}^{m+1}, it suffices to make one recalculation with subsequent interpolation.

Application of the implicit scheme may lead to excessive expansion of the domain of dependence. If the Courant condition is violated only on comparatively few zones of the mesh, then in the remaining part where it is observed explicit approximation should be applied. The following scheme of explicit-implicit approximation satisfies this requirement:

$$\left.\begin{aligned}\frac{s_i^{m+1}-s_i^m}{\tau}-a_i^m\frac{s_{i+1}^m-s_i^m}{h}&=F_{1i}^m,\\ \frac{r_i^{m+1}-r_i^m}{\tau}+a_i^m\frac{r_i^m-r_{i-1}^m}{h}&=F_{2i}^m,\qquad \kappa_i^m\le 1,\end{aligned}\right\}\tag{10}$$

$$\left.\begin{aligned}\frac{s_i^{m+1}-s_i^m}{\tau}-a_i^m\frac{s_{i+1}^{m+1}-s_i^{m+1}}{h}&=F_{1i}^m,\\ \frac{r_i^{m+1}-r_i^m}{\tau}+a_i^m\frac{r_i^{m+1}-r_{i-1}^{m+1}}{h}&=F_{i2}^m,\qquad \kappa_i^m> 1.\end{aligned}\right\}\tag{11}$$

In cases where on one of the boundaries the condition $\kappa\le 1$ is satisfied an explicit-implicit approximation makes possible a simple solution of the boundary conditions. To be specific, suppose that the left boundary is such a boundary. Formula (9) is then stable and makes it possible to start recurrent computation of r_i^{m+1} according to (10) and (6). In the reverse recurrent computation of s_i^{m+1}, on arriving at the left boundary, we again go over to an explicit scheme which does not contradict (9). This running-count scheme has first-order accuracy.

To the scheme (5) there corresponds the scheme of second-order accuracy

$$\frac{s_i^{m+1}-s_i^m}{\tau}+\frac{s_{i+1}^{m+1}-s_{i+1}^m}{\tau}-a_i^{m+1/2}\left[\frac{s_{i+1}^{m+1}-s_i^{m+1}}{h}+\frac{s_{i+1}^m-s_i^m}{h}\right]=2F_{1i}^{m+1/2},$$

$$\frac{r_i^{m+1}-r_i^m}{\tau}+\frac{r_{i-1}^{m+1}-r_{i-1}^m}{\tau}+a_i^{m+1/2}\left[\frac{r_i^{m+1}-r_{i-1}^{m+1}}{h}+\frac{r_i^m-r_{i-1}^m}{h}\right]=2F_{2i}^{m+1/2},$$

$$a_i^{m+1/2}=\frac{a_i^{m+1}+a_i^m}{2},\qquad F_{\alpha i}^{m+1/2}=\frac{F_{\alpha i}^{m+1}+F_{\alpha i}^m}{2}\quad(\alpha=1,2),\tag{12}$$

which requires linearization of the cofficients a_i^{m+1} and $F_{\alpha i}^{m+1}$ and subsequent iteration (see (25)–(29), below). It is also possible to indicate a scheme of second-order accuracy based on the predictor-corrector technique. An auxiliary half step is first made by the formulas

$$\left.\begin{aligned}\frac{s_i^{m+1/2}-s_i^m}{\tau/2}-a_i^m\frac{s_{i+1}^{m+1/2}-s_i^{m+1/2}}{h}&=F_{1i}^m,\\ \frac{r_i^{m+1/2}-r_i^m}{\tau/2}+a_i^m\frac{r_i^{m+1/2}-r_{i-1}^{m+1/2}}{h}&=F_{2i}^m.\end{aligned}\right\}\tag{13}$$

After this a correcting full step is made which improves the accuracy:

$$\left.\begin{aligned} &\frac{1}{2}\left[\frac{s_i^{m+1}-s_i^m}{\tau}+\frac{s_{i+1}^{m+1}-s_{i+1}^m}{\tau}\right]-a_i^{m+1/2}\frac{s_{i+1}^{m+1/2}-s_i^{m+1/2}}{h}=F_{1i}^{m+1/2},\\ &\frac{1}{2}\left[\frac{r_i^{m+1}-r_i^m}{\tau}+\frac{r_{i-1}^{m+1}-r_{i-1}^m}{\tau}\right]+a_i^{m+1/2}\frac{r_i^{m+1/2}-r_{i-1}^{m+1/2}}{h}=F_{2i}^{m+1/2}, \end{aligned}\right\} \quad (14)$$

where $a_i^{m+1/2}$ and $F_{\alpha i}^{m+1/2}$ are computed on the basis of $r_i^{m+1/2}$ and $s_i^{m+1/2}$ from (13). The scheme (13) has first-order accuracy but also has reserve stability, which suffices in order that the scheme (14) having second-order accuracy be stable.

We shall illustrate the application of an implicit schemes for algorithms explicitly distinguishing shock waves with the example of a plane, isothermic flow with a single shock wave propagating from left to right. If the computational formulas depend explicitly on the position of the shock wave relative to the mesh, then replacement of some computational formulas by others occurs when the shock wave passes from one calculation interval to another. This leads to a jump in the parameters of the shock wave and the hydrodynamic quantities which can be eliminated by certain operations or by applying formulas of high accuracy. A convenient method for eliminating such jumps is the choice of a mesh depending on the motion of the shock wave. We assume[22] that at each step the shock wave passes from one mesh point q_i to the next point q_{i+1}(Figure 2.34).

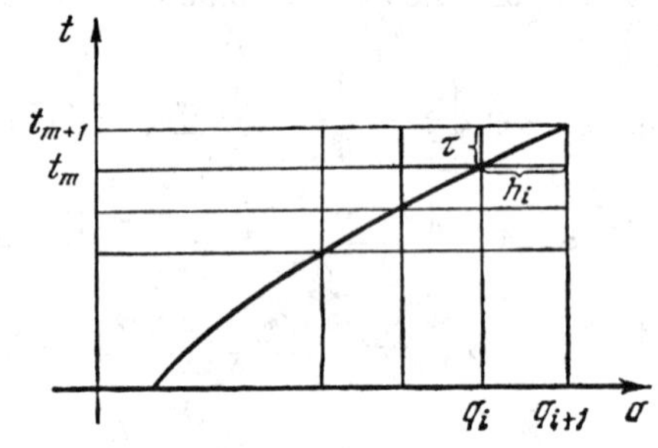

FIGURE 3.24

As we have noted above, in this case it follows from Zemplén's theorem that behind the front of the shock wave a stable scheme cannot be explicit. On the other hand, from the same Zemplén theorem it follows that ahead of the shock wave the Courant stability condition is satisfied. Therefore, an explicit-implicit scheme is convenient.

Suppose that at the point q_i at time t_m there is a shock wave with mass velocity

[22]This was proposed by M. V. Keldyš. A scheme of this type was realized by Ohocimskiĭ et al. (1957).

$\tilde{D}_i^m$. Then the quantities $h_i = q_{i+1} - q_i$ and $\tau_m = t_{m+1} - t_m$ are connected by the relation

$$\tilde{D}_i^m = \frac{h_i}{\tau_m}.$$

Here two variants are possible: either for a fixed spatial mesh $\{q_i\}$ a step size τ_m is chosen or for a fixed step size τ the shock wave itself marks out the mesh.

At the points q_i and q_{i+1} the computational formulas have the form

$$\frac{(r_{i+1}^{m+1})^+ - r_{i+1}^m}{\tau_m} + a_{i+1}^m \frac{r_{i+1}^m - (r_i^m)^+}{h_i} = 0, \tag{15}$$

$$\frac{t_i^{m+1} - (r_i^m)^-}{\tau_m} + (a_i^m)^- \frac{(r_{i+1}^{m+1})^- - r_i^{m+1}}{h_i} = 0, \tag{16}$$

$$\frac{(s_{i+1}^{m+1})^+ - s_{i+1}^m}{\tau_m} - a_{i+1}^m \frac{s_{i+2}^m - s_{i+1}^m}{h_{i+1}} = 0, \tag{17}$$

where the signs $\pm$ correspond to the quantities ahead of and behind the wave front.

Formulas (15) and (16) are stable and make it possible to determine $(r_{i+1}^{m+1})^+$ and $(r_{i+1}^{m+1})^-$, i.e., the invariant r ahead of and behind the front at time t_{m+1}. The formula (17) determines $(s_{i+1}^{m+1})^+$, and hence $(u_{i+1}^{m+1})^+$ is known.

After this, from the Hugoniot conditions at the point (q_{i+1}, t_{m+1})

$$\left.\begin{aligned} &(r_{i+1}^{m+1})^+ - (r_{i+1}^{m+1})^- = \varphi(\theta_{i+1}^{m+1}),\\ &(s_{i+1}^{m+1})^+ - (s_{i+1}^{m+1})^- = \psi(\theta_{i+1}^{m+1}),\\ &\varphi(\theta) = c\left(\theta - \frac{1}{\theta} + \ln \theta^2\right), \qquad \psi(\theta) = c\left(\theta - \frac{1}{\theta} - \ln \theta^2\right),\\ &\theta_{i+1}^{m+1} = \frac{D_{i+1}^{m+1} - (u_{i+1}^{m+1})^+}{c} = \frac{1/(\rho_{i+1}^{m+1})^- \cdot \tilde{D}_{i+1}^{m+1} - (u_{i+1}^{m+1})^+}{c} \end{aligned}\right\} \tag{18}$$

it is possible to determine θ_{i+1}^{m+1}, $\tilde{D}_{i+1}^{m+1}$ and $(\rho_{i+1}^{m+1})^-$, and continue the computation. To refine the trajectory it is possible to recalculate the point q_{i+1} according to the formula

$$q_{i+1} = q_i + \frac{\tilde{D}_i^m + \tilde{D}_{i+1}^{m+1}}{2} \tau_m.$$

This scheme was realized by Janenko and Jaušev (1960).

Implicit schemes of computation are almost unavoidable in the case where thermal conductivity and viscosity are taken into account in the equations of hydrodynamics. We shall present several examples of difference schemes for computing flows with viscosity and thermal conductivity.

We consider the system of equations of gas dynamics with thermal conductivity and viscosity (artificial or physical) in Lagrange coordinates:

$$\frac{\partial u}{\partial t} + \frac{\partial \bar{p}}{\partial q} = 0, \tag{19}$$

$$\frac{\partial v}{\partial t} - \frac{\partial u}{\partial q} = 0, \tag{20}$$

$$\frac{\partial \varepsilon}{\partial t} + \bar{p}\,\frac{\partial u}{\partial q} = \frac{\partial}{\partial q}\left(\kappa\,\frac{\partial T}{\partial q}\right), \tag{21}$$

$$\bar{p} = p + \omega, \qquad \omega = -\,\mu\,\frac{\partial u}{\partial q}, \tag{22}$$

$$p = p(v, T), \qquad \varepsilon = \varepsilon(v, T), \tag{23}$$

$$\mu = \mu(v, T), \qquad \kappa = \kappa(v, T). \tag{24}$$

To the equations (19)–(21) we apply an implicit approximation in which the gradient with respect to q is taken with weight α at the upper step and with weight $\beta = 1 - \alpha$ at the lower step. As a result, we obtain the scheme

$$\begin{aligned}
&\frac{u_i^{m+1}}{\tau} + \frac{\alpha(p_{i+1/2}^{m+1} - p_{i-1/2}^{m+1})}{h} \\
&\quad - \alpha\,\frac{\mu_{i+1/2}^{m+1}(u_{i+1}^{m+1} - u_i^{m+1}) - \mu_{i-1/2}^{m+1}(u_i^{m+1} - u_{i-1}^{m+1})}{h^2} \\
&= \frac{u_i^m}{\tau} - \beta\,\frac{p_{i+1/2}^m - p_{i-1/2}^m}{h} \\
&\quad + \beta\,\frac{\mu_{i+1/2}^m(u_{i+1}^m - u_i^m) - \mu_{i-1/2}^m(u_i^m - u_{i-1}^m)}{h^2} = F_{1i}^m,
\end{aligned} \tag{25}$$

$$\frac{v_{i+1/2}^{m+1}}{\tau} - \alpha\,\frac{u_{i+1}^{m+1} - u_i^{m+1}}{h} = \frac{v_{i+1/2}^m}{\tau} + \beta\,\frac{u_{i+1}^m - u_i^m}{h} = F_{2i}^m, \tag{26}$$

$$\begin{aligned}
&\frac{\varepsilon_{i+1/2}^{m+1}}{\tau} + \alpha(\alpha\bar{p}_{i+1/2}^{m+1} + \beta\bar{p}_{i+1/2}^m)\,\frac{u_{i+1}^{m+1} - u_i^{m+1}}{h} \\
&\quad + \beta\,\frac{u_{i+1}^m - u_i^m}{h}\,(\alpha\bar{p}_{i+1/2}^{m+1} + \beta\bar{p}_{i+1/2}^m) \\
&\quad - \alpha\,\frac{\kappa_{i+1}^{m+1}(T_{i+3/2}^{m+1} - T_{i+1/2}^{m+1}) - \kappa_i^{m+1}(T_{i+1/2}^{m+1} - T_{i-1/2}^{m+1})}{h^2} \\
&= \frac{\varepsilon_{i+1/2}^m}{\tau} + \beta\,\frac{\kappa_{i+1}^m(T_{i+3/2}^m - T_{i+1/2}^m) - \kappa_i^m(T_{i+1/2}^m - T_{i-1/2}^m)}{h^2} = F_{3i}^m.
\end{aligned} \tag{27}$$

This is a nonlinear algebraic system for u_i^{m+1}, $v_{i+1/2}^{m+1}$ and $T_{i+1/2}^{m+1}$ with known right sides F_{si}^m $(s = 1, 2, 3)$.

To find u_i^{m+1}, $v_{i+1/2}^{m+1}$ and $T_{i+1/2}^{m+1}$ we apply the method of successive approximations, linearize equations (25)–(27), and solve the system of linear equations obtained by the method of vectorial double sweep. For this we set

$$f_{i+1/2}^{m+1} = f_{i+1/2}^{(k+1)} + \left(\frac{\partial f}{\partial v}\right)_{i+1/2}^{(k)} (v_{i+1/2}^{(k+1)} - v_{i+1/2}^{(k)}) + \left(\frac{\partial f}{\partial t}\right)_{i+1/2}^{(k)} (T_{i+1/2}^{(k+1)} - T_{i+1/2}^{(k)}), \quad (28)$$

where f denotes any of the functions p, ε, μ or κ; k is the iterative index.

Substituting (28) into (25)-(27) and expressing $v_{i+1/2}^{m+1}$ in terms of $u_{i+1}^{(k+1)}$, u_i^k and the known quantities at the mth level by means of (26), we obtain equations of the form

$$\overset{(k)}{A}_{si-1}\overset{(k+1)}{u}_{i-1} + \overset{(k)}{B}_{si}\overset{(k+1)}{u}_i + \overset{(k)}{C}_{si+1}\overset{(k+1)}{u}_{i+1} + \overset{(k)}{D}_{si-1}\overset{(k+1)}{T}_{i-1} + \overset{(k)}{E}_{si}\overset{(k+1)}{T}_i + \overset{(k)}{G}_{si+1}\overset{(k+1)}{T}_{i+1} = \overset{(k)}{H}_{si}, \quad (29)$$

where the coefficients A, B, C, D, E, G, and H are taken from the preceding iteration. As the zeroth approximation it is possible to take the values of the quantities at the mth level, i.e., $f_{i+1/2}^{(0)} = f_{i+1/2}^m$. The linear equations (29) are solved by the method of vectorial double sweep while taking the boundary conditions into account.[23]

An implicit scheme of this type was developed by I. M. Gel′fand, O. V. Lokucievskiĭ, and V. F. D′jačenko. Of course, the above scheme is applicable also to the computation of flows without thermal conductivity.

By reducing the accuracy of the scheme, it is possible to go over to an implicit scheme solvable by the usual three-point double sweep method (the successive sweep scheme; see Neuvažaev and Janenko (1966). For this in the approximation of (20) and the subsequent linearization (28) we set

$$p_{i+1/2}^{m+1} = p(v_{i+1/2}^{m+1}, T_{i+1/2}^m), \quad (30)$$

whence it follows that

$$\overset{(k+1)}{p}_{i+1/2} = \overset{(k)}{p}_{i+1/2} + \overset{(k)}{\left(\frac{\partial p}{\partial v}\right)}_{i+1/2} (\overset{(k+1)}{v}_{i+1/2} - \overset{(k)}{v}_{i+1/2}). \quad (31)$$

In this case (29) can be solved by the usual three-point double sweep method.

The implicit schemes we considered in the preceding subsection were nonlinear. To solve them it was necessary to linearize them first, with subsequent iteration based on the nonlinearity. It is clear that this considerably complicates the algorithm.

---

[23] See, for example, Godunov and Rjaben′kiĭ (1973) Richtmyer and Morton (1967).

These difficulties can be avoided by giving up the divergence scheme. For example, for the equations

$$\frac{\partial u}{\partial t} + \frac{\partial p}{\partial v}\frac{\partial v}{\partial q} - \mu\frac{\partial^2 u}{\partial q^2} = \frac{\partial u}{\partial t} - a^2\frac{\partial v}{\partial q} - \mu\frac{\partial^2 u}{\partial q^2} = 0,$$

$$\frac{\partial v}{\partial t} - \frac{\partial u}{\partial q} = 0$$

it is possible to write down a linear, implicit scheme of second-order accuracy

$$\frac{u^{m+1} - u^m}{\tau} - (a^2)^{m+1/2}\frac{\Delta_1 + \Delta_{-1}}{2h}\frac{v^{m+1} + v^m}{2} - \mu\frac{\Delta_1\Delta_{-1}}{h^2}\frac{u^m + u^{m+1}}{2} = 0.$$

$$\frac{v^{m+1} - v^m}{\tau} - \frac{\Delta_1 + \Delta_{-1}}{2h}\frac{u^m + u^{m+1}}{2} = 0,$$

where $(a^2)^{m+1/2}$ is determined, for example, by extrapolation from the values of a at the $(m - 1)$st and mth steps or by means of a single recalculation. However, the loss of divergence of the difference scheme is undesirable.

To overcome these difficulties it is possible to apply the "predictor-corrector" method, which is well known in the theory of integrating ordinary differential equations. For example, for the quasilinear heat equation (see Douglas (1958))

$$\frac{\partial u}{\partial t} = \frac{\partial}{\partial x}\left[\kappa(u)\frac{\partial u}{\partial x}\right]$$

the scheme based on the "predictor-corrector" idea has the following form:

$$\frac{u^{m+1/2} - u^m}{\tau/2} = \frac{\Delta_1}{h}\left[\kappa(u^m)\frac{\Delta_{-1}u^{m+1/2}}{h}\right], \tag{32}$$

$$\frac{u^{m+1} - u^m}{\tau} = \frac{\Delta_1}{h}\left[\kappa(u^{m+1/2})\frac{\Delta_{-1}}{h}u^{m+1/2}\right]. \tag{33}$$

Equation (32) provides a linearized scheme of first-order accuracy but maximal stability; (33) refines the computation by using the reserve stability created by the first equation. The scheme (32), (33) is a divergence scheme which is stable, free of iteration, and has second-order accuracy.

A corrector reestablishing divergence conservation laws was proposed in the work of Godunov and Semendjaev (1962) for the method of characteristics. Godunov proposed a scheme for integrating the equations of gas dynamics based on the "predictor-corrector" idea. We shall illustrate this scheme with the example of the equations of gas dynamics (19)–(21) without viscosity and thermal conductivity.

The first half-step is computed on the basis of a stable, linearized, implicit scheme

$$\left.\begin{aligned} &\frac{u_i^{m+1/2} - u_i^m}{\tau/2} + \frac{p_{i+1/2}^{m+1/2} - p_{i-1/2}^{m+1/2}}{h} = 0, \\ &\frac{v_{i+1/2}^{m+1/2} - v_{i+1/2}^m}{\tau/2} - \frac{u_{i+1}^{m+1/2} - u_i^{m+1/2}}{h} = 0, \\ &\frac{p_{i+1/2}^{m+1/2} - p_{i+1/2}^m}{\tau/2} + (c^2)_{i+1/2}^m \frac{u_{i+1}^{m+1/2} - u_i^{m+1/2}}{h} = 0, \quad c^2 = \frac{\partial\varepsilon/\partial v + p}{\partial\varepsilon/\partial p}, \end{aligned}\right\} \tag{34}$$

which is solved by the method of vectorial double sweep. After this the scheme (34) is refined by divergence equations of second-order accuracy using the reserve stability of (34):

$$\frac{u_i^{m+1} - u_i^m}{\tau} + \frac{p_{i+1/2}^{m+1/2} - p_{i-1/2}^{m+1/2}}{h} = 0, \tag{35}$$

$$\frac{v_{i+1/2}^{m+1} - v_{i+1/2}^m}{\tau} - \frac{u_{i+1}^{m+1/2} - u_i^{m+1/2}}{h} = 0, \tag{36}$$

$$\frac{E_i^{m+1} - E_i^m}{\tau} + \frac{p_{i+1/2}^{m+1/2}\, u_{i+1/2}^{m+1/2} - p_{i-1/2}^{m+1/2}\, u_{i-1/2}^{m+1/2}}{h} = 0, \quad E = \varepsilon + \frac{u^2}{2}, \tag{37}$$

where ε_i^{m+1}, ε_i^m and $u_{i+1/2}^{m+1/2}$ are obtained by the corresponding interpolation. Equation (37) can be replaced by

$$\frac{E_{i+1/2}^{m+1} - E_{i+1/2}^m}{\tau} + \frac{p_{i+1}^{m+1/2}\, u_{i+1}^{m+1/2} - p_i^{m+1/2}\, u_i^{m+1/2}}{h} = 0,$$

where $u_i^{m+1/2}$ and $p_i^{m+1/2}$ are obtained by interpolation.

It is clear that the "predictor-corrector" scheme can be used also for the equations of gas dynamics with thermal conductivity.

The "predictor-corrector" scheme considered, which is free of iteration, is more suitable than the implicit method (25)–(27) which requires iteration on the nonlinearity. However, the algorithm for solving (35)–(37) by means of vectorial double sweep is also rather complicated.

We shall formulate the "predictor-corrector" method in more general form. We consider the system of conservation laws

$$\oint u_k\, dx - \varphi_k(u_1, \ldots, u_n)\, dt = 0, \tag{38}$$

to which there corresponds the divergence system of equations

$$\frac{\partial u_k}{\partial t} + \frac{\partial \varphi_k}{\partial x} = 0. \tag{39}$$

We approximate the relations (38) on a difference mesh

$$\frac{u_{i+1/2}^{m+1} - u_{i+1/2}^{m}}{\tau} + \frac{\varphi_{i+1}^{*} - \varphi_{i}^{*}}{h} = 0, \tag{40}$$

where the quantities φ_i^* refer to some intermediate time $t^* = t_m + \tau^*$ in the interval (t_m, t_{m+1}) (Figure 3.25).

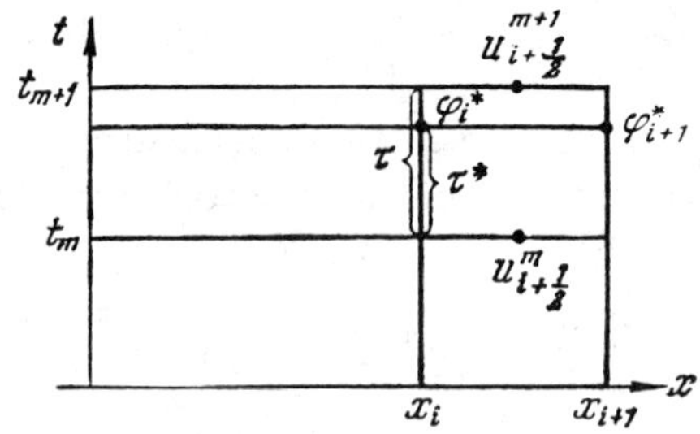

FIGURE 3.25

The integration problem breaks into two steps. At the first step the quantities φ_k^* are computed by means of any maximally stable scheme which is not a divergence scheme as a rule. At the second step (40) is applied. For $t^* = 0.5(t_m + t_{m+1})$ the scheme (40) has second-order accuracy, while the oscillation effects are large. By moving t^* upward, we displace the characteristic roots inside the unit disk, which leads to damping of the oscillations. For $t^* = t_{m+1}$ the scheme (40) has first-order accuracy, but the stability of the scheme and the damping of the oscillations become maximal.

As an illustration, we consider an implicit scheme obtained by applying the "predictor-corrector" technique to running-count schemes (see Janenko and Jaušev (1966)), restricting our attention for simplicity to the case of an isothermic gas. As a predictor for computing $u(t^*)$ it is possible to use the explicit running-count scheme in variants (2.3.11) and (2.3.12) with step size τ^* or, if the Courant stability condition is violated, the implicit running-count scheme (5) in invariants.

As a corrector reestablishing the conservation laws of the equations of gas dynamics, we choose the divergence scheme

$$\left.\begin{aligned} &\frac{u_i^{m+1} - u_i^m}{\tau} + \frac{p_{i+1}^* - p_{i-1}^*}{2h} = 0, \\ &\frac{v_i^{m+1} - v_i^m}{\tau} - \frac{u_{i+1}^* - u_{i-1}^*}{2h} = 0, \quad p^* = c\rho^*, \end{aligned}\right\} \tag{41}$$

where p_i^* and u_i^* are computed from r^* and s^*. For $\tau^* = \tau/2$ the scheme (41) has second-order accuracy and is stable. Here all the characteristic roots lie on the unit circle and oscillate without being damped. For $\tau^* > \tau/2$ the accuracy of the scheme decreases, but damping of the oscillations occurs which becomes stronger with increasing τ^*.

A scheme of this type is considerably simpler in realization, since the equations at the auxiliary step are solved not by double sweep but by a two-point running-count method. Moreover, by applying the explicit or implicit scheme depending on whether the Courant criterion is satisfied, we draw together the domains of dependence of the difference and differential equations. An analogous scheme can also be applied in Euler coordinates.

§10. Singularities of a difference solution

1. ***The behavior of difference solutions near a discontinuity.*** The presence of a discontinuity in the flow parameters leads to a number of singularities in the difference solution.

We first study singularities of the exact solution near a contact boundary. For simplicity we suppose that on both sides of the contact boundary the entropy is constant. On the boundary the usual conditions of continuity of u and p must hold:

$$u_- = u_+ = u, \qquad p_- = p_+ = p. \tag{1}$$

From the Euler equation $\partial u/\partial t + \partial p/\partial q = 0$ and from (1) we have

$$\left(\frac{\partial p}{\partial q}\right)_- = \left(\frac{\partial p}{\partial q}\right)_+, \tag{2}$$

i.e., the function $p(q, t)$ in a neighborhood of the boundary is continuous together with the first derivatives $\partial p/\partial t$ and $\partial p/\partial q$. From the equation of continuity

$$\frac{\partial v}{\partial t} = -\frac{1}{a^2}\frac{\partial p}{\partial t} = \frac{\partial u}{\partial q}, \qquad a^2 = -\frac{\partial p}{\partial v},$$

on the other hand, it follows that $\partial u/\partial q$ is discontinuous:

$$\left[\frac{\partial u}{\partial q}\right] = \left(\frac{\partial u}{\partial q}\right)_+ - \left(\frac{\partial u}{\partial q}\right)_- = -\left[\frac{1}{a^2}\right]\frac{\partial p}{\partial t} = -\left[\frac{1}{a_+^2} - \frac{1}{a_-^2}\right]\frac{\partial p}{\partial t} = f. \tag{3}$$

Suppose that next to the boundary $q = 0$ there are the intervals of the mesh $(-h_0, 0)$ and $(0, h_1)$ characterized by the mean values $p_{-1/2}$, $u_{-1/2}$, and $p_{1/2}$, $u_{1/2}$. Taking into account the relations (2) and (3) for the boundary values u_0 and p_0, we obtain the following interpolation formulas (see Figure 3.26):

$$\left.\begin{aligned} p_0 = \frac{h_1 p_{-1/2} + h_0 p_{1/2}}{h_0 + h_1}, \qquad u_0 = \frac{h_1 u_{-1/2} + h_0 u_{1/2}}{h_0 + h_1} - g, \\ g = \frac{h_1 h_0}{2(h_0 + h_1)} f. \end{aligned}\right\} \tag{4}$$

Hence

$$s_{1/2} - s_H = u_{1/2} - \frac{p_{1/2}}{a_1} - \left(u_0 - \frac{p_0}{a_1}\right) = O(h). \tag{5}$$

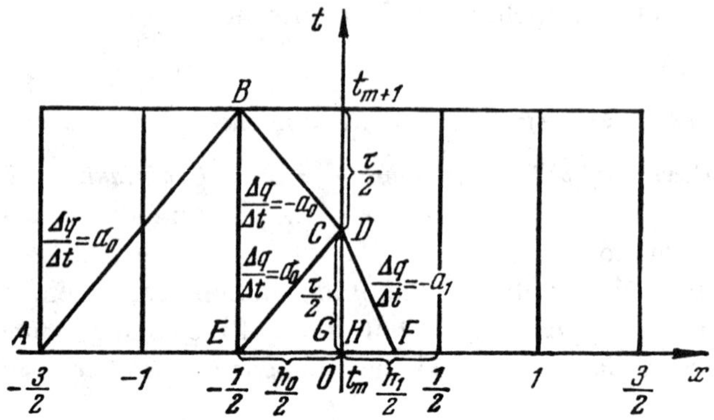

FIGURE 3.26

We now consider the scheme of decay of a discontinuity. It was shown in §8.4 that in a homogeneous medium the scheme of decay of a discontinuity is equivalent in the acoustic approximation to the running-count scheme. It turns out in the presence of a contact boundary equivalence of the schemes holds only if particular conditions on the mesh are satisfied.

We first show how the running-count formulas are changed when quantities are arranged in correspondence with the scheme of decay of a discontinuity. While in the usual running-count scheme the quantities considered are referred to the contact boundary (more precisely, to its left and right sides; see §7), in the scheme of decay of a discontinuity the boundary is located between calculation points. In correspondence with this the formulas matching the left and right quantities on a contact boundary become more complicated. We determine r and s at half-integral points; the boundary is located at the integral point $q = 0$ (Figure 3.26). We assume that the equations of state have the form

$$p = -a_0^2 v, \qquad q < 0, \qquad p = -a_1^2 v, \qquad q > 0, \qquad a_1 > a_0. \tag{6}$$

The quantities p, u, and v are connected with r and s by the relations

$$\left.\begin{aligned} &r = u - av = u + p/a, \qquad s = u + av = u - p/a, \\ &u = \frac{r + s}{2}, \qquad v = \frac{s - r}{2a}, \qquad p = \frac{a}{2}(r - s). \end{aligned}\right\} \tag{7}$$

For simplicity we assume that to the left of the boundary the following condition is satisfied:

$$\kappa_0 = \frac{a_0 \tau}{h_0} = 1. \tag{8}$$

The lines AB and EC are r-characteristics with slope a_0; BC and DF are s-characteristics, the first with slope $-a_0$ and the second with slope $-a_1$. The points C and G of the boundary belong to its left side, and the points D and H to the right.

For the point $-1/2$ the unning-count formulas have the form

$$r^{m+1}_{-1/2} = r^{m}_{-3/2}, \qquad s^{m+1}_{-1/2} = s_C. \tag{9}$$

To determine the quantities s_C we make use of the relations on the contact boundary expressing the continuity of u and p:

$$r_C + s_C = r_D + s_D, \qquad \theta(r_C - s_C) = r_D - s_D, \qquad \theta = a_0/a_1 < 1. \tag{10}$$

From (10) we have

$$s_C = \frac{\theta - 1}{\theta + 1} r_C + \frac{2}{\theta + 1} s_D = \frac{\theta - 1}{\theta + 1} r^m_{-1/2} + \frac{2}{\theta + 1} s_F. \tag{11}$$

As usual, s_F is determined by interpolation between s_H and $s^m_{1/2}$:

$$s_F = (1 - \kappa_1)\, s_H + \kappa_1 s^m_{1/2}, \qquad \kappa_1 = \frac{a_1 \tau}{h_1}. \tag{12}$$

Hence

$$s^{m+1}_{-1/2} = s_C = \frac{\theta - 1}{\theta + 1} r^m_{-1/2} + \frac{2}{\theta + 1} s^m_{1/2} - \frac{2(1 - \kappa_1)}{\theta + 1} (s^m_{1/2} - s_H). \tag{13}$$

We now determine $r^{m+1}_{-1/2}$ and $s^{m+1}_{-1/2}$ from the formulas for decay of a discontinuity (see §8.4):

$$\frac{u_{-1/2}^{m+1} - u_{-1/2}^{m}}{\tau} + \frac{P_0^m - P_{-1}^m}{h_0} = 0, \quad \frac{v_{-1/2}^{m+1} - v_{-1/2}^{m}}{\tau} - \frac{U_0^m - U_{-1}^m}{h_0} = 0,$$

$$P_0^m = \frac{u_{-1/2}^m - u_{1/2}^m + p_{-1/2}^m/a_0 + p_{1/2}^m/a_1}{1/a_0 + 1/a_1} = \frac{r_{-1/2}^m - s_{1/2}^m}{1/a_0 + 1/a_1},$$

$$U_0^m = \frac{p_{-1/2}^m - p_{1/2}^m + a_0 u_{-1/2}^m + a_1 u_{1/2}^m}{a_0 + a_1} = \frac{a_0 r_{-1/2}^m + a_1 s_{1/2}^m}{a_0 + a_1},$$

$$P_{-1}^m = a_0 \frac{r_{-3/2}^m - s_{-1/2}^m}{2}, \qquad U_{-1}^m = \frac{r_{-3/2}^m + s_{-1/2}^m}{2}.$$

Going over to the invariants $r_{-1/2}^m$ and $s_{-1/2}^m$ and noting (10) and (11), we obtain

$$r_{-1/2}^{m+1} = r_{-3/2}^m, \qquad s_{-1/2}^{m+1} = \frac{\theta - 1}{\theta + 1} r_{-1/2}^m + \frac{2}{\theta + 1} s_{1/2}^m. \tag{14}$$

Comparing (9) and (13) with (14), we find that the expressions for $r_{-1/2}^{m+1}$ coincide, while those for $s_{-1/2}^{m+1}$ differ by the quantity

$$\frac{2(1 - \kappa_1)}{1 + \theta}(s_{1/2}^m - s_H). \tag{15}$$

It follows from (5) that the "error" (15) has order $O(h)$, i.e., at the point $-1/2$ the approximation is disrupted. Approximation occurs only under the condition $\kappa_1 = 1$.

A similar analysis shows that for any κ_0 the approximation condition has the form

$$\kappa_1 = \kappa_0, \qquad \frac{h_1}{h_0} = \frac{a_1}{a_0} = \theta. \tag{16}$$

This criterion, which was obtained by Godunov (1959), can be interpreted as the condition of simultaneous arrival of acoustic signals from the points $-1/2$ and $1/2$ at the point 0.

In the nonlinear case the criterion (16) is not satisfied identically, since a_0/a_1 varies in time. This means that in a practical computation there is always a loss of accuracy at the boundary.

The acoustic criterion (16) is easily obtained from dimensional considerations. Indeed, for the wave equation

$$\frac{\partial^2 p}{\partial t^2} = a^2 \frac{\partial^2 p}{\partial q^2} \tag{17}$$

with a discontinuity of the sound speed

$$a = \begin{cases} a_1, & q > 0, \\ a_0, & q \le 0, \end{cases}$$

it is possible to apply the transformation $q = a\bar{q}$. After this (17) takes the form

$$\frac{\partial^2 p}{\partial t^2} = \frac{\partial^2 p}{\partial \bar{q}^2}. \tag{18}$$

It is clear that for (18) a uniform mesh is optimal:

$$\bar{h}_0 = \frac{h_0}{a_0}, \qquad \bar{h}_1 = \frac{h_1}{a_1},$$

and we arrive at the criterion (16).

We now consider the behavior of a difference solution in a neighborhood of a shock front. As the analysis of §2.3 shows, the front of an acoustic shock wave in an explicit running-count scheme is smoothed monotonically (see Figure 3.3). This is explained by the property of monotonicity of the running-count scheme, which is a scheme with positive coefficients. The scheme of decay of a discontinuity, which coincides with the monotone running-count scheme in the acoustic approximation, is not a scheme with positive coefficients and does not possess monotonicity properties, strictly speaking. Therefore, sufficiently strong shock waves are reproduced in the scheme of decay of a discontinuity by nonmonotone profiles.

The Lax scheme, which is analogous to the running-count scheme, is a scheme with symmetric viscosity and does not possess the monotonicity property. Nevertheless, with respect to its asymptotic properties it approaches the running-count scheme; therefore, the lack of monotonicity of the shock front in the Lax scheme is not so sharply expressed as in the von Neumann-Richtmyer scheme. The effects of a difference scheme which find their expression in the lack of monotonicity of the profile of a shock transition are especially evident in the interaction of discontinuities: shock waves with one another, or shock waves with contact boundaries. Major deviations from the true profiles occur in the interaction region. These deviations behave differently depending on the properties of the scheme and the parameters of the scheme and the mesh. Difference schemes here decompose into two basic classes depending on their dispersion properties (see §§2.3, 8.2 and 8.3). To the first class belong schemes admitting contact discontinuities (such as, for example, the von Neumann-Richtmyer scheme). In this case there is a family of standing harmonic waves not affected by the approximation viscosity and which are therefore not damped. To these standing waves there corresponds an "entropy trace" remaining in a neighborhood of the boundary and having the character of a strong, nonmonotone deviation from the true profiles of $\rho(q, t)$ and $s(q, t)$. To the second class belong schemes not admitting contact discontinuities (the Lax scheme, for example). In this case the approximation viscosity acts on all families of solutions of the dispersion equations.

A discontinuity of ρ and S and any nonmonotone profile of ρ and S are smoothed. In such schemes the entropy traces vanish, but the contact boundaries also vanish (are "smeared out"). The magnitude of the entropy trce depends also on the parameters of the scheme. For the wave-train scheme (see §8.4) Kuroptenko (1962, 1966) obtained the following criterion: *when a shock wave passes through the boundary of two media the greatest accuracy is achieved under the condition*

$$\frac{D_0}{h_0} = \frac{D_1}{h_1}. \tag{19}$$

Here D_0 is the speed of the incident shock wave and D_1 the speed of the tians-mitted shock wave: h_0 and h_1 are the lengths of the adjoining intervals. A minimum of the entropy trace on the boundary is observed here (see Figure 3.27, where the graph of the exact solution is shown by a dashed line and that of the numerical solution by a solid line). The theory of the appearance of the entropy trace and other boundary effects has so far not been extensively developed.

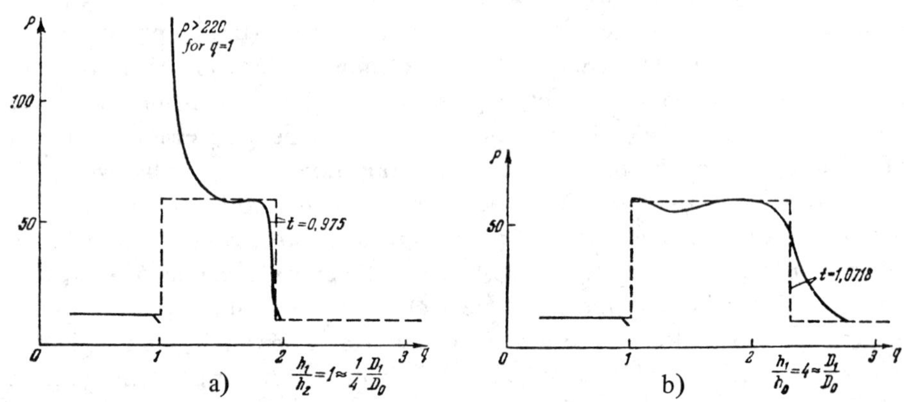

FIGURE 3.27

2. *Remarks.* The analysis of difference schemes we have carried out shows that there are no difference schemes of continuous calculation which are capable of reproducing with high accuracy the entire picture of a hydrodynamic flow containing a large number of singularities.

Methods of continuous calculation replace a shock front by a shock transition and a contact boundary by a contact strip. These transition regions in practice reach widths of several spatial intervals and have, as a rule, nonmonotone profiles. During the interaction of shock transitions with one another or with a contact strip, a transitional process arises which may continue over several time intervals. Suppose, for example, that the width of each of the shock transitions is equal to kh. The local stability condition implies that each front must in time τ pass through a fractional part of the interval h/r $(r > 1)$, while the greater the greater the force of the shock wave. From this it follows that the period of interaction will continue rk time intervals and in practice may reach considerable duration. During the transitional process the hydrodynamic quantities may deviate strongly from their true values. With a large number of discontinuities the transition regions may cover the entire region of computation, which, of course, sharply reduces the accuracy of the computation.

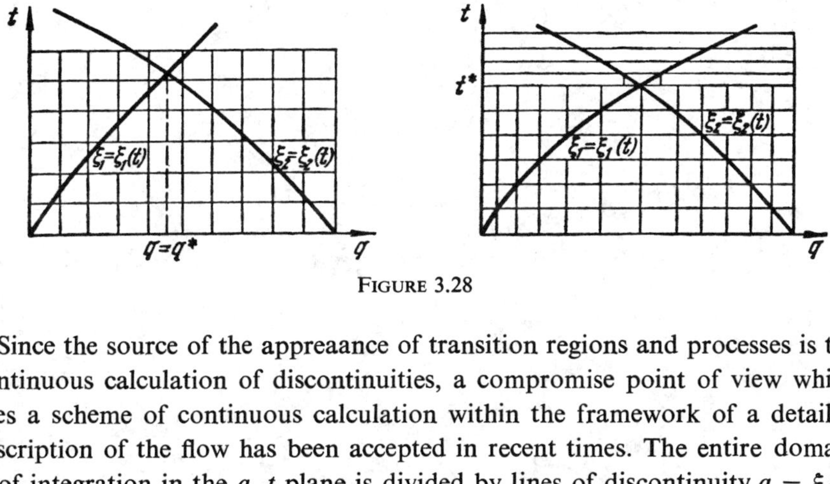

FIGURE 3.28

Since the source of the appreaance of transition regions and processes is the continuous calculation of discontinuities, a compromise point of view which uses a scheme of continuous calculation within the framework of a detailed description of the flow has been accepted in recent times. The entire domain G of integration in the q, t plane is divided by lines of discontinuity $q = \xi_i(t)$ into subdomains G_i, in each of which some scheme of continuous calculation is applied. The mesh here may be either fixed or mobile: in the former case lines of discontinuity $q = \xi_i(t)$, are marked off which need not pass through points of the mesh, while in the latter case the lines of discontinuity $q = \xi_i(t)$ as pass through points of the mesh (Figure 3.28). In both cases the points are arranged on horizontal lines distributed in the q, t plane, generally speaking, with variable step size τ_m. In the first case (a fixed mesh) the intervals (q_i, q_{i+1}) are fixed, while in the second ease (a moving mesh) they depend on time, and not only their length and position but also the number of them may vary.

Thus, in the second case the mesh is determined by the flow itself and its singularities. If newly arising discontinuities are sufficiently strong, then they are included in the boundary lines of the domains G_i, and the mesh is rearranged. Discontinuities of small amplitude are considered by a scheme of continuous calculation.

It is clear that various versions of constructing a mobile mesh are possible. Computational algorithms of this type are rather complicated both arithmetically and logically, which makes them similar to the method of characteristics. In contrast to the latter, the computational front is always a horizontal line, which considerably simplifies the logic of the program.

CHAPTER 4

GENERALIZED SOLUTIONS OF SYSTEMS OF QUASILINEAR EQUATIONS OF HYPERBOLIC TYPE

§1. Formulation of the Cauchy Problem in a Class of Discontinuous Functions

1. ***General remarks.*** In Chapter 2 we saw that the differential equations of gas dynamics are a consequence of the more general integral conservation laws of mass, momentum, and energy. Passage from the integral conservation laws to differential equations is possible only with some smoothness of the flow. If there is no smooth flow, then we must turn to the integral conservation laws to determine the flow (which is discontinuous or does not possess the required smoothness). This is the approach adopted in the theory of discontinuous (generalized) solutions of systems of quasilinear equations of hyperbolic type which has been formed in recent decades.

We consider a conservative system of quasilinear equations

$$\frac{\partial u}{\partial t} + \frac{\partial \varphi(u, x, t)}{\partial x} = f(u, x, t) \tag{1}$$

(u, φ and f are vectors with n components) as a consequence of the system of integral conservation laws

$$\oint_C u\, dx - \varphi(u, x, t)\, dt + \iint_{\mathscr{G}_C} f(u, x, t)\, dx\, dt = 0, \tag{2}$$

which must be satisfied for any piecewise smooth contour C and the region $\mathscr{G}_C$ that it bounds. If a function $u(x, t) \in C_1$ satisfies the integral conservation laws (2) for any smooth contours C and regions $\mathscr{G}_C$, and if $\varphi \in C_1$ and $f \in C_0$, then this implies that $u(x, t)$ satisfies the system of differential equations (1). It is also obvious that any solution $u(x, t)$ of (1) ($u(x, t) \in C_1$) satisfies the integral conservation laws (2). The integral relations (2) are applied to introduce the concept of a generalized solution of (1).

We shall restrict our consideration to the class K of functions $u(x, t)$ satisfying the following conditions:

1. In any finite part of the half plane $t \geq 0$ there are a finite number of lines of discontinuity $x = x(t)$ and a finite number of points of discontinuity; away from these lines and points the function $u(x, t)$ is continuous and possesses continuous first derivatives.

2. On the lines of discontinuity $x = x(t)$ there exist left $u(x - 0, t)$ and right $u(x + 0, t)$ limits; to be specific, we set $u(x, t) = u(x + 0, t)$.

We call a vector-valued function $u(x, t) \in K$ a *generalized solution of the system* (1) if for any piecewise smooth contour C and region $\mathcal{G}_C$ bounded by it the integral conservation laws (2) are satisfied. Other definitions of a generalized solution of the system (1) are also used.

Let $g(x, t) \in C_1$ be a compactly supported function (it vanishes away from a finite part of the x, t plane). We multiply each of the equations in (1) by $g(x, t)$ and integrate the result over the half plane $t \geq 0$. Integrating by parts, we obtain

$$\iint\limits_{t\geq 0}\left[u_i\frac{\partial g}{\partial t} + \varphi_i(u, x, t)\frac{\partial g}{\partial x} + f_i(u, x, t)g(x, t)\right]dx\,dt$$
$$+\int_{-\infty}^{\infty} g(x, 0)u_i(x, 0)\,dx = 0 \qquad (i = 1, 2, \ldots, n). \tag{3}$$

Just like (2), the equations (3) contain no derivatives of $u(x, t)$ and do not become meaningless for discontinuous $u(x, t)$. We call a function $u(x, t) \in K$ a *generalized solution* of (1) if (3) is satisfied for any compactly supported function $g(x, t) \in C_1$.

Finally, using the concept of a potential (Chapter 1, §5), we introduce still another definition of a generalized solution of (1). Suppose that a vector-valued function $\Phi(x, t) = \{\Phi_1(x, t), \ldots, \Phi_n(x, t)\}$ is continuous and possesses first derivatives $\partial\Phi/\partial x \in K$. If at all points at which the derivatives $\partial\Phi/\partial x$ and $\partial\Phi/\partial t$ exist they satisfy the system of nonlinear integrodifferential equations

$$\frac{\partial\Phi_i}{\partial t} + \varphi_i\left(\frac{\partial\Phi}{\partial x}, x, t\right) = \int_0^x f_i\left(\frac{\partial\Phi}{\partial\xi}, \xi, t\right)d\xi \qquad (i = 1, \ldots, n), \tag{4}$$

then we call the function

$$u(x, t) = \partial\Phi/\partial x \tag{5}$$

a *generalized solution* of (1).

It is easy to see that each of these three definitions generalizes in a natural way to broader classes of functions $u(x, t)$. We shall nevertheless limit ourselves to the class K, since more general classes of generalized solutions have so far not been sufficiently studied.

We remark now that a system of quasilinear equations sometimes admits several distinct representations in the form of conservation laws. For example, the equation

$$\frac{\partial u}{\partial t} + u\,\frac{\partial u}{\partial x} = 0 \tag{6}$$

can be represented both in the form

$$\frac{\partial u}{\partial t} + \frac{\partial}{\partial x}\left(\frac{u^2}{2}\right) = 0, \tag{7}$$

and in the form

$$\frac{\partial}{\partial t}\left(\frac{u^2}{2}\right) + \frac{\partial}{\partial x}\left(\frac{u^3}{3}\right) = 0. \tag{8}$$

Thus, any solution of (6) simultaneously satisfies the following integral conservation laws:

$$\oint_C u\,dx - \frac{u^2}{2}\,dt = 0, \tag{9}$$

$$\oint_C \frac{u^2}{2}\,dx - \frac{u^3}{3}\,dt = 0. \tag{10}$$

It is easy to see, however, that a discontinuous function $u(x, t)$ may satisfy one of the equations (9) or (10) and not satisfy the other.

This circumstance reflects an interesting fact that can be interpreted as follows: different processes can be described by the same differential equations but by distinct integral conservation laws. The difference in these processes is thus exhibited only on discontinuous solutions. In introducing a generalized solution of the conservative system (1), we unequivocally fix the integral conservation laws (2).

2. *The Hugoniot conditions.* Let $u(x, t) \in K$, and let $x = x(t)$ be the equation of one of the lines of discontinuity of $u(x, t)$. We set

$$\left.\begin{aligned} D = x'(t), \qquad u_l(t) = u(x(t) - 0, t), \qquad u_r(t) = u(x(t) + 0, t),\\ [u] = [u(x, t)] = u(x + 0, t) - u(x - 0, t),\\ [u(x(t), t)] = u_r(t) - u_l(t),\\ [\varphi(u, x, t)] = \varphi(u(x + 0, t), x, t) - \varphi(u(x - 0, t), x, t). \end{aligned}\right\} \tag{1}$$

In the same way as the Hugoniot conditions on a line of discontinuity of the flow (Chapter 2, §4.1) follow from the integral conservation laws of mass, momentum, and energy, from the integral conservation laws (1.1.2) on the line $x = x(t)$ of discontinuity of the solution $u(x, t)$ we obtain the relations

$$D[u] = [\varphi(u, x, t)] \tag{2}$$

or, in components,

$$D[u_i] = [\varphi_i(u, x, t)].$$

The conditions (2) connect the left and right limit values of the solution on a line of discontinuity.[1] In analogy with the case of gas dynamics we call these equations the *Hugoniot conditions.*

If $u(x, t) \in K$ and everywhere except for the lines of discontinuity it satisifies the system of differential equations (1.1.1) while it satisfies the Hugoniot conditions (2) on the lines of discontinuity, then the integral conservation laws (1.1.2) are obviously satisfied for any closed contour C. In this case $u(x, t)$ is therefore a generalized solution of (1.1.1).

We shall consider the consequence of the Hugoniot conditions for the simplest examples. Suppose that the system (1.1.1) is semilinear. Then

$$\frac{\partial \varphi_i(u, x, t)}{\partial u_j} = a_{ij}(x, t), \tag{3}$$

and therefore the Hugoniot conditions (2) can be transformed to the form

$$D[u_i] = [\varphi_i(u, x, t)] = \sum_{j=1}^{n} a_{ij}(x, t)[u_j]$$

or

$$\sum_{j=1}^{n} (a_{ij}(x, t) - \delta_{ij}D)[u_j] = 0 \qquad (i = 1, 2 \ldots, n). \tag{4}$$

If the solution $u(x, t)$ is discontinuous, then $\sum_{j=1}^{n} [u_j]^2 \neq 0$ and

$$\operatorname{Det}((a_{ij}(x, t) - D\delta_{ij})) = 0,$$

i.e., the quantity $D = x'(t)$ must coincide with an eigenvalue of the matrix $A = ((\partial\varphi_i/\partial u_j))$. Let

$$D = \xi_k(x, t); \tag{5}$$

[1] For piecewise continuous $u(x, t)$ and piecewise continuous lines of discontinuity these same conditions follow also from the definitions of a generalized solution (1.1.3) or (1.1.4), (1.1.5).

then

$$[u] = Ur^k(x, t) \tag{6}$$

or in components

$$[u_i] = Ur_i^k(x, t). \tag{7}$$

Here $r^k(x, t) = \{r_i^k(x, t)\}$ denotes a right eigenvector of the matrix $A(x, t)$ corresponding to the eigenvalue $\xi_k(x, t)$. Thus, for a semilinear system, according to (5),

$$\frac{dx}{dt} = \xi_k(x, t), \tag{8}$$

i.e., the lines of discontinuity of a solution are the characteristics of the system (1.1.1).

In analogy to a weak discontinuity, a strong discontinuity of a solution of a semilinear system of equations also propagates along characteristics of the system. We note that solutions of a weakly nonlinear system of quasilinear qeuations (Chapter 1, §10) have an analogous property. Indeed, it is not hard to verify that the discontinuities of a generalized solution of a weakly nonlinear system of quasilinear equations can be arranged on the characteristics of this system.

In the case of single quasilinear equation ($n = 1$) the Hugoniot conditions (2) can be rewritten in the form

$$x'(t) = D = \frac{[\varphi(u, x, t)]}{[u]} = \frac{\varphi(u_r(t), x(t), t) - \varphi(u_l(t), x(t), t)}{u_r(t) - u_l(t)}. \tag{9}$$

The equality (9) can be interpreted geometrically in the following way. The speed $D = x'(t)$ of the line of discontinuity $x = x(t)$ is equal to the slope (the tangent of the angle α) of the chord AB relative to the u axis (Figure 4.1). For the case shown in Figure 4.1 we have

$$\varphi_u'(u_r(t), x(t), t) < D < \varphi_u'(u_l(t), x(t), t).$$

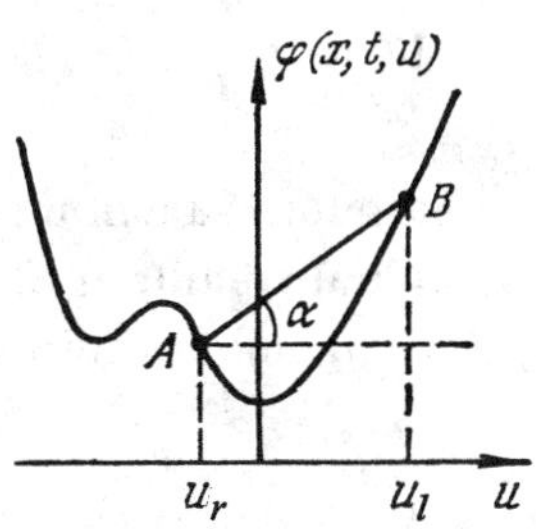

FIGURE 4.1

If $\varphi''_{uu}(u, x, t) \neq 0$, then according to (9) the quantity D is different from both $\varphi'_u(u_r(t), x(t), t)$ and $\varphi'_u(u_l(t), x(t), t)$ provided that $u_r(t) \neq u_l(t)$. Thus, for a nonlinear equation a line of discontinuity $x = x(t)$ is no longer a characteristic.

We observe that to different integral conservation laws corresponding to the same system of quasilinear equations there correspond different Hugoniot conditions. Thus, from the integral conservation law (1.1.9) we obtain the Hugoniot condition

$$D = \frac{[u^2]}{2[u]} = \frac{u_r + u_l}{2}, \tag{10}$$

while from (1.1.10) we obtain

$$D = \frac{2}{3}\frac{[u^3]}{[u^2]} = \frac{2}{3}\frac{u_r^2 + u_r u_l + u_l^2}{u_r + u_l}. \tag{11}$$

It is not hard to produce an example of discontinuous functions $u(x, t)$ satisfying one of the Hugoniot conditions (10) or (11) and not the other.

3. *Stable and unstable discontinuities. Stability conditions.* As we saw in Chapter 1, a (classical) solution of a system of quasilinear equations is uniquely determined in the domain of determinacy by its initial values for $t = 0$. It turns out, however, that satisfying the integral conservation laws and the initial conditions does not guarantee the uniqueness of a discontinuous solution; on the contrary, it is possible to find a set of essentially distinct, discontinuous functions $u(x, t)$ satisfying both the integral conservation laws and the initial conditions. We confirm this with a simple example. For the equation

$$\frac{\partial u}{\partial t} + \frac{\partial}{\partial x}\left(\frac{u^2}{2}\right) = 0 \tag{1}$$

we pose the initial condition

$$u(x, 0) = u_0(x) = \begin{cases} u^- & \text{for } x < 0, \\ u^+ & \text{for } x > 0. \end{cases} \tag{2}$$

We seek a bounded, piecewise continuous function $u(x, t)$ satisfying the integral conservation law (1.1.9) and the initial condition (2).

On the lines of discontinuity the Hugoniot condition (1.2.10) must be satisfied, and therefore the function

$$u_1(x, t) = \begin{cases} u^- & \text{for } x < D_1 t, \\ u^+ & \text{for } x > D_1 t, \end{cases} \qquad D_1 = \frac{u^- + u^+}{2}, \tag{3}$$

is the desired solution. This function is constant to the left and right of the line of discontinuity $x = D_1 t$ on which the Hugoniot condition (1.2.10) is satisfied, and it assumes the initial values (2).

Suppose that $u^- < u^+$. We construct another solution of problem (1), (2):

$$u_2(x, t) = \begin{cases} u^- & \text{for } x \le tu^-, \\ x/t & \text{for } tu^- \le x \le tu^+, \\ u^+ & \text{for } x \ge tu^+. \end{cases} \tag{4}$$

The solution $u = u_2(x, t)$ is continuous for $t > 0$, is continuously differentiable everywhere except on the lines $x = tu^-$ and $x = tu^+$, and it satisfies (1) in the broad sense in the half plane $t \ge 0$.

We have found two solutions $u = u_1(x, t)$ and $u = u_2(x, t)$ of the Cauchy problem (1), (2). Each of these solutions satisfies the integral conservation law (1.1.9) and the initial condition (2). Thus, we here encounter nonuniqueness of a solution. It seems natural, however, that a proper formulation of the Cauchy problem in the class of discontinuous solutions should lead to uniqueness of the solution.

In order to distinguish a unique solution of our problem, we make the following assumptions.

1. Any (classical) solution of a system of quasilinear equations, when it exists, is the "true" solution of this system in the class of generalized (discontinuous) solutions as well.

2. The limits of (classical) solutions of a system of quasilinear equations are the "true" solutions of the integral conservation laws in the class of discontinuous functions.

We shall explain this in somewhat more detail.

Condition 1 is the natural requirement that the class of generalized solutions of the Cauchy problem for a system of quasilinear equations be an extension of the class of classical solutions. If we were to give up this condition, then classical solutions would have no practical significance in considering the Cauchy problem in the class of discontinuous functions.[2]

[2] If we give up condition 1, then the Cauchy problem for equation (1) with the initial condition $u(x, 0) \equiv 0$ has an infinite set of discontinuous solutions. For example,

$$u(x, t) = \begin{cases} 0 & \text{for} \quad x + t < 0, \\ -2 & \text{for} \quad -t < x < 0, \\ 2 & \text{for} \quad 0 < x < t, \\ 0 & \text{for} \quad x - t > 0 \end{cases}$$

is a solution of this problem.

Condition 2 is a natural consequence of condition 1 and the assumption of continuous dependence of generalized solutions on the initial data of the Cauchy problem. The continuous depencence of solutions on the initial data of the Cauchy problem is one of the conditions for correctness of the Cauchy problem.

Thus, the usual idea of a correct formulation of the Cauchy problem leads us to conditions 1 and 2. A more complete formulation of the requirements for a generalized solution will be given below. For the time being we apply conditions 1 and 2 in order to further study the properties of discontinuous solutions of systems of quasilinear equations and, in particular, to distinguish a unique ("true") solution $u = u(x, t)$ of the Cauchy problem (1), (2).

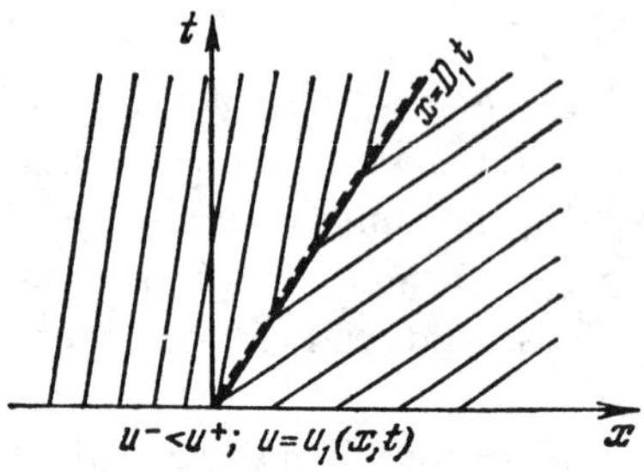

FIGURE 4.2

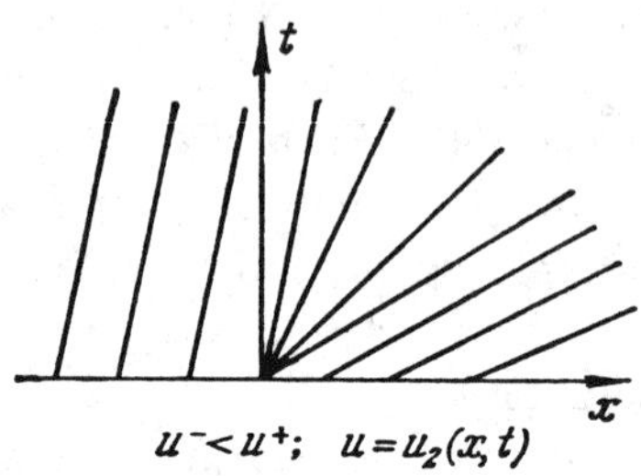

FIGURE 4.3

Figure 4.2 and 4.3 show the behavior of the characteristics $x = x_0 + u(x, t)t$ for the solutions $u = u_1(x, t)$ and $u = u_2(x, t)$. We consider the solution $u = u_\delta(x, t)$ of (1) satisfying the condition

$$u_\delta(x, 0) = u_\delta^0(x), \tag{5}$$

where $u^0(x)$ is a monotonically increasing, continuous function of the variable x which coincides with $u_0(x)$ outside the segment $|x| \leq \delta$ (Figure 4.4). Figure 4.5 shows the picture of the characteristics for the solution $u_\delta(x, t)$.

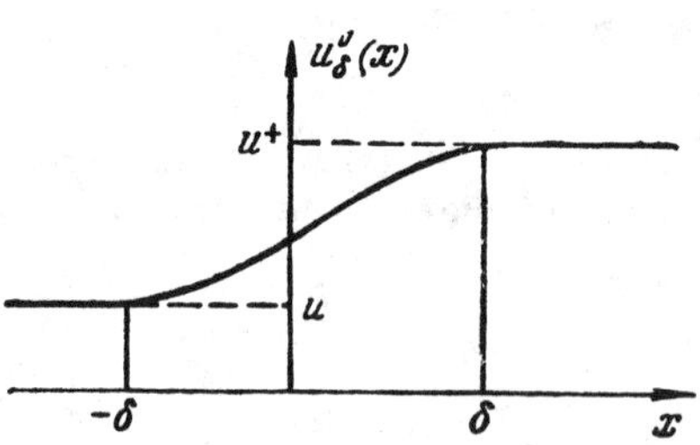

FIGURE 4.4

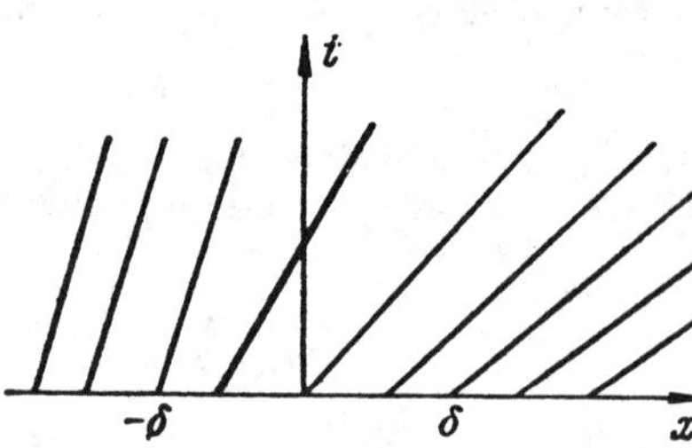

FIGURE 4.5

Comparing Figures 4.3 and 4.5, we see that

$$\lim_{\delta \to 0} u_\delta(x, t) = u_2(x, t), \tag{6}$$

and this holds for any x and t except the point (0, 0), where the limit of $u_\delta(x, t)$ as $\delta \to 0$ does not exist. Therefore, according to condition 2, the "true" solution of the Cauchy problem (1), (2) is $u = u_2(x, t)$. We call $u = u_1(x, t)$ an *unstable discontinuous solution*, since the smoothing of the initial function on an arbitrarily small segment leads to a classical ("true") solution far from $u_1(x, t)$.

From a comparison of Figures 4.2 and 4.3 we conclude that the reason for the instability of $u_1(x, t)$ is that on the line of discontinuity $x = D_1 t$ of the solution $u_1(x, t)$ characteristics issuing from points of the line intersect, rather than characteristics issuing from points of the initial axis $t = 0$ (which hence carry initial data). In this sense it may be said that the discontinuity of the solution shown in Figure 4.1 is "contrived" and is not caused by the intersection of characteristics carrying initial values. Such a discontinuity we call an *unstable discontinuity.*

If $u^- > u^+$ in the conditions of the Cauchy problem (1), (2) then the solution $u_1(x, t)$ given by (3) has the characteristics shown in Figure 4.6. In this case on the line of discontinuity $x = D_1 t$ characteristics carrying initial values (*incoming characteristics*) intersect. Now there is no continuous solution of this Cauchy problem, and the smoothed solution $u_\delta(x, t)$ is also discontinuous. Therefore, the solution $u = u_1(x, t)$ and the discontinuity on the line $x = D_1 t$ for $u^- > u^+$ are called *stable*.

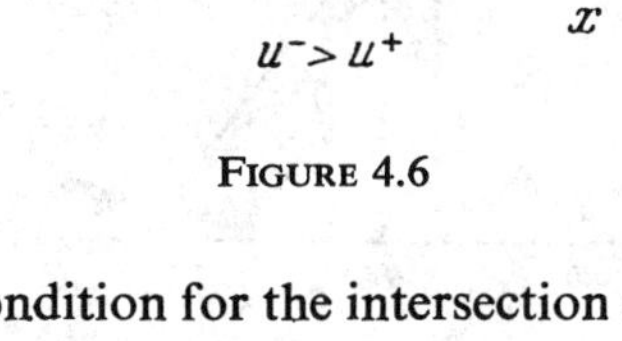

FIGURE 4.6

For equation (1) the condition for the intersection of incoming characteristics on the line of discontinuity can be described by the inequalities

$$u_l(t) > D > u_r(t), \qquad D = x'(t). \tag{7}$$

For the more general quasilinear equation

$$\frac{\partial u}{\partial t} + \frac{\partial \varphi(u, x, t)}{\partial x} = (u, x, t) \tag{8}$$

the speed of the characteristics is $\varphi_u'(u, x, t)$; therefore, the intersection condition on the line $x = x(t)$ for incoming characteristics can be written in analogy with (7):

$$\varphi_u'(u_l(t), x(t), t) > D > \varphi_u'(u_r(t), x(t), t); \qquad D = x'(t). \tag{9}$$

For a semilinear equation (8) ($\varphi_{uu}''(u, x, t) \equiv 0$) the lines of discontinuity of solutions are characteristics. Therefore, if we seek more general stability conditions valid also for linear equations, they should be written in the form

$$\varphi_u'(u_l(t), x(t), t) \geq D \geq \varphi_u'(u_r(t), x(t), t). \tag{10}$$

As we shall see below, the conditions (10) ensure uniqueness and admit the existence of a generalized solution of (8) in the case where $\varphi_{uu}''(u, x, t)$ does not change sign, i.e., for the case in which $\varphi(u, x, t)$ is convex in u. We call (10) the stability conditions for a generalized solution in the case of φ_{uu}'' of constant sign.

If $\varphi_{uu}''(u, x, t)$ is a function of variable sign, then the conditions (10) do not guarantee either the uniqueness or the continuous dependence of a solution on the initial data, i.e., they are not correct conditions. Indeed, suppose that $\varphi = \varphi(u)$ and $f \equiv 0$, and for (8) pose the initial condition

$$u(x, 0) = u_0(x) = \begin{cases} u^- & \text{for } x < 0. \\ u^+ & \text{for } x > 0. \end{cases} \tag{11}$$

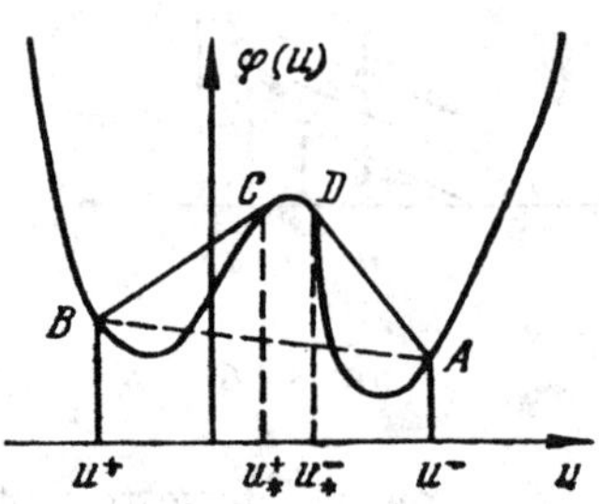

FIGURE 4.7

Suppose that the graph of $\varphi(u)$ and the positions of the points u^- and u^+ are as shown in Figure 4.7. The function

$$u = u_1(x, t) = \begin{cases} u^- & \text{for } x < Dt, \\ u^+ & \text{for } x > Dt, \end{cases} \qquad D = \frac{\varphi(u^+) - \varphi(u^-)}{u^+ - u^-}, \tag{12}$$

satisfies the integral conservation law

$$\oint u\,dx - \varphi(u)\,dt = 0$$

of (8), the initial condition, and the proposed "stability condition" (10). The latter is clear from Figure 4.7 since $\varphi'_u(u^-) > D > \varphi'_u(u^+)$; D is the slope of the chord AB relative to the u axis.

We now construct a second solution $u = u_2(x, t)$ of our Cauchy problem. We define a function $u_2(x, t)$ as follows:

$$u_2(x, t) = \begin{cases} u^- & \text{for } x < D_1^* t, \\ f(x/t) & \text{for } D_1^* t < x < D_2^* t, \\ u^+ & \text{for } x > D_2^* t, \end{cases} \tag{13}$$

where

$$D_1^* = \frac{\varphi(u^-) - \varphi(u_*^-)}{u^- - u_*^-}, \qquad D_2^* = \frac{\varphi(u_*^+) - \varphi(u^+)}{u_*^+ - u^+}; \tag{14}$$

$f(\xi)$ is determined from the equation

$$\xi = \varphi'_u[f(\xi)], \tag{15}$$

at the points u_*^+ and u_*^- (the abscissae of the points C and D); that is, from the condition that the line BC be tangent to the graph of $\varphi(u)$ at C and the line AD be tangent at D; moreover, it is assumed that $\varphi''_{uu}(u) \neq 0$ for $u_*^+ \leq u \leq u_*^-$ (see Figure 4.7).

The solution $u = u_2(x, t)$ defined by (13)–(15) satisfies the integral conservation law for (8) and the conditions (10). Thus, if $\varphi''_{uu}(u, x, t)$ is a function of u with variable sign, then the "stability conditions" (10) are satisfied by two solutions, $u_1(x, t)$ and $u_2(x, t)$.

Figures 4.8 and 4.9 show the fields of characteristics for the solutions $u = u_1(x, t)$ and $u = u_2(x, t)$. We mention an interesting property of $u_2(x, t)$. The line of discontinuity $x = D_2^* t$ is the characteristic of $u_2(x, t)$ computed on the basis of the right limits of the solution $u_r(t)$ on this line; similarly, the line of discontinuity $x = D_1^* t$ is the characteristic computed on the basis of the values $u_l(t)$ on the line $x = D_1^* r$.

In order to give preference to one of these two solutions, we again apply the

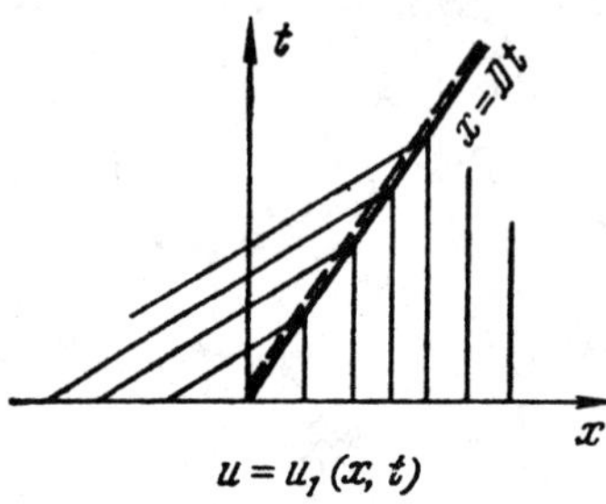

FIGURE 4.8

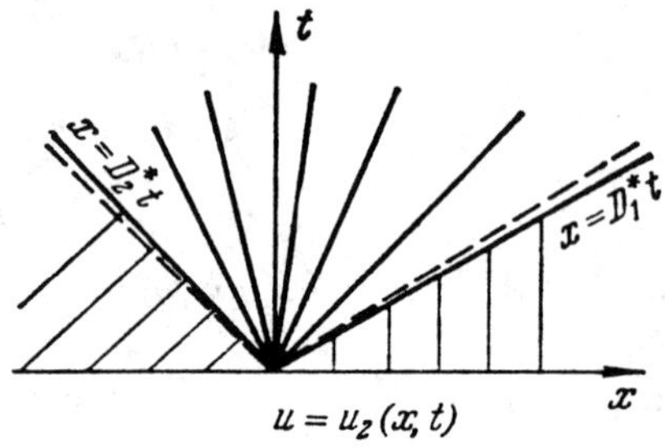

FIGURE 4.9

method of verifying stability of these solutions by smoothing the initial function. We smooth the initial function $u_0(x)$ by means of the montone function $u_0^\delta(x)$ shown in Figure 4.10. The points (x_*^-, u_*^-) and (x_*^+, u_*^+) correspond to the points D and C of Figure 4.7, while u_1^- and u_1^+ are inflection points of the graph of $\varphi(u)$, i.e., $\varphi''_{uu}(u_1^-) = \varphi''_{uu}(u_1^+) = 0$. Figure 4.11 shows the characteristics of the solution $u = u_\delta(x, t)$.

It is evident from Figure 4.7 that $\varphi'_u(u)$ is a monotonically increasing function of u on $[u_1^-, u^-]$, and hence the slope of the characteristic $\varphi'_u(u_0^\delta(x))$ at the initial axis decreases monotonically on $[-\delta, x_1^-]$; similarly, $\varphi'_u(u_0^\delta(x))$ increases monotonically on $[x_1^-, x_1^+]$ and decreases on $[x_1^+, \delta]$.

Careful consideration of the picture of the characteristics (Figure 4.11) leads us to the conclusion that a discontinuity of a solution for which the left value is u^- and the right value is u^+ cannot arise for the smoothed initial function $u_\delta^0(x)$. Indeed, it is evident from Figure 4.11 that there must be two discontinuities, of which the discontinuity I located to the left of the corresponding values u^- and u_*^- has speed $D_1^* = \varphi'_u(u_*^-) < \varphi'_u(u_*^+)$, while the discontinuity II corresponding to the values u_*^+ and u^+ has speed $D_2^* = \varphi'_u(u_*^+)$. Thus, the lines of

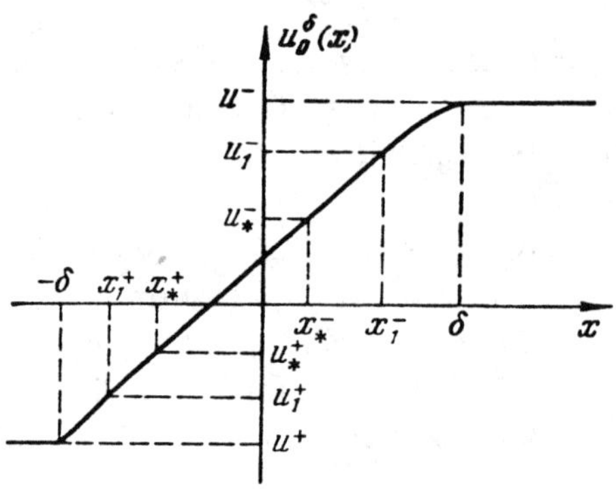

FIGURE 4.10

the discontinuities formed for the initial function $u_0^\delta(x)$ (they are shown in Figure 4.11 by heavy lines and denoted by the numbers I and II) behave as follows. The line of discontinuity I never intersects the characteristic issuing from the point x_*^- of the initial axis and approaches it asymptotically as $t \to \infty$. This obviously follows from the inequality

$$\frac{\varphi(u) - \varphi(u^-)}{u - u^-} < \varphi_u'(u_*^-) \quad \text{for } u_*^- < u < u_1^-.$$

In the same way we conclude that the line of discontinuity II asymptotically approaches the characteristic issuing from the point x_*^+ of the initial axis and never intersects it.

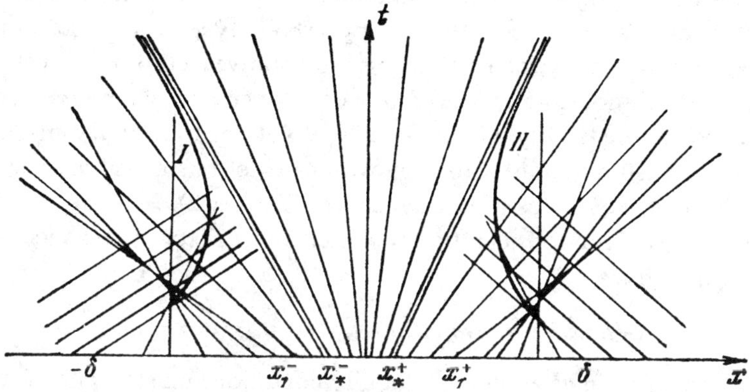

FIGURE 4.11

It now beomes obvious that if we let δ tend to zero, then the solution $u_\delta(x, t)$ tends to $u_2(x, t)$. Thus, the solution $u_1(x, t)$ and the line of discontinuity $x = Dt$ are unstable. Condition (10) therefore does not guarantee stability of a generalized discontinuous solution of the quasilinear equation (8) if $\varphi_{uu}''(u, x, t)$ has variable sign.

The correct generalization of the stability conditions for a discontinuous solution of (8) in the case where φ_{uu}'' has variable sign can be obtained from a careful consideration of solutions obtained by smoothing the discontinuities. Thus, a solution of the type $u_1(x, t)$ (Figure 4.8) would be stable if the lines of discontinuities I and II (Figure 4.11) were to merge, i.e., the picture of the motion of the discontinuities I and II were schematically as shown in Figure 4.12.

These conditions lead us to the following stability condition. Suppose that on a line of discontinuity $x = x(t)$ of a solution of (8) we have $u_l(t) = u^-$ and $u_r(t) = u^+$, and let $x'(t) = D$. The discontinuity is *stable* if the inequalities are satisfied for any values u^* in (u^-, u^+).

$$\frac{\varphi(u^*, x, t) - \varphi(u^-, x, t)}{u^* - u^-} \geq D \geq \frac{\varphi(u^*, x, t) - \varphi(u^+, x, t)}{u^* - u^+},$$

$$u^* \in (u^-, u^+), \qquad x = x(t), \tag{16}$$

It is easy to see that the discontinuity of $u_1(x, t)$ does not satisfy condition (16). Indeed, if we choose, for example, for u^* any value from the interval (u_*^+, u_*^-), then the inequalities (16) are violated. The discontinuities of $u_2(x, t)$ obviously satisfy these inequalities, as is clear from Figure 4.7.

Finally, we observe that the stability conditions (10) automatically follow from (16) if it is assumed that $\varphi''_{uu}(u, x, t) \neq 0$, i.e., for convex $\varphi(u, x, t)$. The stability condltlons (16) were first introduced by Oleĭnik (1958).

The question of stability of discontinuous solutions is still more involved for a system of quasilinear equations of hyperbolic type. The situation is that for complicated systems we actually have no illustrative methods of constructing discontinuous solutions. It is therefore very difficult to determine the correctness of eventual stability conditions for a solution. In formulating stability conditions we shall rely on the analogy with cases of one quasilinear equation and the system of equations of gas dynamics (Chapter 2).

We shall restrict our consideration to systems of strictly hyperbolic systems, i.e., we assume that

$$\xi_1(u, x, t) < \xi_2(u, x, t) < \cdots < \xi_n(u, x, t). \tag{17}$$

As in the case of a single equation, stability conditions must require that on a line of discontinuity $x = x(t)$ incoming characteristics of the same family intersect. Suppose, for example, that on the line $x = x(t)$ incoming characteristics of the kth family intersect; then

$$\xi_k(u_l(t), x(t), t) > D > \xi_k(u_r(t), x(t), t), \qquad D = x'(t). \tag{18}$$

If nothing more is required, it is possible that on the line of discontinuity $x = x(t)$ incoming characteristics of other families will also intersect, while for certain families there will be no incoming characteristics on the line of discontinuity. To (18) we therefore add the two inequalities

$$\xi_{k-1}(u_l(t), x(t), t) < D < \xi_{k+1}(u_r(t), x(t), t), \tag{19}$$

which in conjunction with (17) lead to the picture of the characteristics shown in Figure 4.13 in a neighborhood of the line of discontinuity $x = x(t)$.

Through the point A of the line of discontinuity in Figure 4.13 there pass integral curves of the equations

$$\frac{dx_i}{dt} = \xi_i(u(x_i, t), x_i, t) \tag{20}$$

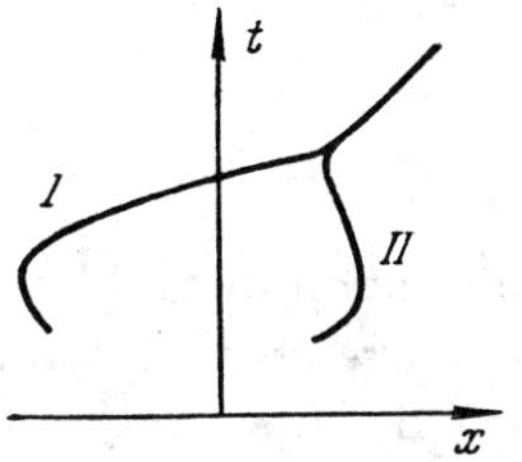

FIGURE 4.12

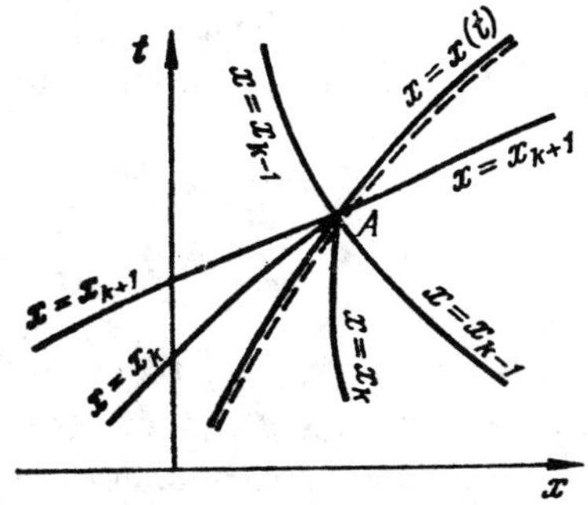

FIGURE 4.13

for indices $i = k - 1, k, k + 1$. According to (17)–(19) $n + 1$ incoming characteristics arrive at A (one from each family and two from the kth family), and $n - 1$ outgoing characteristics leave A (one from each family except the kth). We thus say that on a line of discontinuity $x = x(t)$ of a piecewise continuous solution $u(x, t)$ of a strictly hyperbolic system of quasilinear equations the stability conditions are satisfied provided that the following inequalities hold:

$$\left.\begin{aligned} \xi_k(u_l(t), x(t), t) > D > \xi_k(u_r(t), x(t), t), \\ \xi_{k-1}(u_l(t), x(t), t) < D < \xi_{k+1}(u_r(t), x(t), t). \end{aligned}\right\} \tag{21}$$

The conditions (21) were published by Lax (1957). The index k for which conditions (21) are satisfied is called the *index of the line of discontinuity*.

From the analogy with the case of a single equation it is clear that the stability conditions (21) can ensure uniqueness and continuous dependence of solutions on initial data only in some, possibly narrow, class of systems of quasilinear equations. However, this class has so far not been found. It is possible that the conditions (21) guarantee uniqueness of a discontinuous solution for systems of quasilinear equations satisfying the condition

$$r^k(u, x, t)\,\mathrm{grad}_u\,\xi_k(u, x, t) \neq 0 \qquad (k = 1, 2, \ldots, n) \tag{22}$$

(cf. §10 of Chapter 1), where $r^k(u, x, t)$ is a right eigenvector of the matrix $A = ((\partial\varphi_i/\partial u_j))$.

4. *Irreversibility of processes described by discontinuous solutions of systems of quasilinear equations.* Suppose that for $0 \le t \le t_1$ we have constructed a solution $u = u_1(x, t)$ of the Cauchy problem

$$\frac{\partial u}{\partial t} + \frac{\partial\varphi(u, x, t)}{\partial x} = f(u, x, t), \qquad u(x, 0) = u_0(x). \tag{1}$$

We say that the solution $u = u_1(x, t)$ *describes a reversible process* if the solution of the Cauchy problem with initial condition posed at $t = t_1$

$$\left.\begin{aligned} &\frac{\partial u}{\partial t} + \frac{\partial \varphi(u, x, t)}{\partial x} = f(u, x, t), \\ &u(x, t_1) = u_1(x, t_1), \qquad 0 \le t \le t_1, \end{aligned}\right\} \tag{2}$$

coincides with $u_1(x, t)$ in the strip $0 \le t \le t_1$; if the solution of problem (2) is distinct from $u_1(x, t)$, we say that $u_1(x, t)$ *describes an irreversible process.*

If $u_1(x, t)$ is a classical solution of (1), then it clearly describes a reversible process. Indeed, a smooth function $u_1(x, t)$ is the unique smooth solution both of (1) and of (2), as follows from the uniqueness theorem for a classical solution of a system of quasilinear equations proved in Chapter 1.

In passing to the question of reversibility of a process described by a discontinuous solution of (1), we must recall that we call only a stable solution a solution of this problem. Let $u_1(x, t)$ be a stable, discontinuous solution of (1), and let $x = x(t)$ be its line of discontinuity. Thus, the stability conditions are satisfied on the line $x = x(t)$:

$$\xi_k(u_l(t), x(t), t) > D > \xi_k(u_r(t), x(t), t). \tag{3}$$

We will restrict the phrase *solutions of the inverse Cauchy problem* (2), applying it only to those solutions which are stable with respect to variation of the initial data. It is easy to see, however, that a change of the direction of the time t leads to stability conditions which are the resverse of (3), i.e.,

$$\xi_k(u_l(t), x(t), t) < D < \xi_k(u_r(t), x(t), t), \tag{4}$$

since the right and left positions here change places.

Thus, if $u_1(x, t)$ is a stable, discontinuous solution of problem (1), then $u = u_1(x, t)$ is *not* a stable solution of the inverse Cauchy problem (2), since it does not satisfy conditions (4). Hence, a stable solution of the inverse Cauchy problem (2) is a function $u = u_2(x, t)$ that is clearly distinct from $u_1(x, t)$, and the solution $u_1(x, t)$ thus describes an irreversible process.

We illustrate our conclusions with a simple example. For the equation

$$\frac{\partial u}{\partial t} + \frac{\partial}{\partial x}\frac{u^2}{2} = 0 \tag{5}$$

with the initial condition

$$u(x, 0) = u_0(x) = \begin{cases} u^- & \text{for } x < 0, \\ u^+ & \text{for } x > 0, \end{cases} \tag{6}$$

as we have seen above, in the case $u^- > u^+$ the function

$$u_1(x, t) = \begin{cases} u^- & \text{for } x < Dt, \\ u^+ & \text{for } x > Dt, \end{cases}$$

$$D = \frac{u^- + u^+}{2} \tag{7}$$

is a stable, discontinuous solution.

If we choose some time $t = t_1 > 0$, solve the inverse Cauchy problem for (5), i.e., the problem with the initial condition

$$u(x, t) = u_1(x, t_1) = \begin{cases} u^- & \text{for } x < x_1, \\ u^+ & \text{for } x > x_1, \end{cases} \quad x_1 = Dt_1, \tag{8}$$

given at $t = t_1$, and seek a solution for $0 \le t \le t_1$, then a stable solution of this problem is the function

$$u_2(x, t) = \begin{cases} u^- & \text{for } x - x_1 \le u^-(t_1 - t), \\ \dfrac{x - x_1}{t - t_1} & \text{for } u^-(t_1 - t) \le x - x_1 \le x^+(t_1 - t), \\ u^+ & \text{for } x - x_1 \ge u^+(t_1 - t). \end{cases} \tag{9}$$

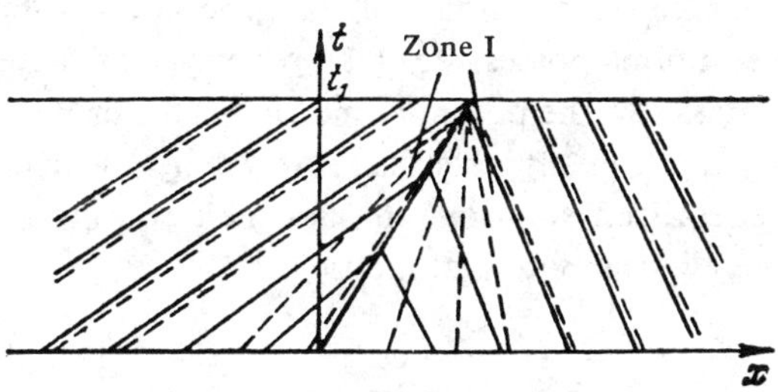

FIGURE 4.14

Figure 4.14 shows the picture of the characteristics for the solutions $u_1(x, t)$ and $u_2(x, t)$. The characteristics of $u_1(x, t)$ are represented by solid lines, and those of $u_2(x, t)$ by dotted lines. Thus, $u_1(x, t)$ and $u_2(x, t)$ are distinct in the zone I.

Thus, a stable, discontinuous solution of a system of quasilinear equations of hyperbolic type describes an irreversible process. This conclusion is true only for hyperbolic equations, since for equations of other types the inverse Cauchy problem may not be well posed, i.e., irreversibility of a process has another character there: in particular, smooth solutions may also describe irreversible processes.

Finally, we point out also that this conclusion is true only for essentially nonlinear systems of quasilinear equations of hyperbolic type. Indeed, discontinuous solutions of linear, semilinear, and weakly nonlinear systems of hyperbolic type describe reversible processes. This follows from the fact that a uniqueness theorem for discontinuous solutions for these systems can be proved on the sole basis of the integral conservation laws, i.e, for these systems there are no unstable solutions.

Finally, we present some considerations regarding the character of continuous dependence of generalized solutions of a system of quasilinear equations on the initial values. From the examples considered above it is evident that the norm of the space C cannot serve as a measure of closeness of generalized solutions. We shall see in this chapter that in the case of a single quasilinear equations stable generalized solutions possess the property that

$$\sup_x |\Phi(x, t) - \bar{\Phi}(x, t)| \to 0, \tag{10}$$

if

$$\sup_x |\Phi(x, 0) - \bar{\Phi}(x, 0)| \to 0. \tag{11}$$

Here $\Phi(x, t)$ and $\bar{\Phi}(x, t)$ are potentials of the generalized solutions $u(x, t)$ and $\bar{u}(x, t)$. We say of solutions possessing this property that they depend continuously on the initial data "in the potential metric". This property of the stability of solutions of one equation suggests that for systems of quasilinear equations also stability of generalized solutions implies their continuous dependence on the initial conditions "in the potential metric".

§2. One quasilinear equation

1. ***Survey of results.*** A single quasilinear equation is the simplest case of a system of quasilinear equations and in this case the details are considerably simpler. It is thus natural that the first results in the study of discontinuous solutions of the Cauchy problem were obtained for a single quasilinear equation.

In his classical paper Hopf (1950) constructed a discontinuous solution of the Cauchy problem

$$\frac{\partial u}{\partial t} + \frac{\partial}{\partial x}\left(\frac{u^2}{2}\right) = 0, \qquad u(x, 0) = u_0(x). \tag{1}$$

Hopf's method is as follows. In place of the Cauchy problem (1) he considers another Cauchy problem

$$\frac{\partial u_\mu}{\partial t} + \frac{\partial}{\partial x}\frac{u_\mu^2}{2} = \mu\frac{\partial^2 u_\mu}{\partial x^2}, \qquad u_\mu(x, 0) = u_0(x),\ \mu > 0, \tag{2}$$

with a solution which can be written out explicitly. The generalized solution of the Cauchy problem (1) is defined as the limit of $u_\mu(x, t)$ as $\mu \to 0$:

$$u(x, t) = \lim_{\mu\to 0} u_\mu(x, t). \tag{3}$$

The explicit expression for $u_\mu(x, t)$ and formula (3) for the generalized solution $u(x, t)$ of the Cauchy problem (1) make it possible to study in detail the properties of a discontinuous solution of problem (1). Thus, Hopf's construction may be considered the first result establishing the existence and uniqueness of a generalized solution of the Cauchy problem for a single quasilinear equation in a rather broad class of initial functions (for Hopf's construction it suffices to require only that the initial function $u_0(x)$ be bounded and measurable).

We note that the Cauchy poblem (2) was also considered by Burgers (1940, 1948) and Cole (1951). This Cauchy problem is well known in the theory of turbulence, and equation (2) itself for $u_\mu(x, t)$ is often called the *Burgers equation*: Burgers studied some aspects of the theory of turbulence with the example of the Cauchy problem (2) (and related problems).

Germain and Bader (1953) established that a solution of problem (1) is unique if at the points of discontinuity of $u(x, t)$ the stability condition $u_l > u_r$ is satisfied. However, rigorous justification of the solution of the Cauchy problem (1) was first given by Hopf. In particular, Hopf first showed that the function $u(x, t)$ given by (3) satisfies (1) in the sense that the integral equality

$$\iint\limits_{t\geq 0}\left(u(x, t)\frac{\partial g}{\partial t} + \frac{u^2(x, t)}{2}\frac{\partial g}{\partial x}\right)dx\,dt + \int_{-\infty}^{\infty} g(x, 0)u_0(x)\,dx = 0 \tag{4}$$

is satisfied for an arbitrary smooth, compactly supported function $g(x, t)$.

In 1954 O. A. Oleĭnik considered the Cauchy problem

$$\frac{\partial u}{\partial t} + \frac{\partial\varphi(u, x, t)}{\partial x} = 0, \tag{5}$$

$$u(x\ \ 0) = u_0(x) \tag{6}$$

in the class of piecewise continuous and piecewise differentiable solutions $u(x, t)$ under the condition $\varphi''_{uu}(u\ \ x\ \ t) > 0$ ($\varphi''_{uu} < 0$) and proved the existence and uniqueness of a generalized solution (see Oleĭnik (1954a, b)). Tihonov and

Samarskiĭ (1954) considered the Cauchy problem for the inhomogeneous conservation law

$$\frac{\partial u}{\partial t} + \frac{\partial \varphi(u, x, t)}{\partial x} = f(u, x, t) \tag{7}$$

in the class $u(x, t) \in K$, also assuming the convexity of $\varphi(u, x, t)$ ($\varphi''_{uu} \neq 0$). Oleĭnik (1956b, 1957b) then extended her results to the class of bounded, measurable solutions $u(x, t)$ of the conservation law (5).

The properties of solutions of the Cauchy problem for equations (5) and (7) were later studied in more detail by a number of authors. The Cauchy problem (5), (6) in the case of a nonconvex function $\varphi(u, x, t)$ when $\varphi''_{uu}(u, x, t)$ has variable sign was considered by Gel'fand (1959), Oleĭnik (1958, 1959b), A. S. Kalašnikov (1959a), and others. As we have seen in §1, the formulation of the problem in this case becomes more complicated, and its solutions have a wider variety of properties. The central feature is the correct generalization of the stability conditions for a generalized solution. The requirements (1.3.16) obtained in the preceding section from graphical considerations are such conditions.

The basic difference in the properties of solutions of the Cauchy problem (5), (6) in the case of a nonconvex function $\varphi(u, x, t)$ from the case of a convex function is that now a solution $u(x, t)$ may have a singularity of the type $u(x, t) = g((x - x_0)/(t - t_0)) + \tilde{u}(x, t)$, where g and $\tilde{u}$ are smooth functions not only at $t_0 = 0$ (which is also the case for $\varphi''_{uu} \neq 0$) but also for $t_0 > 0$. Such singularities of the type of a centered rarefaction wave may occur for $t > 0$ at the points of intersection of two lines of discontinuity. This possibility relates the case of a single equation to systems of equations of hyperbolic type ($n \geq 2$) for which the solutions, as a rule, have singularities of this type. Of course, the case $n = 1$ is far removed from the case $n \geq 2$ both with respect to the properties of solutions and the degree of difficulty of the problems that occur. It is for just this reason that many of the methods used successfully in the case of a single conservation law are not applicable or, more frequently, are difficult to apply in the case of a system of conservation laws. Nevertheless, the case $n = 1$ always serves as a "touchstone" for any new method of studying discontinuous solutions of a system of nonlinear conservation laws. Although in the case when $\varphi''_{uu}(u, x, t)$ is of variable sign the Cauchy problem (5), (6) has been studied in considerably less detail than in the case $\varphi''_{uu} > 0$, it may nevertheless be assumed that the basic facts and theorems have already been established.

We mention also that the case of a single quasilinear equation is the only case of systems of quasilinear equations for which theorems on convergence of solutions of difference equations to a generalized solution of the Cauchy problem have been proved (in this regard see Bahvalov (1961), Kuznecov (1977), Kuznecov and Vološin (1976), and Razumeĭko (1973)).

From what has been said above it is apparent that the theory of generalized solutions of a single quasilinear conservation law (5) or (7) was constructed in general outline during the 1950s. In subsequent years there neverthelass appeared many interesting works illuminating various aspects of this theory. Several new methods of solving the Cauchy problem (5), (6) and new definitions of a solution of this problem have been proposed. We mention the "method of the potential" and the "method of potential smoothing" of a solution of the Cauchy problem (5), (6), which possess great generality and descriptiveness (see Kuznecov and Roždestvenskiĭ (1959a), Roždestvenskiĭ (1961), and Douglis (1961, 1972)). A number of papers have been devoted to more general and descriptive definitions of generalized solutions of a single conservation law and their properties and to the concept of "entropy" for a single quasilinear equation (see Douglis (1959), Vol'pert (1967), Hopf (1969), and Lax (1971)).

The extension of this problem to the case of several space variables led to the construction during the period 1960–1975 of a relatively complete theory of generalized solutions of a single quasilinear equation (conservation law)

$$\frac{\partial u}{\partial t} + \sum_{i=1}^{n} \frac{\partial \varphi_i(u, x, t)}{\partial x_i} = 0 \tag{8}$$

(see papers by N. N. Kuznecov, S. N. Kružkov, A. I. Vol'pert, J. A. Smoller, J. L. Johnson, and others).

Thus, we see that although the basic contours of a theory of generalized solutions of a single quasilinear conservation law have already been created, this problem is still far from exhausted.

We say several words regarding the nature of the exposition in §2. Here we consider almost exclusively the case of a convex function $\varphi(u, x, t)$ ($\varphi''_{uu} > 0$); the case of a nonconvex $\varphi(u)$ is treated only in order to demonstrate the difficulties that arise. For simplicity we restrict ourselves, as a rule, to the class of solutions $u(x, t) \in K$, although the majority of results carry over without notable complication to the class of bounded, measurable solutions.

2. *Hopf's construction.* Hopf considers a solution $u(x, t)$ of the equation

$$\frac{\partial u}{\partial t} + \frac{\partial}{\partial x}\frac{u^2}{2} = 0 \tag{1}$$

with the initial condition

$$u(x, 0) = u_0(x) \tag{2}$$

as the limit as $\mu \to 0$ of solutions $u_\mu(x, t)$ of another Cauchy problem:

$$\frac{\partial u_\mu}{\partial t} + \frac{\partial}{\partial x}\frac{u_\mu^2}{2} = \mu\frac{\partial^2 u_\mu}{\partial x^2}, \qquad \mu > 0, \tag{3}$$

$$u_\mu(x, 0) = u_0(x). \tag{4}$$

We shall assume that the function $u_0(x)$ is bounded on the entire axis $|x| < \infty$, possesses a piecewise continuous first derivative, and has points of discontinuity of first kind.

We suppose that we know a solution $u_\mu(x, t)$ of the Cauchy problem (3), (4) which is continuously differentiable for $t > 0$. Then according to (3), the contour integral

$$\Phi_\mu(x, t) = \int_{(0,0)}^{(x,t)} u_\mu\, dx + \left[\mu\frac{\partial u_\mu}{\partial x} - \frac{u_\mu^2}{2}\right] dt \tag{5}$$

does not depend on the path of integration; here

$$\frac{\partial \Phi_\mu}{\partial x} = u_\mu, \qquad \frac{\partial \Phi_\mu}{\partial t} = \mu\frac{\partial u_\mu}{\partial x} - \frac{u_\mu^2}{2}. \tag{6}$$

Eliminating u_μ from (6), we obtain an equation which $\Phi_\mu(x, t)$ satisfies:

$$\frac{\partial \Phi_\mu}{\partial t} + \frac{1}{2}\left(\frac{\partial \Phi_\mu}{\partial x}\right)^2 = \mu\frac{\partial^2 \Phi_\mu}{\partial x^2}. \tag{7}$$

Formulas (4) and (5) lead to the initial condition for

$$\Phi_\mu(x, 0) = \Phi_0(x) = \int_0^x u_0(\xi)\, d\xi. \tag{8}$$

According to our assumptions,

$$u_0(x) = o(|x|) \quad \text{as } x \to \pm\infty. \tag{9}$$

From this it follows that

$$\Phi_0(x) = o(x^2) \quad \text{as } x \to \pm\infty. \tag{10}$$

Obviously $\Phi_0(x)$ is a continuous function of x and has a piecewise continuous first derivative.

We now solve the Cauchy problem (7), (8) for Φ_μ, find $u_\mu(x, t)$ from (6), and then, letting $\mu \to 0$, obtain in the limit a function $u(x, t)$ which we call a generalized solution of the Cauchy problem (1), (2). The substitution

$$\Phi_\mu(x, t) = -2\mu \ln \Psi_\mu(x, t) \tag{11}$$

reduces the Cauchy problem (7), (8) to the Cauchy problem for the heat equation

$$\frac{\partial \Psi_\mu}{\partial t} = \mu \frac{\partial^2 \Psi_\mu}{\partial x^2} \tag{12}$$

with the initial condition

$$\Psi_\mu(x, 0) = \Psi_u^0(x) = \exp\left\{-\frac{1}{2\mu}\Phi_0(x)\right\} = \exp\left\{-\frac{1}{2\mu}\int_0^x u_0(\xi)\, d\xi\right\}. \tag{13}$$

According to (10),

$$\Psi_\mu^0(x) = o(e^{x^2/2\mu}) \quad \text{as } x \to \pm\infty, \tag{14}$$

and therefore the Cauchy problem (12), (13) has a unique solution given by

$$\begin{aligned}\Psi_\mu(x, t) &= \frac{1}{2\sqrt{\pi\mu t}}\int_{-\infty}^{\infty} \exp\left\{-\frac{(x-\xi)^2}{4\mu t}\right\}\Psi_0^0(\xi)\, d\xi \\ &= \frac{1}{2\sqrt{\pi\mu t}}\int_{-\infty}^{\infty} \exp\left\{-\frac{1}{2\mu}\left[\Phi_0(\xi) + \frac{(x-\xi)^2}{2t}\right]\right\} d\xi.\end{aligned} \tag{15}$$

From (11) and (6) we now obtain an expression for $u_\mu(x, t)$:

$$u_\mu(x, t) = \frac{\int_{-\infty}^{\infty}((x-\xi)/t)\exp\{-\lambda(t, x, \xi)/2\mu\}\, d\xi}{\int_{-\infty}^{\infty}\exp\{-\lambda(t, x, \xi)/2\mu\}\, d\xi}, \tag{16}$$

where

$$\lambda(t, x, \xi) = \Phi_0(\xi) + (x-\xi)^2/2t. \tag{17}$$

It follows from (17) that $\lambda(t, x, \xi)$ is a continuous function of all its arguments for $t > 0$. On the basis of (10) we can assert that for any fixed x, $t \geq 0$

$$\lambda(t, x, \xi) \to +\infty \quad \text{as } \xi \to \pm\infty. \tag{18}$$

This implies that the continuous function $\lambda(t, x, \xi)$ for fixed x, $t > 0$ assumes its minimum value $\lambda_{\min}(x, t)$ on some bounded set of values of ξ. We denote this set by $m(x, t)$.

We introduce the function $\nu(t, x, \xi)$:

$$\nu(t, x, \xi) = \lambda(t, x, \xi) - \lambda_{\min}(x, t) = \Phi_0(\xi) + \frac{(x-\xi)^2}{2t} - \lambda_{\min}(x, t). \tag{19}$$

According to the definition of $\lambda_{\min}(x, t)$ as the absolute minimum of $\lambda(t, x, \xi)$ with respect to ξ, we can obviously conclude that

$$\nu(t, x, \xi) \geq 0; \tag{20}$$

while the equality sign in (20) holds only in the case where ξ belongs to $m(x, t)$. Multiplying the numerator and denominator of (16) by $\exp\{+\lambda_{\min}(x, t)/2\mu\}$, we bring it to the form

$$u_\mu(x, t) = \frac{\int_{-\infty}^{\infty}((x - \xi)/t)\exp\{-\nu(t, x, \xi)/2\mu\}\, d\xi}{\int_{-\infty}^{\infty}\exp\{-\nu(t, x, \xi)/2\mu\}\, d\xi}. \tag{21}$$

Let $\xi^-(x, t)$ and $\xi^+(x, t)$, respectively, denote the infimum and supremum of $m(x, t)$:

$$\xi^-(x,t) = \inf m(x,t), \qquad \xi^+(x,t) = \sup m(x,t), \qquad \xi^-(x,t) \leq \xi^+(x,t). \tag{22}$$

We shall illustrate graphically the definition of the set $m(x, t)$ and its extrema ξ^-, ξ^+. For this we introduce the new function $\eta(t, x, \xi)$:

$$\eta(t, x, \xi) = \lambda(t, x, \xi) - x^2/2t = \eta(t, 0, \xi) - \xi x/t. \tag{23}$$

The functions $\lambda(t, x, \xi)$, $\nu(t, x, \xi)$, and $\eta(t, x, \xi)$ for fixed x, $t > 0$ assume their least value at the same points $\xi \in m(x, t)$.

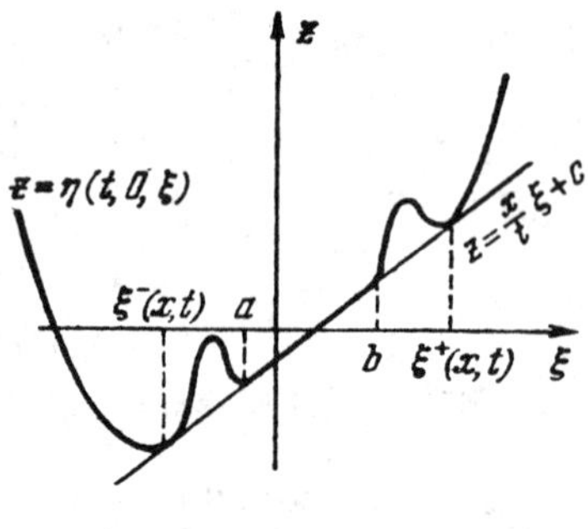

FIGURE 4.15

Figure 4.15 shows in the ξ, z plane the curve $z = \eta(t, 0, \xi)$ and a line $z = (x/t)\xi + c$. For fixed x, $t > 0$, the slope of this straight line is x/t. We now choose the number c so that this line is tangent to the curve $z = \eta(t, 0, \xi)$ from below without intersecting it. The set of points ξ at which $\xi(x/t) = \eta(t, 0, \xi)$ constitutes the set $m(x, t)$. Indeed, if the curve $z = \eta(t, 0, \xi)$ "rests" on the line $z = (x/t)\xi + c$, then this implies that $\eta(t, 0, \xi) - (x/t)\xi = \eta(t, x, \xi)$ at the points at which $\xi(x/t) + c = \eta(t, 0, \xi)$ assumes its least value. We observe that the number c coincides with $\eta_{\min}(x, t)$. In Figure 4.15 the set $m(x, t)$ contains the segment $a \le \xi \le b$ in addition to the extrema $\xi^-(x, t)$ and $\xi^+(x, t)$.

We now establish the following properties of ξ^- and ξ^+:

$$\xi^+(x, t) \le \xi^-(x', t) \quad \text{for } x < x', \tag{24}$$

$$\xi^-(x - 0, t) = \xi^-(x, t), \qquad \xi^+(x + 0, t) = \xi^+(x, t), \tag{25}$$

$$\xi^-(+\infty, t) = +\infty, \qquad \xi^+(-\infty, t) = -\infty. \tag{26}$$

To prove property (24) we consider the difference

$$\begin{aligned} &\eta(t, x + \Delta x, \xi) - \eta(t, x + \Delta x, \xi^+(x, t)) \\ &\quad = \eta(t, x, \xi) - \eta(t, x, \xi^+(x, t)) - \frac{\Delta x}{t}\,[\xi - \xi^+(x, t)]. \end{aligned} \tag{27}$$

By the definition of the supremum, we have

$$\eta(t, x, \xi) - \eta(t, x, \xi^+(x, t)) \begin{cases} \ge 0 & \text{for } \xi \le \xi^+(x, t), \\ > 0 & \text{for } \xi > \xi^+(x, t). \end{cases} \tag{28}$$

Suppose that $\Delta x > 0$. From (27) and (28) it follows that

$$\eta(t, x + \Delta x, \xi) - \eta(t, x + \Delta x, \xi^+(x, t)) \begin{cases} > 0 & \text{for } \xi < \xi^+(x, t), \\ = 0 & \text{for } \xi = \xi^+(x, t). \end{cases} \tag{29}$$

According to the definition of $\xi^-(x, t)$,

$$\eta(t, x + \Delta x, \xi^-(x + \Delta x, t)) - \eta(t, x + \Delta x, \xi^+(x, t)) \le 0 \quad (\Delta x > 0). \tag{30}$$

Comparing (30) and (29), we see that $\xi^-(x + \Delta x, t) \ge \xi^+(x, t)$. This proves (24).

The proof of inequality (24) can be illustrated geometrically. Indeed, if $x' > x$, then the slope of the support line $z = \xi(x'/t) + c'$ is greater than the slope of the line $z = \xi(x/t) + c$, and from Figure 4.15 we immediately obtain (24).

The properties (25) and (26) are established in a similar way.

We note that since $\lambda(t, x, \xi)$ is a continuous function of all its variables for $t > 0$, its absolute minimum $\lambda_{\min}(x, t)$ with respect to ξ is a continuous function of x and t.

We return to (21) and show that for any $x, t > 0$

$$\frac{x - \xi^+(x,t)}{t} \le \liminf_{\substack{\mu\to 0 \\ x'\to x \\ t'\to t}} u_\mu(x', t') \le \limsup_{\substack{\mu\to 0 \\ x'\to x \\ t'\to t}} u_\mu(x', t') \le \frac{x - \xi^-(x, t)}{t}. \tag{31}$$

In particular, it follows from (31) that if at the point x, t ($t > 0$) the condition $\xi^-(x, t) = \xi^+(x, t)$ is satisfied, then at this point there exists the limit

$$\lim_{\substack{\mu\to 0\\ x'\to x\\ t'\to t}} u_\mu(x', t') = \frac{x - \xi^-(x, t)}{t} = \frac{x - \xi^+(x, t)}{t}. \tag{32}$$

Proceeding to the proof of (31), for brevity we set $\xi^+(x, t) = \xi^+$ and $\xi^-(x, t) = \xi^-$. Let $\varepsilon > 0$ be an arbitrary small positive number. We choose numbers a and b so small that for all values of ξ, x' and t' satisfying the conditions

$$|x' - x| + |t' - t| < a, \qquad \xi^- - 2b < \xi < \xi^+ + 2b, \quad 0 < a < t, \tag{33}$$

we have

$$l = \frac{x - \xi^+}{t} - \varepsilon < \frac{x' - \xi}{t'} < \frac{x - \xi^-}{t} + \varepsilon = L. \tag{34}$$

The functions $\xi^-(x, t)$ and $\xi^+(x, t)$ are, respectively, left and right semicontinuous; therefore, if a is sufficiently small and (33) and (34) are satisfied, then it may be assumed that the following inequalities are also satisfied:

$$\xi^- - b < \xi^-(x', t') \le \xi^+(x', t') < \xi^+ + b. \tag{35}$$

We write (21) at the point x', t':

$$u_\mu(x', t') = \frac{\int_{-\infty}^{\infty}((x' - \xi)/t')\exp\{-\nu(t', x', \xi)/2\mu\}\,d\xi}{\int_{-\infty}^{\infty}\exp\{-\nu(t', x', \xi)/2\mu\}\,d\xi}. \tag{36}$$

In (36) we estimate the numerator:

$$\begin{aligned}
&\int_{-\infty}^{\infty}\frac{x' - \xi}{t'}\exp\left\{-\frac{\nu(t', x', \xi)}{2\mu}\right\}d\xi \\
&= l\int_{-\infty}^{\infty}\exp\left\{-\frac{\nu}{2\mu}\right\}d\xi + \int_{-\infty}^{\infty}\left[\frac{x' - \xi}{t'} - l\right]\exp\left\{-\frac{\nu}{2\mu}\right\}d\xi \\
&> l\int_{-\infty}^{\infty}\exp\left\{-\frac{\nu}{2\mu}\right\}d\xi + \int_{-\infty}^{\xi^- - 2b}\left[\frac{x' - \xi}{t'} - l\right]\exp\left\{-\frac{\nu}{2\mu}\right\}d\xi \\
&\quad + \int_{\xi^+ + 2b}^{\infty}\left[\frac{x' - \xi}{t'} - l\right]\exp\left\{-\frac{\nu}{2\mu}\right\}d\xi \\
&= \int_{-\infty}^{\infty}\frac{x' - \xi}{t'}\exp\left\{-\frac{\nu}{2\mu}\right\}d\xi - \int_{\xi - 2b}^{\xi + 2b}\left[\frac{x' - \xi}{t'} - l\right]\exp\left\{-\frac{\nu}{2\mu}\right\}d\xi.
\end{aligned} \tag{37}$$

In obtaining (37) we have used (34). In a similar way we obtain an upper bound for the numerator:

$$\int_{-\infty}^{\infty} \frac{x' - \xi}{t'} \exp\left\{- \frac{\nu(t', x', \xi)}{2\mu}\right\} d\xi$$

$$< L \int_{-\infty}^{\infty} \exp\left\{- \frac{\nu}{2\mu}\right\} d\xi + \int_{-\infty}^{\xi^- - 2b} \left[\frac{x' - \xi}{t'} - L\right] \exp\left\{- \frac{\nu}{2\mu}\right\} d\xi$$

$$+ \int_{\xi^+ + 2b}^{\infty} \left[\frac{x' - \xi}{t'} - L\right] \exp\left\{- \frac{\nu}{2\mu}\right\} d\xi. \tag{38}$$

Dividing (35) by the denominator in (36), we obviously have

$$u_\mu(x', t') > l + \frac{\int_{-\infty}^{\xi^- - 2b} [(x' - \xi)/t' - l] \exp\{- \nu(t', x', \xi)/2\mu\}\, d\xi}{\int_{-\infty}^{\infty} \exp\{- \nu(t', x', \xi)/2\mu\}\, d\xi}$$

$$+ \frac{\int_{\xi^+ + 2b}^{\infty} [(x' - \xi)/t' - l] \exp\{- \nu/2\mu\}\, d\xi}{\int_{-\infty}^{\infty} \exp\{- \nu(t', x', \xi)/2\mu\}\, d\xi}. \tag{39}$$

We shall show that the last two terms in (39) tend to zero as $\mu \to 0$ uniformly with respect to x' and t' if the latter satisfy the condition (33). Noting that the linear function $(x' - \xi)/t' - l$ on the half intervals $(-\infty, \xi^- - 2b]$ and $[\xi^+ + 2b, +\infty)$ can be bounded in modulus by the respective quantities $\omega(\xi^- - \xi)$ and $\omega(\xi - \xi^+)$ while ω may be chosen independent of x' and t' if the latter satisfy (33), we write

$$\frac{\left|\int_{-\infty}^{\xi^- - 2b} [(x' - \xi)/t' - l] \exp\{- \nu/2\mu\}\, d\xi\right|}{\int_{-\infty}^{\infty} \exp\{- \nu/2\mu\}\, d\xi} < \frac{\int_{-\infty}^{\xi^- - 2b} \omega(\xi^- - \xi) \exp\{-\nu/2\mu\}\, d\xi}{\int_{-\infty}^{\xi^-(x', t')} \exp\{-\nu/2\mu\}\, d\xi}, \tag{40}$$

$$\frac{\left|\int_{\xi^+ + 2b}^{\infty} [(x' - \xi)/t' - l] \exp\{-\nu/2\mu\}\, d\xi\right|}{\int_{-\infty}^{\infty} \exp\{- \nu/2\mu\}\, d\xi} < \frac{\int_{\xi^+ + 2b}^{\infty} \omega[\xi - \xi^+] \exp\{-\nu/2\mu\}\, d\xi}{\int_{\xi^+(x', t')}^{\infty} \exp\{- \nu/2\mu\}\, d\xi}. \tag{41}$$

For all x' and t' satisfying (33) we have $\nu(t', x', \xi) > 0$ for $\xi < \xi^- - 2b$ and $\nu(t', x', \xi) = o(\xi^2/2t')$ as $\xi \to -\infty$. Therefore, there exists a number $A > 0$ such that

$$\nu(t', x', \xi) > A(\xi - \xi^-)^2 \quad \text{for } \xi \leq \xi^- - 2b.$$

Using this estimate, we write for the numerator in (40)

$$\int_{-\infty}^{\xi^- - 2b} \omega[\xi^- - \xi] \exp\left\{- \frac{\nu(t', x', \xi)}{2\mu}\right\} d\xi$$

$$< \omega \int_{-\infty}^{\xi^- - 2b} (\xi^- - \xi) \exp\left\{- \frac{A}{2\mu}(\xi - \xi^-)^2\right\} d\xi = \frac{\omega\mu}{A} \exp\left\{- \frac{2Ab^2}{\mu}\right\}. \tag{42}$$

On the other hand, since $\nu(t', x', \xi^-(t', x')) = 0$, there is a $\delta > 0$ such that $\nu(t', x', \xi) < 4Ab^2$ for all ξ in the interval $\xi^-(x', t') - \delta \le \xi \le \xi^-(x', t')$. Therefore, the denominator in (40) can be estimated as follows:

$$\int_{-\infty}^{\xi^-(x',t')} \exp\left\{-\frac{\nu(t', x', \xi)}{2\mu}\right\} d\xi > \int_{\xi^-(x',t')-\delta}^{\xi^-(x',t')} \exp\left\{-\frac{2Ab^2}{\mu}\right\} d\xi$$

$$= \delta \exp\left\{-\frac{2Ab^2}{\mu}\right\}. \tag{43}$$

The right side of (40) does not exceed the ratio of the right side of (42) to (43), i.e., it does not exceed $\omega\mu/A\delta$, and this estimate holds uniformly for all x' and t' satisfying (33). Since an analogous estimate is easily obtained for the second term in (39) as well, we can write

$$u_\mu(x', t') > l + O(\mu). \tag{44}$$

Similar computations for (38) lead to the inequality

$$u_\mu(x', t') < L + O(\mu). \tag{45}$$

The estimates (44) amd (45) hold uniformly for x' and t' satisfying (33). From the definition of the numbers l and L in (34) and the fact that ε is arbitrary it follows that (44) and (45) prove (31) and (32).

We define a generalized solution of the Cauchy problem (1), (2) by the formula

$$u(x, t) = \lim_{\mu\to 0} u_\mu(x, t) \tag{46}$$

at all points x, t at which this limit exists, i.e., where $\xi^-(x, t) = \xi^+(x, t)$. At these points $u(x, t)$ is continuous in x and t. At points where $\xi^+(x, t) \ne \xi^-(x, t)$, to be specific we set

$$u(x, t) = u(x - 0, t) = \frac{x - \xi^-(x, t)}{t}.$$

Suppose that $\xi^-(x, t) \ne \xi^+(x, t)$ at some point (x, t). From (31) and (34) it then follows that

$$u(x - 0, t) > u(x + 0, t), \tag{47}$$

since

$$u(x - 0, t) = \frac{x - \xi^-(x, t)}{t}, \qquad u(x + 0, t) = \frac{x - \xi^+(x, t)}{t}. \tag{48}$$

The inequality (47) shows that the generalized solution $u(x, t)$ defined by (46) satisfies the stability condition introduced in §1.

Suppose now that $\xi^-(x, t) = \xi^+(x, t) = \xi(x, t)$. If $\xi = \xi(x, t)$ is a point of continuity of the initial function $u_0(\xi)$, then, since $\lambda(t, x, \xi)$ assumes a minimum at this point, it follows that

$$u_0(\xi(x, t)) = \frac{x - \xi(x, t)}{t}, \tag{49}$$

$$u_0'(\xi(x, t) - 0) \geq -\frac{1}{t}, \qquad u_0'(\xi(x, t) + 0) \geq -\frac{1}{t} \tag{50}$$

(we assume that $\xi = \xi(x, t)$ may be a point of discontinuity of the derivative of $u_0(\xi)$). If the point $\xi = \xi(x, t) = \xi^-(x, t) = \xi^+(x, t)$ is a point of discontinuity of the initial function $u_0(\xi)$, then

$$u_0(\xi(x, t) - 0) < u_0(\xi(x, t) + 0). \tag{51}$$

Suppose now that $\xi^-(x, t) \neq \xi^+(x, t)$. Then

$$u_0(\xi^-(x, t) - 0) = u(x - 0, t) \leq u_0(\xi^-(x, t) + 0), \tag{52}$$

$$u_0(\xi^+(x, t) - 0) \leq u(x + 0, t) = u_0(\xi^+(x, t) + 0). \tag{53}$$

Finally, we note one further property of the solution $u(x, t)$. Let $x = x(t)$ be a line of discontinuity of $u(x, t)$. Then

$$\lim_{\substack{t' \to t+0 \\ t'' \to t+0}} \frac{x(t'') - x(t')}{t'' - t'} = \frac{1}{2}[u(x(t) - 0, t) + u(x(t) + 0, t)]. \tag{54}$$

This property shows that piecewise continuous solutions $u(x, t)$ of the Cauchy problem (1), (2) satisfy the Hugoniot condition for equation (1).

We now proceed to clarify the question of the sense in which the function $u(x, t) = \lim_{\mu \to 0} u_\mu(x, t)$ satisfies (1) and (2). We consider the function

$$\Phi(x, t) = \lim_{\mu \to 0} \Phi_\mu(x, t) = \lim_{\mu \to 0} \frac{\ln 2\sqrt{\pi \mu t}}{1/2\mu} - \lim_{\mu \to 0} \frac{\ln \int_{-\infty}^{\infty} \exp\{-\lambda(t, x, \xi)/2\mu\}\, d\xi}{1/2\mu}.$$

Applying l'Hôpital's rule, we obtain

$$\Phi(x, t) = \lim_{\mu \to 0} \frac{\int_{-\infty}^{\infty} \lambda(t, x, \xi) \exp\{-\lambda(t, x, \xi)/2\mu\}\, d\xi}{\int_{-\infty}^{\infty} \exp\{-\lambda(t, x, \xi)/2\mu\}\, d\xi}. \tag{55}$$

From (55) it follows that

$$\Phi(x, t) = \lambda_{\min}(x, t) = \lambda(t, x, \xi^-(x, t)) = \lambda(t, x, \xi^+(x, t))$$

$$= \inf_\xi \left[\Phi_0(\xi) + \frac{(x - \xi)^2}{2t} \right] = \inf_\xi \mathscr{G}(t, x, \xi),$$

$$\mathscr{G}(t, x, \xi) = \Phi_0(\xi) + \frac{(x - \xi)^2}{2t}. \tag{56}$$

The proof of (56) is analogous to the proof of (31). It should be noted that $\Phi(x, t)$ is a continuous function of x and t in the half plane $t > 0$. Since $\xi^-(x, t) \to x$ and $\xi^+(x, t) \to x$ as $t \to 0$, it follows from (56) that $\Phi(x, t)$ is also continuous for $t = 0$, i.e., it is continuous for $t \geq 0$.

We consider an arbitrary point (x, t) of the half plane $t > 0$ at which $\xi^-(x, t) = \xi^+(x, t)$. Obviously there exists a neighborhood of (x, t) in which $\xi^-(x', t') = \xi^+(x', t')$ for all points (x', t') in this neighborhood. According to (6).

$$\frac{\partial \Phi_\mu}{\partial x} = u_\mu, \qquad \frac{\partial \Phi_\mu}{\partial t} = \mu \frac{\partial u_\mu}{\partial x} - \frac{u_\mu^2}{2}. \tag{57}$$

Since the sequence u_μ converges uniformly in this neighborhood to $u(x, t)$ and $\Phi_\mu(x, t) \Rightarrow \Phi(x, t)$ by what has been proved above, it is obvious that $\Phi(x, t)$ is differentiable at (x, t), and

$$\frac{\partial \Phi(x, t)}{\partial x} = u(x, t). \tag{58}$$

Differentiating (16) with respect to x, we find

$$\mu \frac{\partial u_\mu}{\partial x} = \frac{1}{2}[u_\mu^2 - \overline{u_\mu^2}] + \frac{\mu}{t}, \tag{59}$$

where $\overline{u_\mu^2}$ denotes the quantity

$$\overline{u_\mu^2} = \frac{\int_{-\infty}^{\infty} ((x - \xi)/t)^2 \exp\{-\lambda(t, x, \xi)/2\mu\}\, d\xi}{\int_{-\infty}^{\infty} \exp\{-\lambda(t, x, \xi)/2\mu\}\, d\xi}. \tag{60}$$

At points (x, t) at which $\xi^-(x, t) = \xi^+(x, t)$ we have

$$\lim_{\mu \to 0} u_\mu^2 = \lim_{\mu \to 0} \overline{u_\mu^2} = \left[\frac{x - \xi(x, t)}{r} \right]^2. \tag{61}$$

Therefore, in the neighborhood of (x, t) considered, according to (57), the sequence $\partial \Phi_\mu / \partial t$ converges uniformly as $\mu \to 0$:

$$\lim_{\mu \to 0} \frac{\partial \Phi_\mu}{\partial t} = \lim_{\mu \to 0} \left(-\frac{u_\mu^2}{2} \right) = -\frac{u^2(x, t)}{2}. \tag{62}$$

Hence, $\Phi(x, t)$ is differentiable at (x, t) with respect to t, and

$$\frac{\partial \Phi(x, t)}{\partial t} = -\frac{u^2(x, t)}{2}. \tag{63}$$

Thus, at all points (x, t) where $\xi^-(x, t) = \xi^+(x, t)$ the function $\Phi(x, t)$ is continuously differentiable and satisfies the equation

$$\frac{\partial \Phi}{\partial t} + \frac{1}{2}\left(\frac{\partial \Phi}{\partial x}\right)^2 = 0. \tag{64}$$

According to (25), $\xi^-(x, t)$ and $\xi^+(x, t)$ are monotonically increasing functions of x, are semicontinuous in x, and are bounded for finite x and t. This implies that on any segment of the line $t =$ const the set of points at which $\xi^-(x, t) \neq \xi^+(x, t)$ is no more than countable. From this it further follows that in any region G of the variables (x, t) there exist no more than a countable number of lines away from which $\Phi(x, t)$ is continuously differentiable and satisfies (64).

Since $\xi^-(x, t) \to x$ and $\xi^+(x, t) \to x$ as $t \to 0$, it follows that

$$\Phi(x, 0) = \Phi_0(x) = \int_0^x u_0(\xi)\, d\xi.$$

Because of the continuity of $\Phi(x, t)$, we conclude that

$$\int_0^x u(\xi, t)\, d\xi \to \int_0^x u_0(\xi)\, d\xi \quad \text{as } t \to 0. \tag{65}$$

The relations (64) and (65) show that the function $u(x, t)$ defined by (46) is a generalized solution of the Cauchy problem (1), (2).

We shall show that the function $u(x, t)$ defined by (46) is a generalized solution of the Cauchy problem (1), (2) in still another sense. Integrating (3) over a region G of the half plane $t > 0$ bounded by the closed contour C we obviously obtain

$$\oint_C u_\mu\, dx - \frac{u_\mu^2}{2}\, dt = -\oint_C u \frac{\partial u_\mu}{\partial x}\, dt. \tag{66}$$

According to (59), $\mu(\partial u_\mu / \partial x) \to 0$ at points where $\xi^-(x, t) = \xi^+(x, t)$. Therefore, if we assume that on the contour C the measure of the set of points (x, t) at which $\xi^-(x, t) \neq \xi^+(x, t)$ is equal to zero, then as $\mu \to 0$ the integral $\oint_C \mu(\partial u_\mu / \partial x)\, dt$ on the right side of (66) tends to zero. Under this same assumption the integral on the left side of (66) remains meaningful as $\mu \to 0$, since $u_\mu(x, t) \to u(x, t)$ almost everywhere on C. Therefore passing to the limit in (66) as $\mu \to 0$, we obtain

$$\oint_C u(x, t)\, dx - \frac{u^2(x, t)}{2}\, dt = 0. \tag{67}$$

Thus, for any closed contour C on which the measure of the set of points of discontinuity of $u(x, t)$ is equal to zero the function $u(x, t)$ satisfies the integral conservation law (67) for equation (1). Since $u(x, t)$ assumes the initial value (2) in the sense of (65), it follows that $u(x, t)$ is a stable generalized solution of the Cauchy problem (1), (2) also in the sense of the integral conservation law (67).

We note now that if at points of discontinuity the function $u(x, t)$ is assigned a particular value, for example,

$$u(x, t) = u(x - 0, t), \tag{68}$$

then (67) holds generally for any closed contour C lying in the half plane $t \geq 0$. Indeed, C may be considered as the limit of contours C' on which the measure of the set of points of discontinuity of $u(x, t)$ is equal to zero. For each such contour C' equality (67) is satisfied according to the foregoing. If at each point of discontinuity (x, t) of $u(x, t)$ the corresponding point (x', t') of C' tends to it from the left, i.e., $x' < x$, $x' \to x$, then $u(x', t') \to u(x, t)$ according to (68), and, passing to the limit in (67), we conclude that this equality is satisfied for C.

Multiplying (3) by an arbitrary smooth, compactly supported function $g(x, t)$ and integrating the results over the half plane $t \geq 0$, we obtain

$$\iint_{t\geq 0} \left[\frac{\partial g}{\partial t} u_\mu + \frac{\partial g}{\partial x}\frac{u_\mu^2}{2}\right] dx\, dt + \int_{-\infty}^{\infty} g(x, 0) u_0(x)\, dx = \iint_{t\geq 0} \mu \frac{\partial u_\mu}{\partial x}\frac{\partial g}{\partial x}\, dx\, dt. \tag{69}$$

S nce $u_\mu \to u(x, t)$ and $u_\mu^2 \to u^2(x, t)$ almost everywhere and, according to (64), $\mu(\partial u_\mu/\partial x) \to 0$, on passing to the limit as $\mu \to 0$ in (69) we obtain

$$\iint_{t\geq 0} \left[\frac{\partial g}{\partial t} u(x, t) + \frac{\partial g}{\partial x}\frac{u^2(x, t)}{2}\right] dx\, dt + \int_{-\infty}^{\infty} g(x, 0) u_0(x)\, dx = 0. \tag{70}$$

This shows that $u(x, t)$ is a generalized solution of the Cauchy problem (1), (2) also in the sense of the last of the three definitions introduced in §1. This definition of a generalized solution of quasilinear equations was first proposed by Hopf.

3. ***The Cauchy problem for the equation*** $u_t + \varphi_x = 0$ ***under the condition*** $\varphi''_{uu} > 0$. Suppose that the function $\varphi(u, x, t)$ possesses two derivatives with respect to all its variables for $t \geq 0$, $-\infty < x < \infty$ and any bounded u. We assume that $\varphi''_{uu}(u, x, t) > 0$ in this region of the variables x, t, u (the case $\varphi''_{uu}(u, x, t) < 0$ is considered in an entirely similar way).

For the equation

$$\frac{\partial u}{\partial t} + \frac{\partial \varphi(u, x, t)}{\partial x} = 0 \tag{1}$$

we pose the initial condition

$$u(x, 0) = u_0(x), \tag{2}$$

assuming that $u_0(x)$ is piecewise continuous and possesses a piecewise continuous first derivative for any bounded values of x.

The Cauchy problem (1), (2) we first considered and solved by O. A. Oleĭnik in the even broader class of initial functions consisting of bounded, measurable functions $u_0(x)$. Our treatment of the Cauchy problem (1), (2) will be less detailed and somewhat different than Oleĭnik's.

For the characteristic system of equation (1)

$$\frac{dx}{dt} = \varphi'_u(u, x, t), \qquad \frac{du}{dt} = -\varphi'_x(u, x, t) \tag{3}$$

we consider the Cauchy problem with initial conditions

$$x(0) = x_0, \qquad u(0) = u_0. \tag{4}$$

We denote the solution of this Cauchy problem by the letters

$$x(t) = X(t, x_0, u_0), \qquad u(t) = U(t, x_0, u_0). \tag{5}$$

By definition, these functions satisfy conditions (4), which we now rewrite in the form

$$X(0, x_0, u_0) \equiv x_0, \qquad U(0, x_0, u_0) \equiv u_0. \tag{6}$$

We assume that the functions X and U remain bounded for any finite x_0, u_0 and $t > 0$. Conditions under which this holds are relatively complicated, and we shall not present them.

If $X(t, x_0, u_0)$ and $U(t, x_0, u_0)$ are bounded, then they have continuous first derivatives with respect to all their variables. This follows from the continuous differentiability of the right sides of the characteristic system (3). Finally, we further assume that

$$X(t, x_0, u_0) \to \pm\infty \quad \text{as } x_0 \to \pm\infty. \tag{7}$$

The second problem that we consider for the characteristic system (3) is the boundary value problem with the following conditions. It is required to find a solution $x(t) = \bar{X}(t, \tau, \xi, x_0)$, $u(t) = \bar{U}(t, \tau, \xi, x_0)$ of the characteristic system (3) satisfying the boundary conditions

$$x(0) = \bar{X}(0, \tau, \xi, x_0) \equiv x_0, \quad x(\tau) = \bar{X}(\tau, \tau, \xi, x_0) \equiv \xi, \tag{8}$$

where x_0 and ξ are arbitrary numbers and $\tau > 0$. As before, we shall assume that this problem has a unique, bounded solution for any x_0, ξ and $\tau > 0$, without writing down the complicated sufficient conditions under which this holds. However, if this is assumed, then $\bar{X}$ and $\bar{U}$ are continuously differentiable with respect to all their arguments.

We list the following obvious relations:

$$\bar{X}(0, \tau, \xi, x_0) \equiv x_0, \quad \bar{X}(\tau, \tau, \xi, x_0) \equiv \xi, \tag{9}$$

$$X(t, x_0, \bar{U}(0, \tau, \xi, x_0)) \equiv \bar{X}(t, \tau, \xi, x_0), \tag{10}$$

$$U(t, x_0, \bar{U}(0, \tau, \xi, x_0)) \equiv \bar{U}(t, \tau, \xi, x_0). \tag{11}$$

For Hopf's equation considered in the preceding subsection we have $\varphi(u, x, t) = u^2/2$ and

$$X(t, x_0, u_0) = x_0 + u_0 t, \quad U(t, x_0, u_0) = u_0, \tag{12}$$

$$\bar{X}(t, \tau, \xi, x_0) = x_0 + \frac{\xi - x_0}{\tau} t, \quad \bar{U}(t, \tau, \xi, x_0) = \frac{\xi - x_0}{\tau}. \tag{13}$$

It is evident from (13) that the solution $u(x, t)$ of the Cauchy problem (2.2.1), (2.2.2) given by (2.2.48) can also be written in the form

$$\left.\begin{aligned} u(x - 0, t) &= \bar{U}(t, t, x, \xi^-(x, t)), \\ u(x + 0, t) &= \bar{U}(t, t, x, \xi^+(x, t)), \end{aligned}\right\} \tag{14}$$

where $\xi^-(x, t)$ and $\xi^+(x, t)$ are defined (see subsection 2) as the infimum and supremum of the set of values ξ on which the function

$$\lambda(t, x, \xi) = \Phi_0(\xi) + \frac{(x - \xi)^2}{2t} = \int_0^\xi u_0(\eta)\, d\eta + \frac{(x - \xi)^2}{2t}$$

achieves a minimum value for fixed x and t. Noting that for Hopf's equation

$$I(t, x, \xi) = \lambda(t, x, \xi) - \frac{x^2}{2t} = \int_0^\xi [u_0(\eta) - \bar{U}(0\ t, x, \eta)]\, d\eta$$

$$= \int_0^\xi \left[u_0(\eta) + \frac{\eta - x}{t}\right] d\eta, \tag{15}$$

we can assert that in this special case the quantities $\xi^-(x,t)$ and $\xi^+(x,t)$ can be defined also as the infimum and supremum of the set of values ξ on which the function

$$I(t, x, \xi) = \int_0^{\xi} [u_0(\eta) - \bar{U}(0, t, x, \eta)]\, d\eta \tag{16}$$

assumes a minimum for fixed values of x, $t > 0$. Oleĭnik showed that (14) gives the solution of the Cauchy problem (1), (2) under the conditions we imposed on $\varphi(u, x, t)$, provided that $\xi^-(x, t)$ and $\xi^+(x, t)$ are now understood as the infimum and supremum of the set of values ξ on which $I(t, x, \xi)$ assumes a minimum.

From the conditions imposed on $\varphi(u, x, t)$ it follows that

$$\bar{U}(0, t, x, \xi) \to \pm\infty \quad \text{as } \xi \to \pm\infty, \tag{17}$$

and, since we assume that the initial function $u_0(x)$ is bounded, it follows that

$$I(t, x, \xi) \to +\infty \quad \text{as } |\xi| \to \infty. \tag{18}$$

This implies that the continuous function $I(t, x, \xi)$ for fixed values of x, $t > 0$ assumes a minimum $I_{\min}(x, t)$ on some bounded set $m(x, t)$ of values of ξ. The infimum and supremum of this set we denote by $\xi^-(x, t)$ and $\xi^+(x, t)$ respectively; these quantities satisfy (2.2.24), (2.2.25), and (2.2.26). The proof is similar to the proof is subsection 2, and is omitted. A generalized solution $u(x, t)$ of the Cauchy problem (1), (2) defined by (14) satisfies (2.2.47).

We introduce the function

$$\mathscr{G}(t, x, \xi) = \int_0^{\xi} u_0(\eta)\, d\eta + \int_0^t [\bar{U}\varphi_u'(\bar{U}, \bar{X}, \tau) - \varphi(\bar{U}, \bar{X}, \tau)]\, d\tau, \tag{19}$$

where for brevity the arguments of the functions $\bar{U} = \bar{U}(\tau, t, x, \xi)$ and $X = \bar{X}(\tau, t, x, \xi)$ have been omitted. We differentiate the integrand in the second term of (19) with respect to ξ. We obtain

$$\begin{aligned}
[\bar{U}\varphi_u'(\bar{U}, \bar{X}, \tau) - \varphi(\bar{U}, \bar{X}, \tau)]_\xi' &= \bar{U}_\xi'\varphi_u'(\bar{U}, \bar{X}, \tau) + \bar{U}[\varphi_u'(\bar{U}, \bar{X}, \tau)]_\xi' \\
&\quad - \varphi_u'(\bar{U}, \bar{X}, \tau)\bar{U}_\xi' - \varphi_x'(\bar{U}, \bar{X}, \tau)\bar{X}_\xi' \\
&= \bar{U}[\varphi_u'(\bar{U}, \bar{X}, \tau)]_\xi' - \varphi_x'(\bar{U}, \bar{X}, \tau)\bar{X}_\xi'.
\end{aligned} \tag{20}$$

Now $\bar{U}$, $\bar{X}$ is a solution of the characteristic system (3), and therefore

$$\varphi_u'(\bar{U}, \bar{X}, \tau) = \bar{X}_\tau', \qquad -\varphi_x'(\bar{U}, \bar{X}, \tau) = \bar{U}_\tau'. \tag{21}$$

Substituting (21) into (20), we obtain

$$[\bar{U}\varphi_u' - \varphi]_\xi' = \bar{U}\bar{X}_{\tau\xi}'' + \bar{U}_\tau' U_\xi' = \frac{\partial}{\partial \tau}[\bar{U}(\tau, t, x, \xi)\bar{X}_\xi'(\tau, t, x, \xi)].$$

Substituting this into (19), we find that

$$\mathscr{G}(t, x, \xi) = \int_0^\xi u_0(\eta)\,d\eta + \int_0^t\int_0^\xi \frac{\partial}{\partial \tau}[\bar{U}(\tau, t, x, \eta)\bar{X}_\eta'(\tau, t, x, \eta)]\,d\tau\,d\eta$$
$$+ \int_0^t [\bar{U}_0\varphi_u'((\bar{U}_0, \bar{X}_0, \tau) - \varphi(\bar{U}_0, \bar{X}_0, \tau)]\,d\tau, \tag{22}$$

where $\bar{U}_0 = \bar{U}(\tau, t, x, 0)$ and $\bar{X}_0 = \bar{X}(\tau, t, x, 0)$.

We consider the second term in (22):

$$\int_0^t\int_0^\xi \frac{\partial}{\partial \tau}[\bar{U}(\tau, t, x, \eta)\bar{X}_\eta'(\tau, t, x, \eta)]\,d\tau\,d\eta$$
$$= \int_0^\xi \{\bar{U}(t, t, x, \eta)\bar{X}_\eta'(t, t, x, \eta) - \bar{U}(0, t, x, \eta)\bar{X}_\eta'(0, t, x, \eta)\}\,d\eta.$$

Now, according to (8), $\bar{X}_\eta'(0, t, x, \eta) = 1$ and $\bar{X}_\eta'(t, t, x, \eta) = 0$. Therefore,

$$\mathscr{G}(t, x, \xi) = \int_0^\xi [u_0(\eta) - \bar{U}(0, t, x, \eta)]\,d\eta + F(x, t), \tag{23}$$

where

$$F(x, t) = \int_0^t [\bar{U}_0\varphi_u'(\bar{U}_0, \bar{X}, \tau) - \varphi(\bar{U}_0, \bar{X}, \tau)]\,d\tau. \tag{24}$$

Formulas (23) and (19) give a new representation for $I(t, x, \xi)$, from which we conclude that for fixed x, $t > 0$ the continuous function $\mathscr{G}(t, x, \xi)$ assumes a minimum on the same set $m(x, t)$ of the variable ξ as the function $I(t, x, \xi)$. Hence,

$$\Phi(x, t) = \mathscr{G}(t, x, \xi^-(x, t)) = \mathscr{G}(t, x, \xi^+(x, t)) = \inf_\xi \mathscr{G}(t, x, \xi) \tag{25}$$

is a well defined and continuous function of x, t.

From (25) and (19) we conclude that the continuous function $\Phi(x, t)$ assumes the following values at $t = 0$:

$$\Phi(x, 0) = \int_0^x u_0(\xi)\,d\xi = \Phi_0(x). \tag{26}$$

Since $\mathscr{G}(t, x, \xi)$ is a Lipschitz continuous function of all its arguments in any bounded domain of the variables $t > 0$, x, ξ, i.e.,

$$|\mathscr{G}(t+\Delta t, x+\Delta x, \xi+\Delta\xi) - \mathscr{G}(t, x, \xi)| \leq M\{|\Delta t| + |\Delta x| + |\Delta\xi|\}, \tag{27}$$

its absolute minimum, the function $\Phi(x, t)$, with respect to ξ possesses the same property, i.e.,

$$|\Phi(x + \Delta x, t + \Delta t) - \Phi(x, t)| \leq \bar{M}\{|\Delta x| + |\Delta t|\}. \tag{28}$$

The Lipschitz continuous function $\Phi(x, t)$ possesses continuous derivatives with respect to x and t almost everywhere. We compute these derivatives formally for the time being. We have

$$\Phi(x, t) = \int_0^{\xi(x,t)} u_0(\eta)\, d\eta + \int_0^t [\bar{U}\varphi_u'(\bar{U}, \bar{X}, \tau) - \varphi(\bar{U}, \bar{X}, \tau)]\, d\tau, \tag{29}$$

where, as before, we have set $\xi(x, t) = \xi^-(x, t)$ and $\bar{U} = \bar{U}(\tau, t, x, \xi(x, t))$. Differentiating (29) with respect to x, we obtain

$$\frac{\partial\Phi}{\partial x} = u_0(\xi(x, t))\frac{\partial\xi(x, t)}{\partial x} + \int_0^t \frac{\partial}{\partial x}[\bar{U}\varphi_u' - \varphi]\, d\tau. \tag{30}$$

In analogy to the foregoing,

$$\frac{\partial}{\partial x}[\bar{U}\varphi_u'(\bar{U}, \bar{X}, \tau) - \varphi(\bar{U}, \bar{X}, \tau)] = \frac{\partial}{\partial\tau}(\bar{U}\bar{X}_x'), \tag{31}$$

where

$$\bar{X}_x' = \frac{\partial}{\partial x}\bar{X}(\tau, t, x, \xi(x, t)).$$

Substituting (31) into (30), we find that

$$\begin{aligned}\frac{\partial\Phi}{\partial x} = u_0(\xi(x, t))\frac{\partial\xi(x, t)}{\partial x} &+ \bar{U}(t, t, x, \xi(x, t))\,\bar{X}_x'(t, t, x, \xi(x, t)) \\ &- \bar{U}(0, t, x, \xi(x, t))\bar{X}_x'(0, t, x\, \xi(x, t)).\end{aligned} \tag{32}$$

Now from (8) it follows that

$$\bar{X}_x'(0, t, x, \xi(x, t)) = \frac{\partial\xi(x, t)}{\partial x}, \qquad X_x'(t, t, x, \xi(x, t)) = 1, \tag{33}$$

and therefore (32) takes the form

$$\frac{\partial \Phi}{\partial x} = [u_0(\xi(x, t)) - \bar{U}(0, t, x, \xi(x, t))] \frac{\partial \xi(x, t)}{\partial x} + \bar{U}(t, t, x, \xi(x, t))$$

$$= [u_0(\xi(x, t)) - U(0, t, x, \xi(x, t))] \frac{\partial \xi(x, t)}{\partial x} + u(x, t). \tag{34}$$

Analogous computations lead to the formula

$$\frac{\partial \Phi}{\partial t} = [u_0(\xi(x, t)) - \bar{U}(0, t, x, \xi(x, t))] \frac{\partial \xi(x, t)}{\partial t} - \varphi(\bar{U}, \bar{X}, t)\,|_{\tau=t}$$

$$= [u_0(\xi(x, t)) - \bar{U}(0, t, x, \xi(x, t))] \frac{\partial \xi(x, t)}{\partial t} - \varphi(u(x. t), x, t). \tag{35}$$

We have obtained (34) formally, and in this connection we have assumed the existence of the derivation $\partial \xi(x, t)/\partial x$ and $\partial \xi(x, t)/\partial t$. Now these derivatives exist at points (x, t) for which $\xi^-(x, t) = \xi^+(x, t)$ regardless of whether the initial function $u_0(\xi)$ is differentiable at $\xi = \xi^+(x, t)$. A rigorous justification of these formulas is not hard, but it is messy For example, (34) and (35) can be justified if $u_0(x)$ is considered the limit of differentiable functions. Now it is easy to see that if $\xi^-(x, t) = \xi^+(x, t)$ then the expressions

$$[u_0(\xi(x, t)) - \bar{U}(0, t, \xi(x, t))] \frac{\partial \xi(x, t)}{\partial x},$$

$$[u_0(\xi(x, t)) - \bar{U}(0, t, x, \xi(x, t))] \frac{\partial \xi(x, t)}{\partial t}$$

vanish,[3] and (34) and (35) under this condition simplify to

$$\frac{\partial \Phi}{\partial x} = u(x, t), \qquad \frac{\partial \Phi}{\partial t} = -\varphi(u(x, t), x, t). \tag{36}$$

If $\xi^-(x, t) \neq \xi^+(x, t)$, then

$$\frac{\partial \Phi}{\partial x}(x - 0, t) = u(x - 0, t), \qquad \frac{\partial \Phi}{\partial x}(x + 0, t) = u(x + 0, t). \tag{37}$$

Thus, $\partial \Phi/\partial x$ and $\partial \Phi/\partial t$ exist almost everywhere, and they can be computed according to (36). This implies that almost everywhere the continuous function $\Phi(x, t)$ satisfies

$$\frac{\partial \Phi}{\partial t} + \varphi\left(\frac{\partial \Phi}{\partial x}, x, t\right) = 0 \tag{38}$$

and assumes the initial values (26).

---

[3] If $\xi^-(x, t) = \xi^+(x, t)$ and $u_0(\xi)$ is continuous at $\xi = \xi(x, t)$, then

$$u_0(\xi(x, t)) = \bar{U}(0, t, x, \xi(x, t));$$

if $\xi(x, t)$ is a point of discontinuity of $u_0(\xi)$, then

$$\partial \xi(x, t)/\partial x = \partial \xi(x, t)/\partial t = 0.$$

We thus conclude that the solution $u(x, t)$ given by (14) for any closed contour C satisfies the identity

$$\oint_C u(x, t)\, dx - \varphi(u(x, t), x, t)\, dt = 0. \tag{39}$$

Since, moreover, $u(x, t)$ assumes the initial values (2) in the sense that

$$\int_0^x u(x, t)\, dx \to \int_0^x u_0(x)\, dx \quad \text{as } t \to 0, \tag{40}$$

and satisfies the stability condition $u(x - 0, t) \geq u(x + 0, t)$, it follows that the discontinuous function $u(x, t)$ defined by (14) is a stable generalized solution of the Cauchy problem (1), (2).

We proved above that (14) defines a stable generalized solution of the Cauchy problem (1), (2). However, this proof does not show how we arrived at (14). We therefore present another method of constructing a discontinuous solution of the Cauchy problem (1), (2) which automatically leads to formula (14) for the generalized solution. This method is applicable under the same assumptions regarding problem (1), (2) as those made above, and even without some of them. However, in the interest of clarity and simplicity we first treat the case of a smooth initial function $u_0(x)$ and then consider the case of a piecewise smooth and piecewise continuous function $u_0(x)$.

Thus, we suppose that $\varphi(u, x, t)$ satisfies our requirements and that the initial function $u_0(x)$ possesses a continuous first derivative and is bounded for any finite values of x. As long as the equation

$$x = X(t, x_0, u_0(x_0)) \tag{41}$$

has a unique solution $x_0 = x_0(x, t)$ for x_0, as we have seen in Chapter 1, the solution is continuous and is given by

$$u(x, t) = U(t, x_0, u_0(x_0(x, t))). \tag{42}$$

If (41) has more than one solution for x_0 at certain points or in certain regions of the variables x and t, then (42) defines a multivalued function of x and t from the branches of which the generalized solution must be constructed.

The existence of a generalized solution $u(x, t)$ of the Cauchy problem (1), (2) assumes the existence of a Lipschitz continuous potential $\Phi(x, t)$ which satisfies the nonlinear equation

$$\frac{\partial \Phi}{\partial t} + \varphi\left(\frac{\partial \Phi}{\partial x}, x, t\right) = 0 \tag{43}$$

almost everywhere and the initial condition

$$\Phi(x, 0) = \Phi_0(x) = \int_0^x u_0(\xi)\, d\xi. \tag{44}$$

We denote by $\tilde{\Phi} = \tilde{\Phi}(t, x_0)$ the geometric integral of the Cauchy problem (43), (44), i.e., the equation of the integral surface of this problem expressed in terms of the parameters t and x_0, where x_0 is a point through which the characteristic (41) passes. This surface, generally speaking, does not project injectively onto the plane of the variables x, t. According to §9.2 of Chapter 1, $\tilde{\Phi}(t, x_0)$ is obtained by integrating the "strip equation"

$$\frac{d\tilde{\Phi}(t, x_0)}{dt} = U\varphi'_u(U, X, t) - \varphi(U, X, t), \tag{45}$$

where $U = U(t, x_0, u(x_0))$ and $X = X(t, x_0, u(x_0))$. Recalling the initial condition (44), we integrate (45):

$$\tilde{\Phi}(t, x_0) = \Phi_0(x_0) + \int_0^t [U\varphi'_u(U, X, t) - \varphi(U, X, t)]_{t=\tau}\, d\tau. \tag{46}$$

In the case where (41) is uniquely solvable for x_0 the potential $\Phi(x, t)$ is given by

$$\Phi(X(t, x_0, u_0(x_0)), t) = \tilde{\Phi}(t, x_0) \tag{47}$$

or by the explicit formula

$$\Phi(x, t) = \tilde{\Phi}(t, x_0(x, t)). \tag{48}$$

If (41) is not uniquely solvable for x_0, (47) and (48) define a multivalued function $\tilde{\Phi}(x, t)$ from the branches of which a single-valued and continuous function $\Phi(x, t)$ must be constructed—the potential of the generalized solution.

We fix any $t > 0$ and consider the behavior of the curve

$$x = X(t, x_0, u_0(x_0)) \tag{49}$$

in the plane of the variables x and x_0 (Figure 4.16). The relation (49) can obviously be represented by a continuous curve which, however, for a sufficiently large value of the argument $t > 0$ does not project injectively onto the line $x_0 = 0$. The curve (49) always projects injectively onto the line $x = 0$. We may therefore assume that to each point of this curve there corresponds a particular value of x_0 for a fixed value of t. It may thus be assumed that on the continuous curve (49) continuous functions $u(t, x_0, u_0(x_0))$ and $\tilde{\Phi}(t, x_0)$a re defined (as functions of x_0).

Our problem is to choose from the branches of the multivalued function (48)

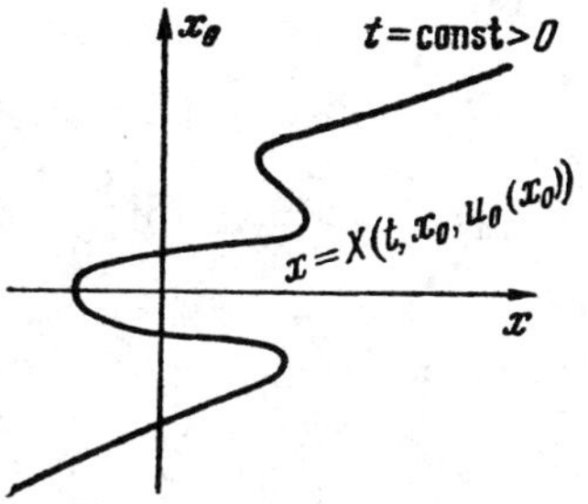

FIGURE 4.16

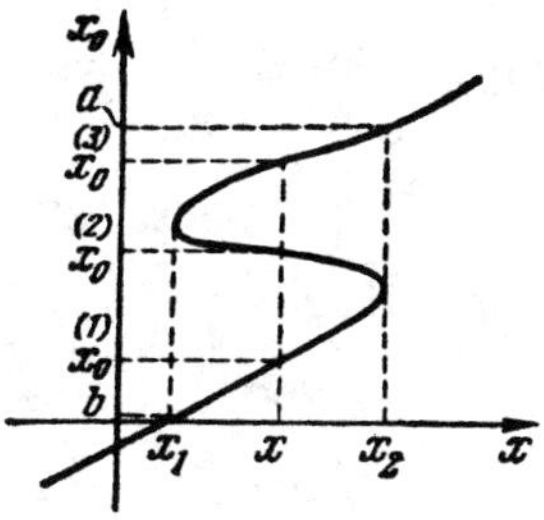

FIGURE 4.17

a single-valued and continuous function—the potential $\Phi(x, t)$. We shall show how to make this selection and, in so doing, establish that it is unique.

Suppose that on some segment $b < x_0 < a$ of the curve (49) there are three intersections with each line $x = c$ for $x_1 < c < x_2$, i.e., the curve projects in a threefold manner onto the segment $x_1 \leq x \leq x_2$ of the axis $x_0 = 0$ (Figure 4.17). Let $x_1 < x < x_2$, and let $\overset{(1)}{x_0} < \overset{(2)}{x_0} < \overset{(3)}{x_0}$ be the values of the variable x_0 satisfying (49).[4] We denote by $\overset{(1)}{u}(x, t)$, $\overset{(2)}{u}(x, t)$, $\overset{(3)}{u}(x, t)$ and $\overset{(1)}{\Phi}(x, t)$, $\overset{(2)}{\Phi}(x, t)$, $\overset{(3)}{\Phi}(x, t)$ the corresponding values of the functions $U(t, x_0, u_0(x_0))$ and $\tilde{\Phi}(t, x_0)$, i.e.,

$$\overset{(i)}{u}(x, t) = U(t, \overset{(i)}{x_0}, u_0(\overset{(i)}{x_0})),$$
$$\overset{(i)}{\Phi}(x, t) = \tilde{\Phi}(t, \overset{(i)}{x_0}).$$

For each of these branches the equality

$$\frac{\partial \overset{(i)}{\Phi}(x, t)}{\partial x} = \overset{(i)}{u}(x, t). \tag{50}$$

is satisfied because of the "strip condition" (45). We assumed previously that the boundary value problem of the characteristic system is uniquely solvable. This assumption implies that

$$\varphi_u'(\overset{(1)}{u}(x, t), x, t) > \varphi_u'(\overset{(2)}{u}(x, t), x, t) > \varphi_u'(\overset{(3)}{u}(x, t), x, t), \tag{51}$$

and since $\varphi_{uu}''(u, x, t) > 0$, it follows that also

[4] The curve (49) may have vertical segments. In this case an infinite set of its points may correspond to some x. As we shall see, this does not affect our constructions.

$$\overset{(1)}{u}(x, t) > \overset{(2)}{u}(x, t) > \overset{(3)}{u}(x, t). \tag{52}$$

According to (50), we can write

$$\frac{\partial\overset{(1)}{\Phi}(x, t)}{\partial x} > \frac{\partial\overset{(2)}{\Phi}(x, t)}{\partial x} > \frac{\partial\overset{(3)}{\Phi}(x, t)}{\partial x}. \tag{53}$$

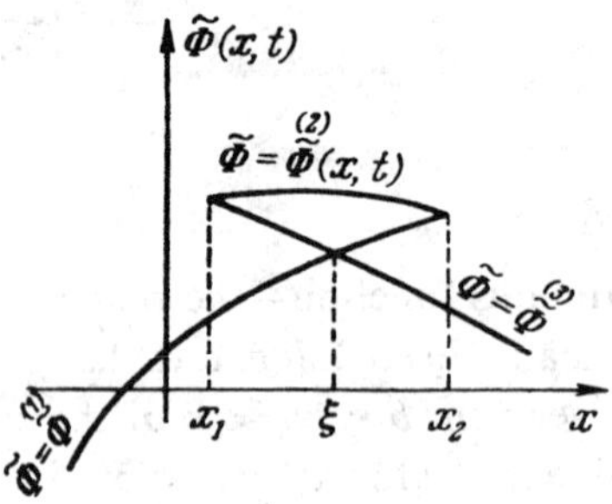

FIGURE 4.18

Figure 4.18 shows a representation of the relation (48) on the segment $[x_1, x_2]$ which, according to our assumption, is three-valued. The branch $\overset{(1)}{\Phi}(x, t)$ is defined for $x \le x_2$ and the branch $\overset{(2)}{\Phi}(x, t)$ for $x_1 \le x \le x_2$; further, in view of the continuity of $\tilde{\Phi}(t, x_0)$ in x_0 we have

$$\overset{(1)}{\Phi}(x_2, t) = \overset{(2)}{\Phi}(x_2, t),$$
$$\overset{(2)}{\Phi}(x_1, t) = \overset{(3)}{\Phi}(x_1, t).$$

Recalling, finally, (53), we conclude that the graph of the relation (48) has the form shown in Figure 4.18.

Thus, on the segment $x_1 \le x \le x_2$ on which $\tilde{\Phi}(x, t)$ is three-valued the selection of a well-defined and continuous function, the potential $\Phi(x, t)$, can obviously be made in a unique manner and is given by

$$\Phi(x, t) = \min \tilde{\Phi}(x, t) \tag{54}$$

where the minimum is taken over all values $\overset{(i)}{\Phi}(x, t)$ at the point x, t. Figure 4.18 shows that this minimum is achieved on the first and third branches of $\tilde{\Phi}(x, t)$, i.e.,

$$\Phi(x, t) = \overset{(1)}{\Phi}(x, t) \quad \text{for } x \le \xi, \qquad \Phi(x, t) = \overset{(3)}{\Phi}(x, t) \quad \text{for } x \ge \xi. \tag{55}$$

The function $u(x, t)$ is defined correspondingly:

$$u(x, t) = \overset{(1)}{u}(x, t) \quad \text{for } x \leq \xi, \qquad u(x, t) = \overset{(3)}{u}(x, t) \quad \text{for } x \geq \xi. \tag{56}$$

If the curve (49) has segments which project in more than a threefold way onto the axis $x_0 = 0$, then, by decomposing it into segments on which $\tilde{\Phi}(x, t)$ is three-valued, we reduce the entire procedure of constructing a continuous potential $\Phi(x, t)$ to the case just considered.

Thus, the generalized solution $u(x,t)$ of the Cauchy problem (1), (2) is given by

$$u(x, t) = U(t, \xi(x, t), u_0(\ (x, t))), \tag{57}$$

where $\xi(x, t)$ is the value of x_0 for which

$$\Phi(x, t) = \tilde{\Phi}(t, \xi(x, t)) \tag{58}$$

is the least value of $\tilde{\Phi}(t_0, x_0)$ for values of x_0 subject to the restriction (49), i.e., $x = X(t, x_0, u_0(x_0))$. Considering (46), (19), and (23), we easily establish that (54) is equivalent to the requirement of a minimum of $I(t, x, \xi)$. Thus, we again arrive at formula (14) for the generalized solution $u(x, t)$.

We have so far considered only the case where the initial function $u_0(x)$ is a smooth and, in particular, continuous function of x. We shall now show that nothing significantly changes in the case of a piecewise smooth and piecewise continuous initial function $u_0(x)$.

If we consider a discontinuous initial function $u_0(x)$ as the limit of continuous functions, then this leads only to the situation that in solving the characteristic system at points of discontinuity of $u_0(x)$ it is necessary to assign to $u_0(x)$ all intermediate va ues between the left and right limits $u_0(x - 0)$ and $u_0(x + 0)$. The relation (49) is then represented by a continuous curve in the plane of the variables x and x_0, as before. The difference is that at points of discontinuity this curve has horizontal segments (Figure 4.19). Thus, for example, in Figure 4.19 to the point $\overset{(1)}{x}_0$ there corresponds the case in which

$$u_0(\overset{(1)}{x}_0 - 0) > u_0(\overset{(1)}{x}_0 + 0),$$

while to the point $\overset{(2)}{x}_0$ there corresponds the case where

$$u_0(\overset{(2)}{x}_0 - 0) < u_0(\overset{(2)}{x}_0 + 0).$$

A pencil of characteristics $x = X(t, x_0, u_0(x_0))$ issues from a point of discontinuity of the initial function $u_0(x)$ which cannot, of course, be described by a

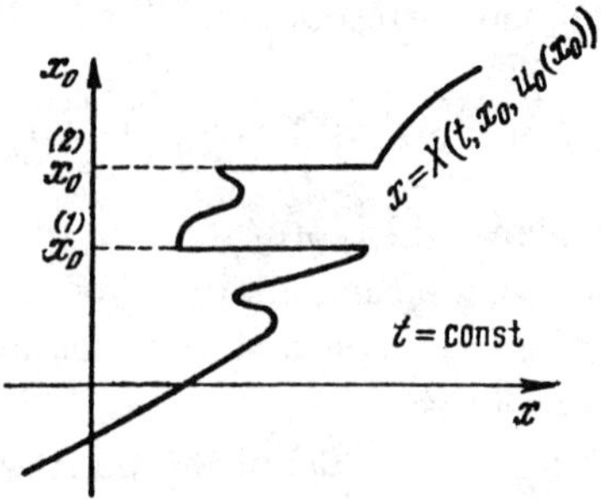

FIGURE 4.19

single parameter x_0. At a point of discontinuity of $u_0(x)$ we therefore introduce another parameter α $(0 \leq \alpha \leq 1)$ and define the functions $X(t, x_0, u_0(x_0), \alpha)$, $U(t, x_0, u_0(x_0), \alpha)$ and $\tilde{\Phi}(t, x_0, \alpha)$ as follows:

$$\left.\begin{aligned} X(t, x_0, u_0(x_0), \alpha) &= X(t, x_0, \alpha u_0(x_0 - 0) + (1 - \alpha)u_0(x_0 + 0)), \\ U(t, x_0, u_0(x_0), \alpha) &= U(t, x_0, \alpha u_0(x_0 - 0) + (1 - \alpha)u_0(x_0 + 0)), \end{aligned}\right\} \tag{59}$$

$$\tilde{\Phi}(t, x_0, \alpha) = \Phi_0(x_0) + \int_0^t [U\varphi'_u(U, X, \tau) - \varphi(U, X, \tau)]\, d\tau. \tag{60}$$

In (60) U and X denote the functions (59) at $t = \tau$. If the parameter x_0 runs through the values from $-\infty$ to ∞ and α varies from 0 to 1, then the curve $x = X(t, x_0, u_0(x_0), \alpha)$ is continuous in the plane of the variables x and x_0; $\tilde{\Phi}(t, x_0, \alpha)$ varies in a continuous manner along this continuous curve. Therefore, by repeating the previous computations, we fird that also in the case of discontinuous initial function $u_0(x)$ formulas (57) and (58) for the generalized solution remain valid.

These formulas now have the appearance

$$u(x, t) = U(t, \xi(x, t), u_0(\xi(x, t)), \alpha(x, t)),$$

where $\xi(x, t)$ and $\alpha(x, t)$ are the values of the parameters x_0 and α for which

$$\Phi(x, t) = \tilde{\Phi}(t, \xi(x, t), \alpha(x, t))$$

is the least value of the function $\tilde{\Phi}(t, x_0, \alpha)$ over the parameters x_0 and α subject to the condition $x = X(t, x_0, u_0(x_0), \alpha)$.

For an inhomogeneous conservation law

$$\frac{\partial u}{\partial t} + \frac{\partial}{\partial x}\varphi(u, x, t) = f(u, x, t) \tag{61}$$

with the initial condition (2) formulas (14) and (54) define a generalized solution as before in the case where

$$f(u, x, t) = f_1(x, t) + f_2(t)u. \tag{62}$$

Now $U(t, x_0, u_0)$, $X(t, x_0, u_0)$, $\bar{U}$ and $\bar{X}$ denote a solution of the characteristic system for equation (61):

$$\frac{dx}{dt} = \varphi_u'(u, x, t), \qquad \frac{du}{dt} = f(u, x, t) - \varphi_x'(u, x, t), \tag{63}$$

and $\tilde{\Phi}(t, x_0)$ is a solution of the "strip condition" for (61):

$$\frac{d\tilde{\Phi}}{dt} = U\varphi_u'(U, X, t) - \varphi(U, X, t) + F(x, t) + f_2(t)\tilde{\Phi}, \tag{64}$$

where $F_x'(x, t) = f_1(x, t)$. In order to see this, we transform (61) to the form

$$\frac{\partial v}{\partial t} + \frac{\partial}{\partial x}[\bar{\varphi}(v, x, t)] = 0, \tag{65}$$

where

$$v = u \exp\left\{-\int_0^t f_2(\tau)\, d\tau\right\},$$

$$\bar{\varphi}(v, x, t) = \exp\left\{-\int_0^t f_2(\tau)\, d\tau\right\}\left[\varphi\left(v\cdot\exp\left\{\int_0^t f_2(\tau)\, d\tau\right\}, x, t\right) - F(x, t)\right].$$

The method of constructing a generalized solution of the Cauchy problem presented above is applicable to (65), and, in particular, (14) and (54) are valid here. Returning again to the variable $u = v\exp(\int_0^t f_2(\tau)\, d\tau)$, we find that the latter is given by (14). This solution satisfies the integral conservation law

$$\oint_C u\, dx - \varphi(u, x, t)\, dt + \iint_{G_C} [f_1(x, t) + f_2(t)u]\, dx\, dt = 0,$$

which follows from (61) under the condition (62).

4. *The Cauchy problem for an inhomogeneous conservation law.* The Cauchy problem for the inhomogeneous conservation law

$$\frac{\partial u}{\partial t} + \frac{\partial\varphi(u, x, t)}{\partial x} = f(u, x, t) \qquad (\varphi_{uu}''(u, x, t) > 0) \tag{1}$$

with the initial condition

$$u(x, 0) = u_0(x) \tag{2}$$

was first considered by Tihonov and Samarskiĭ (1954) assuming that φ and f are twice continuously differentiable and the initial function $u_0(x)$ is piecewise continuous and piecewise differentiable. Their method may be called the *method of integrating the Hugoniot condition.*

The solution of the characteristic system

$$\frac{dx}{dt} = \varphi_u'(u, x, t), \qquad \frac{du}{dt} = f(u, x, t) - \varphi_x'(u, x, t), \tag{3}$$

satisfying the initial conditions

$$x(0) = x_0, \qquad u(0) = u_0, \tag{4}$$

we denote, as before, by

$$x = X(t, x_0, u_0), \qquad u = U(t, x_0, u_0). \tag{5}$$

We shall assume that X and U remain bounded in the region of variation of t, x_0 and u_0.

It suffices to consider the case in which the initial function $u_0(x)$ is defined on a finite segment $|x| \leq a$ and to solve problem (1), (2) in the domain of determinacy of this segment. We choose a so small that the initial function $u_0(x)$ has a single point of discontinuity of first kind on the segment $|x| \leq a$ (which with no loss of generality may be taken to be $x = 0$). We consider two possible cases.

1) Suppose that $u_0(-0) < u_0(+0)$. Since $\varphi_{uu}'' > 0$, this implies that

$$\varphi_u'(u_0(-0), 0, 0) < \varphi_u'(u_0(+0), 0, 0). \tag{6}$$

Through the point (0, 0) we draw the two characteristics $x = x^-(t)$ and $x = x^+(t)$ defined by

$$x^-(t) = X(t, 0, u_0(-0)), \qquad x^+(t) = X(t, 0, u_0(+0)). \tag{7}$$

It follows from (6) that at least for sufficiently small $t > 0$ the condition

$$x^-(t) < x^+(t) \tag{8}$$

is satisfied; and if it is assumed, as previously, that for the characteristic system (3) the boundary value problem (2.3.8) is uniquely solvable, then inequalities (8) are satisfied for all $t > 0$.

Figure 4.20 shows the domain of determinacy of the segment $|x| \leq a$ and the characteristics (7). According to Chapter 1, in zones I and II (Figure 4.20) for

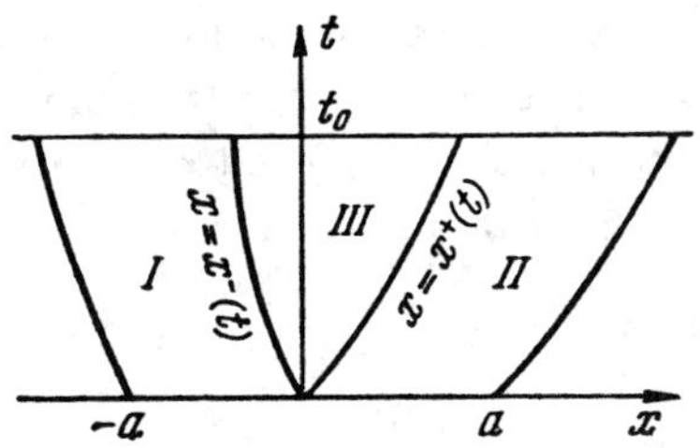

FIGURE 4.20

sufficiently small $t_0 > 0$ there exists a unique classical solution of (1) assuming the initial values (2). This solution is given by

$$u(X(t, x_0, u_0(x_0)), t) = U(t, x_0, u_0(x_0)) \tag{9}$$

implicitly by means of the parameter x_0 or, if we succeed in solving the relation

$$x = X(t, x_0, u_0(x_0)) \tag{10}$$

for x_0,

$$x_0 = x_0(x, t), \tag{11}$$

by the explicit formula

$$u(x, t) = U(t, x_0(x, t), u_0(x_0(x, t))). \tag{12}$$

The formulas (9) and (12) define the solution of the Cauchy problem in zones I and II. It remains to determine the solution in zone III.

Through the point (0, 0) we draw the characteristics $x = X^\alpha$ defined by

$$\left.\begin{aligned} x = X^\alpha = X(t, 0, \alpha u_0(-0) + 1 - \alpha)u_0(+0)), \quad 0 \le \alpha \le 1, \\ X^0 = x^+(t), \qquad X^1 = x^-(t). \end{aligned}\right\} \tag{13}$$

The equation

$$x = X^\alpha \tag{14}$$

is uniquely solvable in zone III for the parameter α:

$$\alpha = \alpha(x, t). \tag{15}$$

Therefore, the solution $u(x, t)$ is given in zone III by

$$u(X^\alpha, t) = U(t, 0, \alpha u_0(-0) + (1 - \alpha)u_0(+0)) \tag{16}$$

or, if (15) is known, by

$$u(x, t) = U(t, 0, \alpha(x, t)u_0(-0) + [1 - \alpha(x, t)]u_0(+0)). \tag{17}$$

The formulas (9) and (16) determine a solution of the Cauchy problem (1), (2) in zones I, II, and III which is continuous for $t > 0$, possesses discontinuities of the first derivatives on the lines $x = x^-(t)$ and $x = x^+(t)$, and has a singularity at (0, 0). To determine the character of this singularity we observe that

$$\lim_{t\to 0} u(X^\alpha(t), t) = u_0^\alpha = \alpha u_0(-0) + (1 - \alpha)u_0(+0), \tag{18}$$

$$\lim_{t\to 0} \frac{dX^\alpha(t)}{dt} = \lim_{t\to 0} \varphi_u'(u(X^\alpha, t), X^\alpha, t) = \varphi_u'(u_0^\alpha, 0, 0). \tag{19}$$

This implies that in zone III the solution (17) has a singularity of the form

$$u(x, t) = g\left(\frac{x}{t}\right) + O(t), \qquad g' = \frac{1}{\varphi_{uu}''(u, 0, 0)}. \tag{20}$$

In gas dynamics a singularity of this type is called a *centered rarefaction wave*. Thus, if (6) is satisfied, then there exists a continuous solution given by (9) and (16).

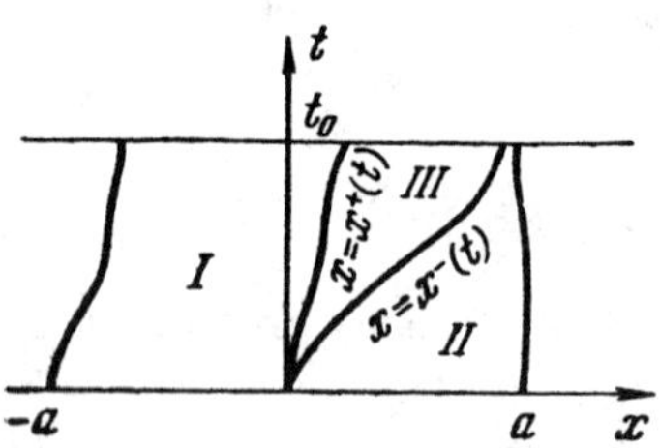

FIGURE 4.21

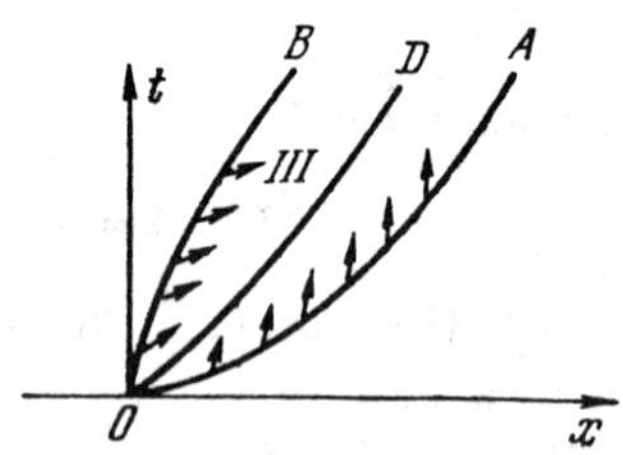

FIGURE 4.22

2) We now consider the second case in which $u_0(-0) > u_0(+0)$, i.e.,

$$\varphi_u'(u_0(-0), 0, 0) > \varphi_u'(u_0(+0), 0, 0). \tag{21}$$

In this case (8) becomes the opposite, and zones I and II overlap one another (Figure 4.21). We call the intersection of zones I and II *zone* III (Figure 4.21). In this case (9) defines a function $u(x, t)$ in zone III twice: one function, which we denote by $u^-(x, t)$, is defined for $x_0 < 0$, and the other, $u^+(x, t)$, for $x_0 > 0$. In this case the solution $u(x, t)$ is discontinuous. We shall assume that through the point $O(0, 0)$ in zone III there passes a line of discontinuity OD (Figure 4.22) with an equation which we assume given in the form $x = x(t)$, $x(0) = 0$;

to the left of the line of discontinuity $u(x, t) = u^-(x, t)$, while to the right of it $u(x, t) = u^+(x, t)$.

In this case on the line of discontinuity $x = x(t)$ the Hugoniot conditions (§1) must be satisfied; they can be written in the form

$$\frac{dx(t)}{dt} = D = \frac{\varphi(u^+(x, t), x, t) - \varphi(u^-(x, t), x, t)}{u^+(x, t) - u^-(x, t)} = A(x, t). \tag{22}$$

According to the existence theorem established in Chapter 1, $u^-(x, t)$ and $u^+(x, t)$ possess bounded first derivatives for sufficiently small t_0. Therefore, the function $A(x, t)$ contained in (22) has a bounded first derivative.

We consider some point $\mathscr{P}$ on the line OB. Let $D_{\mathscr{P}}$ denote the slope of the line OB relative to the axis $t = 0$ at the point $\mathscr{P}$, i.e.,

$$D_{\mathscr{P}} = \left.\frac{dx^+(t)}{dt}\right|_{t=\tau_{\mathscr{P}}}.$$

Through $\mathscr{P}$ there passes one characteristic $x = X(t, x_0, u_0(x_0))$ corresponding to the value $x_0 < 0$. From this we conclude that

$$\begin{aligned}\varphi'_u(u^-(x_{\mathscr{P}}, t_{\mathscr{P}}), x_{\mathscr{P}}, t_{\mathscr{P}}) &= X'_t(t_{\mathscr{P}}, x_0^-, u_0(x_0^-)) \\ &> \varphi'_u(u^+(x_{\mathscr{P}}, t_{\mathscr{P}}), x_{\mathscr{P}}, t_{\mathscr{P}}) = D_{\mathscr{P}} \\ &= X'_t(t_{\mathscr{P}}, 0, u_0(+0)).\end{aligned} \tag{23}$$

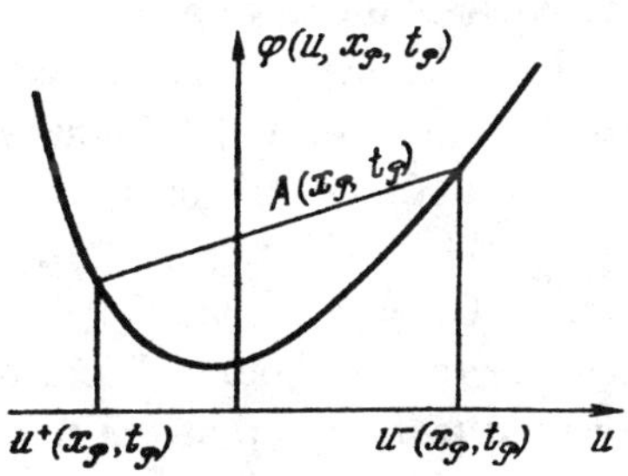

FIGURE 4.23

Figure 4.23 shows the graph of the relation $\varphi = \varphi(u, x_{\mathscr{P}}, t_{\mathscr{P}})$, taking into that $\varphi''_{uu} > 0$. From (23) we conclude that $u^-(x_{\mathscr{P}}, t_{\mathscr{P}}) > u^+(x_{\mathscr{P}}, t_{\mathscr{P}})$. The quantity $A(x_{\mathscr{P}}, t_{\mathscr{P}})$ is obviously the slope of the chord joining the points $\{u^+(x_{\mathscr{P}}, r_{\mathscr{P}}), \varphi(u^+(x_{\mathscr{P}}, t_{\mathscr{P}}), x_{\mathscr{P}}, t_{\mathscr{P}})\}$ and $\{u^-(x_{\mathscr{P}}, t_{\mathscr{P}}), \varphi(u^-(x_{\mathscr{P}}, t_{\mathscr{P}}), x_{\mathscr{P}}, t_{\mathscr{P}})\}$. It is evident from Figure 4.23 that the following inequality is a consequence of (23):

$$A(x_{\mathscr{P}}, t_{\mathscr{P}}) > \varphi'_u(u^+(x_{\mathscr{P}}, t_{\mathscr{P}}), x_{\mathscr{P}}, t_{\mathscr{P}}) = D_{\mathscr{P}}. \tag{24}$$

Thus, on OB we have $A(x_{\mathscr{P}}, t_{\mathscr{P}}) > D_{\mathscr{P}}$, i.e., the speed of the integral curves of the differential equation (22) on the line OB is greater than the speed of the line OB. More briefly, the direction field of equation (22) on OB has the form shown in Figure 4.22. Since in zone III, as we mentioned above, $A(x, t)$ is continuously differentiable in all arguments, it follows that there exists a unique integral curve OD of the ordinary differential equation (22) lying entirely in zone III.

Having determined the line OD, we define the solution in zones I, II and III:

$$u(x, t) = \begin{cases} u^-(x, t) \text{ to the left of } OD, \\ u^+(x, t) \text{ to the right of } OD. \end{cases} \tag{25}$$

This solution is continuous everywhere except on OD and possesses a bounded first derivative. On OD

$$u(x - 0, t) > u(x + 0, t),$$

i.e., this solution satisfies the stability condition. On the line of discontinuity OD the Hugoniot condition is satisfied because of (22); therefore, (25) defines a stable generalized solution of the Cauchy problem (1), (2).

Thus, in the case where the initial function has an isolated discontinuity of first kind we construct in this way a generalized solution of the Cauchy problem in some neighborhood of the point of discontinuity. If the initial function has several points of discontinuity, then, by decompositing the initial segments into parts, we reduce the problem to the case considered.

Finally, we note one further circumstance. The quantity t_0 by which we have in the foregoing restricted the domain of determinacy is bounded above by the condition that $u^-(x, t)$ and $u^+(x, t)$ must have bounded first derivatives in zones I and II.

We have seen, however, in Chapter 1 that the derivatives of solutions of quasilinear equations need not remain bounded and may increase to infinity in absolute value. An essential feature is that the most characteristic case of the behavior of solutions of quasilinear equations is that in which the derivatives of a solution become infinite at some point, while the solution itself remains continuous. Thus, in order that it be possible to apply successively, step-by-step, the method presented above of explicitly distinguishing singularities, it is still necessary to consider the case where the initial function $u_0(x)$ has an unbounded derivative at $x = 0$, while it itself remains continuous (Figure 4.24). Here it suffices to consider the case where in a neighborhood of $x = 0$

$$\frac{\partial u_0(x)}{\partial x} < 0, \tag{26}$$

since the derivative $p = \partial u/\partial x$ satisfies the equation

$$\frac{\partial p}{\partial t} + \varphi_u'(u, x, t)\frac{\partial p}{\partial x} = -\varphi_{uu}''p^2 + [f_u' - \varphi_{ux}'']p + f_x' \tag{27}$$

and remains bounded above: $p(x, t) < A$ ($\varphi_{uu}'' > 0$) if the derivative of $u_0(x)$ is bounded.

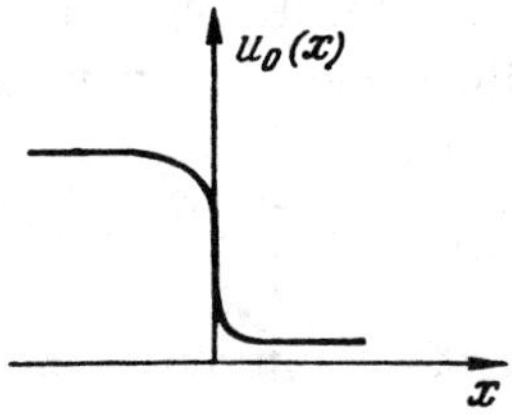

FIGURE 4.24

Thus, we assume that for $|x| \leq a$ condition (26) is satisfied and

$$\frac{\partial u_0(x)}{\partial x} \to -\infty \quad \text{as } x \to 0. \tag{28}$$

In this case we limit ourselves to remarks only, since the construction of the solution in general outline is similar to the case (21). The main difference of this case from the preceding case is that the Cauchy problem for equation (1) considered separately on $[-a, 0]$ and $[0, a]$ has a solution possessing a bounded derivative for any $t_0 > 0$. Therefore, the line OA (Figure 4.22) is now to be understood as the envelope of the family of characteristics $x = X(t, x_0, u(x_0))$, $x_0 < 0$, i.e., characteristics starting to the left of the point O; the line OB is understood to be the envelope of the family of characteristics $x = X(t, x_0, u_0(x_0))$, $x_0 > 0$.

In this case the lines OA and OB are tangent to one another at $t = 0$, the solutions $u^-(x, t)$ and $u^+(x, t)$ have unbounded derivatives on OA and OB respectively, and otherwise the picture is altogether similar to that considered above. Thus, the direction field for the differential equation (22) has the form shown in Figure 4.22, while the uniqueness of the line of discontinuity OD under the conditions of unboundedness of the derivatives of $A(x, t)$ follows from the tangency of the lines OA and OB at $t = 0$. Thus, by applying the method of explicit integration of the Hugoniot condition, we can determine the solution $u(x, t)$ of the Cauchy problem (1), (2) in any region of interest to us in which the number of lines of discontinuity remains bounded, as is most frequently the case in practical problems.

We note further that in the case where $f(u, x, t) \equiv 0$, after introducing the potential of a generalized solution $\Phi(x, t)$, we can write

$$\left.\begin{aligned} u^-(x, t) &= \frac{\partial \Phi^-(x, t)}{\partial x}, & u^+(x, t) &= \frac{\partial \Phi^+(x, t)}{\partial x}, \\ \varphi(u^-(x,t), x, t) &= -\frac{\partial \Phi^-(x, t)}{\partial t}, & \varphi(u^+(x,t), x, t) &= -\frac{\partial \Phi^+(x,t)}{\partial t}, \end{aligned}\right\} \tag{29}$$

where $\Phi^-(x, t)$ and $\Phi^+(x, t)$ are uniquely determined in the region AOB by the "strip equation". Therefore, the Hugoniot condition takes the form

$$\frac{\partial}{\partial t}[\Phi^+(x, t) - \Phi^-(x, t)]\, dt + \frac{\partial}{\partial x}[\Phi^+(x, t) - \Phi^-(x, t)]\, dx = 0, \tag{30}$$

whence it follows that the line of discontinuity OD $x = x(t)$ is a line where

$$\Phi^+(x, t) = \Phi^-(x, t). \tag{31}$$

This shows that in the case of piecewise smooth solutions $u(x, t)$ the method of explicitly distinguishing lines of discontinuity is equivalent for homogeneous conservation laws to the analytic method of finding a solution presented in subsection 3.

In a certain sense the Cauchy problem (1), (2) can be reduced to the problem for a homogeneous conservation law by using the method of successive approximations. We set $\overset{(0)}{u}(x, t) = u_0(x)$. Assuming that the approximation $\overset{(s-1)}{u}(x, t)$ is known, we define $\overset{(s)}{u}(x, t)$ as a solution of the Cauchy problem

$$\frac{\partial \overset{(s)}{u}}{\partial t} + \frac{\partial}{\partial x}\varphi(\overset{(s)}{u}, x, t) = f(\overset{(s-1)}{u}(x, t), x, t), \tag{32}$$

$$\overset{(s)}{u}(x, 0) = u_0(x). \tag{33}$$

Introducing the function

$$\overset{(s)}{F}(x, t) = \int_0^x f(\overset{(s-1)}{u}(\xi, t), \xi, t)\, d\xi, \tag{34}$$

we write (32) in the form

$$\frac{\partial \overset{(s)}{u}}{\partial t} + \frac{\partial}{\partial x}[\varphi(\overset{(s)}{u}, x, t) - \overset{(s)}{F}(x, t)] = 0. \tag{35}$$

The theory of constructing discontinuous solutions developed in subsection 3 is applicable to (35), but we must consider some details.

The characteristic system for (35) can be written in the form

$$\frac{d\overset{(s)}{X}}{dt} = \varphi_u'(\overset{(s)}{U}, \overset{(s)}{X}, t), \quad \frac{d\overset{(s)}{U}}{dt} = -\varphi_x(\overset{(s)}{U}, \overset{(s)}{X}, t) + f(\overset{(s-1)}{u}(\overset{(s)}{X}, t), \overset{(s)}{X}, t), \tag{36}$$

where $f(\overset{(s-1)}{u}(x, t), x, t)$ is a discontinuous function of x and t. At points of discontinuity of f we impose the condition that $\overset{(s)}{x}$ and $\overset{(s)}{U}$ be continuous.

In analogy to the foregoing the solution of problem (35) with the initial condition (33) is given by

$$\overset{(s)}{\Phi}(x, t) = \min \overset{(s)}{\tilde{\Phi}}(t, x_0), \tag{37}$$

where $\overset{(s)}{\tilde{\Phi}}(t, x_0)$ is defined by the quadrature

$$\overset{(s)}{\Phi}(t, x_0) = \Phi_0(x_0) + \int_0^t [\overset{(s)}{U}\varphi_u'(\overset{(s)}{U}, \overset{(s)}{X}, \tau) - \varphi(\overset{(s)}{U}, \overset{(s)}{X}, \tau) + \overset{(s)}{F}(\overset{(s)}{X}, \tau)]\, d\tau \tag{38}$$

(here $\overset{(s)}{U} = \overset{(s)}{U}(\tau, x_0, u_0, (x_0))$.

The sequence $\overset{(s)}{\Phi}(x, t)$ converges uniformly in any bounded region of the variables x and t. Therefore, passing to the limit in (37), we obtain a potential $\Phi(x, t)$ of a generalized solution of the Cauchy problem (1), (2).

5. ***Uniqueness of a generalized solution under the condition*** $\varphi_{uu}'' > 0$. We now show that *a generalized solution of the Cauchy problem*

$$\frac{\partial u}{\partial t} + \frac{\partial \varphi(u, x, t)}{\partial x} = f(u, x, t) \qquad (\varphi_{uu}'' > 0), \tag{1}$$

$$u(x, 0) = u_0(x), \tag{2}$$

satisfying the stability condition

$$u(x - 0, t) \geq u(x + 0, t), \tag{3}$$

is unique.

We shall prove this theorem in the class of piecewise continuous and piecewise differentiable solutions. We suppose that for[5] $t > 0$ problem (1), (2) has two bounded, piecewise continuous and piecewise differentiable solutions, $u(x, t)$ and $\bar{u}(x, t)$, each of which satisfies (1) away from lines of discontinuity and satisfies the stability condition (3) on lines of discontinuity.

Let $\Phi(x, t)$ and $\bar{\Phi}(x, t)$ be the potentials corresponding to these two solutions.

[5] We consider solutions which may have singularities of the type $g((x - x_0)/t, t)$ and are not continuous at $t = 0$.

These functions are continuous, and except on the lines of discontinuity of u and $\bar{u}$ they satisfy the equation

$$\frac{\partial \Phi}{\partial t} + \varphi\left(\frac{\partial \Phi}{\partial x}, x, t\right) = \int_0^x f\left(\frac{\partial \Phi}{\partial \xi}, \xi, t\right) d\xi, \tag{4}$$

while for $t = 0$ they satisfy the initial condition

$$\Phi(x, 0) = \Phi_0(x) = \int_0^x u_0(\xi)\, d\xi. \tag{5}$$

Except on the lines of discontinuity of u and $\bar{u}$ the difference $v = \Phi - \bar{\Phi}$ satisfies the equation

$$\begin{aligned} \frac{\partial v}{\partial t} + A(x, t)\frac{\partial v}{\partial x} &= \int_0^x B(\xi, t)\frac{\partial v(\xi, t)}{\partial \xi} d\xi \\ &= -\int_0^x \frac{\partial B(\xi, t)}{\partial \xi} v(\xi, t)\, d\xi + B(x, t)v(x, t) - B(0, t)v(0, t) \\ &\quad + \sum_i v(x_i(t), t)[B(x_i(t) + 0, t) - B(x_i(t) - 0, t)]. \end{aligned} \tag{6}$$

The summation in (6) goes over all lines of discontinuity $x = x_i(t)$ of the solutions $u(x, t)$ and $\bar{u}(x, t)$ such that $0 < x_i(t) < x$, while

$$A(x, t) = \begin{cases} \dfrac{\varphi(u(x, t), x, t) - \varphi(\bar{u}(x, t), x, t)}{u(x, t) - \bar{u}(x, t)} & \text{for } u \neq \bar{u}, \\ \varphi'_u(u(x, t), x, t) \quad \text{for } u = \bar{u}, \end{cases} \tag{7}$$

$$B(x, t) = \begin{cases} \dfrac{f(u(x, t), x, t) - f(\bar{u}(x, t), x, t)}{u(x, t) - \bar{u}(x, t)} & \text{for } u = \bar{u}, \\ f'_u(u(x, t), x, t) \quad \text{for } u = \bar{u} \end{cases} \tag{8}$$

and

$$v(x, 0) \equiv 0. \tag{9}$$

Let

$$\left.\begin{aligned} &|u(x, t)| < M_0, \qquad |\bar{u}(x, t)| < M_0, \\ &|\varphi'_u(u, x, t)| \leq M_1 \quad \text{for } |u| \leq M_0. \end{aligned}\right\} \tag{10}$$

We shall prove that $v(x, t) \equiv 0$ for any $a > 0$ in the trapezoid

$$x + a \geq M_1 t, \qquad x - a \leq -M_1 t, \quad 0 \leq t \leq T. \tag{11}$$

Under the assumptions made regarding u and $\bar{u}$ and from the conditions of subsection 4 on φ and f, the function $B(x, t)$ obviously has bounded variation in the variable x, while the number of lines of discontinuity $x = x_i(t)$ of the functions u and $\bar{u}$ (they are also lines of discontinuity of $A(x, t)$ and $B(x, t)$) is finite in the trapezoid (11). Therefore, the right side (6) can be estimated by $M_2 V(t)$, where

$$V(t) = \max |v(\xi\ \tau)|, \tag{12}$$

and the maximum in formula (12) is taken over the intersection of the trapezoid (11) with the strip $0 \leq \tau \leq t$.

Thus, in the trapezoid (11) the function $v(x, t)$ satisfies the condition

$$\left| \frac{\partial v}{\partial t} + A(x, t) \frac{\partial v}{\partial x} \right| < M_2 V(t), \tag{13}$$

while for $t = 0$ it satisfies (9).

We consider the ordinary differential equation

$$\frac{dx}{dt} = A(x, t), \tag{14}$$

in which the right side is discontinuous on the lines $x = x_i(t)$. An integral curve of (14) is a continuous curve $x = x(t)$ which satisfies (14) at all points of continuity of $A(x, t)$.

We suppose that through each point of the trapezoid (11) there passes an integral curve of (14) which intersects the axis $t = 0$ on the base of the trapezoid $-a \leq x \leq a$. Integrating (13) along the integral curve passing through (x, t) and using (9), we obtain

$$|v(x, t)| \leq \int_0^t M_2 V(\tau)\, d\tau. \tag{15}$$

This implies that

$$V(t) \leq M_2 \int_0^t V(\tau)\, d\tau \tag{16}$$

and, on the basis of Lemma 1 of Chapter 1, §6.5,

$$v(x, t) \equiv 0. \tag{17}$$

Since $u - \bar{u} = \partial v(x, t)/\partial x$, the proof of our theorem follows from (17).

Thus, the proof of the theorem reduces to a proof of the following fact: through any point (x, t) of the trapezoid (11) there passes an integral curve of equation (14) intersecting the axis $t = 0$. We shall show that if u and $\bar{u}$ satisfy the stability condition (3), then this is actually the case. We observe that through a point (ξ, τ) that is a point of continuity of $A(x, t)$ there passes an integral curve of (14) defined for $t \leq \tau$. If this integral curve does not intersect a line of discontinuity $x = x_i(t)$ for $t < \tau$, then it intersects the axis $t = 0$ on the segment $-a \leq x \leq a$, since $|A(x, t)| \leq M_1$. Therefore, if through any point (ξ, τ) lying on a line of discontinuity $x = x_i(t)$ there passes an integral curve of (14) defined for $t < \tau$, then through any point of the trapezoid (11) there passes an integral curve intersecting the line $t = 0$.

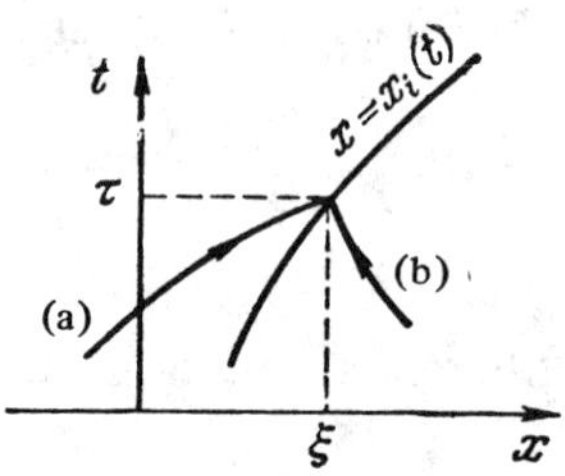

FIGURE 4.25

We consider the case where the point (ξ, τ) lies on a line of discontinuity $x = x_i(t)$ of the function $A(x, t)$ (Figure 4.25). According to our assumptions, at the point (ξ, τ) right and left limit values $A(\xi - 0, \tau)$ and $A(\xi + 0, \tau)$ exist. If

$$A(\xi - 0, \tau) > A(\xi + 0, \tau), \tag{18}$$

then through the point (ξ, τ) there necessarily passes an integral curve of (14) defined for $t < \tau$. Indeed, if we set $D_i = x_i'(t)$, then in this case one of the following two inequalities is satisfied:

$$\text{a)} \quad A(\xi - 0, \tau) > D_i, \qquad \text{b)} \quad A(\xi + 0, \tau) < D_i. \tag{19}$$

Figure 4.25 shows the directions of the integral curves of (14) in cases a) and b). It is evident from these cases that through (ξ, τ) there passes an integral curve defined for $t < \tau$.

We have thus reduced the proof of the theorem to establishing the inequality (18). Since (ξ, τ) is a point of discontinuity of $A(x, t)$, at this point either $u(x, t)$ or $\bar{u}(x, t)$ is discontinuous, or they are discontinuous simultaneously. We may therefore assume that

$$u(\xi - 0, \tau) > u(\xi + 0, \tau), \qquad \bar{u}(\xi - 0, \tau) \geq \bar{u}(\xi + 0, \tau). \tag{20}$$

Figure 4.26 shows the graph of $\varphi(u, \xi, \tau)$ for fixed ξ and τ. Since $\varphi''_{uu} > 0$, this curve is convex upward.

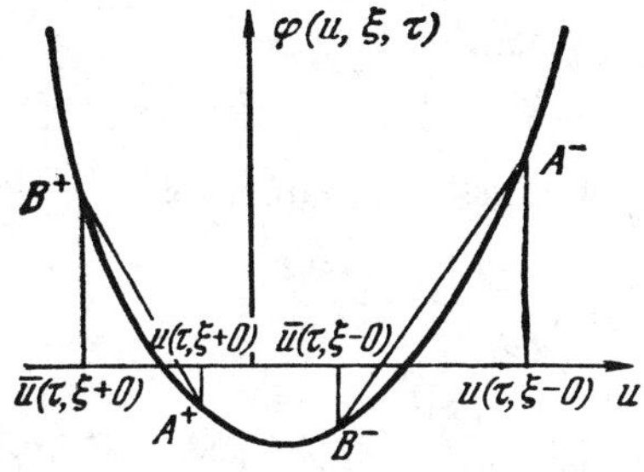

FIGURE 4.26

According to (7), the quantity $A(\xi - 0, \tau)$ is equal to the slope of the chord joining the points A^- and B^-, while $A(\xi + 0, \tau)$ is the slope of the chord A^+B^+. It follows from (20) that each end point of the interval $[\bar{u}(\xi - 0, \tau), u(\xi - 0, \tau)]$ lies to the right of the corresponding end points of the interval $[\bar{u}(\xi + 0, \tau), u(\xi + 0, \tau)]$. For convex curves ($\varphi''_{uu} > 0$) this implies that the slope of A^-B^- is greater than the slope of A^+B^+. This can be seen from Figure 4.26.

Thus, at each point of discontinuity of $A(x, t)$ the condition (18) is satisfied. Hence, through each point of the trapezoid (11) there passes at least one integral curve of (14) intersecting the axis $t = 0$. This means that (17) holds at this point, and $u(x, t) \equiv \bar{u}(x, t)$.

We consider the question of continuous dependence of generalized solutions on the initial data. Suppose that $u(x, t)$ and $\bar{u}(x, t)$ are piecewise smooth, stable generalized solutions of quasilinear equations defined by the conditions

$$\frac{\partial u}{\partial t} + \frac{\partial \varphi(u, x, t)}{dx} = f(u, x, t), \quad u(x, 0) = u_0(x), \quad \varphi''_{uu} > 0, \tag{21}$$

$$\frac{\partial \bar{u}}{\partial t} + \frac{\partial \bar{\varphi}(\bar{u}, x, t)}{\partial x} = \bar{f}(\bar{u}, x, t), \quad \bar{u}(x, 0) = \bar{u}_0(x), \quad \bar{\varphi}''_{uu} > 0. \tag{22}$$

We shall assume that φ, $\bar{\varphi}$, f, $\bar{f}$, u_0 and $\bar{u}_0$ satisfy the conditions imposed in subsections 3 and 4.

To the solutions u and $\bar{u}$ there correspond potentials Φ and $\bar{\Phi}$ satisfying the equations

$$\frac{\partial \Phi}{\partial t} + \varphi\left(\frac{\partial \Phi}{\partial x}, x, t\right) = \int_0^x f\left(\frac{\partial \Phi}{\partial \xi}, \xi, t\right) d\xi, \tag{23}$$

$$\frac{\partial \bar{\Phi}}{\partial t} + \bar{\varphi}\left(\frac{\partial \bar{\Phi}}{\partial x}, x, t\right) = \int_0^x \bar{f}\left(\frac{\partial \bar{\Phi}}{\partial \xi}. \xi, t\right) d\xi \tag{24}$$

and the initial conditions

$$\Phi(x, 0) = \Phi_0(x) = \int_0^x u_0(\xi)\, d\xi, \quad \bar{\Phi}(x, 0) = \bar{\Phi}_0(x) = \int_0^x \bar{u}_0(\xi)\, d\xi. \tag{25}$$

The difference $v = \Phi - \bar{\Phi}$ satisfies the equation

$$\begin{aligned}\frac{\partial v}{\partial t} &+ A(x, t)\frac{\partial v}{\partial x}\\ &= \int_0^x B(\xi, t)\frac{\partial v}{\partial \xi}\, d\xi + [\bar{\varphi}(\bar{u}(x, t), x, t) - \varphi(\bar{u}(x, t), x, t)]\\ &+ \int_0^x [f(\bar{u}(\xi, t), \xi, t)] - \bar{f}(\bar{u}(\xi, t), \xi, t)]\, d\xi\end{aligned} \tag{26}$$

and the initial condition

$$v(x, 0) = v_0(x) = \Phi_0(x) - \bar{\Phi}_0(x) = \int_0^x [u_0(\xi) - \bar{u}_0(\xi)]\, d\xi, \tag{27}$$

while A and B are defined by the (7) and (8).

Suppose that in the trapezoid (11) the conditions $|u|$, $|\bar{u}| \leq M_0$ are satisfied, and suppose that for $|u| \leq M_0$

$$|\varphi_u'(u, x, t)| \leq M_1, \qquad |\varphi(u, x, t) - \bar{\varphi}(u, x, t)| \leq \Delta\varphi,$$
$$|f(u, x, t) - \bar{f}(u, \dot{x}, t)| \leq \Delta f.$$

From (26) we then obtain

$$\left|\frac{\partial v}{\partial t} + A(x, t)\frac{\partial v}{\partial x}\right| \leq \left|\int_0^x B(\xi, t)\frac{\partial v}{\partial \xi}\, d\xi\right| + \Delta\varphi + a\Delta f.$$

In analogy to the foregoing, from this we get the new estimate

$$\left|\frac{\partial v}{\partial t} + A(x, t)\frac{\partial v}{\partial x}\right| \leq MV(t) + \Delta\varphi + a\Delta f, \tag{28}$$

where the constant M depends, in particular, on the number of lines of discontinuity of the solutions u and $\bar{u}$.

We assume that each of the solutions u and $\bar{u}$ satisfies the stability condition, i.e.,

$$u(x-0,t) \geq u(x+0,t), \qquad \bar{u}(x-0,t) \geq \bar{u}(x+0,t).$$

As before, this implies that $A(x-0,t) \geq A(x+0,t)$ at points of discontinuity of $A(x,t)$. Hence through each point of the trapezoid (11) there passes at least one integral curve of (14) intersecting the axis $t=0$. Integrating (28) along this curve and applying Lemma 1 of Chapter 1, §6.5, we obtain

$$|v(x,t)| \leq v_0 e^{Mt} + [\Delta\varphi + a\Delta f]\frac{e^{Mt}-1}{M} \tag{29}$$

where

$$v_0 = \max_{|x|\leq a} |\Phi_0(x) - \bar{\Phi}_0(x)|.$$

Recalling that

$$v(x,t) = \int_0^x [u(\xi,t) - \bar{u}(\xi,t)]\,d\xi,$$

we conclude that (29) establishes the continuous dependence of generalized solutions of quasilinear equations on the initial data of the Cauchy problem in the potential metric. Unfortunately (29) contains the constant M which depends on the number of lines of discontinuity of the solutions u and $\bar{u}$. In this regard we note that this quantity can be estimated in terms of the variation of the functions $u(x,t)$ and $\bar{u}(x,t)$, and the latter can be estimated in terms of the initial conditions.

For homogeneous conservation laws in which $f \equiv \bar{f} \equiv 0$ the estimate (29) simplifies to $|v(x,t)| \leq v_0 + \Delta\varphi t$. In particular if we consider only the dependence of solutions of the Cauchy problem for a single quasilinear equation on the initial data (i.e. we assume that $\Delta\varphi = 0$) then this estimate shows that the solutions constructed above satisfy the principle of continuous dependence (1.5.4) and (1.5.5) of Chapter 1.

There is a method of constructing discontinuous solutions related to the question of continuous dependence of generalized solutions on the initial data, which we call the "method of potential smoothing". For simplicity we consider a homogeneous conservation law

$$\frac{\partial u}{\partial t} + \frac{\partial \varphi(u,x,t)}{\partial x} = 0 \qquad \varphi''_{uu} > 0, \tag{30}$$

with the initial condition

$$u(x, 0) = u_0(x). \tag{31}$$

We assume that the function φ satisfies the previous requirements, and $u_0(x)$ is defined on the segment $|x| \leq a$, has a single point of discontinuity of first kind at $x = 0$, and at all remaining points of the segment possesses a bounded Lipschitz constant. We set $u^- = u_0(-0)$ and $u^+ = u_0(+0)$. As we have seen above, if $u^- \leq u^+$, then the solution of problem (30), (31) is continuous for $t \geq 0$ in some neighborhood $0 \leq t \leq T$ of the initial axis. In particular, it can be obtained as the limit of classical solutions $u_\delta(x, t)$ as $\delta \to 0$, where $u_\delta(x, t)$ is a solution of (30) with the initial condition

$$u_\delta(x, 0) = u_\delta^0(x); \tag{32}$$

here $u_\delta^0(x) \equiv u_0(x)$ for $|x| \geq \delta$ and $u_\delta^0(x)$ is monotone for $|x| \leq \delta$.

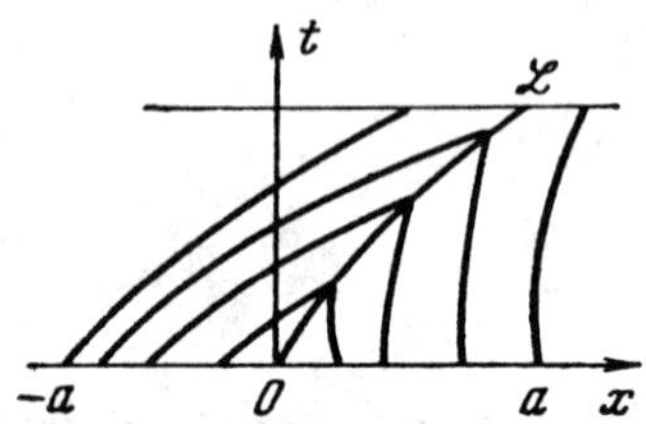

FIGURE 4.27

We therefore consider only the second case, where $u^- > u^+$. In this case for sufficiently small T the solution $u(x, t)$ has in the strip $0 \leq t \leq T$ a unique line of discontinuity $O\mathscr{L}$ issuing from the point $(0, 0)$ of the initial axis (Figure 4.27). Figure 4.27 shows the line of discontinuity $O\mathscr{L}$ and the characteristics of the solution $u(x, t)$. In place of problem (30), (31) we shall solve problem (30), (32), assuming that $u_\delta^0(x) \equiv u_0(x)$ for $|x| \geq \delta$, the function $u_\delta^0(x)$ is monotone for $|x| \leq \delta$ and possesses a Lipschitz constant on the segment $|x| \leq \delta$ not exceeding M/δ, and, finally, that

$$\int_{-\delta}^{\delta} [u_0(x) - u_\delta^0(x)]\, dx = 0. \tag{33}$$

We introduce the potentials of the solutions u and u_δ by the relations

$$\Phi(x, t) = \int_{(-a, 0)}^{(x, t)} u\, dx - \varphi(u, x, t)\, dx,$$

$$\Phi_\delta(x, t) = \int_{(-a, 0)}^{(x, t)} u_\delta\, dx - \varphi(u_\delta, x, t)\, dt.$$

From (33) it then follows that

$$\Phi(x, 0) \equiv \Phi_\delta(x, 0) \quad \text{for } |x| \geq \delta. \tag{34}$$

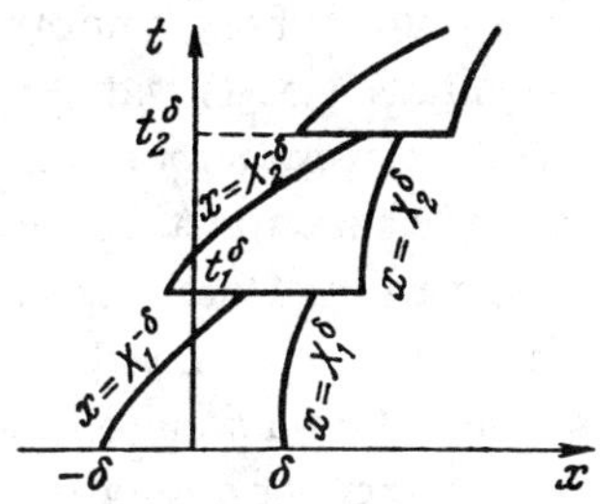

FIGURE 4.28

Under the conditions on $u_\delta^0(x)$ there exists a solution $u_\delta(x, t)$ in the broad sense of problem (30), (32) which is continuous and can be constructed by the classical method of characteristics for $0 \leq t \leq t_1^\delta$, where $t_1^\delta \sim 1/\delta$. We denote by $x = X_1^{-\delta}$ and $x = X_1^\delta$ the equations of the characteristics of this problem issuing from the points $x = -\delta$ and $x = \delta$ of the initial axis respectively (Figure 4.28). It is obvious in view of (34) that $u(x, t) \equiv u_\delta(x, t)$ and $\Phi(x, t) \equiv \Phi_\delta(x, t)$ for $[x - X_1^{-\delta}(t)]\,[x - X_\delta^1(t)] > 0$ and $0 \leq t \leq t_1^\delta$, i.e., $u(x, t)$ and $\Phi(x, t)$ coincide with $u_\delta(x, t)$ and $\Phi_\delta(x, t)$ outside the curvilinear trapezoid formed by the lines $t = 0$ and $t = t_1^\delta$, and the segments of the characteristics $x = X_1^{-\delta}$ and $x = X_1^\delta$.

Suppose that for $t = t_1^\delta$ the solution $u_\delta(x, t)$ has a singularity (an unbounded Lipschitz constant or a discontinuity of first kind) on the segment $[X_1^{-\delta}(t_1^\delta), X_1^\delta(t_1^\delta)]$ of the line $t = t_1^\delta$. It is obvious that

$$X_1^\delta(t_1^\delta) - X_1^{-\delta}(t_1^\delta) < 2\delta.$$

We choose a segment of the line $t = t_1^\delta$ of length 2δ entirely containing $[X_1^{-\delta}(t_1^\delta), X_1^\delta(t_1^\delta)]$, and we again the smooth the function $u_\delta(x, t_1^\delta)$ on this segment, i.e., we introduce a function $u_\delta^1(x, t_1^\delta)$ which satisfies the same conditions as $u_\delta^0(x)$, namely, $u_\delta^1(x, t_1^\delta) \equiv u_\delta(x, t_1^\delta)$ outside the smoothing segment and

$$\int_{a_1}^{b_1} [u_\delta(x, t_1^\delta) - u_\delta^1(x, t_1^\delta)]\, dx = 0,$$

where $[a_1, b_1]$ is the smoothing segment. As before, we require that $u_\delta^1(x, t_1^\delta)$ possess a Lipschitz constant bounded by the quantity M/δ and be monotone on the smoothing segment $[a_1, b_1]$. We seek a solution $u_\delta(x, t)$ for $t_1^\delta \leq t \leq t_2^\delta$, assuming that $u_\delta(x, t)$ satisfies (30) and the initial condition

$$u_\delta(x, t_1^\delta) = u_\delta^1(x, t_1^\delta).$$

To construct $u_\delta(x, t)$ in the strip $t_1^\delta \le t \le t_2^\delta$ we again apply the classical method of characteristics. Once more we find that outside the trapezoid formed by the lines $t = t_1^\delta$ and $t = t_2^\delta$ and segments of the characteristics $x = x_2^{-\delta}(t)$ and $x = X_2^\delta(t)$ issuing from the end points of the smoothing segment the solutions $u(x, t)$ and $u_\delta(x, t)$ and the potentials $\Phi(x, t)$ and $\Phi_\delta(x, t)$ coincide.

Continuing this process of successively smoothing and solving Cauchy problems by the classical method of characteristics, we reach the line $t = T$, i.e., we exhaust the region of interest to us. Here it is necessary to solve, approximately T/δ times, a Cauchy problem with smooth initial conditions.

As a result of this process, we obtain a solution $u_\delta(x, t)$ which is continuous everywhere with the exception of the smoothing segments on the lines $t = t_k^\delta$. The segments of characteristics issuing from the end points of the smoothing segments form a "band" containing the singularities and discontinuities of $u_\delta(x, t)$. As δ decreases, this "band" contracts to the line of discontinuity $O\mathscr{L}$.

Since the potentials $\Phi(x, t)$ and $\Phi_\delta(x, t)$ coincide outside the "band" and the solutions $u(x, t)$ and $u_\delta(x, t)$ are assumed bounded, it follows that

$$|\Phi(x, t) - \Phi_\delta(x, t)| < 2M\delta.$$

This implies the convergence of the "method of potential smoothing" as $\delta \to 0$ in the potential metric. It is clear that outside the "band" the solutions $u(x, t)$ and $u_\delta(x, t)$ coincide.

The method presented above makes it possible to approximately construct discontinuous solutions of quasilinear equations by approximating them by continuous solutions. However, the application of this method to the case of a system of quasilinear equations is of greatest interest. Unfortunately, definitive results have so far not been obtained in this manner.

6. *The asymptotic behavior of generalized solutions as* $t \to \infty$. Let $u(x, t)$ be a generalized solution of the homogeneous conservation law

$$\frac{\partial u}{\partial t} + \frac{\partial \varphi(u)}{\partial x} = 0, \qquad \varphi''_{uu}(u) > 0, \tag{1}$$

with coefficients which do not depend on the independent variables x and t. Suppose further that

$$u(x, 0) = u_0(x) = \begin{cases} u^- & \text{for } x < 0, \\ u^+ & \text{for } x > b > 0, \end{cases} \tag{2}$$

and that, in the segment $(0, b)$, $u_0(x)$ assumes any bounded values. We suppose that $u_0(x)$ is piecewise continuous. Thus,

$$|u_0(x)| \leq M, \qquad |u(x, t)| \leq M. \tag{3}$$

We set

$$\left.\begin{aligned} \xi &= \max_{|u|\leq M} |\varphi_u'(u)|, & A &= \max_{|u|\leq M} \varphi_{uu}''(u), \\ a &= \min_{|u|\leq M} \varphi_{uu}''(u), & 0 &< a \leq A. \end{aligned}\right\} \tag{4}$$

and

$$\int_0^x u_0(\xi)\, d\xi = \Phi_0(x). \tag{5}$$

To the solution $u(x, t)$ we assign the potential

$$\Phi(x, t) = \int_{(0,0)}^{(x,t)} u\, dx - \varphi(u)\, dt, \Phi(x, 0) = \Phi_0(x). \tag{6}$$

In view of conditions (2) and (3) we have

$$\left.\begin{aligned} \Phi_0(x) &= u^- x \quad \text{for } x \leq 0, \\ \Phi_0(x) &= \Phi_0(b) + (x - b)u^+ \quad \text{for } x \geq b, \\ |\Phi_0(x)| &\leq Mx. \end{aligned}\right\} \tag{7}$$

We shall study the behavior of the solution $u(x, t)$ as $t \to \infty$.

If the initial data (1) were prescribed only for $x < 0$ ($u_0(x) \equiv u^-$), then clearly the solution of the Cauchy problem (1). (2) would be defined only for $x < \varphi_u'(u^-)t$ and would coincide there with u^-. We shall estimate the width of the region to the left of the line $x = \varphi_u'(u^-)t$ in which the solution $u(x, t)$ of problem (1), (2) does not coincide with u^-, i.e., $u(x, t) \neq u^-$; here $u(x, t) \neq u^-$ because of characteristics issuing from the segment $[0, b]$. It is obvious that no characteristic $x = X(t, \tilde{x}_0, u_0(\tilde{x}_0))$ for $0 \leq \tilde{x}_0 \leq b$ can lie to the left of the line $x = -\xi t$ because of the boundedness of the initial function $u_0(x)$; however, this estimate of the domain of influence of the segment $[0, b]$ is too rough. Suppose that $u(x, t) \neq u^-$, $x - \varphi_u'(u^-)t < 0$, at the point (x, t) there arrives a characteristic issuing from a point $\tilde{x}_0$ of the segment $0 \leq \tilde{x}_0 \leq b$, and $u(x, t) = u_0(\tilde{x}_0)$ (Figure 4.29). For simplicity we set $u_0(\tilde{x}_0) = \tilde{u}$ and $\varphi_u'(u^-)t - x = h$. We have[6]

$$h = \varphi_u'(u^-)t - x, \tag{8}$$

[6] If $\tilde{x}_0$ is a point of discontinuity of the initial function $u_0(x)$, there u is understood to mean $u_0^\alpha(\tilde{x}_0)$ with an appropriate value of the parameter α.

$$x - \tilde{x}_0 = \varphi_u'(\tilde{u})t \quad \text{or} \quad h = t[\varphi_u'(u^-) - \varphi_u'(\tilde{u})] - \tilde{x}_0. \tag{9}$$

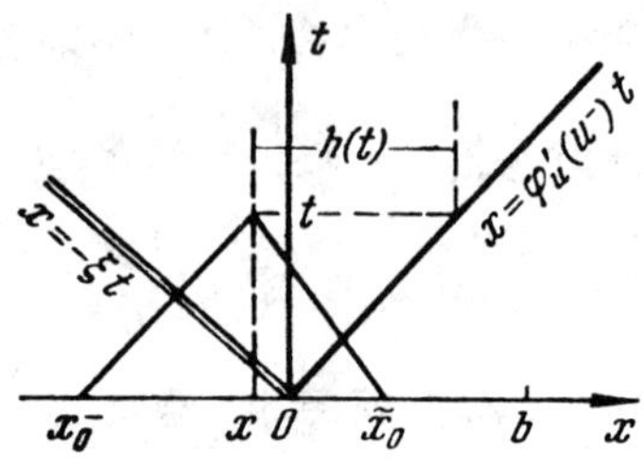

FIGURE 4.29

We compute $\tilde{\Phi}(t, \tilde{x}_0)$ by (2.3.46):

$$\tilde{\Phi}(t, \tilde{x}_0) = \Phi_0(\tilde{x}_0) + t[\tilde{u}\varphi_u'(\tilde{u}) - \varphi(\tilde{u})]. \tag{10}$$

We eliminate $t\tilde{u}\varphi_u'(\tilde{u})$ from (10) by means of (9):

$$t\tilde{u}\varphi_u'(\tilde{u}) = \tilde{u}t\varphi_u'(u^-) - \tilde{u}\tilde{x}_0 - \tilde{u}h, \tag{11}$$

$$\tilde{\Phi}(t, \tilde{x}_0) = \Phi_0(\tilde{x}_0) - \tilde{u}\tilde{x}_0 - \tilde{u}h + t[\tilde{u}\varphi_u'(u^-) - \varphi(\tilde{u})]. \tag{12}$$

On the other hand, at this same point (x, t) for $x - \varphi_u'(u^-)t < 0$ there arrives a characteristic $x = X(t, x_0^-, u^-)$ with $x_0^- < 0$ (Figure 4.29). This characteristic carries the value $u = u^-$ and the value of the potential $\tilde{\Phi}(t, x_0^-)$ to the point (x, t). Thus,

$$x - x_0^- = \varphi_u'(u^-)t, \quad \text{or} \quad x_0^- = -h, \tag{13}$$

while by (2.3.46)

$$\begin{aligned}\tilde{\Phi}(t, x_0^-) &= \Phi_0(x_0^-) + t[u^-\varphi_u'(u^-) - \varphi(u^-)] \\ &= u^-x_0^- + t[u^-\varphi_u'(u^-) - \varphi(u^-)] \\ &= t[u^-\varphi_u'(u^-) - \varphi(u^-)] - hu^-.\end{aligned} \tag{14}$$

According to (2.3.57) and (2.3.58), in order that $u(x, t) = u_0(\tilde{x}^0) = \tilde{u} \neq u^-$ it is necessary that

$$\tilde{\Phi}(t, \tilde{x}_0) \leq \tilde{\Phi}(t, x_0^-) \quad \text{or} \quad \tilde{\Phi}(t, \tilde{x}_0) - \tilde{\Phi}(t, x_0^-) \leq 0. \tag{15}$$

Subtracting (14) from (12), we obtain

$$\begin{aligned}\tilde{\Phi}(t, \tilde{x}_0) - \tilde{\Phi}(t, x_0^-) = {} & [\Phi_0(\tilde{x}_0) - \tilde{u}\tilde{x}_0] + h(u^- - \tilde{u}) \\ & + t[\varphi_u'(u^-)(\tilde{u} - u^-) + \varphi(u^-) - \varphi(\tilde{u})].\end{aligned} \tag{16}$$

We now estimate the last term in (16). Recalling that $\tilde{u} < u^-$ and $\varphi''_{uu}(u) > 0$, by Taylor's formula we obtain

$$\varphi(u^-) - \varphi(\tilde{u}) \geq \varphi'_u(u^-)(u^- - \tilde{u}) + \frac{a}{2}(u^- - \tilde{u})^2. \tag{17}$$

Substituting this into (16), we obtain

$$\tilde{\Phi}(t, \tilde{x}_0) - \tilde{\Phi}(t, x_0^-) \geq [\Phi_0(\tilde{x}_0) - \tilde{u}\tilde{x}_0] + h(u^- - \tilde{u}) + \frac{ta}{2}(u^- - \tilde{u})^2. \tag{18}$$

Recalling that $u^- > \tilde{u}$, from (9) we easily find that

$$\varphi'_u(u^-) - \varphi'_u(\tilde{u}) = \frac{h + \tilde{x}_0}{t} \quad \text{and} \quad (u^- - \tilde{u}) \geq \frac{h + \tilde{x}_0}{At}. \tag{19}$$

This makes it possible to write (18) in the form

$$\tilde{\Phi}(t, \tilde{x}_0) - \Phi(t, x_0^-) \geq [\Phi_0(\tilde{x}_0) - \tilde{u}\tilde{x}_0] + \frac{h(h + \tilde{x}_0)}{At} + \frac{a(h + \tilde{x}_0)^2}{2A^2t}. \tag{20}$$

Thus, in order that (15) be satisfied it is necessary that

$$[\Phi_0(\tilde{x}_0) - \tilde{u}\tilde{x}_0] + \frac{h(h + \tilde{x}_0)}{At} + \frac{a(h + \tilde{x}_0)^2}{2A^2t} \leq 0 \tag{21}$$

or

$$\frac{h(h + \tilde{x}_0)}{At} + \frac{a(h + \tilde{x}_0)^2}{2A^2t} \leq [\tilde{u}\tilde{x}_0 - \Phi_0(\tilde{x}_0)]. \tag{22}$$

The inequality (22) restricts the range of variables x and t lying to the left of the line $x = \varphi'_u(u^-)t$ in the half plane $t \geq 0$. since $h = \varphi'_u(u^-)t - x$. In extending the domain of influence of the segment $[0, b]$, we assume that a point (x, t) belongs to it if there exists at least one value $\tilde{x}_0$ $(0 \leq \tilde{x}_0 \leq b)$ for which (22) holds for the given x and t. Since $|\tilde{u}| \leq M$ and $|\Phi_0(\tilde{x}_0)| \leq M\tilde{x}_0$, strengthening (22)[7] we write

$$h + \frac{(h + \tilde{x}_0)a}{2A} \leq \frac{2M\tilde{x}_0 At}{h + \tilde{x}_0} < 2AMt \tag{23}$$

and

$$h < 2AMt. \tag{24}$$

The inequality (24) shows that for small t the quantity h increases linearly with increasing t. Conversely, we clearly increase the domain of influence of

[7] That is, by increasing the domain of influence of the segment $[0, b]$.

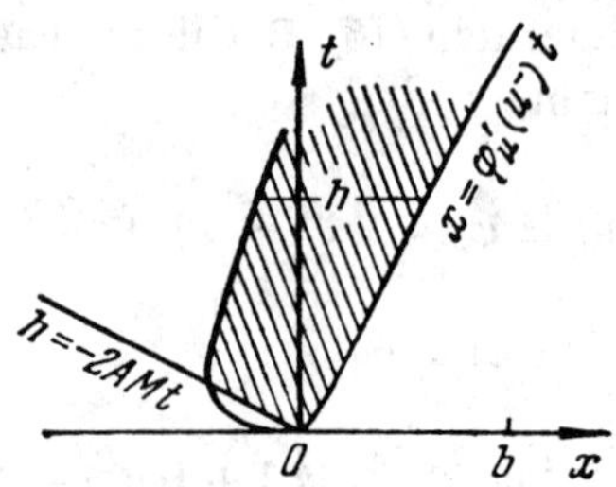

FIGURE 4.30

[0, b] if in (22) we set $\tilde{x}_0 = 0$, while on the right side we take the maximum for all $\tilde{x}_0$. We then write

$$h^2\left(\frac{2A + a}{2A}\right) \le 2MbAt \quad \text{or} \quad h^2 < \frac{4A^2Mbt}{2A + a} < 2AMbt. \tag{25}$$

We see that for large t the quantity h increases like $\sqrt{t}$. Thus, the domain of influence lies inside the hatched region in Figure 4.30:

$$h < \sqrt{2AMbt} = h(t). \tag{26}$$

The domain of influence of the segment [0, b] to the right of the line $x = b + \varphi_u'(u^+)t$ can be estimated in the same way. The inequality (22) remains in force if for h we take the quantity

$$h = x - b - \varphi_u'(u^+)t, \tag{27}$$

and we take the right side with the reverse sign. As we do this, (26) remains completely unchanged.

In the case $u^- > u^+$ there exists $t_1 > 0$ such that for $t \ge t_1$

$$u(x, t) = \begin{cases} u^- & \text{for } x - c < Dt, \\ u^+ & \text{for } x - c > Dt, \end{cases} \tag{28}$$

where

$$D = \frac{\varphi(u^+) - \varphi(u^-)}{u^+ - u^-}. \tag{29}$$

Indeed, if $u^- > u^+$, then the lines $x = \varphi_u'(u^-)t$ and $x = b + \varphi_u'(u^+)t$ are situated as shown in Figure 4.31. Since the distance between these lines grows in proportion to t while the boundaries of the domains of influence of the segment [0, b] are a distance of the order $\sqrt{t}$ from these lines, there exists a time $t = t_1$ at which the domain of influence of [0, b] vanishes (Figure 4.31), i.e., the curves

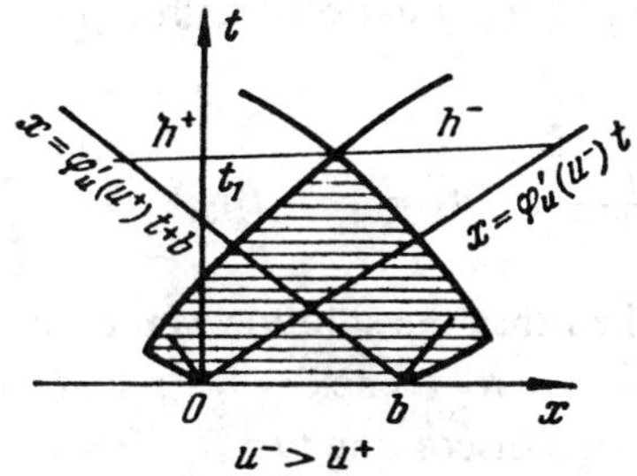

FIGURE 4.31

$$h_1 = \varphi_u'(u^-)t - x = \sqrt{2AMbt},$$
$$h_2 = -b - \varphi_u'(u^+)t + x = \sqrt{2AMbt}$$

intersect. To determine the magnitude of t_1 we have the equation

$$2\sqrt{2AMbt_1} + b = t_1[\varphi_u'(u^-) - \varphi_u'(u^+)]. \tag{30}$$

Suppose that $t_1[\varphi_u'(u^-) - \varphi_u'(u^+)] > 2b$. From (30) we then obtain

$$t_1 \leq \frac{4\sqrt{2AMbt_1}}{\varphi_u'(u^-) - \varphi_u'(u^+)} \quad \text{or} \quad t_1 \leq \frac{32AMb}{a^2[u^- - u^+]^2} = T. \tag{31}$$

We can now obviously conclude that (28) holds for $t_1 = T$. Indeed, since the range of values of x and t prescribed by the conditions

$$b + \varphi_u'(u^+)t < x < \varphi_u'(u^-)t,$$

lies to the left of the line $x = \varphi_u'(u^-)t$ and to the right of the line $x = \varphi_u'(u^+)t + b$, in order that in this region $u(x, t) \neq u^-$ and $u(x, t) \neq u^+$ it is necessary that the point (x, t) belong to the domain of influence of $[0, b]$ for $t \geq T$, which is impossible.

To determine the magnitude of c in (28) we consider a point x_A on the line $t = t_1$. We determine this point from the conditions

$$x_A - \varphi_u'(u^-)t_1 = x_0^-, \qquad x_A - \varphi_u'(u^+)t_1 = x_0^+, \tag{32}$$

$$\tilde{\Phi}(t_1, x_0^-) = \tilde{\Phi}(t_1, x_0^+). \tag{33}$$

The condition (33) can be rewritten in the form

$$u^- x_0^- + t_1[u^- \varphi_u'(u^-) - \varphi(u^-)]$$
$$= \Phi_0(b) + u^+(x_0^+ - b) + t_1(u^+\varphi_u'(u^+) - \varphi(u^+)]. \tag{34}$$

Considering (32) and (34) as equations defining x_0^-, x_0^+ and x_A, we find x_A and c:

$$x_A = \frac{\Phi_0(b) - u^+ b}{u^- - u^+} + Dt_1 = c + Dt_1, \qquad c = \frac{\Phi_0(b) - u^+ b}{u^- - u^+}. \tag{35}$$

We have thus established that the solution $u(x, t)$ of the Cauchy problem (1), (2) under the condition $u^- > u^+$ coincides for $t \geq T$ with the solution $u^0(x, t)$ of the problem on decay of a discontinuity

$$\frac{\partial u^0}{\partial t} + \frac{\partial \varphi(u^0)}{\partial x} = 0, \qquad u^0(x, 0) = \begin{cases} u^-, & x < c, \\ u^+, & x > c. \end{cases} \tag{36}$$

This result can be formulated in another way. Let u and $\bar{u}$ be generalized solutions of (1), and suppose that

$$u(x, 0) \equiv \bar{u}(x, 0) \quad \text{for } |x| \geq b. \tag{37}$$

If, further,

$$u(x, 0) \equiv \bar{u}\,(x, 0) = \begin{cases} u^- & \text{for } x \leq -b, \\ u^+ & \text{for } x \geq b \end{cases} \tag{38}$$

and $u^- > u^+$, then there exists $t_1 > 0$ such that for $t \geq t_1$

$$u(x, t) \equiv \bar{u}(x + x, t). \tag{39}$$

where

$$c = \frac{1}{u^- - u^+} \int_{-b}^{b} [\bar{u}(x, 0) - u(x, 0)]\, dx. \tag{40}$$

The property of generalized solutions we have proved expresses the fact that in the case $u^- > u^+$ generalized solutions of the Cauchy problem (1), (2) "up to translation" do not depend on the initial values on any finite segment of the initial axis $t = 0$ for sufficiently large times.

We remark that the result obtained can be strengthened in two directions:

1) for more general equations where $\varphi = \varphi(u, x, t)$ under certain conditions on φ;

2) the condition (38) can be relaxed:

$$|\bar{u}(x, 0) - u(x, 0)| \to 0 \quad \text{as } x \to \pm\infty,$$

$$u(x, 0) \to \begin{cases} u^- & \text{as } x \to -\infty, \\ u^+ & \text{as } x \to +\infty; \end{cases}$$

here a particular order of passing to the limit is required.

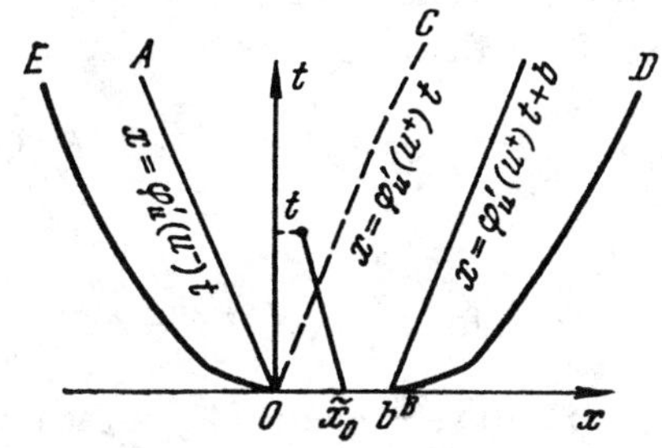

FIGURE 4.32

We shall study the asymptotic behavior of the solution $u(x, t)$ in the case $u^- < u^+$. In this case, as is readily apparent from Figure 4.32, the domain of influence of the segment $[0, b]$ does not vanish as $t \to \infty$, but, on the contrary, becomes unbounded. In this case it cannot be asserted that for sufficiently large t

$$u(x, t) = u^0(x, t),$$

where $u^0(x, t)$ is a solution of the decay problem. However, in this case as well, the solution $u(x, t)$ is close to the solution $u^0(x, t)$ for sufficiently large t. Namely, we shall prove that

$$|u(x, t) - u^0(x, t)| \rightrightarrows 0 \quad \text{as } t \to \infty. \tag{41}$$

Let u^0 be the solution of the decomposition problem

$$\left.\begin{aligned} &\frac{\partial u_0}{\partial t} + \frac{\partial \varphi(u^0)}{\partial x} = 0, \\ &u^0(x, 0) = \begin{cases} u^- & \text{for } x < 0, \\ u^+ & \text{for } x > 0, \end{cases} \qquad u^- \le u^+. \end{aligned}\right\} \tag{42}$$

The function u^0 is given by

$$u^0(x, t) = \begin{cases} u^- & \text{for } x \le \varphi_u'(u^-)t, \\ f(x/t) & \text{for } \varphi_u'(u^-)t \le x \le \varphi_u'(u^+)t, \\ u^+ & \text{for } x \ge \varphi_u'(u^+)t, \end{cases} \tag{43}$$

where

$$\varphi_u'(f(\xi) = \xi, \quad \text{i.e. } f(\xi) = [\varphi_u']^{-1}(\xi). \tag{44}$$

Thus, $u^0(x, t) \equiv u^-$ to the left of the ray OA, $u^0(x, t) \equiv u^+$ to the right of the ray OC, and $u^0 = f(x/t)$ in the zone AOC (Figure 4.32). We see that outside the domain of influence of $[0, b]$ this also hods for $u(x, t)$, i.e.,

$$u(x, t) \equiv u^- \quad \text{for } \varphi_u(u^-)t - x \geq h(t), \tag{45}$$

while

$$u(x, t) \equiv u^+ \quad \text{for } x - \varphi_u'(u^+)t - b \geq h(t). \tag{46}$$

Hence, it is only necessary to prove property (41) in the domain of influence of the segment $[0, b]$—the zone $EOBD$. Let (x, t) be any point lying inside $EOBD$. Since through this point there passes no characteristic $x = X(t, x_0, u_0(x_0))$ or $x_0 \notin [0, b]$, it follows that

$$u(x, t) = u_0(\tilde{x}_0); \qquad x - \varphi_u'(u_0(\tilde{x}_0))t = \tilde{x}_0, \tag{47}$$

where

$$0 \leq \tilde{x}_0 \leq b. \tag{48}$$

From (47) we find that

$$\varphi_u'(u_0(\tilde{x}_0)) = \varphi_u'(u(x, t)) = \frac{x}{t} - \frac{\tilde{x}_0}{t}. \tag{49}$$

Comparing (49) with (44), we conclude that

$$u(x, t) = u_0(\tilde{x}_0) = f\left(\frac{x}{t} - \frac{\tilde{x}_0}{t}\right). \tag{50}$$

Thus, in the zone AOC

$$u(x, t) - u^0(x, t)| = \left|f\left(\frac{x}{t} - \frac{\tilde{x}_0}{t}\right) - f\left(\frac{x}{t}\right)\right| \leq \max_{|\eta| \leq \xi} |f'(\eta)| \left|\frac{\tilde{x}_0}{t}\right|. \tag{51}$$

Since $f_\eta' = 1/\varphi_{uu}''$, it follows from (51) that

$$|u(x, t) - u^0(x, t)| \leq \frac{\tilde{x}_0}{at} \leq \frac{b}{at}. \tag{52}$$

Thus, property (41) has been proved in AOC. Suppose now that (x, t) is any point in $EOBD$; for example, suppose that this point lies to the right of the line $x = \varphi_u'(u^+)t$, i.e., in $COBD$ (Figure 4.32). Then at this point $u^0 = u^+$, and $u(x, t)$ is determined from (47) and (48), as before. According to the definition of the domain of influence of the segment $[0, b]$, we have

$$x - \varphi_u'(u^+)t \le b + h(t) = b + \sqrt{2AMbt}. \tag{53}$$

Hence,

$$\begin{aligned} |u(x,t)-u^0(x,t)| &= \left|f\left(\frac{x-\tilde{x}_0}{t}\right) - u^+\right| \\ &= \left|f\left(\frac{x-\tilde{x}_0}{t}\right) - f(\varphi_u'(u^+))\right| \le \frac{|x-\tilde{x}_0-\varphi_u'(u^+)t|}{at} \\ &\le \frac{|b-\tilde{x}_0+h(t)|}{at} = \frac{b}{at} + \frac{h(t)}{at} = \frac{b}{at} + \frac{\sqrt{2AMb}}{a\sqrt{t}}. \end{aligned} \tag{54}$$

Since to the left of the ray OA inequality (54) can be proved in an altogether similar way, we have thus proved property (41) of the generalized solution and so established the order with which $u(x, t)$ tends to $u^0(x, t)$, namely,

$$|u(x, t) - u^0(x, t)| < \frac{4b}{at} + \frac{8}{a\sqrt{t}}\sqrt{AMb}. \tag{55}$$

Combining the result of the cases $u^- < u^+$ and $u^- > u^+$, we can strengthen and reformulate it as follows.

Suppose that $u(x, t)$ and $\bar{u}(x, t)$ are bounded generalized solutions of (1) with initial values which differ only on a finite segment of the initial axis, i.e.,

$$|u(x, 0) - \bar{u}(x, 0)| \equiv 0 \quad \text{for } |x| \ge b. \tag{56}$$

Then there exist constants c, D and t_1 such that for $t \ge t_1$

$$|u(x, t) - u(x + c, t)| \le D/\sqrt{t}. \tag{57}$$

We emphasize that in this formulation it is not assumed that $u(x,0) \to u^-$ and u^+ as $x \to -\infty$ and $+\infty$. The asymptotic properties of generalized solutions of a single quasilinear equation proved above are connected in an essential way with the nonlinearity of the equation. Indeed, solutions of linear equations with different initial values differ from one another for all values of $t > 0$, and this "remainder" does not, in general, tend to zero as $t \to \infty$.

7. *The method of viscosity.* In subsection 1 of this section we already considered one application of the method of viscosity: a generalized solution of the problem

$$\frac{\partial u}{\partial t} + \frac{\partial}{\partial x}\left(\frac{u^2}{2}\right) = 0, \qquad u(x, 0) = u_0(x)$$

was obtained as the limit as $\mu \to 0$ of solutions $u_\mu(x, t)$ of the Cauchy problem

$$\frac{\partial u_\mu}{\partial t} + \frac{\partial}{\partial x}\left(\frac{u_\mu^2}{2}\right) = \mu\frac{\partial^2 u_\mu}{\partial x^2}, \qquad u_\mu(x, 0) = u_0(x), \quad \mu > 0.$$

An explicit formula for u_μ was thus constructed, whieh made it possible to pass to the limit as $\mu \to 0$ and study the properties of the generalized solution $u(x, t)$. Of course, for more complicated equations one can hardly count on obtaining analytic formulas for the solution. However, in order to acquaint the reader with applications of the method of viscosity, we consider some simple cases from which it is possible to form a conclusion regarding the possibilities of this method.

We consider the Cauchy problem

$$\frac{\partial u}{\partial t} + \frac{\partial \varphi(u)}{\partial x} = 0, \qquad \varphi''_{uu}(u) > 0, \tag{1}$$

$$u(x, 0) = u_0(x), \qquad |u_0(x)| \le M. \tag{2}$$

Together with this we consider another problem:

$$\frac{\partial u_\mu}{\partial t} + \frac{\partial \varphi(u_\mu)}{\partial x} = \mu \frac{\partial^2 u_\mu}{\partial x^2} \qquad (\mu > 0), \tag{3}$$

$$u_\mu(x, 0) = u_0(x). \tag{4}$$

We shall assume that the initial function $u_0(x)$ possesses a continuous first derivative, and that

$$|u_0'(x)| \le K \qquad (-\infty < x < \infty). \tag{5}$$

The very first question that arises is the question of the existence of a solution u_μ and its properties. The study of this question would take us far beyond the framework of our present topic, which is primarily that of quasilinear equations of hyperbolic type and the equations of gas dynamics. Therefore, referring the reader to special investigations on this question (see, for example, Ventcel′ (1957), Ladyženskaja (1956), and Oleĭnik (1955)), we shall assume that the following is know.

1. A bounded solution $u_\mu(x, t)$ exists for any $t > 0$ and is unique.

2. For any initial function $u_0(x)$ satisfying conditions (2) and (5) the derivatives of u_μ exist and remain bounded for all $t > 0$. In particular, for $t > 0$ the continuous derivatives contained in (3) exist.

3. The solution u_μ and its derivative $p_\mu = \partial u_\mu/\partial x$ satisfy the *maximum principle*, which can be formulated in the form of the inequalities

$$|u_\mu| \le M, \tag{6}$$

$$p_\mu = \frac{\partial u_\mu}{\partial x} \le K \tag{7}$$

valid for any x, $t \ge 0$.

We shall illustrate the maximum principle (6) with the following graphic arguments. Suppose that $u_\mu(x, t_1)$, considered as a function of x on the line $t = t_1$, has at $x = x_1$ a relative maximum (minimum). Then at this point

$$\frac{\partial u_\mu}{\partial x}(x_1, t_1) = 0, \qquad \frac{\partial^2 u_\mu}{\partial x^2}(x_1, t_1) \le 0 \quad (\ge 0).$$

From (3) we conclude that

$$\frac{\partial u_\mu(x_1, t_1)}{\partial t} = \left[\mu \frac{\partial^2 u_\mu}{\partial x^2} - \varphi_u'(u_\mu)\frac{\partial u_\mu}{\partial x}\right]_{\substack{x=x_1\\ t=t_1}} \le 0 \qquad (\ge 0).$$

This shows that each relative maximum in x of the function $u_\mu(x, t)$ does not increase with increasing t. However, this still does not imply (6), since u need not assume its maximum (minimum) value at any finite point. A more detailed proof of the maximum principle takes into account that the function $u_0(x)$ is bounded.

As concerns the maximum principle for the derivative p_μ, we point out hat the inequality bounds the derivative p_μ only from one side, namely, from above. This is related to the fact that the equation

$$\frac{\partial p_\mu}{\partial t} + \varphi_u'(u_\mu)\frac{\partial p_\mu}{\partial x} = \mu \frac{\partial^2 p_\mu}{\partial x^2} - \varphi_{uu}''(u_\mu)p_\mu^2, \qquad \varphi_{uu}' > 0,$$

obtained by differentiating (3) makes it possible to assert only that a relative maximum of $p_\mu(x, t)$ does not increase with increasing time t. As concerns the minima of p_μ, they may decrease.

4. For all $\mu > 0$ and any $x, t \ge 0$ we have

$$\left|\mu \frac{\partial u_\mu}{\partial x}\right| \le C, \tag{8}$$

where C is a positive quantity not depending on μ.

We shall prove this.[8] Setting

$$z = \mu \frac{\partial x_\mu}{\partial x} - \varphi(u_\mu), \tag{9}$$

we write (3) in the form $\partial u_\mu/\partial t = \partial z/\partial x$. Differentiating (9) with respect to t, we find that

[8] We were informed of this proof in 1964 by V. G. Susko, then a student at Moscow State University.

$$\frac{\partial z}{\partial t} = \mu \frac{\partial}{\partial x}\left(\frac{\partial u_\mu}{\partial t}\right) - \varphi_u'(u_\mu)\frac{\partial u_\mu}{\partial t}.$$

Substituting $\partial u_\mu/\partial t = \partial z/\partial x$, we obtain

$$\partial z/\partial t + \varphi_u'(u_\mu)(\partial z/\partial x) = \mu(\partial^2 z/\partial x^2).$$

Thus, z satisfies a nonlinear heat equation. Therefore, from the maximum principle we obtain

$$|z(x, t)| \leq \max_x |z(x, 0)|.$$

This implies (8) and, in particular, the more precise inequality

$$\min_x\left[\mu\frac{du_0(x)}{dx} - \varphi(u_0(x))\right] + \varphi(n_\mu)$$
$$\leq \mu\frac{\partial u_\mu}{\partial x} \leq \max_x\left[\mu\frac{du_0(x)}{dx} - \varphi(u_0(x))\right] + \varphi(u_\mu).$$

If conditions 1–4 are satisfied it is also possible to prove some further properties of the solutions $u_\mu(x, t)$. In particular, we shall show that $\mu(\partial u_\mu/\partial x)$ tends to zero in mean as $\mu \to 0$ on any finite domain of the variables x and t. Multiplying (3) by u_μ, we write the result in the form

$$\frac{\partial}{\partial t}\left(\frac{u_\mu^2}{2}\right) + \frac{\partial}{\partial x}F(u_\mu) = \mu\frac{\partial^2}{\partial x^2}\frac{u_u^2}{2} - \mu\left(\frac{\partial u_\mu}{\partial x}\right)^2, \tag{10}$$

where $F'(u_\mu) = u_\mu\varphi'(u_\mu)$. We integrate (10) over the rectangle $0 \leq t \leq t_1$, $x_1 \leq x \leq x_2$:

$$\int_0^{t_1}\int_{x_1}^{x_2}\left[\frac{\partial}{\partial t}\frac{u_\mu^2}{2} + \frac{\partial}{\partial x}F(u_\mu)\right]dt\,dx = \int_0^{t_1}\int_{x_1}^{x_2}\left[\mu\frac{\partial^2}{\partial x^2}\frac{u_\mu^2}{2} - \mu\left(\frac{\partial u_\mu}{\partial x}\right)^2\right]dt\,dx.$$

We rewrite this relation in the form

$$\mu\int_0^{t_1}\int_{x_1}^{x_2}\left(\frac{\partial u_\mu}{\partial x}\right)^2 dt\,dx = \int_{x_1}^{x_2}\left[\frac{u_\mu^2(x, 0) - u_\mu^2(x, t_1)}{2}\right]dx$$
$$+ \int_0^{t_1}\left\{\mu u_\mu(x_2, t)\frac{\partial u_\mu}{\partial x}(x_2, t) - F(u_\mu(x_2, t))\right.$$
$$\left. - \mu u_\mu(x_1, t)\frac{\partial u_\mu(x_1, t)}{\partial x} + F(u_\mu(x_1, t))\right\}dt.$$

Because of (6) and (8) the right side can be bounded above by

$$\mu \int_0^{t_1}\int_{x_1}^{x_2} \left(\frac{\partial u_\mu}{\partial x}\right)^2 dt\, dx \le \frac{1}{2} M^2(x_2 - x_1) + 2Ft_1 + 2MCt_1$$

$$= \frac{M^2}{2}(x_2 - x_1) + 2t_1(F + MC), \tag{11}$$

where the constant F is defined so that $|F| \ge |F(u)|$ for $|u| \le M$. Since the right side of (11) is independent of μ, from this it follows that for any finite region $\mathscr{G}$ of the half plane $t \ge 0$

$$\mu \iint_{\mathscr{G}} \left(\frac{\partial u_\mu}{\partial x}\right)^2 dx\, dt \le D,$$

where the constant D depends only on $\mathscr{G}$. We apply the Schwarz-Bunjakovskiĭ inequality

$$\iint_{\mathscr{G}} |uv| \; dt\, dx \le \left(\iint_{\mathscr{G}} u^2\, dt\, dx \iint_{\mathscr{G}} v^2\, dt\, dx\right)^{1/2},$$

in which we set $u = \mu(\partial u_\mu/\partial x)$ and $v \equiv 1$. We then obtain

$$\mu \iint_{\mathscr{G}} \left|\frac{\partial u_\mu}{\partial x}\right| dt\, dx \le \left(S_{\mathscr{G}}\mu^2 \iint_{\mathscr{G}} \left(\frac{\partial u_\mu}{\partial x}\right)^2 dt\, dx\right)^{1/2} \le \sqrt{\mu}\;\sqrt{S_{\mathscr{G}} D}, \tag{12}$$

where $S_{\mathscr{G}}$ is the area of $\mathscr{G}$. The inequality (12) shows that $\iint_{\mathscr{G}} \mu|\partial u_\mu/\partial x|\, dt\, dx$ tends to zero as $\mu \to 0$ like $\sqrt{\mu}$.

We now outline a method of proving the convergence in mean of the solutions $u_\mu(x, t)$ as $\mu \to 0$ to the solution $u(x, t)$ of the Cauchy problem (1), (2). We introduce the potential $\Phi_\mu(x, t)$. According to (3), the contour integral

$$\Phi_\mu(x, t) = \int_{(0,0)}^{(x,t)} u_\mu\, dx + \left[\mu \frac{\partial u_\mu}{\partial x} - \varphi(u_\mu)\right] dt$$

does not depend on the path of integration and defines a continuous and differentiable function with derivatives that satisfy

$$\frac{\partial \Phi_\mu}{\partial x} = u_\mu, \qquad \frac{\partial \Phi_\mu}{\partial t} = \mu \frac{\partial u_\mu}{\partial x} - \varphi(u_\mu).$$

Eliminating u_μ from these inequalities, we obtain

$$\frac{\partial \Phi_\mu}{\partial t} + \varphi\left(\frac{\partial \Phi_\mu}{\partial x}\right) = \mu \frac{\partial^2 \Phi_\mu}{\partial x^2}. \tag{13}$$

The potential Φ_μ assumes the following initial values:

$$\Phi_\mu(x, 0) = \Phi_0(x) = \int_0^x u_0(\eta)\, d\eta.$$

From (6) and (8) we conclude that

$$|\Phi_\mu(x, t)| \leq M|x| + [C + \varphi]t, \tag{14}$$

where $|\varphi(u)| \leq \varphi$ for $|u| \leq M$, and the family of functions $\Phi_\mu(x, t)$ is bounded for all $\mu > 0$ at any point (x, t). Finally, we write the inequalities

$$\left|\frac{\partial \Phi_\mu}{\partial x}\right| \leq M, \qquad \left|\frac{\partial \Phi_\mu}{\partial t}\right| \leq +\varphi.$$

We shall prove the uniform convergence in any finite part of the half plane $t \geq 0$ of the potential $\Phi_\mu(x, t)$ as $\mu \to 0$. To this end we differentiate (13) with respect to μ. Setting $\Phi_\mu' = \partial\Phi_\mu/\partial_\mu$, we obtain

$$\frac{\partial \Phi_\mu'}{\partial t} + \varphi'(u_\mu)\frac{\partial \Phi_\mu'}{\partial x} = \mu\frac{\partial^2 \Phi_\mu'}{\partial x^2} + \frac{\partial^2 \Phi_\mu}{\partial x^2} = \mu\frac{\partial^2 \Phi_\mu'}{\partial x^2} + \frac{\partial u_\mu}{\partial x}, \tag{15}$$

while according to the initial condition $\Phi_\mu'(x, 0) \equiv 0$. Assuming that u_μ is given, we see that Φ_μ' satisfies the linear heat equation (15) and has zero initial value. According to (14), $\Phi_\mu(x, t)$ grows no faster than a linear function as $x \to \pm\infty$; Φ_μ' thus satisfies the maximum principle for (15):

$$\Phi_\mu'(x, t) \leq t \max_{(\xi, \tau)} \frac{\partial u_\mu(\xi, \tau)}{\partial \xi}.$$

Since $\partial u_\mu/\partial\xi \leq K$ by (7), it follows that $\Phi_\mu'(x, t) \leq Kt$. Therefore, the function $\Psi_\mu = \Phi_\mu - 2Kt_\mu$ satisfies the inequality

$$\frac{\partial \Psi_\mu}{\partial \mu} \leq -Kt,$$

so that the sequence of functions $\Psi_\mu(x, t)$ decreases monotonically as $\mu \to 0$ and therefore converges uniformly with respect to x and t (since it is bounded below). There thus exists the limit

$$\Phi(x, t) = \lim_{\mu\to 0} \Phi_\mu(x, t).$$

We shall establish the simplest properties of the potential $\Phi(x, t)$. Since $|u_\mu| \leq M$ and $\partial\Phi_\mu/\partial x = u_\mu$, it follows that

$$|\Phi(x + \Delta x, t) - \Phi(x, t)| \leq M\,|\Delta x|.$$

Similarly,

$$|\Phi(x, t + \Delta t) - \Phi(x, t)| \leq [C' + \varphi]\,|\Delta t|.$$

Thus, the potential $\Phi(x, t)$ is a Lipschitz continuous function of its variables that assumes the initial values

$$\Phi(x, 0) = \Phi_0(x) = \int_0^x u_0(\eta)\, d\eta$$

and, as any Lipschitz continuous function, it possesses derivatives $\partial\Phi/\partial x$ and $\partial\Phi/\partial t$ almost everywhere. We shall show that almost everywhere these derivatives satisfy the equation

$$\frac{\partial\Phi}{\partial t} + \varphi\left(\frac{\partial\Phi}{\partial x}\right) = 0, \tag{16}$$

and the sequence u_μ tends to $u(x, t)$ as $\mu \to 0$ almost everywhere, where $u(x, t) = \partial\Phi(x, t)/\partial x$. For this we observe that the uniform boundedness of u_μ (see (6)) and the fact that the derivative p_μ is bounded above (7) imply the bounded variation of the functions u_μ on any finite segment of lines $t = \text{const}$. Using this fact, we can show that almost everywhere

$$\frac{\partial\Phi_\mu}{\partial x} \to \frac{\partial\Phi}{\partial x}, \qquad \frac{\partial\Phi_\mu}{\partial t} \to -\varphi\left(\frac{\partial\Phi}{\partial x}\right)$$

as $\mu \to 0$, so that the potential $\Phi(x, t)$ satisfies (16) almost everywhere. Thus, the potential $\Phi(x, t)$ is Lipschitz continuous, satisfies (16) almost everywhere, and assumes the required initial values at $t = 0$. Therefore, the function $u = \partial\Phi/\partial x$ is defined almost everywhere, is bounded and measurable, and because of (7)

$$\frac{\Phi(x + \Delta x, t) - 2\Phi(x, t) + \Phi(x - \Delta x, t)}{\Delta x^2} < K,$$

so that it satisfies the stability condition and is thus a generalized solution of the Cauchy problem (1), (2)

We mention further that it is also possible to consider the Cauchy problem in the case of any bounded, measurable initial function $u_0(x)$ by the method of viscosity (see Oleĭnik (1957)).

We shall indicate still another application of the method of viscosity. Following I. M. Gel'fand, we obtain stability conditions for a generalized solution of a single quasilinear equation with a nonconvex function $\varphi(u)$ (see also §1).

Suppose that the piecewise continuous solution $u(x, t)$ is the limit of solutions $u_\mu(x, t)$ of the equation with viscosity

$$\frac{\partial u_\mu}{\partial t} + \frac{\partial \varphi(u_\mu)}{\partial x} = \mu \frac{\partial}{\partial x}\left(B(u_\mu)\frac{\partial u_\mu}{\partial x}\right), \qquad B(u_\mu) > 0,\ \mu > 0. \tag{17}$$

Suppose at the point $x_1, t_1 > 0$ the solution $u(x, t)$ is discontinuous, $u(x_1 - 0, t_1) = u^-$, $u(x_1 + 0, t_1) = u^+$, and the discontinuity propagates in the x, t plane along the curve $x = x(t)$ with $x'(t_1) = D$.

It is natural to suppose that in a small neighborhood of x_1, t_1 the solution $u(x, t)$ can be represented approximately in the form $u(x, t) = \bar{u}(x - Dt)$, where $\bar{u}(x) = u^-$ for $x < 0$ and $\bar{u}(x) = u^+$ for $x > 0$. We also assume that the solution $u_\mu(x, t)$ in a small neighborhood of x_1, t_1 is a travelling wave, i.e., $u_\mu(x, t) = \bar{u}_\mu((x - Dt)/\mu)$, where $\bar{u}_\mu(x) \to u^-$ as $x \to -\infty$ and $\bar{u}_\mu(x) \to u^+$ as $x \to \infty$ (we here assume that the parameter μ is sufficiently small).

We thus seek a stationary solution $u_\mu(x, t)$ of (17) in the form

$$u_\mu(x, t) = \bar{u}_\mu\left(\frac{x - Dt}{\mu}\right). \tag{18}$$

Setting $\xi = (x - Dt)/\mu$, from (17) we obtain an ordinary differential equation for $\bar{u}_\mu(\xi)$:

$$[\varphi'_u(\bar{u}_\mu(\xi)) - D]\,\frac{d\bar{u}_\mu(\xi)}{d\xi} = \frac{d}{d\xi}\left(B(\bar{u}_\mu)\frac{d\bar{u}_\mu(\xi)}{d\xi}\right); \tag{19}$$

moreover, from (18) we obtain the following boundary conditions for (19):

$$\bar{u}_\mu(\xi) \to \begin{cases} u^- & \text{as } \xi \to -\infty, \\ u^+ & \text{as } \xi \to +\infty. \end{cases} \tag{20}$$

Integrating from $\xi = -\infty$ to ξ and assuming that $d\bar{u}_\mu/d\xi \to 0$ as $\xi \to \pm\infty$, we obtain

$$B(\bar{u}_\mu)\frac{d\bar{u}_\mu}{d\xi} = \varphi(\bar{u}_\mu) - \varphi(u^-) - D(\bar{u}_\mu - u^-) = F(\bar{u}_\mu). \tag{21}$$

In order that $d\bar{u}_\mu/d\xi \to 0$ as $\bar{u}_\mu \to u^+$ it is also necessary that $F(u^+) = 0$, i.e.,

$$D = \frac{\varphi(u^+) - \varphi(u^-)}{u^+ - u^-}, \tag{22}$$

which follows also from the Hugoniot conditions, as we have seen in §1. Thus, for the existence of an integral curve of problem (19), (20) it is necessary that $F(u^-) = F(u^+) = 0$. However, this is not sufficient.

Another necessary condition for the existence of $\bar{u}_\mu$ is that the sign of $F(\bar{u}_\mu)$ not change on the interval (u^-, u^+). Indeed, if in this interval there exists a point

u^* such that $F(u)$ has different signs to the left and right of u^*, then there obviously exists no integral curve $\bar{u}_\mu = \bar{u}_\mu(\xi)$.

Multiplying (21) by $2(\bar{u}_\mu - u^-)$, we bring it to the form

$$B(\underline{n}_\mu)\frac{d}{d\xi}(\bar{u}_\mu - u^-)^2 = 2(\bar{u}_\mu - u^-)F(\bar{u}_\mu). \tag{23}$$

It is now obvious that for the existence of $\bar{u}_\mu(\xi)$ it is also necessary that the right side of (23) be nonnegative, i.e.,

$$2(\bar{u}_\mu - u^-)F(\bar{u}_\mu) \geq 0.$$

Indeed, since $B(\bar{u}_\mu) > 0$, the existence of $\bar{u}_\mu(\xi)$ implies that $(\bar{u}_\mu - u^-)^2$ does not decrease with increasing ξ, since otherwise in (u^-, u^+) the function $F(u)$ would change sign. Similarly, it is easy to see that $(\bar{u}_\mu - u^+)^2$ does not increase with increasing ξ, and hence $2(\bar{u}_\mu - u^+)F(\bar{u}_\mu) \leq 0$. Dividing each of the last two inequalities by the positive quantities $2(\bar{u}_\mu - u^-)^2$ and $2(\bar{u}_\mu - u^+)^2$, we bring them to the form

$$\frac{\varphi(\bar{u}_\mu)-\varphi(u^-)}{\bar{u}_\mu - u^-} \geq D = \frac{\varphi(u^+)-\varphi(u^-)}{u^+ - u^-} \geq \frac{\varphi(\bar{u}_\mu)-\varphi(u^+)}{\bar{u}_\mu - u^+}; \tag{24}$$

moreover, it is obvious that $\bar{u}_\mu$ is any number in (u^-, u^+). We thus conclude that a discontinuous solution $u(x, t)$ may be considered as the limit of solutions $u_\alpha(x, t)$ as $\mu \to 0$ only if at each point of the line of discontinuity $x = x(t)$ the condition (24) is satisfied. It is easy to see that the condition (24) obtained here coincides with the stability condition we obtained earlier in §1 from completely different considerations.

In conclusion we mention that at the present time existence and uniqueness of a generalized solution of the Cauchy problem have been proved in the case of a nonconvex function $\varphi(u)$. Here the stability of the generalized solution means that conditions (24) are satisfied.

§3. A system of quasilinear equations

1. *Introductory remarks.* We now consider a conservative system of quasilinear equations

$$\frac{\partial u_i}{\partial t} + \frac{\partial \varphi_i(u, x, t)}{\partial x} = f_i(u, x, t) \qquad (i = 1, \ldots, n), \tag{1}$$

which we write in the abbreviated form

$$\frac{\partial u}{\partial t} + \frac{\partial \varphi(u, x, t)}{\partial x} = f(u, x, t), \tag{2}$$

where u, φ, and f are understood to be vector-valued functions with n components. We henceforth assume that $\varphi(u, x, t)$ and $f(u, x, t)$ possess two continuous derivatives with respect to all variables in the region of variation of u, x and t considered.

The system of differential equations (2) in the hyperbolic case can be reduced to the form

$$l_\alpha^k (u, x, t)\left[\frac{\partial u_\alpha}{\partial t} + \xi_k(u, x, t)\frac{\partial u_\alpha}{\partial x}\right] = g_k(u, x, t), \tag{3}$$

where

$$\operatorname{Det}((l_\alpha^k(u, x, t))) \neq 0, \tag{4}$$

as we have seen in Chapter 1. The vector $l^k(u, x, t) = \{l_\alpha^k(u, x, t)\}$ is a left eigenvecotor of the matrix $A(u, x, t) = ((\partial\varphi_i/\partial u_j))$, and $\xi_k(u, x, t)$ is the corresponding eigenvalue, i.e.,

$$l^k(u, x, t)A(u, x, t) = \xi_k(u, x, t)l^k(u, x, .t). \tag{5}$$

Throughout this section we shall consider only systems (2), (3) which are strictly hyperbolic, i.e., we assume that

$$\xi_1(u, x, t) < \xi_2(u, x, t) < \cdots < \xi_n(u, x, t). \tag{6}$$

The problem of constructing classical solutions (2) was considered in Chapter 1; here we shall study generalized (discontinuous) solutions. A generalized solution of (2) satisfies the integral conservation laws

$$\oint_C u\, dx - \varphi(u, x, t)\, dt + \iint_{\mathscr{G}_C} f(u, x, t)\, dx\, dt = 0, \tag{7}$$

and in the region where the first derivatives $\partial u/\partial t$ and $\partial u/\partial x$ exist it satisfies the differential equations (2) or (3); the latter we write more briefly as

$$l^k\left(\frac{\partial u}{\partial t} + \xi_k\frac{\partial u}{\partial x}\right) = g_k. \tag{8}$$

For the system of quasilinear equations (2) we pose the Cauchy problem with the initial conditions

$$u(x, 0) = u_0(x), \tag{9}$$

where $u_0(x)$ is, generally speaking, a discontinuous, bounded, vector-valued function.

As we saw in §1, the proper formulation of the Cauchy problem leads to certain stability conditions which generalized solutions of this problem must satisfy. In this section we mainly consider generalized solutions having piecewise smooth lines of discontinuity $x = x(t)$ away from which they are classical solutions of (2). In correspondence with §1 we shall understand stability conditions in the following sense. To each line of discontinuity $x = x(t)$ there is assigned a number k, called the *index* of the line of discontinuity. For this index k on the line $x = x(t)$ the following inequalities are satisfied:

$$\xi_k(u(x(t) - 0, t), x(t), t) > D > \xi_k(u(x(t) + 0, t), x(t), t), \tag{10}$$

$$\xi_{k-1}(u(x(t)-0, t), x(t), t) < D < \xi_{k+1}(u(x(t)+0, t), x(t), t), \; D=x'(t). \tag{11}$$

In their simplest form the properties of generalized solutions of systems of quasilinear equations can be studied with the example of a homogeneous, nonlinear system of equations

$$\frac{\partial u}{\partial t} + \frac{\partial \varphi(u)}{\partial x} = 0, \tag{12}$$

i.e., in the case where the vector φ does not depend on the variables x and t, and $f \equiv 0$.

2. *Self-similar solutions of a system of quasilinear equations.* Solutions of the system of equations (3.1.12) depending only on the single variable $y = (x - x_0)/(t - t_0)$ are called *self-similar*. We assume with no loss of generality that $t_0 = x_0 = 0$. We seek a solution $u(y)$ of (3.1.12) defined for $t \geq 0$ and depending only on the variable $y = x/t$. Making the substitution

$$\frac{\partial}{\partial t} = -\frac{x}{t^2}\frac{d}{dy} = -\frac{y}{t}\frac{d}{dy}, \qquad \frac{\partial}{\partial x} = \frac{1}{t}\frac{d}{dy}$$

in (3.1.12), we find that a self-similar solution $u = u(y)$ satisfies the system of ordinary differential equations

$$[A(u) - yE]\frac{du}{dy} = 0, \qquad A(u) = \left(\left(\frac{\partial \varphi_i(u)}{\partial u_j}\right)\right), \tag{1}$$

where E is the identity matrix of nth order.

From (1) if follows that if $y \neq \xi_k(u)$, $k = 1, \ldots, n$, then the vector du/dy is identically zero, and the solution $u(y) \equiv \text{const}$. We therefore assume that on some interval of y the equality

$$y = \xi_k(u) \tag{2}$$

is satisfied identically, where k is one of the indices $1, \ldots, n$. Then (1) has a nontrivial solution for the vector of derivatives du/dy:

$$\frac{du}{dy} = \lambda r^k(u), \tag{3}$$

where $r^k(u)$, as usual, denotes a right eigenvector of the matrix $A(u)$ (Chapter 1), i.e.,

$$A(u)r^k(u) = \xi_j(u)r^k(u).$$

The system (3) still does not make it possible to determine $u(y)$ by integration, since these equations contain the unknown factor λ. To determine this factor we differentiate (2) with respect to y. We obtain

$$\sum_{j=1}^{n} \frac{\partial \xi_k(u)}{\partial u_j} \frac{du_j}{dy} = 1,$$

which we rewrite in the form

$$\frac{du}{dy} \operatorname{grad} \xi_k(u) = 1, \tag{4}$$

having introduced the notation

$$\operatorname{grad} \xi_k(u) = \left\{ \frac{\partial \xi_k(u)}{\partial u_1}, \quad \ldots, \frac{\partial \xi_k(u)}{\partial u_n} \right\}.$$

Finally, substituting the expression (3) for du/dy into (4), we obtain an equation which serves to determine λ:

$$\lambda r^k(u) \operatorname{grad} \xi_k(u) = \lambda \sum_{j=1}^{n} \frac{\partial \xi_k(u)}{\partial u_j} r_j^k = 1. \tag{5}$$

We consider the following two possibilities.

a) On the interval of variation of y the following inequality is satisfied:

$$r^k(u) \operatorname{grad} \xi_k(u) \neq 0. \tag{6}$$

Systems for which for all values $k = 1, \ldots, n$ and any $u = \{u_1, \ldots, u_n\}$ the inequalities (6) are satisfied were called *genuinely nonlinear* by Lax (1957).

If (6) is satisfied, then λ is uniquely determined from (5), which when substituted into (3) yields the system of ordinary differential equations

$$\frac{du}{dy} = \frac{r^k(u)}{r^k(u) \operatorname{grad} \xi_k(u)}, \tag{7}$$

from which the solution $u(y)$ can be determined by integration if a value

$u_0 = u(y_0)$ is known for which (2) is satisfied, i.e, $y_0 = \xi_k(u_0) = \xi_k(u(y_0))$. We write the solution of the system of ordinary differential equations (7) passing through the point $u_0 = \{u_1^0, \dots, u_n^0\}$ as follows:

$$u(y) = U^k(y, u_0), \qquad U^k(\xi_k(u_0), u_0) \equiv u_0. \tag{8}$$

b) The second possibility occurs when at a given point $y = y_0$

$$r^k(u(y)) \operatorname{grad} \xi_k(u(y)) = 0. \tag{9}$$

In this case (5) is not satisfied for $\lambda \neq 0$, and the solution $u(y)$ is either constant in some neighborhood of y_0 or discontinuous at y_0. The case of a discontinuity of a self-similar solution will be discussed below.

In general, unless otherwise mentioned, we shall restrict our attention to the case where conditions (6) are satisfied for all $k = 1, \dots, n$ and any u. Limiting ourselves to this case, we now study in detail the solution $u(y)$ of (7). Following the analogy with the equations of gas dynamics, we call the solution (8) a "centered rarefaction wave" or a centered traveling wave.

In the space of the variables $u_1, \dots, u_n$ we consider the partial differential equation for one (scalar) function $v = v(u) = v(u_1, \dots, u_n)$:

$$r^k(u) \operatorname{grad} v(u) = \sum_{\alpha=1}^{n} r_\alpha^k(u) \frac{\partial v}{\partial u_\alpha} = 0. \tag{10}$$

Let $v = v(u)$ be some solution of (10). We shall show that

$$v(U^k(y, u_0)) = \text{const}. \tag{11}$$

Indeed,

$$\begin{aligned} dv(U^k(y, u_0)) &= \sum_{\alpha=1}^{n} \frac{\partial v}{\partial u_\alpha} \frac{\partial U_\alpha^k}{\partial y} dy \\ &= \sum_{\alpha=1}^{n} \frac{\partial v}{\partial u_\alpha} \lambda r_\alpha^k(U)\, dy = \lambda\, dy\, r^k(u) \operatorname{grad} v = 0. \end{aligned}$$

Thus, each solution $v(u)$ of (10) is constant on the rarefaction wave (8). This can be formulated in the following manner: an integral curve (8) of the system (7) lies on the hypersurface $v(u) = \text{const}$ if $v(u)$ satisfies (10).

The equation (10) has $n - 1$ independent solutions, which we denote by

$$v = v_1^k(u), \qquad v = v_2^k(u), \qquad \dots, \qquad v = v_{n-1}^k(u):$$

the system of n vectors

$$r^k(u), \qquad \operatorname{grad} v_1^k(u), \qquad \dots, \qquad \operatorname{grad} v_{n-1}^k(u)$$

is thus linearly independent.

It follows from (11) that on the travelling wave $u = u(y)$ defined by (8) the $(n-1)$-dimensional vector $v^k(u) = \{v_1^k(u), \ldots, v_{n-1}^k(u)\}$ is constant, i.e., $v^k(U^k(y, u_0)) = c^k$, where c^k is a constant $(n-1)$-dimensional vector. Thus, the integral curve (8) of equation (7) can also be defined in another manner,

$$v^k(u) = \text{const}, \qquad \xi_k(u) = y,$$

and the problem of integrating the system (7) reduces to determining $n-1$ independent solutions of (10).

We call the vector $v^k(u) = \{v_1^k(u), \ldots, v_{n-1}^k(u)\}$ an $(n-1)$-*dimensional Riemann invariant*. It is easy to show that in cases where the usual Riemann invariants $r_k = r_k(u)$ exist the vector invariant $v^k(u)$ can be represented in the form

$$v^k(u) = \{r_1(u), r_2(u), \ldots, r_{k-1}(u), r_{k+1}(u), \ldots, r_n(u)\}.$$

Indeed, according to §3 of Chapter 1,

$$dr_i(u) = \mu_i(u) l^i(u) du,$$

i.e.,

$$\operatorname{grad} r_i(u) = \mu_i(u) l^i(u),$$

and hence $v_i^k(u) = r_i(u)$ for $i \neq k$ is a solution of (10), since right and left eigenvectors $r^k(u)$ and $l^k(u)$ of the matrix $A(u)$ are biorthogonal: $l^i(u) r^k(u) = 0$ for $i \neq k$. In particular, for $n = 2$ the Riemann invariants always exist, and therefore for $n = 2$

$$v^k(u) = r_i(u), \qquad i \neq k.$$

A solution (8) of the system (7) is a twice continuously differentiable curve situated in the space of the variables $u_1, \ldots, u_n$ and y. If y is taken as the parameter, then this is a curve passing through u_0 in the space of the variables u.

In view of condition (6) the quantity $\xi_k(u(y))$ varies monotonically along this curve. Since $y = x/t$ and $t > 0$, we conclude from (2) that the part of this curve defined by the condition $\xi_k(u(y)) > \xi_k(u_0)$ corresponds to rays $y = x/t$ lying to the right of $y = y_0$, while the part of the curve on which $\xi_k(u(y)) < \xi_k(u_0)$ corresponds to rays $y = x/t$ lying in the x, t plane to the left of $x/t = y_0$.

By letting the index k in (2) run through all values from 1 to n, we arrive at the following conclusion. Through the point u_0 of the space E_n of the variables $u_1, \ldots, u_n$ there pass n smooth curves which are solutions of (3) for $k = 1, \ldots, n$ and are nowhere tangent to one another. Along each of these curves there is a direction of monotonic increase of the variable $y = x/t$.

We now consider the case of a discontinuity of a self-similar solution. A self-similar soluton $u(y) = u(x/t)$ may have discontinuities only on lines $x/t = y = \text{const}$. Suppose that at the point y a solution $u(y)$ is discontinuous. Then on the line $x = yt$ the Hugoniot conditions (§1) must be satisfied:

$$y[u(y-0) - u(y+0)] = \varphi(u(y-0)) - \varphi(u(y+0)), \qquad D = y.$$

We assume that one of the values $u(y+0)$ or $u(y-0)$ is fixed and denote it by u_0; this system then serves to determine the other quantity, which we denote by u. We rewrite this system in the form

$$D(u - u_0) = \varphi(u) - \varphi(u_0), \qquad D = y. \tag{12}$$

We assume that in the space of the variables u the equations (12) define n smooth curves passing through $u = u_0$. We consider one of these curves and introduce some parameter ξ so that a solution of (12) is expressed parametrically in terms of ξ as $u = u(\xi)$, $D = D(\xi)$ and $u(\xi_0) = u_0$. Differentiating (12) with respect to ξ, we find that

$$D\dot{u} + \dot{D}[u(\xi) - u_0] = A(u(\xi))\dot{u},$$

where $\dot{u}$ and $\dot{D}$ denote the derivatives of the corresponding functions with respect to ξ. Setting $\xi = \xi_0$, we obtain

$$[A(u_0) - DE]\dot{u}(\xi_0) = 0, \tag{13}$$

since $u(\xi_0) = u_0$. The quantity $\dot{u}(\xi_0)$ is only nonzero in the case

$$\operatorname{Det}((A(u_0) - D(\xi_0)E)) = 0.$$

We therefore set $D(\xi_0) = D_k(\xi_0) = \xi_k(u_0)$, and denote this branch of the curve determined by the system (12) by $u = \bar{U}^k(\xi)$.

We now choose the parameter ξ in a particular manner:

$$\xi = \xi_k(u). \tag{14}$$

Equations (13) have the corollary

$$\dot{\bar{U}}^k(\xi_0) = \lambda r^k(u_0), \tag{15}$$

and from (14), differentiating with respect to ξ, we obtain

$$\dot{\bar{U}}^k(\xi_0) \operatorname{grad} \xi_k(u_0) = 1. \tag{16}$$

A comparison of (15) and (16) with (3) and (4) leads to the conclusion that

$$\dot{\bar{U}}^k(\xi_0) = \frac{r^k(u_0)}{r^k(u_0) \operatorname{grad} \xi_k(u_0)}, \tag{17}$$

i.e., the derivative $\dot{U}^k(\xi_0)$ has the same value as in the rarefaction wave (7) We shall now show that the second derivatives $\ddot{U}^k(\xi_0)$ also coincide with the second derivatives computed on a rarefaction wave. Differentiating (12) twice with respect to $\xi = \xi_k(u)$ and then setting $\xi_k(u) = \xi_k(u_0)$, we obtain

$$[A(u_0)-\xi_k(u_0)E]\ddot{U}^k(\xi_0)=2\dot{D}_k(\xi_0)\dot{U}^k(\xi_0)-[\nabla A(u_0)\dot{U}^k(\xi_0)]\dot{U}^k(\xi_0), \qquad (18)$$

where $[\nabla A\dot{U}^k]\,\dot{U}^k$ denotes the vector

$$[\nabla A\dot{U}^k]\dot{U}^k = \sum_{j=1}^{n}\sum_{l=1}^{n}\frac{\partial^2\varphi}{\partial u_j\partial u_l}\,\dot{U}_j^k\dot{U}_l^k. \qquad (19)$$

For comparison we differentiate (10) with respect to y; we obtain equations similar to (18) if we set $y = y_0$:

$$[A(u_0) - \xi_k(u_0)E]\frac{d^2u}{dy^2} = \frac{du}{dy} - \left[\nabla A(u_0)\frac{du}{dy}\right]\frac{du}{dy}. \qquad (20)$$

The matrix of coefficients on the left side of (18) and (20) is degenerate, since $\xi_k(u_0)$ is an eigenvalue of the matrix $A(u_0)$. Therefore, for the solvability of (20) with respect to d^2u/dy^2 it is necessary that the right side be orthogonal to the left eigenvector $l^k(u_0)$ of the matrix $A(u_0)$, since the left side of the equality is clearly orthogonal to this vector. Since by (7) the function $u(y)$ has completely determined second derivatives, this condition is clearly satisfied. Hence, writing down the condition that the right side of (20) be orthogonal to $l^k(u_0)$, we obtain

$$\frac{du(y_0)}{dy}\,l^k(u_0) = \frac{r^k(u_0)\,l^k(u_0)}{r^k(u_0)\text{grad }\xi_k(u_0)} = \frac{l^k(u_0)[\nabla A(u_0)\,r^k(u_0)]\,r^k(u_0)}{[r^k(u_0)\text{ grad }\xi_k(u_0)]^2}. \qquad (21)$$

Recalling (17), we write the condition that the right side of (18) be orthogonal to $l^k(u_0)$ in the form

$$\dot{D}_k(\xi_0) = \frac{1}{2}\,\frac{l^k(u_0)\,[\nabla A(u_0)\,r^k(u_0)]\,r^k(u_0)}{(r^k(u_0)\text{ grad }\xi_k(u_0))\,(r^k(u_0)\,l^k(u_0))}. \qquad (22)$$

Using (21), from (22) we obtain

$$\dot{D}_k(\xi_0) = \frac{1}{2}. \qquad (23)$$

From a comparison of the system (18) with (20) we now conlcude that

$$\ddot{U}^k(\xi_0) = \frac{d^2U^k}{dy^2}\,(y_0, u_0). \qquad (24)$$

We denote by $u = \bar{U}^k(\xi_k(u), u_0)$ the solution of (12). From (17) and (24) it then follows that

$$\bar{U}^k(\xi_k(u), u_0) = U^k(\xi_k(u), u_0) + O[(\xi_k(u) - \xi_k(u_0))^3] \tag{25}$$

and

$$\begin{aligned} D_k(\xi_k(u), u_0) &= \xi_k(u_0) + \frac{1}{2}[\xi_k(u) - \xi_k(u_0)] + O[(\xi_k(u) - \xi_k(u_0))^2] \\ &= \frac{\xi_k(u) + \xi_k(u_0)}{2} + O[(\xi_k(u) - \xi_k(u_0))^2]. \end{aligned} \tag{26}$$

The formulas (25) reflect the fact that the state $\bar{U}^k$, which is connected to the state u_0 by means of a shock wave of index k, differs from the state U^k connected with u_0 by a rarefaction wave by a quantity of third-order smallness in the difference $[U^k - u_0]$. For general systems of quasilinear equations this fact was first proved by Lax (1953). In gas dynamics this property of weak shock waves is well known.

Thus, through the point $u = u_0$ there pass two smooth curves $u = U^k(\xi_k(u), u_0)$ and $u = \bar{U}^k(\xi_k(u), u_0)$. The first curve represents the family of states which can be joined with u_0 by means of a rarefaction wave of kth type. The second curve is formed by states which can be connected with u_0 by means of a shock wave of index k. These two curves have second-order tangency at u_0.

It follows from (25) that in a small neighborhood of $u = u_0$ the curves of the shock transition $u = \bar{U}^k(\xi_k, u_0)$ are close to the curves $u = U^k$ describing transitions through rarefaction waves. However, globally (i.e., far from $u = u_0$) the structure of the curves of the shock transition may be rather complicated and not at all similar to the structure of the curves $u = U^k$. Thus, the curves $u = U^k$ can be extended without self-intersection through any point of the region in which the conditions (6) are satisfied right up to the boundary of the region. On the other hand, the curves of a shock transition defined by means of the Hugoniot conditions (12) may behave altogether differently. The example of a strictly hyperbolic system of two equations

$$u_t + (3 \ln u + v)_x = 0, \qquad v_t + \left(\frac{2}{u}\right)_x = 0$$

(see Borovikov (1969)) for $u > 0$ and satisfying the conditions (6) in this half plane corroboratex this.

The set of values (u, v) forming a solution of the Hugoniot conditions (12) for this system is situated on a bounded, closed curve having the form of a figure eight. Thus, for this system even the partition of this set into curves $u = \bar{U}^1$ and $u = \bar{U}^2$ is meaningful only near the point $u = u_0$; the boundedness of this curve leads to major difficulties in solving the problem of the decay of an arbitrary discontinuity.

If we assume that u_0 is the left state, then for $y > y_0$, as we have mentioned above, only the half of the curve $u = U^k(\xi_k(u), u_0)$ defined by the condition $\xi_k(u) > \xi_k(u_0)$ is meaningful. For a shock wave $u = \bar{U}^k(\xi_k(u), u_0)$ the stability conditions (3.1.10) and (3.1.11) require that

$$\xi_k(\bar{U}^k(\xi_k(u_0), (u_0)) > D_k(\xi_k(u), u_0) > \xi_k(\bar{U}^k(\xi_k(u), (u_0)),$$
$$\xi_{k-1}(\bar{U}^k(\xi_k(u_0), u_0)) < D_k(\xi_k(u), u_0) < \xi_{k+1}(\bar{U}^k(\xi_k(u), u_0)),$$

i.e., if $u = u_0$ is the left state, then to the right states there corresponds only the half of the curve $u = \bar{U}^k(\xi_k(u), u_0)$ defined by $\xi_k(u) < \xi_k(u_0)$. Therefore, the curve

$$u = \begin{cases} U^k(\xi_k(u), u_0) & \text{for } u = \xi_k(u) \geq \xi_k(u_0), \\ \bar{U}^k(\xi_k(u), u_0) & \text{for } \xi_k(u) \leq \xi_k(u_0) \end{cases} \tag{27}$$

possesses two continuous derivatives because of (25), passes through u_0, and represents the family of states which can be joined with u_0 considered as the left state relative to u given by (27) by means of a rarefaction wave or a shock wave. On the other hand, if u_0 is considered the right state, then an analogous curve is given by

$$u = \begin{cases} U^k(\xi_k(u), u_0) & \text{for } \xi_k(u) \leq \xi_k(u_0), \\ \bar{U}^k(\xi_k(u), u_0) & \text{for } \xi_k(u) \geq \xi_k(u_0). \end{cases} \tag{28}$$

The curves (27) and (28) consist of all states u which can be connected with u_0 by a rarefaction wave of kth type or a shock transition of kth type; here in the case (27) the state u_0 lies to the left relative to u, while in the case (28) it lies to the right. The curves (27) and (28) are called *wave adiabats* of the system of conservation laws (3.1.12) of kth type.

Finally, we consider the case where for some $k = 1, \ldots, n$

$$r^k(u) \operatorname{grad} \xi_k(u) \equiv 0, \tag{29}$$

A characteristic $\xi = \xi_k(u)$ satisfying (29) we call a *contact characteristic*.

It is obvious that along the line

$$v^k(u) = \{v_1^k(u), v_2^k(u), \ldots, v_{n-1}^k(u)\} = c^k \tag{30}$$

the quantity $\xi_k(u)$ is constant by (10). Therefore, $\xi_k(u)$ cannot be chosen as above as the parameter defining a point on this curve.

We introduce some other parameter on the curve (30), for example, the arc length of this curve computed from any point of the curve. Then (30) is an integral curve of the system $du/ds = r^k(u)$, since we assume that

$\|r^k(u)\| = \Sigma_1^n(r_i^k)^2 = 1$. Multiplying this equation on the left by the matrix $A(u) = \partial\varphi/\partial u$, we obtain

$$A(u)\frac{du}{ds} = \frac{\partial\varphi}{\partial u}\frac{du}{ds} = \frac{d\varphi}{ds} = A(u)r^k(u) = \xi_k(u)r^k(u) = \xi_k(u)\frac{du}{ds}.$$

Now along the curve (30) the quantity $\xi_k(u)$ is constant, and hence these equations can be integrated. We integrate these equations from $u = u_0$ to an arbitrary point $u = u(s)$; then

$$\varphi(u(s)) - \varphi(u_0) = \xi_k(u)[u(s) - u_0] = \xi_k(u_0)[u(s) - u_0]. \tag{31}$$

We thus see that any two points of the curve (30) satisfy the Hugoniot conditions for $D = \xi_k(u_0) = \xi_k(u)$. Discontinuities of this are called *contact discontinuities* in gas dynamics.

Finally, we observe that if the condition (29) is satisfied, then there exists no centered rarefaction wave of kth type. This rarefaction wave goes over into the shock wave (30).

The concept of a wave adiabat of kth type can be generalized, in particular, to allow for the possibility that condition (6) is violated. Any generalization of the concept of the wave adiabat should be based on a generalization of the stability conditions for a discontinuity of a solution.

In analogy with the case of a single equation and the system of equations of gas dynamics (see Chapter 2, §§5.7 and 6.9) for the stability of a discontinuity (u_0, u) with left state u_0 and right state u it is necessary to require that

$$\xi_k(u_0) \geq D \geq \xi_k(u); \tag{32}$$

here $x = Dt$ is the line of discontinuity. However, these conditions are not sufficient even in the simple cases indicated above. Apparently, in the general case also it is necessary to require that u_0 and u be joined by a continuous cure $H(u_0, \bar{u}) = 0$ each point $\bar{u}$ of which corresponds to a solution of the Hugoniot conditions (12), i.e., $\varphi(\bar{u}) - \varphi(u_0) = D(\bar{u} - u_0)$, where $D = D(u_0, u)$ and

$$D(u_0, \bar{u}) \geq D(u_0, u) \tag{33}$$

for any point $\bar{u}$ of the curve $H(u_0, \bar{u}) = 0$ between u_0 and u.

We thus arrive at a generalization of the concept of a wave adiabat of kth type constructed for a point u_0 considered as the left state relative to the stage u. It should be a *continuous* curve in the space of the variables $u = \{u_1, \ldots, u_n\}$ and consist of the set of values u which can be joined with u_0 (considered as the left state) by rarefaction waves $y = \xi_k(u)$ and admissible discontinuities on which conditions (32) and (33) are satisfied.

A similar concept of a wave adiabat was actually used for the equations of gas dynamics by Galin (1958); such a concept was defined by Sidorenko (1968) and was studied by Wendroff (1972) and Liu (1975a).

For more general systems of n quasilinear conservation laws the concept of a wave adiabat was studied by Liu (1975a); using conditions on the conservation laws under which the wave adiabat is uniquely determined.

3. ***The problem of decay of an arbitrary discontinuity.*** The Cauchy problem

$$\frac{\partial u}{\partial t} + \frac{\partial \varphi(u)}{\partial x} = 0, \tag{1}$$

$$u(x, 0) = \begin{cases} u^-, & x < 0, \\ u^+, & x > 0, \end{cases} \tag{2}$$

is called the *problem of decay of an arbitrary discontinuity*. It is easy to see that this problem is invariant under the similarity transformation[9]

$$t = kt', \qquad x = kx', \quad k = \text{const} > 0 \tag{3}$$

Therefore, if the uniqueness of the solution of the Cauchy problem (1), (2) is assumed, this implies that the solution is self-similar. Indeed, suppose

$$u = u(x, t) \tag{4}$$

is the solution of problem (1), (2). By making the similarity transformation (3) we see that the solution $u(x', t')$ coincides with the solution (4) by the uniqueness assumption, i.e.,

$$u\left(\frac{x}{k}, \frac{t}{k}\right) = u(x, t). \tag{5}$$

The equality (5) is satisfied for any values of the parameter $k > 0$. Therefore, setting $k = 1/t$, we obtain

$$u(x, t) = u(1, x/t) = u(1, y) = u_0(y). \tag{6}$$

It is easy to see that if uniqueness of the solution of the Cauchy problem (1), (2) is not assumed, then it cannot be asserted that all solutions of this problem are self-similar, i.e., depend only on the variable $y = x/t$. Therefore, the proof of the uniqueness of a self-similar solution of problem (1), (2) does not enable us to assert that this Cauchy problem has a unique solution. Nevertheless, the question of uniqueness of a self-similar solution of problem (1), (2) has definite

[9] Only $k > 0$ is admissible. This is explained by the fact that stability conditions for the lines of discontinuity enter the formulation of the Cauchy problem (1), (2). These conditions contain concepts of left and right values of a solution on a discontinuity which are not invariant under the transformation (3) for $k < 0$.

independent interest, first of all, because in a number of cases it is possible to prove directly that any stable generalized solution of this problem is self-similar, and, secondly, because in the process of proving the uniqueness of a self-similar solution the basic difficulties are revealed that are encountered in the study of the general Cauchy problem (1), (2).

We shall give a geometric interpretation of the problem of decay of an arbitrary discontinuity. We consider a self-similar solution $u(y)$ of problem (1), (2), assuming that it exists. In the regions of smoothness of the vector-valued function $u(y)$ it satisfies the system (3.2.1)

$$A(u)\frac{du}{dy} = y\frac{du}{dy}, \tag{7}$$

and at points of discontinuity it satisfies the Hugoniot conditions

$$y[u(u+0) - u(y-0)] = \varphi(u(u(y+0)) - \varphi(u(y-0)). \tag{8}$$

We shall assume that the solution $u(y)$ is bounded. It can then be asserted that the vector $u(y)$ is nonconstant only in a bounded interval of values of y. Indeed, suppose that $|u(y)| \leq M$ and $|\xi_k(u)| \leq M$; if $y = y_0$ is a point of continuous variation of $u(y)$, then (7) is satisfied at this point, and, as we saw in subsection 2,

$$|y_0| = |\xi_k(u(y_0))| \leq M. \tag{9}$$

If $y = y_0$ is a point of discontinuity of $u(y)$, then the stability conditions imply that

$$|D| = |y_0| \leq \max_{k=1,\ldots,n} \max_{\substack{y=y_0+0\\ y=y_0-0}} |\xi_k(u(y))| \leq M. \tag{10}$$

Thus, if $|\xi_k(u(y))| \leq M$ $(k = 1, \ldots, n)$, then outside the interval $[-M, M]$ the function $u(y)$ is obviously constant. Condition (2) for $u(y)$ takes the form

$$u(y) \to u^- \quad \text{as} \quad y \to -\infty, \qquad u(y) \to u^+ \quad \text{as } y \to \infty, \tag{11}$$

and because of the boundedness of $u(y)$

$$u(y) = \begin{cases} u^- & \text{for } y < -M, \\ u^+ & \text{for } y > M. \end{cases} \tag{12}$$

Suppose that at a point $y \geq -M$ the solution $u(y) = u^-$. We consider the possible variation of $u(y)$. If $u(y)$ varies by forming a rarefaction wave of kth type, then

$$y = \xi_k(u(y)).$$

The smallest of the quantities ξ_k is ξ_1. We therefore consider the segment of the variable y on which

$$y = \xi_1(u(u)). \tag{13}$$

Since $u(y) = u^-$ for $y < -M_1$, a solution of the type (13) may occur starting with values y_1^-, where

$$y_1^- = \xi_1(u^-). \tag{14}$$

Suppose that on $[y_1^-, y_1^+]$ the inequality (13) is satisfied, i.e., in the segment the solution $u(y)$ forms a rarefaction wave corresponding to the first eigenvalue ξ_1.

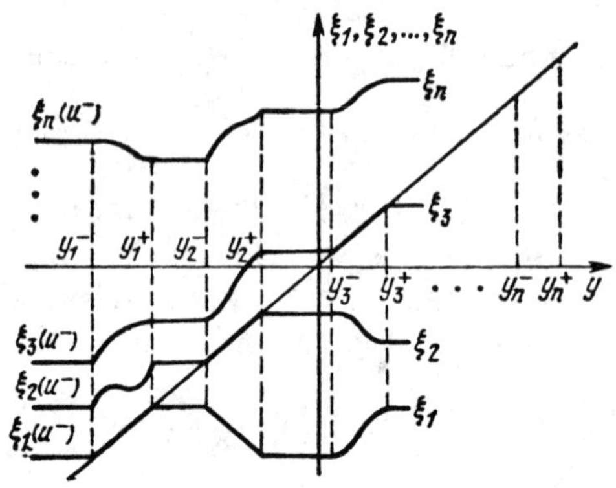

FIGURE 4.33

In Figure 4.33 the value of y is plotted along the axis of the abscissa, the values of $\xi_1(u(y)), \ldots, \xi_n(u(y))$ along the ordinate axis, and the bisectrix $\xi = y$ is drawn. On the segment $y_1^- \le y \le y_1^+$ we have $\xi_1(u(y)) = y$, and the remaining eigenvalues $\xi_2(u), \ldots, \xi_n(u)$ are greater than y. The states $u = u(y)$ on $[y_1^-, y_1^+]$ are connected with u^- as a left state by the rarefaction wave corresponding to the eigenvalue $\xi = \xi_1(u)$. The graph of the dependence $\xi_1 = \xi_1(u(y))$ for $y_1^- \le y \le y_1^+$ lies on the line $\xi = y$.

Suppose that $\xi_2(u(y)) = y$ on $[y_2^-, y_2^+]$ i.e., the solution $u(y)$ on this segment forms a rarefaction wave corresponding to $\xi = \xi_2(u)$. It is then altogether obvious and, in particular, evident from Figure 4.33 that

$$y_2^- > y_1^+, \qquad \xi_3(u(y)) > y, \qquad \xi_1(u(y)) < y. \tag{15}$$

The states $u(y)$ on $[y_2^-, y_2^+]$ are connected with $u(y_1^+)$ as a left state by a rarefaction wave of second type ($y = \xi_2(u)$).

It is now clear from Figure 4.33 that each rarefaction wave corresponding to the kth eigenvalue ($y = \xi_k(u(y))$ lies to the right of another rarefaction wave corresponding to an eigenvalue ξ_l of one lower index l and to the left of any rarefaction wave with one greater index l. From this we conclude that a continuous solution $u(y)$ of problem (7), (12) contains no more than n ordered rarefaction waves.

If it is possible to choose y_k^- and y_k^+ so that $u(y_1^-) = u^-$ and $u(y_n^+) = u^+$, then the function $u(y)$ defined on the segments $[y_k^-, y_k^+]$ as a rarefaction wave, i.e.,

$$u(y) = U^k(y, U_+^{k-1}) \quad \text{for } y_k^- \leq u \leq y_k^+, \tag{16}$$

$$u(y) = U_+^k \quad \text{for } y_k^+ \leq y \leq y_{k+1}^-, \tag{17}$$

where

$$U_+^k = U^k(y_k^+, U_+^{k-1}),\ U_+^0 = u^-,$$

gives a self-similar solution of the problem of decay of a discontinuity.

We consider another possibility, in which at the point $y = y_k$ the solution $u(y)$ is discontinuous. According to the stability conditions, there then exists an index k such that

$$\xi_k(u(y_k - 0)) > y_k > \xi_k(u(y_k + 0)), \tag{18}$$

$$\xi_{k+1}(u(y_k + 0)) > y_k > \xi_{k-1}(u(y_k - 0)). \tag{19}$$

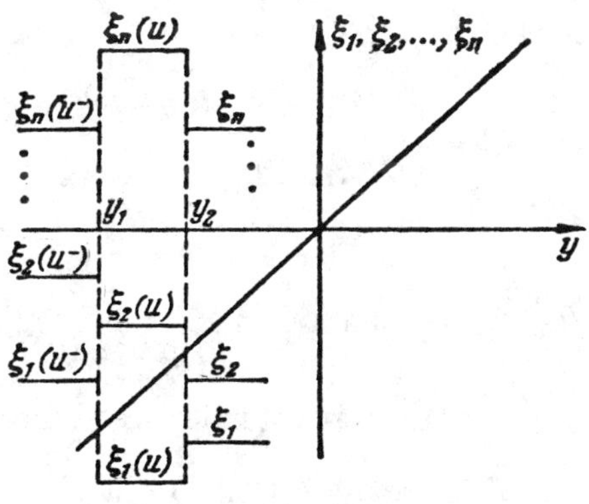

FIGURE 4.34

Suppose that the index of the line of discontinuity at the point y_1 is $k = 1$. Figure 4.34 shows the functional dependence of $\xi_k(u(y))$ for $k = 1, \ldots, n$. In this case at the point y_1

$$\xi_1(y_1 - 0) = \xi_1(u^-) > y_1 > \xi_1(u(y_1 + 0)), \tag{20}$$

$$\xi_k(u(y_1^- + 0)) > y_1 \quad \text{for } k > 1. \tag{21}$$

From this it follows immediately that if at $y = y_1$ there is a shock wave with index $k = 1$, then the solution $u = u(y)$ contains no rarefaction wave corresponding to the value $\xi = \xi_1(u) = y$.

In general, from a comparison of Figures 4.33 and 4.34 we conclude that a stable (satisfying the stability conditions (18) and (19)) generalized self-similar solution $u(y)$ contains no more than n travelling waves (rarefaction waves or shock waves) which are ordered by their indices, since the presence of a shock wave of index k precludes the possibility of a rarefaction wave of index k, and conversely.

Thus, in solving the decay problem (1), (2) in the class of self-similar solutions it is possible to construct the solution as n travelling waves with amplitudes chosen in such a way as to satisfy (11). As we have seen in the preceding section, if the left state is known, then for a wave of kth type the family of states to which it is possible to pass by means of this wave (a shock wave or a rarefaction wave of kth type) is described by a single parameter. Thus, the problem consists in satisfying the conditions (11) by choosing n such parameters.

On the basis of a given value u^-, which is a left state for the solution $u = u(y)$, we determine the solution in the (shock or rarefaction) wave of index 1 as a function of one parameter $\xi_1 = \xi_1(u)$,

$$u^1 = F^1(\xi_1, u^-), \tag{22}$$

where F^1 is given, according to (3.2.27), by

$$F^1(\xi_1, u^-) = \begin{cases} U^1(\xi_1, u^-) & \text{for } \xi_1 \geq \xi_1(u^-), \\ \bar{U}^1(\xi_1, u^-) & \text{for } \xi_1 \leq \xi_1(u^-); \end{cases} \tag{23}$$

and

$$\frac{\partial F^1}{\partial \xi_1}(\xi_1(u^-), u^-) = \frac{r^1(u^-)}{r^1(u^-)\,\text{grad}\,\xi_1(u^-)}. \tag{24}$$

The state (22) is a left state for the wave of index 2. We therefore introduce

$$u^2 = F^2(\xi_2, u^1) = F^2(\xi_2, F^1(\xi_1, u^-)), \tag{25}$$

where F^2 is determined from (3.2.27) in analogy to (23). Continuing our argument, we find that

$$\begin{aligned} u^n &= F^n(\xi_n, u^{n-1}) \\ &= F^n(\xi_n, F^{n-1}(\xi_{n-1}, F^{n-2}(\ldots(\xi_2, F^1(\xi_1, u^-))\ldots))) \\ &= \Phi(\xi_1, \xi_2, \ldots, \xi_n, u^-). \end{aligned} \tag{26}$$

Solution of the decay problem in the class of self-similar solutions reduces to determining $\xi_1, \ldots, \xi_n$ from the system of n equations

$$\Phi(\xi_1, \xi_2, \ldots, \xi_n, u^-) = u^+. \tag{27}$$

We shall assume that the states $u^1, \ldots, u^n = u^+$ lie in a sufficiently small neighborhood of u^-. Since the Jacobian

$$\left.\frac{\partial\Phi(\xi_1, \ldots \xi_n, u^-)}{\partial(\xi_1, \ldots, \xi_n)}\right|_{\substack{\xi_1=\xi_1(u^-)\\ \cdots\cdots\\ \xi_n=\xi_n(u^-)}} = \operatorname{Det}((r_i^k(u^-))) \prod_{k=1}^{n} \frac{1}{(r^k(u^-) \operatorname{grad} \xi_k(u^-))} \tag{28}$$

is nonzero because of the assumed hyperbolicity of the system (1) and the fact that conditions (3.2.6) are satisfied, there exists some neighborhood $|u - u^-| < \delta$ of u^- such that for $u^1, \ldots, u^n$ in this neighborhood the Jacobian

$$\frac{\partial\Phi(\xi_1, \ldots, \xi_n, u^-)}{\partial(\xi_1, \ldots, \xi_n)}$$

is nonzero. Moreover, the equations (27) are compatible for $u^+ = u^-$, $\xi_k = \xi_k(u^-)$. Therefore, on the basis of the implicit function theorem in this neighborhood there exists a unique solution $\xi_1, \ldots, \xi_n$ of the system (27) to which there corresponds the self-similar solution $u = u(y)$. This theorem on the existence of a solution of the problem of decay of a small discontinuity was proved by Lax (1957).

We remark now that existence and uniqueness of a self-similar solution $u(y)$ has now been proved not only under the assumptions that the points u^- and u^+ are close but that the quantities $u^-, u^1, \ldots, u^n = u^+$ are close. Therefore, the question of uniqueness of the solution for sufficiently close u^- and u^+ has still not been solved, since it is entirely possible that for arbitrarily close u^- and u^+ there exists another solution $\bar{u}(y)$ of the decay problem for which the intermediate states $u^1, \ldots, u^{n-1}$ are far from u^- and u^+.

4. *An ex imple of nonuniqueness of a self-similar solution of the decay problem.* We shall now show that if it is not assumed that the vectors u^- and u^+ are close, then without restrictions on the systems of quasilinear equations considered it is not possible to count on the uniqueness of a self-similar solution of this problem.

We remark, first of all, that we have introduced the concept of a discontinuous self-similar solution of the decay problem only for conservative systems of the form (3.3.1); however, a continuous solution $u(y)$, if it exists, is defined also for systems not written as conservation laws. We therefore now consider a strictly hyperbolic system of three quasilinear equations

$$l^k(u)\left[\frac{\partial u}{\partial t} + \xi_k(u)\frac{\partial u}{\partial x}\right] = 0,\ u = \{u_1, u_2, u_3\} \qquad (k = 1, 2, 3), \tag{1}$$

leaving aside the question of whether this system can be written in the form of conservation laws, and we shall show that the decay problem for this system may have several continuous, self-similar solutions $u = u(y)$. We remark that this situation occurs also for conservative systems of quasilinear equations (see D'jačenko (1963)); we consider a nonconservative system, since in this case the example is simpler.

We set

$$l^1 = \{\cos u_2, 0, \sin u_2\}, \qquad l^2 = \{0, 1, 0\}, \qquad l^3 = \{-\sin u_2, 0, \cos u_2\}, \tag{2}$$

and suppose that

$$\xi_1(u) < \xi_2(u) \equiv u_2 < \xi_3(u). \tag{3}$$

The vectors l^1, l^2, l^3 are mutually orthogonal, and we may therefore assume that

$$r^k(u) = l^k(u). \tag{4}$$

Conditions (3.2.6) for the system (1) take the form

$$\left.\begin{aligned} k = 1: &\qquad \frac{\partial \xi_1}{\partial u_1}\cos u_2 + \frac{\partial \xi_1}{\partial u_3}\sin u_2 \neq 0, \\ k = 2: &\qquad \frac{\partial \xi_2}{\partial u_2} = 1 > 0, \\ k = 3: &\qquad -\frac{\partial \xi_3}{\partial u_1}\sin u_2 + \frac{\partial \xi_3}{\partial u_3}\cos u_2 \neq 0. \end{aligned}\right\} \tag{5}$$

It is easy to see that the functions ξ_1 and ξ_3 can always be chosen so that conditions (5) and (3) are satisfied; for example,

$$\xi_1 = u_2 - \alpha(u_1 \cos u_2 + u_3 \sin u_2), \qquad \alpha > 0,\ \alpha' \neq 0, \tag{6}$$

$$\xi_3 = u_2 + \beta(-u_1 \sin u_2 + u_3 \cos u_2), \qquad \beta > 0,\ \beta' \neq 0. \tag{7}$$

The curves $u = U^k(y, u_0)$ describing states that can be joined by means of rarefaction waves for the system (1) are lines:

$$U^1(y, u_0) = u_0 + l^1(u_0)s, \qquad \frac{dy}{ds} = l^1(u_0)\operatorname{grad}\xi_1(U^1), \tag{8}$$

$$U^2(y, u_0) = u_0 + l^2(u_0)(y - u_2^0), \tag{9}$$

$$U^3(y, u_0) = u_0 + l^3(u_0)s, \qquad \frac{dy}{ds} = l^3(u_0)\operatorname{grad}\xi_3(U^3). \tag{10}$$

The line $u = U^1$ lies in the plane $u_2 = u_2^0$ and has the direction of the vector $l^1(u_0)$; the line $u = U^3$ also in the plane $u_2 = u_2^0$ and has the direction of the vector $l^3(u_0)$; the $u = U^2$ is the line $u = u_1^0$, $u_2 = u_2^0$ (Figure 4.35). The arrows on the lines $u = U^k$ denote the direction of increase of the quantities $\xi_k(u)$.

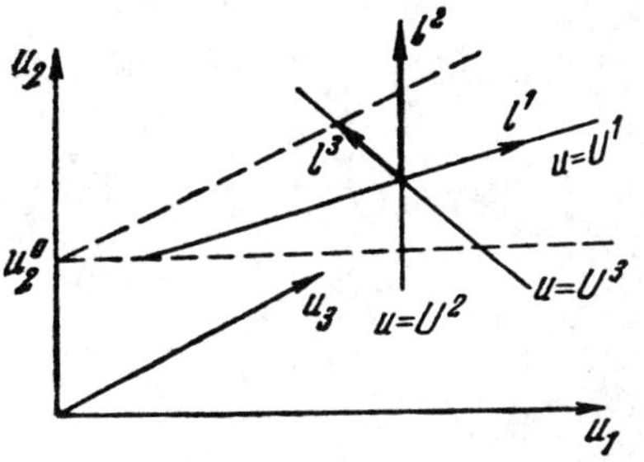

FIGURE 4.35

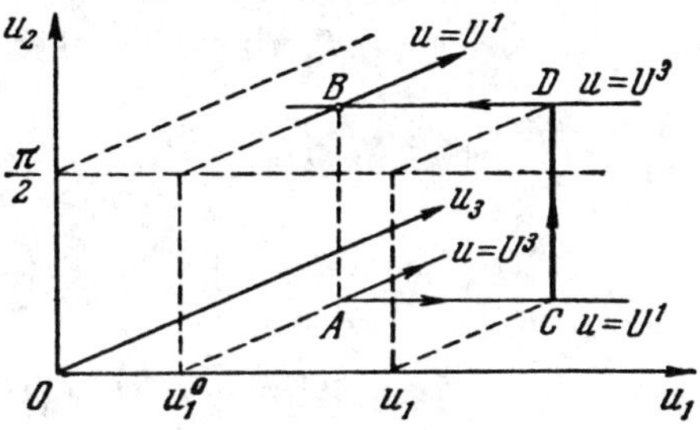

FIGURE 4.36

We consider any two planes $u_2 = c$ and $u_2 = c + \pi/2$, for example, $u_2 = 0$ and $u_2 = \pi/2$ (Figure 4.36). Since $l^1(u_1, 0, u_3) = l^3(u_1, \pi/2\ u_3)$, the lines $u = U^1$ and $u = U^3$ lying in these two planes project onto one another. The arrows on these lines $u = U^k$ indicate the direction of increase of the quantity $\xi_k(u)$, as before. For the system (1) we now consider the decay problem for the case where

$$u^- = \{u_1^0, 0, u_3^0\}, \qquad u^+ = \left\{u_1^0, \frac{\pi}{2}, u_3^0\right\}. \tag{11}$$

We construct one of the solutions of this problem,

$$u(y) = \begin{cases} u^- & \text{for } y \leq 0, \\ U^2(y, u^-) & \text{for } 0 \leq y \leq \pi/2, \\ u^+ & \text{for } y \geq \pi/2, \end{cases} \tag{12}$$

which is obviously continuous in y. We shall show that for this problem there exists an infinite set of other self-similar solutions $u(y)$. Suppose, for example, that the points C and D lie on a line parallel to the u_2 axis: $C(u_1, 0, u_3^0)$ and $D(u_1, \pi/2, u_3^0)$ (Figure 4.36). Because of our assumptions,

$$\xi_1(u_1^0, 0, u_3^0) > \xi_1(u_1, 0, u_3^0), \qquad \xi_3\left(u_1, \frac{\pi}{2}, u_3^0\right) > \xi_3\left(u_1^0, \frac{\pi}{2}, u_3^0\right). \tag{13}$$

We can therefore write out another self-similar and continuous solution of this decay problem:

$$
u(y) = \begin{cases}
u^- & \text{for } y \le \xi_1(u^-) < 0, \\
U^1(y, u^-) & \text{for } \xi_1(u_1^0, 0, u_3^0) \le y \le \xi_1(u_1, 0, u_3^0), \\
U^1(\xi_1(u_1, 0, u_3^0), u^-) & \text{for } \xi_1(u_1, 0, u_3^0) \le y \le 0, \\
U^2(y, (u_1, 0, u_3^0)) & \text{for } 0 \le y \le \frac{\pi}{2}, \\
U^2\left(\frac{\pi}{2}, (u_1\, 0, u_3^0)\right) & \text{for } \frac{\pi}{2} \le y \le \xi_3\left(u_1, \frac{\pi}{2}, u_3^0\right), \\
U^3\left(y, \left(u_1, \frac{\pi}{2}, u_3^0\right)\right) & \text{for } \xi_3\left(u_1, \frac{\pi}{2}, u_3^0\right) \le y \le \xi_3\left(u_1^0, \frac{\pi}{2}, u_3^0\right), \\
u^+ = U^3\left(\xi_3\left(u_1^0, \frac{\pi}{2}, u_3^0\right), \left(u_1, \frac{\pi}{2}, u,\right)\right) & \text{for } y \ge \xi_3\left(u_1^0, \frac{\pi}{2}, u_3^0\right),
\end{cases} \tag{14}
$$

which is obviously distinct from the solution (12). By varying the quantity $u > u_1^0$ in an arbitrary way, we obtain an infinite set of self-similar solutions.

This example shows that for a hyperbolic system of three quasilinear equations the decay problem may have an infinite set of self-similar solutions.

Finally, we observe that it is possible that the nonuniqueness of the solution of the problem on the decay of an aribtrary discontinuity for a system of the type (1) is somehow related to a general property of systems of three and more quasilinear equations of hyperblic type which consisst in the following. For a system

$$
\frac{\partial u}{\partial t} + A(u)\frac{\partial u}{\partial x} = 0 \tag{15}
$$

of three or more quasilinear equations the growth of a solution $u(x, t)$ of the Cauchy problem with initial data $u(x, 0) = u_0(x)$, in contrast to systems of two quasilinear equations, is determined not only by the quantity $U_0 = \max_x \|u_0(x)\|$, but also by the magnitude of the derivatives $\|\partial u_0/\partial x\|$. Therefore, a solution of (15) for $n \ge 3$, in general, becomes unbounded for some $t > 0$. This shows that, generally speaking, the problem of decay of an arbitary discontinuity for such systems becomes meaningless, since in this case a solution of the problem cannot be considered the limit of solutions with smoothed initial data.

The question of the uniqueness of a continuous solution of the problem on the decay of a discontinuity for systems of three and more equations has been considered in more detail by Tupčiev (1972, 1973). He showed that for $n \ge 3$ in any neighborhood of a given continuous solution of the problem on decay of a discontinuity there may be another solution of the same problem. The solution of the decay problem in this case is called *nonisolated.* Conditions are considered under which any continuous solution of the decay problem is isolated; here both conservative and nonconservative systems of equations are considered. It is shown that for conservative systems the investigation of isolation and uniqueness of a solution of the decay problem is easier.

We note other complications which may arise in the problem of decay of a discontinuity. In subsection 2 we presented a simple system of two quasilinear equations for which the curves $u = \bar{U}_k$ $(k = 1, 2)$ lie in a bounded region of the half plane $u > 0$. From this it follows, in particular, that the decay problem does not have a solution if u^- and u^+ are sufficiently far from one another, while for sufficiently close u^- and u^+ there may be several distinct solutions (see Borovikov (1969)). Conditions for the uniqueness of a solution of the problem of decay of an arbitrary discontinuity for $n = 2$ and the case of "convex" systems ($r^k \nabla \xi_k \neq 0$, $k = 1, 2$) were studied by Borovikov (1972).

5. *The decay problem for a system of two quasilinear equations.* In the case $n = 2$ a system of quasilinear equations reduces to Riemann invariants (Chapter 1, §3) and can be written in the form

$$\frac{\partial r_k}{\partial t} + \xi_k(r)\frac{\partial r_k}{\partial x} = 0 \qquad (k = 1, 2). \tag{1}$$

We shall assume that condition (3.2.6) is satisfied, which for the system (1) can be written in the form

$$\frac{\partial \xi_k}{\partial r_k} \neq 0.$$

The curve $u = U^k(y, u_0)$ in the plane of the variables u_1, u_2 goes over into the line $r_j = \text{const}$ $(j \neq k)$ in the plane of the variables r_1, r_2 (Figure 4.37).

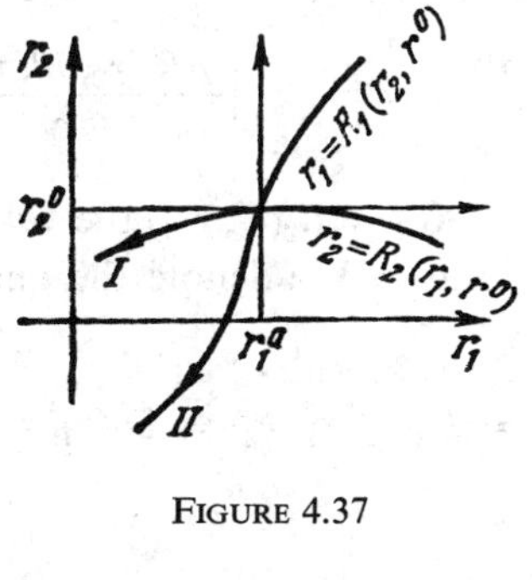

FIGURE 4.37

Thus, a rarefaction wave for the system (1) corresponds to a segment of the line $r_j = \text{const}$. According to the foregoing, through each point $r = r^0$ there pass two curves $u = \bar{U}^k(y, u_0)$ representing families of states which can be connected with u_0 by means of a shock transition. As we have seen above, at the point u_0 these curves have second-order tangency with the lines $r_j = r_j^0$; therefore, at least in some neighborhood of $r = r^0$ the equations of these curves can be written in the form

$$\left.\begin{aligned} k_1 = 1; \qquad r_2 = R_2(r_1, r_1^0, r_2^0) = R_2(r_1, r^0), \\ k_2 = 2; \qquad r_1 = R_1(r_2, r_1^0, r_2^0) = R_1(r_2, r^0), \end{aligned}\right\} \tag{3}$$

i.e., these curves project injectively onto the lines $r_2 = \text{const}$ and $r_1 = \text{const}$. In Figure 4.37 the arrows indicate the direction of increase of the variable $y = x/t$ in the rarefaction waves $r_j = \text{const}$ and in the shock waves $r_j = R_j$; here we require that

$$\frac{\partial \xi_k(r)}{\partial r_k} > 0, \tag{4}$$

which may always be assumed satisfied because of (2).

The question of unique solvability of the decay problem in the class of self-similar solutions depends in an essential way on the behavior of the curves $r_2 = R_2$ and $r_1 = R_1$ globally, i.e., for sufficiently large values of $|r - r_0|$. However, it is difficult to study the global behavior of these curves, since they are determined from essentially nonlinear equations. We therefore now indicate some sufficient conditions under which the problem of decay of an aribtrary solution.

Under the assumptions made above we assume, in addition, that the Hugoniot conditions in the form (3) can be solved for any r_1, r_2, r_1^0 and r_2^0, i.e., the corresponding curves probject injectively onto the axes $r_2 = \text{const}$ and $r_1 = \text{const}$ for any r and r^0. Suppose further that

$$\left|\frac{\partial R_2(r_1, r_1^0, r_2^0)}{\partial r_1}\right| < 1, \qquad \left|\frac{\partial R_1(r_2, r_1^0, r_2^0)}{\partial r_2}\right| < 1. \tag{5}$$

At each point of the curve I ($r_2 = R_2(r_1, r^0)$) the value of $D = y$ contained in the Hugoniot conditions is known. We denote the quantity D along the curve I by

$$D = D_1(r_1, r_1^0, r_2^0) = D_1(r_1, r^0), \tag{6}$$

and on the curve II by

$$D - D_2(r_2, r_1^0, r_2^0) - D_2(r_2, t^0). \tag{7}$$

As we saw in subsection 2, at $r - r^0$

$$\left.\begin{aligned} D_1(r_1^0, r^0) &= \xi_1(r^0), \\ D_2(r_2^0, r^0) &= \xi_2(r^0) \end{aligned}\right\} \tag{8}$$

and

$$\left.\frac{\partial D_1(r_1, r^0)}{\partial r_1}\right|_{r_1=r_1^0} = \frac{1}{2}\frac{\partial \xi_1(r^0)}{\partial r_1} > 0, \qquad \left.\frac{\partial D_2(r_2, r^0)}{\partial r_2}\right|_{r_2=r_2^0} = \frac{1}{2}\frac{\partial \xi_2(r^0)}{\partial r_2} > 0, \tag{9}$$

$$\left.\frac{\partial R_2(r_1, r^0)}{\partial r_1}\right|_{r_1=r_1^0} = 0, \qquad \left.\frac{\partial R_1(r_2, r^0)}{\partial r_2}\right|_{r_2=r_2^0} = 0. \tag{10}$$

In addition to (5), we assume that for any r_1^0, r_2^0 and $r_1 < r_1^0$, $r_2 > r_2^0$ the following conditions are satisfied:

$$\left.\begin{aligned} &\xi_1(r_1, R_2(r_1, r^0)) < D_1(r_1, r^0) < \xi_1(r^0), \\ &\xi_2(r_1, R_2(r_1, r^0)) > D_1(r_1, r^0), \qquad r_1 < r_1^0, \end{aligned}\right\} \tag{11}$$

$$\left.\begin{aligned} &\xi_2(R_1(r_2, r^0), r_2) > D_2(r_2, r^0) > \xi_2(r^0), \\ &\xi_1(R_1(r_2, r^0), r_2) < D_2(r_2, r^0), \qquad r_2 > r_2^0. \end{aligned}\right\} \tag{12}$$

Conditions (11) and (12) obviously imply that the solutions $r_2 - R_2$, $D = D_1$ and $r_1 = R_1$, $D = D_2$ of the Hugoniot conditions satisfy the stability conditions.

If (11) and (12) are satisfied for any r^0 and $r_1 < r_1^0$, $r_2 > r_2^0$, then for $r_1 > r_1^0$ and $r_2 < r_2^0$ in these inequalities the signs are replaced by the opposite signs. Indeed, suppose that $r_1 > r_1^0$. Through the point $(r_1, R_2(r_1, r^0))$ there passes the curve $r_2 = R_2$ which, according to the Hugoniot conditions, satisfies the equations

$$R_2(r_1^0, r_1, R_2(r_1, r^0)) = r_2^0, \qquad D_1(r_1^0, r_1, R_2(r_1, r^0)) = D_1(r_1, r^0). \tag{13}$$

These equations express the obvious fact that the Hugoniot conditions do not change if the left and right values of the solution change places. From (13) the validity of our assertion now follows easily.

Conditions (5), (11), and (12) are somewhat complicated. We therefore verify that they are satisfied for a simple example. We consider the system of two quasilinear equations describing the motion of an isothermic gas in Lagrange variables (Chapter 2, §2.7):

$$\frac{\partial V}{\partial t} - \frac{\partial u}{\partial x} = 0, \quad \frac{\partial u}{\partial t} + \frac{\partial p(V)}{\partial x} = 0, \quad p'(V) < 0, \quad p''(V) > 0, \tag{14}$$

which can be written in Riemann invariants as

$$\frac{\partial r_1}{\partial t} - c(V)\frac{\partial r_1}{\partial x} = 0, \qquad \frac{\partial r_2}{\partial t} + c(V)\frac{\partial r_2}{\partial x} = 0, \tag{15}$$

where

$$r_1 = u - \int_V^\infty c(\eta)\,d\eta, \quad r_2 = u + \int_V^\infty c(\eta)\,d\eta, \quad c^2(V) = -\frac{\partial p}{\partial V}. \tag{16}$$

Equations (16) make it possible to express V in terms of the difference $r_2 - r_1$. To verify (2) we compute the derivatives $\partial\xi_k/\partial r_k$:

$$\frac{\partial\xi_1}{\partial r_1} = \frac{\partial\xi_2}{\partial r_2} = -\frac{c'(V)}{2c(V)} = \frac{p''(V)}{c^2(V)} > 0. \tag{17}$$

Proceeding to the verification of (5), (11), and (12), we write the Hugoniot conditions for the system (14):

$$D(V - V_0) = u_0 - u, \qquad D(u_0 - u) = p(V_0) - p(V), \tag{18}$$

whence

$$(u - u_0)^2 = \{p(V) - p(V_0)\}(V_0 - V), \tag{19}$$

$$D^2 = \frac{p(V) - p(V_0)}{V_0 - V}. \tag{20}$$

From (20) we find that D may have positive and negative values. Suppose that $D_1(r_1, r^0) < 0$ and $D_2(r_2, r^0) > 0$. Since $p'(V) < 0$ and $p''(V) > 0$, if $V_0 > V$, then

$$-c(V) < D_1(r_1, r^0) < -c(V_0), \tag{21}$$

$$c(V_0) < D_2(r_2, r^0) < c(V). \tag{22}$$

To verify that the inequalities (5) are satisfied we differentiate the Hugoniot conditions (18) assuming that V, u, and D depend on r_1 and $D = (r_1, r^0) < 0$. Denoting the derivatives with respect to r_1 by a prime, we obtain from (18)

$$DV' + D'(V - V_0) = -u', \qquad Du' + D'(u - u_0) = -c^2(V)V' \tag{23}$$

and

$$\frac{dr_1}{dr_1} = u' + c(V)V' = 1. \tag{24}$$

The values of V', u', and D' are determined from (23) and (24). Since

$$\frac{\partial R_2(r, r_0)}{\partial r_1} = u' - c(V)V' = \frac{[D_1(r_1, r^0) + c(V)]^2}{[D_1(r_1, r^0) - c(V)]^2} \tag{25}$$

and $D_1(r_1, r^0) < 0$, it follows that

$$0 \le \frac{\partial R_2(r_1, r^0)}{\partial r_1} < 1. \tag{26}$$

Analogous computations lead to the result

$$\frac{\partial R_1(r_2, r^0)}{\partial r_2} = \frac{[D_2(r_2, r^0) - c(V)]^2}{[D_2(r_2, r^0) + c(V)]^2}, \tag{27}$$

and we thus see that for the system of two equations (14) the requirements (5) are satisfied. Since along the curve $r_2 = R_2(r_1, r^0)$

$$V' = \frac{dV}{dr_1} = \frac{-2D_1(r_1, r^0)}{[D_1(r_1, r^0) - c(V)]^2} > 0, \tag{28}$$

while $V' < 0$ along the curve $r_1 = R_1(r_2, r^0)$, the inequalities (21) and (22) imply that (11) and (12) are satisfied.

Returning to the general case, we shall show that if a system of two quasilinear equations satisfies our conditions, then the problem of decay of a discontinuity for this system has no more than one self-similar solution.[10] Through the point r^- we draw the segment of the curve $r_2 = R_2(r_1, r^-)$ for $r_1 \le r_1^-$ and the ray $r_2 = r_2^-$ for $r_1 \ge r_1^-$ (Figure 4.38), while through r^+ we draw the segment of the curve $r_1 = R_1(r_2, r^+)$ for $r_2 \ge r_2^+$ and the ray $r_1 = r_1^+$ for $r_2 \le r_2^+$. The curve I represents the family of states which can be connected with r^- if the latter is considered the left state; the curve II represents the states which can be connected with r^+ considered as the right state. It follows from (5) that curves I and II can intersect at only one point. This implies the uniqueness of the self-similar solution of the decay problem.

It is easy that if $r_1^- \le r_1^+$ and $r_2^- \le r_2^+$, then the self-similar solution consists

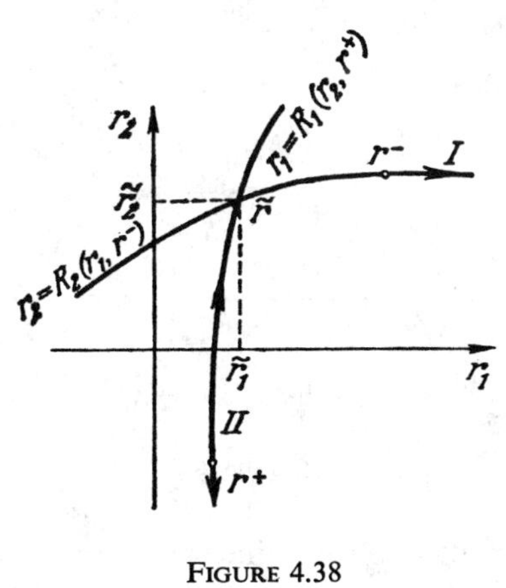

FIGURE 4.38

[10] If in (5) it is required that $|\partial R_i/\partial r_j| < q < 1$, it can then be asserted also that there exists a self-similar solution of the decay problem.

of two rarefaction waves (Figure 4.39, a); if $r_1^- \leq r_1^+$ and $r_2^- > r_2^+$, then the solution consists of either a rarefaction wave and a shock wave (Figure 4.39, b) or of two shock waves (Figure 4.39, c); if $r_1^- > r_1^+$ and $r_2^- > r_2^+$, there is also necessarily at least one shock wave. If we succeed in determining the functions R_2, D_1, R_1, and D_2, then with the exception of the case where the self-similar solution under consideration has two shock waves the solution $r(y)$ can be explicitly determined. In the case of two shock waves (Figure 4.38) the construction of a self-similar solution $r(y)$ reduces to solving the system of two equations

$$R_2(\tilde{r}_1, r^-) = \tilde{r}_2, \qquad R_1(\tilde{r}_2, r^+) = \tilde{r}_1, \tag{29}$$

where $\tilde{r}_1$, $\tilde{r}_2$ is the value of the self-similar solution between the two shock waves. The system (29) can be solved by the method of successive approximations:

$$\overset{(s+1)}{\tilde{r}_1} = R_1(\overset{(s)}{\tilde{r}_2}, r^+), \qquad \overset{(s+1)}{\tilde{r}_2} = R_2(\overset{(s)}{\tilde{r}_1}, r^-); \tag{30}$$

here we can set $\overset{(0)}{\tilde{r}_1} = r_1^-$ and $\overset{(0)}{\tilde{r}_2} = r_2^+$. The condition for convergence of the successive approximations

$$\left| \frac{\partial R_2(\tilde{r}_1, r^-)}{\partial \tilde{r}_1} \frac{\partial R_1(\tilde{r}_2, r^+)}{\partial \tilde{r}_2} \right| < 1, \tag{31}$$

is obviously satisfied if (5) is satisfied. We now observe that condition (31) is broader than (5) and suffices for the uniqueness of the self-similar solution. Moreover, this condition is invariant under change of the dependent variables, in contrast to (5). We have used condition (5) for simplicity; actually condition (31) suffices.

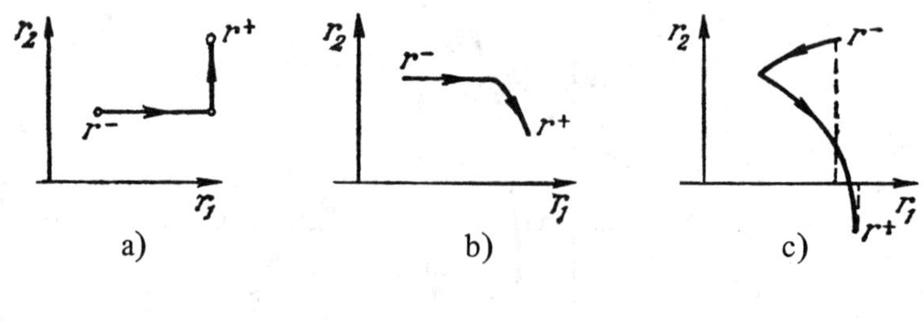

FIGURE 4.39

Suppose that a system of two quasilinear equations satisfies conditions (2), (11), (12) and (31). Then the solution of the decay problem for this system is self-similar. Indeed, suppose for the system of quasilinear equations (1) we pose the Cauchy problem

$$r(x, 0) = \begin{cases} r^-, & x < 0, \\ r^+, & x > 0. \end{cases} \tag{32}$$

We denote a bounded, stable generalized solution of this problem by $r(x, t)$ and prove that it is self-similar, i.e., depends only on the variable $y = x/t$.

Suppose that the solution $r = r(x, t)$ of the Cauchy problem (1), (32) is continuous for $t > 0$. This is possible only in the case $\xi_1(r^-) < \xi_2(r^+)$. Through the origin (0, 0) we draw the two characteristics $x = \xi_1(r^-)$ and $x = \xi_2(r^+)t$ (Figure 4.40).

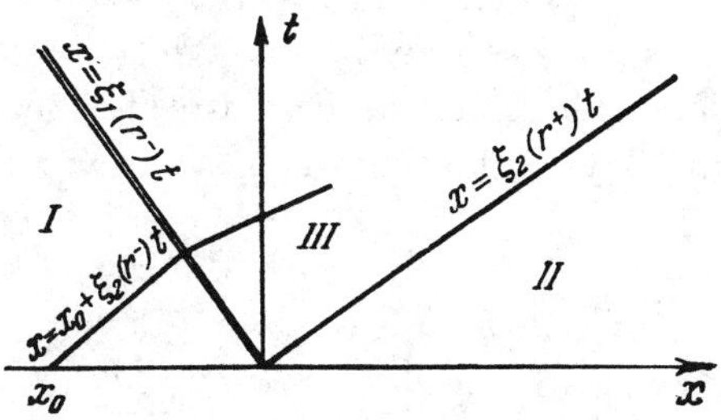

FIGURE 4.40

It is obvious that in zones I and II the solution is constant. We suppose that in some neighborhood of the line $x = \xi_1(r^-)t$ in zone III the solution $r(x, t)$ is variable. Now in this neighborhood $r_2(x, t) = r_2^-$, since $\xi_2(r^-) > \xi_1(r^-)$. Therefore, in a neighborhood of the characteristic $x = \xi_1(r^-)t$ the function $r_1(x, t)$ satisfies the equation

$$\frac{\partial r_1}{\partial t} + \xi_1(r_1, r_2^-)\frac{\partial r_1}{\partial x} = 0. \tag{33}$$

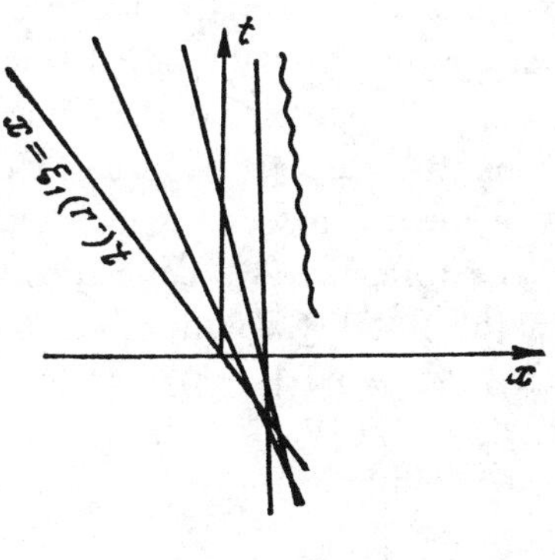

FIGURE 4.41

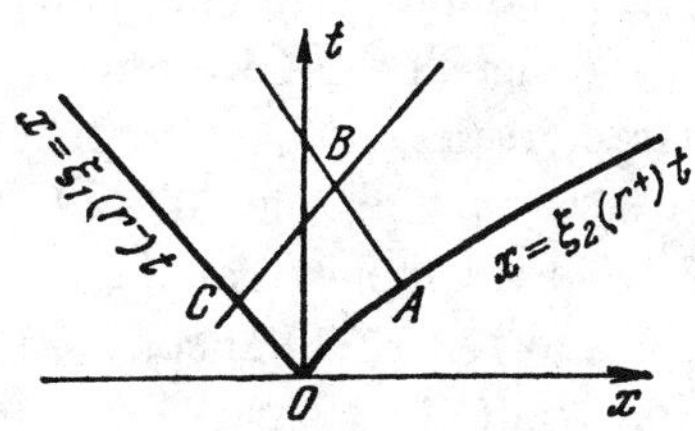

FIGURE 4.42

From this we conclude that near the line $x = \xi_1(r^-)t$ in zone III the function $r_1(x, t)$ is constant along the straight lines $x = \xi_1(r_1, r_2^-)\, t + \alpha$. It is obvious that these lines can intersect only for $t \le 0$ (Figure 4.41). In the same way we conclude that in zone III in a neighborhood of the characteristic $x = \xi_2(r^+)t$ we have $r_1(x, t) = r_1^+$, and $r_2(x, t)$ is constant along lines which can intersect only for $t \le 0$. Thus, $r_1(x, t)$ is constant along lines $x = \xi_1(r_1, r_2^-)t + \alpha(r_1)$, $\alpha \ge 0$, while $r_2(x, t)$ is constant along lines $x = \xi_2(r_1^+, r_2)t + \beta(r_2)$, $\beta \le 0$. We shall show that $\alpha = \beta = 0$. If this is not so, then these two families of lines intersect for $t > 0$. At the point of intersection we have $r_2(x, t) = r_2^-$ and $r_1(x, t) = r_1^+$. From this we conclude that in the region $OABC$ (Figure 4.42) $r(x, t) = \text{const}$, which is possible only in the case $r^- = r^+$. In this case the solution is constant altogether, and hence it is self-similar. If $r^- \ne r^+$, then $\alpha = \beta = 0$. If $\alpha = \beta = 0$, then the continuous solution $r(x, t)$ is self-similar.

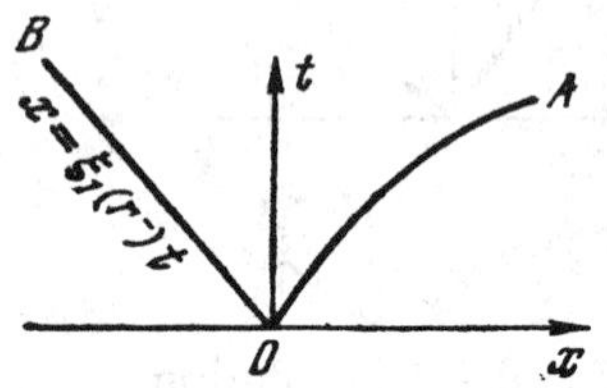

FIGURE 4.43

We now consider the case of discontinuous solution $r(x, t)$ of the decay problem. We first note that from the conditions (3.1.10) and (3.1.11) for stability of a generalized solution it follows that a stable generalized solution which is bounded and piecewise continuous for $t > 0$ cannot have more than two lines of discontinuity issuing from a single point of the domain $t > 0$. It therefore sufficies to consider the cases where the solution has one and two lines of discontinuity issuing from the point (0, 0) of discontinuity of the initial values. We consider the case of one line of discontinuity. Suppose that the solution $r(x, t)$ has one line of discontinuity OA of index 2 (Figure 4.43). It is obvious that to the right of OA we have $r(x, t) = r^+$, while $r(x, t) = r^-$ to the left of the characteristic $x = \xi_1(r^-)t$. In analogy to the foregoing we conclude that $r_2(x, t) = r_2^-$ in the zone BOA. Indeed, from the stability conditions (3.1.10) and (3.1.11) it follows that the characteristics $dx/dt = \xi_2(r(x, t))$ intersect the lines OB and OA simultaneously, and hence $r_2(x - 0, t)|_{OA} = r_2^- = r_2(x, t)|_{OB}$. Therefore, on OA we have

$$r_2(x - 0, t) = r^-, \qquad r(x + 0, t) = r^+. \tag{34}$$

According to our assumptions regarding the system of two quasilinear equations, the speed D of the line of discontinuity OA and the value $r(x - 0, t)$ are determined on the basis of these data, and these quantities are constant:

$$D = D_2(r_2^-, r^+), \qquad r_1(x - 0, t) = R_1(r_2^-, r^+). \tag{35}$$

Thus, the line of discontinuity OA is a straight line. After this, just as above, we establish that $r(x, t) = r(y)$ in the zone BOA, i.e., the self-similar solution of the decay problem which has one line of discontinuity is self-similar.

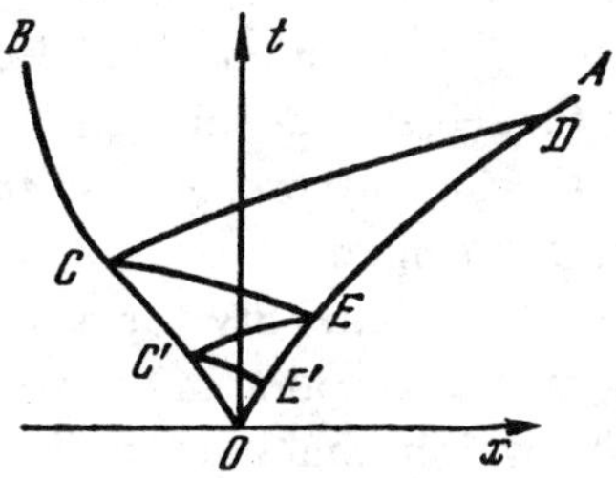

FIGURE 4.44

Finally, we consider the case where the solution $r = r(x, t)$ has two lines of discontinuity OA and OB (Figure 4.44) on which the stability conditions (3.1.10) and (3.1.11) are satisfied. Omitting some details, we note that the integral curves of the equation

$$\frac{dx}{dt} = \xi_1(r(x, t)) \tag{36}$$

in the zone BOA intersect the lines OA and OB simultaneously. Figure 4.44 shows an integral curve of equation (36) (the curve CE) with $t_C > t_E$. Similarly, an integral curve CD of the equation

$$\frac{dx}{dt} = \xi_2(r(x, t)) \tag{37}$$

intersects OB and OA simultaneously, and $t_D > t_C$. Thus, we can write

$$r_2(D) = r_2(C), \qquad r_2(C) = R_2(r_1(C), r^-) \tag{38}$$

and

$$r_1(C) = r_1(E), \qquad r_1(E) = R_1(r_2(E), r^+), \tag{39}$$

where $r(D)$, $r(C)$, and $r(E)$ denote the values of the solution in zone III at the corresponding points. Substituting (39) into (38), we find that

$$r_2(D) = R_2(R_1(r_2(E), r^+), r^-). \tag{40}$$

Analogously we could obtain

$$r_2(D) = R_2(R_1(R_2(R_1(r_2(E'), r^+), r^-), r^+), r^-). \qquad (41)$$

Continuing this process and comparing it with the process of successive approximations (3), we conclude that $r_2(D) = \bar{r}_2$. Since the point D is arbitrary, we conclude that $r_2(x - 0, t)$ is constant on the line OA and equal to $\bar{r}_2$; similarly, $r_1(x + 0, t)$ on OB is equal to $\bar{r}_1$. Thus, the solution $r(x, t)$ is constant in BOA, and OA and OB are straight lines. This implies the self-similarity of the solution $r(x, t) = r(y)$ containing two lines of discontinuity.

The proof of self-similarity of the solution of the decay problem in combination with the uniqueness of the self-similar solution of this problem allow us to assert that the solution of the Cauchy problem on the decay of an arbitrary discontinuity is unique and self-similar.

The problem of decay of a discontinuity for a system of two quasilinear equations was studied under the assumption of "convexity" of the conservation laws, i.e., conditions (2), in by Johnson and Smoller (1969), Smoller (1969b) and a number of others. The structure of a solution and conditions for uniqueness (and existence) of a solution were studied.

Questions of existence and uniqueness of the solution of the decay problem for a system of two quasilinear equations have also been studied without the convexity conditions (see, for example, Tupčiev (1966), Wendroff (1972), Dafermos (1973a, 1974), Liu (1974), and Leibovich (1974)). In this case there arise lines of discontinuity of the solution which coincide with characteristics along one or both sides of the line of discontinuity of the type considered for the equations of gas dynamics in §6.9 of Chapter 2. The conditions (3.2.32) and (3.2.33) are taken as the stability conditions; (3.2.33) must be satisfied along the wave adiabat of index k between the points u_l and u_r, where u_l and u_r are the values of the solution of opposite sides of the line of discontinuity It should be kept in mind that for an arbitrary system of two quasilinear equations these stability conditions do not suffice to distinguish a unique, stable solution of the problem on the deacy of an arbitrary discontinuity.

6. *A Goursat problem for a system of quasilinear equations.* For a system of two quasilinear equations

$$\frac{\partial r_k}{\partial t} + \xi_k(r, x, t)\frac{\partial r_k}{\partial x} = f_k(r, x, t) \qquad (k = 1, 2) \qquad (1)$$

we consider some problems more general than the problem of decay of a discontinuity. Let $\xi_k, f_k \in C_2$, and suppose that

$$\xi_1(r, x, t) < \xi_2(r, x, t), \qquad \frac{\partial \xi_2(r, x, t)}{\partial r_k} > 0 \quad (k = 1, 2). \qquad (2)$$

Suppose that the curve $O\mathscr{L}_2$ with equation $x = x_2(t)$ is a characteristic of the system (1). On $O\mathscr{L}_2$ the differentiable functions $r_1^0(x, t)$ and $r_2^0(x, t)$ are known. The curve $O\mathscr{L}_2$ possesses a continuous tangent and is an integral curve of the equation

$$\frac{dx}{dt} = \xi_2(r^0(x, t), x, t),$$

while $r^0(x, t)$ satisfies on $O\mathscr{L}_2$ the compatibility condition

$$\frac{d}{dt} r_2^0(x_2(t), t) = f_2(r^0(x_2(t), t), x, t).$$

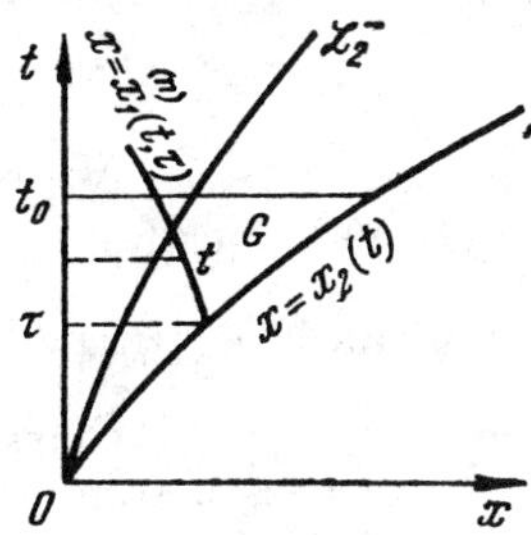

FIGURE 4.45

In a region $\mathscr{L}_2 O \mathscr{L}_2^-$ (Figure 4.45) we seek a solution $r(x, t)$ of the system (1) satisfying the following conditions.

1. The solution $r(x, t)$ on $O\mathscr{L}_2$ assumes the given values $r^0(x, t)$:

$$r(x, t)\,|_{O\mathscr{L}_2} = r^0(x, t)\,|_{O\mathscr{L}_2}. \tag{3}$$

2. The function $r_2(x, t)$ has a singularity at the point (0, 0) of the type of a self-similar rarefaction wave. Analytically this condition can be written

$$\lim_{t \to 0} r_2(t\xi_2(r_1^0(0, 0), \beta, 0, 0), t) = \beta, \tag{4}$$

where the parameter β varies over some interval

$$\beta_1 \le \beta \le \beta_0 = r_2^0(0, 0).$$

Condition (4) implies that at (0, 0) the function $r_2(x, t)$ must possess a singularity of the form

$$r_2(x, t) = g_2\left(\frac{x}{t}, t\right). \tag{5}$$

We seek a solution of problem (1), (3), (4) such that $r_2(x, t)$ can be represented in the form (5); moreover, $g_2(y, t)$ should possess continuous first derivatives with respect to y and t. We seek a solution of the problem by a method of successive approximations which differs from that applied in Chapter 1 in connection with a singularity of the solution $r_2(x,t)$. Here we must observe that boundedness of all successive approximations can be guaranteed if we consider only a sufficiently small strip $0 \leq t \leq T$ in the variable t.

We suppose that in the zone $\mathscr{L}_2^- O \mathscr{L}_2$ an approximation $\overset{(n-1)}{r}(x, t)$ is known which satisfies conditions (3) and (4). We define $\overset{(n)}{r_1}(x, t)$ as a solution of the single quasilinear equation

$$\frac{\partial \overset{(n)}{r_1}}{\partial t} + \xi_1(\overset{(n)}{r_1}, \overset{(n-1)}{r_2}(x, t), x, t) \frac{\partial \overset{(n)}{r_1}}{\partial x} = f_1(\overset{(n)}{r_1}, \overset{(n-1)}{r_2}(x, t), x, t), \tag{6}$$

satisfying (3), i.e.,

$$\overset{(n)}{r_1}(x_2(t), t) = r_1^0(x_2(t), t). \tag{7}$$

This Cauchy problem is normal, since $O\mathscr{L}_2$ is a characteristic of the second family.

The function $\overset{(n)}{r_2}(x, t)$ is also defined as a solution of the single quasilinear equation

$$\frac{\partial \overset{(n)}{r_2}}{\partial t} + \xi_2(\overset{(n-1)}{r_1}(x, t), \overset{(n)}{r_2}, x, t) \frac{\partial \overset{(n)}{r_2}}{\partial x} = f_2(\overset{(n-1)}{r_1}(x, t), \overset{(n)}{r_2}, x, t), \tag{8}$$

satisfying condition (4):

$$\lim_{t \to 0} \overset{(n)}{r_2}(t\xi_2(r_1^0(0, 0), \beta, 0, 0), t) = \beta.$$

We seek a solution of these two problems in the region G formed by the intersection of $\mathscr{L}_2^- O \mathscr{L}_2$ with the strip $0 \leq t \leq t_0$. We determine the magnitude of t_0 below from the condition that all successive approximations $\overset{(n)}{r_1}$ and $\overset{(n)}{g_2}(y, t) = \overset{(n)}{r_2}(yt, t)$ posses bounded first derivatives in G: $\overset{(n)}{r_1}$ with respect to x and t, and $\overset{(n)}{g_2}(y, t)$ with respect to y and t.

The Cauchy problem (6), (7) is solved by the usual method of characteristics. As concerns the problem (8), (4), here we have a problem with a singularity at (0, 0). We denote by $x = \overset{(n)}{x_2}(t, \beta)$ the equation of the characteristics along which as $t \to 0$ the function $\overset{(n)}{r_2}(x, t)$ assumes the values β. It is obvious that by definition

$$\overset{(n)}{x_2}(0, \beta) = 0 \quad \text{and} \quad \lim_{t \to 0} \overset{(n)}{r_2}(\overset{(n)}{x_2}(t, \beta), t) = \beta.$$

To simplify notation we set

$$\overset{(n)}{r}_2(\overset{(n)}{x}_2(t,\ \beta),\ t) = \overset{(n)}{\bar{r}}_2(t,\ \beta).$$

The quantities $\overset{(n)}{\bar{r}}_2(t,\ \beta)$ and $\overset{(n)}{x}_2(t,\ \beta)$ satisfy the characteristic system of equations (8)

$$\left.\begin{aligned} \frac{d\overset{(n)}{x}_2}{dt} &= \xi_2(\overset{(n-1)}{r_1}(\overset{(n)}{x}_2,\ t),\ \overset{(n)}{\bar{r}}_2,\ \overset{(n)}{x}_2,\ t), \\ \frac{d\overset{(n)}{\bar{r}}_2}{dt} &= f_2(\overset{(n-1)}{r_1}(\overset{(n)}{x}_2,\ t),\ \overset{(n)}{\bar{r}}_2,\ \overset{(n)}{x}_2,\ t) \end{aligned}\right\} \tag{9}$$

and the initial conditions

$$\overset{(n)}{x}_2(0,\ \beta) = 0, \qquad \overset{(n)}{\bar{r}}_2(0,\ \beta) = \beta. \tag{10}$$

We shall assume that t_0 is so small that all the successive approximations $\overset{(n)}{r}_1$ and $\overset{(n)}{r}_2$ remain bounded in G:

$$|\overset{(n)}{r}_i(x,\ t)| \le R, \qquad i = 1,\ 2,\ n = 1,\ 2,\ \ldots$$

We shall show that for sufficiently small t_0 the derivatives of $\overset{(n)}{r}_1$ and $\overset{(n)}{g}_2(y, t)$ remain bounded for all $n = 1,\ 2,\ \ldots$

We denote by E_1^0 a quantity exceeding the derivatives of $r_1^0(x,\ t)$ in modulus:

$$\left|\frac{\partial r_1^0}{\partial x}\right|,\ \left|\frac{\partial r_1^0}{\partial t}\right| \le E_1^0,$$

and similarly

$$\left|\frac{\partial \overset{(n-1)}{r_1}(x,\ t)}{\partial x}\right| \le \overset{(n-1)}{E_1}, \qquad \left|\frac{\partial \overset{(n-1)}{g_2}(y,\ t)}{\partial y}\right| \le \overset{(n-1)}{E_2},$$

and we estimate $|\partial \overset{(n)}{r}_1(x,\ t)/\partial x|$. To this end we differentiate (6) with respect to x. We obtain

$$\begin{aligned} \frac{\partial \overset{(n)}{p}_1}{\partial t} + \xi_1 \frac{\partial \overset{(n)}{p}_1}{\partial x} &= -\frac{\partial \xi_1}{\partial r_1}(\overset{(n)}{p}_1)^2 + \overset{(n)}{p}_1\left[\frac{\partial f_1}{\partial r_1} - \frac{\partial \xi_1}{\partial x} - \frac{1}{t}\frac{\partial \xi_1}{\partial r_2}\frac{\partial \overset{(n-1)}{g_2}}{\partial y}\right] \\ &\quad + \left[\frac{\partial f_1}{\partial x} + \frac{\partial f_1}{\partial r_2}\frac{\partial \overset{(n-1)}{g_2}}{\partial y}\frac{1}{t}\right], \qquad \overset{(n)}{p}_1 = \frac{\partial \overset{(1)}{r}_1(x,\ t)}{\partial x}. \end{aligned} \tag{11}$$

We compute the initial value of the derivative $\overset{(n)}{p_1}$ on the line $O\mathscr{L}_2$. For this we determine $\overset{(n)}{p_1}$ from two conditions: equation (6)

$$\overset{(n)}{q_1} + \xi_1 \overset{(n)}{p_1} = f_1,$$

and the differential corollary of (7)

$$\overset{(n)}{p_1} \frac{dx_2}{dt} + \overset{(n)}{q_1} = \frac{\partial r_1^0}{\partial x} \frac{dx_2}{dt} + \frac{\partial r_1^0}{\partial t}.$$

Recalling that $dx_2/dt = \xi_2$, from these two conditions we find that

$$\overset{(n)}{p_1}\big|_{O\mathscr{L}_2} = \frac{(\partial r_1^0/\partial x)\xi_2 + \partial r_1^0/\partial t - f_1}{\xi_2 - \xi_1}\bigg|_{O\mathscr{L}_2}.$$

Since $\xi_2 - \xi_1 > \varepsilon > 0$, this implies that there exists a number C such that

$$|\overset{(n)}{p_1}|\big|_{O\mathscr{L}_2} \leq C.$$

The growth of $\overset{(n)}{p_1}$ is estimated by means of the solution of the ordinary differential equation

$$\frac{dp_1}{dt} = Mp_1^2 + M\left[2 + \frac{\overset{(n-1)}{E_2}}{t}\right]p_1 + M\left[1 + \frac{\overset{(n-1)}{E_2}}{t}\right]$$

$$\leq M(p_1 + 1)^2 + \frac{M\overset{(n-1)}{E_2}}{t}(p_1 + 1), \tag{12}$$

assuming the value C at $t = \tau$.

We denote by τ the value of t on the characteristic of (11) at the intersection with the line $O\mathscr{L}_2$ (Figure 4.45); the quantity M is chosen so that for $|r_i| \leq R$ and $(x, t) \in G$

$$|\xi_i| \leq M, \quad |f_i| \leq M, \quad \left|\frac{\partial \xi_i}{\partial r_j}\right| \leq M, \quad \left|\frac{\partial \xi_i}{\partial x}\right| \leq M, \quad \left|\frac{\partial f_i}{\partial r_j}\right| \leq M,$$

$$\left|\frac{\partial f_i}{\partial x}\right| \leq M \qquad (i, j = 1, 2).$$

The solution of the equation

$$\frac{dp_1}{dt} = M(p_1 + 1)^2 + \frac{M\overset{(n-1)}{E_2}}{t}(p + 1) \tag{13}$$

dominates the solution of (12). Rewriting (13) in the form

$$\frac{d}{dt}(p_1+1)\left(\frac{\tau}{t}\right)^{M\overset{(n-1)}{E_2}} = M(p_1+1)^2\left(\frac{\tau}{t}\right)^{M\overset{(n-1)}{E_2}},$$

we integrate it while recalling the condition $p_1(\tau) = C$:

$$p_1 + 1 = \frac{(C+1)\,(t/\tau)^{M\overset{(n-1)}{E_2}}}{1 - ((C+1)\,\tau M/(M\overset{(n-1)}{E_2}+1))\,[(t/\tau)^{M\overset{(n-1)}{E_2}+1} - 1]}. \tag{14}$$

Thus, the solution $\overset{(n)}{p_1}(x, t)$ of (11) can be bounded above:

$$|\overset{(n)}{p_1}(x\ t)| < \frac{(C+\tau)\,(t/\tau)^{M\overset{(n-1)}{E_2}}}{1 - ((C+1)\,\tau M/(M\overset{(n-1)}{E_2}+1))[(t/\tau)^{M\overset{(n-1)}{E_2}+1} - 1]}. \tag{15}$$

The ratio t/τ contained in (15) is the ratio of the variable t on the characteristic of equations (11) to the value of this same variable at the point of intersection of the characteristic with the line $O\mathscr{L}_2$ (Figure 4.45). If the ratio t/τ is bounded in G, then for sufficiently small t_0 the estimate (15) implies that the derivatives $\overset{(u)}{p_1}$ are bounded.

The magnitude of $\partial\overset{(u)}{g_2}/\partial y$ is easily estimated on the basis of the characteristic system (9). Differentiating (9) and the initial conditions (10) with respect to the parameter β, we obtain

$$\left.\begin{aligned}
&\frac{d}{dt}\overset{(n)}{x_2'} = \frac{\partial\xi_2}{\partial r_2}\overset{(n)}{\bar r_2'} + \left[\frac{\partial\xi_2}{\partial r_1}\frac{\partial\overset{(n-1)}{r_1}}{\partial x} + \frac{\partial\xi_2}{\partial x}\right]\overset{(n)}{x_2'},\\
&\frac{d}{dt}\overset{(n)}{\bar r_2'} = \frac{\partial f_2}{\partial r_2}\overset{(n)}{\bar r_2'} + \left[\frac{\partial f_2}{\partial r_1}\frac{\partial\overset{(n-1)}{r_1}}{\partial x} + \frac{\partial f_2}{\partial x}\right]\overset{(n)}{x_2'},\\
&\overset{(n)}{x_2'}(0,\beta) = 0, \qquad \overset{(n)}{\bar r_2'}(0,\beta) = 1,
\end{aligned}\right\} \tag{16}$$

where we have introduced the notation

$$\overset{(n)}{x_2'} = \frac{\partial}{\partial\beta}\overset{(n)}{x_2}(t,\beta), \qquad \overset{(n)}{\bar r_2'} = \frac{\partial}{\partial\beta}\overset{(n)}{\bar r_2}(t,\beta).$$

From (16) it follows that

$$\left.\begin{aligned}
\overset{(n)}{x_2'}(t,\beta) &= \int_0^t \frac{\partial\xi_2}{\partial r_2}\overset{(u)}{\bar r_2'}(\tau,\beta)\exp\left\{\int_\tau^t\left[\frac{\partial\xi_2}{\partial r_1}\frac{\partial\overset{(n-1)}{r_1}}{\partial x} + \frac{\partial\xi_2}{\partial x}\right]d\zeta\right\}d\tau,\\
\overset{(n)}{\bar r_2'}(t,\beta) &= \exp\left\{\int_0^t\frac{\partial f_2}{\partial r_2}\,d\tau\right\}\\
&\quad + \int_0^t\left[\frac{\partial f_2}{\partial r_1}\frac{\partial\overset{(n-1)}{r_1}}{\partial x} + \frac{\partial f_2}{\partial x}\right]\overset{(n)}{x_2'}(\tau,\beta)\exp\left\{\int_\tau^t\frac{\partial f_2}{\partial r_2}\,d\zeta\right\}d\tau.
\end{aligned}\right\} \tag{17}$$

The usual estimates of (17) lead to the inequalities

$$t\delta \min_{0\le\tau\le t} \overset{(n)}{\bar{r}}{}_2'(\tau, \beta)\, e^{-[\overset{(n-1)}{E_1}+1]\, Mt} \le \overset{(n)}{x}{}_2'(t, \beta) \le Mt \max_{0\le\tau\le t} \overset{(n)}{\bar{r}}{}_2'(\tau, \beta)\, e^{[\overset{(n-1)}{E_1}+1]\, Mt}. \tag{18}$$

Here δ denotes the quantity

$$\delta = \min_{|r_i|\le R,\, (x,t)\in G} \frac{\partial \xi_2}{\partial r_2}(r, x, t).$$

which is greater than zero according to (2).

Substituting (18) into the second formula of (17), we obtain

$$\begin{aligned} e^{-Mt} - [\overset{(n-1)}{E_1} + 1]\frac{M^2t^2}{2} \max_{0\le\tau\le t} \overset{(n)}{\bar{r}}{}_2'(\tau, \beta) \exp\{(\overset{(n-1)}{E_1} + 2)Mt\} &\le \overset{(n)}{\bar{r}}{}_2'(t, \beta) \\ \le e^{Mt} + [\overset{(n-1)}{E_1} + 1)\frac{M^2t^2}{2} \max_{0\le\tau\le t} \overset{(n)}{\bar{r}}{}_2'(\tau, \beta) \exp\{(\overset{(n-1)}{E_1} + 2)\, Mt\}&. \end{aligned} \tag{19}$$

Let t_0 be so small that

$$\frac{M^2t_0^2}{2} [\overset{(n-1)}{E_1} + 1] \exp\{(\overset{(n-1)}{E_1} + 2)\, Mt_0\} < \frac{1}{2}. \tag{20}$$

From (19) we then obtain the simpler estimates

$$\frac{2}{3} e^{-Mt} < \overset{(n)}{\bar{r}}{}_2'(t, \beta) < 2e^{Mt}. \tag{21}$$

Substituting (21) into (18), we obtain

$$\frac{2}{3}\, t\delta e^{-[\overset{(n-1)}{E_1}+2]Mt} \le \overset{(n)}{x}{}_2'(t, \beta) \le 2Mte^{[\overset{(n-1)}{E_1}+2]Mt}. \tag{22}$$

Since

$$\frac{\partial \overset{(n)}{g}{}_2}{\partial y} = t\frac{\partial \overset{(n)}{r}{}_2}{\partial x} \quad\text{and}\quad \frac{\partial \overset{(n)}{r}{}_2}{\partial x} = \frac{\overset{(n)}{\bar{r}}{}_2'(t, \beta)}{\overset{(n)}{x}{}_2'(t, \beta)},$$

from (21) and (22) we have

$$\left|\frac{\partial \overset{(n)}{g}{}_2(y, t)}{\partial y}\right| < \frac{2e^{Mt}t}{2t\delta \exp(-[\overset{(n-1)}{E_1} + 2]\, Mt)/3} = \frac{3}{\delta} e^{[\overset{(n-1)}{E_1}+3]Mt}. \tag{23}$$

The derivatives $\partial \overset{(n)}{r}{}_1/\partial t$ and $\partial \overset{(n)}{g}{}_2/\partial t$ can be estimated in a similar way. Assuming that the ratio t/τ is bounded in G, $t/\tau < K$, from (15) we obtain

$$|\overset{(n)}{p_1}(x, t)| < \frac{(C+1)\, K^{M \overset{(n-1)}{E_1}}}{1 - ((C+1)\tau M/(M \overset{(n-1)}{E_2} + 1))[K^{M \overset{(n-1)}{E_2}+1} - 1]}. \tag{24}$$

Suppose that in G

$$[K^{M \overset{(n-1)}{E_2}+1} - 1]\, \frac{(C+1)\, t_0 M}{M \overset{(n-1)}{E_2} + 1} < \frac{1}{2}. \tag{25}$$

Then from (24) we have

$$|\overset{(n)}{p_1}(x, t)| < 2(C+1) K^{M \overset{(n-1)}{E_1}}. \tag{26}$$

If, further more,

$$[\overset{(n-1)}{E_1} + 3] M t_0 < \alpha, \tag{27}$$

then from (23) we obtain

$$\left| \frac{\partial \overset{(n)}{g_2}}{\partial y} \right| < \frac{3}{\delta} e^{\alpha} = E_2 \tag{28}$$

and, according to (26),

$$|\overset{(n)}{p_1}(x, t)| < 2(C+1) K(3Me^{\alpha}/\delta) = E_1. \tag{29}$$

Thus, if in G the inequalities

$$\frac{t}{\tau} < K, \tag{30}$$

$$M^2 \frac{t_0^2}{2} [E_1 + 1]\, e^{[E_1+2]\, M t_0} < \frac{1}{2}, \tag{31}$$

$$[K^{ME_2+1} - 1] \frac{(C+1)\, t_0 M}{M E_2 + 1} < \frac{1}{2}, \tag{32}$$

$$(E_1 + 3)\, M t_0 < \alpha \tag{33}$$

are simultaneously satisfied, then (20), (25), and (27) are simultaneously satisfied, and hence

$$\overset{(n-1)}{E_1} < E_1, \qquad \overset{(n-1)}{E_2} < E_2.$$

Therefore, if the conditions (30)–(33) are satisfied in G we have

$$|\overset{(n)}{p}_1(x, t)| = \left|\frac{\partial \overset{(n)}{r}_1(x, t)}{\partial x}\right| < E_1, \qquad \left|\frac{\partial \overset{(n)}{g}_2(y, t)}{\partial y}\right| < E_2$$
$$(n = 1, 2, 3 \ldots), \tag{34}$$

and all the successive approximations $\overset{(n)}{r}_1$ and $\overset{(n)}{g}_2(y, t)$ possess bounded first derivatives in G.

We consider the question of the possibility of satisfying the inequalities (30)–(33) in G. We prescribe any finite values $\alpha > 0$ and $K > 1$. After this the quantities E_1 and E_2 are computed from (29) and (28). The inequalities (31)–(33) can obviously be satisfied if the parameter t_0 is chosen sufficiently small. As for (30), it provides a lower bound for the range of the values β. Thus, since $K > 1$ is arbitrary, we can assume that (30) is satisfied in G. Thus, there exists a region G of the type indicated in Figure 4.45 in which all the successive approximations $\overset{(n)}{r}_1(x, t)$ and $\overset{(n)}{g}_2(y, t)$ possess bounded first derivatives.

We remark that if the initial data of the Goursat problem are Lipschitz continuous, then the uniform Lipschitz continuity of the successive approximations can be proved in an analogous way.

We shall prove the convergence of the successive approximations. From (6) we have the corollary

$$\left|\frac{\partial}{\partial t}(\overset{(n)}{r}_1 - \overset{(n-1)}{r}_1) + \xi_1(\overset{(n)}{r}_1, \overset{(n-1)}{r}_2, x, t)\frac{\partial}{\partial x}(\overset{(n)}{r}_1 - \overset{(n-1)}{r}_1)\right|$$
$$< \left[\max\left|\frac{\partial f_1}{\partial r_2}\right| + \max\left|\frac{\partial \xi_1}{\partial r_1}\right| E_1\right]|\overset{(n)}{r}_1 - \overset{(n-1)}{r}_1|$$
$$+ \left[\max\left|\frac{\partial f_1}{\partial r_2}\right| + E_1\left|\frac{\partial \xi_1}{\partial r_2}\right|\right]|\overset{(n-1)}{r}_2 - \overset{(n-2)}{r}_2|$$
$$\le M[E_1 + 1]\{|\overset{(n)}{r}_1 - \overset{(n-1)}{r}_1| + \max_G|\overset{(n-1)}{r}_2 - \overset{(n-2)}{r}_2|\}. \tag{35}$$

By virtue of (7), this implies that there exists a $B > 0$ such that

$$|\overset{(n)}{r}_1(x, t) - \overset{(n-1)}{r}_1(x, t)| < B(t - \tau)\max_G|\overset{(n-1)}{r}_2 - \overset{(n-2)}{r}_2|$$
$$< Bt\max_G|\overset{(n-1)}{r}_2 - \overset{(n-2)}{r}_2|. \tag{36}$$

It is easier for us to estimate $|\overset{(n)}{r}_2 - \overset{(n-1)}{r}_2|$ on the basis of the characteristic system (9). From (9) and the initial conditions (10) it follows that

$$\left.\begin{aligned} |\overset{(n)}{x}_2(t,\beta) - \overset{(n-1)}{x}_2(t,\beta)| &\le Bt \max_G |\overset{(n-1)}{r}_1 - \overset{(n-2)}{r}_1|, \\ |\overset{(n)}{\bar{r}}_2(t,\beta) - \overset{(n-1)}{\bar{r}}_2(t,\beta)| &\le Bt \max_G |\overset{(n-1)}{r}_1 - \overset{(n-2)}{r}_1|. \end{aligned}\right\} \tag{37}$$

In accordance with our notation,

$$\begin{aligned} |\overset{(n)}{r}_2(x,t) - \overset{(n-1)}{r}_2(x,t)| &= |\overset{(n)}{r}_2(\overset{(n)}{x}_2(t,\beta),t) - \overset{(n-1)}{r}_2(\overset{(n)}{x}_2(t,\beta),t)| \\ &\le |\overset{(n)}{r}_2(\overset{(n)}{x}_2(t,\beta),t) - \overset{(n-1)}{r}_2(\overset{(n-1)}{x}_2(t,\beta),t)| \\ &\quad + |\overset{(n-1)}{r}_2(\overset{(n-1)}{x}_2(t,\beta),t) - \overset{(n-1)}{r}_2(\overset{(n-1)}{x}_2(t,\beta),t)| \\ &< |\overset{(n)}{\bar{r}}_2(t,\beta) - \overset{(n-1)}{\bar{r}}_2(t,\beta)| \\ &\quad + \frac{E_2}{t}|\overset{(n)}{x}_2(t,\beta) - \overset{(n-1)}{x}_2(t,\beta)|. \end{aligned}$$

Substituting (37), we obtain

$$|\overset{(n)}{r}_2(x,t) - \overset{(n-1)}{r}_2(x,t)| < B \max_G |\overset{(n-1)}{r}_1 - \overset{(n-2)}{r}_1|\{t + E_2\}. \tag{38}$$

The estimates (36) and (38) prove the uniform convergence in G of the successive approximations for sufficiently small t_0.

We have thus proved the existence of a solution to the Goursat problem in question for a system of two quasilinear equations. It is interesting to observe that in the case $n \ge 3$ neither the question of the existence of a solution nor even the formulation of the Goursat problem is clear.

7. *Construction of discontinuous solutions of a system of two quasilinear equations.* We consider now several cases of the construction of discontinuous solutions of a system of two quasilinear equations which we write in invariants,

$$\frac{\partial r_k}{\partial t} + \xi_k(r,x,t)\frac{\partial r_k}{\partial x} = f_k(r,x,t) \qquad (k = 1, 2), \tag{1}$$

and also in the form of conservation laws,

$$\frac{\partial u_i}{\partial t} + \frac{\partial \varphi_i(x,x,t)}{\partial x} = g_i(u,x,t) \qquad (i = 1, 2). \tag{2}$$

We shall assune that the system (2) satisfies conditions (3.5.4)–(3.5.12), only now the functions ξ_k, R_1, R_2, D_1, and D_2 contained in these conditions depend on x and t as well as on the arguments indicated in subsection 5.

The construction of piecewise smooth, discontinuous solutions of a system of two conservation laws (2) was carried out independently by a large group of Chinese researchers (see Gu, Li, Hou et al. (1961–1962)) and by Roždestvenskiĭ (1962b, c, 1963). The local construction of discontinuous solutions was carried out under almost the same assumptions on the system (2), but the methods of construction in these two sets of papers differ somewhat. The exposition in this subsection follows Roždestvenskiĭ (1962c, 1963), since this approach is more closely connected with the exposition of the methods of solving systems of quasilinear equations of hyperbolic type adopted in this book.

We shall seek a generalized solution $r = r(x, t)$ of the system (1) assuming the initial values

$$r(x, 0) = r_0(x). \tag{3}$$

We assume that $r_0(x)$ has a discontinuity of first kind at the point $x = 0$; with the exception of this point the function $r_0(x)$ is assumed to be continuously differentiable for $|x| \leq a$.

In the case where

$$r_1^0(-0) \leq r_1^0(+0), \qquad r_2^0(-0) \leq r_2^0(+0),$$

the solution $r(x, t)$ will contain only centered rarefaction waves and will have no lines of discontinuity (shock waves). In this case a solution can be constructed by solving the two Goursat problems considered in the preceding subsection. We therefore consider here the case of discontinuous solutions, and of the initial function $r_0(x)$ we require that, for example,

$$r_1^0(-0) > r_1^0(+0). \tag{4}$$

In analogy to the case of the problem on the decay of an arbitrary discontinuity considered in subsection 5, problem (1), (3) with condition (4) can be broken down into three mutually exclusive cases:

a) $$r_1^0(+0) < r_1^0(-0), \quad R_2(r_1^0(+0), r_1^0(-0), r_2^0(-0), 0, 0) \leq r_2^0(+0); \tag{5}$$

b) $$r_2^0(+0) < r_2^0(-0), \quad R_1(r_2^0(-0), r_1^0(+0), r_2^0(+0), 0, 0) \geq r_1^0(-0); \tag{6}$$

c) $$\left.\begin{aligned} &R_2(r_1^0(+0), r_1^0(-0), r_2^0(-0), 0, 0) > r_2^0(+0), \\ &R_1(r_2^0(-0), r_1^0(+0), r_2^0(+0), 0, 0) < r_1^0(-0). \end{aligned}\right\} \tag{7}$$

If (5) is satisfied, the solution $r(x, t)$ of problem (1), (3) has a line of discontinuity $O\mathscr{L}_{D_1}$ of index 1 issuing from the point (0, 0) and a rarefaction wave $\mathscr{L}_2^- O \mathscr{L}_2^+$ (Figure 4.46, a); if (6) is satisfied, a rarefaction wave $\mathscr{L}_1^- O \mathscr{L}_1^+$ and a line of discontinuity $O\mathscr{L}_{D_2}$ of index 2 issue from the point (0, 0) (Figure 4.46, b); and if (7) is satisfed, two lines of discontinuity $O\mathscr{L}_{D_1}$ of index 1 and $O\mathscr{L}_{D_2}$ of index 2 emanate from the point (0, 0) (Figure 4.46, c).

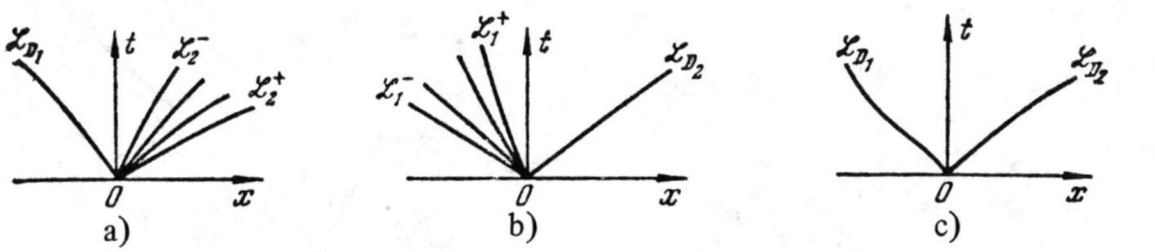

FIGURE 4.46

The construction of a solution $r(x, t)$ is different in each of these cases, but cases a) and b) differ only in the indices of the lines of discontinuity and the rarefaction waves. It therefore suffices for us to consider the question of constructing a solution only when inequalities (5) or (7) are satisfied. We shall present a method of constructing a solution $r(x, t)$ in each of these two cases.

The common feature of these two cases consists in solving the Cauchy problem for equation (1) with the initial conditions

$$r(x, 0) = r_0(x), \qquad -a \leq x < 0,$$

given only to the left of the point $x = 0$ and the initial conditions

$$r(x, 0) = r_0(x), \qquad 0 < x \leq a,$$

given only to the right of $x = 0$. The solution of each of these two Cauchy problems is continuously differentiable and can be determined in the domain on definition of each of the problems by the method of characteristics presented in Chapter 1. The solution of the first problem is determined in the region I bounded on the right by a characteristic of the first family $O\mathscr{L}_1^-$; the solution of the second problem is determined in the region II bounded on the left by a characteristic of the second family which we denote by $O\mathscr{L}_2^+$ (Figure 4.47). We observe that in some cases regions I and II may overlap, i.e., the line $O\mathscr{L}_2^+$ may lie to the left of $O\mathscr{L}_1^-$, but this is of no basic significance for our subsequent argument. We henceforth denote the solutions of these two problems in regions I and II by $r_0(x, t)$. According to the results of Chapter 1, $r_0(x, t)$ possesses bounded first derivatives with respect to x and t; we shall asume that these derivatives are bounded in modulus by a number $C > 0$.

In the case where conditions (5) are satisfied the construction of a solution begins with the solution of a Goursat problem for the system of equations (1) with conditions given on the characteristic $O\mathscr{L}_2^+$

$$r_1(x, t)|_{O\mathscr{L}_2^+} = r_1^0(x, t),$$

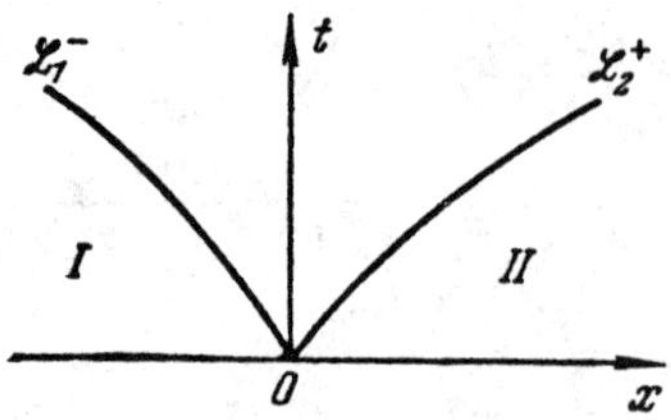

FIGURE 4.47

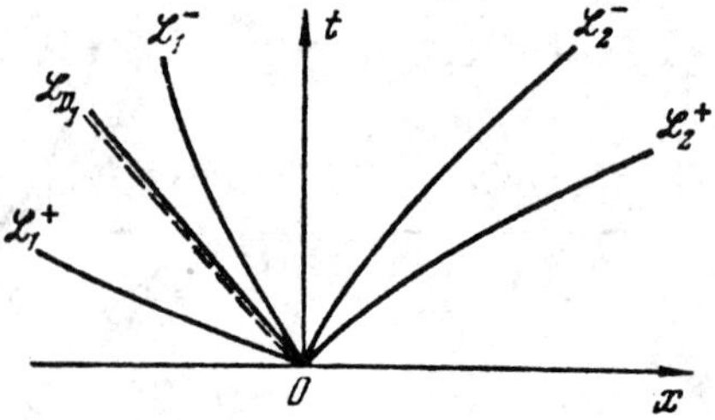

FIGURE 4.48

and at the point (0, 0)

$$\lim_{t \to 0} r_2(t\xi_2(r_1^0(+0), \beta, 0, 0), t) = \beta,$$

where

$$R_2(r_1^0(+0), r_1^0(-0), r_2^0(-0), 0, 0) \leq \beta \leq r_2^0(+0).$$

A solution of this problem is constructed by the method set forth in subsection 6. Suppose that for $0 \leq t \leq T$ in the zone $\mathscr{L}_2^- O \mathscr{L}_2^+$ (Figure. 4.48) a solution of this problem has been constructed. According to subsection 6, the solution $r(x, t)$ of this Goursat problem is a smooth function and, in particular, is differentiable along $O\mathscr{L}_2^-$. The solution $r(x, t)$ in the zone $\mathscr{L}_2^- O \mathscr{L}_2^+$ we also denote by $r_0(x, t)$.

The next constructions are aimed at determining functions $\tilde{r}_1(x, t)$ and $\tilde{r}_2(x, t)$ satisfying the following conditions.

1) The functions $\tilde{r}_1$ and $\tilde{r}_2$ are defined in the zone $\mathscr{L}_1^+ O \mathscr{L}_2^-$ containing $\mathscr{L}_1^- O \mathscr{L}_2^-$ and satisfy (1) in $\mathscr{L}_1^+ O \mathscr{L}_2^-$.

2) The integral curve $O\mathscr{L}_{D_1}$ of the equation

$$\frac{dx}{dt} = D_1(\tilde{r}_1(x, t), r_0(x, t), x, t), \tag{8}$$

passing through the point (0, 0),

$$x(0) = 0, \tag{9}$$

lies entirely within the zone $\mathscr{L}_1^+ O \mathscr{L}_1^-$ for $0 \leq t \leq T$.

3) On the line $O\mathscr{L}_2^-$ the condition

$$\tilde{r}_1(x, t)\big|_{O\mathscr{L}_2^-} = r_1^0(x, t) \tag{10}$$

is satisfied, while on $O\mathscr{L}_{D_1}$ we have

$$\tilde{r}_2(x, t)|^{O\mathscr{L}_{D_1}} = R_2(\tilde{r}_1(x, t), r_0(x, t), x, t). \tag{11}$$

It is easy to see that if such functions $\tilde{r}_1$ and $\tilde{r}_2$ and the line $O\mathscr{L}_{D_1}$ have been found, then the generalized solution is given by

$$r(x, t) = \begin{cases} \tilde{r}(x, t) & \text{in} \quad \mathscr{L}_{D_1}O\mathscr{L}_2^-, \\ r_0(x, t) & \text{outside } \mathscr{L}_{D_1}O\mathscr{L}_2^-. \end{cases} \tag{12}$$

Indeed, (12) defines a function discontinuous on $O\mathscr{L}_{D_1}$ which satisfies (1) everywhere except on the line of discontinuity. On the line of discontinuity $O\mathscr{L}_{D_1}$ the solution $r(x, t)$ satisfies the Hugoniot conditions and the stability conditions. Thus, the problem has been reduced to the construction of functions $\tilde{r}_1$ and $\tilde{r}_2$ satisfying conditions (1)–(3). We now indicate a method of successive approximations by means of which these functions can be constructed.

We determine the function $\overset{(0)}{\tilde{r}_2}(x, t)$ to the left of the line $O\mathscr{L}_2^-$. We shall assume that $\overset{(0)}{\tilde{r}_2}(x, t)$ does not depend on x and on the line $O\mathscr{L}_2^-$ it assumes the same values as $r_2^0(x, t)$. Thus, $\overset{(0)}{\tilde{r}_2}(x, t)$ has been determined to the left of $O\mathscr{L}_2^-$ and is given by

$$\overset{(0)}{\tilde{r}_2}(x, t) = r_2^0(x(t), t),$$

where $x = x(t)$ is the equation of $O\mathscr{L}_2^-$.

After this we determine $\overset{(1)}{r_1}(x, t)$ as a solution of the equation

$$\frac{\partial \overset{(1)}{\tilde{r}_1}}{\partial t} + \xi_1(\overset{(1)}{\tilde{r}_1}, \overset{(0)}{\tilde{r}_2}(x, t), x, t)\frac{\partial \overset{(1)}{\tilde{r}_1}}{\partial x} = f_1(\overset{(1)}{\tilde{r}_1}, \overset{(0)}{\tilde{r}_2}(x, t), x, t), \tag{13}$$

satisfying on $O\mathscr{L}_2^-$ the condition

$$\overset{(1)}{\tilde{r}_1}|_{O\mathscr{L}_2^-} = r_1^0(x, t).$$

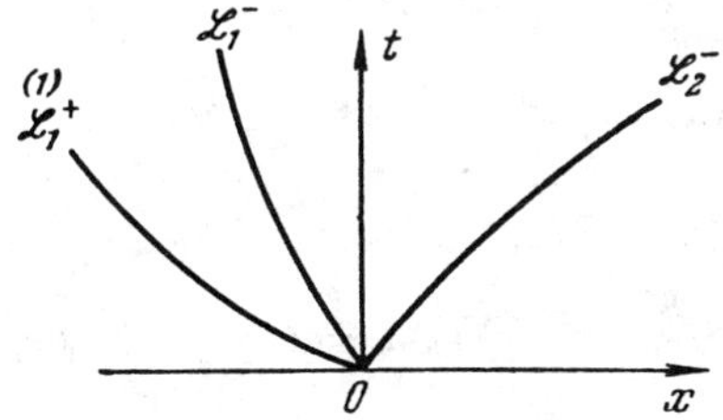

FIGURE 4.49

The solution $\overset{(1)}{\tilde{r}_1}$ is uniquely determined in $\overset{(1)}{\mathscr{L}_1^+}O\mathscr{L}_2^-$ (Figure 4.49), where $O\overset{(1)}{\mathscr{L}_1^+}$ is the characteristic of (13) passing through (0, 0). According to our conditions $r_1^0(+0) < r_1^0(-0)$, and therefore for sufficiently small T in $\overset{(1)}{\mathscr{L}_1^+}O\mathscr{L}_1^-$ we have

$$\overset{(1)}{\bar{r}}_1(x, t) < r_1^0(x, t).$$

From (3.5.11) it follows that

$$\begin{aligned}
&\xi_1(\overset{(1)}{\bar{r}}_1(0, 0), \overset{(0)}{\bar{r}}_2(0, 0), 0, 0) \\
&\qquad = \xi_1(r_1^0(+0), R_2(r_1^0(+0), r_0(-0), 0, 0), 0, 0) \\
&\qquad < \xi_1(r_1^0(-0), r_2^0(-0), 0, 0).
\end{aligned} \tag{14}$$

Here the left side contains the slope of $O\overset{(1)}{\mathscr{L}_1^+}$ at (0, 0), while the right side contains the slope of $O\mathscr{L}_1^-$ at (0, 0). Therefore, this inequality implies that for sufficiently small T the curve $O\overset{(1)}{\mathscr{L}_1^+}$ lies to the left of $O\mathscr{L}_1^-$, as shown in Figure 4.49.

In $\overset{(1)}{\mathscr{L}_1^+}O\mathscr{L}_1^-$ we consider the ordinary differential equations

$$\frac{dx}{dt} = D_1(\overset{(1)}{\bar{r}}_1(x, t), r_0(x, t), x, t), \tag{15}$$

for which we pose the initial condition $x(0) = 0$. Noting that $\overset{(1)}{\bar{r}}_1(x, t)$ is a differentiable function of its variables for sufficiently small T, we conclude that the right side of (15) is also a differentiable function of its variables. At the point (0, 0) it follows from (3.5.11) that

$$\begin{aligned}
\xi_1(\overset{(1)}{\bar{r}}_1(0, 0), \overset{(0)}{\bar{r}}_2(0, 0), 0, 0) &< D_1(\overset{(1)}{\bar{r}}_1(0, 0), r_0(0, 0), 0, 0) \\
&< \xi_1(r_0(0, 0), 0, 0).
\end{aligned}$$

By the continuity of these functions we now conclude that for sufficiently small T on the line $O\mathscr{L}_1^-$ we have

$$D_1(\overset{(1)}{\bar{r}}_1(x, t), r_0(x, t), x, t) < \xi_1(r_0(x, t), x, t), \tag{16}$$

while on $O\overset{(1)}{\mathscr{L}_1^+}$ we have

$$\xi_1(\overset{(1)}{\bar{r}}_1(x, t), \overset{(0)}{\bar{r}}_2(x, t), x, t) < D_1(\overset{(1)}{\bar{r}}_1(x, t), r_0(x, t), x, t). \tag{17}$$

The inequalities (16) and (17) imply that on $O\mathscr{L}_1^-$ and $O\overset{(1)}{\mathscr{L}_1^+}$ the direction field for the differential equation (15) has the form shown in Figure 4.50. We conclude that there exists a unique integral curve of equation (15) passing through (0, 0). We denote this curve by $O\overset{(1)}{\mathscr{L}}_{D_1}$. It is obvious that $O\overset{(1)}{\mathscr{L}}_{D_1}$ is a smooth and, in particular, differentiable curve.

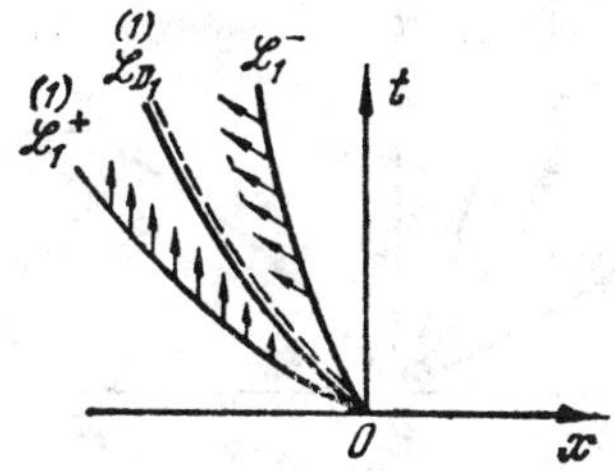

FIGURE 4.50

After determining $O\overset{(1)}{\mathscr{L}}_{D_1}$ we define $\overset{(1)}{\tilde{r}}_2(x, t)$ in $\overset{(1)}{\mathscr{L}_1^+}O\mathscr{L}_2^-$ as a solution of the equation

$$\frac{\partial \overset{(1)}{\tilde{r}}_2}{\partial t} + \xi_2(\overset{(1)}{\tilde{r}}_1(x, t), \overset{(1)}{\tilde{r}}_2, x, t), \frac{\partial \overset{(1)}{\tilde{r}}_2}{\partial x} = f_2(\overset{(1)}{\tilde{r}}_1(x, t), \overset{(1)}{\tilde{r}}_2, x, t) \tag{18}$$

(in which $\overset{(1)}{r}_1(x, t)$ is a known function) satisfying the initial condition posed on $O\overset{(1)}{\mathscr{L}}_{D_1}$

$$\overset{(1)}{\tilde{r}}_2(x, t)\big|_{O\overset{(1)}{\mathscr{L}}_{D_1}} = R_2(\overset{(1)}{\tilde{r}}_1(x, t), r_0(x, t), x, t).$$

This problem obviously has a solution $\overset{(1)}{\tilde{r}}_2(x, t)$ which is uniquely determined in the entire zone $\overset{(1)}{\mathscr{L}_1^+}O\mathscr{L}_2^-$ and is differentiable in this zone for sufficiently small T; the values of the solution in $\overset{(1)}{\mathscr{L}_1^+}O\mathscr{L}_2^-$ here match the values of $\overset{(1)}{r_2^0}(x, t)$ on $O\mathscr{L}_2^-$ in continuous fashion.

To the left of $O\overset{(1)}{\mathscr{L}_1^+}$ we assume that $\overset{(1)}{\tilde{r}}_2(x, t)$ does not depend on x and assumes the same values as on $O\mathscr{L}_1^+$. After this the process of successive approximations is repeated.

Suppose that the Lipschitz continuous function $\overset{(n-1)}{\tilde{r}_2}(x, t)$ is known to the left of $O\mathscr{L}_2^-$. In the zone $\overset{(n)}{\mathscr{L}_1^+}O\mathscr{L}_2^-$ we define $\overset{(n)}{\tilde{r}}_1(x, t)$ as a solution of the problem

$$\frac{\partial \overset{(n)}{\tilde{r}}_1}{\partial t} + \xi_1(\overset{(n)}{\tilde{r}}_1, \overset{(n-1)}{r_2}(x, t), x, t)\frac{\partial \overset{(n)}{\tilde{r}}_2}{\partial x} = f_1(\overset{(n)}{\tilde{r}}_1, \overset{(n-1)}{\tilde{r}_2}(x, t), x, t), \tag{19}$$

$$\overset{(n)}{\tilde{r}}_1(x, t)\big|_{O\mathscr{L}_2^-} = r_1^0(x, t). \tag{20}$$

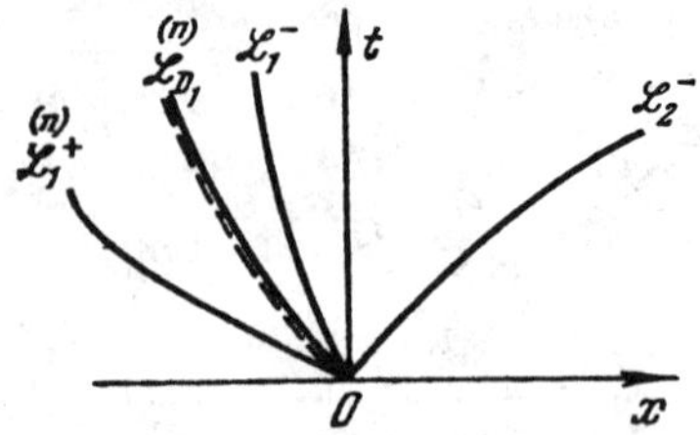

FIGURE 4.51

In analogy to the foregoing the line $O\overset{(n)}{\mathscr{L}}_1^+$ is situated to the left of $O\mathscr{L}_1^-$, and in $\overset{(n)}{\mathscr{L}}_1^+ O\mathscr{L}_1^-$ (Figure 4.51) there exists a unique integral curve $O\overset{(1)}{\mathscr{L}}_{D_1}$ of the equation

$$\frac{dx}{dt} = D_1(\overset{(n)}{\bar{r}}_1(x, t), r_0(x, t), x, t), \qquad x(0) = 0. \tag{21}$$

After this the function $\overset{(n)}{\bar{r}}_2(x, t)$ is determined in $\overset{(n)}{\mathscr{L}}_1^+ O\mathscr{L}_2^-$ as a solution of the problem

$$\frac{\partial \overset{(n)}{\bar{r}}_2}{\partial t} + \xi_2(\overset{(n)}{\bar{r}}_1(x, t), \overset{(n)}{\bar{r}}_2, x, t)\frac{\partial \overset{(n)}{\bar{r}}_2}{\partial x} = f_2(\overset{(n)}{\bar{r}}_1(x, t), \overset{(n)}{\bar{r}}_2, x, t), \tag{22}$$

$$\overset{(n)}{\bar{r}}_2\Big|_{O\overset{(n)}{\mathscr{L}}D_1} = R_2(\overset{(n)}{\bar{r}}_1(x, t), r_0(x, t), x, t) \tag{23}$$

and it is assumed to be independent of x to the left of $O\overset{(n)}{\mathscr{L}}_1^+$ and continuous on $O\overset{(n)}{\mathscr{L}}_1^+$.

In analogy to subsection 6, to prove the convergence of the successive approximations, we estimate the derivatives $\overset{(n)}{\bar{p}}_1 = \partial\overset{(n)}{\bar{r}}_1/\partial x$, $\overset{(n)}{\bar{p}}_2 = \partial\overset{(n)}{\bar{r}}_2/\partial x$.

Suppose that

$$|\overset{(n)}{\bar{r}}_i(x, t)| \leq R \qquad (i = 1, 2),\ n = 1, 2, \ldots,$$
$$|\overset{(n-1)}{p_i}(x, t)| \leq \overset{(n-1)}{E_i}, \qquad |\overset{(n)}{\bar{p}}_1|\,|_{O\mathscr{L}_2^-} \leq C,$$

and M has the same meaning as in subsection 6. From (19) we obtain an equation analogous to (3.6.11):

$$\frac{\partial \overset{(n)}{\tilde{p}}_1}{\partial t} + \xi_1 \frac{\partial \overset{(u)}{\tilde{p}}_1}{\partial x} = -\frac{\partial \xi_1}{\partial r_1}(\overset{(n)}{\rho}_1)^2$$
$$+\left[\frac{\partial f_1}{\partial r_1} - \frac{\partial \xi_1}{\partial x} - \frac{\partial \xi_1}{\partial r_2}\frac{\partial \overset{(n-1)}{\tilde{r}_2}}{\partial x}\right]\overset{(n)}{p}_1 + \left[\frac{\partial f_1}{\partial x} + \frac{\partial f_1}{\partial r_2}\frac{\partial \overset{(n-1)}{\tilde{r}_2}}{\partial x}\right].$$

In analogy to (12) we have an equation for a majorant p_1:

$$\frac{dp_1}{dt} = Mp_1^2 + M[\overset{(n-1)}{E_2} + 2]p_1 + M[E_2 + 1] < M(E_2 + 1)(p_1 + 1)^2.$$

As in the computations in subsection 6, this easily implies that

$$|\overset{(n)}{\tilde{p}}_1(x, t)| < \frac{C + 1}{1 - (C + 1)M(\overset{(n-1)}{E_2} + 1)(t - \tau)}. \tag{24}$$

If

$$T \leq \frac{1}{2(C+1)M(\overset{(n-1)}{E_2} + 1)},$$

then for $0 \leq \tau \leq t < T$

$$|\overset{(n)}{\tilde{p}}_1(x, t)| < 2(C + 1). \tag{25}$$

The equation for $\overset{(n)}{\tilde{p}}_2$ is found similarly; the initial value $\overset{(n)}{\tilde{p}}_2(x, t)$ is found on the line $O\overset{(n)}{\mathscr{L}}_{D_1}$ from condition (23):

$$\overset{(n)}{\tilde{p}}_2(x, t) = \frac{\partial R_2/\partial t + D_1(\partial R_2/\partial x) - f_2}{\xi_2(\overset{(n)}{\tilde{r}}_1, \overset{(n)}{\tilde{r}}_2, x, t) - D_1(\overset{(n)}{\tilde{r}}_1, r_0(x, t), x, t)}. \tag{26}$$

Here $\partial R_2/\partial t$ and $\partial R_2/\partial x$ denote the corresponding derivatives on the right side of (23). According to (3.5.12),

$$\xi_2(\overset{(n)}{\tilde{r}}_1, \overset{(n)}{\tilde{r}}_2, x, t) - D_1(\overset{(n)}{\tilde{r}}_1, r_0(x, t), x, t) > \varepsilon > 0,$$

and so for sufficiently small T the denominator in (26) does not vanish, and because of (25) it may be assumed that there exist constants D and E such that on $O\mathscr{L}_{D_1}$

$$|\overset{(n)}{\tilde{p}}_2(x, t)| < DC + E = D_0.$$

In analogy to the foregoing, for

$$T \leq \frac{1}{2(D_0 + 1)M(\overset{(n)}{E}_1 + 1)}, \tag{27}$$

we find that

$$|\overset{(n)}{\bar{p}}(x, t)| < 2D_0. \tag{28}$$

Choosing now

$$T = \min\left\{\frac{1}{2(C+1)M(2D_0+1)}, \frac{1}{2(D_0+1)M(2C+3)}\right\}, \tag{29}$$

we find that for $0 \le t \le T$ the estimates (25) and (28) are simultaneously satisfied for any $n = 1, 2, \ldots$, i.e., all the successive approximations have bounded first derivatives.

We proceed to the proof of convergence of the successive approximations $\overset{(n)}{\bar{r}}(x, t)$, assuming that T is given by (29). We choose the common part of the regions $\overset{(n)}{\mathscr{L}_1^+ O \mathscr{L}_2^-}$ for $n = 1, 2, \ldots$, and restrict it by the condition $0 \le t \le T' \le T$, where T is so small that the common part of the regions $\overset{(n)}{\mathscr{L}_1^+ O \mathscr{L}_2^-}$ always contains in its interior all the lines$\overset{(n)}{O\mathscr{L}_{D_1}}$ for $n = 1, 2, \ldots$ Since the slope of $\overset{(n)}{O\mathscr{L}_1^+}$ at (0, 0) does not depend on the index n, while $\overset{(n)}{\bar{r}_1}$ and $\overset{(n)}{\bar{r}_2}$ have uniformly bounded derivatives, this is always possible.

We denote by G this common part of the regions $\overset{(n)}{\mathscr{L}_1^+ O \mathscr{L}_2^-}$ and introduce the notation

$$\Delta\overset{(n)}{\bar{r}_i}(t) = \max_{G_t} |\overset{(n)}{\bar{r}_i}(\xi, \tau) - \overset{(n-1)}{\bar{r}_i}(\xi, \tau)|,$$

where G_τ is the intersection of G with the strip $0 \le \tau \le t$,

$$\Delta\overset{(n)}{x_{D_1}}(t) = \max_{0 \le \tau \le t} |\overset{(n)}{x_{D_1}}(\tau) - \overset{(n-1)}{x_{D_1}}(\tau)|.$$

Here $x = \overset{(n)}{x_{D_1}}(t)$ denotes the equation of the line $\overset{(n)}{O\mathscr{L}_{D_1}}$. From (19) and (20) we easily find the estimates (we obtained similar estimates in subsection 6)

$$\Delta\overset{(n)}{\bar{r}_1}(t) \le \bar{M}t\Delta\overset{(n-1)}{\bar{r}_2}(t),$$

from (21)

$$\Delta\overset{(n)}{x_{D_1}}(t) \le \bar{M}t\Delta\overset{(n)}{\bar{r}_1}(t) \le \bar{M}^2t^2\Delta\overset{(n-1)}{\bar{r}_2}(t),$$

and, finally, from (22), and (23)

$$\Delta\overset{(n)}{\bar{r}_2}(t) \le \bar{M}\Delta\overset{(n)}{\bar{r}_1}(t) + \bar{M}t\Delta\overset{(n)}{\bar{r}_1}(t) + \bar{M}\Delta\overset{(n)}{x_{D_1}}(t) < \bar{\bar{M}}t\Delta\overset{(n-1)}{\bar{r}_2}(t).$$

Here $\bar{M}$ and $\bar{\bar{M}}$ are bounded. The uniform convergence of the sequences $\{\overset{(n)}{\bar{r}_2}\}$, $\{\overset{(n)}{\bar{r}_1}\}$ and $\{\overset{(n)}{x_{D_1}}\}$ follows easily from these formulas.

Because of the uniform convergence, the quantities

$$\tilde{r}_1(x,\, t) = \lim_{n\to\infty} \overset{(n)}{\tilde{r}_1}(x,\, t), \qquad \tilde{r}_2(x,\, t) = \lim_{n\to\infty} \overset{(n)}{\tilde{r}_2}(x,\, t)$$

satisfy (1) in G, the line $O\mathscr{L}_{D_1}$ $(x_{D_1}(t) = \lim_{n\to\infty} \overset{(n)}{x}_{D_1}(t))$ is an integral curve of (8), and $\tilde{r}_2(x,\, t)$ satisfies (11) on $O\mathscr{L}_{D_1}$.

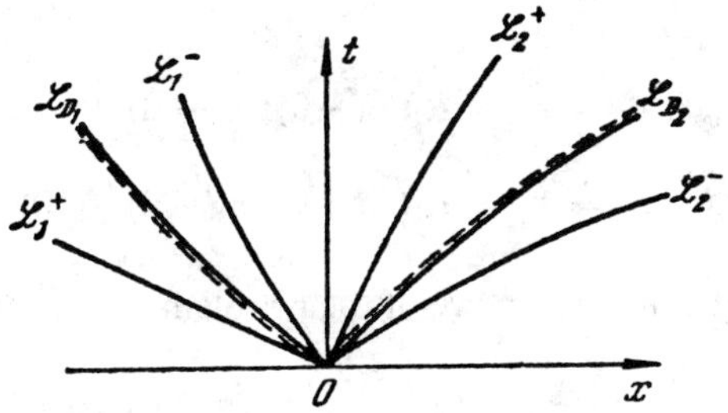

FIGURE 4.52

We now proceed to consider the second case. If the inequalities (7) are satisfied, the solution contains two lines of discontinuity, as we have noted: the line $O\mathscr{L}_{D_1}$ of index 1 and the line $O\mathscr{L}_{D_2}$ of index 2 emanating from the point (0, 0) (Figure 4.52). In analogy to the foregoing we introduce the functions $\tilde{r}_1(x,\, t)$ and $\tilde{r}_2(x,\, t)$, of which we require the following.

1) $\tilde{r}_1$ and $\tilde{r}_2$ are defined in the zone $\mathscr{L}_1^+ O \mathscr{L}_2^-$ containing $\mathscr{L}_{D_1} O \mathscr{L}_{D_2}$; in this zone they possess bounded first derivatives satisfying (1).

2) The integral curve $O\mathscr{L}_{D_1}$ of the equation

$$\frac{dx}{dt} = D_1(\tilde{r}_1(x,\, t),\, r_0(x,\, t),\, x,\, t), \tag{30}$$

passing through (0, 0) for $0 \leq t \leq T$ lies in $\mathscr{L}_1^+ O \mathscr{L}_1^-$, while the integral curve $O\mathscr{L}_{D_2}$ of the equation

$$\frac{dx}{dt} = D_2(\tilde{r}_2(x,\, t),\, r_0(x,\, t),\, x,\, t), \tag{31}$$

issuing from (0, 0) lies in $\mathscr{L}_2^+ O \mathscr{L}_2^-$ for $0 \leq t \leq T$.

3) On $O\mathscr{L}_{D_1}$ and $O\mathscr{L}_{D_2}$ the following conditions are satisfied:

$$\tilde{r}_2(x,\, t)O|_{O\mathscr{L}_{D_1}} = R_2(\tilde{r}_1(x,\, t),\, r_0(x,\, t),\, x,\, t), \tag{32}$$

$$\tilde{r}_1(x,\, t)|O_{O\mathscr{L}_{D_1}} = R_1(\tilde{r}_2(x,\, t),\, r_0(x,\, t),\, x,\, t). \tag{33}$$

If functions $\tilde{r}_1$ and $\tilde{r}_2$ and lines $O\mathscr{L}_{D_1}$ and $O\mathscr{L}_{D_2}$ satisfying these requirements have been constructed, then the solution of the Cauchy problem (1), (3) is given by

$$r(x, t) = \begin{cases} \tilde{r}(x, t) & \text{in } \mathscr{L}_{D_1} O \mathscr{L}_{D_2}, \\ r_0(x, t) & \text{outside } \mathscr{L}_{D_1} O \mathscr{L}_{D_2}. \end{cases} \tag{34}$$

We present a method of successive approximations by means of which $\tilde{r}_1$ and $\tilde{r}_2$ can be constructed; here we omit certain details which are similar to the preceding construction.

We introduce the notation

$$\overset{(0)}{\tilde{r}_1}(x, t) = \overset{(0)}{\tilde{r}_1} = R_1(r_2^0(-0), r_0(+0), 0, 0),$$

$$\overset{(0)}{\tilde{r}_2}(x, t) = \overset{(0)}{\tilde{r}_2} = R_2(r_1^0(+0), r_0(-0), 0, 0)$$

and define $O\overset{(0)}{\mathscr{L}}_{D_1}$ as the integral curve of the problem

$$\frac{dx}{dt} = D_1(\overset{(0)}{\tilde{r}_1}, r_0)(x, t), x, t), \qquad x(0) = 0, \tag{35}$$

and $O\overset{(0)}{\mathscr{L}}_{D_2}$ as the integral curve of the problem

$$\frac{dx}{dt} = D_2(\overset{(0)}{\tilde{r}_2}, r_0(x, t), x, t), \qquad x(0) = 0. \tag{36}$$

If T is sufficiently small, then for $0 \le t \le T$ the curve $O\overset{(1)}{\mathscr{L}}_{D_1}$ lies to the left of $O\overset{(0)}{\mathscr{L}_1^-}$, and $O\overset{(0)}{\mathscr{L}}_{D_2}$ lies to the right of $O\overset{(0)}{\mathscr{L}}_{D_1}$.

After this we define the next approximation $\overset{(1)}{\tilde{r}_1}(x, t)$ as a solution of the Cauchy problem

$$\frac{\partial \overset{(1)}{\tilde{r}_1}}{\partial t} + \xi_1(\overset{(1)}{\tilde{r}_1}, \overset{(0)}{\tilde{r}_2}(x, t), x, t) \frac{\partial \overset{(1)}{\tilde{r}_1}}{\partial x} = f_1(\overset{(1)}{\tilde{r}_1}, \overset{(0)}{r_2}(x, t), x, t), \tag{37}$$

$$\overset{(1)}{\tilde{r}_1}(x, t)|_{O\overset{(0)}{\mathscr{L}}_{D2}} = R_1(\overset{(0)}{\tilde{r}_2}(x, t), r_0(x, t), x, t), \tag{38}$$

and $\overset{(1)}{\tilde{r}_2}(x, t)$ as a solution of the Cauchy problem

$$\frac{\partial \overset{(1)}{\tilde{r}_2}}{\partial t} + \xi_2(\overset{(0)}{\tilde{r}_1}(x, t), \overset{(1)}{\tilde{r}_2}, x, t) \frac{\partial \overset{(1)}{\tilde{r}_2}}{\partial x} = f_2(\overset{(0)}{\tilde{r}_2}(x, t), \overset{(1)}{\tilde{r}_2}, x, t), \tag{39}$$

$$\overset{(1)}{\tilde{r}_2}(x, t)|_{O\overset{(0)}{\mathscr{L}}_{D1}} = R_2(\overset{(0)}{r_1}(x, t), r_0(x, t), x, t). \tag{40}$$

A solution of the problem (37), (38) can be determined in $\overset{(1)}{\mathscr{L}_1^+} O \overset{(1)}{\mathscr{L}_2^-}$; in this same zone we find a solution of the Cauchy problem (39), (40); outside this zone we assume that $\overset{(1)}{\tilde{r}_1}$ and $\overset{(1)}{\tilde{r}_2}$ continuously match their values on the lines $O\overset{(1)}{\mathscr{L}_1^+}$ and $O\overset{(1)}{\mathscr{L}_2^-}$ and do not depend on x.

After this the process of successive approximations becomes standard. Suppose the approximation $\overset{(n-1)}{\tilde{r}}(x, t)$ possessing a bounded derivative is known. We first define the lines $\overset{(n-1)}{O\mathscr{L}_{D_1}}$ and $\overset{(n-1)}{O\mathscr{L}_{D_2}}$ as integral curves of the problems

$$\frac{dx}{dt} = D_1(\overset{(n-1)}{\tilde{r}_1}(x, t), r_0(x, t), x, t), \qquad x(0) = 0,$$

$$\frac{dx}{dt} = D_2(\overset{(n-1)}{\tilde{r}_2}(x, t), r_0(x, t), x, t), \qquad x(0) = 0. \tag{41}$$

For sufficiently small T these integral curves lie in $\overset{(n-1)}{\mathscr{L}_1^+ O \mathscr{L}_1^-}$ and $\overset{(n-1)}{\mathscr{L}_2^+ O \mathscr{L}_2^-}$ respectively. We see this by investigating the direction fields of the differential equations (41) and (42) on the lines $\overset{(n-1)}{O\mathscr{L}_1^+}$, $O\mathscr{L}_1^-$ and $O\mathscr{L}_2^+$, $\overset{(n-1)}{O\mathscr{L}_2^-}$ in exactly the same way as above.

After this we define $\overset{(n)}{\tilde{r}_1}(x, t)$ as the solution of the Cauchy problem

$$\left.\begin{aligned} &\frac{\partial \overset{(n)}{\tilde{r}_1}}{\partial t} + \xi_1(\overset{(n)}{\tilde{r}_1}, \overset{(n-1)}{\tilde{r}_2}(x, t), x, t)\frac{\partial \overset{(n)}{\tilde{r}_1}}{\partial x} = f_1(\overset{(n)}{\tilde{r}_1}, \overset{(n-1)}{\tilde{r}_2}(x, t), x, t), \\ &\overset{(n)}{\tilde{r}_1}(x, t)\big|_{\overset{(n-1)}{O\mathscr{L}_{D2}}} = R_1(\overset{(n-1)}{\tilde{r}_2}(x, t), r_0(x, t), x, t) \end{aligned}\right\} \tag{43}$$

in $\overset{(n)}{\mathscr{L}_1^+} O \overset{(n)}{\mathscr{L}_2^-}$; in this same zone we find $\overset{(n)}{\tilde{r}_2}(x, t)$ from the conditions

$$\left.\begin{aligned} &\frac{\partial \overset{(n)}{\tilde{r}_2}}{\partial t} + \xi_2(\overset{(n-1)}{\tilde{r}_1}(x, t), \overset{(n)}{\tilde{r}_2}, x, t)\frac{\partial \overset{(n)}{\tilde{r}_2}}{\partial x} = f_2(\overset{(n-1)}{\tilde{r}_1}(x, t), \overset{(n)}{\tilde{r}_2}, x, t), \\ &\overset{(n)}{\tilde{r}_2}(x, t)\big|_{\overset{(n-1)}{O\mathscr{L}_{D1}}} = R_2(\overset{(n-1)}{\tilde{r}_1}(x, t), r_0(x, t), x, t) \end{aligned}\right\} \tag{44}$$

and we assume that $\overset{(n)}{\tilde{r}_1}$ and $\overset{(n)}{\tilde{r}_2}$ do not depend on x outside $\overset{(n)}{\mathscr{L}_1^+} O \overset{(n)}{\mathscr{L}_2^-}$ (Figure 4.53).

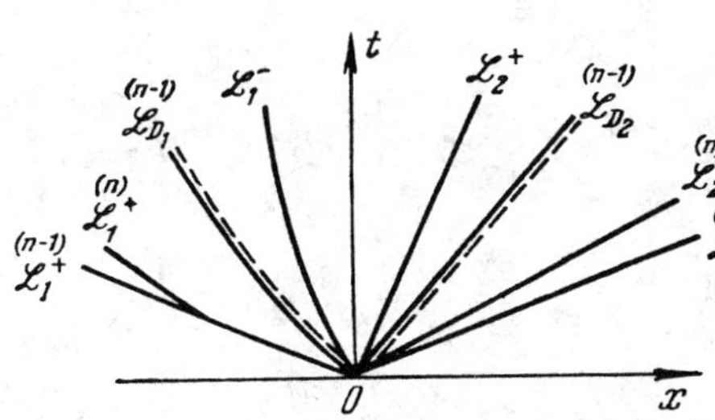

FIGURE 4.53

We shall not carry out here the involved computations connected with estimating the first derivatives of the successive approximations $\overset{(n)}{\bar{r}}_1$ and $\overset{(n)}{\bar{r}}_2$, since they basically repeat earlier computations; we only state the results. If system (1) satisfies the conditions enumerated above and the solutions $\overset{(n)}{\bar{r}}_1$ and $\overset{(n)}{\bar{r}}_2$ remain bounded for any $n = 1, 2, \ldots$, then there exists $T > 0$ such that for $0 \le t \le T$ all the successive approximations $\overset{(n)}{\bar{r}}(x, t)$ $(n = 1, 2, \ldots)$ possess bounded first derivatives.

We denote by $\overset{(n)}{\mathscr{L}_1^+}O\overset{(n)}{\mathscr{L}_2^-}$ the region of the variables x, t in which $\overset{(n)}{\bar{r}}_1(x, t)$ and $\overset{(n)}{\bar{r}}_2(x, t)$ are simultaneously defined as solutions of problems (43) and (44) (Figure 4.53). Outside $\overset{(n)}{\mathscr{L}_1^+}O\overset{(n)}{\mathscr{L}_2^-}$ we complete the definition of $\overset{(n)}{\bar{r}}(x, t)$ so that in the process of the method of successive approximations no shrinking of the regions $\overset{(n)}{\mathscr{L}_1^+}O\overset{(n)}{\mathscr{L}_2^-}$ occurs due to the functions $\overset{(n-1)}{\bar{r}_1}$ and $\overset{(n-1)}{\bar{r}_2}$ being unknown in the domain of determinacy of the solutions of the Cauchy problems (43) and (44).

Finally, we sketch the proof of convergence of the method of successive approximations in order to determine the requirements that guarantee convergence. In the common part of the zones $\overset{(n)}{\mathscr{L}_1^+}O\overset{(n)}{\mathscr{L}_2^-}$ for all $n = 1, 2, \ldots$ (this part exists and for sufficiently small T contains the lines $O\overset{(n)}{\mathscr{L}}_{D_1}$ and $O\overset{(n)}{\mathscr{L}}_{D_2}$ for all $n = 1, 2, \ldots$) in analogy to the foregoing it is not hard to obtain the estimates

$$\Delta\overset{(n)}{\bar{r}}_i(t) \le \max\left|\frac{\partial R_i(\bar{r}_j, r_0, x, t)}{\partial \bar{r}_j}\right| \Delta\overset{(n-1)}{\bar{r}_j}(t) + \bar{M}t\Delta\overset{(n-1)}{\bar{r}_j}(t) + \bar{M}\Delta\overset{(n-1)}{x_{D_j}}(t)$$

$$(i = 1, 2;\ j \ne i), \tag{45}$$

$$\Delta\overset{(n)}{x}_{D_i}(t) \le \bar{M}t\Delta\overset{(n)}{\bar{r}}_i(t). \tag{46}$$

In (45) the expression $\partial R_i/\partial \bar{r}_j$ is evaluated at some value of the first argument which is intermediate for successive approximations. From (45) and (46) it follows that

$$\Delta\overset{(n)}{\bar{r}}_i(t) \le \left\{\max\left|\frac{\partial R_i}{\partial \bar{r}_j}\right| + \bar{\bar{M}}t\right\}\Delta\overset{(n-1)}{\bar{r}_j}(t) \qquad (i = 1, 2;\ j \ne i). \tag{47}$$

This further implies that

$$\Delta\overset{(n)}{\bar{r}}_i(t) \le \left\{\max\left|\frac{\partial R_i}{\partial \bar{r}_j}\right| + \bar{\bar{M}}\right\}\left\{\max\left|\frac{\partial R_j}{\partial r_i}\right| + \bar{\bar{M}}t\right\}\Delta\overset{(n-2)}{r_i}(t)$$

$$(i = 1, 2;\ j \ne i). \tag{48}$$

Since

$$\left|\frac{\partial R_1(\bar r_2, r_0, x, t)}{\partial \bar r_2}\right| \cdot \left|\frac{\partial R_2(\bar r_1, r_0, x, t)}{\partial \bar r_1}\right| < 1$$

for any $\bar r_1$, $\bar r_2$, r_0 and $\bar r_0$, from this it follows that there exists a small $T > 0$ for which (48) can be strengthened so that

$$\Delta \overset{(n)}{\bar r_i}(t) \le \beta \Delta \overset{(n-2)}{\bar r_i}(t), \qquad 0 < \beta < 1 \ (i = 1, 2).$$

This implies the uniform convergence for $0 \le t \le T$ of the successive approximations $\overset{(n)}{\bar r}(x, t)$ and the lines of discontinuity $\overset{(n)}{O\mathscr{L}}_{D_1}$ and $\overset{(n)}{O\mathscr{L}}_{D_2}$. In this way we can show that there exist limit functions $\bar r_i(x, t) = \lim_{n\to\infty} \overset{(n)}{\bar r_i}(x, t)$ ($i = 1, 2$) which satisfy the conditions formulated above. This concludes the proof of the existence of a solution to the Cauchy problem (1), (3) in the case that conditions (7) are satisfied.

8. *Remarks on the uniqueness of a discontinuous solution of a system of two equations.* Here we briefly discuss the question of the uniqueness of the discontinuous solution $r(x, t)$ constructed above.

In subsection 7, on the basis of a known approximation $\overset{(n-1)}{\bar r}(x, t)$ we constructed the next approximation $\overset{(n)}{r}(x, t)$. For brevity we denote the procedure of constructing the approximation $\overset{(n)}{\bar r}(x, t)$ by the equality

$$\overset{(n)}{\bar r}(x, t) = T \overset{(n-1)}{\bar r}(x, t). \tag{1}$$

where T denotes the nonlinear operator that takes $\overset{(n-1)}{\bar r}(x, t)$ into $\overset{(n)}{r}(x, t)$. We established above that for each of the configurations considered the inequality

$$\|\overset{(n)}{\bar r}(x, t) - \overset{(n-1)}{\bar r}(x, t)\| \le \beta \|\overset{(n-1)}{r}(x, t) - \overset{(n-2)}{\bar r}(x, t)\|, \qquad 0 < \beta < 1, \tag{2}$$

holds, where the norm $\|r(x, t)\|$ is understood to be the maximum modulus of $r(x, t)$ in G. The inequality (2) can also be written in the form

$$\|T \overset{(n-1)}{\bar r} - T \overset{(n-2)}{\bar r}\| \le \beta \|\overset{(n-1)}{\bar r} - \overset{(n-2)}{\bar r}\|, \qquad 0 \le \beta < 1. \tag{3}$$

We now suppose that there exist two distinct solutions $r(x, t)$ and $\bar r(x, t)$ of the Cauchy problem in question. There are two possibilities:

1) r and $\bar r$ correspond to two distinct configurations;

2) r and $\bar r$ correspond to the same configuration of lines of discontinuity and rarefaction waves.

Now the first possibility can be eliminated immediately, since it is equivalent to the nonuniqueness of a stable generalized solution of the problem on decay of an arbitrary discontinuity, and, as we saw in subsection 5, this is excluded by the conditions imposed on the system of two quasilinear equations.

It therefore remains for us to consider the case in which the solutions $r(x, t)$ and $\bar{r}(x, t)$ correspond to the same configuration. This means that there exist two solutions $\bar{r}(x, t)$ and $\bar{\bar{r}}(x, t)$ satisfying the conditions formulated in subsection 7 and the same operator equation:

$$\bar{r}(x, t) = T\bar{r}(x, t), \qquad \bar{\bar{r}}(x, t) = T\bar{\bar{r}}(x, t). \tag{4}$$

The method of estimates used above for the successive approximations is applicable to the solutions $\bar{r}$ and $\bar{\bar{r}}$ of (4); therefore, (4) implies the estimate

$$\|\bar{r} - \bar{\bar{r}}\| = \|T\bar{r} - T\bar{\bar{r}}\| \leq \beta\|\bar{r} - \bar{\bar{r}}\|, \qquad 0 < \beta < 1, \tag{5}$$

which is impossible. We thus conclude that $\bar{r} = \bar{\bar{r}}$. The uniqueness of a stable generalized solution of the Cauchy problem (3.71), (3.73) has thus been proved.

Our second remark concerns the region in which a discontinuous solution of a system of two quasilinear equations can be constructed. As long as the number of singularities of the solution (lines of discontinuity and rarefaction waves) remains finite, we can apply the method indicated above for constructing a solution by subdividing the region into parts in which the singularties are isolated. However, singularities of a solution arise even from smooth initial data, and the number of them may multiply, possibly even without bound. This circumstance hampers the construction of discontinuous solutions of a system of two quasilinear equations globally, i.e., for any $t > 0$. We remark, however, that in the majority of practical problems the number of singularities remains bounded.

9. *Glimm's theorem.* Glimm (1965) proved that the Cauchy problem

$$\frac{\partial u}{\partial t} + \frac{\partial \varphi(u)}{\partial x} = 0, \qquad u(x, 0) = u_0(x) \tag{1}$$

is globally solvable (i.e., in the entire half plane $t \leq 0$) if the initial function $u_0(x)$ differs sufficiently little from a constant vector $\tilde{u}$. Regarding the system of conservation laws in (1), we assume only that it is strictly hyperbolic and the "convexity condition"

$$r^k(u) \cdot \nabla\xi_k(u) > 0 \qquad (k = 1, 2, \ldots, n) \tag{2}$$

is satisfied in some domain Ω of the variables $u_1, \ldots, u_n$.

The quantities

$$d_0 = \max_x |u_0(x) - \tilde{u}|(1 + \operatorname*{tot\,var}_{-\infty<x<\infty} u_0(x)),$$

$$d_1 = \max_x |u_0(x) - \tilde{u}| + \operatorname*{tot\,var}_{-\infty<x<\infty} u_0(x)$$

are used as a measure of deviation of the function $u_0(x)$ from a constant vector $\tilde{u}$. The quantity d_1 is used in considering general conservation laws, while d_0 is used for systems (1) which are reducible to Riemann invariants. The smallness of d_0 imposes considerably less stringent conditions on $u_0(x)$ than the smallness of d_1.

A generalized solution of the Cauchy problem (1) is understood to be a bounded, measurable function $u(x, t)$ on the half plane $t \leq 0$ satisfying the integral relations

$$\iint\limits_{\substack{t\geq 0\\ -\infty<x<\infty}} (g_t u + g_x \varphi(u))\,dt\,dx + \int_{-\infty}^{\infty} g(x, 0)u_0(x)\,dx = 0 \tag{3}$$

for any smooth vector-valued function $g(x, t)$ which is nonzero only in a bounded region of this half plane.

THEOREM. *Suppose for the system of conservation laws* (1) *the conditions formulated above are satisfied in some domain* Ω. *Then for any* $\tilde{u} \in \Omega$ *it is possible to find numbers* $\delta > 0$ *and* $K > 0$ *such that if*

$$d_1 < \delta, \tag{4}$$

then the Cauchy problem (1) *has a generalized solution* $u(x, t)$ *in the half plane* $t \leq 0$ *satisfying the estimates*

$$\sup_x |u(x, t) - \tilde{u}| \leq K \sup_x |u_0(x) - \tilde{u}|, \tag{5}$$

$$\operatorname*{tot\,var}_{-\infty<x<\infty} u(x, t) \leq K \operatorname*{tot\,var}_{-\infty<x<\infty} u_0(x), \tag{6}$$

$$\int_{-\infty}^{\infty} |u(x, t_2) - u(x, t_1)|\,dx \leq K|t_2 - t_1| \operatorname*{tot\,var}_{-\infty<x<\infty} u_0(x). \tag{7}$$

If, moreover, the system of conservation laws can be reduced to Riemann invariants, then (4) *can be replaced by the weaker condition* $d_0 \geq \tau$.

We point out tha Glimm's theorem establishes only the existence of a generalized solution. The question of the uniqueness of a solution of the Cauchy problem (1) remains open.

We limit ourselves here to a discussion of the central features of the rather complicated proof of this theorem and refer the reader to Glimm's paper (1965) for details. A certain family of "approximate" solutions of problem (1) is first constructed. On the half plane $t \geq 0$ we introduce a "checkered" mesh of points $(mh, n\ )$, $n = 0, 1, \ldots$, while, for each n, m runs through the set of all integers such that $m + n$ is even.

Suppose that the function $v(x, t)$ is piecewise constant on each level $t = n\tau$, where it takes the constant values v_m^n on each interval $(m-1)h < x < (m+1)h$ (for $m + n$ odd). In Figure 4.54 the boundaries of these intervals are indicated by points, and the constant values corresponding to these intervals by letters.

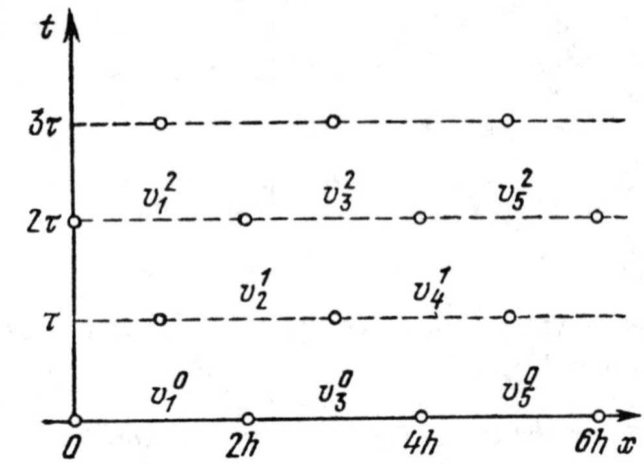

FIGURE 4.54

We shall describe the procedure of constructing, on the basis of known v_m^n, an approximate solution $v(x, t)$ in the strip $n\tau \leq t \leq (n + 1)$, and the next choice of the constant values v_m^{n+1}. We take the piecewise constant values v_m^n at the nth level $t = n\tau$ for the initial conditions for the system of conservation laws (1), and we find a solution $v(x, t)$ of this Cauchy problem in the strip $n\tau \leq t \leq (n + 1)\tau$. Thus, in this strip $v(x, t)$ is an exact generalized solution of (1) with piecewise constant initial values, i.e., finding $v(x, t)$ reduces to successive solution of problems on the decay of a discontinuity concentrated at the points of the mesh at the nth level. The width of this strip (the step size τ) can be chosen so small that the waves arising as a result of the decay of the discontinuity at the points $(mh, n\tau)$ remain for $t \in [n\tau, (n + 1)\tau]$ in the intervals $(m - 1)h \leq x \leq (m + 1)h$. Then during time τ these waves cannot interact with one another. The condition

$$\frac{\tau}{h} \sup_{u \in \Omega} |\xi_k(u)| < 1 \qquad (k = 1, 2, \ldots, n), \tag{8}$$

obviously provides the required smallness of τ.

Because of the closeness of the quantities v^n_{m-1} and v^n_{m+1} (which follows from the smallness of the variation of $u_0(x)$), according to Lax's theorem (see subsection 4), there exists a solution

$$v = V\left(\frac{x - mh}{t - n\tau}; v^n_{m-1}, v^n_{m+1}\right), \qquad t \leq n\tau$$

of the decay problem

$$\frac{\partial v}{\partial t} + \frac{\partial \varphi(v)}{\partial x} = 0, \qquad v(x, n\tau) = \begin{cases} v^n_{m-1}, & x < mh, \\ v^n_{m+1}, & x > mh. \end{cases}$$

Because of (8), $v^n(x, t)$ is determined for all m with $m + n$ even by the formula

$$v^n(x, t) = V\left(\frac{x - mh}{t - n\tau}; v^n_{m-1}, v^n_{m+1}\right),$$
$$n\tau \leq t \leq (n + 1)\tau, \qquad (m - 1)h \leq x \leq (m + 1)h. \tag{9}$$

Suppose that on each mesh interval $t = (n + 1)\tau$, $(m - 1)h \leq x \leq (m + 1)h$ an arbitrary point x^{n+1}_m is selected. We then set

$$v^{n+1}_m = v^n(x^{n+1}_m, (n + 1)\tau) = V\left(\frac{x^{n+1}_m - mh}{\tau}; v^n_{m-1}, v^n_{m+1}\right). \tag{10}$$

This makes it possible to determine recurrently sequences $\{v^n_m\}$, $n = 1, 2, \ldots$, provided, of course, that the sequences $\{x^n_m\}$ are given. We set $v^0_m = u_0(x^0_m)$; then (10) may be considered a finite-difference scheme approximating the Cauchy problem (1) on the mesh $\{x^n_m\}$ (a *Glimm scheme*).

We remark that prescribing the collection of points $\{x^n_m\}$ on the half plane $x, t \geq 0$ is equivalent to prescribing a collection of points $a^n_m = (2h)^{-1}[x^n_m - (m - 1)h]$ of the fixed interval $[0, 1]$. This collection we denote by the letter a. It may be considered a point in the infinite-dimensional space $A = \pi_{m,n}[0, 1]$.

We take the function $v(x, t)$ defined by (9) and (10) as an approximate solution of (1). It depends on the parameters τ and h and the infinite-dimensional parameter $a = \{a^n_m\}$. Of the parameters τ and h we consider only h to be independent, with the ratio τ/h fixed by (8). The parametefs a^n_m we consider independent. Thus, as an approximate solution we take the function $v = v(x, t; h, a)$ defined by (9) and (10), in which $x^n_m = (m - 1)h + 2ha^n_m$. By the construction of the approximate solution the function $v(x, t; h, a)$ is an exact generalized solution of the system of conservation laws inside each strip $n\tau < t < (n + 1)\tau$.

To prove the existence of $v(x, t; h, a)$ it is necessary to justify the applicability of Lax's theorem to the decay of a discontinuity at all points of the mesh, i.e., to prove that v^n_{m-1} and v^n_{m+1} are sufficiently close for all n and m, and that the values $v(x, t; h, a)$ belong to the domain Ω, which is necessary to satisfy (8). For a proof of these details we refer the reader to Glimm's paper.

We also mention without proof that the approximate solution $v(x, h; t, a)$ satisfies estimates analogous to (5)–(7):

$$\sup_x |v(x, t; h, a) - \tilde{u}| \leq K \sup_x |v(x, 0; h, a) - \tilde{u}|, \tag{11}$$

$$\operatorname*{tot\,var}_{-\infty<x<\infty} v(x, t; h, a) \leq K \operatorname*{tot\,var}_{-\infty<x<\infty} v(x, 0; h, a), \tag{12}$$

$$\begin{aligned} \int_{-\infty}^{\infty} &| v(x, t_2; h, a) - v(x, t_1; h, a) |\, dx \\ &\leq K(\tau + |t_2 - t_1|) \operatorname*{tot\,var}_{-\infty<x<\infty} v(x, 0; h, a) \end{aligned} \tag{13}$$

with a constant K not depending on h, a. By well-known theorems of analysis it follows from (11) and (12) that for each fixed t and a the family of functions $v(x, t; h, a)$ contains a sequence $v(x, t; h_i, a)$ $(h_i \to 0)$ which converges almost everywhere and in mean:

$$v(x, t; h_i, a) \to v(x, t; a) \quad \text{for almost all } x,$$

$$\int_{x_1}^{x_2} | v(x, t; h_i, a) - v(x, t; a) |\, dx \to 0 \quad \text{for any } x_1 \text{ and } x_2.$$

It follows from (13) by a simple argument that from this sequence we can extract a subsequence which converges almost everywhere in the half plane $t \leq 0$ and also in mean in an bounded region of this half plane. We shall assume that the sequence $v(x, t; h_i, a) \to v(x, t; a)$ itself possesses these properties (almost everywhere for $t \leq 0$).

The proof of Glimm's theorem is completed by considering the error of the approximate solution $v(x, t; h, a)$; that is, the error in satisfying (3) with a smooth, compactly supported function $g(x, t)$:

$$\delta(g; h; a) = \iint_{\substack{t\geq 0 \\ -\infty<x<\infty}} (g_t v + g_x \varphi(v))\, dt\, dx + \int_{-\infty}^{\infty} g(x, 0)\, u_0(x)\, dx. \tag{14}$$

Consideration of the fact that in the strips $n\tau \leq t \leq (n + 1)\tau$ the approximate solution $v(x, t; h, a)$ is an exact solution leads to the equalities

$$\int_{n\tau}^{(n+1)\tau} dt \int_{-\infty}^{\infty} (g_t v + g_x \varphi(v))\, dx$$

$$= \int_{-\infty}^{\infty} [g(x, (n+1)\tau) v(x, (n+1)\tau - 0; h; a)$$

$$- g\,(x, n\tau) v(x, n\tau; h, a)]\, dx,$$

and hence $\delta(g; h, a) = \sum_0^\infty \delta_n(g; h; a)$, where

$$\delta_0(g; h, a) = \int_{-\infty}^{\infty} g(x, 0)[u_0(x) - v(x, 0; h, a)]\, dx,$$

$$\delta_n(g, h, a) = \int_{-\infty}^{\infty} g(x, n\tau)[v(x, n\tau - 0; h, a) - v(x, n\tau; h, a)]\, dx.$$

It is not possible to prove that $\delta \to 0$ as $h \to 0$ for any fixed a (i.e., everywhere on A). We therefore consider δ as a function of the variable $a = \{a_m^n\}$ and show that the error δ tends to zero in mean square, i.e., in the metric of $\mathscr{L}_2(A)$:

$$\int_A |\delta(g; h, a)|^2\, da \to 0 \quad \text{as } h \to 0. \tag{15}$$

Since

$$\delta^2 = \sum_{n=0}^{\infty} \delta_n^2 + 2 \sum_{n=1}^{\infty} \sum_{k<n} \delta_n \delta_k$$

and the number of values of n for which $\delta_n \neq 0$ is $O(1/h)$ (because g is compactly supported), to prove (15) it suffices to show that

$$\int \delta_n^2\, da \le Ch^2, \qquad \left| \int \delta_n \delta_m\, da \right| \le Ch^2. \tag{16}$$

The first estimate of (16) is a consequence of the uniform estimate on A

$$|\delta_n(g; h, a)| \le Ch \text{ tot var } v(x, t; h, a),$$

which is easily varified. To prove the second estimate of (16) we denote by a^n the set of a_m^n with fixed n and note that

$$\int \delta_n(g; h, a)\, da^n$$

$$= \sum_k \int_{(k-1)h}^{(k+1)h} g(x, \tau n)\, dx \int da^n[v^{n-1}(x, n\tau) - v^{n-1}(x_k^n, n\tau)]$$

$$= \sum_k \int_{(k-1)h}^{(k+1)h} g(x, n\tau)\, dx \int_0^1 da_k^n[v^{n-1}(x, n\tau) - v^{n-1}(x_n^k, n\tau])$$

$$= \sum_k \frac{1}{2h} \int_{(k-1)h}^{(k+1)h} g(x, n\tau)\, dx \int_{(k-1)h}^{(k+1)h} [v^{n-1}(x, n\tau] - v^{n-1}(y, n\tau)]\, dy$$

$$= \sum_k \frac{1}{2h} \int_{(k-1)h}^{(k+1)h} [g(x, n\tau) - g(kh, n\tau)]\, dx \int_{(k-1h)}^{(k+1)h} [v^{n-1}(x, n\tau) - v^{n-1}(y, n\tau)]\, dy.$$

Therefore

$$\left| \int \delta_n(g; h, a) da^n \right|$$

$$\leq C_1 \sum_k \int_{(k-1)h}^{(k+1)h} dx \int_{-2h}^{2h} | v^{n-1}(x, n\tau) - v^{n-1}(x + y, n\tau)|\, dy$$

$$\leq 4h^2 C_1 \text{ tot var } v(x, t; h, a),$$

where C_1 is a constant bounding $|g_x(x, t)|$. Since $\delta_m(g; h, a)$ does not depend on a^n for $m < n$ (it depends only on a^k for $k \leq m$), it follows that

$$\left| \int \delta_n \cdot \delta_m da^n \right| \leq |\delta_m| \cdot O(h^2) = O(h^3).$$

This implies the second estimate of (16), since obviously

$$\left| \int \delta_n \cdot \delta_m\, da \right| \leq \left| \int \delta_n \cdot \delta_m\, da^n \right|.$$

Thus, it has been shown that $\delta(g; h, a)$ tends to zero in the mean-square sense in the space A (15) for any compactly supported, smooth function $g(x, t)$. According to well-known theorems of analysis, there exists a subsequence $\delta(g; h_i, a)$ which tends to zero almost everywhere on A as $h_i \to 0$. A simple argument shows that this subsequence may be chosen to be the same for all test functions $g(x, t)$. We may therefore assume that the sequence $\{h_i\}$ determining $v(x, t; a)$ is this subsequence.

Thus, for any point $a \in B \subset A$, where $\text{meas}(A - B) = 0$, and any test function $g(x, t)$ the error $\delta(g; h_i, a) \to 0$. Passing to the limit as $h_i \to 0$ in (14), we obtain

$$\iint\limits_{\substack{t\geq 0\\ -\infty<x<\infty}} (g_t v(x, t; a) + g_x\varphi(v(x, t; a))\, dt\, dx + \int_{-\infty}^{\infty} g(x, 0)u_0(x)\, dx = 0.$$

Hence, $v(x, t; a)$ is a generalized solution of the Cauchy problem (1), and the theorem is proved.

We emphasize again that Glimm's theorem does not establish uniqueness of the solution of (1). Another basic shortcoming of thes theorem is that it is not constructive: it does not give a concrete algorithm for constructing approximate solutions which tend to the exact solutions as $h_i \to 0$, since the mesh $\{a_m^n\}$ is unknown. Finally, perhaps the most serious drawback of Glimm's method is that even with the a priori assumption of the convergence of the approximate solutions $v(x, t; h, a)$ as $h \to 0$ to some function $v(x, t; a)$ it cannot be asserted that this function is an exact solution of (1), since $\delta(g; h, a)$ need not tend to zero as $h \to 0$. It may therefore be said that Glimm'e method is not "convergent with respect to error" (more precisely, it has not been proved otherwise).

In spite of these shortcomings Glimm's theorem was the first theorem establishing the global existence of a generalized solution of problem (1) for the case of a *system* ($n \geq 2$) of quasilinear equations and for a class of initial functions which, while strongly restricted by the condition (4), is still an interesting class. Glimm's method was applied by a number of mathematicians to more special systems of conservation laws: as a result, more general existence theorems were obtained.

Nishida (1968) proved that for a system of two quasilinear conservation laws

$$\frac{\partial V}{\partial t} - \frac{\partial u}{\partial x} = 0, \qquad \frac{\partial u}{\partial t} + \frac{\partial}{\partial x} p(V) = 0 \tag{17}$$

the Cauchy problem with initial functions

$$V(x, 0) = V_0(x), \qquad u(x, 0) = u_0(x),$$

which are bounded and have locally bounded variation while $V_0(x) > \delta > 0$, also has a generalized solution in the case $p(V) = a^2/V$.

Bahvalov (1970) distinguished a class of systems of two quasilinear equations including, in particular, the system (17) for $p(V) = a^2V^{-\gamma}, 0 < \gamma < 2$, for which Glimm's theorem holds without assuming the smallness of the total variation of the initial function. For these specific systems of two conservation laws, due to their special features, it is possible to estimate the boundedness of the solutions and their variation (i.e., to prove the estimates (5)–(7) without the smallness condition (4)). There is an analogous approach to the problem in the works of DiPerna (1973, 1976b) and in a number of others.

Kuznecov and Tupčiev (1975) generalized Glimm's theorem in another direction: they proved the applicability of this method for more general systems of conservation laws not satisfying the "convexity conditions" (2).

Zhang Tong and Guo Yu-fa (1965) developed another approach to the problem of existence of a solution of problem (1); they considered the Cauchy problem for the system (17) with a pressure $P = p(V)$ satisfying the usual convexity conditions $p'(V) < 0$ and $p''(V) > 0$. However, the initial functions are not arbitrary: they must satisfy certain order conditions which imply, roughly speaking, that all shock waves arising in the solution move in one direction and all rarefaction waves in another. In this case smallness of the variation of the initial functions is not required; it suffices that they be bounded.

Ljapidevskiĭ (1974) proved the uniqueness of the solution of the Cauchy problem for the system (17) under the same assumptions. In a later paper (1975) he proved uniqueness of the solution of the Cauchy problem for a more general system of two conservation laws, but only in a class of solutions having a finite number of lines of discontinuity.

Finally, we note that the technique of Glimm presented here has been generalized to the case of initial boundary value problems for systems of conservation laws (see Doktor (1977)).

10. ***The method of viscosity for a system of quasilinear equations. Phenomena of the method of viscosity.*** In Chapter 2 we saw that shock waves in a gas or liquid may be considered as limits of flows of a viscous and thermally conducting liquid, and we encountered the application of a certain nonlinear viscosity (the von Neumann-Richtmyer viscosity). In §§7.2 and 7.7 of that chapter it was shown that a stable generalized solution of a single quasilinear equation is the limit of solutions of an equation with "visccsity" as the coefficient of viscosity tends to zero.

For systems of quasilinear equations the method of viscosity has still been insufficiently studied. With the example of a homogeneous system of quasilinear equations

$$\frac{\partial u}{\partial t} + \frac{\partial \varphi(u)}{\partial x} = 0, \qquad u = \{u_1, \ldots, u_n\}, \tag{1}$$

which is strictly hyperbolic ($\xi_1(u) < \xi_2(u) < \cdots < \xi_n(u)$), we shall show that the question of the choice of a particular viscosity is critical and requires great care. The system of equations with viscosity corresponding to (1) we write in the form

$$\frac{\partial u_\mu}{\partial t} + \frac{\partial \varphi(u_\mu)}{\partial x} = \mu \frac{\partial}{\partial x} B(u_\mu) \frac{\partial u_\mu}{\partial x}, \tag{2}$$

where $B(u_\mu)$ is a square $b \times n$ matrix. We thus restrict attention to the class of "divergence" viscosities.[11]

We first make some very general remarks. It is obvious that the matrix B must be chosen so that the following conditions are satisfied:

a) correctness of the formulation of the Cauchy problem for the system (2);

b) the solutions u_μ are smooth for $t > 0$ for any piecewise continuous and piecewise smooth initial data;

c) there is convergence (in some norm, for example, in mean) of the solutions u_μ as $\mu \to 0$ to stable generalized solutions $u(x, t)$ of (1).

At present it is not possible to list sufficient conditions that ensure that requirements a)–c) are met, because the systems (2) have not been sufficiently studied. We shall therefore try to restrict the class of matrices B on the basis of some simple analogies.

We consider the case of linear systems (1) and (2) with constant coefficients, where $\varphi = Au$ and A and B are constant matrices. We note first of all that then any solution $u_\mu(x, t)$ can be represented in the form

$$u_\mu(x, t) = u\left(\frac{x}{\mu}, \frac{t}{\mu}\right),$$

where $u(x, t)$ is a solution of the system

$$\frac{\partial u}{\partial t} + A\frac{\partial u}{\partial x} = B\frac{\partial^2 u}{\partial x^2}. \tag{3}$$

This system has particular solutions of the form

$$u = u_0 \exp\{i(\lambda t + \nu x)\}, \tag{4}$$

where u_0 is a (right eigenvector and $-i\lambda/\nu^2$ an eigenvalue of the matrix $B + (i/\nu)A$. If β_j are the eigenvalues of the matrix B, then $\lambda = i\beta_j \nu^2 + O(\nu)$ as $|\nu| \to \infty$. According to condtion a), we wish to ensure that $u(x, t) \to 0$ if $u(x, 0) \to 0$, and we should therefore require that

$$\operatorname{Re} \beta_j > 0 \qquad (j = 1, 2, \ldots, n). \tag{5}$$

Indeed, for $\operatorname{Re} \beta_{j_0} < 0$ it is possible to find a sequence of initial functions, say, $u_\nu(x, 0) = \nu^{-1}u_0 e^{i\nu x}$, which tends to zero as $\nu \to \infty$ but for which the solution $u_\nu(x, t)$, according to (4), tends (in absolute value) to ∞ for any $t > 0$.

Condition (5) is the well-known Hadamard correctness condition for the Cauchy problem for the system of equations (3). It has been pointed out (see

[11] The divergence form of the viscosity $(\partial/\partial x)B(\partial u/\partial x)$ ensures that the Hugoniot conditions are satisfied on the front of a smeared-out shock wave (cf. Chapter 2, §5).

Iskander-zade (1966)) that in a certain sense conditions (5) are not sufficient for correctness of the problem.

We consider the behavior of the numbers λ for small ν. If α_j are the eigenvalues of the matrix A and r_j and l_j are the corresponding eigenvectors (as usual, we assume that the α_j are real and distinct), then it is easily seen that

$$\lambda = -\alpha_j \nu + i\nu^2(l_j B r_j) + O(\nu^3).$$

If $l_{j_0} B r_{j_0} < 0$, for some $j = j_0$, then, according to (4), the system with viscosity will have special solutions of the form

$$u_\mu = (\exp(\nu^2 |(l_{j_0} B r_{j_0})| t)/u) v_\mu,$$

where v_μ is a bounded function (as $\mu \to 0$) which as $\mu \to 0$ will tend to ∞ for any $t > 0$. According to requirement c), we must eliminate this possibility and require that together with (5) the following condition be satisfied:

$$(l_j B r_j) \geq 0 \qquad (j = 1, 2, \ldots, n). \tag{6}$$

We observe that (6) implies that the diagonal elements of the viscosity matrix arising in (3) after reducing it to Riemann invariants must be positive. In analogy with the linear case we shall also require that (5) and (6) be satisfied for a nonlinear system.

We shall consider some simple examples. We seek particular solutions $u_\mu((x - Dt)/\mu)$ of (2) depending only on the variable $y = (x - Dt)/\mu$, i.e.,

$$u_\mu(x, t) = u_\mu(y), \qquad y = \frac{x - Dt}{\mu}.$$

The function $u_\mu(y)$ satisfies the system of equations

$$\frac{d}{dy} B(u_\mu) \frac{du_\mu}{dy} = \frac{d\varphi(u_\mu)}{dy} - D \frac{du_\mu}{dy} = [A(u_\mu) - DE] \frac{du_\mu}{dy}.$$

which admits the integration

$$B(u_\mu) \frac{du_\mu}{dy} = \varphi(u_\mu) - Du_\mu - C = F(u_\mu), \tag{7}$$

where C is any constant vector. We require of the solution u_μ that it tend to constant values as $y \to \pm\infty$, i.e.,

$$u_\mu(y) \to \begin{cases} u^- & \text{as } y \to -\infty, \\ u^+ & \text{as } y \to +\infty. \end{cases}$$

For the existence of such a solution it is necessary that $u_\mu = u^-$ and $u_\mu = u^+$

be stationary points of the system (7), i.e., $F(u^-) = F(u^+) = 0$. These conditions can be rewritten in the form

$$\varphi(u^-) - \varphi(u^+) = D(u^- - u^+).$$

From this we conclude that the states u^- and u^+ must be related by the Hugoniot conditions. However, this is still insufficient for the existence of an integral curve $u_\mu(y)$ of the system (7) passing through u^- at $y = -\infty$ and u^+ at $y = +\infty$.

We shall find some additional necessary conditions; to this end we write (7) in the form

$$\frac{du_\mu}{dy} = B^{-1}(u_\mu)F(u_\mu),$$

where B^{-1} is the matrix inverse to B. We expand the right side of this system in a series in a neighborhood of a stationary point, for example, $u = u^-$. Then

$$\frac{d}{dy}(u_\mu - u^-) = B^{-1}(u^-)\left(\frac{\partial F(u_\mu)}{\partial u}\right)\Big|_{u=u^-}(u_\mu - u^-) + O(|u_\mu - u^-|^2)$$
$$= B^{-1}(u^-)[A(u^-) - DE](u_\mu - u^-) + O(|u_\mu - u^-|^2).$$

Taking the inner product of this system with the vector $(u_\mu - u^-)$, we obtain

$$\frac{d}{dy}\frac{(u_\mu - u^-)^2}{2}$$
$$= (u_\mu - u^-)B^{-1}(u^-)[A(u^-) - DE](u_\mu - u^-) + O(|u_\mu - u^-|^3).$$

With increase of y the quantity $(u_\mu - u^-)^2/2$ does not decrease in a neighborhood of $u = u^-$; therefore, if there exists an integral curve $u_\mu(y)$ of the system (7), then the matrix $B^{-1}(u^-)[A(u^-) - DE]$ cannot be negative definite. In the same way it is established that the matrix $B^{-1}(u^+)[A(u^+) - DE]$ cannot be positive definite.

Suppose that the desired solution $u_\mu(y)$ of (7) exists. Then the limit

$$\lim_{\mu\to 0} u_\mu(x, t) = u(x, t) = \begin{cases} u^- & \text{as } x - Dt < 0, \\ u^+ & \text{as } x - Dt > 0 \end{cases}$$

is a discontinuous function that satisfies the Hugoniot conditions on the line of discontinuity $x = Dt$ and hence the integral conservation laws of the system (1).

However, this solution may be an unstable solution of (1), since the conditions for existence of a solution $u_\mu(y)$ discussed above and the stability conditions for the discontinuity u^-, u^+ are different. We confirm this with a simple example constructed by Byhovskiĭ (1962). Suppose that

$$n = 2; \qquad \varphi(u) = \{-u_2;\, p(u_1)\}; \qquad p'(u_1) < 0,\, p''(u_1) > 0.$$

The (1) is the system of equations for the isothermic flow of a normal gas (see Chapter 2, §2).

Let $u^+ = \{1, 0\}$ and $D > 0$. We shall show that there exists a solution $u_\mu(y)$ with $u_1^- > 1$. Such a solution goes over as $\mu \to 0$ into a rarefaction shock wave which is an unstable solution of the equations of gas dynamics, as we saw in Chapter 2.

We choose a constant, positive definite matrix

$$B = \begin{pmatrix} a & 0 \\ c & b \end{pmatrix}, \qquad a, b > 0,\, c \neq 0.$$

Then

$$B^{-1} = \begin{pmatrix} 1/a & 0 \\ c/ab & 1/b \end{pmatrix}.$$

It is easily verified that the eigenvalues of $B^{-1}(A(u) - DE)$ are roots of the quadratic equation

$$\lambda^2 + \left[\frac{(a+b)D - c}{ab}\right]\lambda + \frac{D^2 + p'(u_1)}{ab} = 0. \tag{8}$$

According to the Hugoniot conditions,

$$D^2 = \frac{p(1) - p(u_1^-)}{u_1^- - 1},$$

and, according to our assumption, $u_1^- > 1$ and $D > 0$. Since $p'(u_1) < 0$ and $p''(u_1) > 0$, this implies that

$$\sqrt{-p'(1)} > D > \sqrt{-p'(u_1^-)}, \tag{9}$$

or

$$p'(1) + D^2 < 0, \qquad p'(u_1^-) + D^2 > 0.$$

Since $a, b > 0$, the roots of (8) for $u_1 = u_1^+ = 1$ have different signs. Thus, the eigenvalues of $B^{-1}(A(u^+) - DE)$ have different signs. This means that $u = u^+$ is a saddle point.

On the other hand, at $u = u^-$ the roots of (8) have the same sign (if they are real). Therefore, for the existence of an integral curve joining u^- and u^+ we must require that the roots of (8) be real and positive, i.e., that the matrix $B^{-1}(u^-)[A(u^-) - DE]$ be positive definite. This occurs if

$$\frac{c - (a + b) D}{ab} > 0, \quad \frac{[c - (a + b)D]^2}{4ab} - D^2 - p'(u_1^-) > 0. \tag{10}$$

It is easy to see that these two inequalities can be satisfied by fixing an arbitrary $c > 0$ and choosing $D > 0$ sufficiently small.

Suppose that the quantities $a, b, c, D > 0$ are such that these inequalities are satisfied. Then $u^- = \{u_1^-, u_2^-\}$ is a node ($u_1^- > 1$, $u_2^- < 0$), and $u^+ = \{1, 0\}$ is a saddle point.

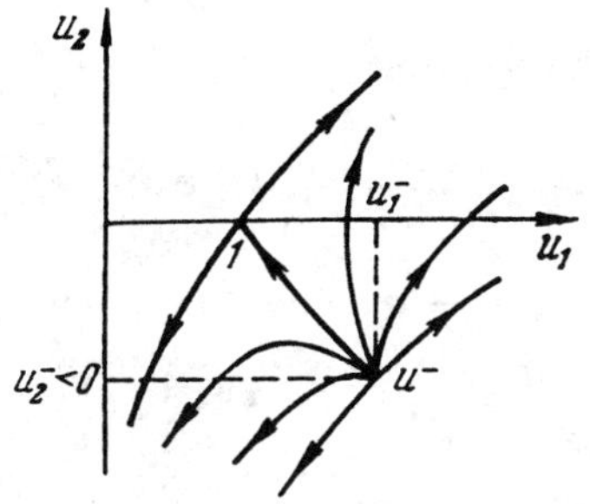

FIGURE 4.55

Thus, the picture of the integral curves of the system (7) in the plane of the variables u_1 and u_2 is as shown in Figure 4.55. The arrows in the picture indicate the direction of increasing y. Thus, under these assumptions there exists an integral curve $u = u_\mu(y)$ passing through u^- and u^+. Letting the parameter μ tend to zero in the solution u_μ, we obtain in the limit a rarefaction shock wave which is unstable.

So far we have not considered condition (6). In our case it is easy to verify that

$$l_1Br_1 = a\sqrt{-p'(u_1)} - c + b\sqrt{-p'(u_1)},$$
$$l_2Br_2 = a\sqrt{-p'(u_1)} + c + b\sqrt{-p'(u_1)}.$$

If the conditions (10), under which we obtained an unstable shock wave, are satisfied, then, according to (9),

$$(l_1Br_1)_{u_1=u_1^-} < (a + b)D - c < 0,$$

i.e., (6) is violated.

A still simpler and more striking example[12] is the system (2) in the case

$$\varphi(u) = \left\{\frac{u_1^2}{2}, \frac{u_2^2}{2}\right\}, \qquad B = \left\| \begin{matrix} \sqrt{2} - \left(\frac{1}{2} + \sqrt{2}\right)\alpha & 1 - \alpha \\ 1 - \alpha & \frac{1}{\sqrt{2}} \end{matrix} \right\|.$$

For $O < \alpha < 1 - 1/2\sqrt{2}$ the matrix B is positive definite; moreover, it is always symmetric and obviously satisfies (6). For $\varphi = \{u_1^2/2, u_2^2/2\}$ the system (1) decomposes into the two equations

$$\frac{\partial u_1}{\partial t} + \frac{\partial}{\partial x}\frac{u_1^2}{2} = 0, \qquad \frac{\partial u_2}{\partial t} + \frac{\partial}{\partial x}\frac{u_2}{2} = 0,$$

and in this case (2) has the solution

$$u_2 = -\sqrt{2}\,u_1, \qquad \frac{u_1 + u_1^0}{u_1 - u_1^0} = \exp\left\{\frac{2u_1^0 x}{\alpha\mu}\right\}, \quad u_0^1 > 0.$$

This implies that, as $\mu \to 0$, $u_1 = u_1(y) = u_1(x/\mu)$ tends to an unstable solution

$$u_1(x) = \begin{cases} -u_1^0, & x < 0, \\ u_1^0, & x > 0, \end{cases} \qquad u_1^0 > 0,$$

of the Hopf equation ($u_2(y)$ tends to a stable solution).

We observe that the integral curve in this case passes through the point $u_1 = u_2 = 0$, at which the eigenvalues $\xi_1 = u_1$ and $\xi_2 = u_2$ of the system coincide (i.e, the system is not strictly hyperbolic). The fact that condition (6) in the present case is insufficient is related to this. It is not known if similar examples are possible with a symmetric viscosity matrix for systems that are strictly hyperbolic.

We consider the viscosity matrix $B = b(u)E$, which clearly satisfies condition (6). We shall show that in this case for $n = 2$ the system (7) cannot have solutions which as $\mu \to 0$ correspond to unstable discontinuities. Indeed, if there exists an integral curve $u_\mu(y)$ passing through the points u^- ($y = -\infty$) and u^+ ($y = \infty$), then it is impossible to satisfy the inequalities

$$\xi_k(u^-) < D < \xi_k(u^+) \tag{11}$$

for any $k = 1, 2$. Indeed, suppose these inequalities hold, for example, for $k = 1$. Since $\xi_2(u) > \xi_1(u)$, it then follows that $\xi_2(u^+) > D$. Therefore the matrix

[12] This example was suggested by N. N. Kuznecov.

$$B^{-1}(u^+)[A(u^+) - DE] = \frac{B[A(u^+) - DE]}{b(u^+)} = \frac{[A(u^+) - DE]}{b(u^+)}$$

has the eigenvalues $[\xi_k(u^+) - D]/b(u^+)$, which are both positive by (11). As we have seen above, in this case there exists no integral curve of (4) joining u^- and u^+, and therefore it is not possible to satisfy (11). Similarly, it can be proved that (11) is impossible for $k = 2$. Thus, in the case $B = b(u)E$ condition (11) cannot be satisfied for any integral curve.

We remark that this still does not imply that as $\mu \to 0$ all solutions $u_\mu(y)$ tend to stable solutions of (1), since for arbitrary systems the stability conditions are not formulated in the form (3.1.10) and (3.1.11). Nevertheless, for systems for which these inequalities guarantee uniqueness, viscosity matrices of the form considered give only stable solutions.

Some advantages of the identity viscosity matrix can also be established in more general cases (see Foy (1964)). General viscosity matrices were considered by Kulikovskiĭ (1962), but for more special systems. The advantages of the identity viscosity matrix do not imply, of course, that it is preferable. Thus, for example, for a viscous and thermal conducting fluid the "viscosity matrix" is not the identity matrix.

The question of admissible discontinuities for general systems of two and more conservation laws has been investigated successfully with the viscosity method. Tupčiev (1964, 1966, 1973a) formulated conditions for the admissibility of a discontinuity for a rather general system of conservation laws. The concept of an admissible viscosity matrix is introduced there, and sufficient conditions for a viscosity matrix to be admissible are established. There is an interesting application of a special form of viscosity in which the matrix $B(u)$ of (2) is replaced by $tB(u)$, and the system (2) has the form

$$\frac{\partial u_\mu}{\partial t} + \frac{\partial \varphi(u_\mu)}{\partial x} = t\mu \frac{\partial}{\partial x} B(u_\mu) \frac{\partial u_\mu}{\partial x}. \tag{12}$$

This form of viscosity was applied by Kalašnikov (1959), Tupčiev (1964, 1972b), and Dafermos (1973a, 1974).

The system (12) is invariant under the similarity transformation $x' = kx$, $t' = kt$ $(k > 0)$. Therefore, the problem of decay of an arbitrary discontinuity for the system (12) has a self-similar solution $u_\mu = u_\mu(y)$, $y = x/t$, analogous to the problem of decay of a discontinuity for a system without viscosity. This solution satisfies the system of ordinary differential equations

$$\mu \frac{d}{dy} B(u_\mu) \frac{du_\mu}{dy} = \frac{d\varphi(u_\mu)}{dy} - y \frac{du_\mu}{dy} \tag{13}$$

and the boundary conditions at infinity

$$u_\mu(y) = \begin{cases} u^- & \text{as } y \to -\infty, \\ u^+ & \text{as } y \to +\infty. \end{cases} \tag{14}$$

The study of the structure of solutions of problem (13), (14) leads to necessary conditions on the discontinuities of the solution $u(y) = \lim_{\mu \to +0} u_\mu(y)$ of the decay problem for a system of hyperbolic type. A system of two conservation laws has been considered in more detail, the structure of the solution $u(y)$ of this system has also been studied for "nonconvex" systems (see the papers cited above). In particular, it has been shown that on admissible discontinuities of the solution $u(y)$ the conditions (3.2.32) and (3.2.33) must be satisfied.

We now consider more general problems for equations with viscosity. The equations of motion of a viscous and thermally conducting gas and closely related systems of equations are of greatest interest. In Chapter 2, §5, we have already considered self-similar soltions of the equations of motion for a viscous and thermally conducting gas which were applied to study the structure of a shock transition. More general problems such as the Cauchy problem and mixed problems have not yet been sufficiently studied.

The mathematical investigation of the system of Navier-Stokes equations for a compressible, viscous, thermally conducting gas was begun by Nash (1962). He proved that, for any smooth initial data, for a sufficiently small time interval there exists a unique classical solution of the Cauchy problem. These results were then repeated and generalized with application of another method by Itaya (1970) and Vol'pert and Hudjaev (1972). Solonnikov (1976) proved the local solvability of a mixed boundary value problem for the equations of a barotropic, compressible, viscous fluid. The behavior of solutions of the equations of a viscous gas globally in time has so far been studied only in the one-dimensional case and also for model equations.

Itaya (1974) considered the simplest model equations

$$\frac{\partial \rho}{\partial t} + \rho^2 \frac{\partial u}{\partial q} = 0, \qquad \frac{\partial u}{\partial t} = \mu \frac{\partial}{\partial q}\left(\rho \frac{\partial u}{\partial q}\right) \tag{15}$$

(the so-called Burgers model), which coincide with the equations of gas dynamics with viscosity in Lagrange coordinates if we set $p \equiv \text{const}$ there. He proved the existence and uniqueness of a classical solution of the Cauchy problem for the system (15) for any $t > 0$.

Kažihov (1975, 1976) proved global theorems for certain mixed boundary value problems for a more general model of a barotropic fluid and for the system of equations of a perfect polytropic gas with viscosity and thermal conductivity

$$\frac{\partial V}{\partial t} - \frac{\partial u}{\partial q} = 0, \qquad \frac{\partial u}{\partial t} = \frac{\partial}{\partial q}\left(\mu V^{-1}\frac{\partial u}{\partial q} - p\right),$$
$$\frac{\partial}{\partial t}\left(\varepsilon + \frac{u^2}{2}\right) + \frac{\partial}{\partial q}\left[u\left(p - \mu V^{-1}\frac{\partial u}{\partial b}\right)\right] = \frac{\partial}{\partial q}\left(\kappa V^{-1}\frac{\partial T}{\partial q}\right) \tag{16}$$

with the equations of state

$$pV = RT, \qquad \varepsilon = c_V T \tag{17}$$

(see (2.5.14) and (2.5.16) in Chapter 2, with $\nu = 0$).

The central feature in the proof of the global solvability of the Cauchy problem for (15) is establishing boundedness and strict positivity of the density $\rho(q, t)$ for any $t \geq 0$; in the case of a boundary value problem for (16) it is also necessary to establish this for the temperature T. Upper and lower bounds for the velocity $u = u(q, t)$ for any $t > 0$ follow from the second equation of (15), after which upper and lower bounds for $\rho(q, t)$ are easily found.

For the system (16), (17) Kažihov (1976) considers the initial conditions

$$u(q,0) = u_0(q), \quad V(q,0) = V_0(q) > 0, \quad T(q,0) = T_0(q) > 0,$$
$$0 \leq q \leq q_1, \tag{18}$$

and the boundary conditions

$$\mu V^{-1}\frac{\partial u}{\partial q} - p = 0, \qquad \kappa\frac{\partial T}{\partial q} = 0 \quad (q = 0, q = q_1). \tag{19}$$

The conditions (19) describe the flow of a thermally insulated mass of gas into a vacuum.

By introducing the new dimensionless variables

$$\tau = \frac{t}{t_1}, \qquad x = \frac{q}{q_1}, \qquad \rho = \frac{V_1}{V}, \qquad v = \frac{u}{u_1}, \qquad \theta = \frac{T}{T_1},$$

where

$$V_1 = \frac{1}{q_1}\int_0^{q_1} V_0(q)\,dq, \qquad u_1 = \frac{\mu}{q_1}, \qquad t_1 = \frac{q_1^2 V_1}{\mu}, \qquad T_1 = \frac{\mu^2}{q_1^2 c_V},$$

we bring the equations (16)–(19) to the form

$$\frac{\partial \rho}{\partial \tau} + \rho^2 \frac{\partial v}{\partial x} = 0, \qquad \frac{\partial v}{\partial \tau} = \frac{\partial}{\partial x}\left(\rho \frac{\partial v}{\partial x}\right) - k \frac{\partial}{\partial x}(\rho\theta),$$

$$\frac{\partial \theta}{\partial \tau} = \lambda \frac{\partial}{\partial x}\left(\rho \frac{\partial \theta}{\partial x}\right) + \rho\left(\partial\left(\frac{\partial v}{\partial x}\right)^2 - k\rho\theta \frac{\partial v}{\partial x}\right.$$

$$(k = R/c_v,\ \lambda = \kappa/\mu c_v) \tag{20}$$

and

$$\frac{\partial v}{\partial x} - k\theta = 0, \qquad \frac{\partial \theta}{\partial x} = 0 \quad \text{for } x = 0,\ x = 1. \tag{21}$$

The initial values of $\rho_0(x)$ and $\theta_0(x)$ are strictly positive and bounded:

$$0 < m \le \rho_0(x), \qquad \theta_0(x) \le M < \infty. \tag{22}$$

We estimate $\rho(x, t)$. Integrating the second of the equations (20), we find that

$$\frac{\partial}{\partial \tau} \int_0^x v(\xi, \tau)\, d\xi = \rho \frac{\partial v}{\partial x} - k\rho\theta.$$

Since it follows from the first equation of (20) that $\rho(\partial v/\partial x) = -(\partial \ln \rho/\partial \tau)$, we obtain from this, after integration with respect to τ,

$$\rho(x, \tau) \exp\left\{k \int_0^\tau \rho(x, s)\theta(x, s)\, ds\right\}$$
$$= \rho_0(x) \exp\left\{\int_0^x [v_0(\xi) - v(\xi, \tau)]\, d\xi\right\}. \tag{23}$$

We multiply (23) by $k\theta(x, \tau)$ and again integrate with respect to τ. We obtain

$$\exp\left\{k \int_0^\tau \rho(x, s)\theta(x, x)\, ds\right\}$$
$$= 1 + k\rho_0(x) \int_0^\tau \theta(x, s) \exp\left\{\int_0^x [v_0(\xi) - v(\xi, s)]\, d\xi\right\} dx. \tag{24}$$

Combining (23) and (24), we get

$$\rho(x, \tau) = \frac{\rho_0(x) \exp\{\int_0^x [v_0(\xi) - v(\xi, \tau)]\, d\xi\}}{1 + k\rho_0(x) \int_0^\tau \theta(x, s) \exp\{\int_0^x [v_0(\xi) - v(\xi, s)]\, d\xi\}\, ds}, \tag{25}$$

which makes it possible to estimate $\rho(x, \tau)$. From (25) we have

$$M_\rho(\tau) \leq N\Big(1 + kN^{-1}\int_0^\tau m_\theta(s)\,ds\Big)^{-1}, \tag{26}$$

$$m_\rho(\tau) \geq N^{-1}\Big(1 + kN\int_0^\tau M_\theta(s)\,ds\Big)^{-1}, \tag{27}$$

where we have introduced the notation

$$m_\rho(\tau) = \min_x \rho(x, \tau), \qquad m_\theta(\tau) = \min_x \theta(x, \tau),$$

$$M_\rho(\tau) = \max_x \rho(x, \tau), \qquad M_\theta(\tau) = \max_x \theta(x, \tau).$$

The constant N in (26) and (27) depends only on the initial data. If $m_\theta(\tau) > 0$, then from (26) we have the upper bound $M\rho(\tau) \leq N$ for the density $\rho(x, \tau)$.

We write the third equation of (20) in the form

$$\frac{\partial\theta}{\partial\tau} = \lambda\frac{\partial}{\partial x}\left(\rho\frac{\partial\theta}{\partial x}\right) + \rho\left(\frac{\partial v}{\partial x} - \frac{k}{2}\theta\right)^2 - \frac{k^2}{4}\rho\theta^2$$

and consider it as a parabolic equation for the temperature $\theta(x, \tau)$. We then obtain

$$m_\theta^{-1}(\tau) \leq m^{-1} + \frac{k^2}{4}\int_0^\tau M_\rho(s)\,ds.$$

Strengthening this estimate by means of (26), we obtain an integral inequality for $m_\theta(\tau)$, from which there follows a lower bound for $m_\theta(\tau)$. From (27) we obtain a lower bound for $\rho(x, \tau)$ in a similar way.

The remaining a priori estimates are obtained from well-known techniques developed in the theory of parabolic equations. Without going into this, we formulate the main result of Kažihov (1976).

THEOREM. *Suppose that the initial data* (18) *satisfy conditions* (22) *and belong to the Sobolev space* W_2^1. *Then one any finite time interval* $[0, T]$ *there exists a unique solution of problem* (16)–(19) *with the following properties*:

$$\max_{0\leq\tau\leq T}\left[\|v(\tau)\|_{W_2^1(\Omega)} + \|\theta(\tau)\|_{W_2^1(\Omega)} + \left\|\frac{\partial\rho}{\partial\tau}(\tau)\right\|_{L_2(\Omega)} + \left\|\frac{\partial\rho}{\partial x}(\tau)\right\|_{L_2(\Omega)}\right] < \infty,$$

$$\|v_\tau\|_{L_2(Q)} + \|v_{xx}\|_{L_2(Q)} + \|\theta_\tau\|_{L_2(Q)} + \|\theta_{xx}\|_{L_2(Q)} < \infty.$$

Here $\Omega = [0, 1]$ and $Q = \Omega \times [0, T]$. *The quantities* $\rho(x, \tau)$ *and* $\theta(x, \tau)$ *are strictly positive and bounded in* Q. *If, moreover, the initial data are sufficiently smooth and the necessary consistency conditions are satisfied, then the solution of this mixed problem is a classical solution.*

§4. Applications of the general theory of systems of quasilinear equations of hyperbolic type

In this section we indicate some problems of physics, chemistry, and mathematics that are related to the theory of systems of quasilinear equations of hyperbolic type.

The best-known application of systems of quasilinear equations of hyperbolic type is the study of one-dimensional flows of compressible gases and liquids without viscosity and thermal conductivity. This question and its relation to the theory of systems of quasilinear equations was studied in detail in Chapter 2.

Other well-known examples of problems connected with systems of quasilinear equations are the motion of incompressible fluids in shallow channels ("shallow water" theory), the supersonic, stationary flow of a gas or liquid in the two-dimensional case, problems of the nonlinear theory of elasticity, the theory of filtration, and some others. Here we shall briefly consider some of these.

1. ***"Shallow water" theory.*** Suppose that a heavy (in a gravitational field), incompressible fluid flows in a channel having the form shown in Figure 4.56. We shall assume that the liquid is free of internal friction and friction with the sides and bottom of the channel, and that the height h of the liquid above the bottom of the channel is a small quantity as compared with the dimensions of irregularities of the bottom, the characteristic dimensions of the flow, etc. We assume that the flow of liquid is characterized by a single space variable x and depends on the time t. We thus assume that the fluid velocity u has a nonzero component u_x, which we denote by u, while the remaining components can be neglected; we moreover assume that the height h depends only on x and t. Under these assumptions we shall derive the equations describing the flow of the liquid.[13]

Let $h(x, t)$ denote the height of the liquid above the bottom of the channel at the point x, let ρ be the fluid density, let l be the width of the channel, and let $u(x, t)$ denote the velocity of the liquid directed along the x axis. The quantity of liquid between the cross-sections $x = x_1$ and $x = x_2$ of the channel at time t is obviously

$$\int_{x_1}^{x_2} \rho l h(x, t)\, dx = \rho l \int_{x_1}^{x_2} h(x, t)\, dx.$$

[13] Our derivation is only a sketch. For a rigorous foundation of "shallow water" theory see Ovsjannikov (1973).

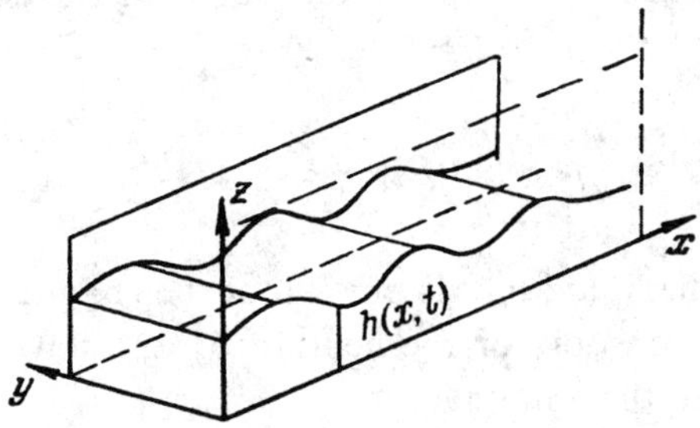

FIGURE 4.56

The change in the amount of liquid in this portion of the channel between times $t = t_1$ and $t = t_2$ is

$$\rho l \int_{x_1}^{x_2} [h(x, t_2) - h(x, t_1)]\, dx, \tag{1}$$

which obviously must be equal to the amount of liquid flowing in through the planes $x = x_1$ and $x = x_2$ during the time from $t = t_1$ to $t = t_2$, i.e., to the quantity

$$-\rho l \int_{t_1}^{t_2} [h(x_2, t)u(x_2, t) - h(x_1, t)u(x_1, t)]\, dt. \tag{2}$$

Equating (1) and (2), we obtain

$$\int_{x_1}^{x_2} \{h(x, t_2) - h(x, t_1)\}\, dx + \int_{t_1}^{t_2} [hu|_{x=x_2} - hu|_{x=x_2}]\, dt = 0, \tag{3}$$

which obviously represents the integral law of conservation of mass of the liquid. As usual, from (3) we obtain the integral conservation law

$$\oint_C h\, dx - hu\, dt = 0, \tag{4}$$

which is valid for any closed contour C in the plane of the variables x, t, and the differential equation

$$\frac{\partial h}{\partial t} + \frac{\partial}{\partial x} hu = 0 \tag{5}$$

in the case of smooth flows.

The change of total momentum of the liquid in this same portion of the channel from the time $t = t_1$ to $t = t_2$ is

$$\rho l \int_{x_1}^{x_2} [hu|_{t=t_2} - hu|_{t=t_1}]\, dx. \tag{6}$$

The momentum of the liquid in this portion of the channel changes due to two effects: due to the transport of momentum by the flow through the planes $x = x_1$ and $x = x_2$ in the amount

$$-\rho l \int_{t_1}^{t_2} [u^2h|_{x=x_2} - u^2h|_{x=x_1}]\, dt, \tag{7}$$

and also due to the momentum of pressure forces in the planes $x = x_1$ and $x = x_2$.

We shall compute the total pressure $p(x, t)$ acting on a cross section of the channel. Assuming that on the free surface $z = h$ of the liquid the pressure is equal to zero, by the barometric formula we have $p = \rho g(h - z)$ (g is the gravitational acceleration) and

$$p(x, t) = l \int_0^h p\, dz = l\rho g \int_0^h (h - z)\, dz = \frac{1}{2} l\rho g h^2. \tag{8}$$

Therefore, the momentum of pressure forces on the sections $x = x_1$ and $x = x_2$ from the time $t = t_1$ to $t = t_2$ is given by

$$-\frac{l\rho g}{2} \int_{t_1}^{t_2} [h^2(x_2, t) - h^2(x_1, t)\, dt. \tag{9}$$

Equating (6) to the sum of (7) and (9), we obtain the integral law for the conservation of momentum of the liquid:

$$\int_{x_1}^{x_2} [hu|_{t=t_2} - hu|_{t=t_1}]\, dx$$
$$+ \int_{t_1}^{t_2} \left\{ \left[[hu^2 + g\frac{h^2}{2} \right] \Big|_{x=x_2} - \left[hu^2 + g\frac{h^2}{2} \right] \Big|_{x=x_1} \right\} dt = 0. \tag{10}$$

Equation (10) can be written as the integral law of conservation of momentum

$$\oint_C hu\, dx - \left(hu^2 + g\frac{h^2}{2} \right) dt = 0, \tag{11}$$

from which for smooth flows we obtain the differential equation

$$\frac{\partial hu}{\partial t} + \frac{\partial}{\partial x}\left(hu^2 + \frac{g}{2}h^2\right) = 0. \tag{12}$$

Combining (5) and (12), we obtain a system of two quasilinear equations for $h(x, t)$ and $u(x, t)$:

$$\frac{\partial h}{\partial t} + \frac{\partial hu}{\partial x} = 0, \qquad \frac{\partial hu}{\partial t} + \frac{\partial}{\partial x}\left[hu^2 + \frac{g}{2}h^2\right] = 0. \tag{13}$$

It is now easy to see that (13) coincides with the system of equations of gas dynamics for the isentropic flow of an ideal gas with an exponent of the adiabat $\gamma = 2$. Indeed, if the quantity h is denoted by ρ and it is assumed that $p = g\rho^2/2$, then (13) goes over into the corresponding system for the case indicated (see Chapter 2).

From this comparison we conclude, in particular, that (13) is a system of quasilinear equations of hyperbolic type and that solutions of this system are, in general, discontinuous. To a discontinuity of a solution of (13) there corresponds a sudden increase of the height $h(x, t)$—a so-called "water jump". The usual Hugoniot conditions and stability conditions must be satisfied on the front of discontinuity.

2. ***Plane stationary flow of a compressible gas.*** Another well-known example of a system of quasilinear equations of hyperbolic type is the system of equations describing the plane, stationary, supersonic flow of a compressible gas. If u and v are the components of the velocity vector $\boldsymbol{q}$, then this system has the form

$$\left.\begin{aligned} u\frac{\partial \rho}{\partial x} + \rho\frac{\partial u}{\partial x} + v\frac{\partial \rho}{\partial y} + \rho\frac{\partial v}{\partial y} &= 0,\\ u\frac{\partial u}{\partial x} + v\frac{\partial u}{\partial y} + \frac{1}{\rho}\frac{\partial p}{\partial x} &= 0,\\ u\frac{\partial v}{\partial x} + v\frac{\partial v}{\partial y} + \frac{1}{\rho}\frac{\partial p}{\partial y} &= 0,\\ u\frac{\partial S}{\partial x} + v\frac{\partial S}{\partial y} &= 0. \end{aligned}\right\} \tag{1}$$

This system describes only smooth flows. The required conservative form of the equations (1) for describing discontinuous flows is presented below.

The characteristic equation (of fourth degree) for the system (1) has the form

$$\begin{vmatrix} v - \xi u & -\xi\rho & \rho & 0 \\ -\xi\dfrac{c^2}{\rho} & v - \xi u & 0 & \dfrac{-\xi p_S'}{\rho} \\ \dfrac{c^2}{\rho} & 0 & v - \xi u & \dfrac{p_S'}{\rho} \\ 0 & 0 & 0 & v - \xi u \end{vmatrix} = 0$$

or

$$(v - \xi u)^2[\xi^2(u^2 - c^2) - 2uv\xi + (v^2 - c^2)] = 0, \tag{2}$$

were $\xi = dy/dx$ is a characteristic direction of (1). The first factor gives the double root $\xi = v/u$. The corresponding characteristic is obviously a flow line. It is thus twofold degenerate. The second factor on the left side of (2) has real roots only for $q^2 = u^2 + v^2 \geq c^2$. In the sonic case ($q = c$) both these roots coincide, and it is not hard to verify that the system (1) is not hyperbolic. On the other hand, in the supersonic case ($q > c$) it is of hyperbolic type. In order to see this, it suffices to verify that to the multiple eigenvalue $\xi = v/u$ there correspond two linearly independent eigenvectors, i.e., two independent equations in the characteristic form of (1). Simple computations lead to the following two equations which contain differentiation only along the flow line:

$$\frac{dS}{dq} = 0, \tag{3}$$

$$\frac{1}{\rho}\frac{dp}{dq} + \frac{d}{dq}\left(\frac{u^2 + v^2}{2}\right) = 0 \tag{4}$$

(here $d/dq = u(\partial/\partial x) + v(\partial/\partial y)$). The relation (3) implies that the entropy is constant on the flow line, and is obviously independent of (4).

Introducing the function $H(\rho, S)$ defined by

$$H(\rho, S) = \int \frac{c^2(\rho, S)d\rho}{\rho},$$

we obtains from (3) and (4)

$$\frac{d}{dq}\left(H(\rho, S) + \frac{q^2}{2}\right) = 0. \tag{5}$$

Thus, to the flow line there correspond the two Riemann invariants: the entropy S and $B = H + q^2/2$. The equality $B = \text{const}$ on the flow line, which follows from (5), is called the *Bernoulli integral*.

The expression for the remaining (so-called sonic) characteristic values of the system (1) are

$$\xi^{\pm} = \frac{uv \pm c\sqrt{q^2 - c^2}}{u^2 - c^2} = \frac{v^2 - c^2}{uv \mp c\sqrt{q^2 - c^2}}.$$

The characteristics $dy/dx = \xi^{\pm}$ form angles α and $-\alpha$ respectively with the flow line, where $\sin\alpha = c/q = 1/M$ and $M = q/c$ (the angle α is called the *Mach angle*, and the function M the *Mach number*).

In conclusion we present the divergence form of (1):

$$\frac{\partial \rho u}{\partial x} + \frac{\partial \rho v}{\partial y} = 0,$$

$$\frac{\partial}{\partial x}(p + \rho u^2) + \frac{\partial}{\partial y}(\rho uv) = 0,$$

$$\frac{\partial}{\partial x}(\rho uv) + \frac{\partial}{\partial y}(p + \rho v^2) = 0,$$

$$\frac{\partial}{\partial x}\left[\rho u\left(\varepsilon + \frac{p}{\rho} + \frac{q^2}{2}\right)\right] + \frac{\partial}{\partial y}\left[\rho v\left(\varepsilon + \frac{p}{\rho} + \frac{q^2}{2}\right)\right] = 0.$$

3. ***Chemical sorption and problems of chromatography.***[14] Suppose that through a pipe containing an active sorption material (the sorbent) there passes a liquid or gaseous mixture of substances to be separated. Ignoring effects connected with the walls of the pipe, we shall assume that the problem is one-dimensional. Let t denote the time, x the coordinate along the axis of the sorption column, u_i the concentration of the ith component in the mixture, a_i the concentration of the ith component in the sorbent, and V the rate of motion of the mixture in the column (assumed to be constant).

Neglecting diffusion flows of substances both in the mixture and in the sorbent, we write the equations for the conservation of mass of each component:

$$\int_{x_1}^{x_2} \{[u_i + a_i]\,|_{t=t_2} - [u_i + a_i]\,|_{t=t_1}\}\,dx + \int_{t_1}^{t_2} V[u_i(x_2, t) - u_i(x_1, t)]\,dx = 0$$

$$(i = 1, 2, \ldots, n), \qquad (1)$$

which for smooth u_i and a_i reduce to the differential equations

$$V\frac{\partial u_i}{\partial x} + \frac{\partial}{\partial t}(u_i + a_i) = 0. \qquad (2)$$

[14] See Kuznecov (1967).

We make an assumption regarding the instantaneous character of sorption, i.e., we atsume that at each point of the pipe and at each moment of time equilibrium previals between the sorption material and the free mixture. Mathematically this is expressed by the fact that the concentartion of sorption material is determined by the composition of the mixture, i.e.,

$$a_i = f_i(u_1, \ldots, u_n) = f_i(u) \qquad (i = 1, \ldots, n). \tag{3}$$

The equations (3) are usually called the *equations of isothermic sorption.* Under this condition the system (2) can be rewritten in the form

$$\frac{\partial}{\partial t}(u + f(u)) + V\frac{\partial u}{\partial x} = 0, \tag{4}$$

where u and f are vectors with n components.

The characteristic values $\xi = \xi(u)$ of (4) are defined from the equation

$$\operatorname{Det}\left(\left((V - \xi)\,\delta_{ij} - \xi\frac{\partial f_i}{\partial u_j}\right)\right) = 0. \tag{5}$$

We denote by $\lambda = \lambda(u)$ an eigenvalue of the matrix $((\partial f_i/\partial u_j))$. From (5) we then obviously have

$$\frac{V - \xi}{\xi} = \lambda, \qquad \xi = \frac{V}{1 + \lambda}. \tag{6}$$

We shall assume (this actually follows from general laws of the character of sorption) that all the eigenvalues $\lambda_k(u)$ of the matrix $((\partial f_i/\partial u_j))$ are positive, i.e., the maxtrix $\partial f/\partial u$ is positive definite. From (6) it then follows that the eigenvalues $\xi = \xi_k(u)$ of (4) satisfy the inequalities

$$\xi_k(u) = \frac{V}{1 + \lambda_k(u)} < V \qquad (V > 0), \tag{7}$$

i.e., the speed $\xi_k(u)$ of the characteristics

$$\frac{dx}{dt} = \xi_k(u) \tag{8}$$

of (4) is less than the speed V of the flow.

We consider in more detail the case of a special sorption isotherm (3) in which

$$a_i = \frac{a_i^\infty k_i u_i}{1 + \sum_{s=1}^n k_s u_s} \qquad (i = 1, 2, \ldots, n), \tag{9}$$

which is usually called the case of *Langmuir sorption.* Here a_i^∞ is the magnitude of adsorption saturation, and k_i is the sorption coefficient.

Introducing the notation $v_i = k_i u_i$ and $\Gamma_i = a_i^\infty k_i$ (the Henry coefficients), we rewrite the equations of the Langmuir isotherm (9) in the form

$$\varphi_i(v) = k_i a_i = \Gamma_i \frac{v_i}{1 + \sum_{s=1}^n v_s} = \frac{\Gamma_i v_i}{p}, \qquad p = 1 + \sum_{s=1}^n v_s, \tag{10}$$

and (4) in the form

$$\frac{\partial}{\partial t}(v + \varphi(v)) + V\frac{\partial v}{\partial x} = 0. \tag{11}$$

We shall assume that all the Henry coefficients Γ_i are distinct (the case when some of the Γ_i coincide can be reduced to this case), and we enumerate them in increasing order:

$$\Gamma_1 < \Gamma_2 < \cdots < \Gamma_n. \tag{12}$$

For the isotherm (10) the eigenvalues λ are determined from the equation

$$\mathrm{Det}(((\Gamma_i p - \lambda)\,\delta_{ij} - \Gamma_i v_i)) = 0. \tag{13}$$

We consider the case where $v_i \neq 0$ for all $i = 1, \ldots, n$. Then (13) can be transformed to the form

$$\mathrm{Det}\left(\left(\frac{\Gamma_i p - \lambda}{\Gamma_i v_i}\delta_{ij} - 1\right)\right) = \left[\prod_{k=1}^n \frac{\Gamma_k p - \lambda}{\Gamma_k v_k}\right]\left(1 - \sum_{k=1}^n \frac{\Gamma_k v_k}{\Gamma_k p - \lambda}\right) = 0. \tag{14}$$

The equation (14) can also be written in the form

$$F(\lambda, v) = 1, \tag{15}$$

where

$$F(\lambda, v) = \sum_{k=1}^n \frac{\Gamma_k v_k}{\Gamma_k p - \lambda}. \tag{16}$$

Since

$$F'_\lambda(\lambda, v) = \sum_{k=1}^n \Gamma_k v_k (\Gamma_k p - \lambda)^{-2} > 0 \qquad (v_k > 0), \tag{17}$$

the function $F(\lambda, v)$ is a monotonically increasing function λ which has poles at the points $\lambda_k^* = \Gamma_k p > 0$ (Figure 4.57). Since the roots $\lambda = \lambda_k(v)$ are the abscissae of the points of intersection of the graph of $F(\lambda, v)$ with the line $F = 1$ (Figure 4.57) and

$$0 < F(0, v) = \sum_{k=1}^n \frac{\Gamma_k v_k}{\Gamma_k p} = 1 - \frac{1}{p} < 1,$$

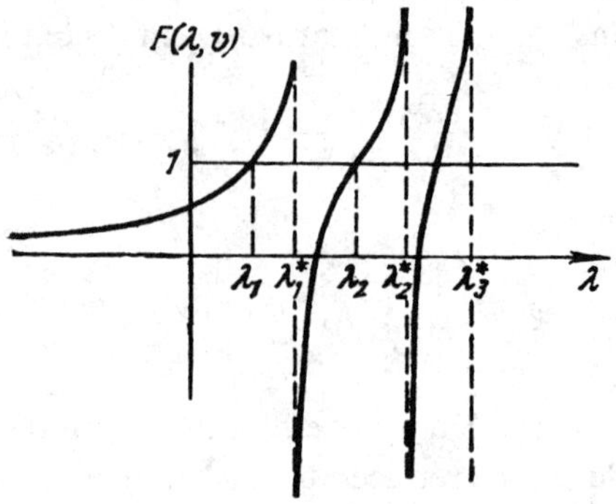

FIGURE 4.57

we conclude that (15) has n real, distinct, positive roots $\lambda = \lambda_k(v)$. These roots satisfy the inequalities

$$0 < \lambda_1(v) < \lambda_1^*, \qquad \lambda_{i-1}^* < \lambda_i(v) < \lambda_i^*, \tag{18}$$

and, according to (6), the characteristic values $\xi = \xi_k(v)$ of (11) are also real, distinct, and positive; moreover,

$$V > \xi_1(v) > \xi_2(v) > \cdots > \xi_n(v) > 0, \tag{19}$$

i.e., under the condition $\Gamma_i v_i \neq 0$ the system (11) is strictly hyperbolic.

Computing the left eigenvectors $l^k = \{l_i^k\}$ of (11), we find that (up to a factor)

$$l_i^k = \frac{1}{\Gamma_i p - \lambda_k(v)}, \tag{20}$$

while the right eigenvectors $r^k = \{r_i^k\}$ are

$$r_i^k = \frac{\Gamma_i v_i}{\Gamma_i p - \lambda_k(v)}. \tag{21}$$

We mention some special features of the system (11). For it there exist n Riemann invariants, so that it can be brought to the form

$$\frac{\partial R_i}{\partial t} + \xi_k(R)\frac{\partial R_i}{\partial x} = 0 \qquad (i = 1, 2, \ldots, n), \tag{22}$$

where $R_i(v) = \lambda_i(v)/p$.

As usual, discontinuities of a solution $v(x, t)$ of (11) satisfy the Hugoniot conditions, which in this case take the form

$$D[v + \varphi(v)] = V[v], \qquad D = \frac{dx}{dt}, \tag{23}$$

and the stability condition

$$\xi_k(v(x-0,t)) > D > \xi_k(v(x+0,t)). \tag{24}$$

It is interesting to note that for the system (11) rarefaction waves (R_i = const for $i \neq k$) of kth type coincide with the Hugoniot adiabat of kth type, i.e., the line $R_i(v)$ = const ($i \neq k$) is a solution of (23).

The difference in the speeds of the characteristics $\xi_k(v)$ is explained by the method of separating the components of the mixture in the sorbent. The method of separating components based on difference of the Henry coefficients is called *chromatography*.

4. ***An application to differential geometry.*** Problems of geometry are connected with nonlinear differential equations. It is for this reason that geometers first began a systematic study of nonlinear differential equations and their solutions It is no accident that the outstanding geometer of the last century, Rieman, obtained basic results in gas dynamics in many respects unsurpassed at the present time.

Here we shall indicate the connection of the theory of quasilinear equations with one branch of differential geometry—the theory of surfaces. Suppose that on some smooth surface in three-dimensional space there is a metric

$$ds^2 = dx^2 + B^2(x,y)dy^2, \tag{1}$$

where the lines y = const are geodesic lines on the surface, while the lines x = const form the family of orthogonal trajectories to them. Such a coordinate system (x, y) on a surface is called a *semigeodesic* system.

The Gaussian curvature $K(x, y)$ of the surface is determined only by the metric (1) by means of the formula

$$B''(x,y) + K(x,y)B(x,y) = 0. \tag{2}$$

If only the metric (1) (the first quadratic form) is given, the question of the existence of a surface in three-dimensional Euclidean space realizing this metric reduces to finding the coefficients $L(x, y)$, $M(x, y)$ and $N(x, y)$ of the second quadratic form. These coefficients must satisfy the basic equations of the theory of surfaces—the Peterson-Condazzi equations (see Blaschke (1924)):

$$M_x - L_y = -\frac{B_x}{B}M, \qquad N_x - M_y = \frac{B_x}{B}(N + B^2L) - \frac{B_y}{B}M. \tag{3}$$

On the other hand, the Gaussian curvature of the surface can also be computed in an exterior fashion in terms of the coefficients of the second quadratic form. For this there is Gauss' formula

$$B^2K = LN - M^2. \tag{4}$$

If we now eliminate L, M or N from (3) by means of (4), we obtain a system of two quasilinear equations in two independent variables x and y for the two unknown functions. On investigating this system of equations, we can easily establish that it is elliptic for $K > 0$, parabolic for $K = 0$, and hyperbolic for $K < 0$.

Thus, in the case of negative Gaussian curvature $K < 0$ the Peterson-Codazzi equations reduce to a system of two quasilinear equations of hyperbolic type. Here the characteristics are integral curves of the equation

$$L\,dx^2 + 2M dx\,dy + N\,dy^2 = 0; \tag{5}$$

in geometry these lines are called the *asymptotic lines of the surface*.

As we saw in Chapter 1, for any system of two quasilinear equations of hyperbolic type it is possible to introduce the Riemann invariants. Simple computations lead to the expressions for the invariants

$$r = B\frac{-M-B\sqrt{-K}}{N}, \qquad s = B\frac{-M+B\sqrt{-K}}{N}, \tag{6}$$

after which (3) reduces to the form

$$\left.\begin{aligned} &\frac{\partial r}{\partial x} + \frac{s}{B}\frac{\partial r}{\partial y} = -s(1+r^2)\frac{B_x}{B} + \frac{(r-s)}{2}\left[\frac{\partial Q}{\partial x} + \frac{r}{B}\frac{\partial Q}{\partial y}\right],\\ &\frac{\partial s}{\partial x} + \frac{r}{B}\frac{\partial s}{\partial y} = -r(1+s^2)\frac{B_x}{B} + \frac{(s-r)}{2}\left[\frac{\partial Q}{\partial x} + \frac{s}{B}\frac{\partial Q}{\partial y}\right],\\ &Q = \ln\sqrt{-K(x, y)}, \qquad B = B(x, y). \end{aligned}\right\} \tag{7}$$

If we set $r = \tan\varphi_1$ and $s = \tan\varphi_2$, then φ_1 and φ_2 are the angles formed on the surface between the direction of the asymptotic lines (characteristics) and the direction of the geodesic lines $y =$ const. The system (7) can also now be written in the form

$$\left.\begin{aligned} &\frac{d\varphi_1}{ds_2} = \cos\varphi_2\frac{\partial\varphi_1}{\partial x} + \frac{\sin\varphi_2}{B}\frac{\partial\varphi_1}{\partial y} = -\sin\varphi_2\frac{B_x}{B} + \frac{1}{2}\sin\omega\frac{dQ}{ds_1},\\ &\frac{d\varphi_2}{ds_1} = \cos\varphi_1\frac{\partial\varphi_2}{\varphi x} + \frac{\sin\varphi_1}{B}\frac{\partial\varphi_2}{\partial y} = -\sin\varphi_1\frac{B_x}{B} - \frac{1}{2}\sin\omega\frac{dQ}{ds_2},\\ &\omega = \varphi_1 - \varphi_2. \end{aligned}\right\} \tag{8}$$

If we now examine (7) or (8), we easily see that (7) is a weakly nonlinear system of quasilinear equations, since

$$\xi_1 = \frac{s}{B(x, y)}, \qquad \xi_2 = \frac{r}{B(x, y)}.$$

A regular surface of negative Gaussian curvature $K(x, y) < 0$ possesses distinct directions of the asymptotic lines (characteristics), so that we may assume that $r \neq s$ at regular points, i.e., $\xi_1 \neq \xi_2$, which means that the systems (7) and (8) are strictly hyperbolic at regular points.

As we saw in Chapter 1, weakly nonlinear systems possess a remarkable property: solutions of such systems remain continuous and smooth as long as the solution itself remains bounded. Solutions of (7) or (8) thus possess an analogous property.

At the beginning of this century Hilbert conjectured that there does not exist a complete regular surface of negative Gaussian curvature $K(x, y) < \varepsilon < 0$ imbedded in three-dimensional Euclidean space. He proved this assertion for the case of constant Gaussian curvature $K = -1$. Since the sysrem (8) is weakly nonlinear, the reason for the nonexistence of a surface with $K(x, y) \leq -1$ is not the formation of a discontinuity of the solution (as might be thought) nor the unboundedness of the solution (it is evident from the form (8) that the solution remains bounded at finite points x, y). Therefore, the reason for the nonexistence of global solutions of (8) is the degeneration of the surface, i.e., the case $\varphi_1 = \varphi_2$, $\omega = 0$. In all known cases it actually turns out that $\omega = 0$ on the boundary of the surface, and the surface cannot be extended in a smooth manner beyond the boundary (edge). We mention that a proof of this conjecture has been obtained in recent years with some restrictions on the derivatives of $K(x, y)$.

5. *The equations of magnetohydrodynamics.* In this subsection we obtain the differential equations[15] describing the one-dimensional motion of an electrically conducting gas in a magnetic field.

If we denote by $\boldsymbol{E}$ and $\boldsymbol{H}$ respectively the electric and magnetic fileds, then it is known (see Tamm (1946)) that the force $\boldsymbol{f}$ which the electromagnetic field exerts on a unit volume of gas can be written in the form

$$\boldsymbol{f} = \rho_e \boldsymbol{E} + \frac{1}{c}[\boldsymbol{j} \times \boldsymbol{B}], \tag{1}$$

where ρ_e is the density of the electric charge, $\boldsymbol{j}$ is the current density, and c is the speed of light. In (1) the Caussian system of units for the electromagnetic quantities has been adopted.

When an electric current $\boldsymbol{j}$ flows in a stationary substance the amount of heat

[15] Integral conservation laws must be derived to study discontinuous solutions. For simplicity we limit ourselves here to smooth soiutions of the equations of magnetohydrodynamics.

energy (Joule heat) evolved per unit time and unit volume is $Q = \boldsymbol{Ej}$. In the case where the conductor moves with velocity $\boldsymbol{u}$ we have $Q = \boldsymbol{E}'\boldsymbol{j}'$, where $\boldsymbol{E}'$ and $\boldsymbol{j}'$ can be written according to the formulas of electrodynamics up to terms of order u^2/c^2 in the form

$$\boldsymbol{E}' = \boldsymbol{E} + \frac{1}{c}[\boldsymbol{u} \times \boldsymbol{B}], \qquad \boldsymbol{j}' = \boldsymbol{j} - \rho_e \boldsymbol{u}, \tag{2}$$

where $\boldsymbol{B}$ is the magnetic induction.

The electromagnetic fields satisfy the system of Maxwell's equations

$$\left.\begin{aligned} \boldsymbol{H} &= \frac{4\pi}{c}\boldsymbol{j} + \frac{1}{c}\frac{\partial \boldsymbol{D}}{\partial t}, \qquad \operatorname{curl} \boldsymbol{E} = -\frac{1}{c}\frac{\partial \boldsymbol{B}}{\partial t}, \\ \operatorname{div} \boldsymbol{D} &= 5\pi\rho_e, \qquad \operatorname{div} \boldsymbol{B} = 0. \end{aligned}\right\} \tag{3}$$

Here $\boldsymbol{D} = \varepsilon\boldsymbol{E}$ and $\boldsymbol{E} = \mu\boldsymbol{H}$ (ε and μ are the electric and magnetic permeabilities of the gases).

We shall assume that the media we consider satisfy the condition of quasi-neutrality. This means that the total electric charge of any elementary volume is equal to zero ($\rho_e = 0$). Then $\boldsymbol{j}' = \boldsymbol{j}$. A medium satisfying this condition of quasineutrality is usually called a *plasma*.

Finally, for a well-ionized plasma it may be assumed with sufficient accuracy that $\boldsymbol{B} = \boldsymbol{H}$ and $\boldsymbol{D} = \boldsymbol{E}$.

To determine the current density $\boldsymbol{j}$ we use Ohm's law:

$$\boldsymbol{j} = \sigma\boldsymbol{E}' = \sigma\left(\boldsymbol{E} + \frac{1}{c}[\boldsymbol{u} \times \boldsymbol{H}]\right). \tag{4}$$

We shall obtain the equations describing the motion of the plasma. They consist of two groups: Maxwell's equations in a moving medium and the equations of hydrodynamics. The latter must take account of the effect of the electromagnetic force (1) and the evolution of Joule heat $Q = \boldsymbol{j}^2/\sigma$ per unit volume.

It is obvious that the equation of continuity remains unchanged:

$$\frac{\partial \rho}{\partial t} + \operatorname{div} \rho\boldsymbol{u} = 0; \tag{5}$$

the equation of motion contains the force $\boldsymbol{f}$:

$$\frac{\partial \boldsymbol{u}}{\partial t} + (\boldsymbol{u}\nabla)\,\boldsymbol{u} + \frac{1}{\rho}\operatorname{grad} p = \frac{\boldsymbol{f}}{\rho}; \tag{6}$$

and the equation of energy balance written for the entropy S of unit mass of the gas contains the Joule heat:

$$\frac{\partial S}{\partial t} + (\boldsymbol{u}\nabla)\, S = \frac{Q}{\rho T} = \frac{j^2}{\rho T\sigma}. \tag{7}$$

Thus, the complete system of equations describing the motion of a plasma in an electromagnetic field is the system (3)–(7). Here the "electromagnetic" equations (3) and (4) are coupled with the "hydrodynamic" equations (5)–(7) only through the right sides: (3) contains the function $\boldsymbol{j}$ depending on the velocity $\boldsymbol{u}$, while (5)–(7) contain the functions $\boldsymbol{j}$ and $\boldsymbol{f}$ depending on $\boldsymbol{E}$ and $\boldsymbol{H}$.

In the majority of cases of practical interest the system (3)–(7) can be simplified somewhat. The situation is that even in the case of a rather cold plasma it is usually possible in (3) to neglect the dislacement current $(1/4\pi)(\partial D/\partial t)$ as compared with the conduction current $\boldsymbol{j}$. In this case it is obviously necessary to drop the equation div $\boldsymbol{D} = 0$. The quantity $\boldsymbol{E}$ can then be eliminated from the system (3)–(7), and the equations (3) take the form

$$\left.\begin{aligned} &\operatorname{div} \boldsymbol{H} = 0, \\ &\frac{\partial \boldsymbol{H}}{\partial t} + \operatorname{curl}\left[\frac{c^2}{3\pi\sigma} \operatorname{curl} \boldsymbol{H} - (\boldsymbol{u} \times \boldsymbol{H})\right] = 0. \end{aligned}\right\} \tag{8}$$

We consider a further simplification of the system by assuming that the electrical conductivity σ of the plasma is infinitely large, i.e., the gas is an ideal conductor. Here Ohm's law (4) is replaced by the condition that the current $\boldsymbol{j}$ be finite, i.e., by the equation $\boldsymbol{E} = -(1/c)[\boldsymbol{u} \times \boldsymbol{H}]$, and the equations (8) take the form

$$\begin{aligned} &\operatorname{div} \boldsymbol{H} = 0, \\ &\frac{\partial \boldsymbol{H}}{\partial t} + \operatorname{curl}\, [\boldsymbol{u} \times \boldsymbol{H}] = 0. \end{aligned} \tag{9}$$

The system (5)–(7) in this case can be simplified very considerably. Since

$$\boldsymbol{j} = \frac{c}{\pi 4} \operatorname{curl} \boldsymbol{H}, \qquad \boldsymbol{f} = \frac{1}{c} [\boldsymbol{j} \times \boldsymbol{H}] = \frac{1}{4\pi} [\operatorname{curl} \boldsymbol{H} \times \boldsymbol{H}]$$

and

$$\operatorname{curl} \boldsymbol{H} \times \boldsymbol{H} = -\operatorname{grad} \frac{H^2}{2} + \frac{\partial}{\partial x}(H_x \boldsymbol{H}) + \frac{\partial}{\partial y}(H_y \boldsymbol{H}) + \frac{\partial}{\partial z}(H_z \boldsymbol{H}),$$

it follows that (6) can be brought to the form

$$\begin{aligned} &\frac{\partial \boldsymbol{u}}{\partial t} + (\boldsymbol{u}\nabla)\boldsymbol{u} + \frac{1}{\rho} \operatorname{grad}\left(p + \frac{H^2}{8\pi}\right) \\ &\quad = \frac{1}{4\pi\rho}\left[\frac{\partial}{\partial x}(H_x \boldsymbol{H}) + \frac{\partial}{\partial y}(H_y \boldsymbol{H}) + \frac{\partial}{\partial z}(H_z \boldsymbol{H})\right], \end{aligned} \tag{10}$$

while (7) goes over into the adiabatic condition

$$\frac{\partial S}{\partial t} + (\boldsymbol{u}\nabla)S = 0. \tag{11}$$

We shall consider one-dimensional motion, i.e., we assume that all quantities depend only on x and t. From (9) it then follows that $H_x = H_0 = \text{const}$.

We write out our system (5), (9), (10), (11):

$$\left.\begin{aligned}
&\frac{\partial H_y}{\partial t} + \frac{\partial}{\partial x}(u_x H_y - u_y(H_0) = 0,\\
&\frac{\partial H_z}{\partial t} + \frac{\partial}{\partial x}(u_x H_z - u_z H_0) = 0,\\
&\frac{\partial u_x}{\partial t} + u_x \frac{\partial u_x}{\partial x} + \frac{1}{\rho}\frac{\partial}{\partial x}\left(p + \frac{H_y^2 + H_z^2}{8\pi}\right) = 0,\\
&\frac{\partial u_y}{\partial t} + u_x \frac{\partial u_y}{\partial x} - \frac{1}{4\pi\rho}\frac{\partial}{\partial x}(H_0 H_y) = 0,\\
&\frac{\partial u_z}{\partial t} + u_x \frac{\partial u_z}{\partial x} - \frac{1}{4\pi\rho}\frac{\partial}{\partial x}(H_0 H_z) = 0,\\
&\frac{\partial \rho}{\partial t} + \frac{\partial \rho u_x}{\partial x} = 0,\\
&\frac{\partial S}{\partial t} + u_x \frac{\partial S}{\partial x} = 0.
\end{aligned}\right\} \tag{12}$$

This system relates the seven variables ρ, S, u_x, u_y, u_z, H_y and H_z by means of seven equations. Writing down the characteristic equation for (12), we obtain

$$(u_x - \xi)\,[(u_x - \xi)^2 - a^2]\Big\{(u_x - \xi)^2\Big[(u_x - \xi)^2 - \frac{H^2}{4\pi\rho}\Big] - c^2[(u_x - \xi)^2 - a^2]\Big\} = 0, \tag{13}$$

where

$$c^2 = (\partial p/\partial\rho)_S, \qquad a^2 = H_0^2/4\pi\rho, \qquad b^2 = (H_y^2 + H_z^2)/4\pi\rho.$$

The first factor gives the usual entropy characteristics (the flow line) $\xi_0 = u_x$, and the second factor gives the so-called Alvén characteristics $\xi_{\pm 1} = u_x \pm a$. Finally, the last factor in (13) given four more real roots

$$\xi_{\pm 2} = u_x \pm a_-; \qquad \xi_{\pm 3} = u_x \pm a_+,$$

where

$$a_- = \frac{1}{\sqrt{2}}\sqrt{(c^2 + a^2 + b^2) - \sqrt{(c^2 + a^2 + b^2)^2 - 4c^2a^2}},$$

$$a_+ = \frac{1}{\sqrt{2}}\sqrt{(c^2 + a^2 + b^2) + \sqrt{(c^2 + a^2 + b^2)^2 - 4c^2a^2}}.$$

These characteristics are called the *slow* and *fast magnetosonic characteristics* respectively.

In the case where there is no longitudinal component of the magnetic field ($H_0 = 0$) we have $a = a_- = 0$, so that (12) has a fivefold degenerate characteristic $\xi = u_x$. Three independent equations on this characteristic are the fourth, fifth, and seventh equations in (12). To obtain two more equations we compute $(H_y/\rho)\partial/\partial t$ and $(H_z/\rho)\partial/\partial t$. From (12) we obtain

$$\frac{\partial}{\partial t}\left(\frac{H_y}{\rho}\right) = -\frac{1}{\rho}\frac{\partial u_x H_y}{\partial x} - \frac{H_y}{\rho^2}\frac{\partial \rho}{\partial t} = -\frac{1}{\rho}\frac{\partial u_x H_y}{\partial x} + \frac{H_y}{\rho^2}\frac{\partial \rho u_x}{\partial x}$$

$$= \frac{-1}{\rho}\left(\rho u_x \frac{\partial (H_y/\rho)}{\partial x} + \frac{H_y}{\rho}\frac{\partial \rho u_x}{\partial x}\right) + \frac{H_y}{\rho^2}\frac{\partial \rho u_x}{\partial x} = -u_x\frac{\partial (H_y/\rho)}{\partial x}.$$

Therefore, the desired relations are

$$\frac{\partial}{\partial t}\left(\frac{H_y}{\rho}\right) + u_x\frac{\partial}{\partial x}\left(\frac{H_y}{\rho}\right) = 0,$$

$$\frac{\partial}{\partial t}\left(\frac{H_z}{\rho}\right) + u_x\frac{\partial}{\partial x}\left(\frac{H_z}{\rho}\right) = 0.$$

From these equations it follows that along a flow line

$$\frac{\boldsymbol{H}}{\rho} = \text{const.} \tag{14}$$

The equality (14) is said to express the "freezing-in" of the magnetic field.

Using the integral (14), in the case $H_0 = 0$ we reduce the system (12) to three equations:

$$\left.\begin{aligned} \frac{\partial \rho}{\partial t} + \frac{\partial \rho u_x}{\partial x} = 0, \qquad \frac{\partial u_2}{\partial t} + u_x\frac{\partial u_x}{\partial x} + \frac{1}{\rho}\frac{\partial}{\partial x}p_e = 0, \\ \frac{\partial S}{\partial t} + u_x\frac{\partial S}{\partial x} = 0, \end{aligned}\right\}$$

where p_e denotes the so-called effective magnetohydrodynamic pressure: $p_e = p(\rho, S) + A\rho^2$. The quantity A as well as the entropy S may vary from one flow line to another, but along any one line they are constant.

In the case considered of $H_0 = 0$ the fast magnetosonic waves propagate relative to the gas with speed $a_+ = \sqrt{c^2 + b^2} > c$.

BIBLIOGRAPHY

S. Abarbanel and D. Gottlieb

(1973) *Higher order accuracy finite difference algorithms for quasi-linear, conservation law hyperbolic systems*, Math. Comp. **27**, 505–523.

V. B. Adamskiĭ

(1956) *Integration of a system of self-similar equations in the problem of a shock of short duration in a cold gas*, Akust. Ž. **2**, 3–9; English transl. in Soviet Phys. Acoustics **2**.

D. N. de G. Allen and R. V. Southwell

(1955) *Relaxation methods applied to determine the motion, in two dimensions, of a viscous fluid past a fixed cylinder*, Quart. J. Mech. Appl. Math. **8**, 129–145.

L. V. Al'tšuler

(1965) *Use of shock waves in high pressure physics*, Uspehi Fiz. Nauk **85**, 197–258; English transl. in Soviet Phys. Uspekhi **8**.

D. A. Anderson

(1974) *A comparison of numerical solutions to the inviscid equations of fluid motion*, J. Comput. Phys. **15**, 1–20.

P. A. Andrejanov

(1975) *The stability of the solutions of the Cauchy problem for first order quasilinear equations*, Mat. Zametki **17**, 79–89; English transl. in Math. Notes **17**.

V. Ja. Arsenin and N. N. Janenko

(1956) *On the interaction of shock waves with simple waves*, Dokl. Akad. Nauk SSSR **109**, 713–716. (Russian)

K. I. Babenko and I. M. Gel'fand

(1958) *Some remarks on hyperbolic systems*, Naučn. Dokl. Vysš. Školy Fiz.-Mat. Nauk, no. 1, 12–18. (Russian)

N. S. Bahvalov

(1961) *Error estimates for numerical integration of quasilinear first-order equations*, Ž. Vyčisl. Mat. i Mat. Fiz. **1**, 771–783; English transl. in USSR Comput. Math. and Math. Phys. **1**.

(1967) *Parabolic systems with small parameters for the leading derivatives*, Dokl. Akad. Nauk SSSR **174**, 263–266; English transl. in Soviet Math. Dokl. **8**.

(1970) *The existence in the large of a regular solution of a quasilinear hyperbolic system*, Ž. Vyčisl. Mat. i Mat. Fiz. **10**, 969–980; English transl. in USSR Comput. Math. and Math. Phys. **10**.

V. B. Balakin

(1970) *Methods of the Runge-Kutta type for gas dynamics*, Ž. Vyčisl. Mat. i Mat. Fiz. **10**, 1512–1519; English transl. in USSR Comput. Math. and Math. Phys. **10**.

Donald P. Ballou

(1970) *Solutions to nonlinear hyperbolic Cauchy problems without convexity conditions*, Trans. Amer. Math. Soc. **152**, 441–460.

N. K. Bari

(1961) *Trigonometric series*, Fizmatgiz, Moscow; English transl., Vols. 1, 2, Pergamon Press, Oxford, and Macmillan, New York, 1964.

Karl Bechert

(1940) *Zur Theorie ebener Störungen in reibungsfreien Gasen*. I, II, Ann. Physik (5) **37(429)**, 89–123; (5) **38(430)**, 1–25.

(1941) *Über die Ausbreitung von Zylinder- und Kugelwellen in reibungsfreien Gasen und Flüssigkeiten*, Ann. Physik (5) **39(431)**, 169–202.

R. Becker

(1921) *Stosswelle und Detonation*, Z. Phys. **8**, 321–362.

Wilhelm Blaschke

(1924) *Vorlesungen über Differentialgeometrie und geometrische Grundlagen von Einsteins Relativitätstheorie*, Band I, 2nd ed., Springer-Verlag, 1924; reprint, Dover, New York, 1945. [Also 5th rev. ed., W. Blaschke and K. Leichtweiss, *Elementare Differentialgeometrie*, Springer-Verlag, 1973.]

Ju. E. Bojarincev

(1966) *On convergence of difference schemas for equations with variable coefficients*, Trudy Mat. Inst. Steklov. **74**, 16–37; English transl. in Proc. Steklov Inst. Math. **74**.

V. A. Borovikov

(1969) *On the problem of discontinuity decay for a system of two quasilinear equations*, Dokl. Akad. Nauk SSSR **185**, 250–252; English transl. in Soviet Math. Dokl. **10**.

(1972) *On the decay of a discontinuity for a system of two quasilinear equations*, Trudy Moskov. Mat. Obšč. **27**, 53–92; English transl. in Trans. Moscow Math. Soc. **27**.

K. V. Brušlinskiĭ and Ja. M. Každan

(1963) *Self-similar solutions of certain problems of gas dynamics*, Uspehi Mat. Nauk **18**, no. 2(110), 3–24; English transl. in Russian Math. Surveys **18**.

J. M. Burgess

(1940) *Application of a model system to illustrate some points of the statistical theory of free turbulence*, Nederl. Akad. Wetensch. Proc. **43**, 2–12.

(1948) *A mathematical model illustrating the theory of turbulence*, Advances in Appl. Mech., Academic Press, pp. 171–199.

Samuel Z. Burstein and Arthur A. Mirin

(1970) *Third order difference methods for hyperbolic equations*, J. Comput. Phys. **5**, 547–571.

È. B. Byhovskiĭ

(1962a) *Nonadmissible viscosity matrices for the equations of isothermal gas motion*, Dokl. Akad. Nauk SSSR **146**, 751–753; English transl. in Soviet Math. Dokl. **3**.

(1962b) *The small-parameter method ("vanishing viscosity") for a system of equations of gas dynamics*, Ž. Vyčisl. Mat. i Mat. Fiz. **2**, 1128–1131; English transl. in USSR Comput. Math. and Math. Phys. **2**.

(1972) *An initial boundary value problem for the equation* $u_t + a_x(u) = 0$, Dokl. Akad. Nauk SSSR **202**, 511–514; English transl. in Soviet Math. Dokl. **13**.

(1974) *Boundary and initial-boundary value problems "in the large" for a quasilinear conservation law*, Dokl. Akad. Nauk SSSR **215**, 17–20; English transl. in Soviet Math. Dokl. **15**.

Élie Cartan

(1945) *Les systèmes différentiels extérieurs et leurs applications géométriques*, Actualités Sci. Indust., No. 994, Hermann, Paris, 1945.

Sydney Chapman and T. G. Cowling

(1952) *The mathematical theory of non-uniform gases*, 2nd ed., Cambridge Univ. Press.

W. Chester

(1950) *The propagation of sound waves in an open-ended channel*, Philos. Mag. (7) **4**, 11–33.

R. F. Chisnell

(1955) *The normal motion of a shock wave through a non-uniform one-dimensional medium*, Proc. Roy. Soc. London Ser. A **232**, 350–370.

(1957) *The motion of a shock wave in a channel, with applications to cylindrical and spherical shock waves*, J. Fluid Mech. **2**, 286–298.

Julian D. Cole

(1951) *On a quasi-linear parabolic equation occurring in aerodynamics*, Quart. Appl. Math. **9**, 225–236.

Lothar Collatz

(1951) *Numerische Behandlung von Differentialgleichungen*, Springer-Verlag.

Charles C. Conley and Joel A. Smoller

(1970) *Viscosity matrices for two-dimensional nonlinear hyperbolic systesm*, Comm. Pure Appl. Math. **23**, 867–884.

(1971) *Shock waves as limits of progressive wave solutions of higher order equations*, Comm. Pure Appl. Math. **24**, 459–472.

E. D. Conway and J. A. Smoller

(1973) *Shocks violating Lax's condition are unstable*, Proc. Amer. Math. Soc. **39**, 353–356.

E. T. Copson

(1953) *On sound waves of finite amplitude*, Proc. Roy Soc. London Ser. A **216**, 539–547.

R. Courant

(1962) *Methods of mathematical physics*. Vol. II: *Partial differential equations*, Interscience.

R. Courant and K. O. Friedrichs

(1948) *Supersonic flow and shock waves*, Interscience.

R. Courant, K. Friedrichs and H. Lewy

(1928) *Über die partiellen Differentialgleichungen der mathematischen Physik*, Math. Ann. **100**, 32–74; English transl., IBM J. Res. Develop. **11** (1967), 215–234.

Richard Courant, Eugene Isaacson and Mina Rees

(1952) *On the solutions of nonlinear hyperbolic differential equations by finite differences*, Comm. Pure Appl. Math. **5**, 243–255.

Richard Courant and Peter Lax

(1949) *On nonlinear partial differential equations with two independent variables*, Comm. Pure Appl. Math. **2**, 255–273.

Constantine M. Dafermos

(1973a) *Solution of the Riemann problem for a class of hyperbolic systems of conservation laws by the viscosity method*, Arch. Rational Mech. Anal. **52**, 1–9.

(1973b) *The entropy rate admissibility criterion for solutions of hyperbolic conservation laws*, J. Differential Equations **14**, 202–212.

(1974) *Structure of solutions of the Riemann problem for hyperbolic systems of conservation laws*, Arch. Rational Mech. Anal. **53**, 203–217.

C. M. Dafermos and R. J. DiPerna

(1976) *The Riemann problem for certain classes of hyperbolic systems of conservation laws*, J. Differential Equations **20**, 90–114.

Ronald J. DiPerna

(1973) *Global solutions to a class of nonlinear hyperbolic systems of equations*, Comm. Pure Appl. Math. **26**, 1–28.

(1975) *Decay and asymptotic behavior of solutions to nonlinear hyperbolic systems of conservation laws*, Indiana Univ. Math. J. **24**, 1047–1071.

(1976a) *Singularities of solutions of nonlinear hyperbolic systems of conservations laws*, Arch. Rational Mech. Anal. **60**, 75–100.

(1976b) *Global existence of solutions to nonlinear hyperbolic systems of conservation laws*, J. Differential Equations **20**, 187–212.

V. F. D'jačenko

(1961a) *Cauchy's problem for quasilinear systems*, Dokl. Akad. Nauk SSSR **136**, 16–17; English transl. in Soviet Math. Dokl. **2**.

(1961b) *On the numerical calculation of discontinuous solutions of quasilinear systems*, Ž. Vyčisl. Mat. i Mat. Fiz. **1**, 1127–1129; English transl. in USSR Comput. Math. and Math. Phys. **1**.

(1963) *Conditions for the uniqueness of a continuous solution of the problem of the decay of the discontinuity for a set of three equations*, Dokl. Akad. Nauk SSSR **153**, 1245–1248; English transl. in Soviet Math. Dokl. **4**.

Alexander Doktor

(1977) *Global solution of mixed problem for a certain system of nonlinear conservation laws*, Czechoslovak Math. J. **27(102)**, 69–95.

G. A. Dombrovskiĭ

(1964) *A method of approximation to an adiabatic curve in the theory of plane gas flows*, "Nauka", Moscow. (Russian)

Jim Douglas, Jr.

(1956) *On the relation between stability and convergence in the numerical solution of linear parabolic and hyperbolic differential equations*, J. Soc. Indust. Appl. Math. **4**, 20–37.

(1958) *The application of stability analysis in the numerical solution of quasilinear parabolic differential equations*, Trans. Amer. Math. Soc. **89**, 484–518.

Avron Douglis

(1952) *Some existence theorems for hyperbolic systems of partial differential equations in two independent variables*, Comm. Pure Appl. Math. **5**, 119–154.

(1959) *An ordering principle and generalized solutions of certain quasi-linear partial differential equations*, Comm. Pure Appl. Math. **12**, 87–112.

(1961) *The continuous dependence of generalized solutions of non-linear partial differential equations upon initial data*, Comm. Pure Appl. Math. **14**, 267–284.

(1972) *Layering methods for nonlinear partial differential equations of first order*, Ann. Inst. Fourier (Grenoble) **22**, fasc. 3, 141–227.

V. G. Dulov

(1958) *Decay of an arbitrary initial discontinuity of the gas parameters across a jump-discontinuity in the cross-sectional area*, Vestnik Leningrad. Univ. No. 19 (Ser. Mat. Met. Astr. vyp. 4), 76–99. (Russian)

K. V. Emel'janov

(1970) *The difference scheme for a differential equation with a small parameter multiplying the highest derivatives*, Čisl. Metody Meh. Splošnoĭ Sredy **1**, no. 5, 20–30. (Russian)

(1973) *The difference scheme for a three-dimensional elliptic equation with a small parameter multiplying the highest derivatives*, Trudy Inst. Mat. i Meh. Ural. Naučn. Centr Akad. Nauk SSSR Vyp. 11, 30–42. (Russian)

Z. I. Fedotova and Ju. I. Šokin

(1975) *Invariant difference schemes with a polynomial viscosity matrix*, Dokl. Akad. Nauk SSSR **222**, 51–53; English transl. in Soviet Math. Dokl. **16**.

S. P. Finikov

(1948) *Cartan's method of exterior forms in differential geometry*, OGIZ, Moscow. (Russian)

Linus Richard Foy

(1964) *Steady state solutions of hyperbolic systems of conservation laws with viscosity terms*, Comm. Pure Appl. Math. **17**, 177–188.

K. O. Friedrichs

(1948) *Nonlinear hyperbolic differential equations for functions of two independent variables*, Amer. J. Math. **70**, 555–589.

(1954) *Symmetric hyperbolic linear differential equations*, Comm. Pure Appl. Math. **7**, 345–392.

G. Ja. Galin

(1958) *Shock waves in media with arbitrary equations of state*, Dokl. Akad. Nauk SSSR **119**, 1106–1109; English transl. in Soviet Phys. Dokl. **3**.

(1959) *A theory of shock waves*, Dokl. Akad. Nauk SSSR **127**, 55–58; English transl. in Soviet Phys. Dokl. **4**.

F. R. Gantmaher

(1966) *The theory of matrices*, 2nd ed., "Nauka", Moscow; English transl. of 1st ed., Vols. 1, 2, Chelsea, New York, 1959.

John Gary and Richard Helgason

(1970) *A matrix method for ordinary differential eigenvalue problems*, J. Comput. Phys. **5**, 169–187.

C. W. Gear

(1967) *The numerical integration of ordinary equations*, Math. Comp. **21**, 146–156.

(1969) *The automatic integration of stiff ordinary equations*, Information Processing 68 (Proc. IFIP Congr., Edinburgh), Vol. I: Math., Software, North-Holland, pp. 187–193.

I. M. Gel'fand

(1959) *Some problems in the theory of quasilinear equations*, Uspehi Mat. Nauk **14**, no. 2(86), 87–158; English transl. in Amer. Math. Soc. Transl. (2) **29** (1963).

I, M. Gel'fand and G. E. Šilov

(1958) *Generalized functions*. Vols. I, II, III, Fizmatgiz, Moscow; English transl., Academic Press, 1964, 1968, 1967.

P. Germain and R. Bader

(1953) *Unicité des écoulements avec chocs dans la méchanique de Bergers*, Office Nat. d'Études et de Recherches Aeronautiques (ONÉRA), Paris.

David Gilbarg

(1951) *The existence and limit behavior of the one-dimensional shock layer*, Amer. J. Math. **73**, 256–274.

James Glimm

(1965) *Solutions in the large for nonlinear hyperbolic systems of equations*, Comm. Pure Appl. Math. **18**, 697–715.

James Glimm and Peter D. Lax

(1970) *Decay of solutions of systems of nonlinear hyperbolic conservation laws*, Mem. Amer. Math. Soc., No. 101, Amer. Math. Soc., Providence, R. I.

S. K. Godunov

(1956) *On uniqueness of the solution of hydrodynamic equations*, Mat. Sb. **40(82)**, 467–478. (Russian)

(1959) *A difference method for numerical calculation of discontinuous solutions of the equations of hydrodynamics*, Mat. Sb. **47(89)**, 271–306.

(1960) *On the concept of generalized solution*, Dokl. Akad. Nauk SSSR **134**, 1279–1282; English transl. in Soviet Math. Dokl. **1**.

(1961a) *An interesting class of quasilinear systems*, Dokl. Akad. Nauk SSSR **139**, 521–523; English transl. in Soviet Math. Dokl. **2**.

(1961b) *On nonunique "smearing out" of discontinuities in solutions of quasilinear systems*, Dokl. Akad. Nauk SSSR **136**, 272–273; English transl. in Soviet Math. Dokl. **2**.

(1962) *The problem of a generalized solution in the theory of quasilinear equations and in gas dynamics*, Uspehi Mat. Nauk **17**, no. 3(105), 147–158; English transl. in Russian Math. Surveys **17**.

S. K. Godunov and V. S. Rjaben'kiĭ

(1973) *Difference schemes (introduction to the theory)*, "Nauka", Moscow, French transl., "Mir", Moscow, 1977.

S. K. Godunov and K. A. Semendjaev

(1962) *Difference methods for numerical solution of the problems of gas dynamics*, Ž. Vyčisl. Mat. i Mat. Fiz. **2**, 3–14; English transl. in USSR Comput. Math. and Math. Phys. **2**.

V. Ja. Gol'din, N. I. Ionkin and N. N. Kalitkin

(1969) *The entropy scheme of computation in gas dynamics*, Ž. Vyčisl. Mat. i Mat. Fiz. **9**, 1411–1413; English transl. in USSR Comput. Math. and Math. Phys. **9**.

Édouard Goursat

(1929) *Cours d'analyse mathématique*. Vol. II, Part 2, 5th ed., Gauthier-Villars, Paris; English transl. of 2nd (1915) ed., Ginn, Boston, Mass., 1917; reprint, Dover, New York, 1959.

J. M. Greenberg

(1973) *Estimates for fully developed shock solutions to the equation* $\partial u/\partial t - \partial v/\partial x = 0$ *and* $\partial v/\partial t - \partial \sigma(u)/\partial x = 0$, Indiana Univ. Math. J. **22**, 989–1003.

Ku [Gu] Chao-hao, Li Ta-chien [Li Da-qian], Houtsung-yi [Hou Zong-yi] and others

(1961/62) *Discontinuous initial value problems for systems of quasilinear hyperbolic equations*. I, II, III, Acta Math. Sinica **11**, 314–323, 324–327; **12**, 132–143; English transl. in Chinese Math. Acta **2**, **3**.

G. Guderley

(1942) *Starke kugelige und zylindrische Verdichtungsstösse in der Nähe des Kugelmittelpunktes bzw. der Zylinderachse*, Luftfahrtforschung **19**, 302–311.

A. R. Hačaturov

(1977) *Spectral characteristics of a family of difference schemes for Navier-Stokes equations*, Čisl. Metody Meh. Splošnoĭ Sredy **8**, no. 6, 108–119. (Russian)

Wolf Häfele

(1954) Z. Naturforsch. 9a, 269.*

G. H. Hardy and W. W. Rogosinski

(1950) *Fourier series*, 2nd ed., Cambridge Univ. Press.

Francis H. Harlow

(1963) *The particle-in-cell method for numerical solution of problems in fluid dynamics*, Proc. Sympos. Appl. Math., vol. XV, Amer. Math. Soc., Providence, R. I., pp. 269–288.

Philip Hartman

(1964) *Ordinary differential equations*, Wiley.

Philip Hartman and Aurel Wintner

(1952) *On hyperbolic partial differential equations*, Amer. J. Math. **74**, 834–864.

Einar Hille

(1948) *Functional analysis and semi-groups*, 1st ed., Amer. Math. Soc. Colloq. Publ., vol. 31, Amer. Math. Soc., Providence, R. I., Chapter XX.

Sebastian v. Hoerner

(1955) *Lösungen der hydrodynamischen Gleichungen mit linearen Verlauf der Geschwindigkeit*, Z. Naturforsch. **10a**, 687–692.

---

* *Editor's note.* This item is reproduced exactly as it appears in the original. On the basis of V. Horner's paper cited below it seems likely that the authors intended one of the following three articles:

Wolf Häfele

(1955a) *Zur analytischen Behandlung ebener, starker, instationärer stosswellen*, Z. Naturforsch. **10a**, 1006–1016.

(1955b) *Über die Stabilität des Stosswellentypus aus der Klasse der Homologie-Lösungen*, Z. Naturforsch. **10a**, 1017–1027.

C. F. v. Weizsäcker

(1954) *Genäherte Darstellung starker instationärer Stosswellen durch Homologie-Lösunger*, Z. Naturforsch. **9a**, 269–275.

Eberhard Hopf

(1950) *The partial differential equation* $u_t + uu_x = \mu u_{xx}$, Comm. Pure Appl. Math. **3**, 201–230.

(1969) *On the right weak solution of the Cauchy problem for a quasilinear equation of first order*, J. Math. Mech. **19**, 483–487.

Lars Hörmander

(1963) *Linear partial differential operators*, Springer-Verlag, Berlin, and Academic Press, New York.

S. A. Hristianovič

(1937) *Le problème de Cauchy pour les équations non linéaires hyperboliques*, Mat. Sb. **2(44)**, 871–899. (Russian; French summary)

A. M. Il'in

(1965) *Stability of difference schemes for the Cauchy problem for systems of partial differential equations*, Dokl. Akad. Nauk SSSR **164**, 491–494; English transl. in Soviet Math. Dokl. **6**.

(1969) *A difference scheme for a differential equation with a small parameter multiplying the highest derivative*, Mat. Zametki **6**, 237–248; English transl. in Math. Notes **6**.

A. M. Il'in and O. A. Oleĭnik

(1958) *Behavior of the solutions of the Cauchy problem for certain quasilinear equations for unbounded increase of the time*, Dokl. Akad. Nauk SSSR **120**, 25–28; English transl. in Amer. Math. Soc. Transl. (2) **42** (1964).

(1960) *Asymptotic behavior of solutions of the Cauchy problem for some quasilinear equations for large values of the time*, Mat. Sb. **51(93)**, 191–216. (Russian)

Z. A. Iskander-Zade

(1966) *On the question of the stability of the trivial solutions of a parabolic system of partial differential equations*, Ž. Vyčisl. Mat. i Mat. Fiz. **6**, 921–927; English transl. in USSR Comput. Math. and Math. Phys. **6**.

Nobutoshi Itaya

(1970) *The existence and uniqueness of the solution of the equations describing compressible viscous fluid flow*, Proc. Japan Acad. **46**, 379–382.

(1974) *On the temporally global problem of the generalized Burgess' equation*, J. Math. Kyoto Univ. **14**, 129–177.

N. N. Janenko

(1955a) *Reduction of a system of quasilinear equations to a quasilinear equation*, Uspehi Mat. Nauk **10**, no. 3 (65), 173–178. (Russian)

(1955b) *On discontinuities of solutions of quasilinear equations*, Uspehi Mat. Nauk **10**, no. 2(64), 195–202. (Russian)

(1964) *Compatibility theory and integration methods for systems of nonlinear partial differential equations*, Proc. Fourth All-Union Math. Congr. (Leningrad, 1961), "Nauka", Leningrad, pp. 247–252. (Russian)

N. N. Janenko and Ju. E. Bojarincev

(1961) *The convergence of difference schemes for the heat conduction equation with variable coefficients*, Dokl. Akad. Nauk SSSR **139**, 1322–1324; English transl. in Soviet Math. Dokl. **2**.

N. N. Janenko and G. V. Demidov

(1971) *On the structure of absolutely approximating and absolutely correct difference schemes*, Problems in Appl. Math. and Mech. (A. A. Dorodnicyn Sixtieth Birthday Volume), "Nauka", Moscow, pp. 137–144. (Russian)

N. N. Janenko, G. V. Demidov and S. A. Kantor

(1972) *Evolutionary double-layer difference schemes*, Čisl. Metody Meh. Splošnoĭ Sredy **3**, no. 5, 95–114. (Russian)

N. N. Janenko and I. K. Jaušev
(1966) *On an absolutely stable schema for integration of the equations of hydrodynamics*, Trudy Mat. Inst. Steklov. **74**, 141–146; English transl. in Proc. Steklov Inst. Math. **74**.
N. N. Janenko and V. E. Neuvažaev
(1966) *A method of computing gas-dynamic motions with nonlinear heat conduction*, Trudy Mat. Inst. Steklov. **74**, 138–140; English transl. in Proc. Steklov Inst. Math. **74**.
N. N. Janenko and Ju. I. Šokin
(1968) *Correctness of first differential approximations of difference schemes*, Dokl. Akad. Nauk SSSR **182**, 776–778; English transl. in Soviet Math. Dokl. **9**.
(1973) *On a group classification of difference schemes for the system of equations of gas dynamics*, Trudy Mat. Inst. Steklov. **122**, 85–97; English transl. in Proc. Steklov Inst. Math. **122**.
N. N. Janenko, E. V. Vorožcov and V. M. Fomin*
(1976a) *Differential analyzers of shock waves*, Dokl. Akad. Nauk SSSR **227**, 50–53; English transl. in Soviet Math. Dokl. **17**.
(1976b) *Differential analyzers of shock waves. Applications ofthe theory*, Čisl. Metody Meh. Splošnoĭ Sredy **7**, no. 6, 8–23. (Russian)
(1976c) *Differential analyzers of shock waves*: *Theory*, Computers & Fluids **4**, 171–183.
I. K. Jaušev
(1967) *Decay of an arbitrary discontinuity in a channel with a jump in the cross-sectional area*, Izv. Sibirsk. Otdel. Akad. Nauk SSSR **1967**, no. 8 (Ser. Tehn. Nauk, Vyp. 2), 109–120. (Russian)
Alan Jeffrey
(1973) *Quasi-linear hyperbolic systems and continuum mechanics*, Math. Balkanica **3**, 166–183.
(1975) *Smooth fronted waves in the shallow water approximation*, Proc. Roy Soc. Edinburgh Sect. A **73**, 107–116.
(1976) *Quasilinear hyperbolic systems and waves*, Res. Notes in Math., No. 5, Pitman.
Fritz John
(1974) *Formation of singularities in one-dimensional nonlinear wave propagation*, Comm. Pure Appl. Math. **27**, 377–405.
A. S. Kalašnikov
(1959a) *Construction of generalized solutions of quasilinear equations of first order without convexity conditions as limits of solutions of parabolic equations with a small parameter*, Dokl. Akad. Nauk SSSR **127**, 27–30. (Russian)
(1959b) *Uniqueness of the solution of the Cauchy problem for a class of quasilinear hyperbolic systems*, Uspehi Mat. Nauk **14**, no. 2(86), 195–202. (Russian)
L. V. Kantorovič and G. P. Akilov
(1959) *Functional analysis in normed spaces*, Fizmatgiz, Moscow; English transl., Macmillan, 1964.
A. V. Kažihov
(1975) *Correctness "in the large" of mixed boundary value problems for a model system of equations of a viscous gas*, Dinamika Splošnoĭ Sredy Vyp. 21, 18–47. (Russian)
(1976) *The global solvability of one-dimensional boundary value problems for the equations of a viscous heat-carrying gas*, Dinamika Splošnoĭ Sredy Vyp. 24, 44–61. (Russian)
J. B. Keller
(1956) *Spherical, cylindrical and one-dimensional gas flows*, Quart. Appl. Math. **14**, 171–184.
N. E. Kotchine [Kočin]
(1926) *Sur la théorie des ondes de choc dans une fluide*, Rend. Circ. Mat. Palermo **50**, 305–344.

**Editor's note*. For b) the actual order of the authors' names is Vorožcov, Fomin and Janenko, and for c) it is Fomin, Vorozhtsov and Yanenko.

I. V. Konoval'cev

(1968) *Stability in C and L_p of double layer difference schemes for parabolic equations with variable coefficients*, Ž. Vyčisl. Mat. i Mat. Fiz. **8**, 894–899; English transl. in USSR Comput. Math. and Math. Phys. **8**.

V. P. Korobeĭnikov, N. S. Mel'nikova and E. V. Rjazonov

(1961) *Theory of point explosions*, Fizmatgiz, Moscow. (Russian)

Heinz-Otto Kreiss

(1958) *Über Sachsgemässe Cauchy probleme für Systeme von linearen partiellen Differentialgleichungen*, Kungl. Tekn. Högskol. Handlingar (Stockholm), No. 127.

(1962) *Über die Stabilitätsdefinition für Differenzengleichungen die partielle Differentialgleichungen approximieren*, Nordisk Tidskr. Informationsbehandlung (BIT) **2**, 153–181.

(1964) *On difference approximations of the dissipative type for hyperbolic differential equations*, Comm. Pure Appl. Math. **17**, 335–353.

S. N. Kružkov

(1960) *The Cauchy problem in the large for certain nonlinear first-order differential equations*, Dokl. Akad. Nauk SSSR **132**, 36–39; English transl. in Soviet Math. Dokl. **1**.

(1964) *Generalized solutions of nonlinear equations of the first order and certain problems for quasilinear parabolic equations*, Vestnik Moskov. Univ. Ser. I Mat. Meh. No. 6, 65–74. (Russian)

(1965) *Methods for constructing generalized solutions for the Cauchy problem for a quasilinear equation of the first order*, Uspehi Mat. Nauk **20**, no. 6(126), 112–118. (Russian)

(1970) *First order quasilinear equations in several independent variables*, Mat. Sb. **81(123)**, 228–255; English transl. in Math. USSR Sb. **10**.

A. G. Kulikovskiĭ

(1962) *The structure of shock waves*, Prikl. Mat. Meh. **26**, 631–641; English transl. in J. Appl. Math. Mech. **26**.

V. F. Kuropatenko

(1962) *A method of constructing difference schemes for the numerical integration of the equations of gas dynamics*, Izv. Vysš. Učebn. Zaved. Matematika no. 3(28), 75–83. (Russian)

(1966) *On difference methods for the equations of hydrodynamics*, Trudy Mat. Inst. Steklov. **74**, 107–137; English transl. in Proc. Steklov Inst. Math. **74**.

Paul Kutler and Harvard Lomax

(1971) *The computation of supersonic flow fields about wing-body combinations by "shock-capturing" finite difference techniques*, Proc. Second Internat. Conf. Numer. Methods Fluid Dynamics (Berkeley, Calif., 1970), Lecture Notes in Physics, Vol. 8, Springer-Verlag, 1971, pp. 24–29.

Paul Kutler, Leonidas Sakell and Gene Aiello

(1975) *Two-dimensional shock-on-shock interaction problem*, AIAA J. **13**, 361–367.

N. N. Kuznecov

(1959) *Some asymptotic properties of the generalized solution of the Cauchy problem for a quasilinear equation of first order*, Uspehi Mat. Nauk **14**, no. 2(86), 203–209. (Russian)

(1960) *Problem of the decay of an arbitrary discontinuity for a system of quasilinear equations of the first order*, Dokl. Akad. Nauk SSSR **131**, 503–506; English transl. in Soviet Math. Dokl. **1**.

(1967a) *The weak solutions of the Cauchy problem for a multidimensional quasilinear equation*, Mat. Zametki **2**, 401–410; English transl. in Math. Notes **2**.

(1967b) *Some mathematical problems of chromatography*, Vyčisl. Metody i Programmovanie Vyp. 6, 242–258. (Russian)

(1972) *The asymptotic behavior of the solutions of the finite difference Cauchy problem*, Ž. Vyčisl. Mat. i Mat. Fiz. **12**, 334–351; English transl. in USSR Comput. Math. and Math. Phys. **12**.

(1973) *Application of the smoothing method to some systems of hyperbolic quasilinear equations*, Ž. Vyčisl. Mat. i Mat. Fiz. **13**, 92–102; English transl. in USSR Comput. Math. and Math. Phys. **13**.

(1975) *On stable methods of solving a quasilinear equation of first order in a class of discontinuous functions*, Dokl. Akad. Nauk SSSR **225**, 1009–1012; English transl. in Soviet Math. Dokl. **16**.

(1977) *On stable methods for solving non-linear first order partial differential equations in the class of discontinuous functions*, Topics in Numer. Anal. III (Proc. Roy. Irish Acad. Conf., Dublin, 1976), Academic Press, pp. 183–197.

N. N. Kuznecov and Ci Čžun-tao [Chi Chung-tao]

(1964a) *A uniqueness theorem in the theory of quasilinear hyperbolic equations*, Vestnik Moskov. Univ. Ser. I Mat. Meh. No. 3, 25–30. (Russian)

(1964b) *On the uniqueness of the generalized solution of the Cauchy problem for a hyperbolic system of two quasilinear equations*, Vestnik Moskov. Univ. Ser. I Mat. Meh. No. 4, 3–6. (Russian)

N. N. Kuznecov and B. L. Roždestvenskiĭ

(1959a) *Construction of the generalized solution of the Cauchy problem for a quasilinear equation*, Uspehi Mat. Nauk **14**, no. 2(86), 211–215. (Russian)

(1959b) *Existence and uniqueness of the generalized solution of the Cauchy problem for a nonhomogeneous law of conservation*, Dokl. Akad. Nauk SSSR **126**, 486–489. (Russian)

(1965) *On the construction of a generalized solution of the Cauchy problem for a quasilinear equation*, Uspehi Mat. Nauk **20**, no. 1(121), 209–212. (Russian)

N. N. Kuznecov and V. A. Tupčiev

(1975) *On an extension of a theorem due to Glimm*, Dokl. Akad. Nauk SSSR **221**, 287–290; English transl. in Soviet Math. Dokl. **16**.

N. N. Kuznecov and S. A. Vološin

(1976) *On monotone difference approximations for a first-order quasilinear equation*, Dokl. Akad. Nauk SSSR **229**, 1317–1320; English transl. in Soviet Math. Dokl. **17**.

O. A. Ladyženskaja

(1952) *Solution of Cauchy's problem for hyperbolic systems by the method of finite differences*, Leningrad. Gos. Univ. Učen. Zap. **144** (Ser. Mat. No. 23), 192–246. (Russian)

(1956) *On the construction of discontinuous solutions of quasilinear hyperbolic equations as limits of solutions of the corresponding parabolic equations when the "coefficient of viscosity tends toward zero*, Dokl. Akad. Nauk SSSR **111**, 291–294. (Russian)

L. D. Landau and E. M. Lifšic

(1953) *Mechanics of continuous media*, 2nd ed., GITTL, Moscow; English transl., Parts 1, 2, *Course of theoretical physics*, Vols. 6, 7, Pergamon Press, Oxford, and Addison-Wesley, Reading, Mass., 1959.

L. D. Landau, N. N. Meĭman and I. M. Halatnikov

(1958) *Numerical methods for the integration of partial differential equations by the method of finite differences*, Proc. Third All-Union Math. Congr. (Moscow, 1956), Vol. III, Izdat. Akad. Nauk SSSR, Moscow, pp. 92–100. (Russian)

Richard Latter

(1955) *Similarity solution for a spherical shock wave*, J. Appl. Phys. **26**, 954–960.

Peter D. Lax

(1953) *Nonlinear hyperbolic equations*, Comm. Pure Appl. Math. **6**, 231–258.

(1954a) *The initial value problem for nonlinear hyperbolic equations in two independent variables*, Contributions to the Theory of Partial Differential Equations, Ann. of Math. Studies, No. 33, Princeton Univ. Press, Princeton, N. J., pp. 211–229.

(1954b) *Weak solutions of nonlinear hyperbolic equations and their numerical computation*, Comm. Pure Appl. Math. **7**, 159–193.
(1957) *Hyperbolic systems of conservation laws*. II, Comm. Pure Appl. Math. **10**, 537–566.
(1960) *The scope of the energy method*, Bull. Amer. Math. Soc. **66**, 32–35.
(1964) *Development of singularities of solutions of nonlinear hyperbolic partial differential equations*, J. Mathematical Phys. **5**, 611–613.
(1971) *Shock waves and entropy*, Contributions to Nonlinear Functional Analysis (Proc. Sympos., Madison, Wisc.), Academic Press, pp. 603–634.

P. D. Lax and R. D. Richtmyer
(1956) *Survey of the stability of linear finite difference equations*, Comm. Pure Appl. Math. **9**, 267–293.

Peter Lax and Burton Wendroff
(1960) *Systems of conservation laws*, Comm. Pure Appl. Math. **13**, 217–237.

Lewis Leibovich
(1974) *Solutions of the Riemann problem for hyperbolic systems of quasilinear equations without convexity conditions*, J. Math. Anal. Appl. **45**, 81–90.

M. A. Leontovič
(1950) *Introduction to thermodynamics*, GITTL, Moscow; German transl., VEB Deutscher Verlag Wiss., Berlin, 1953.

Hans Lewy
(1927) *Über das Anfangswertproblem einer hyperbolischen nichtlinearen partiellen Differentialgleichung zweiter Ordnung mit zwei unabhängigen Veranderlichen*, Math. Ann. **98**, 179–191.

C. C. Lin
(1955) *The theory of hydrodynamic stability*, Cambridge Univ. Press.

Tai-Ping Liu
(1974) *The Riemann problem for general* 2×2 *conservation laws*, Trans. Amer. Math. Soc. **199**, 89–112.
(1975a) *The Riemann problem for general systems of conservation laws*, J. Differential Equations **18**, 218–234.
(1975b) *Existence and uniqueness theorems for Riemann problems*, Trans. Amer. Math. Soc. **212**, 375–382.
(1976) *Uniqueness of weak solutions of the Cauchy problem for general* 2×2 *conservation laws*, J. Differential Equations **20**, 369–388.

V. Ju. Ljapidevskiĭ
(1973) *The global well-posedness of the Cauchy problem for a certain class of nonlinear hyperbolic systems of equations*, Dinamika Splošnoĭ Sredy Vyp. 15, 74–88. (Russian)
(1974a) *The continuous dependence on the initial conditions of generalized solutions of the system of equations of gas dynamics*, Ž. Vyčisl. Mat. i Mat. Fiz. **14**, 982–991; English transl. in USSR Comput. Math. and Math. Phys. **14**.
(1974b) *On the uniqueness of the generalized solution of a system of equations of gas dynamics*, Dokl. Akad. Nauk SSSR **215**, 535–538; English transl. in Soviet Math. Dokl. **15**.
(1975) *On correctness classes for nonlinear hyperbolic systems*, Dokl. Akad. Nauk SSSR **225**, 507–510; English transl. in Soviet Math. Dokl. **16**.

L. A. Ljusternik and V. I. Sobolev
(1951) *Elements of functional analysis*, GITTL, Moscow; English transl., Ungar, New York, 1961.

G. S. S. Ludford
(1955) *Generalised Riemann invariants*, Pacific J. Math. **5**, 441–450.

R. W. MacCormack
(1969) *The effect of viscosity on hypervelocity impact cratering*, AIAA Paper No. 69–354.

G. I. Maičuk

(1961) *Methods of calculation for nuclear reactors*, Gosatomiz dat, Moscow. (Russian)

(1967) *Numerical methods in weather forecasting*, Gidrometeoizdat, Leningrad; English transl., Academic Press, 1974.

(1973) *Methods of numerical mathematics*, "Nauka", Sibirsk. Otdel., Novosibirsk; English transl., Springer-Verlag, 1975.

M. H. Martin

(1953a) *The propagation of a plane shock into a quiet atmosphere*, Canad. J. Math. **5**, 37–39.

(1953b) *The Monge-Ampère partial differential equation* $rt - s^2 + \lambda^2 = 0$, Pacific J. Math. **3**, 165–187.

N. N. Meĭman

(1954) *On the theory of partial differential equations*, Dokl. Akad. Nauk SSSR **97**, 593–596. (Russian)

John Nash

(1962) *Le problème de Cauchy pour les équations différentielles d'un fluide général*, Bull. Soc. Math. France **90**, 487–497.

J. von Neumann and R. D. Richtmyer

(1950) *A method for the numerical calculation of hydrodynamic shocks*, J. Appl. Phys. **21**, 232–237.

V. E. Neuvažaev

(1962) *The propagation of a spherical blast wave in a heat-conducting gas*, Prikl. Mat. Meh. **26**, 1094–1099; English transl. in J. Appl. Math. Mech. **26**.

Takaaki Nishida

(1968) *Global solution for an initial boundary value problem of a quasilinear hyperbolic system*, Proc. Japan Acad. **44**, 642–646.

Takaaki Nishida and Joel A. Smoller

(1973) *Solutions in the large for some nonlinear hyperbolic conservation laws*, Comm. Pure Appl. Math. **26**, 183–200.

Johannes Nitsche

(1953) *Über Unstetigkeit in den Ableitungen von Lösungen quasilinearer hyperbolischer Differentialgleichungenssysteme*, J. Rational Mech. Anal. **2**, 291–297.

Arnold Nordsieck

(1962) *On numerical integration of ordinary differential equations*, Math. Comp. **16**, 22–49.

D. E. Ohocimskiĭ et al.

(1957) *Computation of point explosion taking into account counterpressure*, Trudy Mat. Inst. Steklov. **50**. (Russian)

O. A. Oleĭnik

(1954a) *The Cauchy problem for nonlinear equations in a class of discontinuous functions*, Dokl. Akad. Nauk SSSR **95**, 451–454; English transl. in Amer. Math. Soc. Transl. (2) **42** (1964).

(1954b) *The Cauchy problem for nonlinear equations in a class of discontinuous functions*, Uspehi Mat. Nauk **9**, no. 3(61), 231–233. (Russian)

(1955) *Boundary value problems for partial differential equations with a small parameter multiplying the highest derivatives, and the Cauchy problem for nonlinear equations in the large*, Uspehi Mat. Nauk **10**, no. 3(65), 229–234. (Russian)

(1956a) *The Cauchy problem for nonlinear differential equations of the first order with discontinuous initial conditions*, Trudy Moskov. Mat. Obšč. **5**, 433–454. (Russian)

(1956b) *On discontinuous solutions of nonlinear differential equations*, Dokl. Akad. Nauk SSSR **109**, 1098–1101. (Russian)

(1957a) *On the uniqueness of the generalized solution of the Cauchy problem for a nonlinear system of equations occurring in mechanics*, Uspehi Mat. Nauk **12**, no. 6(78), 169–176. (Russian)

(1957b) *Discontinuous solutions of nonlinear differential equations*, Uspehi Mat. Nauk **12**, no. 3(75), 3–73; English transl. in Amer. Math. Soc. Transl. (2) **26** (1963).

(1958) *On a class of discontinuous solutions of quasilinear equations of the first order*, Naučn. Dokl. Vysš. Školy Fiz.-Mat. Nauki No. 3, 91–98. (Russian)

(1959a) *Construction of a generalized solution of the Cauchy problem for a quasilinear equation of first order by the introduction of "vanishing velocity"*, Uspehi Mat. Nauk **14**, no. 2(86), 159–164; English transl. in Amer. Math. Soc. Transl. (2) **33** (1963).

(1959b) *Uniqueness and stability of the generalized solution of the Cauchy problem for a quasilinear equation*, Uspehi Mat. Nauk **14**, no. 2(86), 165–170; English transl. in Amer. Math. Soc. Transl. (2) **33** (1963).

O. A. Oleĭnik and N. D. Vvedenskaja

(1957) *The solution of the Cauchy problem and a boundary value problem for nonlinear equations in a class of discontinuous functions*, Dokl. Akad. Nauk SSSR **113**, 503–506. (Russian)

L. V. Ovsjannikov

(1960) *Group properties of Čaplygin's equation*, Ž. Prikl. Meh. i Tehn. Fiz. (PMTF) no. 3, 126–145. (Russian)

(1962) *Group properties of differential equations*, Izdat. Sibirsk. Otdel. Akad. Nauk SSSR, Novosibirsk. (Russian)

(1973) *On the foundations of the theory of shallow water*, Dinamika Splošnoĭ Sredy Vyp. 15, 104–125. (Russian)

D. Ju. Panov

(1957) *Numerical solution of quasilinear hyperbolic systems of partial differential equations*, GITTL, Moscow. (Russian)

G. N. Patterson

(1956) *Molecular flow of gases*, Wiley, New York, and Chapman & Hall, London.

B. V. Pavlov and A. Ja. Povzner

(1973) *A method for the numerical integration of systems of ordinary differential equations*, Ž. Vyčisl. Mat. i Mat. Fiz. **13**, 1056–1059; English transl. in USSR Comput. Math. and Math. Phys. **13**.

I. Petrowsky [I. G. Petrovskiĭ]

(1937) *Über das Cauchysche Problem für Systeme von partiellen Differentialgleichungen*, Mat. Sb. **2(44)**, 815–870.

(1961) *Lectures on partial differential equations*, 3rd ed., Fizmatgiz, Moscow; English transl., Saunders, Philadelphia, Pa., 1967.

A. R. Pinskiĭ and A. I. Ruzanov

(1976) *Application of Picard's method to the solution of nonstationary elasto-plastic problems*, Prikl. Probl. Pročnosti i Plastičnosti Vyp. 5, 53–58. (Russian)

G. N. Položiĭ

(1962) *Numerical solution of two- and three-dimensional boundary-value problems of mathematical physics and functions of a discrete argument*, Izdat. Kiev. Univ., Kiev, 1962; English transl., *The method of summary representation for numerical solution of problems of mathematical physics*, Pergamon Press, 1965.

Ju. P. Popov and A. A. Samarskiĭ

(1969) *Totally conservative difference schemes*, Ž. Vyčisl. Mat. i Mat. Fiz. **9**, 953–958; English transl. in USSR Comput. Math. and Math. Phys. **9**.

(1970) *Totally conservative difference schemes for the equations of gas dynamics in Euler's variables*, Ž. Vyčisl. Mat. i Mat. Fiz. **10**, 773–779; English transl. in USSR Comput. Math. and Math. Phys. **10**.

P. K. Raševskiĭ

(1947) *Geometrical theory of partial differential equations*, OGIZ, Moscow. (Russian)

V. E. Raspopov, V. P. Šapeev and N. N. Janenko

(1974) *An application of the method of differential relations to the one-dimensional equations of gas dynamics*, Izv. Vysš. Učebn. Zaved. Math. no. 11(150), 69–74; English transl. in Soviet Math. (Iz. VUZ) **18**.

(1977) *The method of differential relations for the one-dimensional equations of gas dynamics*, Čisl. Metody Meh. Splošnoĭ Sredy **8**, no. 2, 100–105. (Russian)

R. V. Razumeĭko

(1973) *Error estimation for the numerical integration of a first order quasilinear equation*, Mat. Zametki **13**, 207–215; English transl. in Math Notes **13**.

Robert D. Richtmyer and K. W. Morton

(1967) *Difference methods for initial-value problems*, 2nd ed., Interscience.

Bernhard Riemann

(1860) *Über die Fortpflanzung ebener Luftwellen von endlicher Schwingungsweite*, Abh. Königl. Gess. Wiss. Göttingen **8**; reprinted in his *Gesammelte mathematische Werke und wissenschaftlicher Nachlass*, Teubner, Leipzig, 1876 (reprint, Dover, New York, 1953), pp. 145–164.

(1876) *Versuch einer allgemeinen Auffassung der Integration und Differentiation* (manuscript, 1847), first published in his *Gesammelte mathematische Werke und wissenschaftlicher Nachlass*, Teubner, Leipzig, 1876 (reprint, Dover, New York, 1953), pp 331–344.

Charles Riquier

(1910) *Les systèmes d'équations aux dérivees partielles*, Gauthier-Villars, Paris.

V. S. Rjaben'kiĭ

(1952) *On the application of the method of finite differences to the solution of the Cauchy problem*, Dokl. Akad. Nauk SSSR **86**, 1071–1074. (Russian)

V. S. Rjaben'kiĭ and A. F. Filippov

(1956) *On the stability of difference equations*, GITTL, Moscow; German transl., VEB Deutscher Verlag Wiss., Berlin, 1960.

B. L. Roždestvenskiĭ

(1957) *Systems of quasilinear equations*, Dokl. Akad. Nauk SSSR **115**, 454–457; English transl. in Amer. Math. Soc. Transl. (2)**42** (1964).

(1958a) *The Cauchy problem for quasilinear equations*, Dokl. Akad. Nauk SSSR **122**, 551–554; English transl. in Amer. Math. Soc. Transl. (2)**42** (1964).

(1958b) *Uniqueness of the generalized solution of the Cauchy problem for hyperbolic systems of quasilinear equations*, Dokl. Akad. Nauk SSSR **122**, 762–765; English transl. in Amer. Math. Soc. Transl. (2)**42** (1964).

(1959a) *Conservativeness of systems of quasilinear equations*, Uspehi Mat. Nauk **14**, no. 2(86), 217–218; English transl. in Amer. Math. Soc. Transl. (2)**42** (1964).

(1959b) *On the discontinuity of solutions of quasilinear equations*, Mat. Sb. **47(89)**, 485–494; English transl. in Amer. Math. Soc. Transl. (2) **101** (1973).

(1960) *Discontinuous solutions of hyperbolic systems of quasilinear equations*, Uspehi Mat. Nauk **15**, no. 6(96), 59–117; English transl. in Russian Math. Surveys **15**.

(1961) *A new method of solving the Cauchy problem in the large for quasilinear equations*, Dokl. Akad. Nauk SSSR **138**, 309–312; English transl. in Soviet Math. Dokl. **2**.

(1962a) *A system of quasilinear equations in the theory of surfaces*, Dokl. Akad. Nauk SSSR **143**, 50–52; English transl. in Soviet Math. Dokl. **3**.

(1962b) *Construction of discontinuous solutions of systems of two quasilinear equations*, Dokl. Akad. Nauk SSSR **144**, 58–61; English transl. in Soviet Math. Dokl. **3**.

(1962c) *Construction of discontinuous solutions to systems of quasilinear equations*. I, Ž. Vyčisl. Mat. i Mat. Fiz. **2**, 1019–1043; English transl. in USSR Comput. Math. and Math. Phys. **2**.

(1963) *Construction of discontinuous solutions to systems of quasilinear equations*. II, Ž. Vyčisl. Mat. i Mat. Fiz. **3**, 79–98; English transl. in USSR Comput. Math. and Math. Phys. **3**.

(1974) *The Picard method as a method for the numerical solution of problems of mathematical physics*, Čisl. Metody Meh. Splošnoĭ Sredy **5**, no. 2, 96–107. (Russian)

B. L. Roždestvenskiĭ, E. I. Ermakova and V. G. Priĭmak

(1977) *A study of the stability of difference schemes of higher order accuracy*, Preprint No. 14 (1977), Keldyš Inst. Appl. Math. Acad. Sci. USSR, Moscow. (Russian) MR **58** #19207.

B. L. Roždestvenskiĭ and N. N. Janenko

(1968) *Systems of quasilinear equations and their applications to the dynamics of gases*, 1st ed., "Nauka", Moscow. (Russian)

B. L. Rozdestvenskiĭ and A. D. Sidorenko

(1967) *On the impossibility of "gradient catastrophe" for weakly nonlinear systems*, Ž. Vyčisl. Mat. i Mat. Fiz. **7**, 1176–1179; English transl. in USSR Comput. Math. and Math. Phys. **7**.

V. V. Rusanov

(1968) *Difference schemes of the third order of accuracy for a continuous computation of discontinuous solutions*, Dokl. Akad. Nauk SSSR **180**, 1303–1305; English transl. in Soviet Math. Dokl. **9**.

O. S. Ryžov and S. A. Hristianovič

(1958) *On nonlinear reflection of weak shock waves*, Prikl. Mat. Meh. **22**, 586–599; English transl. in J. Appl. Math. Mech. **22**.

A. A. Samarskiĭ

(1971) *Introduction to the theory of difference schemes*, "Nauka", Moscow. (Russian)

(1977) *Theory of difference schemes*, "Nauka", Moscow. (Russian)

A. A. Samarskiĭ and V. Ja. Arsenin

(1961) *On the numerical solution of the equations of gas dynamics with various types of viscosity*, Ž. Vyčisl. Mat. i Mat. Fiz. **1**, 357–360; English transl. in USSR Comput. Math. and Math. Phys. **1**.

A. A. Samarskiĭ and A. V. Gulin

(1973) *Stability of difference schemes*, "Nauka", Moscow. (Russian)

A. A. Samarskiĭ and Ju. P. Popov

(1975) *Difference schemes for gas dynamics*, "Nauka", Moscow. (Russian)

L. Schwartz

(1950) *Théorie des distributions*. Vols. I, II, Actualités Sci. Indust., Nos. 1091, 1122, Hermann, Paris.

L. I. Sedov

(1946) *Le mouvement d'air en cas d'une forte explosion*, C. R. (Dokl.) Acad. Sci. URSS **52**, 17–20.

(1957) *Similarity and dimensional methods in mechanics*, 4th ed., GITTL, Moscow; English transl., Academic Press, 1959.

A. D. Sidorenko

(1968) *Wave adiabats for media with an arbitrary equation of state*, Dokl. Akad. Nauk SSSR **178**, 818–821; English transl. in Soviet Phys. Dokl. **13**.

(1973) *A variant of the problem with a contact discontinuity for systems of three quasilinear equations*, Differencial'nye Uravnenija **9**, 774–777; English transl. in Differential Equations **9**.

A. D. Sidorenko and B. L. Roždestvenskiĭ

(1968) *A problem with contact discontinuity*, Ž. Vyčisl. Mat. i Mat. Fiz. **8**, 1352–1359; English transl. in USSR Comput. Math. and Math. Phys. **8**.

V. I. Smirnov

(1958) *A course in higher mathematics*, Vol. IV, 5th ed., Fizmatgiz, Moscow; English transl. of 3rd ed., Pergamon Press, Oxford, and Addison-Wesley, Reading, Mass., 1964.

(1959) *A course in higher mathematics*, Vol. V, 2nd rev. ed., Fizmatgiz, Moscow; English transl., Pergamon Press, Oxford, and Addison-Wesley, Reading, Mass., 1964.

V. Smirnov and S. Sobolev

(1932) *Sur une méthode nouvelle dans le problème plan des vibrations élastiques*, Trudy Seĭsmolg. Inst. Akad. Nauk SSSR No. 20.

J. A. Smoller

(1969a) *On the solution of the Riemann problem with general step data for an extended class of hyperbolic systems*, Michigan Math. J. **16**, 201–210.

(1969b) *A uniqueness theorem for Riemann problems*, Arch. Rational Mech. Anal. **33**, 110–115.

(1970) *Contact discontinuities in quasi-linear hyperbolic systems*, Comm. Pure Appl. Math. **23**, 791–801.

Joel A. Smoller and Charles C. Conley

(1972) *Viscosity matrices for two-dimensional non-linear hyperbolic systems*. II, Amer. J. Math. **94**, 631–650.

J. A. Smoller and J. L. Johnson

(1969) *Global solutions for an extended class of hyperbolic systems of conservation laws*, Arch. Rational Mech. Anal. **32**, 169–189; erratum, **37** (1970), 399–400.

S. L. Sobolev

(1962) *Applications of functional analysis in mathematical physics*, reprint, Izdat. Sibirsk. Otdel. Akad. Nauk SSSR, Novosibirsk; English transl., Amer. Math. Soc., Providence, R. I., 1963.

Ju. I. Šokin

(1973) *On the first differential approximation method in the theory of difference schemes for hyperbolic systems of equations*, Trudy Mat. Inst. Steklov. **122**, 66–84; English transl. in Proc. Steklov Inst. Math. **122**.

(1976) *Analysis of the properties of approximation viscosity of difference schemes by means of the method of differential approximation*, Proc. Fifth Internat. Conf. Numer. Methods Fluid Dynamics (Enschede, 1976), Lecture Notes in Physics, Vol. 59, Springer-Verlag, pp. 410–414.

V. A. Solonnikov

(1976) *The solvability of the initial-boundary value problem for the equations of motion of a viscous incompressible fluid*, Zap. Naučn. Sem. Leningrad. Otdel. Mat. Inst. Steklov. (LOMI) **56**, 128–142; English transl. in J. Soviet Math. **14** (1980), no. 2.

K. P. Stanjukovič

(1955) *Unsteady motion of continuous media*, GITTL, Moscow; English transl., Pergamon Press, 1960.

V. V. Stepanov

(1958) *Course in differential equations*, Fizmatgiz, Moscow; German transl., VEB Deutscher Verlag Wiss., Berlin, 1963.

Gilbert Strang

(1963) *Accurate partial difference methods*. I, Arch Rational Mech. Anal. **12**, 392–402.

(1964) *Accurate partial difference methods*. II, Numer. Math. **6**, 37–46.

(1966) *Necessary and sufficient conditions for well-posed Cauchy problems*, J. Differential Equations **2**, 107–114.

I. E. Tamm

(1946) *Foundations of the theory of electricity*, 3rd ed., GITTL, Moscow. (Russian)

A. H. Taub

(1946) *Interaction of progressive rarefaction waves*, Ann. of Math. (2) **47**, 811–828.

A. N. Tihonov and A. A. Samarskiĭ

(1954) *Discontinuous solutions of quasilinear equations of the first order*, Dokl. Akad. Nauk SSSR **99**, 27–30; English transl. in Amer. Math. Soc. Transl. (2) **42** (1964).

(1959) *Convergence of difference schemes in the class of discontinuous coefficients*, Dokl. Akad. Nauk SSSR **124**, 529–532. (Russian)

(1961) *Homogeneous difference schemes*, Ž. Vyčisl. Mat. i Mat. Fiz. **1**, 5–63; English transl. in USSR Comput. Math. and Math. Phys. **1**.

(1972) *The equations of mathematical physics*, 4th ed., "Nauka", Moscow; English transl. of 2nd ed., Pergamon Press, Oxford, and Macmillan, New York, 1963; Vols. 1, 2, Holden-Day, San Francisco, Calif., 1964, 1967.

V. A. Titov and G. I. Šiškin

(1976) *A numerical solution of a parabolic equation with small parameters multiplying the derivatives with respect to the space variables*, Trudy Inst. Mat. i Meh. Ural. Naučn. Centr Akad. Nauk SSSR Vyp. 21, 38–43. (Russian)

V. E. Troščiev

(1970) *The divergence of the "cross" scheme for the equations of gas dynamics*, Čisl. Metody Meh. Splošnoĭ Sredy **1**, no. 5, 87–93. (Russian)

H. S. Tsien

(1956) *The Poincaré-Lighthill-Kuo method*, Adv. in Appl. Mech., Vol. IV, Academic Press, pp. 281–349.

V. A. Tupčiev

(1964) *On the decay of an arbitrary discontinuity for a system of two first-order quasilinear equations*, Ž. Vyčisl. Mat. i Mat. Fiz. **4**, 817–825; English transl. in USSR Comput. Math. and Math. Phys. **4**.

(1966) *The problem of decay of an arbitrary discontinuity for a system of quasilinear equations without the convexity condition*, Ž. Vyčisl. Mat. i Mat. Fiz. **6**, 527–547; English transl. in USSR Comput. Math. and Math. Phys. **6**.

(1972a) *On the isolatedness of a solution of the problem of decay of an arbitrary discontinuity*, Inform. Bjull. Cisl. Metody Meh. Splošnoĭ Sredy **1**, no. 2, 82–93. (Russian)

(1972b) *The asymptotic behavior of the solution of the Cauchy problem for the equation* $\varepsilon^2 t u_{xx} = u_t + [\varphi(u)]_x$ *that degenerates for* $\varepsilon = 0$ *into the problem of the decay of an arbitrary discontinuity for the case of a rarefaction wave*, Ž. Vyčisl. Mat. i Mat. Fiz. **12**, 770–775; English transl. in USSR Comput. Math. and Math. Phys. **12**.

(1973a) *The uniqueness of the continuous solution of the problem of the decay of an arbitrary discontinuity for a gradient system*, Mat. Zametki **13**, 251–258; English transl. in Math. Notes **13**.

(1973b) *On the method of introducing viscosity in the study of problems involving the decay of a discontinuity*, Dokl. Akad. Nauk SSSR **211**, 55–58; English transl. in Soviet Math. Dokl. **14**.

T. D. Ventcel′

(1957) *Certain quasilinear parabolic systems*, Dokl. Akad. Nauk SSSR **117**, 21–24. (Russian)

(1963) *Quasilinear parabolic systems with increasing coefficients*, Vestnik Moskov. Univ. Ser. I Mat. Meh. no. 6, 34–44. (Russian)

A. I. Vol′pert

(1967) *The spaces BV and quasilinear equations*, Mat. Sb. **73(115)**, 255–302; English transl. in Math. USSR Sb. **2**.

A. I. Vol′pert and S. I. Hudjaev

(1972) *On the Cauchy problem for composite systems of nonlinear differential equations*, Mat. Sb. **87(129)**, 504–528; English transl. in Math. USSR Sb. **16**.

E. V. Vorožcov

(1977) *Numerical tests of differential analyzers of shock waves*, Čisl. Metody Meh. Splošnoĭ Sredy **8**, no. 2, 12–27. (Russian)

N. D. Vvedenskaja

(1956) *Solution of the Cauchy problem for a nonlinear equation with discontinuous initial conditions by the method of finite differences*, Dokl. Akad. Nauk SSSR **111**, 517–520. (Russian)

(1961) *An example of nonuniqueness of a generalized solution of a quasilinear system of equations*, Dokl. Akad. Nauk SSSR **136**, 532–533; English transl. in Soviet Math. Dokl. **2**.

R. F. Warming, Paul Kutler and Harvard Lomax

(1973) *Second- and third-order noncentered difference schemes for nonlinear hyperbolic equations*, AIAA J. **11**, 189–196.

Burton Wendroff

(1972) *The Riemann problem for materials with nonconvex equations of state*. I, II, J. Math. Anal. Appl. **38**, 454–466, 640–658.

Hermann Weyl

(1949) *Shock waves in arbitrary fluids*, Comm. Pure Appl. Math. **2**, 103–122.

G. B. Whitham

(1958) *On the propagation of shock waves through regions of non-uniform area or flow*, J. Fluid Mech. **4**, 337–360.

Masaya Yamaguti and Takaaki Nishida

(1968) *On some global solutions for quasilinear hyperbolic equations*, Funkcial. Ekvac. **11**, 51–57.

Ju. S. Zav′jalov

(1955) *On some integrals of one-dimensional flow of a gas*, Dokl. Akad. Nauk SSSR **103**, 781–782. (Russian)

(1956) *On the integration of certain equations for the nonisentropic motion of a gas*, Dissertation, Tomsk State Univ., Tomsk. (Russian)

Ja. B. Zel′dovič

(1956) *Motion of a gas under the action of a momentary pressure (shock)*, Akust. Ž. **2**, 28–38; English transl. in Soviet Phys. Acoustics **2**.

Ja. B. Zel′dovič and Ju. P. Raĭzer

(1966) *Physics of shock waves and high-temperature hydrodynamics phenomena*, 2nd ed., "Nauka", Moscow; English transl., Vols. I, II, Academic Press, 1966, 1967.

Zhang Tong and Guo Yu-fa

(1965) *A class of initial-value problems for systems of aerodynamic equations*, Acta Math. Sinica **15**, 386–396; English transl. in Chinese Math. Acta **7**, no. 1.

A. I. Žukov

(1959) *A limit theorem for difference operators*, Uspehi Mat. Nauk **14**, no. 3(87), 129–136; English transl. in Amer. Math. Soc. Transl. (2)**102** (1973).

(1960) *Application of the method of characteristics to the numerical solution of one-dimensional problems in gas dynamics*, Trudy Mat. Inst. Steklov. **58**; English transl., Clearinghouse Federal Sci. Tech. Information, Springfield, Va., 1967.

A. I. Žukov and Ja. M. Každan

(1956) *On the motion of gas under the influence of a short impulse*, Akust. Ž. **2**, 352–357; English transl. in Soviet Phys. Acoustics **2**.

A. Zygmund

(1959) *Trigonometric series*, 2nd rev. ed., Vols. I, II, Cambridge Univ. Press.

ABCDEFGHIJ–EB–89876543